Advance Praise for Investment Decisions and the Logic of Valuation

"The subject of valuation and project selection has a long history originating in multiple disciplines including financial economics, accounting, and engineering. Historically these literatures evolved along three different paths that were sometimes overlapping but often not such that they appear disjointed. This makes it difficult for students and business professionals to take advantages of the unique contributions of each. Carlo Alberto Magni has simplified this problem for us in his book titled *Investment Decisions and the Logic of Valuation*. He has done a masterful job of melding the contributions of each of these disciplines in a single volume. Carlo Alberto has the rare gift of analytical sophistication and superb communication skills. This is a book that every student of finance will want to have on their book shelf."

—John Martin, *Carr P. Collins Chairholder and Professor of Finance, Baylor University, co-author of* Valuation. The Art and Science of Corporate Investment Decisions

"This textbook takes a refreshing approach to engineering economics by integrating traditional engineering topics such as statics and dynamics. As a leading authority on determining the economic value of projects, this book affords Dr. Magni the opportunity to cohesively tie together the various aspects of investment appraisal. It is only a matter of time before his rate of return analysis is the mainstream approach in corporate finance."

—Joseph Hartman, *Provost and Vice Chancellor for Academic Affair, University of Massachusetts Lowell, editor-in-chief of* The Engineering Economist (2004–2012), *author of* Engineering Economy and the Decision-Making Process

"Professor Magni has created a very unique pedagogical framework for describing the complex interrelationships between finance, accounting and economic decision-making using engineering analogies that will appeal to engineering students and to professionals involved in the justification of project investments."

—Thomas Boucher, *Professor of Industrial and Systems Engineering Emeritus, Rutgers University, editor-in-chief of* The Engineering Economist (2013–2016)

"Engineers, economists and accountants have all had a hand in the development of the tools needed for making rational investment decisions. But the results have been scattered across the three disciplines, with each tending to focus on the aspects that most bear on their concerns. Carlo Alberto Magni is a leading authority on how to use rate of return measures properly, measures treated by most economic texts as misguided and therefore quickly dismissed, even though they are used widely in practice. His magnificent book guides the reader through a very difficult area, one that not only provides the necessary tools for the job, but which shows how to tie together the cash flow and accounting forecast inputs in a coherent and systematic fashion when doing so. All three groups will learn much from this novel and important text."

—Ken Peasnell, *Distinguished Professor of Accounting Emeritus, Lancaster University Management School*

"Measurements of capital, income, cash flow, and rates of return are regularly employed in decision-making as well as in analyzing and appraising capital investments. For decades, the relationship between these concepts remained obscure if not largely misunderstood. In this novel book, Carlo Alberto Magni, a virtuous scholar and a leading authority on the correct measurement of rates of return, eloquently presents the golden link that relates them all in a simple, logical, comprehensive and intuitive manner. This important book is an eye-opener, rich with examples and is highly recommended to all who wish to analyze investments and use these basic concepts properly, whether they are economists, accountants, engineers or finance people. The book provides an enormous benefit even to experienced professionals in assessing investment profitability."

—Moshe Ben-Horin, *Professor of Finance and President of Ono Academic College*

"Put in a room a manager (possibly an engineer), an accountant and a financial mathematician. Close the room and throw away the key until they reach an agreement about how to evaluate investments. If the financial mathematician is Carlo Alberto Magni and if the agenda is the index of this book, well, try to recover quickly the key."

—Lorenzo Peccati, *Emeritus Professor of Applied Mathematics, Bocconi University; co-editor-in-chief of the* European Journal of Operational Research (2005–2014)

Carlo Alberto Magni

Investment Decisions and the Logic of Valuation

Linking Finance, Accounting, and Engineering

Springer

Carlo Alberto Magni
Department of Economics “Marco Biagi”
University of Modena and Reggio Emilia
Modena, Italy

ISBN 978-3-030-26775-9 ISBN 978-3-030-27662-1 (eBook)
https://doi.org/10.1007/978-3-030-27662-1

This Springer imprint is published by the registered company Springer Nature Switzerland AG
The registered company address is: Gewerbestrasse 11, 6330 Cham, Switzerland

To Roberta, my capital affective project

About the Author

Carlo Alberto Magni is an Associate Professor at the Department of Economics "Marco Biagi" and the School of Doctorate E4E (Engineering for Economics—Economics for Engineering) of the University of Modena and Reggio Emilia. He has received the national qualification as full professor in Financial Markets and Institution in 2014 and in Mathematical Methods of Economics, Finance, and Actuarial Sciences in 2017.

Graduated in Economics and Business in 1989, he received his Ph.D. in Mathematics for Economic Decisions from the University of Trieste in 1996. He holds a Master in Business Administration from the University of Turin, received in 1991.

His teaching activities include engineering economics and financial management, principles and models for managerial decisions, mathematics for investment and credit, and calculus. His research areas are engineering economics, corporate finance, managerial finance, financial management, management accounting, and financial mathematics. He has written more than 100 papers and published in more than 30 different journals, including *European Journal of Operational Research*, *International Journal of Production Economics*, *Quantitative Finance*, *Journal of the Operational Research Society*, *Journal of Mathematical Economics*, *The Engineering Economist*, *European Accounting Review*, *Theory and Decision*, *Managerial Finance*, *Economic Modelling*, *Insurance: Mathematics and Economics*.

In 2011, he won the "Eugene L. Grant" Award, granted by the Engineering Economy Division of the American Society for Engineering Education (ASEE). Since 2013, he serves as area editor for *The Engineering Economist*, a joint publication of the Engineering Economy Division of ASEE and the Institute of Industrial Engineers (IIE).

Contents

Preface

> Solving a problem simply means representing it so as to make the solution transparent
>
> Simon (1981, p. 153)

This book analyzes valuation and decision-making for capital budgeting projects, also known as capital asset investments. An investment decision is a process which leads to a change in the firm's operations. It starts with the recognition of a problem that needs a solution or an opportunity to take advantage of. After investigating the technical feasibility of the project, input data are estimated which must be aggregated in a suitable *accounting-and-finance* model in order to provide a reliable valuation of the project. Valuation is meant to lead to an economically rational decision, one which is capable of *increasing the wealth* of the firm's shareholders.

Any economic entity and, in particular, a project may be viewed as a system of activities whose basic elements are: *Capital*, *income*, *cash flow*. The book is built upon the mechanics of this system. Specifically, the statics studies the forces that act upon the "economic body" at rest and determine the condition of static equilibrium; the dynamics studies the forces that govern the motion of the economic body over time. The statics is represented by the *law of conservation* (for capital, for income, and for cash flow) while the dynamics is represented by the *law of motion* (linking capital, income, and cash flow). The former has to do with a fundamental principle of accounting (accounting identity); the latter has to do with a fundamental principle of finance (time value of money).

Accounting and financial modeling of projects in this book is grounded upon such a mechanical system. We present a framework which is designed to integrate the law of conservation and the law of motion in a simple way so as to make the resulting accounting and financial modeling intuitive and easy to apply. In particular, no prerequisite of accounting and finance are required nor any knowledge of calculus.

Statics and dynamics are integrated in a matrix (the *split-screen Matrix*), a working device resulting in an innovative system of *accounting-and-finance engineering*. This system will be the beacon of a transparent and robust economic analysis of investments and will serve the purpose of linking theory and practice.

The accounting-and-finance engineering system provides the methodological framework for analyzing *capital asset projects*. Whatever the definition attached to these activities, they mostly involve the work and the expertise of one or more engineers, either for assessing the technical feasibility of some design (electrical engineers, chemical engineers, civil engineers, industrial engineers, environmental engineers, mechanical engineers, computer science engineers, etc.) or for coordinating the work of a team or managing complex systems (engineering managers, systems engineers). Therefore, to deal with capital asset projects means to deal with engineering and industrial projects.

However, this book is about *logic*. While a technical design is discipline-specific (different laws of nature are used for different designs), the *logical* design of valuation and decision is general and is not affected by the kind of activity under consideration. Therefore, since the accounting-and-finance engineering system presented in the book holds for any economic entity, the book provides the tools for modeling *any* kind of corporate decision which has an impact on the project's amount of capital, income, and cash flow, thereby including financial investments as well as real asset investments, tangible and intangible projects, manufacturing schemes as well as human resource initiatives, privately owned projects as well as (the financial aspects of) public investments. Even *firm valuation* is incorporated in the approach, for a firm may be viewed as a portfolio of projects and, symmetrically, a project may be viewed as an incremental firm. And while the estimation of the input data in terms of capital, income, and cash flow may differ across economic entities, the logical reciprocal relationships among the various types of capital, income, and cash flow do not. As a result, it suffices to change "project" with "firm" and a logical approach to firm valuation is obtained.

Logic is a tool, not an end in itself. This book does not deny the major role of informed judgment in decision-making. Quite the contrary, the book aims at providing decision-makers with a logically consistent tool which may be part of a wider process involving technical expertise, social skills, human interactions, creativity, and ethical considerations.[1]

Potential Readers

The book is addressed to three classes of readers: Practitioners, academics, and students. It is hoped that:

- practitioners (engineers, financial and operational managers, engineering managers, advisors, analysts, professionals, accountants) will find

[1]For example, to choose between a program that is expected to reduce the number of work accidents and a program that reduces water pollution and/or greenhouse gas emissions involves an ethical trade-off issue.

  - a helpful guide to the financial modeling of projects
  - user-friendly analytical tools for managing the subtleties of accounting and financial relations
  - a toolkit of techniques for carrying out an in-depth analysis of economic efficiency and for making rational decisions
  - an applicative error-free model easily convertible to spreadsheets

- academics (finance scholars, accounting scholars, engineering economists, management scientists, operations researchers, financial mathematicians) will welcome

  - the unified theory of investment decision-making
  - the robust theoretical apparatus
  - the novel way of linking finance and accounting with engineering decision-making and, more generally, corporate decision-making

- students of relevant courses (corporate finance, engineering economics, managerial finance, financial management, management accounting) will appreciate

  - the agility and easiness of the accounting and financial engineering system
  - the strong correspondence between highly theoretical notions and the ready-to-use metrics and techniques logically derived from the former
  - the use of basic algebra and simple working rules.

Some Methodological Remarks

The book is concerned with both theory and practice. No metric or technique, whether simple or complex, is reliable if it does not derive from a corpus of theoretical principles which are logically deducted from some primitive assumptions. At the same time, no theory of investment decisions is helpful if it is not applicable to the practice of real-life investment analysis and decision-making.

For this reason, the theoretical apparatus of principles, concepts, definitions of this book is in a strict correspondence with a toolkit of applicative techniques and metrics. This implies that every theoretical notion and principle presented in the book leads to one or more applicative tools, metrics, and techniques (i.e., no theoretical principle is explained without a precise correspondence in applicative terms) and every applicative tool, metric, and technique emanates from one or more theoretical principles (i.e., no metric nor technique is presented without a solid theoretical ground, logical rationale, and economic meaningfulness).

Each chapter starts with a summary and a list of skills the reader will acquire and ends with a list of key points where tables and graphs are often used in order to summarize the notions, results, relations, principles, methods, and techniques presented. They are aimed at providing snapshots of the concepts presented in the chapter for a more effective systematization of the learning material.

The book contains more than 180 guided examples. They are essential to the theory as well as the practice and are not designed to be slavish illustrations of formulas. They have several aims:

- guide the reader to a deeper understanding of the notions and methods presented
- show the application of the theory to the practice of financial modeling in a painless way
- provide conceptual insights, trigger new perspectives, introduce new notions, and stimulate comments and remarks on theory or techniques.

As such, they should be considered an essential part of the book.

Some of the worked examples are explicitly inspired by examples already presented in engineering economy textbooks or finance textbooks.

In order not to swamp the reader with excessive accuracy, which would obscure the presentation, numbers in the examples are often rounded. Some rounding errors may then arise, but the prime concern of this book is logic, not decimal place accuracy.

The book avails itself of more than 160 tables to help the reader assimilate concepts, principles, techniques, and appreciate the networks of relations involved.

The book is designed to be self-explaining and makes use of a *low-math-no-economics* teaching approach. This means that no use of calculus is made, and no prior studies of finance and accounting are required. The only prerequisite is basic (undergraduate) algebra.

Terminology

In this book, several disciplines are intertwined, which raises the major issue of using a terminology which may suit the linguistic habits and intellectual perspectives of readers of various backgrounds. This also raises an even more important problem: Different notions are sometimes referred to with the same expression or, vice versa, the same notion is sometimes referred to with different labels. This causes ambiguities, misunderstanding, frictions across scientific domains, and between academics and practitioners.

I had to make a choice capable of encompassing the multitude of notions and concepts stemming from corporate finance, accounting, and engineering economics.

The choice of some terms, expressions, definitions, and labels will meet the expectations of some while frustrating the expectations of others. To the extent that linguistic confusion would not arise and for the purpose of underlining the conceptual equivalence, in some cases I have used some terms or expressions interchangeably (e.g., *return on investment* and *return on capital*) and, in some cases, I have favored one label over other possible ones (e.g., *cash flow from operations* is favored over *capital cash flow*). Whenever labels were unavailable, as in the case of the creation of a new concept or definition, I have tried to use as intuitive a term as possible (e.g., the class of *net operating liabilities*). Several remarks are strewn over the book for clarification of terminological issues. The substantial use of symbols is meant to disambiguate, and a detailed list of notational conventions is provided at the end of the book.

Logical Structure of the Book

To describe the logical structure of the book means to describe the system of accounting-and-finance engineering that governs the process of valuation and decision-making. Figure 1 proposes a sequential chart which provides a conceptual map the reader may always turn to for a better orientation.

The starting point of the process is the determination of project input data and market input data.

Project input data consist of the differential costs (additional cost and/or cost savings), prices and quantities of new products, growth rates of costs and prices, purchases or disposals of plants and equipment, credit policy, tax rates, etc. Market input data are represented by the expected rates of return on financial assets equivalent in risk to the project's constituents (i.e., required return on assets, required return on equity, required return on debt, etc.).

The book does not deal with this first step. It assumes both project data and market data as exogenously given and guides the analyst from this point through the appraising process up to valuation and decision.

Each input belongs to one of three fundamental categories which are typical of any economic entity: *Capital*, *income*, and *cash flow*. They are ascribed to project input data and market input data, and generate two different economic systems: The project system and the benchmark system.

The project system is subjected to forces (customers, suppliers, employees, debtholders, equityholders, etc.) that determine its static equilibrium through a *law of conservation* and its dynamics through a *law of motion*. Each basic element is then broken down into subsystems (e.g., operating system, non-operating system, debt system, equity system), which are in turn broken down in further lower-level subsystems (e.g., fixed-asset system, accounts receivable system, net operating liabilities system, etc.) and gathered in the *Matrix*. The latter is a flexible machine equipped with a quadruple-entry system of working rules (*split-screen technique*) such that the entire mechanics of the project is captured by a sequence of split-screen Matrices.

The benchmark system is itself equipped with a mechanics (statics, dynamics, and associated Matrix) which is governed by the forces acting in the capital markets (demand and offer by market investors). The *Law of One Price* guarantees that all marketed assets equivalent in risk share the same equilibrium, that is, all market assets equivalent in risk have the same expected rate of return. Therefore, the benchmark system represents the expectations of the market's investors for all equivalent-risk assets.

The objective of the firm is to make decisions which increase shareholders' wealth (*rational decision-making*). This is expressed by saying that *value is created*. In order to measure shareholder value creation, the project system is compared with the benchmark system, where investors might alternatively invest their fund if they did not invest in the project system.

The project system and the benchmark system may themselves be in reciprocal equilibrium or disequilibrium. In the former case, they are expected to have equal economic profitability, which implies the firm's shareholders do not receive any additional benefit (nor loss) in undertaking the project as opposed to investing in the benchmark system. If they are in reciprocal disequilibrium, then one of the two systems increases shareholders' wealth more than the other system does. Value is created if and only if the two systems are in reciprocal disequilibrium; in particular, value is created if the disequilibrium is such that the project system's economic profitability is greater than the benchmark system's economic profitability.

The degree of disequilibrium may be measured in absolute terms (values expressed in dollars) or in relative terms (rates of return, profitability indices, benefit-cost ratios, etc.). The value of the model output (either absolute or relative) establishes whether a project should be undertaken or not and provide information on the amount of wealth increase (absolute measure) and the magnitude of the project's economic efficiency (relative measure).

However, the benchmark system is not unambiguously described. There are three ways of benchmarking a project (which refer to three ways of replicating a project by purchasing equivalent-risk assets in the capital market).

To each benchmark system, there corresponds one absolute approach and one relative approach. This brings about three pairs of valuation and decision-making approaches, each associated with a specific benchmark system. In turn, each pair is associated with one of the three basic elements:

Absolute approach	**Relative approach**	**Basic element**
Net present value (NPV)	Internal average rate of return (IARR)	Cash flow
Residual income (RI)	Average internal rate of return (AIRR)	Income
Net future value (NFV)	Aggregate return on investment (AROI)	Capital

(The widely employed internal rate of return approach is a special, rather problematic case of AIRR approach.)

The six approaches are economically rational and, therefore, financially equivalent: They lead to the same valuation and same (accept or reject) decision.

Table of Contents Overview

Chapter		Highlights
1	Dynamics. The law of motion	Presents the basic elements of an economic system (capital, income, and cash flow) and the equation of motion; introduces the income rate; distinguishes invested capital and borrowed capital; defines a project as a differential/incremental system; provides a (minimal) taxonomy for projects in terms of expansion, replacement, abandonment
2	Statics. The law of conservation	Presents the static equilibrium for each basic element; distinguishes the financial structure from the capital structure
3	Financial statements	Introduces the Matrix and the split-screen technique; examines the capital components (fixed assets, working capital, liquid assets, debt, equity), the income components (operating income, interest income, interest expense, net income), and the cash-flow components (operating and non-operating cash flow, cash flow to debt, cash flow to equity)
4	Estimating the cash flows	Employs the split-screen technique for deducting the cash flows from the accounting estimates; presents the related notions of net operating profit after taxes (NOPAT) and free cash flow (FCF); presents the mosaic of a project's cash flows
5	Valuation and value creation	Analyzes the Law of One Price and the notion of value creation; introduces a benchmark system based on replication of the project's prospective cash flows; analyzes the notions of cost of capital, minimum attractive rate of return (MARR), and net present value (NPV); defines (economically) rational decision-making
6	Project appraisal	Presents 12 discounted cash-flow valuation methods, including CFE, CCF, APV, WACC; introduces two other benchmark systems giving rise to the notions of residual income (RI) and value added (VA) (or net future value, NFV); illustrates the RI approach and the NFV approach to valuation and decision
7	The quest for a relative approach	Addresses the issue of measuring economic efficiency; argues in favor of relative measures as complements of absolute measures; anticipates the link between scale and efficiency
8	Average internal rate of return	Sets up the intuition for the notion of rate of return; provides a logical derivation of rate of return; presents the average internal rate of return (AIRR) approach, the relative counterpart of the RI approach
9	Internal rate of return	Presents the internal rate of return (IRR) approach and describes its pitfalls; shows that the IRR is a special case of the AIRR approach; illustrates the modified internal rate of return (MIRR) approach and the Teichroew-Robichek-Montalbano (TRM) model
10	IARR and AROI	Builds a genuinely internal average rate of return (IARR) which is the relative counterpart of the NPV approach; presents the aggregate return on investment (AROI), which is genuinely internal as well but is the relative counterpart of the NFV
11	Ranking projects	Develops rational ranking via relative measures of worth as well as absolute measure of worth; presents the incremental method and the direct method
12	Three decisions	Illustrates three final examples: Two engineering projects (one accept/reject decision and a choice between mutually exclusive alternatives) and a human resource initiative

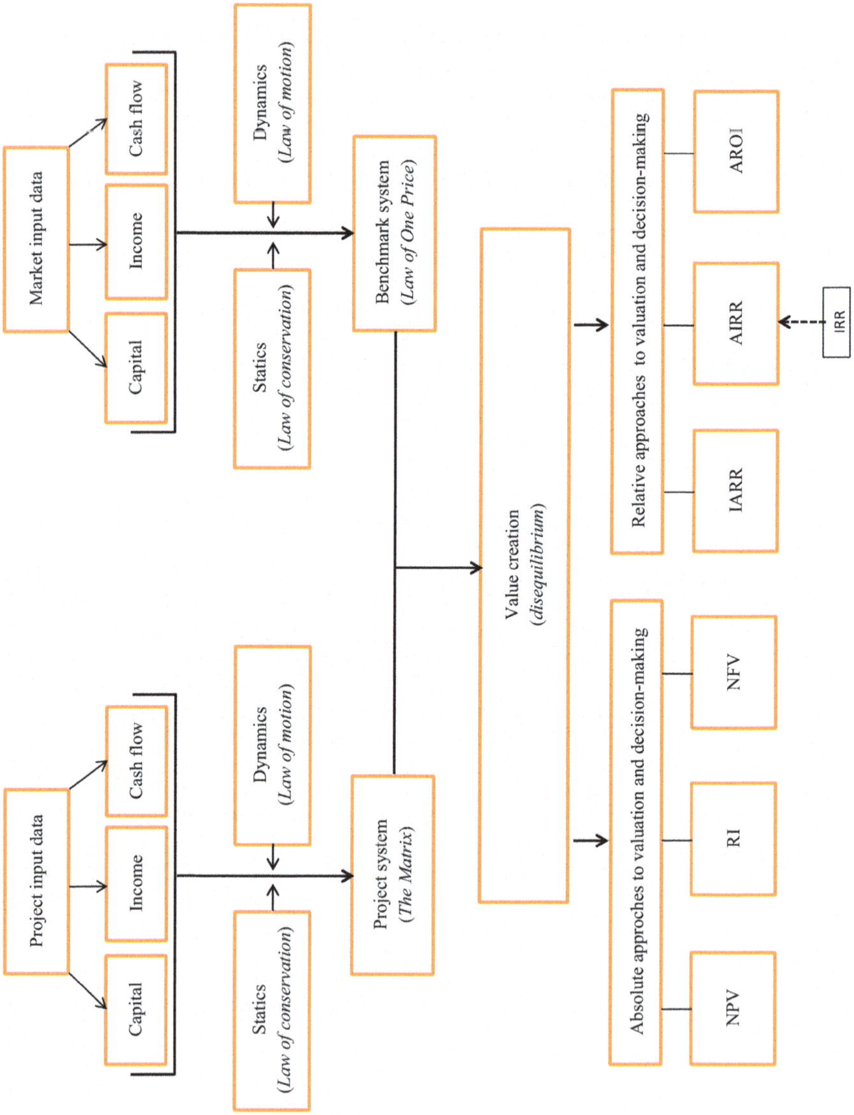

Fig. 1 Accounting-and-finance engineering system

Acknowledgements

This book originates from an invitation by Anthony Doyle, executive editor for Springer. The book would not have been written without his invitation, so I thank him for this opportunity (and for his patience). Actually, I must add that the book would not have been written if I had not attended, 30 years ago, the outstanding lectures by Lorenzo Peccati, who first opened my mind on financial mathematics, a quite obscure world before meeting him. He and his inspiring works have contributed to direct my researches toward capital budgeting analysis. His appreciation of my research results and his constant support in these years have been very important. I am deeply grateful to him.

In recent years, the appreciation of my work by Joseph Hartman has been essential to head towards engineering economics. I would like to express my gratitude to him. I thank Thomas Boucher for giving me the opportunity of gaining experience as an area editor of *The Engineering Economist*. Both Joe and Tom have contributed to my ever-increasing interest in linking engineering and economic decision-making. On the "engineering" part, I also wish to thank Gordon Hazen, who has always been supportive of my approach to financial efficiency.

My inveterate, almost philosophical attitude to symmetry has brought me to accounting, which I see as a means for unifying different valuation methods and decision criteria. This disposition has led me to the authoritative, enlightening papers by Ken Peasnell and has triggered fruitful exchanges with him. I am very grateful to him for his support and encouragement to my effort of establishing links between accounting and finance.

It is precisely the "finance turn" taken by my researches which has set up the conditions to get in touch with various scholars involved in firm valuation as well as project appraisal. In particular, I would like to thank John Martin, whose books have helped me make my way in the world of corporate valuation theory. A special thank goes to Ignacio Vélez-Pareja; we have had long, vibrant discussions about firm and project valuation and fertile exchanges on several topics related to corporate finance and financial modeling. I thank Yoram Kroll and Moshe Ben-Horin for their appreciation of my work and for their hospitality. I am much obliged to

Moshe, who meticulously read, evaluated, and analyzed the whole book. His rigorous and cogent remarks on each and every part of it have been extremely important to improve the presentation and to test the solidity and transparency of my approach to valuation and decision.

I am indebted to "eagle-eyed" Andrea Marchioni, who painstakingly read the book and supplied a great help in editing it, detecting many annoying typos and errors; he examined its preciseness and analytical consistency and we had lively and useful exchanges on the various topics treated in the book.

I thank Franco Leone, accountant, for some clarifications of accounting subtleties and for his unfailing friendship. A last, personal thank is due to my mother-in-law, Carla, for his sympathetic support during these years.

Part I
Accounting-and-Finance Engineering System: The Mechanics

In this part of the book, we present the building blocks of the **accounting-and-finance engineering system**.

Specifically, we illustrate the **mechanics** of an economic "body" or system. The mechanics of an economic system is illustrated in terms of **dynamics** (diachronic perspective) and **statics** (synchronic perspective). The dynamics studies the way the system evolves through time and investigates the relationships among the fundamental elements of an economic system: The capital, the income, and the cash flow. The statics studies the equilibrium relations among various kinds of capital, income, and cash flow at a given point in time.

The two perspectives give rise to two fundamental laws: The **law of motion**, which is described by an intertemporal relation linking **capital**, **income**, and **cash flow**, and the **law of conservation**, which gives voice to a balancing principle between investment forces and financing forces.

Capital, income, and cash flow are absolute amounts of money. A fourth, relative notion is also introduced, which is derived from capital and income: The **income rate** or **rate of return**. The law of motion and the law of conservation are reframed in terms of income rate.

Chapter 1
Dynamics. The Law of Motion

> Dynamics is concerned with the prediction of the motions of the widest possible variety of objects and with the computation of the effects that these motions entail …Its logical structure has long been a model for other branches of science.
>
> Goodman and Warner (2001, p. 9)

> [T]he BS [balance sheet] shows the level of the different reservoirs where the stocks of the various items are kept. The IS [income statement] shows the dynamic part or the movements of those stocks. It tells us the amount of flow of goods expressed in dollars …that flows into or out of the reservoir. Then, we can express the idea of the stock balance equation as follows:
> Level of the reservoir$_n$ = Level of the reservoir$_{n-1}$ + New flow into the reservoir$_n$ − Flow out of the reservoir$_n$
>
> Tham and Vélez-Pareja (2004, p. 111)

> As the saying goes, "Capital is the key"
>
> Pacioli and Cripps (1494/1994, p. 1)

C. A. Magni, *Investment Decisions and the Logic of Valuation*,
https://doi.org/10.1007/978-3-030-27662-1_1

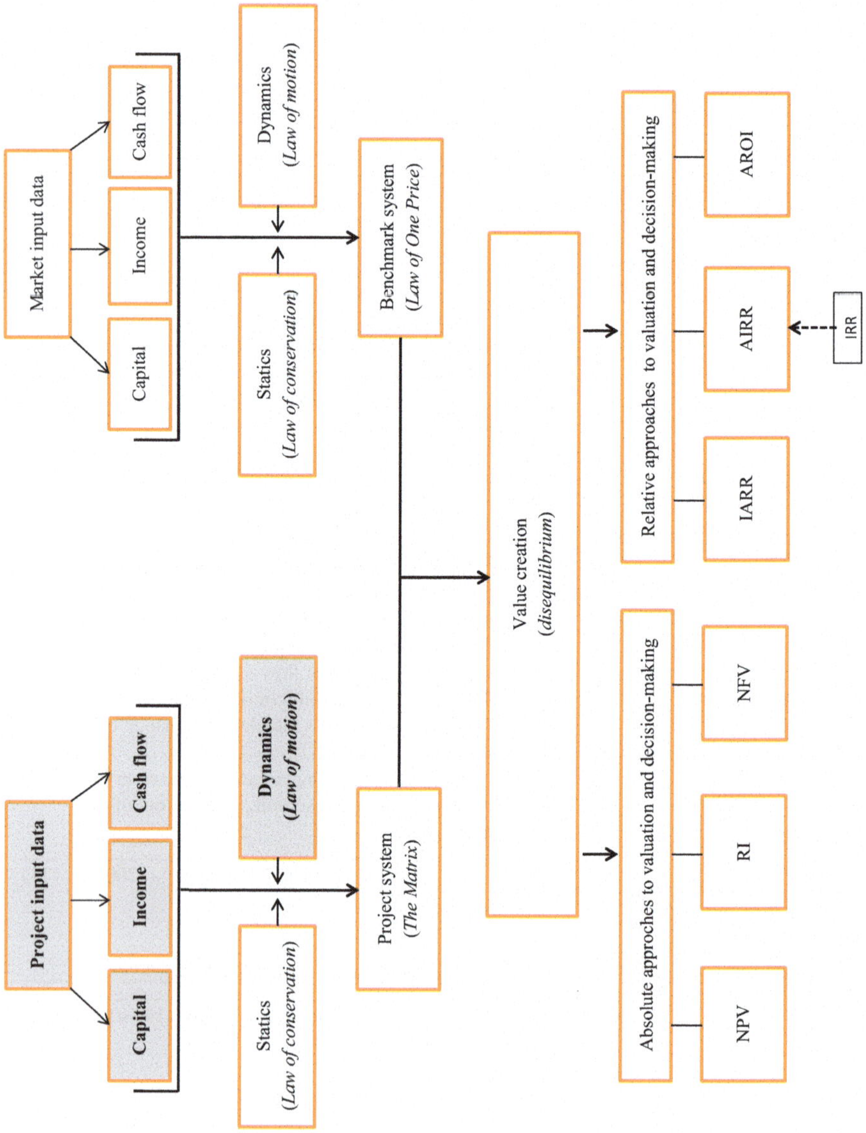

Fig. 1.1 Chapter 1: The dynamics

This chapter introduces the notion of **economic system**. An economic system consists of a **set of activities** put in place by one or more economic agents or parties which involves a system of business, trading, financial transactions and a **system of monetary relations** with several agents such as investors, capital providers, suppliers, customers, employees, government, etc.

The chapter investigates the **dynamics** of an economic system. Specifically, it illustrates the fundamental features underlying an economic system from a *diachronic* perspective; that is, it examines the way the asset's economic structure evolves over time. In particular, it studies how the forces that act upon the system affect the position of the system (i.e., the **capital**) by means of two sources of increase/decrease: **Income and cash flow**.

The dynamics of an economic system is characterized by a fundamental intertemporal, recursive equation. It is called the **Law of Motion**. We analyze this law in some depth.

We introduce a fourth basic notion: The **income rate**, also known as rate of return. The law of motion is reframed, thereby enabling deducting the principle of the **time value of money**, a basic tenet in finance and economics.

We provide a definition of project in terms of **incremental economic system** and classify projects as expansion projects, replacement projects, and abandonment projects.

After completing this chapter, the reader should be able to:

- understand the meaning of the three basic elements of an economic system: Capital, income, cash flow
- understand the *law of motion* of an economic system, its position in terms of capital, and the sources of change in terms of income and cash flow
- interpret the notion of income rate (rate of return) as the relative growth of the system's capital position
- derive the time value of money as a logical consequence of the law of motion
- understand that an economic agent may either be in an investment (i.e., lending) position or a financing (i.e., borrowing) position, being represented by a capital position above zero or below zero, respectively
- differentiate investment/lending rates, occurring whenever capital is being invested in the system (capital position above zero), from financing/borrowing rates, occurring whenever capital is being absorbed from the system (capital position below zero)
- conceptualize a project as an *incremental* economic system which changes the firm's *status quo* (null alternative)
- describe any project as a triplet of incremental capital stream, incremental income stream, and incremental cash-flow stream governed by the law of motion
- distinguish expansion projects, abandonment projects, and replacement projects

1.1 The Three Basic Notions

Any economic activity can be described in terms of three basic elements:

- capital
- income
- cash flow.

These elements constitute a fundamental economic trinity:

- the capital measures the monetary amount which is committed in the economic activity
- the income represents the growth of the capital within a given period as the result of the operations made in the economic activity. It is often conceptualized as the *remuneration* of the capital
- the cash flow expresses the monetary amount that flows out of (or in) the economic activity.

Thus, the time evolution of the capital depends on income and cash flow: It increases (decreases) by a positive (negative) income and decreases (increases) by a positive (negative) cash flow, as the following example illustrates.

Example 1.1 ♦
Suppose an economic agent A invests \$100 in an economic activity B at a given date. The beginning-of-period (BOP) capital is then \$100. This is the initial state of the system. Suppose that, at the end of the first period, the income is \$10. This means that the end-of-period (EOP) capital value is \$110 = \$100 + \$10. Suppose that, at the end of the first period, A withdraws \$30 from B. The capital that remains invested in B at the beginning of the second period is \$80 = \$110 − \$30. Now, suppose that, in the second period, the capital grows by an income equal to \$4; then, the EOP capital is \$84 = \$80 + \$4. Assume that, at the end of this period, the investor injects an additional \$50 capital in the system. The BOP capital of the third period is then \$134 = \$84 + \$50. And so on for the following periods.

In general, consider an economic activity. It may be described as a dynamical system starting at a point in time, denoted as $t = 0$, and stopping at a point in time $t = n \in \mathbb{N}$. The interval $[t-1, t]$ denotes the t-th period or period t, that is, the interval between the two points in time $t-1$ and t. Time $t-1$ is the *beginning of the period* (BOP) and time t is the *end of the period* (EOP).

Let C_{t-1} be the BOP capital at time $t-1$, that is, the capital invested at the beginning of the t-th period, $[t-1, t]$. It represents the state of the system at the beginning of the t-th period. Also, let I_t be the income generated by the project at time t, and let F_t be the cash flow distributed or raised at time t by the firm. Income and cash flow are forces that act upon the capital in the following way:

$$C_t = C_{t-1} + I_t - F_t, \quad t \in \mathbb{N}_n^0 \tag{1.1}$$

where $C_{-1} = 0$ and $\mathbb{N}_n^0 = \{0, 1, \ldots, n\}$ denotes the natural numbers from 0 to n (more generally, $\mathbb{N}_k^j = \{j, j+1, j+2, \ldots, k\}$ denotes the set of natural numbers between j and k). At time 0, $C_0 - C_{-1} = C_0 = I_0 - F_0$ and, if $I_0 = 0$, then $C_0 = -F_0$.[1] Equation (1.1) is a recursive intertertemporal relation. Mathematically, it is a first-order one-dimensional autonomous discrete dynamical system. We call it the ***law of motion*** of an economic entity.[2]

Note that every date t separates two consecutive periods, the t-th period (interval $[t-1, t]$) and the $t+1$-th period (interval $[t, t+1]$). Therefore, time t may be viewed as the end of the t-th period or as the beginning of the $t+1$-th period.

We denote as E_t the capital at the end of period t (EOP capital), $t \in \mathbb{N}_n^0$, as opposed to C_t, the capital at the beginning of period $t+1$ (BOP capital). For every fixed date t, the BOP capital C_t is equal to the sum of the BOP capital at time $t-1$ and the income at time t:

$$E_t = C_{t-1} + I_t. \tag{1.2}$$

Alternatively, C_t is defined as the EOP capital, E_t, netted out of time-t cash flow, F_t:

$$C_t = E_t - F_t \tag{1.3}$$

(see Fig. 1.2).

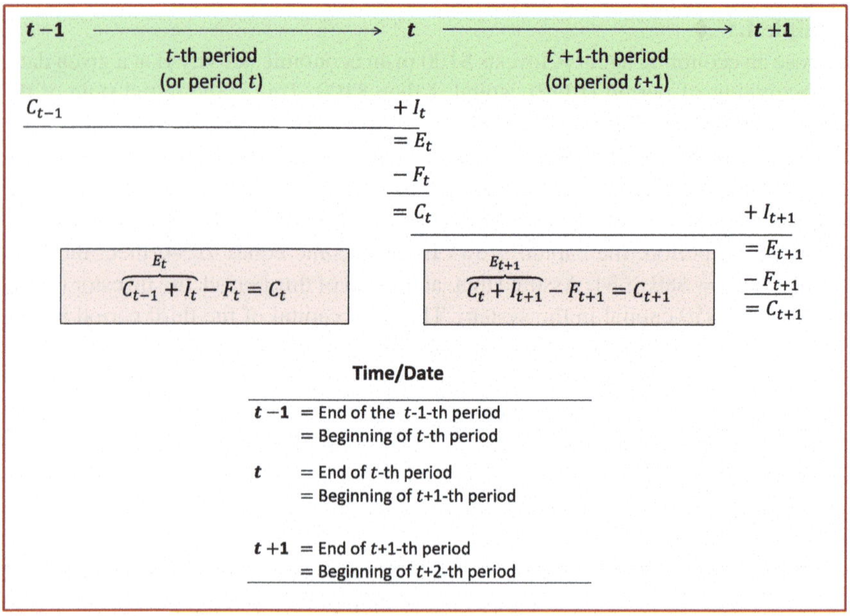

Fig. 1.2 Periods, times, and the law of motion

[1] In Example 1.1, we have

$C_0 = 100$	$C_1 = 80$	$C_2 = 134$
$I_0 = 0$	$I_1 = 10$	$I_2 = 4$
$F_0 = -100$	$F_1 = 30$	$F_2 = -50$

[2] We will refer to an economic "system", "unit", "entity", or "activity" interchangeably.

One can then define the EOP capital in two ways:

- sum of C_{t-1} (BOP capital at time $t-1$) and I_t (income generated at time t),
- sum of C_t (BOP capital at time t) and F_t (cash flow at time t).

In symbols,

$$\overbrace{C_{t-1}}^{\text{BOP capital at time } t-1} + \overbrace{I_t}^{\text{Income at time } t} = \overbrace{E_t}^{\text{EOP capital at time } t} = \overbrace{C_t}^{\text{BOP capital at time } t} + \overbrace{F_t}^{\text{Cash flow at time } t} \tag{1.4}$$

with $E_0 = C_{-1} + I_0 = I_0$. Put it differently, the income is the difference between EOP capital at time t and BOP capital at time $t-1$: $I_t = E_t - C_{t-1}$.

Highlighting the change in BOP capital, one may write

$$\Delta C_t = I_t - F_t \tag{1.5}$$

where $\Delta C_t = C_t - C_{t-1}$ denotes the change in BOP capital between time $t-1$ and time t. The amount $C_{t-1} - C_t = -\Delta C_t = F_t - I_t$ is the so-called *capital depreciation*. A capital depreciates by $-\Delta C_t$ in period t, $t = 1, 2, \ldots, n$. If $\Delta C_t < 0$, then capital increases in period t and $-\Delta C_t$ represents a capital *appreciation*.

Equation (1.5) signifies that the evolution of the capital depends on two key parameters: The income flow (i.e., capital growth within the period) and cash flow (i.e, the monetary amount injected in, or subtracted from, the economic entity at the end of the period). The above identity says that the change in the capital stock equates the net flow. Figure 1.3 portraits the capital system as a reservoir whose level increases by the (assumed positive) income I_t and decreases by the (assumed positive) cash flow F_t.

Remark 1.1 (*Terminology*) Throughout the book, the word "time" refers to a date, a point in the time interval $(-\infty, +\infty)$. Time $t = 0$ denotes the current time, $t > 0$ refers to future dates, $t < 0$ refers to past dates. For example, time $t = -1$ denotes one unit of time ago. If the unit of time is a year it means one year ago.

We will always use discrete time, so cash flows (and incomes) occur at time t, where t is an integer, $t \in \mathbb{Z}$. Mathematically, this is not inappropriate, because if any cash flow occurs between date $t-1$ and date t, then it is possible to change the unit of time in such a way that every monetary amount refers to a given date (and the dates are equidistant). As a result, the intertemporal relation (1.1) holds *sensu stricto*.

Practically, it is usual for the analyst to first select a unit of time that may fit the aim and purpose of the analysis (a year, a quarter, a month, a day, etc., depending on the type of transaction) and then assume that the cash flow and the income generated within a given interval occur at the end of that interval.[3] This simplifies the analysis

[3]For example, suppose a transaction consists of three cash flows available now, in five months, and in one year, respectively. The analyst has two options: (i) choose the year as a unit of time (and, therefore, shift the interim cash flow at the end of the year, so that the transaction is described as a two-date operation), or (ii) choose the month as a unit of time and describe the transaction as an operation with ten equidistant cash flows: Three nonzero cash flows at times 0, 5, and 12, and ten cash flows equal to zero at the remaining dates. The choice depends on the degree of approximation one is willing to work with.

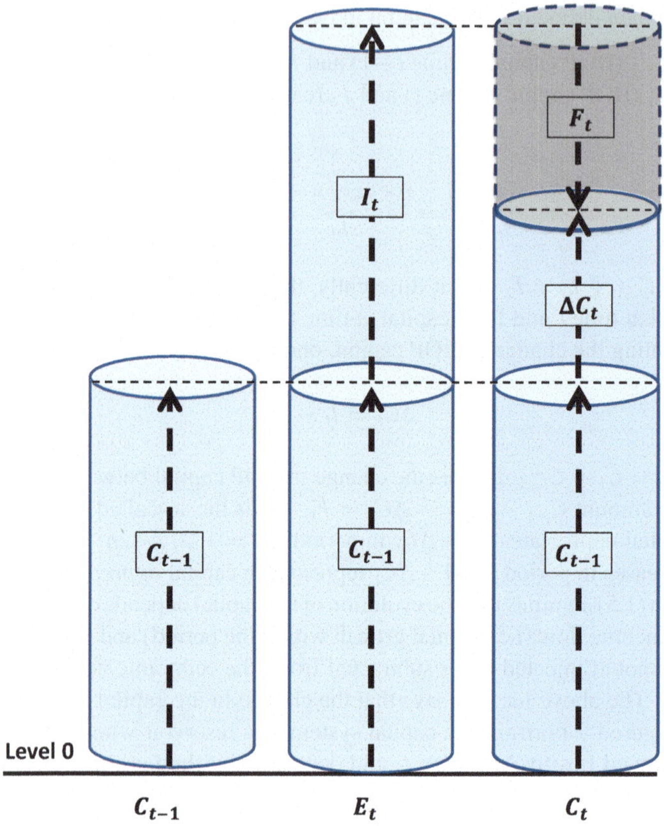

Fig. 1.3 Capital reservoir and the change in level. The BOP capital is increased by the income to get the EOP capital, which is decreased by the cash flow to get the BOP capital of the next period. Notice that the EOP reservoir level, E_t, may be calculated from the left reservoir level, C_{t-1} (by adding I_t) or from the right reservoir level, C_t (by adding F_t). The change in consecutive BOP capitals is $\Delta C_t = I_t - F_t > 0$ which means that the invested capital increases from time $t-1$ to time t

from a practical point of view. Typically, in the financial modelling of real-life applications, a year is often selected as the unit of time for corporate projects (a quarter of year or a month or even a day is used for short-term financial investments).

In tables, we will sometimes use a year-column. A number in that column denotes a point in time, not an interval. ▲

Example 1.2 ♦

Let $C_3 = 100$, $I_4 = 30$, $F_4 = 50$. Let us compute both the EOP capital and the BOP capital at time 4.

The EOP capital at time 4 is $E_4 = 100 + 30 = 130$, while the BOP capital at time 4 is $C_4 = 130 - 50 = 80$. The difference between EOP capital and BOP capital is equal to the cash flow: $E_4 - C_4 = 130 - 80 = 50 = F_4$.

Example 1.3 ♦

An economic agent (Mr. EcAg) injects \$100 in an economic unit. The capital grows by \$10 in the first period; at the end of the first period, the agent withdraws \$20. In the second period, the capital grows by \$30; at the end of the second period, the agent deposits additional \$40. In the third period, the capital decreases by \$50; at the end of that period, the agent withdraws the entire capital and closes off the investment. Table 1.1 describes the evolution of the system. For example, let us consider row 3 of the table, which refers to time $t = 2$. Starting from the capital invested at the beginning of the second period ($C_1 = \$90$), one adds the first force acting on the system, namely, the income ($I_2 = \$30$), so getting the EOP capital at time 2 ($E_2 = \$120$). Then, one subtracts the second force acting on the system, namely, the cash flow ($F_2 = -\$40$) and gets the BOP capital at time $t = 2$ ($C_2 = \$160$).

At time 0, just before income and cash flow act upon the system, the latter is at a zero level. When, at that time, income and cash flow trigger the motion, the system makes a jump equal to $\Delta C_0 = I_0 - F_0$. If $I_0 = 0$, then $\Delta C_0 = -F_0 = C_0$. This implies that the first cash flow is the first capital changed in sign (as in Example 1.1). In many projects, $I_0 = 0$. Unless otherwise stated, we assume that $I_0 = 0$. However, there are important cases where $I_0 \neq 0$ (see Sect. 1.4). At time n, the BOP capital is necessarily zero, $C_n = 0$ (the transaction is over, no capital is put at work at the beginning of the $n+1$-th period). This also implies that the terminal value of an economic system, at time n, is equal to the last cash flow that flows in or out of the system: $E_n = 0 + F_n = F_n$.[4]

Table 1.1 Dynamics of Mr. EcAg's economic unit

t	C_{t-1}	$+I_t$	=	E_t	$-F_t$	=	C_t
0	\$0	+\$0	=	\$0	+\$100	=	\$100
1	\$100	+\$10	=	\$110	−\$20	=	\$90
2	\$90	+\$30	=	\$120	+\$40	=	\$160
3	\$160	−\$50	=	\$110	−\$110	=	\$0

t	C_{t-1}	$+I_t$	=	E_t	=	F_t	$+C_t$
0	\$0	+\$0	=	\$0	=	−\$100	+\$100
1	\$100	+\$10	=	\$110	=	\$20	+\$90
2	\$90	+\$30	=	\$120	=	−\$40	+\$160
3	\$160	−\$50	=	\$110	=	\$110	+\$0

---

[4]Using the fundamental relation (1.1), one may write

$$C_t = \sum_{k=0}^{t} I_k - \sum_{k=0}^{t} F_k. \tag{1.6}$$

Therefore, the law of motion (1.1) describes a dynamical system which is at rest at a zero level, then jumps up or down at time 0 to reach level C_0, then moves up and/or down for n periods until it finally stops at time n, when it reaches the zero level again and keeps at rest in that position permanently.

We will often use the symbols $\boldsymbol{C} = (C_0, C_1, \ldots, C_{n-1}, 0)$, $\boldsymbol{F} = (F_0, F_1, \ldots, F_n)$, and $\boldsymbol{I} = (I_0, I_1, I_2, \ldots, I_n)$ to denote the vector of capital amounts, cash flows, and incomes. The triplet $(\boldsymbol{C}, \boldsymbol{I}, \boldsymbol{F})$ fully describes the dynamics of the system.

The dynamical system implied by the recursive relation (1.1) is not specific to an economic activity; it is a general-but-simple description that accounts for many different phenomena where the state of the system changes owing to two different kinds of forces:

i. [*Demography*] a population increases (decreases) by the life births (deaths) and decreases (increases) by the emigrants (immigrants)
ii. [*Geology*] the water level of a lake increases (decreases) owing to precipitation (evaporation) and decreases (increases) owing to a higher (lower) atmospheric pressure
iii. [*Chemistry*] the volume of a mass of gas increases (decreases) with a higher (lower) temperature and decreases (increases) with a greater (smaller) pressure
iv. [*Economics*] the amount of capital increases (decreases) by the income and decreases (increases) by the cash flows.

Phenomena (i)–(iv) share the same logical structure: A given amount of a variable increases or decreases as a consequence of two individual effects: Life births and emigrants, precipitation and pressure, temperature and pressure, income and cash flow, respectively. Therefore, Eq. (1.1) enables grouping, conceptually and formally, phenomena that would otherwise seem unrelated.

However, (i)–(iii) differs from (iv) in one respect: Each of them describe one *specific* phenomenon of the respective field: Demography, geology, chemistry. In contrast, (iv) describes *any* conceivable phenomenon in economics. In other words, the structure of (1.1) subsumes *every* existing or yet-to-be economic transaction. It then represents a general law in economics.

Therefore, the BOP capital can be viewed as the accumulated difference of incomes and cash flows. For $t = n$, Eq. (1.6) becomes

$$0 = \sum_{k=0}^{n} I_k - \sum_{k=0}^{n} F_k \implies \sum_{k=0}^{n} I_k = \sum_{k=0}^{n} F_k. \tag{1.7}$$

This means that the total income is equal to the net cash flow. In overall terms, income and cash flow coincide: They are the same overall amount which is distributed differently across the various periods. For example, if the income vector is equal to $\boldsymbol{I} = (10, 30, -20, 50, 70)$, then the overall net cash flow is necessarily equal to 140. This result will turn to be important in developing the notion of *internal average rate of return* in Chap. 10.

Thus, the terms "capital", "income" and "cash flow" should be intended as most generic terms that may take on different labels depending on the underlying economic milieu and the specific economic transaction that takes place. In particular, consider the following specific economic activities, where different terminologies are used. Note that their logical structure is the same; in each of these situations, the fundamental law of motion is fulfilled:

Bank account The 'account balance' increases (decreases) by the 'interest' and decreases (increases) by the 'withdrawal' ('deposit').

Loan The 'principal outstanding' (or 'residual debt') increases by the 'interest' and decreases by the 'instalment'.

Firm The 'assets' increase by the 'profit' (or 'income') and decrease by the 'cash flow from assets'. The 'liabilities' increase by the 'interest' and decrease by the 'cash flow to debtholders'. The 'equity' increases (or decreases) by 'net income' ('earnings') and decreases by the 'cash flow to equityholders'

Security The 'price' or 'value' of a security (stock, bond, etc.) increases by the 'price increase' ('capital gain') and decrease by the 'distribution' ('dividend', 'coupon', etc.).

Project The 'invested capital' increases or decreases by the project 'return' or 'income' and decreases (increases) by the 'inflow' ('outflow').

Fund The 'net asset value' increases or decreases by the fund's 'return' and decreases (increases) by a 'distribution' from ('contribution' into) the fund.

Public debt The 'total deficit' increases or decreases by the 'interest expenses' and decreases (increases) by the 'primary surplus' ('primary deficit').[5]

It should be clear now that the above economic units refer to different economic domains, yet they are logically and conceptually equivalent. They share the following structure:

> For any given period, the capital stock increases or decreases owing to a given **flow of income** and increases or decreases owing to a given **flow of cash**. The change in stock equates the difference between the two flows.

Symmetrically, the same set of monetary amounts can be generated by different transactions, which can be conceptualized under the same theoretical framework presented in Eq. (1.1). For example, consider the case where $C_0 = 500$, $C_1 = 360$, $I_1 = 100$, $F_1 = 240$. Following are different transactions that fit these values:

Bank account Mrs. A deposits \$500 in a bank account B, whose interest rate is 20%. The interest in the period is then \$100. At the end of the period, Mrs. A withdraws \$240 from the account. At time 1, the account BOP balance is \$500 + \$100 − \$240 = \$360.

[5]The primary surplus/deficit is the difference between tax revenues and government spending.

Loan Bank A lends \$500 to company B at an interest rate of 20%. The first instalment is due after one period and amounts to \$240. At time 1, the principal outstanding is \$500 + \$100 − \$240 = \$360.

Firm The book value of firm A's assets is \$500. The operating profit is \$100. At the end of the period, firm A distributes \$240 to claimholders (capital providers). At time 1, the book value of assets is \$500 + \$100 − \$240 = \$360.

Security Mr. A purchases 100 bonds of firm B. The bond's price is \$5. At the end of the period, the price increase is \$1 and the coupon paid on each bond is \$2.4. At time 1, the investment value of Mr. A is \$500 + \$100 − \$240 = \$360.

Project Company A invests \$500 in a capital asset investment B. The investment generates a return of \$100 in the first period. The firm distributes \$240 to the capital providers. The capital that remains invested in the project is \$500 + \$100 − \$240 = \$360.

Fund The market value of a private equity investment A undertaken by agent B is \$500. At the end of the period, the value increases by \$100. Agent B decides to withdraw \$240 from investment A. The market value of A at the beginning of the next period is \$500 + \$100 − \$240 = \$360

Public debt The public debt of Government A is \$500. The interest rate on debt is 20%. Therefore, the outstanding debt increases by the deficit, which is equal to the interest expenses (\$100) minus the primary surplus (\$240); the latter is paid to the class B of holders of the government debt. Therefore, the public debt outstanding at the beginning of the next period is \$500 + \$100 − \$240 = \$360.

It is observed that the 7 economic units described above do not necessarily refer to disjoint sets of transaction. For example, a *bond* is contractually evidenced by a security which underlies a *loan*. In a project financing transaction a *firm* is incorporated for the sole purpose of undertaking a *project* (i.e., firm = project). A financial portfolio or a *fund* consist of a bundle of *securities* (fund = set of securities). The equity share of a *firm* listed in the stock market is itself a *security* (firm's share = security). Public debt consists of *bonds* (as well as T-bills and notes, themselves special cases of collective *loans*).

1.2 The Income Rate and the Time Value of Money

In this section we show that the law of motion logically entails the principle of *time value of money*. To this end, we need to first derive a fourth basic notion from the previous elements: The income rate.

As seen, the income, I_t, represents the capital growth occurred between the beginning and the end of a period, *before* any cash movement; that is, it expresses the difference between EOP capital and BOP capital. This capital appreciation/depreciation is an *absolute* measure of growth. A *relative* measure of growth is obtained by dividing the income by the BOP capital, thereby getting the income per unit of committed capital:

$$i_t = \frac{I_t}{C_{t-1}} \tag{1.8}$$

for $t \in \mathbb{N}_n^1$ or, equivalently,

$$i_t = \frac{C_t + F_t - C_{t-1}}{C_{t-1}} = \frac{E_t - C_{t-1}}{C_{t-1}} \tag{1.9}$$

where $C_{t-1} \neq 0$ is assumed. The income rate, i_t, is a *flux*: It represents the rate at which the capital grows (or decreases) period by period. Depending on the context, it is named in various ways, including the expressions ***rate of return*** or *holding period rate*.

In general, there is a plethora of different labels that are used by scholars and practitioners to denote the same concept. Symmetrically, the same label may be used to denote different concepts, depending on the domain and the perspective adopted.[6] This causes semantic ambiguities, not merely between the scholarly community and the business community, but also among scholars and among practitioners themselves. To make matters worse, the use of several kinds of adjectives which are often associated with the nouns increases the degree of confusion, thereby obscuring the conceptual and formal relations among the four basic elements of a given economic activity, to such an extent that the relevance of Eq. (1.1) as a building block of economics is concealed. Table 1.2 collects some expressions for the four basic notions that are variously used in practice and/or found in textbooks of finance, accounting, and engineering.[7]

Noting that $I_t = i_t C_{t-1}$, the law of motion (1.1) can be reframed as

$$C_t = C_{t-1}(1 + i_t) - F_t \quad \text{or, proceeding backwards,} \quad C_{t-1} = \frac{C_t + F_t}{1 + i_t} \quad t \in \mathbb{N}_n^1 \tag{1.10}$$

Hence, $C_0 = -F_0$ and, iterating,

[6] For example, 'revenues' may be used to denote income or cash flows or even a firm's sales; 'return' may be used to denote an income or, rather, an income rate; 'income' itself is sometimes used to denote an inflow payment instead of the remuneration of capital.

[7] The highly specialized jargons of these disciplines may even impede exchanges and communications among scholars of different fields. Paradoxically, some notions, results, and approaches may be well-known in one field long since while they may appear as novel and original in another field (or just known and neglected).

Table 1.2 Synonymous expressions for the basic elements of an economic system

Capital	wealth, value (book value, market value, historic value, fair value, net asset value, present value), equity, assets (financial assets, operating assets, current assets, liquid assets etc.), principal, residual debt, outstanding debt, stock, resources, net worth, outstanding capital, balance (outstanding balance, project balance, bank balance, account balance, etc.), carrying amount, total deficit
Income	(operating income, non-operating income, net income), profit (gross profit, operating profit, net profit), interest, interest on capital, interest income, interest expenses, return, compensation, remuneration, reward, gain/loss, capital gain/loss, earnings (before/after taxes, before interest and taxes, etc.), revenue, appreciation/depreciation, yield, service, margin
Cash flow	instalment, flow, equity cash flow, free cash flow, capital cash flow, project cash flow, dividend, payout, inflow/outflow, income/outcome, contribution/distribution, deposit/withdrawal, investment/disinvestment, payment, disbursement, drawdown, benefit/disbenefit, benefit/cost, cash outlay/receipt, revenue, primary surplus/deficit, service
Income rate	interest rate, return, rate of return, return on equity (ROE), return on assets (ROA), return on net assets (RONA), return on investment (ROI), return on capital (ROC), return on invested capital (ROIC), return on debt (ROD), holding period rate, growth rate, appreciation rate, cost of capital, benchmark return, minimum attractive rate of return, minimum acceptable rate of return, yield, effective rate

$$C_1 = -F_0(1+i_1) - F_1$$
$$C_2 = -F_0(1+i_1)(1+i_2) - F_1(1+i_2) - F_2$$
$$C_3 = -F_0(1+i_1)(1+i_2)(1+i_3) - F_1(1+i_2)(1+i_3) - F_2(1+i_3) - F_3$$

and so on. For a generic date t, the following *retrospective* relation holds:

$$C_t = -\sum_{k=0}^{t} F_k \cdot (1+i_{k+1})(1+i_{k+2}) \cdot \ldots \cdot (1+i_t). \tag{1.11}$$

The above equation says that the amount of capital committed at time t is equal to the accumulated cash flows, changed in sign.[8] Hence, picking $t = n$ and recalling that $C_0 = -F_0$ and $C_n = 0$,

[8]If $I_0 \neq 0$, implying $C_0 = I_0 - F_0 \neq -F_0$, then the relation becomes

$$C_t = I_0(1+i_1)(1+i_2) \cdot \ldots \cdot (1+i_t) - \sum_{k=0}^{t} F_k(1+i_{k+1})(1+i_{k+2}) \cdot \ldots \cdot (1+i_n). \tag{1.12}$$

Considering that $C_n = 0$, the above equation implies

$$(I_0 - F_0)(1+i_1)(1+i_2) \cdot \ldots \cdot (1+i_n) = \sum_{k=1}^{n} F_k(1+i_{k+1})(1+i_{k+2}) \cdot \ldots \cdot (1+i_t).$$

As $\Delta C_0 = C_0 = I_0 - F_0$, the latter is equivalent to (1.13).

$$C_0 \cdot (1+i_1)(1+i_2) \cdot \ldots \cdot (1+i_n) = \sum_{k=1}^{n} F_k \cdot (1+i_{k+1})(1+i_{k+2}) \cdot \ldots \cdot (1+i_n). \tag{1.13}$$

Alternatively, iterating backward,

$$\begin{aligned}
C_n &= 0 \\
C_{n-1} &= \frac{0+F_n}{1+i_n} \\
C_{n-2} &= \frac{C_{n-1}+F_{n-1}}{1+i_{n-1}} = \frac{F_{n-1}}{1+i_{n-1}} + \frac{F_n}{(1+i_{n-1}) \cdot (1+i_n)} \\
C_{n-3} &= \frac{C_{n-2}+F_{n-2}}{1+i_{n-2}} = \frac{F_{n-2}}{1+i_{n-2}} + \frac{F_{n-1}}{(1+i_{n-2}) \cdot (1+i_{n-1})} \\
&\quad + \frac{F_n}{(1+i_{n-2}) \cdot (1+i_{n-1}) \cdot (1+i_n)}
\end{aligned}$$

and so on;[9] for a generic date t, the following *prospective* relation holds:[10]

$$C_t = \sum_{k=t+1}^{n} \frac{F_k}{(1+i_{t+1})(1+i_{t+2}) \cdot \ldots \cdot (1+i_k)}. \tag{1.14}$$

Hence, picking $t = 0$,

$$C_0 = \sum_{k=1}^{n} \frac{F_k}{(1+i_1)(1+i_2) \cdot \ldots \cdot (1+i_k)}. \tag{1.15}$$

The above equation expresses the ***time value of money***, a basic tenet in finance and economics.[11] It states that the capital value of any economic unit is equal to the sum of the cash flows discounted at the return rates $i_1, i_2, \ldots, i_n$. Note that this tenet is a straightforward implication of the law of motion presented in (1.1).

If one assumes that the growth rate for capital is constant, $i_t = i$, Eq. (1.1) becomes

$$C_t = C_{t-1}(1+i) - F_t \quad \text{or, proceeding backwards,} \quad C_{t-1} = \frac{C_t + F_t}{1+i} \tag{1.16}$$

---

[9]We may also frame the equation in terms of ending capitals: $C_t = E_{t+1}/(1+i_{t+1})$.

[10]The adjectives *retrospective* and *prospective* refer to time t as the reference date. Therefore, in the retrospective relation, cash flows prior to time t are *compounded* (i.e., ' moved' forward) to time t; in the prospective relation, cash flows after time t are *discounted* (i.e., 'moved' backward) to time t.

[11]If $I_0 \neq 0$, then (1.15) becomes

$$I_0 - F_0 = \sum_{k=1}^{n} \frac{F_k}{(1+i_1)(1+i_2) \cdot \ldots \cdot (1+i_k)}.$$

which implies, using the initial condition $C_0 = -F_0$,

$$C_t = -\sum_{k=0}^{t} F_k(1+i)^{t-k} \tag{1.17}$$

(retrospective relation) or, proceeding backwards,

$$C_t = \sum_{k=t+1}^{n} \frac{F_k}{(1+i)^{k-t}} \tag{1.18}$$

(prospective relation).

The set of equations from (1.10) to (1.18) show that the *time value of money* is nothing but an iteration of the definition of rate of return. Finance scholars often illustrates it in reference to financial assets, but it is, in fact, a general principle which is valid for any economic entity, be it a project, a firm, a loan, a security, a savings-and-credit account, etc. Any economic unit fulfills the time value of money, which connects the unit's cash flows, the unit's income rates, and the unit's capitals.

Remark 1.2 (*Terminology*) The expression "income rate" is more often replaced in finance by the expression "rate of return", and the expression "income" is often replaced by "return". Unfortunately, the expressions "return" and "rate of return" are often used interchangeably. For example, in investment performance measurement, "return" often means "*rate* of return". In corporate finance and management science, the well-known label "return on investment" refers to a *rate* of return: It stands for "return *divided by* investment". The distinction is often apparent with the context, but it is important to always bear in mind the difference between an absolute measure (money) and a relative one (rate). The symbols i_t and I_t will be used throughout the book to help distinguish the two concepts. ▲

Example 1.4 ♦

Firm Alpha undertakes a project whose cash-flow stream is $\boldsymbol{F} = (-33, 30, -10, 20)$. Assuming the capital invested grows at a constant rate i, then i is necessarily equal to 12.1%. This is confirmed by the following equation:

$$33 = \frac{30}{(1+0.121)} + \frac{-10}{(1+0.121)^2} + \frac{20}{(1+0.121)^3}$$

which is Eq. (1.18) for $t = 0$. The capital values can be found by applying Eq. (1.16),

$$\begin{aligned} C_0 &= 33 \\ C_1 &= 33(1.121) - 30 = 6.99 \\ C_2 &= 6.99(1.121) + 10 = 17.8 \\ C_3 &= 17.8(1.121) - 20 = 0 \end{aligned}$$

or

$$
\begin{aligned}
C_3 &= 0 \\
C_2 &= \frac{0+20}{1.121} = 17.8 \\
C_1 &= \frac{17.8-10}{1.121} = 6.99 \\
C_0 &= \frac{6.99+30}{1.121} = 33.
\end{aligned}
$$

Example 1.5 ♦
Company A is incorporated at time 0. The capital is estimated to grow at time-varying income rates: $i_1 = 5\%$, $i_2 = 10\%$, $i_3 = -2\%$. The cash flow distributed to capital providers are estimated at $F_1 = 30$, $F_2 = -10$, $F_3 = 90.65$. At time 3, the firm will terminate operations. From these figures and using Eq. (1.15), it is straightforward to compute the initial capital raised by the firm from the claimholders:

$$\frac{30}{1.05} + \frac{-10}{1.05 \cdot 1.1} + \frac{90.65}{1.05 \cdot 1.1 \cdot 0.98} = 100.$$

The cash-flow stream is then $\boldsymbol{F} = (-100, 30, -10, 90.65)$.

The dynamical system (1.1) gives expression to the diachronic aspect of an economic system. However, the same aspect may be represented graphically by using a (t, C_t)-plane and a piecewise function. The graph of the function consists of n linear segments (one for each period) which describe the within-period capital growth (the slope of the segments being equal to the income I_t). At every integer date t, a discontinuity occurs (as long as $F_t \neq 0$), which represents an outflow from the system ($F_t > 0$) or an inflow into the system ($F_t < 0$).[12]

Figure 1.4 illustrates the evolution of a system from time 0 to time 9, the date at which the system stops. Using the triplet $(\boldsymbol{C}, \boldsymbol{I}, \boldsymbol{F})$ to describe the system,

Time	0	1	2	3	4	5	6	7	8	9
$\boldsymbol{C}$	16	14	8	3	14	7	4	12	1	0
$\boldsymbol{I}$	0	2	1	−3	8	0	6	−1	−8	1
$\boldsymbol{F}$	−16	4	7	2	−3	7	9	−9	3	2

The segments with a positive (negative) slope describe the periods where the income is positive (negative). The subtraction of capital from the system ($F_t > 0$) are graphically signalled by downward jumps, while the capital injections into the system ($F_t < 0$) are characterized by upwards jumps. For example, in the third period (from

---

[12]As will become apparent later, an outflow from the firm/project represents an inflow for the investor, whereas an inflow into the firm/project represents an outflow for the investor.

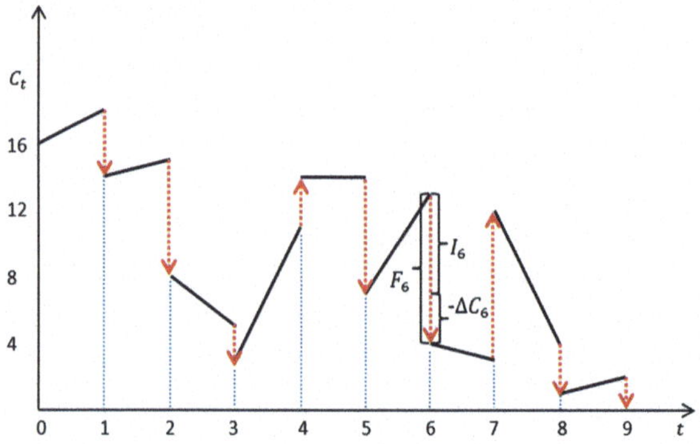

Fig. 1.4 Graphical representation of the dynamical system (example of a 9-year system)

time 2 to time 3) the income is negative. Specifically, a loss of \$3 occurs (from a BOP capital equal to \$8 to an EOP capital equal to \$5). An inflow is received by the investor equal to \$2, so the invested capital at the beginning of the fourth period is equal to \$5 − \$2 = \$3.

Analytically, considering that $C_{-1} = 0$ and assuming $I_0 = 0$,

$$C_0 = 0 + 0 - (-16) = 16$$
$$C_1 = 16 + 2 - 4 = 14$$
$$C_2 = 14 + 1 - 7 = 8$$
$$C_3 = 8 - 3 - 2 = 3$$
$$C_4 = 3 + 8 - (-3) = 14$$
$$C_5 = 14 + 0 - 7 = 7$$
$$C_6 = 7 + 6 - 9 = 4$$
$$C_7 = 4 - 1 - (-9) = 12$$
$$C_8 = 12 - 8 - 3 = 1$$
$$C_9 = 1 + 1 - 2 = 0.$$

Note the upward or downward jump generated by the cash flow can be decomposed into two components representing the income effect and the (change in) capital effect:

$$F_t = I_t + (-\Delta C_t). \tag{1.19}$$

The above relation reframes the law of motion and describes cash flow as a function of income and capital. For example, if $I_6 = 6$ and $\Delta C_6 = -3$, then $F_6 = I_6 - \Delta C_6 = 6 + 3 = 9$. As we will see, this reframing plays a major role in the evaluation of capital asset investments (see Chap. 3).

1.3 Investment *or* Financing

As seen, the capital is the amount of the economic resources committed in the economic activity. The capital can be either *injected* into an economic activity or *subtracted* from an economic activity. Whether it is injected or subtracted depends on the perspective: An economic activity is a transaction (or a set of transactions) involving two parties, say agent A and agent B, such that A injects/invests capital in B while B subtracts/absorbs capital from A (see also Chap. 2). Injecting economic resources or subtracting economic resources is financially equivalent to lending money or borrowing money, respectively. It is *as if* every single economic activity were a transaction made by a lender and a borrower. From this point of view, the expressions 'to invest', 'to inject', 'to use', 'to infuse', 'to lend' are, conceptually, synonymous expressions, as well as the expressions 'to subtract', 'to absorb', 'to raise', 'to borrow'. The former points to a lending position, the latter points to a financing position.[13]

Therefore, at a given date t, agent A is either a lender or a borrower, a user of funds or a raiser of funds. (Agent B takes on the symmetric position of either borrower or lender). This is valid for any economic activity, irrespective of the terminology that may be used.

Table 1.3 details the conceptual equivalence of the 7 economic units introduced in the previous section, highlighting the four basic notions and the position of agents A and B.[14]

One can algebraically distinguish investment and financing, as well as increase or decrease of investment and financing, by checking the sign of C_t, I_t and F_t. Let A and B be two opposite parties:

- a positive capital ($C_t > 0$) for A means that A is *investing* resources in B; a negative capital ($C_t < 0$) for A means that A is *borrowing* resources from B
- a positive income ($I_t > 0$) for A means that the capital invested increases (if $C_{t-1} > 0$) or the capital borrowed decreases (if $C_{t-1} < 0$) by the amount I_t; a negative income ($I_t < 0$) means that capital invested decreases (if $C_{t-1} > 0$) or the capital borrowed increases (if $C_{t-1} < 0$) by the amount $|I_t|$
- a positive cash flow at time t ($F_t > 0$) for A means that a monetary amount is withdrawn from B and distributed to A, thereby decreasing the capital invested in B (if $C_{t-1} > 0$) or increasing the capital borrowed from B (if $C_{t-1} < 0$); a negative cash flow ($F_t < 0$) for A means that monetary amounts are contributed by A and distributed to B, thereby increasing the capital invested in B (if $C_{t-1} > 0$) or decreasing the capital borrowed from B (if $C_{t-1} < 0$).[15]

[13] In financial markets, the expressions 'long position' and 'short position' are also often employed to reflect the distinction between investment and financing (see Chap. 5).

[14] The reader is invited to complete the table with other economic transactions.

[15] For example, dividends are inflows for shareholders and outflows for the firm; capital contributions are inflows for the firm and outflows for shareholders; a project's positive cash flow is an inflow for the firm and an outflow for the project; money injected in a fund by an investor is an inflow for the fund and an outflow for the investor.

Table 1.3 The two-party relation for different economic systems

	Loan	Bank account	Project	Firm	Fund	Security	Public debt
transaction	loan contract	account agreement	supply and sales agreements	company agreement	investment management agreement	purchase agreement	purchase contract
agent A	borrower	bank	asset/service providers and receivers	firm	fund manager	issuer	Government
agent B	lender	customer	firm	claimholders	investor	purchaser	investor
C_t	principal	account balance	capital	assets	net asset value	asset value	outstanding debt
I_t	interest	interest	return, profit, income	earnings, profit, income	return, profit	return	interest
F_t	instalment	deposit or withdrawal	inflow or outflow	cash flow from assets	contribution or distribution	dividend, coupon	primary surplus/deficit
i_t	interest rate	interest rate	rate of return	return on capital	return (rate)	rate of return	interest rate

Table 1.4 Investment rate versus financing rate

	$C_{t-1} > 0$	$C_{t-1} < 0$
$i_t > 0$	investment rate positive income	financing rate negative income
$i_t < 0$	investment rate negative income	financing rate positive income

Merging the two sources of variation of the capital, one may describe all the cases in a concise way as follows:

State	Capital invested $C_{t-1} > 0$		Capital borrowed $C_{t-1} < 0$	
$I_t - F_t$	> 0	< 0	> 0	< 0
Effect	increases (↑)	decreases (↓)	decreases (↓)	increases (↑)

(see also Figs. 1.3 and 1.5).

As for the income rate, its financial nature (investment/lending rate vs. financing/borrowing rate) depends on the financial nature of the capital: If the capital is positive, the holding period rate is an investment rate; if the capital is negative, the holding period rate is a financing rate. In particular, if an investment rate is negative ($i_t < 0$ and $C_{t-1} > 0$), the investor is lending money but part of it is lost; if a financing rate is negative ($i_t < 0$ and $C_{t-1} < 0$), the agent is borrowing from the system and is making money out of that borrowing (see Table 1.4).[16]

Consider a transaction where the sequence of capital amounts $\boldsymbol{C} = (C_0, C_1, \ldots, C_{n-1}, 0)$ is such that some amounts are positive and some other amounts are negative. That is, there exists some j, k such that $C_j \cdot C_k < 0$. In this case, the agent alternates lending and borrowing position; all considered, the economic entity may not be said to be an investment nor a financing; its financial nature alternates across periods. We may then supply the following definition.

[16]In recent years, the government bonds of some countries have been negative for investors. Among others, Germany Switzerland, Sweden, Denmark, and Japan have charged depositors instead of remunerating them. In these (rather extreme) situations, the borrower is remunerated for borrowing and the lender is penalized for lending. In corporate and engineering contexts, negative capitals with negative rates often occur whenever the firm retires from an activity or abandons a given course of action. This may conceptualized by saying that the firm borrows from the project and is remunerated for this borrowing (see Sect. 1.4).

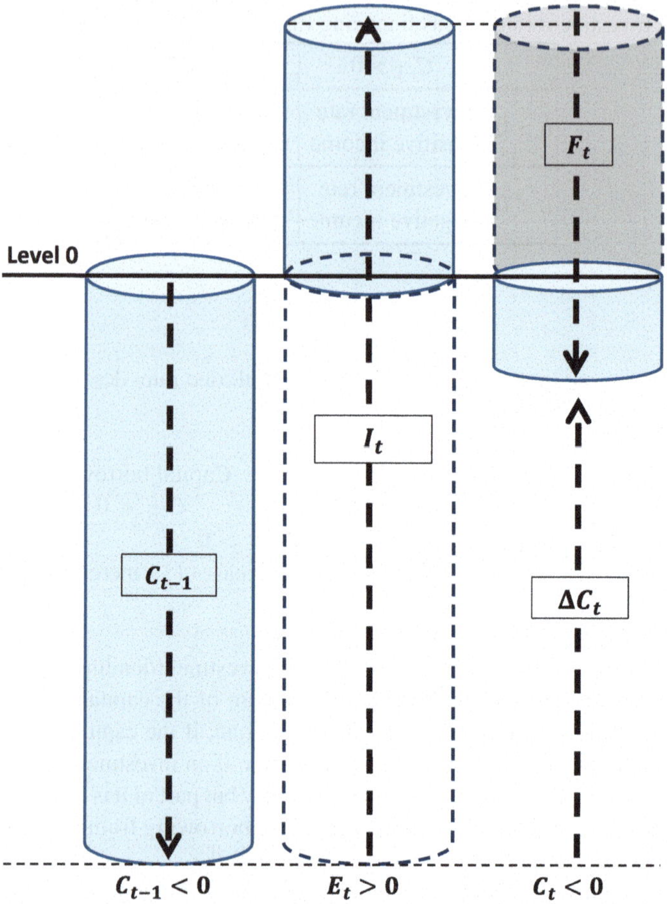

Fig. 1.5 Capital reservoir and the change in level. The BOP capital in $t-1$ is negative (financing position) and is increased by the income to such an extent that the EOP capital is positive. Cash flow is distributed to such an extent that the capital turns negative again. However, $\Delta C_t = I_t - F_t > 0$ and, therefore, the borrowed capital decreases

Definition 1.1 Consider a vector of capital amounts $\boldsymbol{C} = (C_0, C_1, \ldots, C_{n-1}, 0)$. If $C_t \geq 0$ for all t with $C_j > 0$ for some $j \in \mathbb{N}_n^0$, the economic system is a *pure investment transaction*; if $C_t \leq 0$ for all t with $C_j < 0$ for some $j \in \mathbb{N}_n^0$, the system is a *pure financing (or borrowing) transaction*; if $C_j \cdot C_k < 0$ for some $j, k \in \mathbb{N}_{n-1}^0$, the system is a *mixed transaction*.

In a mixed transaction, the financial nature of the income rate i_t changes across periods.

In terms of dynamics, if the system is above zero level (i.e., $C_t > 0$), the agent is investing (investment position), whereas if the system is below zero level (i.e., $C_t < 0$), the agent is financing (financing position). At a zero level, the agent is neither investing nor financing. The jump of the system is obtained as ΔC_t and is the effect of income and cash flow upon the system: $\Delta C_t = I_t - F_t$. If the state of the system is never below zero in the interval $[0, n]$, a pure investment occurs; if the state of the system is never above zero in the interval $[0, n]$ then a pure financing occurs.

Example 1.6 ♦
Example 1.1 describes a pure investment, that is, a case where A invests or 'lends' money in each period and the (positive) cash flows represent distributions to A (contributions from B). Alternatively, suppose agent A invests a capital amount equal to \$100 in an economic activity B and suppose that income, at the end of the first period, is \$10. This means that the EOP capital is \$110=\$100 + \$10. Suppose A withdraws \$130 from the economic activity. The capital decreases to such an extent that the BOP capital committed in B drops to −\$20 = \$110 − \$130. This means that, in the second period, A subtracts resources or 'borrows' \$20 from B. Now, suppose that the income in that period is −\$4. This means that the 'financed' amount has increased by \$4 and the EOP capital is −\$24 = −\$20 − \$4. Assume that, at that point, agent A injects an additional \$150 in B, thereby increasing the capital to such an extent that it grows to a positive \$126 = −\$24 + \$150. This means that, at the beginning of the third period, agent A invests or 'lends' \$126 to asset B. The income of the third period is, say, −\$19 and agent A terminates the operation by withdrawing the EOP value, equal to \$107 = \$126 − \$19. To sum up,

Time	Capital	Position	Income	Cash flow
0	100	investment		−100
1	−20	financing	10	130
2	126	investment	−4	−150
3	0		−19	107

This transaction is a mixed transaction and the holding period rates are $i_1 = 10/100 = 0.1$ (investment rate), $i_2 = -4/-20 = 0.2$ (financing rate), $i_3 = -19/126 = -0.151$ (investment rate). Note that, by Eq. (1.15),

$$100 = \frac{130}{1+0.1} + \frac{-150}{(1+0.1)(1+0.2)} + \frac{107}{(1+0.1)(1+0.2)(1-0.151)}.$$

Example 1.7 ♦
A bank offers a 'savings-and-credit' account that warrants a 1% interest whenever the account balance is positive and charges a 2% interest whenever the account

balance is negative. Suppose agent A injects \$150 in the account at time 0, withdraws \$160 at time 1, deposits \$50 at time 2 and withdraws the entire ending balance at time 3. The BOP account balances are $C_0 = 150$, $C_1 = 150(1.01) - 160 = -8.5$, $C_2 = -8.5(1.02) + 50 = 41.3$. At the end of the third period, the account balance is $41.3(1.01) = 41.7$. Below are the values for BOP capital, income, cash flow at each date:

Time	Capital Position	Interest	Cash flow
0	150 investment		−150
1	−8.5 financing	1.5	160
2	41.3 investment	−0.17	−50
3	0	0.413	41.7

This is a mixed transaction, alternating investment and borrowing positions. Also, by Eq. (1.15),

$$150 = \frac{160}{1+0.01} + \frac{-50}{(1+0.01)(1+0.02)} + \frac{41.7}{(1+0.01)^2(1+0.02)}.$$

Given that any economic entity can be viewed as a transaction between a lender and a borrower, the notion of *loan* deserves a privileged theoretical status. The reader may think of any economic unit as a loan where capital is the outstanding principal, income is the interest, and cash flow is the instalment.

Far from being a mere metaphor, the equivalence between any economic entity and the notion of loan is evidenced by the very term 'capital' which is historically connected with loans and, in general, with a two-party relation between a lender and a borrower. More precisely, the term 'capital' derives from the medieval latin expression 'capitalis pars' (from 'caput', head), which was referred to the principal sum of a money loan (Fetter 1937, p. 5). The term capital thus originated in a lending/borrowing context and was only later extended to include the

> monetary wares sold on credit, and still more generally the worth of any other credit (receivable) expressed in terms of money. The next inevitable expansion of the meaning of capital made it include the estimated value of a merchant's stock of goods and of agents (such as tools, shops, ships, land, etc.) employed in his business by himself as well as when loaned to another for an agreed interest or rental…All these were resources, or assets (to use a later terms) which might be sold for money, and which were thus alternative forms of business investment, the equivalents in their money's worth of a principal sum loaned at interest. (Fetter 1937, pp. 5–6).

Finally, the term was expanded to include the worth of any kind of economic activity (Fetter 1937, p. 6). For example, in a firm, the lending party is represented by two classes of investors, debtholders and shareholders. Indeed, equity represents the residual amount the company owes to shareholders: "The corporation *owes* the capital, it does not *own* it. The shareholders own it" (Fetter 1937, p. 9). (See also Cannan 1921 on the early history of this term). And since the use of interest as the remuneration of the lender is known since ancient times (Van de Mieroop 2005), it is then not wonder that the notions of income and interest may be assimilated, something that is acknowledged explicitly by some authors: "the profit is equal to interest on the capital value existing at the beginning of the period" (Hansen 1972, p. 15).

Even more compelling (and more general) is the interpretation of any economic system as a set of financial positions, determined by (1.1), potentially changing, period by period, from investment to financing, and vice-versa. This implies that any multiperiod economic entity may be conceived as if it were a ***savings-and-credit account***, that is, an account that can be used for a mixed transaction, one which includes both investing (savings) and borrowing (credit). The capital (invested or borrowed) of the economic activity under consideration corresponds to the account balance (positive or negative), the income (positive or negative) corresponds to the interest (earned or charged), and the inflow/outflow corresponds to the deposit/withdrawal.[17]

The metaphor of a savings-and-credit is useful for cognizing the dynamics of any economic system and, in particular, a capital asset investment, making the economic analysis of the project under consideration simple and intuitive.

Example 1.8 ♦

A firm is incorporated with \$100 equity. The net income in the first year is \$5; \$10 are distributed to shareholders at the end of the first period. In the second period, net income is \$38 and an additional \$25 is contributed by shareholders and infused in the firm at the end of the second period. The BOP capital at time 2 is

$$\$158 = \overbrace{100}^{\text{Contributed equity}} + \overbrace{5-10}^{\text{Change in equity at time 1}} + \overbrace{38+25.}^{\text{Change in equity at time 2}}$$

This scenario is financially equivalent to one where an investor deposits \$100 in a savings account whose interest rate in the first period is $5/100 = 5\%$, withdraws \$10 from the account at the end of the first period, so that the balance becomes \$95$= 105(1 + 0.05) - 10$. In the second period, the interest rate is $38/95 = 40\%$ and the investor additional injects \$25 in the account so that the account balance increases to \$158$= 95(1 + 0.4) + 25$.

Remark 1.3 (*Terminology*) In this book, we will use diverse terms such as return, income, profit, earnings interchangeably. Likewise, we will use terms such as profit rates, income rates, return rates interchangeably. The latter refer to the same conceptual entity: The relative growth of the capital, abstracting from any cash withdrawals or deposits in the economic system. Whether 10% is the interest rate on a loan or the increase in a share price or the ratio of net income to equity book value, etc. they all represent an increase of a unit of invested or borrowed capital. Whether the capital is derived from transactions in the security market or from a loan contract or from supply agreements or from purchases of a firm's shares, and whether the capital

[17]A loan is then a less general concept than savings-and-credit account.

is invested or borrowed, the engineering of the system is the same. In any case, an economic system is one where the income/profit/return rate expresses the relative growth of the system's capital.

Rather, a relevant distinction should be made between income and cash flow. If the BOP capital is $C_{t-1} = \$100$ and the EOP capital is $E_t = \$120$, the income rate is 20% regardless of the cash flow injected or subtracted from the system at the end of the period. The distributed or contributed cash flow does not affect an income/profit/return rate. Failing to grasp the distinction between profit and cash flow means failing to understand the concept of rate of return. Rate of return has to do with income and capital, not with cash flows.[18] ▲

Remark 1.4 The distinction among capital, income, and cash flow is important but, at the same time, slippery and tricky. On one hand, project appraisal (as well as firm appraisal) requires a clear-cut separation among the three notions, and, diachronically, between the economic position of the system (the capital) and the sources of change in the economic position (the income and the cash flow). This enables the analyst to keep logical consistency and avoid errors in the financial modelling. On the other hand, the three notions are crucially intertwined and chained to one another, so that any one may turn into the other one. A negative cash flow turns into capital as soon as it is injected in the economic system; a positive cash flow is a distribution of capital to investors; income is itself a 'surplus' capital.[19] Therefore, one dollar might be viewed as an income, a capital or a cash flow depending on the purpose, the perspective, and the moment when the analysis is carried out. The dynamical system of the capital resembles the flow of a river, with incomes and cash flows playing the part of tributaries and distributaries. Once some amount of water flows from the tributary into the river, it becomes water of the river; once some amount of water in the river flows into the distributary, it becomes water of the distributary. Equation (1.4) testifies to the oneness of the notion of capital: Income and cash flow are part of it. The triplet $(\boldsymbol{C}, \boldsymbol{I}, \boldsymbol{F})$ (capital, income, cash flow) is three in one. ▲

1.4 The Project as an Incremental System

While the analytical toolkit derived from the law of motion presented in (1.1) is valid for any economic activity, this book especially deals with projects. Roughly speaking, a project is often viewed as an economic system which is generated by a

[18]An income/profit/return rate is unambiguously defined as the ratio $i_t = I_t/C_{t-1}$. The ratio F_t/C_{t-1} is a rate, but not an income/profit/return rate. However, an income rate may well be expressed in terms of cash flow: From $i_t = I_t/C_{t-1}$ and considering that I_t is equal to $F_t + \Delta C_t$, i_t may be written as $i_t = (F_t + \Delta C_t)/C_{t-1} = (F_t + C_t - C_{t-1})/C_{t-1}$ (see Eq. (1.9)).

[19]Historically, income has been viewed as a 'surplus capital': "income is the surplus net worth (proprietor's interest) accruing during the period after maintaining opening capital (net worth) intact." (Whittington 2017, p. 43). Traces of it can still be found in the accounting expression "clean surplus relation" which is the name accountants and accounting scholars give to Eq. (1.1), where I_t is called "comprehensive income" (see, for example, Brief and Peasnell 1996).

firm whenever incremental funds are raised to undertake incremental investments. More precisely, a project is an economic system which increments (or decrements) the firm's income, capital, and cash flow with respect to the ***null alternative*** or ***status quo*** (i.e., the state where the existing operations remain unchanged and no operations are added or subtracted). A project generates incremental incomes, incremental capital amounts and, hence, incremental cash flows (the project's cash flows). The following definition highlights the differential character of a project.

Definition 1.2 (*Project as incremental system*) An n-period project is a set of capital amounts, incomes, and cash flows summarized in the triplet $(\boldsymbol{C}, \boldsymbol{I}, \boldsymbol{F})$. For every given date t, the cash flow, income, and capital of a project is equal to the difference between the cash flow, income, and capital of the firm-if-the-project-is-undertaken and the cash flow, income, and capital of the firm-if-the-project-is-not-undertaken.

Whenever a firm deviates from the status quo, incremental capitals, incremental incomes, and incremental cash flows are generated. The firm-with-the-project is then economically equivalent to a portfolio of the firm-without-the-project and the project. In what follows, the reader should never forget that capital, income, and cash flow of a project represent incremental amounts with respect to the status quo. In other words, the triplet $(\boldsymbol{C}, \boldsymbol{I}, \boldsymbol{F})$ is an incremental triplet.

Practically, the triplet $(\boldsymbol{C}, \boldsymbol{I}, \boldsymbol{F})$ may be calculated directly by estimating the increments in capital, income, and cash flow generated by the project or, indirectly, by calculating the difference between the triplet of the firm-with-the-project and the triplet of the firm-without-the-project:[20]

$$\text{Firm-with-the-project} = \text{Firm-without-the-project} + \text{Project}$$

The above definition makes it clear that the project measures a change, a variation, a deviation from the firm's *status quo* system:

$$\text{Project} = \text{Firm-with-the-project} - \text{Firm-without-the-project}$$

Such a deviation may be accomplished in three ways:

- by adding an activity to the existing operations of the firm (*expansion* project)

[20]This difference must take account of any direct or indirect effect arising as a consequence of the project undertaking: Externalities, cannibalism, spillover effects, etc. (e.g., see Example 9.19 and Rainbow program in Chap. 12).

- by substituting an existing operation with a different (or equivalent) operation (*replacement* project)
- by subtracting an economic activity from the existing operations of the firm (*abandonment* project).

1.4.1 Expansion Projects

Expansion projects add an activity or a set of activities to the firm. It may consist in expanding the production capacity, developing and launching new products, constructing a new facility, purchasing additional equipment, and, in general, expanding the firm's assets (major repairs and overhaul of existing equipment may be included in this class). In many of these projects, no extra revenues/costs usually occurs at time 0. This implies $I_0 = 0$ and $F_0 = -C_0$. However, in some other cases, a nonzero initial income may occur (e.g., research and development costs occur and are expensed immediately; a prior sale of existing assets is required in order to expand the assets; clean-up costs must be incurred for using an unused space; etc.). In these cases, $I_0 \neq 0$ and $F_0 = I_0 - \Delta C_0 = I_0 - C_0$.[21]

Example 1.9 ♦
(*Expansion project*) Suppose the firm faces the opportunity of launching a new product. The firm's incomes and capital values with and without the project are reported in Table 1.5. The project represents the difference between the two alternatives. The year-0 income of the firm under the two alternative scenarios is the same, which implies that the project's income in year 0 is zero: $I_0 = 0$, and the first cash flow equates the first capital changed in sign: $F_0 = -C_0 = -10$. Starting from the project's incomes and capital amounts one may compute the project's cash flows by applying (1.19):

$$\begin{aligned}
F_0 &= 0 - (0 + 10) = -10 \\
F_1 &= 25 - (50 - 10) = -15 \\
F_2 &= 20 - (-20 - 50) = 90 \\
F_3 &= 20 - (60 + 20) = -60 \\
F_4 &= 10 - (20 - 60) = 50 \\
F_5 &= 5 - (0 - 20) = 25
\end{aligned}$$

That is, $\boldsymbol{C} = (10, 50, -20, 60, 20, 0)$, $\boldsymbol{I} = (0, 25, 20, 20, 10, 5)$, $\boldsymbol{F} = (-10, -15, 90, -60, 50, 25)$.

[21] The incomes and cash flows of the following examples (and, in general, of the first two chapters of the book) may be considered after-tax amounts. Alternatively, one may well assume that they are pre-tax amounts (or that the projects are undertaken in a tax-free world). Taxes will be introduced in Chap. 3.

Table 1.5 Expansion project

	Firm with project			Firm without project			Expansion project			
Year	Capital	Income	Cash flow	Capital	Income	Cash flow	Capital C_t	Income I_t	Cash flow F_t	Income rate i_t
−1	130			130			0			
0	110	70	90	100	70	100	10	0	−10	
1	130	50	30	80	25	45	50	25	−15	250.0%
2	40	40	130	60	20	40	−20	20	90	40.0%
3	90	−10	-60	30	−30	0	60	20	−60	−100.0%
4	70	0	20	50	−10	−30	20	10	50	16.7%
5	25	15	60	25	10	35	0	5	25	25.0%

Alternatively, one may apply the law of motion separately to the firm with the project and the firm without the project to get the respective cash flows, and then compute the difference between the two alternative sets of cash flows. In this case, assuming the invested capital at time −1 is 130 and applying (1.19), one gets $f_0^{\text{with}} = 70 - (110 - 130) = 90$ and $f_0^{\text{without}} = 70 - (100 - 130) = 100$. The cash-flow vectors generated by the firm with the project and by the firm without the project are $\boldsymbol{f}^{\text{with}} = (90, 30, 130, -60, 20, 60)$ and $\boldsymbol{f}^{\text{without}} = (100, 45, 40, 0, -30, 35)$, respectively. Hence, $\boldsymbol{F} = \boldsymbol{f}^{\text{with}} - \boldsymbol{f}^{\text{without}} = (-10, -15, 90, -60, 50, 25)$. The project is a mixed project, according to Definition 1.1, for the capital amounts do not have the same sign. In particular $C_2 = -20 < 0$. This means that $i_3 = -100\%$ is a financing rate. As the rate is negative, in the third period the firm makes money by subtracting resources from the project (see Table 1.4). More precisely, the firm injects capital in the first and second year and the income rates represent single-period rate of return on invested amounts. In the third year, the firm subtracts \$20 from the project. On that amount, it earns a 100% return which means that the EOP capital is $E_3 = 0$: The capital has jumped from \$20 below zero to exactly \$0. The BOP capital of year 3 is $C_3 = 0 + 60 > 0$, which signals that the firm switches back to an investment position at time 3. At time 4, the firm remains in a lending position.

Expansion projects are, usually, investment projects, that is, pure investment transactions according to Definition 1.1. However, they may also be financing projects (pure financing transactions), if they subtract resources from the project in every period. This may occur, for example, whenever upfront payments arise. Following is an example of a financing expansion project.

Example 1.10 ♦
(*Expansion financing project*) A customer of firm A makes a purchase order for the production of a good whose price is \$130. The customer is ready to make an upfront payment of \$100 at time 0. The relevant data for the firm with and without the project are collected in Table 1.6. If the firm does not accept the purchase order and the upfront payment, its assets will amount to \$5,000; if it accepts the

Table 1.6 Expansion financing project

	Firm with project			Firm without project			Expansion project			
Year	Capital	Income	Cash flow	Capital	Income	Cash flow	Capital C_t	Income I_t	Cash flow F_t	Income rate i_t
−1	4,000			4,000			0			
0	4,900	1,200	300	5,000	1,200	200	−100	0	100	
1	5,900	1,240	240	6,000	1,300	300	−100	−60	−60	60%
2	6,900	1,460	460	7,000	1,500	500	−100	−40	−40	40%
3	8,000	1,695	595	8,000	1,600	600	0	95	−5	−95%

purchase order and the upfront payment, a deferred revenue will occur, representing a non-interest bearing liability (the customer loans to the firm). Assuming the money amount will be distributed to the firm's equityholders, the firm's assets will decrease by \$100. Suppose that the production of the good entails additional operational costs amounting to \$60, \$40, and \$35 at time 1, 2, and 3, respectively. At time 3, the customer settles the account by paying the remaining \$30. Note that, by undertaking the project, the firm raises funds from the customer, thereby getting an incremental financing. The project's income rates are then financing rates. In the third and last period, the project's income rate is negative. This is due to the fact that the firm makes an additional income equal to $\$95 = \$130 - \$35$ out of the additional financing position, which amounts to \$100, whence $i_3 = 95/(-100) = -95\%$

Expansion projects may be ***corporate financed*** or ***project financed***. In the latter case, a firm is incorporated for the sole purpose of undertaking a project. This situation is known as a *project financing transaction* and the firm is often called *special purpose vehicle* (SPV) (see Gatti 2013; Finnerty 2013). The equity capital of a SPV is raised from other firms (the sponsors) which are then the equityholders of the SPV. Therefore, in project financing transactions, the project is itself a firm. In other words, a project financed transaction is an off-balance sheet project, and the debt which finances the project's investments is a non-recourse debt (or the recourse is limited). In contrast, a corporate financed project is on-balance sheet and the debtholders have full recourse to the firm's other assets (see Example 2.8).

Project finance is a form of no-recourse debt, whereby a legal entity is incorporated with the intention that such a special purpose vehicle (SPV) or project company will be used to undertake a specific project with a limited (usually long-term) life. A group of sponsoring firms, which constitute the SPV's equity holders, and a group of banks, which constitute the SPV's debt holders, typically supply the funds needed for designing, building, and operating the new project (Gatti 2013). *Loans are guaranteed by the SPV's net assets, so project finance transactions enable equity holders to limit their financial exposure to the fund committed in the project:*

> *By segregating risky assets in a project company, managers can prevent a failing project from dragging the parent firm into default. Project finance allows the firm to isolate asset risk in a separate entity where it has limited ability to inflict collateral damage on the sponsoring firm.* (Esty 2004, p. 217)

At the same time, debtholders impose a large set of covenants aimed at limiting, to some extent, the flexibility of the use of funds by the SPV's shareholders (Smith and Warner 1979). *Project finance originated in the energy generation sector and is now used to fund large-scale projects such as power generation facilities, oil and natural gas pipelines, electric utilities, chemical plants, water and wastewater treatment facilities, renewable energy and green technologies, and so forth* (Scannella 2012), *and recently, Internet and e-commerce projects. The creation of an SPV is associated with a careful economic analysis of the capacity of the entity to generate sufficient cash to meet the expectations of the capital providers. For this reason, considerable time and resources are devoted to financial modeling and financial accounting. In particular, extensive negotiations occur between equity holders and debt holders for estimating cash flows. The estimation process comprises forecasting revenues and costs* [. . .] *constructing pro forma financial statements that collect the forecasts about economic and accounting variables. By constructing a sequence of income statements and a sequence of balance sheets for every period of the project's life, the analysts try to describe the future actual reported statements of the project company; thus, full consideration is given to accounting and fiscal rules in the country or region where the SPV operates.*

Magni (2016a). An average-based accounting approach to capital asset investment: The case of project finance. *European Accounting Review*, 25(2), 275–286 (p. 277), https://doi.org/10.1080/09638180.2015.1009143

1.4.2 Abandonment Projects

Abandonment projects are also called retirement projects. They consist of termination (in full or in part) of a given set of operations. They may consist in closing facility, terminating the production of a given product, retiring from a market, stopping a process, etc.

Contrary to expansion projects, to terminate a group of operations from the firm's current operations often means that some fixed assets are sold (e.g., a piece of equipment, machinery, plant, etc.). If an asset is sold and the disposal value is greater (smaller) than the asset's carrying amount, then a gain (loss) on disposed assets must be recorded as an income accruing at time 0. To abandon or suspend an economic activity may also imply payments of fees, suspension costs, penalties, environmental costs, etc. For this reason, abandonment projects are typical examples of projects where the initial income is nonzero, $I_0 \neq 0$.

Given that they involve disinvestment, it is often the case that abandonment projects represent financing transactions, as the following example shows.

Example 1.11 ♦
(*Abandonment project*) Suppose a firm's management team must decide whether it should discontinue the production of a product or keep the operations for additional five years. The income generated by the operations in year 0 is equal to \$600 and, if operations are continued, it is expected that the income will decrease and be equal to \$400, \$200, \$100, \$50, \$10 in 1, 2, 3, 4, and 5 years respectively. If, instead, operations are discontinued now, the firm will dispose of the relative plant and equipment. The capital invested in the on-going operations (carrying amount of plant and equipment) is \$6,000 and it will depreciate evenly by $-\Delta C_t = \$1,200$ a year for the following 5 years if operations are continued. Plant and equipment may be sold at \$6,500, that is, \$500 above the carrying amount.

Discontinuing production will entail foregoing the above mentioned prospective incomes but will generate a \$500 incremental income (gains on disposed asset).[22]

Table 1.7 collects

- the capital, income, and cash-flow streams of the firm if operations are continued
- the capital, income, and cash-flow streams of the firm if operations are discontinued
- the capital, income, and cash-flow streams of the project.

This project is a pure financing, for $C_t \leq 0$ for all $t \in \mathbb{N}_5^0$. That is, the firm subtracts capital; specifically, it borrows from the project an amount of \$6,000. This is equal to the current carrying amount changed in sign. This financing immediately yields an income of \$500 (= 6,500 − 6,000), caused by the disposal of plant and equipment. Therefore, the initial project's cash flow is positive and equal to

$$6{,}500 = \overbrace{500}^{I_0} - \overbrace{(-6000)}^{\Delta C_0}.$$

[22] A disposal of assets is often associated with a gain or (more often) a loss on disposal, whenever the resale price is different from the carrying amount (see treatment in Sects. 3.6.2 and 4.4).

Table 1.7 Abandonment project

	Continuing operations			Discontinuing operations			Abandonment project			
Year	Capital	Income	Cash flow	Capital	Income	Cash flow	Capital C_t	Income I_t	Cash flow F_t	Income rate i_t
−1	7,200			7,200						
0	6,000	600	1,800	0	1,100	8,300	−6,000	500	6,500	
1	4,800	400	1,600	0	0	0	−4,800	−400	−1,600	6.7%
2	3,600	200	1,400	0	0	0	−3,600	−200	−1,400	4.2%
3	2,400	100	1,300	0	0	0	−2,400	−100	−1,300	2.8%
4	1,200	50	1,250	0	0	0	−1,200	−50	−1,250	2.1%
5	0	10	1,210	0	0	0	0	−10	−1,210	0.8%

The nonzero year-0 income is obtained as the difference between the income that would be generated in case of discontinued operations ($1,100) and the income that would be generated in case of continued operations ($600): $I_0 = 500$. It expresses the incremental income over and above the status quo's at time 0.

The cash flows emerge as incremental cash flows or with the law of motion, $F_t = I_t - \Delta C_t$ (income + capital depreciation).[23]

Note also that the income rates are financing rates, given that $C_t \leq 0$ for all t (see Table 1.4). They are positive, which means that the firm pays a cost for absorbing resources (see also Sect. 1.3). The financed amount increases by 6.7% in the first year, rising to $|E_1| = \$6,400$, but decreases by $|F_1| = \$1,600$, so lowering the financed amount to $|C_1| = 4,800$. Analogously in the next periods, until the system reaches the zero level at time 5.

It is easy that to verify that (1.15) holds. Specifically, discounting the project's prospective cash flow at the project's income rate, one gets the initial capital:

$$-6,000 = -\frac{1,600}{1.067} - \frac{1,400}{1.067 \cdot 1.042} - \frac{1,300}{1.067 \cdot 1.042 \cdot 1.028} - \frac{1,250}{1.067 \cdot 1.042 \cdot 1.028 \cdot 1.021} - \frac{1,210}{1.067 \cdot 1.042 \cdot 1.028 \cdot 1.021 \cdot 1.008}.$$

The metaphor project = loan is particularly compelling for describing this project. Discontinuing operations is financially equivalent to accepting a loan whose financed amount is $6,000 and is gradually reimbursed by the firm with 5 instalments equal to $1,600, $1,400, $1,300, $1,250, $1,210, respectively.

---

[23] Note that the capital at time $t = -1$, which we assume to be equal to $7,200, is irrelevant: The difference between the cash flows of the alternatives does not depend on it and is, invariably, equal to $6,500.

1.4.3 Replacement Projects

A replacement project expresses a change in a set of activities. It may consist in replacing equipment, changing process or technology, changing suppliers, etc. It may also consist of a continuation of the same activities by changing processes, equipment, locations (e.g., replacement of equivalent assets).

In replacement projects, the status quo is often called the *defender*. Thus, the defender represents the current operations, which are challenged by the new course of action, often called the *challenger*.[24] The replacement project is then the incremental system arisen from the difference between the challenger and the defender.

As with abandonment projects, the adoption of the challenger often implies the disposal of assets. If the challenger is accepted, the asset is sold and a gain or loss on the disposed asset must be recorded as an income as a result of the acceptance of the challenger (as long as the disposal value is different from the carrying amount). The initial income may or may not be zero, depending on the situation.

Replacement analysis often comes along with (not one but) two or more alternative projects. For example, if a firm must decide whether to overhaul equipment by additionally investing funds or replace it with new equipment, the decision process is not a replacement project, for it is a decision about three courses of action: (i) Keep the status quo, (ii) overhaul old equipment, (iii) purchase new equipment and sell old equipment. The difference between (ii) and (i) is an expansion project, the difference between (iii) and (i) is a replacement project. In this case, the decision maker will have to establish a rational ranking among the three alternatives to make a decision about whether or not the status quo should be kept and, if not, make a decision about which one of the two projects is preferably undertaken (see Chap. 11 on choices between two mutually exclusive projects and, in general, project ranking. See Bumps example in Sect. 12.1 for a relevant case).

Example 1.12 ♦

(*Replacement project*) A firm purchased a piece of equipment three years ago at a price of \$180 and the firm expected that the asset would depreciate by a constant \$36 every year (this means that the carrying amount at time 0 is \$72). The firm is considering substituting the old equipment (defender) with a high-tech one (challenger), whose current price is \$100 and is expected to depreciate evenly by a constant \$20, being therefore worthless in five years, after which it will not be usable. If the challenger is selected, the old equipment is sold at a market price of \$80, which means that the income at time 0 becomes \$25 + \$80 − \$72 = \$33. Owing to the replacement, cost savings will occur, which are expected to increase incomes by 10% each year. Table 1.8 reports the data for the defender, the challenger, and the project. The latter's capitals and incomes are obtained as the difference between the capitals and the incomes of challenger and defender, respectively. The project's cash flows are obtained via (1.19). It is worth noting that the project is a pure investment transaction ($C_t \geq 0$ for all $t \geq 0$) and all income rates are investment rates. The first cash-flow is

[24] This terminology might well be extended to expansion projects and abandonment projects.

Table 1.8 Replacement project

	Defender			Challenger			Replacement project			
Year	Capital	Income	Cash Flow	Capital	Income	Cash Flow	Capital C_t	Income I_t	Cash flow F_t	Income rate i_t
−3	180			180						
−2	144			144						
−1	108			108						
0	72	25	61	100	33.0	41.0	28	8.0	−20.0	
1	36	20	56	80	22.0	42.0	44	2.0	−14.0	7.1%
2	0	15	51	60	16.5	36.5	60	1.5	−14.5	3.4%
3	0	10	10	40	11.0	31.0	40	1.0	21.0	1.7%
4	0	5	5	20	5.5	25.5	20	0.5	20.5	1.3%
5	0	2	2	0	2.2	22.2	0	0.2	20.2	1.0%

$F_0 = I_0 - \Delta C_0 = 8 - 28 = -20$. This is caused by an investment of \$100 in new equipment accompanied by a sale of old equipment at \$80, which means that the firm only needs to raise funds by $\$20 = 80 - 100$ to undertake the project. The incremental invested capital is $C_0 = 100 - 72 = 28$ and, using (1.15), it is equal to the value of the prospective cash flows discounted at the investment rates 7.1%, 3.4%, 1.7%, 1.3%, 1%, respectively:

$$28 = \frac{-14}{1.071} + \frac{-14.5}{1.071 \cdot 1.034} + \frac{21}{1.071 \cdot 1.034 \cdot 1.017} + \frac{20.5}{1.071 \cdot 1.034 \cdot 1.017 \cdot 1.013} + \frac{20.2}{1.071 \cdot 1.034 \cdot 1.017 \cdot 1.013 \cdot 1.01}.$$

Note that F_1 and F_2 are negative, owing to the fact that the increase in invested capital is greater than the respective income in both the first and second period.

The project's cash flows can also be found as the difference between the cash flows of the challenger and the cash flows of the defender. Specifically, the latter are

$$\begin{aligned}
f_0^{\text{def}} &= 25 - (72 - 108) = 61\\
f_1^{\text{def}} &= 20 - (36 - 72) = 56\\
f_2^{\text{def}} &= 15 - (0 - 36) = 51\\
f_3^{\text{def}} &= 10 - (0 - 0) = 10\\
f_4^{\text{def}} &= 5 - (0 - 0) = 5\\
f_5^{\text{def}} &= 2 - (0 - 0) = 2;
\end{aligned}$$

the former are

$$\begin{aligned} f_0^{\text{ch}} &= 33 - (100 - 108) \;=\; 41 \\ f_1^{\text{ch}} &= 22 - (80 - 100) \;=\; 42 \\ f_2^{\text{ch}} &= 16.5 - (60 - 80) \;=\; 36.5 \\ f_3^{\text{ch}} &= 11 - (40 - 60) \;=\; 31 \\ f_4^{\text{ch}} &= 5.5 - (20 - 40) \;=\; 25.5 \\ f_5^{\text{ch}} &= 2.2 - (0 - 20) \;=\; 22.2. \end{aligned}$$

The difference between the two cash-flow streams is equal to the project's cash-flow stream:

$$\begin{aligned} F_0 &= f_0^{\text{ch}} - f_0^{\text{def}} = 41 - 61 = -20 \\ F_1 &= f_1^{\text{ch}} - f_1^{\text{def}} = 42 - 56 = -14 \\ F_2 &= f_2^{\text{ch}} - f_2^{\text{def}} = 36.5 - 51 = -14.5 \\ F_3 &= f_3^{\text{ch}} - f_3^{\text{def}} = 31 - 10 = 21 \\ F_4 &= f_4^{\text{ch}} - f_4^{\text{def}} = 25.5 - 5 = 20.5 \\ F_5 &= f_5^{\text{ch}} - f_5^{\text{def}} = 22.2 - 2 = 20.2. \end{aligned}$$

Example 1.13 ♦

(*Solar PhV system*) Consider the case where a firm currently purchases electric power through a conventional system (retail electricity from a utility), which represents the defender. It faces the opportunity of purchasing and operating a standalone solar photovoltaic (PhV) system sold by a provider, which represents the challenger. Suppose the utility bill, to be paid periodically, is equal to b and the standalone plant implies a leasing contract whereby a lease payment equal to l is made periodically. After several years, at time t^* (expiration date) the lessee may pay a lump sum P to acquire the plant, and the system will keep on generating electric power for some years. Assuming that no other change in revenues and costs occur, we may abstract from the firm's costs and revenues which do not pertain to the choice as well as from the related cash flows, which are equal in both scenarios. The relevant cash flows of the defender may be represented by the period payments of the utility bills, denoted as b: $\boldsymbol{f}^{\text{retail}} = (-b, -b, \ldots, -b)$. In contrast, if the project is undertaken, the relevant cash flows of challenger amount to lease payments l for $t < t^*$ and to P for $t = t^*$; afterwards, no outflow related to the solar PhV system is incurred.[25] Therefore, $\boldsymbol{f}^{\text{PhV}} = (-l, -l, \ldots, -l, -P, 0, 0, \ldots, 0)$. As a result, the cash-flow stream of the solar PhV system project are as follows:

$$\boldsymbol{F} = \boldsymbol{f}^{\text{PhV}} - \boldsymbol{f}^{\text{retail}} = (s, s, \ldots, s, -P + b, b, b, \ldots, b)$$

where $s = b - l$ is the annual cash saving (we assume $l < b$).

[25] In principle, one might consider an additional terminal outflow for dismantling the plant.

For example, suppose $n = 20$, $l = 800$, $b = 900$, $P = 8,000$. Let $t^* = 12$ be the expiration date, when the lessee purchases the PhV system. Then, the relevant outflows of the firm with the PhV system are

$$f_0^{\text{PhV}} = f_1^{\text{PhV}} = \cdots = f_{11}^{\text{PhV}} = -800, \quad f_{12}^{\text{PhV}} = -8,000, \quad f_{13}^{\text{PhV}} = f_{14}^{\text{PhV}} = \cdots = f_{20}^{\text{PhV}} = 0.$$

Conversely, the relevant outflows of the firm without the PhV system are

$$f_0^{\text{retail}} = f_1^{\text{retail}} = \cdots = f_{20}^{\text{retail}} = -900.$$

The difference between the relevant firm's outflows in the two alternative states (with or without the solar PhV system) provides the project's cash flows. Specifically,

$$F_0 = F_1 = \cdots = F_{11} = 100, \quad F_{12} = -7,100, \quad F_{13} = F_{14} = \cdots = F_{20} = 900.$$

Table 1.9 illustrates the project in more detail, showcasing the capital components and the income components as well as the cash flows. As anticipated, we assume that the firm's revenues (and other costs) are not affected by the decision made. Thus, we may disregard revenues and other costs from the analysis and focus on the electricity costs.

We assume that the solar PhV system decreases in value by a constant \$1,000 and that it has no salvage value, such that the carrying amounts will be \$8,000, \$7,000, \$6,000, \$5,000, \$4,000, \$3,000, \$2,000, \$1,000, in year 12–19. This implies that the firm's income component will be -900 if the retail system is maintained; if, instead, the solar PhV system is purchased, the income component will be -800 for $t \in \mathbb{N}_{11}^{0}$ and $-1,000$ for $t \in \mathbb{N}_{20}^{13}$. The difference between the two alternative income components is a cost saving, which represents the project's income, I_t.

This replacement project does not entail any disposal of asset at time 0. However, the year-0 income is nonzero, owing to the cost saving equal to the difference between utility bill and lease payment: $I_0 = 900 - 800 = 100$. This is also equal to the year-0 cash flow, since no change in capital occurs. At time 12, when the solar PhV system is purchased, the project's income is $I_{12} = 900$ (no lease payment is due any more). After that date, an additional cost is paid, equal to the difference between the capital depreciation and the ceased utility bill, so the project's income is $I_t = 900 - 1,000 = -100$ for $t > 12$.

The project's cash flows are found with the usual relation $F_t = I_t - \Delta C_t$, which conveys the same cash flows that we have previously found with the relation $F_t = f_t^{\text{PhV}} - f_t^{\text{retail}}$.

This project is a pure investment transaction, for $C_t \geq 0$ for all t. The income rates are investment rates. In the first 12 periods, the rate of return is $+\infty$.[26] In the last 8 periods, the investment rate is negative, meaning that some of the capital invested in those periods is lost.[27]

---

[26] More precisely, the income rate is not defined for the capital is zero. As the income is positive, one might use the concept of limit to conventionally define it to be equal to $i_t = +\infty$.

[27] See Magni amd Marchioni (2019) for description of a more realistic example of PhV plant.

Table 1.9 Solar PhV versus Retail

	Retail (defender)			**Solar PhV (challenger)**			**Replacement project**			
Year	Capital	Income	Cash flow	Capital	Income	Cash flow	Capital C_t	Income I_t	Cash flow F_t	Income rate i_t
0	0	−900	−900	0	−800	−800	0	100	100	
1	0	−900	−900	0	−800	−800	0	100	100	$+\infty$
2	0	−900	−900	0	−800	−800	0	100	100	$+\infty$
3	0	−900	−900	0	−800	−800	0	100	100	$+\infty$
4	0	−900	−900	0	−800	−800	0	100	100	$+\infty$
5	0	−900	−900	0	−800	−800	0	100	100	$+\infty$
6	0	−900	−900	0	−800	−800	0	100	100	$+\infty$
7	0	−900	−900	0	−800	−800	0	100	100	$+\infty$
8	0	−900	−900	0	−800	−800	0	100	100	$+\infty$
9	0	−900	−900	0	−800	−800	0	100	100	$+\infty$
10	0	−900	−900	0	−800	−800	0	100	100	$+\infty$
11	0	−900	−900	0	−800	−800	0	100	100	$+\infty$
12	0	−900	−900	8,000	0	−8,000	8,000	900	−7,100	$+\infty$
13	0	−900	−900	7,000	−1,000	0	7,000	−100	900	−1.3%
14	0	−900	−900	6,000	−1,000	0	6,000	−100	900	−1.4%
15	0	−900	−900	5,000	−1,000	0	5,000	−100	900	−1.7%
16	0	−900	−900	4,000	−1,000	0	4,000	−100	900	−2.0%
17	0	−900	−900	3,000	−1,000	0	3,000	−100	900	−2.5%
18	0	−900	−900	2,000	−1,000	0	2,000	−100	900	−3.3%
19	0	−900	−900	1,000	−1,000	0	1,000	−100	900	−5.0%
20	0	−900	−900	0	−1,000	0	0	−100	900	−10.0%

1.5 Key Points

- An economic system is an economic activity (or a set of economic activities) involving business, trading and/or financial transactions with other agents in the guise of suppliers, customers, employees, raisers of funds, investors or other economic roles
- Any economic entity consists of three basic elements: Capital, income, and cash flow
- Capital, income, and cash flow are interrelated dynamically via a *law of motion* (see Eq. (1.1)) which defines the evolution of the capital position. Income and cash flow represent the sources of motion for the system. The law of motion defines the dynamics of any economic system
- Any economic activity can be viewed as a two-party relation between a lender and a borrower, a raiser of funds and a user of funds, an investing party and a financing party. From the point of view of a single agent, the economic system is above zero level whenever the capital is positive at some time t ($C_t > 0$), signalling that the agent is investing money; the system is below zero level whenever the capital is negative ($C_t < 0$), signalling that the agent is borrowing money, that is, subtracting resources from the system
- As a consequence of the previous point, any economic system may conveniently be interpreted as a *savings-and-credit account*, one where deposits and withdrawals

can be made and where the account balance is allowed to be positive or negative, thereby signalling lending or borrowing position

- The income rate expresses the amount of income/profit per unit of invested (or borrowed) capital. The income rate is either an investment rate (if capital is invested in the system) or a financing rate (if capital is absorbed from the system)
- Dynamically, a project may be described as a mechanical system such that its opening capital is at rest at a zero level, then starts jumping up and/or down the zero level owing to perturbations caused by incomes and cash flow, and then finally gets back to zero level at time n, the date where the project is completed. We summarize the dynamics of an economic activity and, in particular, of a project, as follows:

Dynamics of an economic system: Law of motion

Any economic activity is a dynamical system representing the capital position of an economic agent. The sources of perturbation of the capital position are the income and the cash flow, both of which may either be negative or positive. Formally, the capital position evolves according to the recursive intertemporal relation Eq. (1.1), reported below:

$$C_t = C_{t-1} + I_t - F_t$$

or, employing income rates,

$$C_t = C_{t-1} \cdot (1 + i_t) - F_t$$

Economically, a project may conveniently be likened to a savings-and-credit account. The capital amount, the income, and the cash flow of a project play the same role as the account balance, the interest, and the deposit/withdrawal of a savings-and-credit account. Likewise, the income rate plays the role of the account's interest rate.

- The law of motion gives rise to a most important principle in economics: The *time-value of money*, according to which the capital is equal to the discounted sum of the prospective cash flows or, equivalently, equal to the compounded sum of the prior cash flows (changed in sign), calculated at the income rates i_t
- A project is an economic system which is incremental with respect to the *status quo* of a firm, which is the current course of action pursued by the firm. So, although one writes of capital, income, cash flow (and income rate) one always refers to incremental (differential) capital, income, cash flow (and income rate)
- Projects are characterized by a triplet $(\boldsymbol{C}, \boldsymbol{I}, \boldsymbol{F})$ of capitals, incomes, and cash flows. Projects may be classified as (i) expansion projects, (ii) abandonment projects, (iii) replacement projects. In the first case, the time-0 income is usually zero; in the other two cases, gains or losses on disposal of assets may occur, which implies that the time-0 income is nonzero. A project may be corporate financed (on-balance sheet project) or project financed (off-balance sheet project).

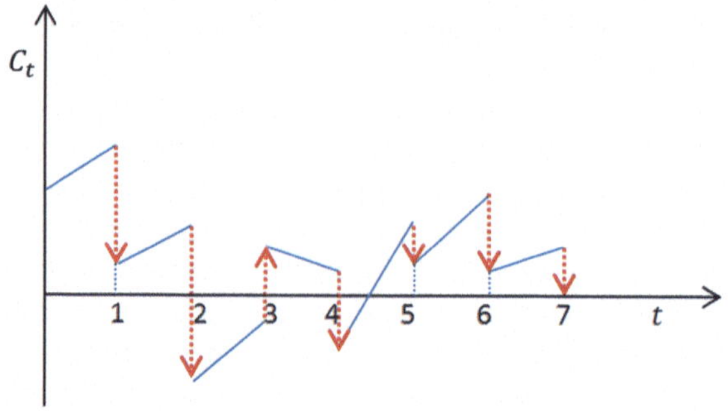

Fig. 1.6 Dynamics of a 7-year project

The dynamics of a project may be fully unfolded graphically as in Fig. 1.6, where the t-axis represents the time and the C-axis represents the capital. The example presents the capital amounts of the economic agent: There are 5 positive capital values (investments), 2 negative capital values (borrowings), 6 positive incomes and 1 negative income (see segments), 6 inflows and 2 outflows (see arrows).[28] The system starts from zero, jumps up and down, also reaching negative values, and finally gets back to zero level at time 7.

[28] $I_0 = 0$ is assumed.

Chapter 2
Statics. The Law of Conservation

> Statics deals with the equilibrium of bodies ... To maintain equilibrium ... requires the resultant force acting on a particle to be equal to zero ... This condition is stated by the *equation of equilibrium* ... For equilibrium, these forces must sum to produce a zero force resultant
>
> Hibbeler (2016, p. 87 and p. 91, italics in original)

> Just as we established that the value of a firm's assets is always equal to the combined value of the liabilities and the value of the equity, the cash flows received from the firm's assets ..., CF(A), must equal the cash flows to the firm's creditors, CF(B), and equity investors, CF(S):
>
> $$\mathrm{CF}(A) \equiv \mathrm{CF}(B) + \mathrm{CF}(S)$$
>
> Ross, Westerfield and Jaffe (2013, p. 29)

> The accounting equation applies to all economic entities regardless of size, nature of business, or form of business organizations.
>
> Weygandt, Kimmel and Kieso (2012, p. 12)

C. A. Magni, *Investment Decisions and the Logic of Valuation*,
https://doi.org/10.1007/978-3-030-27662-1_2

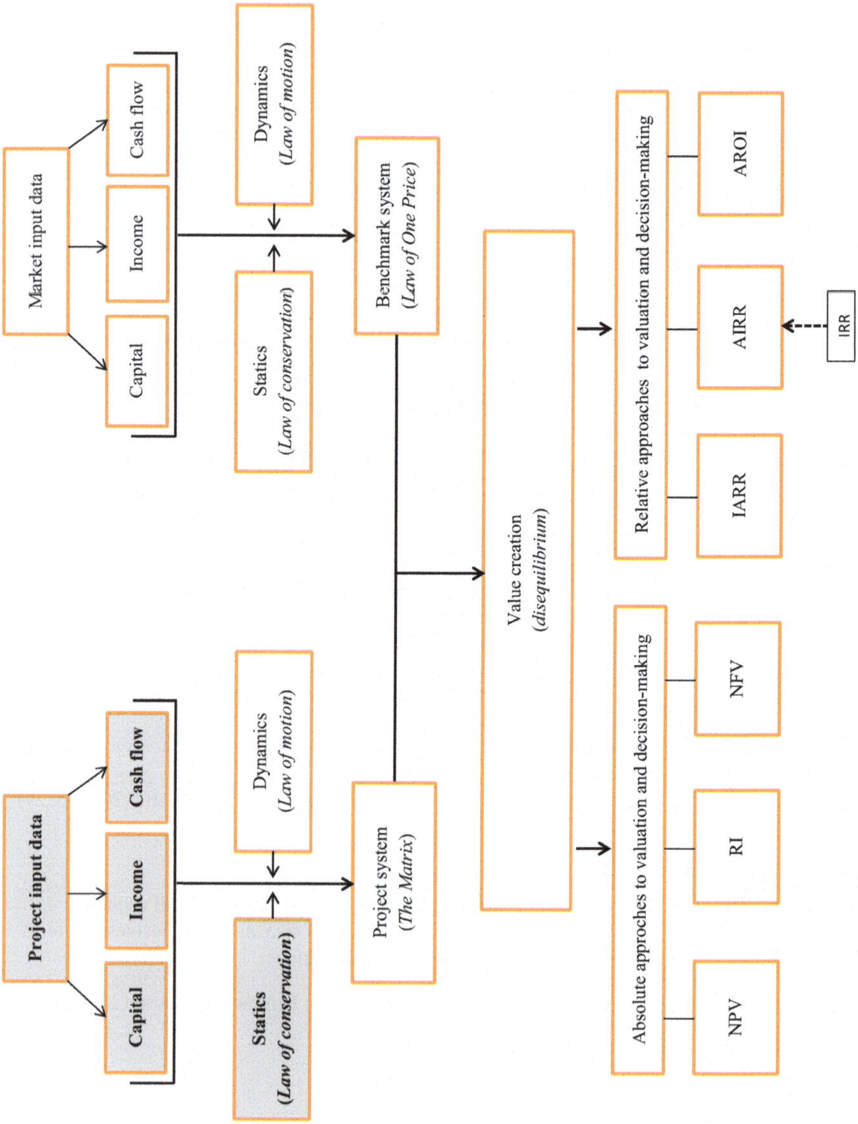

Fig. 2.1 Chapter 2: The statics

The chapter investigates the **statics** of an economic system. Specifically, we adopt a *synchronic* perspective, that is, we analyze the forces that act upon an economic system at a **given point in time**. We introduce the **Law of Conservation**, according to which each of the basic elements of an economic system (capital, income, cash flow, income rate) is subjected to **opposite forces that balance**.

To understand the resulting effect of opposite forces, it is crucial to conceptualize an economic agent as being both a lender and a borrower, a user of funds and a raiser of funds. In particular, for any fixed date, a firm raises funds from capital providers and, synchronically, invests funds in some assets. The two symmetric positions offset each other. Therefore, the law of conservation manifests itself in the shape of four **balancing laws** holding for (i) capital, (ii) income, (ii) cash flow, and (iv) income rate. The four laws logically describe the **static equilibrium** of the system.

We break down an economic system into four subsystems representing the **operating** capital, **non-operating** capital, **debt** capital, and **equity** capital, governed by four laws of motion and interlocked by the laws of conservation. We also clarify the important distinction between two kinds of borrowing: **Borrowing from debtholders** (increasing the amount of the firm's loans, bonds, etc.) and **borrowing from the project** (decreasing the amount of the firm's assets).

After completing this chapter, the reader should be able to:

- grasp the fundamental idea that any economic system (and, in particular, a project) has two sides: The investment side and the financing side. Specifically, an economic agent is, at any given point in time, a raiser of funds *and* a user of funds, a borrower *and* a lender. For a firm, this means that, at any given point in time, it is raising capital from capital providers *and* investing the resources raised in some assets
- understand the *law of conservation*, which logically requires that the forces acting on the system at a given point in time balance, so that the following equilibrium conditions hold:
  - the capital *invested* in some investing activities equals the capital *financed* by capital providers
  - the income generated by investments equals the income accrued to capital providers
  - the cash flow generated by investments equals the cash flow distributed to capital providers
  - the return on investments is equal to the return on financings
- break down the system into four subsystems and familiarize with the resultant four classes of capital (operating capital, non-operating capital, debt capital, equity capital), the respective classes of income (operating income, non-operating income, interest expense, net income), and the respective classes of cash flow (cash flow from operations, cash flow from liquid assets, cash flow to debt, cash flow to equity), as well as the respective classes of income rates (return on operating investment, return on non-operating investment, return on debt, return on equity)
- understand that the equilibrium condition for rates of return means that the rate of return on investments and the rate of return on financings are *average* rates: The return on investments equals a weighted average of the return on operating assets and the return on non-operating assets; the return of financings equals the weighted average of the return on debt and the return on equity
- appreciate the difference between financial structure and capital structure and, consistently, differentiate the (gross) debt from the net debt.

2.1 Investment Side and Financing Side

Consider a scenario where, for a given date t, agent A finds itself in an investment position. This means that the agent must raise funds from somewhere/someone to invest in B. We may use the following account-form representation.

A invests funds in	A raises funds from
B $100	?

The question mark highlights the fact that information is missing about the sources of funds of A. Consider now a scenario where A raises funds from C:

A invests funds in	A raises funds from
?	C $150

The question mark means that information is missing about the uses of funds by A.

In other words, if agent A invests in B (first case), then it must also raise funds from some source; if agent A raises funds from C (second case), then it must invest those funds somewhere. In general, every investment is matched by a financing of equal amount (and vice versa), which means that any economic agent A is a lender with respect to some agent and a borrower with respect to some other agent. For example, consider a very simple economic unit consisting of an individual (Mr. A) who raises funds from his personal wealth at time 0 in order to purchase a portfolio of 100 bonds at a price of $15 each. A simple description of this economic unit at time 0 is as follows.

A invests funds in	A raises funds from
Bond $1,500	Personal wealth $1,500

Investments and financings balance. More generally, agent A may undertake, at the same time, multiple investments by some given amount and raise funds from multiple sources by the same amount. For example, suppose Mr. A raises funds from personal wealth ($500) and a loan ($600) and uses them to purchase an apartment house ($900) and deposit the residual funds ($200) in a savings account. Mr. A's economic position can be graphically described as follows.

A invests funds in		A raises funds from	
Apartment house	$900	Personal wealth	$500
Savings account	$200	Loan	$600
TOTAL	$1,100	TOTAL	$1,100

The same situation might also be described in a report form:

A invests funds in	
Apartment house	\$900
Savings account	\$200
TOTAL	\$1,100

A raises funds from	
Personal wealth	\$500
Loan	\$600
TOTAL	\$1,100

The investments of a firm or a project are often called assets. Assets can be conveniently divided into two main classes:

- Operating assets (assets held for operating activities)
- Non-operating or liquid assets (assets held for financial activities).[1]

The financings of a firm or a project are also called liabilities. They can be conveniently divided into two classes:

- Debt (financings raised from economic agents who are entitled to payments which are fixed in advance according to specific schedules)
- Equity (financings raised from economic agents whose payments are not fixed in advance and who are entitled to payments only if debtholders are satisfied in full).

Suppose now agent A is a firm, which undertakes a project by raising incremental funds from equityholders and debtholders. Suppose that, at time t, equityholders claim rights by \$1,000 while debtholders claims rights by \$400. Also, the project's operating assets are \$1,200 while the liquid assets are \$200. The capital position of A may be equivalently described in account form or report form:

A invests funds in		**A raises funds from**	
Operating assets	\$1,200	Debt capital	\$400
Non-operating assets	\$200	Equity capital	\$1,000
TOTAL	\$1,400	TOTAL	\$1,400

[1] Strictly speaking, non-operating assets do not coincide with liquid assets. For example, the firm may own some real estate assets for investment (not operational) purposes. The real asset is, from the point of view of the firm, a non-operating asset although it is less liquid than cash or marketable securities (it does not have a high degree of liquidity). For simplicity, we will make use of the terms 'liquid assets' and 'non-operating assets' as synonyms, with the understanding that any income related to non-operating assets must be added to the interest income of liquid assets (e.g., see rental income in HomeNet project in Chap. 12).

A invests funds in	
Operating assets	\$1,200
Non-operating assets	\$200
TOTAL	\$1,400

A raises funds from	
Debt capital	\$400
Equity capital	\$1,000
TOTAL	\$1,400

The firm subtracts funds (borrows) from the lower side of the sheet, where financings are listed, and invests in (lends to) the upper side, where investments are listed.[2] Note that investments and financings balance.

In general, for any fixed date $t \in \mathbb{N}$, any project may be broken down into $4 \times 3 = 12$ basic elements. Specifically, the capital is broken up into four classes:

- Invested capital (C_t^{inv})
  1. Capital invested in the operations associated with the project (C_t^o).[3]
  2. Capital invested in non-operating assets such as marketable securities, financial investments and other liquid assets (C_t^l).[4]
- Borrowed capital (C_t^{fin})
  3. Capital raised from debtholders (C_t^d)[5]
  4. Capital contributed by equityholders (C_t^e).[6]

Correspondingly, the income is broken up into four types of income:

- Income generated by the investments (I_t^{inv})
  5. Operating income (or profit), generated by the operations (I_t^o)
  6. Non-operating income (interest income), generated by the non-operating assets (I_t^l)
- Income accrued to capital providers (I_t^{fin})
  7. Interest expense, generated by the debt capital (I_t^d)
  8. Net income, generated by the equity capital (I_t^e)

---

[2]If an account form is used, replace "lower" with "right" and "upper" with "left".

[3]This capital quantifies the operating assets, resources that are associated with transactions with customers and suppliers, as well as with workers and employees involved in the business and with the government for tax payments.

[4]These liquid assets correspond to transactions made in financial markets. Other non-operating assets may include real estate assets, rental property, vacant land, etc.

[5]The debt has to do with financial transactions with the creditors.

[6]Equity has to do with economic transactions with the firm's equityholders.

Finally, the cash flow is broken up into

- Cash flow generated by the investments (F_t^{inv})

9. Cash flow generated by the operations, which flows out of the operating assets or, if negative, flows into the operating assets (F_t^o)
10. Cash flow generated by the non-operating assets, which flows out of the liquid assets or, if negative, flows into the liquid assets (F_t^l)

- Cash flow distributed to capital providers (F_t^{fin})

11. Cash flow distributed (or raised from) the debtholders (F_t^d)
12. Cash flow distributed to (or contributed by) the firm's equityholders (F_t^e)

Both investments and financings abide by their own law of motion fulfilling the fundamental intertemporal Eq. (1.1):

$$\begin{aligned} \text{Investments:} \quad & C_t^{\text{inv}} = C_{t-1}^{\text{inv}} + I_t^{\text{inv}} - F_t^{\text{inv}} \\ \text{Financings:} \quad & C_t^{\text{fin}} = C_{t-1}^{\text{fin}} + I_t^{\text{fin}} - F_t^{\text{fin}}. \end{aligned} \tag{2.1}$$

Equation (2.1) expresses the law of motion for investments and the law of motion for financings, respectively.

In turn, each of the economic activities that make up investments and financings abides by its own law of motion:

$$\begin{aligned} \text{Operating assets:} \quad & C_t^o = C_{t-1}^o + I_t^o - F_t^o \\ \text{Liquid assets:} \quad & C_t^l = C_{t-1}^l + I_t^l - F_t^l \\ \text{Debt:} \quad & C_t^d = C_{t-1}^d + I_t^d - F_t^d \\ \text{Equity:} \quad & C_t^e = C_{t-1}^e + I_t^e - F_t^e. \end{aligned} \tag{2.2}$$

The four classes of income and capital give rise to four classes of income rates:

- Income rate generated by the investments (i_t^{inv})

13. Return on operating investment, which measures the relative growth of the operating assets (i_t^o)
14. Return on non-operating investment, which measures the relative growth of the non-operating assets (i_t^l)

- Income rate earned by capital providers (i_t^{fin})

15. Return on debt, which measures the relative growth of the debt capital (i_t^d)
16. Return on equity, which measures the relative growth of the equity capital (i_t^e).

Hence, using Eq. (1.10), one may frame (2.2) as

$$\begin{aligned} \text{Operating assets:} \quad & C_t^o = C_{t-1}^o(1 + i_t^o) - F_t^o \\ \text{Liquid assets:} \quad & C_t^l = C_{t-1}^l(1 + i_t^l) - F_t^l \\ \text{Debt:} \quad & C_t^d = C_{t-1}^d(1 + i_t^d) - F_t^d \\ \text{Equity:} \quad & C_t^e = C_{t-1}^e(1 + i_t^e) - F_t^e \end{aligned} \tag{2.3}$$

It should be clear that, for any given date t, the investment side and the financing side are always balanced. This means that, from a static perspective, an economic entity (and, in particular, a project) is subjected to forces that act upon the system whose net amount is zero. We will refer to it as the ***law of conservation***. Every element of an economic system fulfills its own law of conservation, which expresses the static equilibrium determined by the opposite forces triggered by investment and financing. We illustrate below these equilibria in some detail (see also Fig. 2.2).

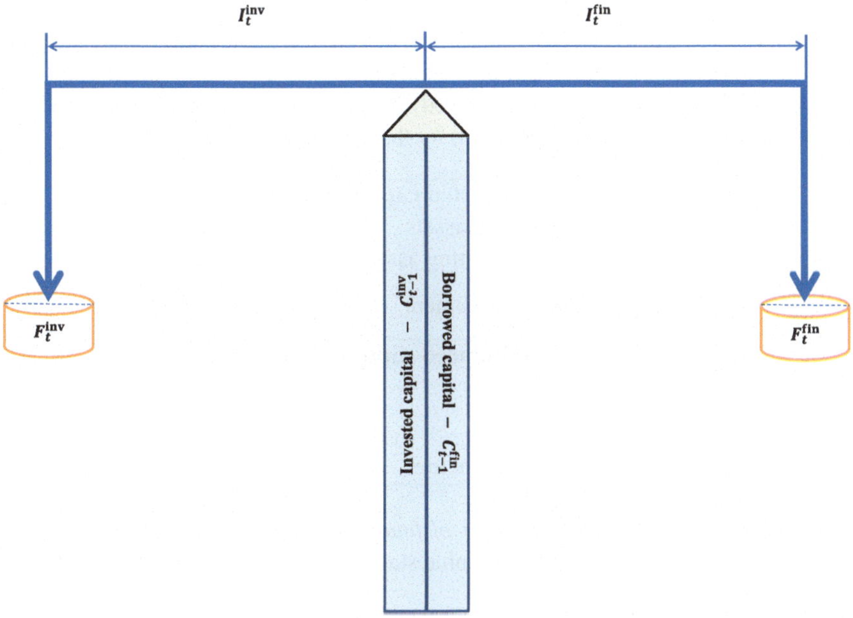

Fig. 2.2 The statics of a project and its laws of conservation may be represented with a weighing scale, consisting of a frame composed of two equal-height columns, the capital invested in the project and the capital raised from capital providers: $C_{t-1}^{\text{inv}} = C_{t-1}^{\text{fin}}$. The columns hold up a horizontal lever (the beam) which has two arms of equal length, the income from investments and the income to capital providers: $I_t^{\text{inv}} = I_t^{\text{fin}}$. Two equal loads are placed on the suspended pans, the cash flow from investments and the cash flow to capital providers: $F_t^{\text{inv}} = F_t^{\text{fin}}$. The system is in equilibrium at every date $t = 0, 1, 2, \ldots, n$

2.1.1 *Equilibrium of Capital Values*

Mathematically, the law of conservation is translated by the following balancing identity:

$$C_t^{\text{inv}} = C_t = C_t^{\text{fin}} \iff C_t^{\text{inv}} - C_t^{\text{fin}} = 0 \tag{2.4}$$

where C_t denotes the common value of C_t^{inv} and C_t^{fin}. In other words, netting out investments and financings, the balance is zero for every t.[7]

In terms of the four main classes above described, in symbols,

$$\overbrace{C_t^o + C_t^l}^{C_t^{\text{inv}}} = C_t = \overbrace{C_t^d + C_t^e}^{C_t^{\text{fin}}}. \tag{2.5}$$

or, equivalently,

$$C_t^e = C_t^o + C_t^l - C_t^d. \tag{2.6}$$

Equation (2.7) below is a typical report form description of the equilibrium between the investment side of capital and the financing side of capital. *Invested* capital is the capital which is invested (i.e., injected) into the project, *financed* capital is the capital which is financed (i.e., supplied) by the capital providers.

Invested capital	
Operating assets	(C_t^o)
Non-operating assets	(C_t^l)
Total Investment	(C_t^{inv})
Financed capital	
Debt	(C_t^d)
Equity	(C_t^e)
Total Financing	(C_t^{fin})

(2.7)

This statement is called the *statement of financial position* or *balance sheet* (see Chap. 3). Using an account form, the same statement may be reframed as

Invested capital		**Financed capital**	
Operating capital	(C_t^o)	Debt capital	(C_t^d)
Non-operating capital	(C_t^l)	Equity capital	(C_t^e)
Total Investment	(C_t^{inv})	Total Financing	(C_t^{fin})

Example 2.1 ♦
The capital invested in a project is $C_t = 460$, all of which is operating capital (i.e., $C_t^l = 0$). One fourth of the investment is financed by debt (e.g., loans and bonds).

[7] In accounting and finance, investments are called "assets" and financings are called "liabilities" (debt) and "net worth" (equity). Hence, the balancing identity may also be written as

$$\text{Assets} = \text{Liabilities} + \text{Net Worth}.$$

Strictly speaking, the expression "liabilities" might be used as a synecdoche, that is, including equity, because both debtholders and equityholders are, financially speaking, in a lending position; in particular, equityholders have indeed a legal claim on equity (albeit residual), so the firm is *liable* to pay the residual capital to equityholders when the asset (project, firm) is liquidated.

Using the balancing principle (specifically, Eq. (2.6)), it is straightforward to compute the equity invested in the project:

$$C_t^e = 460 - 115 = 345.$$

Example 2.2 ♦
Assume the ratio of debt to equity is 40% and equity is equal to $C_t^e = 1{,}250$ (i.e., $C_t^d = 0.4 \cdot 1{,}250 = 500$). Assuming there are no liquid assets involved in the project, one may find the operating capital by using the law of conservation: From (2.5),

$$C_t^o = C_t = 1{,}250 + 500 = 1{,}750.$$

2.1.2 *Equilibrium of Incomes*

The law of conservation holds for incomes as well. The incomes generated by the investments necessarily match with the incomes accrued to capital providers:

$$I_t^{\text{inv}} = I_t = I_t^{\text{fin}} \iff I_t^{\text{inv}} - I_t^{\text{fin}} = 0. \tag{2.8}$$

In symbols,

$$\overbrace{I_t^o + I_t^l}^{I_t^{\text{inv}}} = I_t = \overbrace{I_t^d + I_t^e}^{I_t^{\text{fin}}}. \tag{2.9}$$

Equation (2.10) below is a report form description of the equilibrium of invested side and financing side.

Incomes from investments	
Operating income	(I_t^o)
Non-operating income	(I_t^l)
Total income from assets	(I_t^{inv})
Incomes to financings	
Interest expenses	(I_t^d)
Net income	(I_t^e)
Total income to capital providers	(I_t^{fin})

(2.10)

Example 2.3 ♦
Suppose the operating profit from a project at time t is $I_t^o = 190$, the interest payments amount to $I_t^d = 120$ while the net income is $I_t^e = 80$. What is the interest income? Using the balancing Eq. (2.9), $I_t^l = 120 + 80 - 190 = 10$.

2.1.3 Equilibrium of Cash Flows

From an investment perspective, the sum of the cash flow from operations (CFO) and the cash flow from liquid assets (CFL) is the cash flow from assets (CFA). In symbols, $F_t^{\text{inv}} = F_t^o + F_t^l$. It represents the total cash flow that flows out of the project's assets at time t. From a financing perspective, the sum of the cash flow to equity (CFE) and the cash flow to debt (CFD) is the cash flow distributed to capital providers (CFP). In symbols, $F_t^{\text{fin}} = F_t^e + F_t^d$.

The law of conservation for cash flows says that, whenever cash flow is withdrawn from the investments, then it is necessarily distributed to the capital providers (debtholders and/or equityholders), so cash flow from assets and cash flow to capital providers balance:

$$F_t^{\text{inv}} = F_t = F_t^{\text{fin}} \iff F_t^{\text{inv}} - F_t^{\text{fin}} = 0. \tag{2.11}$$

In terms of the four classes of assets and financings,

$$\overbrace{F_t^o + F_t^l}^{F_t^{\text{inv}}} = F_t = \overbrace{F_t^e + F_t^d}^{F_t^{\text{fin}}} \tag{2.12}$$

(see report form description of the equilibrium between investment side and financing side in (2.13) below.[8]

Cash flow from investments	
Cash flow from operating assets	(F_t^o)
Cash flow from non-operating assets	(F_t^l)
Cash flow from assets	(F_t^{inv})
Cash flows to financings	
Cash flow to debt	(F_t^d)
Cash flow to equity	(F_t^e)
Cash flow to capital providers	(F_t^{fin})

(2.13)

Remark 2.1 (***Capital Cash Flow***) The CFO is often referred to as the *capital cash flow* (CCF) (Ruback 1994, 2002). However, the latter label does not make it clear that this cash flow is not the cash flow which is distributed to the capital providers (CFP) but, rather, the cash flow which is distributed to the capital providers *net of* the cash flow generated by the non-operating assets:

[8]It is worth pointing out that, if $F_t^l < 0$ (i.e., $F_t^o > F_t^e + F_t^d$), then an amount smaller than the CFO is distributed to capital providers and an additional investment in financial activities occurs. In this case, the CFL represents *retained* or *undistributed* cash, which flows in the liquid assets (see Sect. 4.10). Conversely, if $F_t^l > 0$ (i.e., $F_t^o < F_t^e + F_t^d$), an amount greater than CFO is distributed to debtholders and equityholders.

$$\text{CCF} = \text{CFO} = \text{CFP} - \text{CFL} = (\text{CFE} + \text{CFD}) - \text{CFL}$$

or, in symbols,

$$\text{CCF} = F_t^o = (F_t^e + F_t^d) - F_t^l.$$

If one excludes the cash flow generated by the non-operating assets, F_t^l, then one gets a cash flow generated by the operating assets. In other words, the *capital* cash flow only refers to the operating component of the invested capital. For this reason, the expression *cash flow from operations* or *operating cash flow* (CFO) more properly fits the financial nature of F_t^o. As the CFO is the cash flow available for distribution to investors, an appropriate term would be *free cash flow* (FCF). However, this term is used in finance to denote a different concept (we will tackle this notion in Sect. 4.7). Throughout the book, we will prefer the term CFO over CCF but, given that CCF is well-known in finance, we will often remind that they are the same notion. ▲

$$\overbrace{\text{CCF} + \text{CFL}}^{\text{CFA}} = \underbrace{\text{CFE} + \text{CFD}}_{\text{CFP}}$$

Example 2.4 ♦
Consider a project whose CFO/CCF is, at a certain date t, $F_t^o = \$340$. Assume no cash flow is distributed to equityholders and the CFO/CCF is used for paying the instalment of a loan associated with the project (\$210) and paying the coupon of a corporate bond previously issued (\$81). We can compute the cash flow from liquid assets, CFL, by using (2.12):

$$F_t^l = (210 + 81) - 340 = -49.$$

The CFL is negative, meaning that it flows into the liquid assets. More precisely, the \$340 cash flow generated by the operations is split up into two parts: \$291 are distributed to debtholders while \$49 are retained in the business and invested in liquid assets.

Example 2.5 ♦
The stream of CFOs of a 5-period project is $\boldsymbol{F}^o = (-70, 100, 200, 50, -80, 90)$. Suppose the CFA stream is $\boldsymbol{F} = (-70, 90, 265, 20, 5, 95)$. Then, from (2.12), one finds the CFLs (see (2.14) below). This means that injections are made in liquid assets at time 1 and 3 while withdrawals are made at time 2, 4, and 5. At time 0, no cash flow is either contributed or distributed in the account (i.e., the initial operating investment is not financed with liquid assets).

$$\begin{aligned}
F_0^l &= -70 - (-70) = 0\\
F_1^l &= 90 - 100 \quad = -10\\
F_2^l &= 265 - 200 \quad = 65\\
F_3^l &= 20 - 50 \quad = -30\\
F_4^l &= 5 - (-80) \quad = 85\\
F_5^l &= 95 - 90 \quad = 5.
\end{aligned} \tag{2.14}$$

2.1.4 Equilibrium of Income Rates

The equilibrium between investments and financings holds for income rates as well:

$$i_t^{\text{inv}} = i_t = i_t^{\text{fin}}. \tag{2.15}$$

However, unlike capital, income and cash flow, the project income rate, i_t, enters (2.3) in a non-additive way. This implies that the law of conservation takes the form of an equality of means. This can be seen as follows. On the investment side, using Eqs. (2.3) and (2.5), the relation $C_t = C_t^o + C_t^l$ becomes

$$C_{t-1}^{\text{inv}}(1 + i_t^{\text{inv}}) - F_t^{\text{inv}} = C_{t-1}^o(1 + i_t^o) - F_t^o + C_{t-1}^l(1 + i_t^l) - F_t^l$$

whence, remembering that $F_t^o + F_t^l = F_t^{\text{inv}}$,

$$i_t^{\text{inv}} = \frac{i_t^o C_{t-1}^o + i_t^l C_{t-1}^l}{C_{t-1}^o + C_{t-1}^l}. \tag{2.16}$$

Analogously, on the financing side,

$$C_{t-1}^{\text{fin}}(1 + i_t^{\text{fin}}) - F_t^{\text{fin}} = C_{t-1}^d(1 + i_t^d) - F_t^d + C_{t-1}^e(1 + i_t^e) - F_t^e$$

whence, remembering that $F_t^d + F_t^e = F_t^{\text{fin}}$,

$$i_t^{\text{fin}} = \frac{i_t^d C_{t-1}^d + i_t^e C_{t-1}^e}{C_{t-1}^d + C_{t-1}^e}. \tag{2.17}$$

This result implies that the income rate i_t may be alternatively interpreted under the investment perspective or under the financing perspective. Specifically, it can be computed as a weighted arithmetic mean of the return on liquid assets, i_t^l, and the return on operating assets, i_t^o; alternatively, it can be computed as a weighted arithmetic mean of return on debt, i_t^d, and the return on equity, i_t^e. The law of conservation for income rates can then be written as follows:

$$\overbrace{\frac{i_t^o \cdot C_{t-1}^o + i_t^l \cdot C_{t-1}^l}{C_{t-1}^o + C_{t-1}^l}}^{i_t^{\text{inv}}} = i_t = \overbrace{\frac{i_t^d \cdot C_{t-1}^d + i_t^e \cdot C_{t-1}^e}{C_{t-1}^d + C_{t-1}^e}}^{i_t^{\text{fin}}}. \tag{2.18}$$

Evidently, as the income is the product of income rate and capital, one can compute the project's rate of return in a given period ($i_t = I_t/C_{t-1}$) as the ratio of operating income (I_t^o) plus interest income (I_t^l) to assets (investment perspective) or as the ratio of net income (I_t^e) plus interest expenses (I_t^d) to liabilities (financing perspective):

$$\frac{\overbrace{I_t^o + I_t^l}^{I_t^{\text{inv}}}}{\underbrace{C_{t-1}^o + C_{t-1}^l}_{C_{t-1}^{\text{inv}}}} = i_t = \frac{\overbrace{I_t^d + I_t^e}^{I_t^{\text{fin}}}}{\underbrace{C_{t-1}^d + C_{t-1}^e}_{C_{t-1}^{\text{fin}}}} \tag{2.19}$$

(as long as the denominator is nonzero). A third way to compute i_t is to use the ratio of EOP capital to BOP capital:

$$i_t = \frac{E_t}{C_{t-1}} - 1 = \frac{C_t + F_t}{C_{t-1}} - 1$$

(see Eqs. (1.8)–(1.9)).

Remark 2.2 (*Terminology*) In academia and in practice, there is no accepted standard to label the project's holding period rate i_t; several different expressions are available in textbooks and papers, including return on capital (ROC), return on assets (ROA), return on net assets (RONA), return on investment (ROI), return on invested capital (ROIC). ROC is often used in engineering economics; ROI and ROA seems to be preferred acronyms in finance; ROIC seems to be used in both finance and engineering economics; ROI seems to be rather widespread in operations research and production economics (as well as in human resource management and public policy). Unfortunately, the same expressions are often used to mean i_t^o rather than i_t (i.e., a return on operating capital) or other variants of i_t^o, thereby causing some semantic confusion. We will use the acronyms ROC or ROI interchangeably and the symbols i_t^o or i_t will help disambiguate the perspective adopted. Linguistically, we will disambiguate (whenever necessary) by adding the qualification *operating* (which refers to the operating assets) or *asset* (which refer to both operating and non-operating assets). Correspondingly, the return on non-operating assets will be called *non-operating* ROC or ROI.[9] To sum up,

[9]Splitting hairs, the rate i_t is a return on capital (ROC) and i_t^{inv} is a return on investments/assets (ROI/ROA). Given the law of conservation (2.15), ROC and ROI/ROA are equal. However, the ROC/ROI/ROA is, at the same time, a cost of financing (from the point of view of the firm): $i_t = i_t^{\text{fin}}$. Unfortunately, no particular expression or acronym is used in finance for this financing rate.

- i_t: (asset) ROC/ROI
- i_t^o: (operating) ROC/ROI.
- i_t^l: (non-operating) ROC/ROI.

(If $C_t^l = 0$, then $C_t^o = C_t$ and $i_t^o = i_t$, so ROI is unambiguous and no other specification is needed.) The income rate for debt is the return on debt (ROD) and the income rate for equity is the return on equity (ROE). ▲

The equilibrium of income rates may also be framed in a tabular (report) form as in (2.20) below (with the understanding that rates do not *sum up* but, rather, *average out*).

Income rates from investments	
Operating ROI/ROC	(i_t^o)
Non-operating ROI/ROC	(i_t^l)
Rate of return from investments (ROI/ROC)	(i_t^{inv})

Income rates to financings	
ROD	(i_t^d)
ROE	(i_t^e)
Rate of return to capital providers (ROI/ROC)	(i_t^{fin})

(2.20)

Table 2.1 summarizes the relations between the law of conservation and the 16 basic constituents of an economic system. Equilibria of forces acting synchronically upon the four elements is displayed by the balancing of upper and lower lines, respectively (the sum of the elements of each line in the investment side is equal to the sum of the elements of the corresponding line in the financing side), except for the income-rate rows, where sum must be replaced by average.

Remark 2.3 Operating assets, financial assets, debt, and equity may themselves be classes composed of several accounts, each one associated with its own law of motion. In this case, $i_t^o, i_t^l, i_t^d, i_t^e$ are themselves weighted average rates of return. For example, suppose debt is composed of two loans and the respective loan rates are i_t^{d1} and i_t^{d2}, then the ROD is

$$i_t^d = \frac{i_t^{d1} \cdot C_{t-1}^{d1} + i_t^{d2} \cdot C_{t-1}^{d2}}{C_{t-1}^{d1} + C_{t-1}^{d2}}.$$

Analogously with a class of $m > 2$ accounts. For example, if the liquid assets consist of m financial investments,

$$i_t^l = \frac{i_t^{l1} \cdot C_{t-1}^{l1} + i_t^{l2} \cdot C_{t-1}^{l2} + \cdots + i_t^{lm} \cdot C_{t-1}^{lm}}{C_{t-1}^{l1} + C_{t-1}^{l2} + \cdots + C_{t-1}^{lm}}$$

with obvious meaning of the symbols. ▲

Table 2.1 The four laws of conservation

Investment side	**Operating assets**	**Liquid assets**
$\boldsymbol{C_t^{\text{inv}}}$	capital invested in operating assets (C_t^o)	capital invested in liquid assets (C_t^l)
$\boldsymbol{I_t^{\text{inv}}}$	operating profit (I_t^o)	interest income (I_t^l)
$\boldsymbol{F_t^{\text{inv}}}$	cash flow from operations (F_t^o)	cash flow from liquid assets (F_t^l)
$\boldsymbol{i_t^{\text{inv}}}$	return on operating assets (i_t^o)	return on liquid assets (i_t^l)
Financing side	**Debt**	**Equity**
$\boldsymbol{C_t^{\text{fin}}}$	capital raised from debtholders (C_t^d)	capital raised from equityholders (C_t^e)
$\boldsymbol{I_t^{\text{fin}}}$	interest expense (I_t^d)	net income (I_t^e)
$\boldsymbol{F_t^{\text{fin}}}$	cash flow to debt (F_t^d)	cash flow to equity (F_t^e)
$\boldsymbol{i_t^{\text{fin}}}$	return on debt (i_t^d)	return on equity (i_t^e)

Remark 2.4 (*Terminology*) The conventional prepositions used for signalling the direction of the cash flows ("from" and "to") reflect the implicit assumption that, after the initial investment at time 0, the firm withdraws cash flows *from* the assets and distributes it *to* capital providers. However, it may well occur that one or more cash flows have a different direction, whenever the firm raises cash flow *from* the capital providers and distributes it *to* the assets. For example, the cash flow to debt, F_t^d, and the cash flow to equity, F_t^e, are, in actual facts, cash flow *from* debt and cash flow *from* equity whenever additional financing is raised from debtholders and additional equity is issued from equityholders. Likewise, the cash flow from operations, F_t^o, may turn out to be a cash flow *to* operations whenever additional funds are injected in the operating assets (e.g., capital expenditures).

The case with CFL is particularly important. The cash flow from liquid assets, F_t^l, is, in actual facts, a cash flow *to* liquid assets whenever part or all of the CFO/CCF is not distributed to the investors and is invested in liquid assets. In this case, we will also call it *undistributed cash* or *retained cash* (see Sect. 4.10).

For any one of the four accounts (operating, non-operating, debt, equity) the sign of a cash flow signals whether cash flows are withdrawn from the account or contributed to the account: A positive sign means that a cash flow is withdrawn from the account, a negative sign denotes a cash flow injected into the account. For example, $F_t^o < 0$ means that the operating cash flow is injected into the operating assets, and, hence, this is cash flow received by the firm and distributed *to* the operating assets. Likewise, $F_t^d < 0$ means that cash flow is injected in the debt account and, therefore, this is cash flow distributed from debtholders to the firm. Tables 2.2 and 2.3 summarize the eight possible cases.[10] ▲

Example 2.6 ♦
Given a fixed date t, suppose that the capital invested in a project is partially financed with debt. In particular, two loans are active on the project, one with residual debt equal to \$50 and a second one with residual debt equal to \$460. The interest rates are, in the given period, 5% and 7%, respectively. Thus, the ROD is

$$i_t^d = \frac{0.05 \cdot 50 + 0.07 \cdot 460}{50 + 460} = 0.068 \implies i_t^d = 6.8\%.$$

Example 2.7 ♦
Consider a firm undertaking a capital asset investment. The project's financial position (balance sheet) at time $t-1$ is reported in (2.21) below. Suppose $I_t^o = 200$, $i_t^l = 1\%$, $i_t^d = 3\%$. This implies $I_t^e = 200 - 0.03 \cdot 290 + 0.01 \cdot 250 = 193.8$. Suppose also that $F_t^o = 54$, $F_t^d = 85$, $F_t^e = 100$. This implies $F_t^l = 185 - 54 = 131$.

[10] See also Sect. 1.3 on the sign of cash flows, incomes, capital amounts, and income rates.

$$
\begin{array}{lll}
\multicolumn{3}{c}{\textbf{Investments}} \\
\text{Operating assets} & C^o_{t-1} & = 650 \\
\text{Liquid assets} & C^l_{t-1} & = 250 \\
\text{TOTAL} & C_{t-1} & = 900 \\
\multicolumn{3}{c}{\textbf{Financings}} \\
\text{Debt} & C^d_{t-1} & = 290 \\
\text{Equity} & C^e_{t-1} & = 610 \\
\text{TOTAL} & C_{t-1} & = 900
\end{array}
\tag{2.21}
$$

Applying (2.2)–(2.3),

$$
\begin{aligned}
C^o_t &= 650 + 200 - 54 = 796 \\
C^l_t &= 250(1 + 0.01) - 131 = 121.5 \\
C^d_t &= 290(1 + 0.03) - 85 = 213.7 \\
C^e_t &= 610 + 193.8 - 100 = 703.8
\end{aligned}
$$

The financial position of the system at time t is represented in (2.22) below.

Table 2.2 Cash flow "from" or "to"—Framing (a)

	Cash flow is withdrawn from the account	Cash flow is injected in the account
Operating assets	$F^o_t > 0$	$F^o_t < 0$
Liquid assets	$F^l_t > 0$	$F^l_t < 0$
Debt	$F^d_t > 0$	$F^d_t < 0$
Equity	$F^e_t > 0$	$F^e_t < 0$

Table 2.3 Cash flow "from" or "to"—Framing (b)

Sign	F^o_t	F^l_t	F^d_t	F^e_t
> 0	Cash flow from operating assets	Cash flow from liquid assets	Cash flow to debt	Cash flow to equity
< 0	Cash flow to operating assets	Cash flow to liquid assets	Cash flow from debt	Cash flow from equity

Investments		
Operating assets	C_t^o	$= 796.0$
Liquid assets	C_t^l	$= 121.5$
TOTAL	C_t	$= 917.5$
Financings		
Debt	C_t^d	$= 213.7$
Equity	C_t^e	$= 703.8$
TOTAL	C_t	$= 917.5$

(2.22)

The capital has grown by $\Delta C_t = C_t - C_{t-1} = 17.5$ as a result of an increase in equity ($\Delta C_t^e = 93.8$) and a decrease in debt ($\Delta C_t^d = -76.3$). The net increase in financings is matched by an equal net increase in investments, which is the result of an increase in operations ($\Delta C_t^o = 146$) and a decrease in liquid assets ($\Delta C_t^l = -128.5$). The asset ROI computed in an investment perspective is $i_t = (200 + 0.01 \cdot 250)/900 = 22.5\%$; the same ROI, computed in a financing perspective, is $i_t = (193.8 + 0.03 \cdot 290)/900 = 22.5\%$. As a third equivalent computation, consider that the CFA is $F_t = 54 + 131 = 185$, whence the EOP capital is $E_t = C_t + F_t = 917.5 + 185 = 1{,}102.5$, which implies $i_t = (1{,}102.5 - 900)/900 = 22.5\%$.

Example 2.8 ♦
(*Project financing versus corporate financing*) A project financing transaction may be considered a special case of an expansion project where the assets of a firm are increased by the equity of a legal entity, the SPV (see Sect. 1.4). To better appreciate the difference between project financing and corporate financing, consider a simple example where company XYZ faces the opportunity of undertaking a large scale engineering project, called EngPro, whose cost is \$250 million. Let us focus on date $t = 0$. The equity contributed by company XYZ is \$40 million while debt financing is \$210 million. Company XYZ may undertake the project with recourse debt (corporate financing) or non-recourse debt (project financing).

Suppose the financial position of company XYZ without EngPro consists of \$1,175 assets and liabilities, with a structure as described in Table 2.4.

Let us now describe how the balance sheet of XYZ is affected in the two cases of corporate financing and project financing. To this end, we will use an account

Table 2.4 Firm XYZ—without EngPro

Investments		Financings	
Operating assets	1,050	Debt	560
Liquid assets	125	Equity	615
TOTAL	1,175	TOTAL	1,175

form representation, with investments on the left-hand side and financings on the right-hand side.[11]

If corporate financing is used, both assets and liabilities of XYZ will be increased by $250; in particular, the debt will be increased by $210 and the equity will be increased by $40. If project financing is used, the assets and the equity of XYZ will be increased only by $40 (see Tables 2.5 and 2.6. See also Titman and Martin 2011, pp. 163–164 on project financing).

Table 2.5 Firm XYZ—with EngPro (corporate financing)

Investments		Financings	
Operating assets	1,050	Debt	770
EngPro	250	Equity	655
Liquid assets	125		
TOTAL	1,425	TOTAL	1,425

Table 2.6 Firm XYZ—with EngPro (project financing)

Investments		Financings	
Operating assets	1,050	Debt	560
Equity of EngPro	40	Equity	655
Liquid assets	125		
TOTAL	1,215	TOTAL	1,215

Table 2.7 EngPro

Investments		Financings	
Operating assets	250	Debt	210
		Equity	40
TOTAL	250	TOTAL	250

As for EngPro, its financial position is the same regardless of the financing, with the caveat that, in case of corporate financing, the financial position of EngPro is embedded in XYZ's balance sheet, whereas in case of project financing the financial

[11] The choice between an account form representation or a report form representation is only a matter of taste. However, we will often use a report form representation, and we will build a technique of analysis grounded on it.

position of EngPro is equal to the financial position of a separate firm (see Table 2.7). It should be noted that, in case of corporate financing, the transaction is financed by XYZ with $210 additional debt whereas in case of project financing, XYZ undertakes an equity-financed transaction.[12]

If project financing is selected, XYZ's shareholders will incrementally contribute $40 at time 0 and will receive the stream of CFEs distributed by EngPro. This project is equity financed and the project's prospective cash flows coincide with EngPro's prospective CFEs (assuming no cash will be retained in the firm). The following section specifies the difference between equity financed projects, levered projects and unlevered projects.

2.2 Financial Structure and Capital Structure

Debt and liquid assets may be conveniently matched against each other and mingled so as to give rise to the notion of ***net debt*** or *net financial obligations* or *net financial position*. This is obtained as the difference between debt and liquid assets. We will use the symbol $C_t^{\mathcal{D}} = C_t^d - C_t^l$ to denote the net debt. The reason why debt and liquid assets are often netted out in project analysis is that liquid assets and debt may be interpreted as financially homogeneous in the following sense: A firm could use cash and cash equivalents not used for operations to pay off the debt immediately.[13] If the firm has liquid assets in excess of liquid liabilities, it may conceptually pay all of its debt in a short time and still hold extra liquid assets in an amount that equals to the negative position: $C_t^{\mathcal{D}} = -C_t^l < 0$.

Under this reframing, we may define the net interest expense as the difference between interest expense and interest income: $I_t^{\mathcal{D}} = I_t^d - I_t^l$. Correspondingly, the CFD may be netted out of the CFL to get the cash flow to net debt (CF$\mathcal{D}$). We use the symbol $F_t^{\mathcal{D}} = F_t^d - F_t^l$ to formally denote the CF$\mathcal{D}$. Diachronically, $C_t^{\mathcal{D}} = C_{t-1}^{\mathcal{D}} + I_t^{\mathcal{D}} - F_t^{\mathcal{D}}$. Synchronically, the laws of conservation for capital, income, and cash flow may be reframed as follows:

$$C_t^o = C_t^{\mathcal{D}} + C_t^e \tag{2.23}$$

$$I_t^o = I_t^{\mathcal{D}} + I_t^e \tag{2.24}$$

$$F_t^o = F_t^{\mathcal{D}} + F_t^e \tag{2.25}$$

(see also the representation in (2.26) below).

---

[12] As for EngPro, it is a *levered* firm/project (see Sect. 2.2 on the notion of levered/unlevered project).

[13] See, for example, also Berk and DeMarzo (2014, p. 40 and p. 410).

Operating investments		**Incomes from operations**		**Cash flows from operations**	
Operating assets	(C_t^o)	Operating income	(I_t^o)	Operating cash flow	(F_t^o)
Net financings		**Incomes to net financing**		**Cash flows to net financings**	
Net financial obligations	$(C_t^{\mathcal{D}})$	Net interest	$(I_t^{\mathcal{D}})$	Cash flow to net debt	$(F_t^{\mathcal{D}})$
Equity	(C_t^e)	Net income	(I_t^e)	Cash flow to equity	(F_t^e)

(2.26)

As for the income rate, the operating ROC may be expressed as

$$i_t^o = \frac{i_t^e \cdot C_{t-1}^e + i_t^{\mathcal{D}} \cdot C_{t-1}^{\mathcal{D}}}{C_{t-1}^e + C_{t-1}^{\mathcal{D}}} \tag{2.27}$$

where $i_t^{\mathcal{D}}$ is the return on net debt (RO$\mathcal{D}$):

$$i_t^{\mathcal{D}} = \frac{I_t^{\mathcal{D}}}{C_{t-1}^{\mathcal{D}}} = \frac{I_t^d - I_t^l}{C_{t-1}^d - C_{t-1}^l} = \frac{i_t^d C_{t-1}^d - i_t^e C_{t-1}^l}{C_{t-1}^d - C_{t-1}^l}.$$

The decision about whether debt or net debt should be used to describe a project's investments and financings boils down to a decision about whether liquid assets should be placed on the asset side with positive sign or on the financing side with negative sign. This choice is a matter of convention and has to do with the conceptualization of the liquid assets as an activity that goes along with the operations of the project/firm or as a negative investment that serves the purpose of reducing the financial exposition of the firm. In the former case, gross investments are matched against gross financings, whereas in the latter case net investments (i.e., operating investments) are matched against net financings.[14] The composition of the liabilities in the gross-financing structure is referred to as the ***financial structure***, whereas the composition of the liabilities in the net-financing structure is referred to as the ***capital structure*** (see also Titman et al. 2017). The former represents the sources of financing of the project's assets, the latter represents the sources of financing of the project's operating assets.

The financial structure and capital structure are affected by the so-called financing decisions, that is, decisions regarding borrowing, payout policy, and issuance of equity (see also Remark 4.2 on financing policy)

The notion of financial/capital structure triggers the notion of ***leverage ratio***: It refers to the extent to which a project is financed with debt. The definition of leverage ratio depends on the framing that is adopted. One may define it as the ratio of gross debt to gross financings, $C_t^d/(C_t^d + C_t^e)$ (financial-structure perspective) or as the ratio of net debt to net financings, $C_t^{\mathcal{D}}/(C_t^{\mathcal{D}} + C_t^e)$ $(= C_t^{\mathcal{D}}/C_t^o)$ (capital-structure perspective). The leverage may also be framed in terms of debt-equity ratio: C_t^d/C_t^e (financial-structure perspective) or $C_t^{\mathcal{D}}/C_t^e$ (capital-structure perspective).

[14]The convention of netting liquid assets out of debt is analogous to the convention of netting accounts payable out of accounts receivable to give rise to the notion of net operating working capital (see also Sect. 3 and Table 3.11).

Consequently, a project may be said to be unlevered if and only if $C_t^d = 0$ for all t (financial-structure perspective) or, alternatively, if and only if $C_t^{\mathcal{D}} = 0$ for all t (capital-structure perspective). The distinction of the two framings vanishes whenever a full payout policy is employed (and, at time 0, the capital raised is invested in operating assets alone). In this case, every cash flow generated by the project's operations is distributed to investors (i.e., no fund is retained in the business) so the balance of liquid assets is zero: $C_t^l = 0$ for every t; this means that the project's assets are made of operating assets only (i.e., $C_t = C_t^o$ for all t as well as $C_t^{\mathcal{D}} = C_t^d$ for all t). We adopt a third, more stringent definition of unlevered project.

Definition 2.1 (*Unlevered project*) A project is (operationally) unlevered if and only if $C_t^d = C_t^l = 0$ for every $t \in \mathbb{N}_n^0$

A *levered* project is a project which is not unlevered, that is, a project where either liquid assets or debt (or both) are nonzero for some $t \in \mathbb{N}_t^0$. We also introduce the following definition of equity-financed project.

Definition 2.2 (*Equity financed*) A project is equity financed if and only if $C_t^d = 0$ for every $t \in \mathbb{N}_n^0$ and there exists some t such that $C_t^e > 0$.

According to the above definitions, a project which is equity financed is (negatively) levered if $C_t^l > 0$ for some t; namely, $C_t^{\mathcal{D}}/C_t^e = -C_t^l/C_t^e < 0$. For example, if the stream of non-operating assets is $\boldsymbol{C}^l = -\boldsymbol{C}^{\mathcal{D}} = (0, 10, 20, 30)$, then the project is negatively levered. If, in this case, $\boldsymbol{C}^e$=(100, 90, 80, 70), then $\boldsymbol{C}^o$=$\boldsymbol{C}^e + \boldsymbol{C}^{\mathcal{D}}$=(100, 80, 60, 40) and the leverage (net-debt-to-equity) ratios are, respectively, 0%, −11.1%, −25%, −42.9%. A negative amount of liquid assets, $C_t^l < 0$, may be conceptualized as an ***internal financing*** (the firm raises funds internally from the liquid assets).

Remark 2.5 According to Definition 2.1, a project is not necessarily unlevered if $C_t^{\mathcal{D}} = 0$ for every $t \in \mathbb{N}_n^0$, because the condition $C_t^l = C_t^d = 0$ is not equivalent to $C_t^{\mathcal{D}} = C_t^d - C_t^l = 0$: The former implies the latter, but the latter does not imply the former. For example, consider a two-period project such that $C_0^l = C_0^d = 0$, $C_1^l = C_1^d = 100$. This implies $C_1^{\mathcal{D}} = 0$ notwithstanding the fact that both debt and liquid assets are nonzero. Note that, while the condition $C_t^d = C_t^l = 0$ for all t guarantees $I_t^{\mathcal{D}} = 0$ for all t, the condition $C_t^{\mathcal{D}} = 0$ for all t does not guarantee $I_t^{\mathcal{D}} = 0$ for all t. For example, assume that the return on non-operating assets is $i^l = 1\%$ and the ROD is constant and equal to $i^d = 3\%$. This implies $I_2^{\mathcal{D}} = 3\% \cdot 100 - 1\% \cdot 100 = 2$. Now, the presence or absence of net interest has important implications on valuation, for it determines whether an interest tax shield exists or not (see Chap. 3). For this reason, we prefer to define a project *unlevered* only when both debt and non-operating assets are zero in each period, a condition which logically implies that net interest is zero in each period. ▲

Table 2.8 Operationally unlevered project (Definition 2.1) and financially unlevered project (Definition 2.2)

Assumptions	**unlevered** or *operationally* **unlevered**	**equity financed** or *financially* **unlevered**
$C_t^d \neq 0, C_t^l \neq 0$	NO	NO
$C_t^d \neq 0, C_t^l = 0$	NO	NO
$C_t^d = 0, C_t^l \neq 0$	NO	YES
$C_t^d = 0, C_t^l = 0$	YES	YES

Remark 2.6 (***Operationally unlevered, financially unlevered***) The notions of levered project and equity financed project reflect the difference between capital structure and financial structure. From a capital-structure point of view, investment is made of operations only and financing consists of debt, liquid assets, and equity; hence, "unlevered" means *operationally unlevered* (operations are supported by equity, not by debt capital nor by internal financing). From a financial-structure point of view, investment is made of operations and liquid assets; hence, financing consists of debt and equity. Therefore, "equity financed" means, so to say, *financially unlevered* (operations and liquid assets are supported by equity, not by debt). A project which is operationally unlevered is, necessarily, financially unlevered (operating assets and equity coincide). Vice versa, a financially unlevered project may or may not be operationally unlevered. Table 2.8 clarifies the relationships between the two notions (under the natural assumption $C_t^e > 0$ for some t). The difference between the two notions shows up only if $C_t^d = 0$ and $C_t^l \neq 0$. ▲

To nip possible misunderstandings in the bud, the reader should bear in mind the following special cases:[15]

- Assume $C_t^d = 0$ for all t, then $I_t = I_t^e$ and $F_t = F_t^e$ for all t. Hence,

$$i_t = i_t^e = \frac{i_t^o C_{t-1}^o + i_t^l C_{t-1}^l}{C_{t-1}^o + C_{t-1}^l} \quad \text{for all } t.$$

- Assume $C_t^l = 0$ for all t. Then, $I_t^o = I_t$ and $F_t^o = F_t$ for all t. Hence,

$$i_t = i_t^o = \frac{i_t^e C_{t-1}^e + i_t^d C_{t-1}^d}{C_{t-1}^e + C_{t-1}^d} \quad \text{for all } t.$$

- Assume $C_t^d = C_t^l = 0$ for all t (project is unlevered). Then, $F_t = F_t^o = F_t^e$ and $I_t = I_t^o = I_t^e$ for all t. Hence,

$$i_t = i_t^e = i_t^o \quad \text{for all } t.$$

[15] Proof is straightforward by using the law of conservation.

In corporate finance texts, it is usual to reason in terms of net financial position and, therefore, adopt a capital-structure perspective rather than a financial-structure perspective. However, the notion of potential dividends (i.e., the cash flow that is available for distribution to equityholders) is conceptually easier to understand if one separates liquid assets and debt (see Sect. 4.10). Also, internal financing is rather common, especially for small-scale projects or for firms which have a substantial amount of liquid assets. Therefore, while capital structure is useful whenever the operating value of the project is at stake, financial structure is preferable when dealing with potential dividends or internal financing. The two perspectives are logically equivalent and lead to the same valuations.

Remark 2.7 Financial investments, as opposed to real investments, are well served by a financial-structure perspective, for, in those cases, no operating assets are present (i.e., no goods are produced or sold nor a service is delivered). A financial investment changes the firm's status quo in such a way that $C_t^o = 0$ for all t. If one employed a capital-structure perspective, one would be left with a zero amount of investments and a zero amount of financings:

$$\overbrace{C_t^o}^{\text{invested capital}} = 0 = \overbrace{C_t^e + C_t^d - C_t^l}^{\text{financed capital}}.$$

In order to avoid to work with a zero capital, it may be preferable to consider C_t^l as an investment (rather than a negative financing) so that C_t^l will be properly placed on the investment side while C_t^d and C_t^e will be placed on the financing side:

$$\overbrace{C_t^l}^{\text{invested capital}} = \overbrace{C_t^d + C_t^e}^{\text{financed capital}}.$$

Further, while it is true that a financial investment may be viewed as a way of reducing the net debt, in other cases financial investments are made for increasing the equitholders' wealth, which makes the positive-investment interpretation more compelling than the negative-financing one. ▲

Description of a project or firm may be made with a financial-structure perspective or a capital-structure perspective. This book manages both. However, from a graphical point of view, **projects will be illustrated with a financial-structure perspective**, in order to better disentangle the roles of debt and liquid assets in the financing/distribution policy. (If the analyst prefers the capital structure perspective, it will suffice to remove the liquid-asset account from the asset side and determine the net debt in the financing side by subtracting the liquid assets from the gross debt: $C_t^{\mathcal{D}} = C_t^d - C_t^l$.)

However, we adopt a capital-structure perspective for the definition of unlevered project. Specifically, a project will be said to be ***unlevered*** if it is operationally unlevered (capital structure) and ***equity financed*** if it is financially unlevered (financial structure).

Example 2.9 ♦
Consider Example 2.7. If net debt is used, one may describe the financial position as follows:

Financial position at time $t-1$		Financial position at time t	
Operating Investments		**Operating Investments**	
Operating assets	$C^o_{t-1} = 650.0$	Operating assets	$C^o_{t-1} = 796.0$
TOTAL	$C^o_{t-1} = 650.0$	TOTAL	$C_{t-1} = 796.0$
Net Financings		**Net Financings**	
Net financial position	$C^{\mathcal{D}}_{t-1} = 40.0$	Net financial position	$C^{\mathcal{D}}_{t-1} = 92.2$
Equity	$C^e_{t-1} = 610.0$	Equity	$C^e_{t-1} = 703.8$
TOTAL	$C^o_{t-1} = 650.0$	TOTAL	$C^o_{t-1} = 796.0$

(2.28)

The operating ROC is $i^o_t = 200/650 = 30.77\%$; the BOP net debt is $C^{\mathcal{D}}_{t-1} = 40$, the net interest payment is $I^{\mathcal{D}}_t = 0.03 \cdot 290 - 0.01 \cdot 250 = 6.2$, and the RO$\mathcal{D}$ is

$$i^{\mathcal{D}}_t = \frac{6.2}{40} = 15.5\%.$$

As seen, the ROE is $i^e_t = 31.77\%$ and i^o_t may be written as the weighted average of ROE and RO$\mathcal{D}$:

$$i^o_t = \frac{0.3177 \cdot 610 + 0.155 \cdot 40}{610 + 40} = 30.77\%.$$

Example 2.10 ♦
(*Borrowing from the project: Positive or negative framing*) There are circumstances where the project may be a financing project, which means that the firm absorbs resources from the assets. Typically, abandonment projects are financing projects (see Sect. 1.4.2). However, replacement projects and expansion projects may also imply that the firm borrows from the project, at least in some periods (see Sect. 1.4.1 and, specifically, Table 1.6). While we crossrefer the reader to Sect. 2.3 for a detailed treatment, we here sketch a situation where the firm absorbs resources from the operating assets.

Consider that operating assets consist of net operating working capital (NOWC) and fixed assets, net of depreciation (NFA).[16] Both classes are usually positive; however, there may be circumstances where they become negative. For example, consider a firm which, at time t, employs a small amount of fixed assets, whose book

[16] See Sect. 3.1 for details.

value is $\text{NFA}_t = \$50$, and suppose the firm's customers purchase the goods produced by the firm in cash, while the firm's trade credit agreements enable the firm to pay suppliers after 90 days. This means that money is received from customers before money is paid to suppliers, which implies that NOWC is negative (cash cycle is negative).[17] Specifically, suppose $\text{NOWC}_t = -\$300$. Assume that the cash received from the customers is invested in short-term financial assets and that, currently, the balance amounts to $C_t^l = \$1{,}200$. Equity is assumed to be equal to \$200 and debt is assumed to be equal to \$750.

The situation may be described in terms of financial structure or in terms of capital structure. In terms of financial structure, one may say that the firm absorbs

(i) \$300 from customers and suppliers
(ii) \$200 from the equityholders
(iii) \$750 from debtholders

while investing

(i) \$50 in NFA
(ii) \$1,200 in financial assets.

(left-hand side of Eq. (2.30)); in terms of capital structure, the firm absorbs

(i) \$300 from customers and suppliers
(ii) \$200 from the equityholders

while investing

(i) \$50 in NFA
(ii) \$450 in net financial assets.

(right-hand side of Eq. (2.30)). Overall, in the latter case, the firm subtracts capital from the project (net assets are negative) by an amount of \$250 ($C_t^o = \text{NOWC}_t + \text{NFA}_t = -300 + 50 = -250$), which is invested in the net financings (net financings are negative) ($C_t^e + C_t^d - C_t^l = 200 + 750 - 1{,}200 = -250$). The reason is that a negative investment is a ... financing, and a negative financing is an ... investment.[18] Therefore, in some cases, whether a firm is in an investment position or in a financing

[17] See treatment of NOWC in Sect. 3.2 and definition of cash cycle in Sect. 3.2.4. See also Sect. 2.3.

[18] A further framing may get rid of negative numbers by placing investments and financings on the same side:

Borrowing from the project – positive numbers

Investments		
Net fixed assets	NFA_t	$= +50$
Net financial asset	$-C_t^{\mathcal{D}}$	$= +450$
TOTAL		$= +500$
Financings		
Net operating working capital	$-\text{NOWC}_t$	$= +300$
Equity	C_t^e	$= +200$
TOTAL		$= +500$

(2.29)

position is a matter of how the situation is conventionally represented by the analyst. Logical equivalence holds whatever the framing, but the economic interpretation is different.

Borrowing from the project – Financial structure			Borrowing from the project – Capital structure		
Gross Investments			**Net Investments**		
Net fixed assets	NFA_t	$= +50$	Net fixed assets	NFA_t	$= +50$
Net operating working capital	$NOWC_t$	$= -300$	Net operating working capital	$NOWC_t$	$= -300$
Liquid assets	C_t^l	$= +1{,}200$			
TOTAL	C_t	$= 950$	TOTAL	C_t^o	$= -250$
Gross Financings			**Net Financings**		
Debt	C_t^d	$= +750$	Net financial position	$C_t^{\mathcal{D}}$	$= -450$
Equity	C_t^e	$= +200$	Equity	C_t^e	$= +200$
TOTAL	C_t	$= 950$	TOTAL	C_t^o	$= -250$

(2.30)

Example 2.11 ♦
(*Equity-invested project*) The opposite of "equity-financed" is not "unlevered", but "equity-invested". An equity-invested project is a project where the equity coincides with the assets (debt is zero) and the latter are negative: $C_t = C_t^e < 0$ for all t.

This is often the case with abandonment projects. For example, consider the abandonment project described in Example 1.11. As already seen, Table 1.7 shows that the firm is in a financing position: It absorbs resources from the operating assets. As terminating the operations does not entail any deviation in the firm's debt outstanding, the project's debt is zero: $C_t^d = 0$ for all t. We also assume $C_t^l = 0$ for all t. Hence, the equity capital is $C_t^e = C_t^o$ for all t. As $C_t^o = C_t^e < 0$, discontinuing the operations implies that the firm's equityholders borrow from the firm (which in turn borrows from the project). Precisely, they absorb resources amounting to $|C_0^e| = 6{,}000$ from the firm. However, the subtraction of such resources generates an immediate income of $I_0 = I_0^e = 500$ for equityholders. In the next periods, equityholders will pay back monetary amounts to the firm equal to $|F_1^e| = 1{,}600$, $|F_2^e| = 1{,}400$, $|F_3^e| = 1{,}300$, $|F_4^e| = 1{,}250$, and $|F_5^e| = 1{,}210$.[19] Equityholders keep their financing position until the end of the project: $C_5^e = 0$. Borrowing from the firm entails that they bear financing costs equal to $|I_1^e| = 400$, $|I_2^e| = 200$, $|I_3^e| = 100$, $|I_4^e| = 50$, and and $|I_5^e| = 10$.

The ratio of equity to assets is 100% in every period. However, this is obtained by dividing a negative number (C_t^e) by a negative number (C_t) (for example, at time 0,

We will keep the convention according to which the symbol C_t^{inv} denotes the sum of operating and non-operating assets, whatever the sign. Accordingly, we will keep the symbol C_t^{fin} to denote the sum of equity and debt, regardless of the sign.

[19]More precisely, the firm will distribute to equityholders an amount of money which will be smaller by $|F_t^e|$ than the amount that would be distributed if the project were rejected (however, if the latter is nonnegative and smaller than $|F_t^e|$, then equityholders will indeed make actual payments to the firm).

the ratio is −6,000/(−6,000) = 1). This is interpretable by saying that the project is equity-invested; that is, 100% of the capital subtracted by the firm from the assets is invested in equity. (Symmetrically, one might say that it is …100% asset-financed).

The project	loans $\lvert C_t \rvert$	to the firm
The firm	borrows $\lvert C_t \rvert$	from the project
The firm	loans $\lvert C_t^e \rvert = \lvert C_t \rvert$	to equityholders
The equityholders	borrow $\lvert C_t^e \rvert = \lvert C_t \rvert$	from the firm

At time 0, the statement of the project's financial position is as in Table 2.9. Investments are negative and financings are negative. As already noted, a negative investment is a financing and a negative financing is an investment. Therefore, one might switch headings and change signs, as in Table 2.10. The same structure holds in the following periods.

2.3 Borrowing from the Project versus Borrowing from Debtholders

Examples 2.10 and 2.11 bring attention to those situations where the firm borrows from the project. These cases should not be confused with borrowing from debtholders. Borrowing from a debtholder means to receive funds from an economic agent who, by the enforcement of a legal contract, is entitled to receive payments of principal and interest. As seen, such financings are measured with positive sign in the liability side of the balance sheet.

In contrast, borrowing from the project means that an asset is used as a *source* of funds instead of a *use* of funds or, in other words, that the firm subtracts resources from the assets. For example, an upfront payment from a customer means that the firm is borrowing from the customer. In general, whenever the firm succeeds in receiving cash from customers before payments to suppliers are due (a situation often described in terms of *negative cash cycle*), then the NOWC is financing the firm:

> Some firms are able to operate with zero or even negative net working capital. Dell Computer and Amazon are examples. When customers order computers from Dell's Web site or books from Amazon, they must provide a credit card number. Dell and Amazon then receive next-day cash, even before the product is shipped and even before they have paid their own suppliers. This results in a negative CCC [Cash Conversion Cycle], which means that working capital *provides* cash rather than *uses* it. In order to grow, companies normally need cash for working capital. However, if the CCC is negative, then growth in sales *provides* cash rather than *uses* it. This cash can be invested in plant and equipment, research and development, or for any other corporate purpose. Analysts recognize this point when they value Dell and Amazon, and it certainly helps their stock prices. (Ehrhardt and Brigham 2016, p. 667)

Table 2.9 Equity-invested project (a)

Investments		Financings	
Operating assets	−$6,000	Debt	$0
Liquid assets	$0	Equity	−$6,000
TOTAL	−$6,000	TOTAL	−$6,000

Table 2.10 Equity-invested project (b)

Financings		Investments	
Operating assets	$6,000	Liquid assets	$0
Debt	$0	Equity	$6,000
TOTAL	$6,000	TOTAL	$6,000

(See also Fig. 3.3 in Sect. 3.2.4) The case of Dell is particularly interesting, because its business strategy precisely consists of absorbing resources from operating assets and investing them in financial assets. Penman (2010) explains this strategy with the following words:

> [T]he striking feature of Dell's strategic balance sheet is the negative net operating assets …How can it be? …Keep operating assets low with just-in-time inventory, require a credit card before shipping retail customer sales (thus keeping accounts receivable low), outsource production (reducing investment in plant and equipment), require cash upfront for servicing contracts (and thus amass large deferred revenues) and, importantly, require suppliers to carry Dell's payables and supply operating credit. (Penman 2010, p. 299)

We will henceforth denote such financings with negative signs in the investment side of the balance sheet. However, consider that, when speaking of a *project*, a negative NOWC or a negative operating capital do not necessarily mean that the *firm*'s NOWC or operating capital are negative. To understand this, remember that a project represents an incremental system (see Sect. 1.4); by definition, the project's capital is the difference between the firm's capital with the project and the firm's capital without the project. Therefore, a negative project's operating capital may well mean that the firm's operating capital is positive in both cases (with and without the project) but the firm's operating capital in case the project is undertaken is smaller than the firm's operating capital in case the project is rejected.

Likewise, NFA may turn out to be negative: In some periods, the firm's NFA with the project may be smaller than the firm's NFA without the project. Replacement projects and abandonment projects are examples of situations where the project's NFA may be negative.

Even liquid assets may well turn out to be negative in some periods, as a result of their being an incremental asset. This occurs, for example, whenever the project is financed with the firm's liquidity (internal financing). For example, in the Rainbow

Educational Program presented in Sect. 12.3, the project's capital is negative (as well as the equity capital), as a result of internal financing, meaning that the firm subtracts resources from the assets and invests them in equity (in particular, in this project, operating assets are negative as well as liquid assets).

In other words, borrowing from a project implies that, overall, the firm will use less resources if the project is undertaken than it will use if the project is not undertaken. This is reflected in the negative sign of the project's assets, which signals that the project undertaking frees out resources. By law of conservation, such resources will be necessarily invested somewhere. Specifically, if resources are wholly absorbed by the firm from the asset side, they are necessarily invested in the liability side.

Regardless of the kind of financing, one should always bear in mind that the firm is, at any given point in time, in a lending position with respect to some agent and in a financing position for the same amount with respect to some other agent. If the whole project's capital is negative at a given time (i.e., $C_t = C_t^o + C_t^l < 0$), the capital subtracted is used for reducing the claimholders' (equityholders+debtholders) committed capital. Which means that, in differential terms, the amount invested in the firm by the claimholders with the project is less than the amount invested in the firm by the claimholders if the project is not undertaken:

$$\overbrace{C_t^d + C_t^e}^{C_t^{\text{inv}}} = \underbrace{C_t}_{<0} = \overbrace{C_t^o + C_t^l}^{C_t^{\text{fin}}}$$

Remark 2.8 (*Terminology*) It is worth mentioning that C_t^{inv} refers to the capital associated with the project's assets, which are, usually, investments, which is the reason why we use the superscript "inv" to denote the project's assets. Accordingly, the superscript "fin" of C_t^{fin} refers to the project's liabilities, which are usually financings. However, when C_t is negative, then C_t^{inv} and C_t^{fin} are negative as well. Thus, C_t^{inv} denotes a borrowed amount (assets become "financings"), while C_t^{fin} denotes an invested amount (liabilities becomes "investments"). ▲

A firm may even raise, at the same time, some funds from the project's assets *and* some funds from debtholders. In this case, by the law of conservation, equity capital will be negative:

$$\underbrace{C_t}_{<0} - \underbrace{C_t^d}_{>0} = \underbrace{C_t^e}_{<0}.$$

(that is, the project is equity-invested at time t. See Example 2.11).

It should then be clear that the cost of subtracting resources from the assets is the ROI, not the ROD. More precisely, adopting the point of view of the firm,

- if $C_{t-1}^o < 0$, the operating ROI, i_t^o, is the cost of raising funds from the operating assets (NOWC and/or NFA)
- if $C_{t-1}^l < 0$, the non-operating ROI, i_t^l, is the cost of raising funds from the liquid assets

- if $C^d_{t-1} > 0$, the ROD, i^d_t, is the cost of raising funds from debtholders
- if $C^e_{t-1} > 0$, the ROE, i^e_t, is the cost of raising funds from equityholders
- if $C_{t-1} < 0$, the asset ROI, i_t, is the cost of raising funds from the assets

(see also Sect. 1.3).

Borrowing from the project is economically different from raising funds from lenders such as a bank (via loans) or investors in the capital market (via bonds or issuance of new shares). While the latter implies that the project undertaking increases the amount of debt in the firm (or the amount of equity, in case of stock issuance), the former means that **the project undertaking decreases the amount of assets** in the firm. More precisely, the firm borrows from the economic agents that are representative of the project (suppliers, customers, etc.). The cost of subtracting such funds is not equal to the interest rate on debt; it has to do with the law of motion of the project's assets (and, therefore, with the business transactions involving customers and suppliers as well as the withdrawals/deposits of cash and sales/purchases of financial assets). The cost of raising debt from debtholders is the ROD, i^d_t; the cost of subtracting funds from the project is the ROC, i_t, that is, the ratio of the project's income to the (negative) project's assets.

2.4 Key Points

- For any given date t, an economic system presents a financing side as well as an investment side. In particular, a firm always finds itself in a financing position *and* in an investment position: It is financed by capital providers for some capital amount and, at the same time, it invests that amount in some economic activities. Investments are also called assets and financings are also called liabilities (including net worth)
- Assets are divided into two classes: Operating assets and non-operating (liquid) assets. Financings are divided into two classes: Debt capital and equity capital
- The law of conservation establishes the equilibrium between the forces acting on the economic system synchronically. This law takes the form of 4 balancing principles according to which (i) the capital invested in the assets is equal to the capital raised from debtholders and equityholders, (ii) the income generated by the assets is equal to the income accrued to debtholders and equityholders, (iii) the cash flow flowing out of the assets is distributed to debtholders and equityholders, and (iv) the rate of return generated by the assets is equal to the rate of cost paid by the firm to debtholders and equityholders
- Point (iv) above implies that the income rate is a weighted mean. Specifically, it is the capital-weighted mean of the return rates generated by the operating assets and the liquid assets and, at the same time, the capital-weighted mean of the return

rates earned by debtholders and equityholders, respectively. We summarize the static equilibrium of an economic entity and, in particular, a project, as follows:

Statics of an economic system: Law of conservation

Any economic system presents an investment side and a financing side which always match in four senses:

- capital invested in the economic activity = capital raised from capital providers:

$$\overbrace{C_t^o + C_t^l}^{C_t^{\text{inv}}} = C_t = \overbrace{C_t^d + C_t^e}^{C_t^{\text{fin}}}$$

- income generated by the economic activity = income accrued to capital providers:

$$\overbrace{I_t^o + I_t^l}^{I_t^{\text{inv}}} = I_t = \overbrace{I_t^d + I_t^e}^{I_t^{\text{fin}}}$$

- cash flow contributed by the economic activity = cash flow distributed to capital providers:

$$\overbrace{F_t^o + F_t^l}^{F_t^{\text{inv}}} = F_t = \overbrace{F_t^d + F_t^e}^{F_t^{\text{fin}}}$$

- rate of return generated from the assets = rate of return accrued to capital providers:

$$\overbrace{\frac{i_t^o C_{t-1}^o + i_t^l C_{t-1}^l}{C_{t-1}^o + C_{t-1}^l}}^{i_t^{\text{inv}}} = i_t = \overbrace{\frac{i_t^e C_{t-1}^e + i_t^d C_{t-1}^d}{C_{t-1}^e + C_{t-1}^d}}^{i_t^{\text{fin}}}$$

- The liquid assets may be merged with debt to give rise to the net financial position or net debt. The financial structure is the proportion of debt and equity; the capital structure is the proportion of net debt and equity
- A project is unlevered if both debt and liquid assets are zero. A project is negatively levered whenever liquid assets exceed debt
- In some (or all) periods, the project may be such that the firm absorbs resources from it. In this case, the project is a source of funds rather than a use of funds. By the law of conservation, this implies that the firm invests those resources in the financing side. Typical situations are replacement projects and abandonment projects. In these cases, the ROI represents the cost of financing *for* the claimholders. It is important to distinguish the situation where the firm is borrowing from the project (C_t is negative) and the situation where the firm is borrowing from debtholders (C_t^d is positive)

- In the light of what seen above, "assets" are often "investments" and "liabilities" are often "financings". However, they are not synonyms. In some cases, when dealing (with firms and, more compellingly) with capital asset projects, assets and liabilities may well be negative, acting as financings and investments, respectively.

The law of conservation for the capital amounts is often visualized in tabular format, either in a report form or an account form, either using a financial structure or a capital structure. The former lists liquid assets as an investment, the latter lists liquid assets as a (negative) financing, deducted from the debt to form the net debt (i.e., net financial position). Using an account form statement,

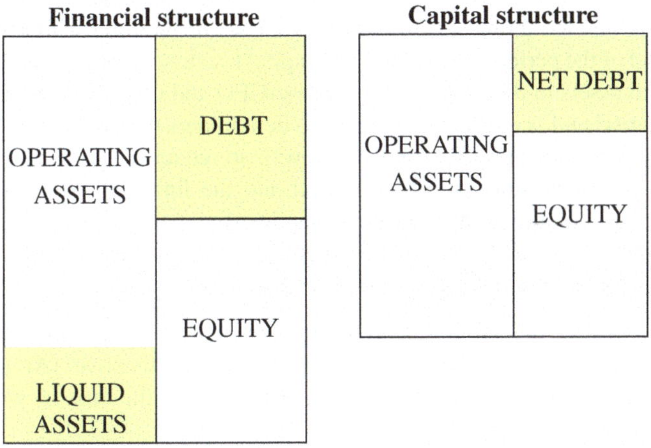

Using a report form statement,

Financial structure		*Capital structure*	
Gross Investments		**Net Investments**	
OPERATING ASSETS	C_t^o	OPERATING ASSETS	C_t^o
LIQUID ASSETS	C_t^l		
TOTAL	C_t	TOTAL	C_t^o
Gross Financings		**Net Financings**	
DEBT	C_t^d	NET DEBT	$C_t^{\mathcal{D}}$
EQUITY	C_t^e	EQUITY	C_t^e
TOTAL	C_t	TOTAL	C_t^o

The statics and the dynamics of an economic system constitute the mechanics, that is, the economic process triggered by the laws of motion and conservation. The mechanics may also be grasped by inspecting Fig. 2.3 which provides, at the same time, the diachronic perspective and the synchronic perspective.

Specifically, inspecting the figure, and considering a standard situation:

1. At the beginning of any given period, the firm is financed by debt capital and equity capital
2. The capital raised is entirely invested by the firm in operating assets and liquid assets [**law of conservation for capital**]
3. Operating assets and liquid assets generate incomes which increase the capital invested (or decrease if incomes are, overall, negative) [**law of motion for investments**]. The total income generated by the assets (operating income and interest income) exactly offsets the total income accrued to debtholders and equityholders (i.e., interest expense and net income) [**law of conservation for incomes**] so that the capital borrowed by the firm is increased [**law of motion for financings**]
4. At the end of the period, part of the EOP capital is extracted from operating assets and liquid assets in the shape of cash flows (CFO and CFL) [**law of motion for investments**] and is entirely distributed to debtholders (CFD) and equityholders (CFE) [**law of conservation for cash flows**], so reducing the firm's committed capital [**law of motion for investments**] and the financial exposition toward capital providers [**law of motion for financings**]
5. In the following period, the same loop is observed, and so on until investment and financing are gradually reduced to zero at time n.

Figure 2.4 describes the same diachronic-synchronic structure using the capital-structure perspective instead of the financial-structure perspective. (An alternative graphical representation with a strong operational power, the *Matrix*, will be presented in Chap. 3.)

Taken together, the two fundamental laws presented in the first two chapters constitute

- the primary sources for developing a reliable logic of investment valuation and decision-making
- the first part of the accounting-and-finance engineering system
- the basis for unifying economics, finance, and accounting
- the ground for elaborating a unified (theoretical and practical) methodology for analyzing any kind of project, either real or financial, and designing reciprocally consistent (absolute and relative) measures of worth

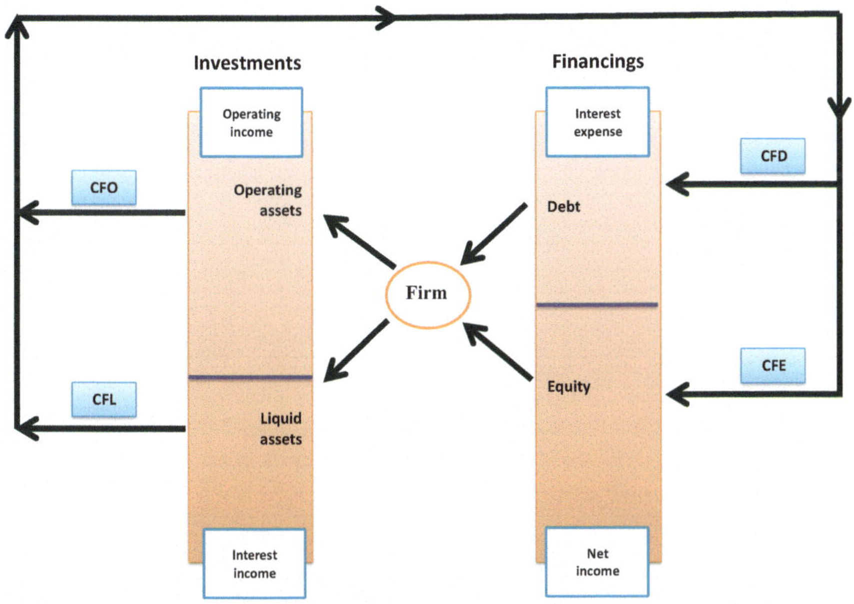

Fig. 2.3 Statics and dynamics (financial-structure perspective)

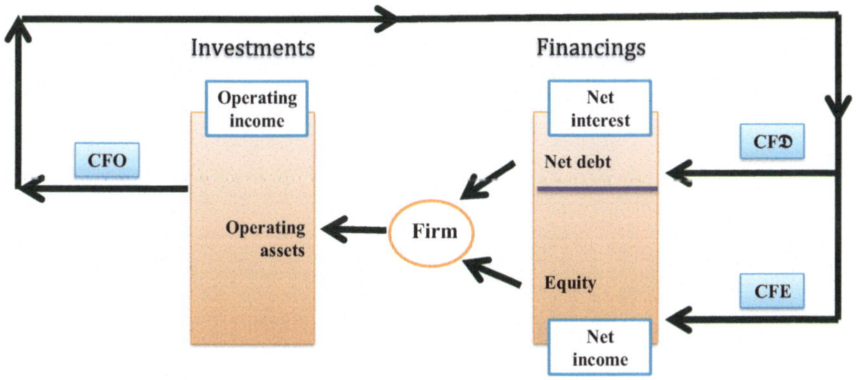

Fig. 2.4 Statics and dynamics (capital-structure perspective)

Part II
Accounting-and-Finance Engineering System: The Matrix

Building upon the mechanics of an economic system, as illustrated in Part I, this part is devoted to **designing a logical structure** which engineers, so to say, accounting and financial data in a systematic, rigorous way, so that a refined economic analysis may be carried out and the valuation and decision-making process may follow in a natural and intuitive way.

The task is accomplished by building a *diachronic-synchronic* **Matrix**, which embodies both statics relations (vertical dimension) and dynamics relations (horizontal dimension).

The Matrix is then gradually expanded to incorporate several subsystems, each one equipped with its own mechanics: After breaking down the system into the subsystems of operating assets, non-operating assets, debt financings, equity financings, we disaggregate the four classes into lower-level subsystems, including, among others, the subsystems of **fixed assets, working capital, net operating liabilities, excess cash, marketable securities, loans, bonds**. They all fulfill their own laws of motion and are reciprocally interlocked by the laws of conservation.

The Matrix is equipped with a *split-screen* technique, giving rise to a **quadruple-entry system** which enables

- visualizing, at the same time, the **statics** and the **dynamics** of an economic system in an intuitive tabular format and, consequently,
- engineering the "**accounting and finance**" of a project
- condensing into one single visualization the **pro forma (projected) financial statements**
- deriving complex notions and relationships with **simple working rules**
- analyzing the project under several **different perspectives** (levered vs. unlevered, investment vs. financing, financial structure vs. capital structure, actual cash flow vs. potential cash flow)
- acting as a **diagnostic tool** for revealing logical inconsistencies and minimizing the risk of modelling errors.

Chapter 3
Financial Statements

SCREEN
Do you want to know what the Matrix is, Neo?
[...]
NEO
The Matrix. What is the Matrix?

The Matrix (DVD). Produced by J. Silver. Script by Larry and Andy Wachowski, April 1996.

C. A. Magni, *Investment Decisions and the Logic of Valuation*,
https://doi.org/10.1007/978-3-030-27662-1_3

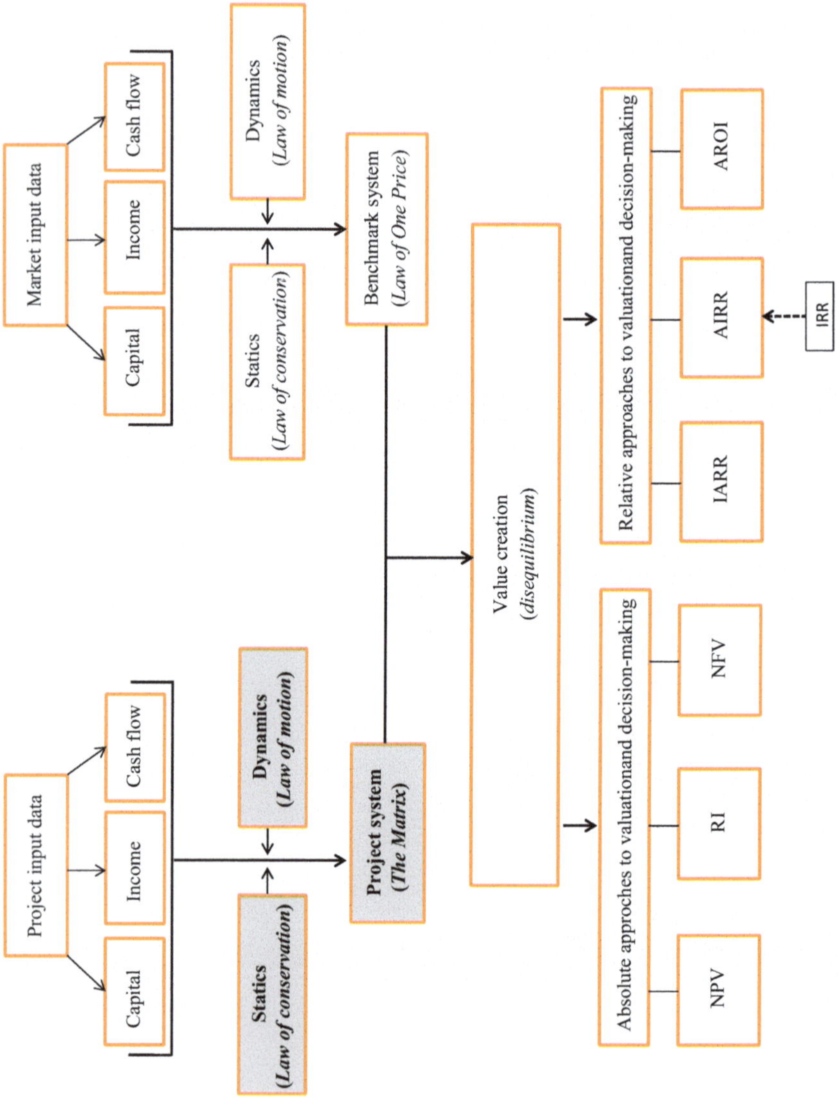

Fig. 3.1 Chapter 3: The Matrix (financial statements)

This chapter delves into the details of the project system.

Building upon Chap. 2, the system is broken down into four subsystems, namely the classes of operating assets, non-operating assets, debt, and equity. They are in turn broken down into lower-level subsystems, including the subsystems of **fixed assets**, **working capital**, **operating liabilities**, **loans**, **bonds**, **excess cash**, **marketable securities**, which may in turn be broken down into further lower-level subsystems, up to the desired level of detail. Each subsystem fulfills its own law of motion and, therefore, has its own income components and cash-flow components, which are reciprocally interlocked with the other subsystems by the laws of conservation.

We embed this nested network into a diachronic-synchronic matrix, called **the Matrix**, which collects, in one single visual representation, both the statics and the dynamics of a project system. Economically, the Matrix turns the triplet of the basic notions $(\boldsymbol{C}, \boldsymbol{I}, \boldsymbol{F})$ alongside its mechanics into a triplet of fundamental statements:

- the statement of financial (i.e., capital) position or **Balance Sheet**
- the statement of incomes or **Income Statement**
- the statement of cash flows or **Cash Flow Statement**.

We use a simple and intuitive technique, called **split-screen technique**, upon the Matrix, in order to effectively manage the labyrinth of intricate relations implied by the three statements. Among other things, this quadruple-entry system enables deriving, with minimum cognitive effort, a network of tricky relations among subtly different notions of income, capital, cash flow, and income rate.

The Matrix and its associated split-screen technique will also enable one to easily understand the various notions of costs and the various notions of income, as well as extract significant relations that are relevant for project appraisal and decision-making while, at the same time, minimizing the risk of making errors that would result in model inconsistencies.

After completing this chapter, the reader should be able to:

- break down the capital into operating assets, non-operating assets, debt capital, equity capital and, in turn, break them down into further subsystems:
  - net fixed assets (property, plant and equipment)
  - net operating working capital (accounts receivable, accounts payable, inventories, taxes payable, salaries and wages payable)
  - excess cash, marketable securities and other financial assets
  - loans, bonds, and notes payable
  - equity
- blend the statics and the dynamics into the *Matrix* of a project, which horizontally displays the laws of motion and vertically displays the laws of conservation
- use the Matrix for analyzing the capital components, the income components, and the cash-flow components. These components are summarized in the Matrix's columns, called, respectively,
  - balance sheet
  - income statement
  - cash flow statement
- split the Matrix into four screens and use a related *split-screen technique* in order to easily conceptualize the (fourfold) interrelations among the various types of capital, income, and cash flow at any level of complexity
- understand in depth such notions as gross profit (GP), earnings before interest, taxes, depreciation and amortization (EBITDA), earnings before interest and taxes (EBIT), operating income (OI), net income
- differentiate, on one hand, between manufacturing and nonmanufacturing costs and, on the other hand, between variable and fixed costs
- calculate the *cost of goods sold* in either a financial or a managerial accounting perspective
- reframe the income statement in terms of a cost-volume-profit analysis.

3.1 The Breakdown and the Matrix: Starting Steps

The fundamental principles describing the statics and the dynamics of an economic system have been presented in Chaps. 1–2 and visualized in Fig. 2.3. In some sense, the statics of an economic system may be seen as consisting in breaking down the law of motion (1.1) into four laws of motion, corresponding to the four main classes of an economic system: Operating assets, non-operating assets, debt, and equity (see Eq. (2.2)). In the following sections we proceed to split up each such class into subclasses, so reaching a greater level of details. We show that each subclass enjoys its own law of motion. The breakdown will be described with the aid of an effective and logically robust representation in the format of a matrix structure:

$$\begin{array}{c|ccc} +C_t^o & +C_{t-1}^o & +I_t^o & -F_t^o \\ +C_t^l & +C_{t-1}^l & +I_t^l & -F_t^l \\ \hline +C_t^d & +C_{t-1}^d & +I_t^d & -F_t^d \\ +C_t^e & +C_{t-1}^e & +I_t^e & -F_t^e \end{array} \tag{3.1}$$

We call this representation the ***split-screen Matrix*** or, simply, ***the Matrix***. It is partitioned into four blocks, delimited by a horizontal bar and a vertical bar. The bars act as equality signs. For any given row or column, the elements of opposite partitions sum up to the same amount. Therefore, the rows of the Matrix represents the diachronic aspect (laws of motion), the columns of the Matrix represent the synchronic aspect (laws of conservation). This Matrix is the germ of an enlarged Matrix that may be obtained by breaking down the rows. More precisely, we will show that each row of the germ Matrix may be disaggregated into several multiple rows expressing the laws of motion of several subsystems, thereby generating an expanded Matrix.

We show that such an expanded Matrix naturally encompasses three statements, one for each basic element: The statement of financial position (or ***balance sheet***, BS) for the capital, the statement of incomes (or ***income statement***, IS) for the income, and the statement of cash flows (or ***cash flow statement***, CFS) for the cash flow. Afterwards, we show that the Matrix is gifted with interesting invariance properties that make it a powerful working tool for financial and accounting analysis.

3.2 Operating Assets

The operating assets are needed for supporting the operations related to the project/firm. They are typically divided into net fixed assets (NFA) and net operating working capital (NOWC).

Net Fixed Assets (NFA)

The NFA are long-term assets consisting of property, plant, and equipment. The two sources of change of this class are:

i. Noncash expenses for **depreciation**, amortization and depletion (Dep_t). They decrease or increase NFA depending on whether the decrease in value of some assets is more than compensated by the increase in some others (typically, real estate assets, land, precious metals, artwork)
ii. Cash for **capital expenditures**, net of disposal of fixed assets (F_t^{capex}).[1]

One may then write a law of motion for net fixed assets:

$$\text{NFA}_t = \text{NFA}_{t-1} - \text{Dep}_t + F_t^{capex}. \tag{3.2}$$

Conceptually, depreciation reflects the fact that, while fixed assets do not physically consume like supplies, they decline in value owing to wear and tear. Depreciation expense is the portion of the historic cost of a fixed asset that is allocated to a given period or, in more accounting terms, it "is the allocation of the cost of a tangible plant asset to expense in the periods in which services are received from the asset" (Williams et al. 2010, p. 395). The appropriate amount of depreciation is an estimate, so it is a matter of informed judgment, and it depends on various factors, including the years the asset is expected to be in use. There are several different methods for depreciating fixed assets, the most common being straight-line depreciation (SL), where an equal portion of the asset's cost is depreciated each year (thus, the depreciation expense is the ratio of the cost of the asset to the estimated useful life). Other common depreciation patterns consist in accelerated depreciation, where higher depreciation expenses are recorded in the early years. A general and simple definition of accelerated depreciation is: "A depreciation method that writes off more of the asset's cost near the start of its useful life than the straight-line method does." (Horngren et al. 2012, p. 479). The accelerated methods include, among others, double-declining balance, modified accelerated cost recovery system, sum-of-the-years' digits, double-declining balance switching to straight-line depreciation. We focus on three different depreciation patterns:

- straight-line (SL)
- double-declining balance (DDB)
- double-declining-balance switching to straight-line (DDB-SL).

Let Q_0 be the amount invested by the firm in a fixed asset at time 0 and Q_t be its carrying amount (depreciated amount); let N be the depreciable life.

The SL depreciation pattern applies the depreciation rate $1/N$ to the initial investment Q_0, in order to get a uniform charge, $\text{Dep}_t^{SL} = (1/N) \cdot Q_0$ so that the depreciated amount is

$$Q_t^{SL} = Q_{t-1}^{SL} - \text{Dep}_t^{SL} = Q_{t-1}^{SL} - Q_0/N$$

or, in terms of initial investment, $Q_t^{SL} = Q_0(1 - t/N)$.

The double-declining-balance (DDB) method makes use of a depreciation rate which is twice the depreciation rate used in the SL depreciation. However, the depreciation rate is not applied to Q_0 but to the carrying amount. Formally, the carry-

[1]Capital expenditures include major repairs, major maintenance, and improvements.

ing amount at time t is recursively obtained as $Q_t^{DDB} = Q_{t-1}^{DDB} - \text{Dep}_t^{DDB}$ where $\text{Dep}_t^{DDB} = (2/N) \cdot Q_{t-1}^{DDB}$. Equivalently,

$$Q_t^{DDB} = Q_{t-1}^{DDB}\left(1 - \frac{2}{N}\right) = Q_0\left(1 - \frac{2}{N}\right)^t \tag{3.3}$$

for $t < n$. For $t = n$, $\text{Dep}_t^{DDB} = Q_{n-1}^{DDB}$.[2]

The two above methods are sometimes combined into the so-called *double-declining balance-switching-to-straight-line* (DDB-SL) depreciation method. The DDB-SL depreciation is defined as

$$\text{Dep}_t^{DDB\text{-}SL} = \begin{cases} Q_{t-1}^{DDB} \cdot \dfrac{2}{N}, & \text{for } t < t^*; \\ Q_{t^*-1}^{DDB} \cdot \dfrac{1}{N-(t^*-1)} & \text{for } t \geq t^* \end{cases} \tag{3.4}$$

where t^* is the first date for which $\dfrac{1}{N-(t^*-1)} \geq \dfrac{2}{N}$.[3] In other words, the depreciation charge is calculated as in the DDB depreciation pattern from time 1 to time t^*-1, which is the last date at which the corresponding adjusted SL depreciation charge $Q_{t-1}^{DDB}/(N-(t-1))$ is smaller than the DDB charge, $2Q_{t-1}^{DDB}/N$. At time t^* (switching time) the adjusted SL charge exceeds (or is equal to) the DDB charge, so the DDB-SL charge is constant and thereafter equal to $Q_{t^*-1}^{DDB}/(N-(t^*-1))$, $t \in \mathbb{N}_N^{t^*}$.

Time	1	2	...	t^*-1	t^*	t^*+1	...	N
Depreciation method	DDB	DDB	...	DDB	SL	SL	...	SL
Depreciation charge	$\frac{2Q_0^{DDB}}{N}$	$\frac{2Q_1^{DDB}}{N}$	...	$\frac{2Q_{t^*-2}^{DDB}}{N}$	$\frac{Q_{t^*-1}^{DDB}}{N-(t^*-1)}$	$\frac{Q_{t^*-1}^{DDB}}{N-(t^*-1)}$	...	$\frac{Q_{t^*-1}^{DDB}}{N-(t^*-1)}$

Example 3.1 ♦
(*DDB-SL depreciation*) Consider a piece of equipment which is purchased at $Q_0 =$ \$10,000 and let $N = 15$ be it useful life. If SL depreciation were used, then the depreciation rate would be $1/15 = 0.06667$ and the depreciation charge would be $Q_t^{SL} = 0.06667 \cdot Q_0 = 0.06667 \cdot 10{,}000 = 666.67$. Suppose, instead, that the firm depreciates the fixed asset via the DDB-SL pattern. Table 3.1 displays all the relevant data. Following we describe the procedure for determining the depreciation charge in both formal terms and descriptive terms.

---

[2]See an example of DDB depreciation in Chap. 12 (Bumps example).

[3]Note that $t^* > 1$, for $t^* = 1$ implies $1/N < 2/N$. Therefore, the first depreciation charge is necessarily equal to $Q_0(2/N)$.

Formally, the depreciation rate for DDB depreciation is $2/15 = 0.1333$. The depreciation charge is obtained, for every period, by application of (3.4). Therefore, the first depreciation charge is

$$\text{Dep}_1^{DDB\text{-}SL} = 10{,}000 \cdot 0.1333 = 1{,}333.33$$

so the residual amount at time 1 is

$$Q_1^{DDB\text{-}SL} = 10{,}000 - 1{,}333.33 = 8{,}666.67.$$

The second depreciation charge is

$$\text{Dep}_2^{DDB\text{-}SL} = 8{,}666.67 \cdot 0.1333 = 1{,}155.56$$

so the carrying amount at time 2 is

$$Q_2^{DDB\text{-}SL} = 8{,}666.67 - 1{,}155.56 = 7{,}511.11,$$

and so on until time $t^* = 9$; at that date, $1/(15 - 8) = 1/7 = 0.1429 > 0.1333$, which means that depreciation switches to straight-line depreciation. As $Q_8^{DDB\text{-}SL} = 3{,}182.85$, the depreciation charge is

$$\text{Dep}_9^{DDB\text{-}SL} = 0.1429 \cdot 3{,}182.85 = 454.69.$$

From time $t = 9$ to time $t = 15$, the depreciation charge remains constant: $\text{Dep}_t^{DDB\text{-}SL} = \text{Dep}_9^{DDB\text{-}SL} = 454.69$ for $t \in \mathbb{N}_{15}^9$.

Conceptually, the DDB-SL depreciation method requires, for application, that two sets of numbers be compared, which are reported in column 3 (DDB depreciation) and column 4 (Adjusted SL depreciation); such a comparison determines the actual depreciation charge, which is reported in column 5 (DDB-SL depreciation). More precisely, for every date, the number in column 3 is obtained as the product of 0.1333 ($= 2/15$) and the previous-date carrying amount, while the number in column 4 is obtained by dividing the previous-date carrying amount by the number of residual periods. Starting from time $t = 1$, the actual depreciation charge is the greater between the two numbers. Inspecting Table 3.1, the number in column 3 is greater than the number in column 4 from time $t = 1$ to $t = 8$. From time $t = 9$ to time $t = 15$, the number in column 4 exceeds the number in column 3. Therefore, from time $t = 1$ to time $t = 8$, the DDB depreciation is employed, whereas from time $t = 9$ to time $t = 15$ SL is employed. Date $t^* = 9$ is then the switching time. The amount \$454.69 is the depreciation charge of the asset from time $t = 9$ to time $t = 15$. This amount is calculated on the basis of the carrying amount at time $t = 8$, not the carrying amount at time 0:

$$454.69 = \frac{3{,}182.85}{15 - 8}.$$

Table 3.1 Double-declining balance switching to straight-line

Time	Carrying amount	DDB depreciation	Adjusted SL depreciation	DDB-SL depreciation
0	$10,000.00			
1	$8,666.67	$1,333.33	$666.67	$1,333.33 (DDB)
2	$7,511.11	$1,155.56	$619.05	$1,155.56 (DDB)
3	$6,509.63	$1,001.48	$577.78	$1,001.48 (DDB)
4	$5,641.68	$867.95	$542.47	$867.95 (DDB)
5	$4,889.46	$752.22	$512.88	$752.22 (DDB)
6	$4,237.53	$651.93	$488.95	$651.93 (DDB)
7	$3,672.52	$565.00	$470.84	$565.00 (DDB)
8	$3,182.85	$489.67	$459.07	$489.67 (DDB)
9	$2,728.16	$424.38	$454.69	$454.69 (SL)
10	$2,273.47	$367.80	$459.75	$454.69 (SL)
11	$1,818.77	$318.76	$478.14	$454.69 (SL)
12	$1,364.08	$276.26	$517.98	$454.69 (SL)
13	$909.39	$239.42	$598.55	$454.69 (SL)
14	$454.69	$207.50	$778.12	$454.69 (SL)
15	$0.00	$1,384.74	$1,348.74	$454.69 (SL)
Total		$10,000.00		$10,000.00

In essence, it is as if, at time $t = 8$, the counter is set again to 0, and the asset is depreciated evenly for 7 periods.[4]

Net Operating Working Capital (NOWC)

The NOWC represents current (i.e., short-term) operating assets net of current operating liabilities. We may classify this class as follows:

- Current Operating Assets
  - *Inventories* (Inv_t): This subclass may be itself broken down into three items: Finished goods inventory, Work in process inventory, Raw materials inventory.
  - *Accounts Receivable* (AR_t): This item refers to the amount customers owe to the firm for purchases of goods made on account.

[4]As the switch occurs at time $t^* = 9$, the remaining numbers in columns 3 and 4 are irrelevant, so they are reported in grey font. If the depreciation method used were the DDB method, then the depreciation charges from time $t^* = 9$ to time $t = 15$ would precisely be represented by the grey numbers in column 3.

- Current Operating Liabilities
  - *Accounts Payable* (AP_t): It is the amount owed to suppliers for purchases made on account.
  - *Salaries and Wages Payable* (SWP_t): It is the amount owed to employees.
  - *Income taxes payable* (TP_t): This class represents the obligation to pay the income taxes accrued as a consequence of the profit generated by the operations.

Therefore, synchronically, the NOWC is obtained as

$$\text{NOWC}_t = \overbrace{\text{Inv}_t + \text{AR}_t}^{\text{Current Assets}} - \overbrace{(\text{AP}_t + \text{SWP}_t + \text{TP}_t)}^{\text{Current liabilities}}. \tag{3.5}$$

The operating capital is the sum of NOWC and NFA:

$$C_t^o = \text{NFA}_t + \text{NOWC}_t = \overbrace{\text{NFA}_t}^{\text{Long-term assets}} + \overbrace{\text{Inv}_t + \text{AR}_t}^{\text{Current Assets}} - \overbrace{(\text{AP}_t + \text{SWP}_t + \text{TP}_t)}^{\text{Current liabilities}}. \tag{3.6}$$

3.2.1 Decomposing Operating Assets, Income, and Cash Flow

We now introduce the class of ***net operating liabilities*** (NOL_t) defined as the class of liabilities that are increased by the cost for running the operations and decreased by the respective payments. Essentially, NOL consists of legal obligations towards suppliers and employees for the operating costs arising from the purchase-and-manufacturing process (or the purchase-and-merchandising process), net of the inventories. The NOL measures the exposition of the firm toward suppliers and employees. Such an exposition is decreased by the inventories (they may be converted into cash that may be used to reduce the liabilities toward suppliers and employees): $\text{NOL}_t = \text{AP}_t + \text{SWP}_t - \text{Inv}_t$ (see also Sect. 3.2.3). The operating capital may then be reframed as

$$C_t^o = \text{AR}_t + \text{NFA}_t - \text{NOL}_t - \text{TP}_t. \tag{3.7}$$

Each of the four classes that make up the operating capital enjoys its own law of motion, according to which each capital amount is increased/decreased owing to two different sources of change: An income component and a cash-flow component. Specifically,

- Accounts receivable increase by the sales revenues (S_t) and decrease by the cash received from customers (F_t^c):[5]

[5]More precisely, F_t^c indicates the cash received from customers for revenues generated by sales of goods or by gains/losses on disposal of assets (see also Sects. 3.6.2 and 4.4).

$$\text{AR}_t = \text{AR}_{t-1} + S_t - F_t^c \tag{3.8}$$

- Net operating liabilities increase by the (pre-tax) operating costs (OpC_t) while decrease by the associated payments to suppliers and employees (F_t^{opc}).

$$\text{NOL}_t = \text{NOL}_{t-1} + \text{OpC}_t - F_t^{opc} \tag{3.9}$$

- Net fixed assets decrease by the depreciation charge (Dep_t) and increase by the capital expenditures (F_t^{capex}):

$$\text{NFA}_t = \text{NFA}_{t-1} - \text{Dep}_t + F_t^{capex} \tag{3.10}$$

- Taxes payable increase by the income taxes due (T_t) and decrease by the tax payments (F_t^T):

$$\text{TP}_t = \text{TP}_{t-1} + \text{T}_t - F_t^T. \tag{3.11}$$

Summing algebraically the four laws of motion, one gets a decomposition of the operating capital's law of motion:

$$C_t^o = \overbrace{\underbrace{\text{AR}_{t-1} - \text{NOL}_{t-1} - \text{TP}_{t-1}}_{\text{NOWC}_{t-1}} + \text{NFA}_{t-1}}^{C_{t-1}^o} + \overbrace{S_t - \text{OpC}_t - \text{Dep}_t - \text{T}_t}^{I_t^o} - \overbrace{(F_t^c - F_t^{opc} - F_t^{capex} - F_t^T)}^{F_t^o}. \tag{3.12}$$

We have disaggregated the operating capital into four capital components, the operating income into four income components, the operating cash flow into four cash-flow components (see also Table 3.2, where parentheses denote negative sign).

Equation (3.13) is the matrix form of such a breakdown. The first column says that the operating assets are obtained from the accounts receivable (which collect accumulated sales on credit) by deducting the (pre-tax) operating liabilities net of the inventories (which collect the accumulated costs of the goods sold and the accumulated selling, general, and administrative expenses), adding the net fixed assets and deducting the income taxes payable (which collect the accumulated taxes accrued). The second column says that the operating income is obtained by deducting operat-

Table 3.2 The breakdown of the operating assets (a)

Capital component	Income component	Cash-flow component
Accounts receivable	Sales	Cash receipts from customers
(Net operating liabilities)	(Operating costs)	(Payments for operating costs)
Net fixed assets	(Depreciation charges)	(Capital expenditures)
(Taxes payable)	(Income taxes)	(Tax payments)
Operating assets	**Operating profit**	**Cash flow from operations**

ing costs, depreciation and taxes from sales revenues (analogously, the third column regards the calculation of CFO/CCF, changed in sign).

The matrix of the operating assets (a)

$$\begin{array}{|l|llll|}
\hline
+\mathrm{AR}_t & +\mathrm{AR}_{t-1} & +S_t & -F_t^c \\
-\mathrm{NOL}_t & -\mathrm{NOL}_{t-1} & -\mathrm{OpC}_t & +F_t^{opc} \\
+\mathrm{NFA}_t & +\mathrm{NFA}_{t-1} & -\mathrm{Dep}_t & +F_t^{capex} \\
-\mathrm{TP}_t & -\mathrm{TP}_{t-1} & -\mathrm{T}_t & +F_t^T \\
\hline
+C_t^o & +C_{t-1}^o & +I_t^o & -F_t^o \\
\hline
\end{array} \tag{3.13}$$

3.2.2 *Cost of Goods Sold*

The pre-tax operational costs (OpC_t) may be divided into two classes:

$$\mathrm{OpC}_t = \mathrm{COGS}_t + \mathrm{SGA}_t. \tag{3.14}$$

The COGS are ***product costs***, that is, they represent costs that are tied to the products sold in the period; the SGA are ***period costs***, that is they are expensed in the period in which they occur.[6] The COGS are manufacturing costs, that is, they are associated with the manufacturing process, whereas the SGA are nonmanufacturing costs related to the management of the business.[7]

Corresponding to this decomposition, one splits up the payments for operating costs into the respective payments, F_t^{cogs}, F_t^{sga}:

$$F_t^{opc} = F_t^{cogs} + F_t^{sga}. \tag{3.15}$$

One may then write

$$\begin{aligned}
C_t^o = \underbrace{\overbrace{\mathrm{AR}_{t-1} - \mathrm{NOL}_{t-1} - \mathrm{TP}_{t-1}}^{}_{\mathrm{NOWC}_{t-1}} + \mathrm{NFA}_{t-1}}_{}\ \ &\overset{C_{t-1}^o}{} + \underbrace{(S_t - \overbrace{(\mathrm{COGS}_t + \mathrm{SGA}_t)}^{\mathrm{OpC}_t} - \mathrm{Dep}_t - \mathrm{T}_t)}_{I_t^o} \\
&- \underbrace{(F_t^c - \overbrace{(F_t^{cogs} + F_t^{sga})}^{F_t^{opc}} - F_t^{capex} - F_t^T)}_{F_t^o}
\end{aligned} \tag{3.16}$$

---

[6]The amount OpC_t refers to pre-tax *cash* operating costs, that is, operating costs that will be converted in cash outflows. Therefore, it does not include noncash operating costs (Dep_t).

[7]In a merchandising firm, no transformation process is necessary, thus COGS refers to the purchase activity alone.

Table 3.3 The breakdown of the operating assets (b)

Capital component	Income component	Cash-flow component
Accounts receivable	Sales	Cash receipts from customers
(Net operating liabilities)	(COGS and SGA)	(Payments for COGS and SGA)
Net fixed assets	(Depreciation charges)	(Capital expenditures)
(Taxes payable)	(Income taxes)	(Tax payments)
Operating assets	**Operating profit**	**Cash flow from operations**

(see also Table 3.3 and Eq. (3.17)).

The matrix of the operating assets (b)

$$\begin{array}{|l|llr|}\hline +\text{AR}_t & +\text{AR}_{t-1} & +S_t & -F_t^c \\ -\text{NOL}_t & -\text{NOL}_{t-1} & -\text{COGS}_t - \text{SGA}_t & +F_t^{cogs} + F_t^{sga} \\ +\text{NFA}_t & +\text{NFA}_{t-1} & -\text{Dep}_t & +F_t^{capex} \\ -\text{TP}_t & -\text{TP}_{t-1} & -\text{T}_t & +F_t^T \\ \hline +C_t^o & +C_{t-1}^o & +I_t^o & -F_t^o \\ \hline \end{array} \tag{3.17}$$

In a manufacturing firm, three types of inventory exist: Raw materials inventory (RM), Work in process inventory (WIP), Finished goods inventory (FG). The amount of inventory held by the firm is then

$$\text{Inv}_t = \text{RM}_t + \text{WIP}_t + \text{FG}_t. \tag{3.18}$$

The COGS is the cost incurred by the firm for producing the goods sold in the period. It is defined as

$$\begin{aligned} \text{COGS}_t = &+\text{FG}_{t-1} +\text{WIP}_{t-1} +\text{RM}_{t-1} + \text{COM}_t + \text{DL}_t + \text{MOH}_t \\ &-\text{FG}_t \quad -\text{WIP}_t \quad -\text{RM}_t \end{aligned} \tag{3.19}$$

or, equivalently,

$$\text{COGS}_t = -\Delta\text{Inv}_t + \text{COM}_t + \text{DL}_t + \text{MOH}_t \tag{3.20}$$

where

$$\begin{aligned} \text{COM}_t &= \text{Cost of material purchased} \\ \text{DL}_t &= \text{Direct labor costs} \\ \text{MOH}_t &= \text{Manufacturing overhead.} \end{aligned}$$

MOH_t consists of indirect labor, indirect material and other operating (manufacturing) costs (see Table 3.5 in Sect. 3.2.3 for a breakdown of operating costs).

To explain (3.19), consider first COM_t. It is the cost of direct material *purchased* in the *period*. In contrast, the amount $\text{DM}_t = \text{RM}_{t-1} + \text{COM}_t - \text{RM}_t$ represents the cost of direct material *used* in the period. Adding DL_t and MOH_t to DM_t one gets the total manufacturing costs. Deducting the change in WIP one finds the cost of goods *manufactured* in the period. Finally, deducting the change in FG, one finds the costs of goods *sold* in the period. In tabular form,

$+\text{RM}_{t-1} + \text{COM}_t$
$-\text{RM}_t$
$=$ **Direct Material Used** (DM_t)
$+\text{DL}_t$
$+\text{MOH}_t$
$=$ **Total Manufacturing Costs**
$+\text{WIP}_{t-1}$
$-\text{WIP}_t$
$=$ **Cost of Goods Manufactured**
$+\text{FG}_{t-1}$
$-\text{FG}_t$
$=$ **Cost of Goods Sold**

From (3.18) and (3.19), one can write

$$\text{Inv}_t = \text{Inv}_{t-1} + \text{COM}_t + \text{DL}_t + \text{MOH}_t - \text{COGS}_t. \tag{3.21}$$

This means that inventory increases by the cost of material purchased, direct labor, and manufacturing overhead, while decreases by the COGS. Therefore, the net operating liabilities increase by the COGS and SGA ($= \text{OpC}_t$, an income component) and decreases by the payments to suppliers and employees (F_t^{opc}, a cash-flow component). The matrix (3.17) becomes

The matrix of the operating assets (b) (detailed)

$+\text{AR}_t$	$+\text{AR}_{t-1} \quad +S_t$	$-F_t^c$
$-\text{NOL}_t$	$-\text{NOL}_{t-1} \quad +\Delta\text{Inv}_t - \text{COM}_t - \text{DL}_t - \text{MOH}_t - \text{SGA}_t$	$+F_t^{cogs} + F_t^{sga}$
$+\text{NFA}_t$	$+\text{NFA}_{t-1} \quad -\text{Dep}_t$	$+F_t^{capex}$
$-\text{TP}_t$	$-\text{TP}_{t-1} \quad -\text{T}_t$	$+F_t^T$
$+C_t^o$	$+C_{t-1}^o \quad +I_t^o$	$-F_t^o$

(3.22)

For projects where no manufacturing process exists (e.g., merchandising firms), only the Finished goods inventory exists ($\text{Inv}_t = \text{FG}_t$), no manufacturing costs exists, and accounts payable only refer to nonmanufacturing activities. Therefore, the COGS is equal to $\text{COGS} = -\Delta\text{FG}_t + \text{COM}_t$.

Equation (3.21) implies that the change in the inventory level is equal to the difference between the manufacturing period costs (CM, DL, and MOH) and the product costs

(COGS). Such a difference may be interpreted, if positive, as the income generated by inventories in a given period and, if negative, as the cost generated by inventories. The amount $\mathrm{COM}_t + \mathrm{DL}_t + \mathrm{MOH}_t$ is an expense that is incurred in a given period owing to the manufacturing process. In contrast, the COGS is a product cost; precisely, COGS does not refer to the expenses that are incurred in a given period, but to the costs that have been incurred for manufacturing the goods that are sold in that period. If one were to accept an accounting system where all costs are period costs, then inventory would not exist. Inventory is an item that precisely serves the scope of passing from manufacturing period costs to manufacturing product costs (so-called *matching principle*). The difference

$$\overbrace{\mathrm{COM}_t + \mathrm{DL}_t + \mathrm{MOH}_t}^{\text{manufacturing period costs}} - \overbrace{\mathrm{COGS}_t}^{\text{manufacturing product costs}} = \Delta\mathrm{Inv}_t \tag{3.23}$$

expresses the income which is generated (if positive) by passing from a perspective on period costs to a perspective on product costs. Inventory is the item that does the job: It takes period costs and turns (part of) them into capital. For example, if $\mathrm{COM}_t + \mathrm{DL}_t + \mathrm{MOH}_t = \$100$ and no product is sold in the period, then the matching principle says that no cost is charged to expense (i.e., the COGS is zero), for there are no goods sold; in other words, that cost is, so to say, suspended. In accounting terminology: the cost is *capitalized*, that is, it is turned to invested capital (inventory). In the next period, assuming all the goods manufactured in the previous period are sold, the change in inventory ($100) is charged to expense as the COGS incurred in that period.

3.2.3 Splitting Up Operating Costs and Liabilities

As anticipated, the NOL class consists of Accounts Payable (AP) and Salaries and Wages Payable (SWP), net of inventories (Inv):

$$\mathrm{NOL}_t = \mathrm{AP}_t + \mathrm{SWP}_t - \mathrm{Inv}_t. \tag{3.24}$$

Financially, AP and SWP are financings and the amount $\mathrm{AP}_t + \mathrm{SWP}_t$ measures the exposition of the firm toward suppliers and employees. As noted above, this exposition is reduced by the amount of inventories held by the firm.

Equation (3.21) may be interpreted as the law of motion for inventory, and the difference between the manufacturing period costs and the manufacturing product costs presented in (3.23) (i.e., the change in inventory) is an income component which may be called *inventory income* (if positive) or *inventory cost* (if negative). Accordingly, one may write the law of motion for accounts payable and salaries and wages payable as well, along with the law of motion for inventory:

Table 3.4 The breakdown of the operating assets (c)

Capital component	Income component	Cash-flow component
Accounts receivable	Sales	Cash receipts from customers
Inventory	Change in inventory	
(Accounts payable)	(Cost of purchases)	(Payments to suppliers)
(Salaries and wages payable)	(Cost of labor)	(Payments to employees)
Net Fixed Assets	(Depreciation charges)	(Capital expenditures)
(Taxes Payable)	(Income taxes)	(Tax payments)
Operating assets	**Operating profit**	**Cash flow from operations**

$$
\begin{aligned}
\mathrm{AP}_t &= \mathrm{AP}_{t-1} + \mathrm{COP}_t - F_t^{cop} \\
\mathrm{SWP}_t &= \mathrm{SWP}_{t-1} + \mathrm{LC}_t - F_t^{lc} \\
\mathrm{Inv}_t &= \mathrm{Inv}_{t-1} + \Delta\mathrm{Inv}_t - 0
\end{aligned}
\tag{3.25}
$$

where

$$
\begin{aligned}
\mathrm{COP}_t &= \text{Cost of purchases (manufacturing and nonmanufacturing)} \\
\mathrm{LC}_t &= \text{Labor costs (manufacturing and nonmanufacturing)} \\
F_t^{cop} &= \text{Payments to suppliers} \\
F_t^{lc} &= \text{Payments to employees}
\end{aligned}
$$

(note that the cash-flow component of inventory is zero).

Equation (3.25) splits up the NOL system presented in Eq. (3.9) into its three constituents, each one abiding by its own law of motion. Hence, the breakdown of the operating assets becomes as depicted in Table 3.4. The matrix of the operating assets may then be reframed as in Eq. (3.26).

The matrix of the operating assets (c)

$$
\begin{array}{|l|llll|}
\hline
+\mathrm{AR}_t & +\mathrm{AR}_{t-1} & +S_t & -F_t^c \\
+\mathrm{Inv}_t & +\mathrm{Inv}_{t-1} & +\Delta\mathrm{Inv}_t & \\
-\mathrm{AP}_t & -\mathrm{AP}_{t-1} & -\mathrm{COP}_t & +F_t^{cop} \\
-\mathrm{SWP}_t & -\mathrm{SWP}_{t-1} & -\mathrm{LC}_t & +F_t^{lc} \\
+\mathrm{NFA}_t & +\mathrm{NFA}_{t-1} & -\mathrm{Dep}_t & +F_t^{capex} \\
-\mathrm{TP}_t & -\mathrm{TP}_{t-1} & -\mathrm{T}_t & +F_t^T \\
\hline
+C_t^o & +C_{t-1}^o & +I_t^o & -F_t^o \\
\hline
\end{array}
\tag{3.26}
$$

In project appraisal, it is usual to assume that salaries and wages are paid in the year where they accrue (the same for taxes payable). In this case $\mathrm{SWP}_t = \mathrm{TP}_t = 0$ and $\mathrm{NOL}_t = \mathrm{AP}_t - \mathrm{Inv}_t$ for every t so that $\mathrm{T}_t = F_t^T$ and $F_t^{lc} = \mathrm{LC}_t$. Hence, the matrix becomes

$+\text{AR}_t$	$+\text{AR}_{t-1}$	$+S_t$	$-F_t^c$
$+\text{Inv}_t$	$+\text{Inv}_{t-1}$	$+\Delta\text{Inv}_t$	
$-\text{AP}_t$	$-\text{AP}_{t-1}$	$-\text{COP}_t$	$+F_t^{cop}$
		$-\text{LC}_t$	$+\text{LC}_t$
$+\text{NFA}_t$	$+\text{NFA}_{t-1}$	$-\text{Dep}_t$	$+F_t^{capex}$
		$-\text{T}_t$	$+\text{T}_t$
$+C_t^o$	$+C_{t-1}^o$	$+I_t^o$	$-F_t^o$

(do not cancel rows 4 and 6, otherwise the law of conservation is infringed!).

Comparing Eqs. (3.22) and (3.26), it is straightforward that

$$\text{OpC}_t = \overbrace{\text{COM}_t + \text{DL}_t + \text{MOH}_t - \Delta\text{Inv}_t}^{\text{COGS}_t} + \text{SGA}_t \tag{3.27}$$

$$\text{OpC}_t = \text{COP}_t + \text{LC}_t - \Delta\text{Inv}_t \tag{3.28}$$

Table 3.5 is a tabular description of the above pair of equations: Operating costs are split up, on one hand, according to their relevance to the process (manufacturing or nonmanufacturing) and, on the other hand, according to their relevance to suppliers or employees.

Likewise, splitting up the operating costs into two dimensions logically implies splitting up the net operating liabilities into the same two dimensions. Specifically, let NOL^m denote the manufacturing liabilities and NOL^{nm} denote the nonmanufacturing liabilities. The former consist of manufacturing supplies and manufacturing salaries and wages, net of inventories; the latter consist of nonmanufacturing supplies and nonmanufacturing salaries and wages. Using Table 3.5 as a template, one may consistently describe the twofold decomposition of net operating liabilities replacing the income components with the respective capital components (compare Tables 3.5 and 3.6). The manufacturing liabilities are increased by the COGS (and decreased by the respective payments), while the nonmanufacturing liabilities are increased by the SGA (and decreased by the respective payments):

$$\begin{aligned} \text{NOL}_t^m &= \text{NOL}_{t-1}^m + \text{COGS}_t - F_t^{cogs} \\ \text{NOL}_t^{nm} &= \text{NOL}_{t-1}^{nm} + \text{SGA}_t - F_t^{sga}. \end{aligned} \tag{3.29}$$

As a result, the NOL may be computed in two ways:

$$\text{NOL}_t = \text{NOL}_t^m + \text{NOL}_t^{nm} \tag{3.30}$$

$$\text{NOL}_t = \text{AP}_t + \text{SWP}_t - \text{Inv}_t. \tag{3.31}$$

The operating matrix may then be reframed in terms of manufacturing and nonmanufacturing components:

Table 3.5 Breakdown of operating costs

	Cost of purchases	Cost of labor	Cost of inventory	Operating costs
Manufacturing costs	Direct and indirect material, plant insurance, utilities repair and maintenance, etc.	Direct and indirect labor, manufacturing plant managers' salaries, etc.	– Change in inventories (raw materials, work in process, finished goods)	$\overbrace{+\text{COM}_t + \text{DL}_t + \text{MOH}_t - \Delta\text{Inv}_t}^{\text{COGS}_t}$
Nonmanufacturing costs	Office supplies, office insurance, utilities, etc.	Sales and marketing expenses, accounting and administration expenses, delivery expense, rent expense, executive compensation, etc.		$+\text{SGA}_t$
Operating costs	$+\text{COP}_t$	$+\text{LC}_t$	$-\Delta\text{Inv}_t$	OpC_t

Table 3.6 Breakdown of net operating liabilities

	Accounts Payable	Salaries and Wages Payable	Inventory	Net Operating Liabilities
Manufacturing NOL	Accounts payable for manufacturing supplies	Salaries and wages payable for manufacturing labor costs	−Inventory	$+NOL_t^m$
Nonmanufacturing NOL	Accounts payable for nonmanufacturing supplies	Salaries and wages payable for nonmanufacturing labor costs		$+NOL_t^{nm}$
Net Operating Liabilities	$+AP_t$	$+SWP_t$	$-Inv_t$	NOL_t

The matrix of the operating assets (d)

$$\begin{array}{|l|llll|}\hline +\text{AR}_t & +\text{AR}_{t-1} & +S_t & & -F_t^c \\ -\text{NOL}_t^m & -\text{NOL}_{t-1}^m & -\text{COGS}_t & & +F_t^{cogs} \\ -\text{NOL}_t^{nm} & -\text{NOL}_{t-1}^{nm} & -\text{SGA}_t & & +F_t^{sga} \\ +\text{NFA}_t & +\text{NFA}_{t-1} & -\text{Dep}_t & & +F_t^{capex} \\ -\text{TP}_t & -\text{TP}_{t-1} & -\text{T}_t & & +F_t^T \\ \hline +C_t^o & +C_{t-1}^o & +I_t^o & & -F_t^o \\ \hline \end{array} \tag{3.32}$$

where F_t^{cogs} are the payments for manufacturing costs and F_t^{sga} are payments for nonmanufacturing costs. (See also the related breakdown tables in Fig. 3.7 in Sect. 3.10.)

3.2.4 *Operating Cycle and Cash Cycle*

To manage NOWC means to manage AR and NOL (i.e., Inv, AP, and SWP). This has to do with the so-called operating cycle. The operating cycle is the interval of time elapsing from the time when the raw material is received and cash from customers is collected. It may be divided into accounts payable period and cash cycle. The accounts payable period is the time between when the raw material is purchased and the time when suppliers are paid by the firm; the cash cycle is the time between when the suppliers are paid and the time when customers pay the goods sold by the firm.

Under another perspective, the operating cycle may be broken down to inventory period and accounts receivable period. The inventory period is the lapse of time between when the raw material is purchased and the finished goods are sold. It therefore includes the manufacturing period, which is the interval of time from when the raw material is withdrawn from the raw materials inventory and the good is stored in the finished good inventory, waiting for being sold. The accounts receivable period is the lapse of time between when the finished good is sold and the cash is received from customers. Figure 3.2 is a friendly representation of the operating and cash cycles (see also Ross et al. 2011, Fig. 16.1, pp. 500–506; Berk and DeMarzo, p. 887, Fig. 26.1).

From left to right, the four vertical arrows denote the points in time when

i. inventory is purchased,
ii. cash is paid to suppliers,
iii. inventory is sold to customers,
iv. cash is received from customers.

Operating cycle and cash cycle are equal if and only if inventory purchased from suppliers is paid on a cash-only basis (i.e., accounts payable period is zero). Cash cycle is zero if the cash is paid to suppliers in the same moment when cash is received from customers. If cash is received from customers whenever the inventory is sold, then the inventory period coincides with the operating cycle (accounts receivable period is zero). If cash from customers is received before cash is paid to suppliers

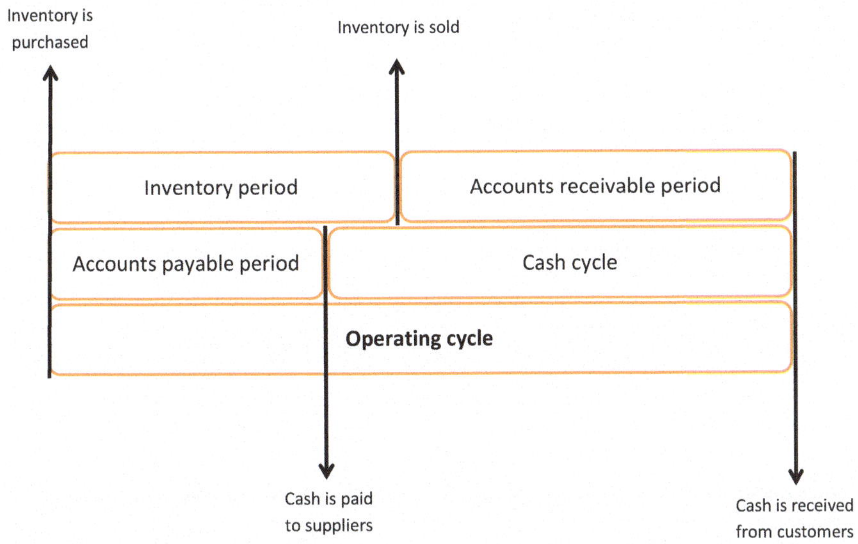

Fig. 3.2 Operating cycle and (positive) cash cycle

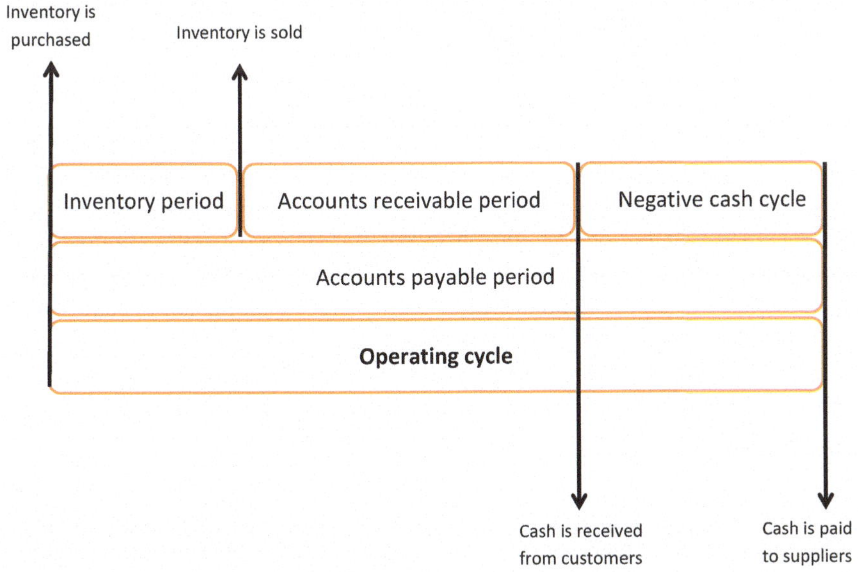

Fig. 3.3 Negative cash cycle: cash from customers precedes payments to suppliers

(e.g., upfront payments from customers), the cash cycle is negative. In this case, the sum of the inventory period and the accounts receivable period is smaller than the accounts payable period (see Fig. 3.3).

3.3 Non-operating Assets

The non-operating assets are liquid assets represented by excess cash (EC_t), marketable securities and other financial assets (FA_t)[8]:

$$C_t^l = \mathrm{EC}_t + \mathrm{FA}_t. \tag{3.33}$$

Table 3.7 breaks down the law of motion for the non-operating assets.

Table 3.7 The breakdown of the non-operating assets

Capital component	Income component	Cash-flow component
Excess cash Financial Assets	Interest on excess cash Interest on financial assets	Net withdrawal from excess cash Net cash flow from financial assets
Non-operating assets	**Interest income**	**Cash flow from non-operating assets**

The matrix of the non-operating assets

$$\begin{array}{|l|llll|} \hline +\mathrm{EC}_t & +\mathrm{EC}_{t-1} & +I_t^{ec} & -F_t^{ec} \\ +\mathrm{FA}_t & +\mathrm{FA}_{t-1} & +I_t^{fa} & -F_t^{fa} \\ \hline +C_t^l & +C_{t-1}^l & +I_t^l & -F_t^l \\ \hline \end{array} \tag{3.34}$$

The income components of these classes consist of interest on excess cash (if any) (I_t^{ec}) and interest income generated by the marketable securities and other financial assets (I_t^{fa}), so that the non-operating income is $I_t^l = I_t^{fa} + I_t^{ec}$. The cash-flow component consists of sales of financial assets (net of purchases) (F_t^{fa}) and withdrawal from excess cash (net of additional deposits) (F_t^{ec}), so that the non-operating cash flow is $F_t^l = F_t^{fa} + F_t^{ec}$.

3.4 Debt

The debt can be represented by loans (LO_t), bonds (BO_t), and notes payable (NP_t):

$$C_t^d = \mathrm{LO}_t + \mathrm{BO}_t + \mathrm{NP}_t. \tag{3.35}$$

The income components consist of the interest paid for notes payable (I_t^{np}) as well as bonds (I_t^{bo}) and loans (I_t^{lo}), so that $I_t^d = I_t^{np} + I_t^{bo} + I_t^{lo}$. The cash-flow component of loans is the loan *instalment* (F_t^{lo}) and the cash flow paid to bondholders (F_t^{bo}) consists of *coupons* and, at the expiring date, of face value. Holders of notes payable

[8] As mentioned earlier, if other non-operating assets exist, they should be included in this class. See footnote 27 in Chap. 4.

Table 3.8 Breakdown of the debt

Capital component	Income component	Cash-flow component
Loans	Interest on loans	Loan instalments
Bonds	Interest on bonds	Cash to bondholders
Notes payable	Interest on notes payable	Cash to notesholders
Debt	**Interest expense**	**Cash flow to debt**

receive the *interest payable* (and the face value at the expiring date) (F_t^{np}). Hence, $F_t^d = F_t^{np} + F_t^{bo} + F_t^{lo}$. Table 3.8 breaks down the law of motion for debt. Equation (3.36) is the corresponding Matrix.

The matrix of the debt

$$\begin{array}{|l|llll|} \hline +\text{LO}_t & +\text{LO}_{t-1} & +I_t^{lo} & -F_t^{lo} \\ +\text{BO}_t & +\text{BO}_{t-1} & +I_t^{bo} & -F_t^{bo} \\ +\text{NP}_t & +\text{NP}_{t-1} & +I_t^{np} & -F_t^{np} \\ \hline +C_t^d & +C_{t-1}^d & +I_t^d & -F_t^d \\ \hline \end{array} \tag{3.36}$$

3.5 Equity

We have thus far broken down the law of motion for operating assets, non-operating assets, and debt, giving rise to three matrices which encompass both the dynamics (reading by row) and the statics (reading by column) of the three subsystems. Equity is the fourth subsystem of the project system.[9] From the law of conservation, equity may be found as a residue from the other three subsystem. Specifically, we remind that

- $C_t^e = C_t^o + C_t^l - C_t^d$: Equity capital is equal to the difference between all the operating and non-operating asset balances and all the debt balances
- $I_t^e = I_t^o + I_t^l - I_t^d$: Net income is computed as the difference between all the incomes from the operating and non-operating assets and all the incomes from the debt balances
- $F_t^e = F_t^o + F_t^l - F_t^d$: The CFE distributed to equityholders is equal to the sum of all the cash flow generated by the operations and the cash flows received by the investment in financial assets, minus the payments made to debtholders.

We then take the three breakdown tables (Tables 3.2, 3.7, and 3.8) and combine them to get the table of the equity breakdown. Table 3.9 is the resulting breakdown of

[9]Equity can be divided into contributed capital and accumulated retained earnings. Contributed capital (C_0^e) is the incremental paid-in capital raised from equityholders at time 0 for undertaking the project (or incorporating the firm). Retained earnings is defined as that part of the project's (or firm's net income) which is not distributed to shareholders: $I_t^e - F_t^e = \Delta C_t^e$. Accumulated (retained) earnings are the sum of such earnings, $\sum_{k=1}^t (I_k^e - F_k^e) = \sum_{k=1}^t \Delta C_k^e$. Therefore, one may write $C_t^e = C_0^e + \sum_{k=1}^t (I_k^e - F_k^e)$. The distinction between contributed capital and accumulated earnings is not useful for our purposes, so we will not deal with it.

Table 3.9 Breakdown of the equity

Capital component	Income component	Cash-flow component
Accounts receivable	Sales	Cash receipts from customers
(Net operating Liabilities)	(Operating costs)	(Payments for operating costs)
Net Fixed Assets	(Depreciation charges)	(Capital expenditures)
(Taxes payable)	(Income taxes)	(Tax payments)
Excess cash	Interest on excess cash	Net withdrawal from excess cash
Financial assets	Interest on financial assets	Net cash flow from financial assets
(Loans)	(Interest on loans)	(Loan instalments)
(Bonds)	(Interest on bonds)	(Cash to bondholders)
(Notes payable)	(Interest on notes payable)	(Cash to notesholders)
Equity	**Net income**	**Cash flow to equity**

equity in terms of the other classes. The first column of the table includes the items of the ***balance sheet*** (BS), the second column displays the items of the ***income statements*** (IS), the third column contains the elements of the ***cash-flow statement*** (CFS). Reading the table by row, the reader catches a view of the fundamental law of motion applied to all aspects of the economic activity; reading the table by column, the reader catches a view of the fundamental law of conservation applied to all aspects of the project's (or firm's) economic activity. (Parentheses signify negative sign).

The corresponding matrix form is Eq. (3.37), which puts in full use the dynamical-static structure presented in Part I of the book: The dynamical perspective is grasped by rows, the static equilibrium is captured by columns.

The Matrix

$$\mathbf{M}_t = \left[\begin{array}{c|ccc} +\mathrm{AR}_t & +\mathrm{AR}_{t-1} & +S_t & -F_t^c \\ -\mathrm{NOL}_t & -\mathrm{NOL}_{t-1} & -\mathrm{OpC}_t & +F_t^{opc} \\ +\mathrm{NFA}_t & +\mathrm{NFA}_{t-1} & -\mathrm{Dep}_t & +F_t^{capex} \\ -\mathrm{TP}_t & -\mathrm{TP}_{t-1} & -\mathrm{T}_t & +F_t^T \\ +\mathrm{EC}_t & +\mathrm{EC}_{t-1} & +I_t^{ec} & -F_t^{ec} \\ +\mathrm{FA}_t & +\mathrm{FA}_{t-1} & +I_t^{fa} & -F_t^{fa} \\ -\mathrm{LO}_t & -\mathrm{LO}_{t-1} & -I_t^{lo} & +F_t^{lo} \\ -\mathrm{BO}_t & -\mathrm{BO}_{t-1} & -I_t^{bo} & +F_t^{bo} \\ -\mathrm{NP}_t & -\mathrm{NP}_{t-1} & -I_t^{np} & +F_t^{np} \\ \hline +C_t^e & +C_{t-1}^e & +I_t^e & -F_t^e \end{array}\right] \tag{3.37}$$

Table 3.10 Breakdown of the equity (reframed)

Capital component	Income component	Cash-flow component
Accounts receivable	Sales	Cash receipts from customers
Inventory	Change in inventory	
(Accounts payable)	(Cost of purchases)	(Payments to suppliers)
(Salaries and wages payable)	(Cost of labor)	(Payments to employees)
Net Fixed Assets	(Depreciation charges)	(Capital expenditures)
(Taxes payable)	(Income taxes)	(Tax payments)
Excess cash	Interest on excess cash	Net withdrawal from excess cash
Financial assets	Interest on financial assets	Net cash flow from financial assets
(Loans)	(Interest on loans)	(Loan instalments)
(Bonds)	(Interest on bonds)	(Cash to bondholders)
(Notes payable)	(Interest on notes payable)	(Cash to notesholders)
Equity	**Net income**	**Cash flow to equity**

Remembering Table 3.4 and Eq. (3.26), the breakdown table and the Matrix may be reframed as Table 3.10 and Eq. (3.38), respectively.

The Matrix – reframed

$$\mathbf{M}_t = \begin{array}{|l|llll|}
\hline
+\text{AR}_t & +\text{AR}_{t-1} & +S_t & -F_t^c \\
+\text{Inv}_t & +\text{Inv}_{t-1} & +\Delta\text{Inv}_t & \\
-\text{AP}_t & -\text{AP}_{t-1} & -\text{COP}_t & +F_t^{cop} \\
-\text{SWP}_t & -\text{SWP}_{t-1} & -\text{LC}_t & +F_t^{lc} \\
+\text{NFA}_t & +\text{NFA}_{t-1} & -\text{Dep}_t & +F_t^{capex} \\
-\text{TP}_t & -\text{TP}_{t-1} & -\text{T}_t & +F_t^T \\
+\text{EC}_t & +\text{EC}_{t-1} & +I_t^{ec} & -F_t^{ec} \\
+\text{FA}_t & +\text{FA}_{t-1} & +I_t^{fa} & -F_t^{fa} \\
-\text{LO}_t & -\text{LO}_{t-1} & -I_t^{lo} & +F_t^{lo} \\
-\text{BO}_t & -\text{BO}_{t-1} & -I_t^{bo} & +F_t^{bo} \\
-\text{NP}_t & -\text{NP}_{t-1} & -I_t^{np} & +F_t^{np} \\
\hline
+C_t^e & +C_{t-1}^e & +I_t^e & -F_t^e \\
\hline
\end{array} \tag{3.38}$$

Equations (3.37) and (3.38) may also be reframed by separating investments from financings:

The Matrix

$$
\mathbf{M}_t = \begin{array}{|l|llll|}
\hline
+\mathrm{AR}_t & +\mathrm{AR}_{t-1} & +S_t & -F_t^c \\
-\mathrm{NOL}_t & -\mathrm{NOL}_{t-1} & -\mathrm{OpC}_t & +F_t^{opc} \\
+\mathrm{NFA}_t & +\mathrm{NFA}_{t-1} & -\mathrm{Dep}_t & +F_t^{capex} \\
-\mathrm{TP}_t & -\mathrm{TP}_{t-1} & -\mathrm{T}_t & +F_t^T \\
+\mathrm{EC}_t & +\mathrm{EC}_{t-1} & +I_t^{ec} & -F_t^{ec} \\
+\mathrm{FA}_t & +\mathrm{FA}_{t-1} & +I_t^{fa} & -F_t^{fa} \\
\hline
+\mathrm{LO}_t & +\mathrm{LO}_{t-1} & +I_t^{lo} & -F_t^{lo} \\
+\mathrm{BO}_t & +\mathrm{BO}_{t-1} & +I_t^{bo} & -F_t^{bo} \\
+\mathrm{NP}_t & +\mathrm{NP}_{t-1} & +I_t^{np} & -F_t^{np} \\
+C_t^e & +C_{t-1}^e & +I_t^e & -F_t^e \\
\hline
\end{array} \quad (3.39)
$$

The Matrix

$$
\mathbf{M}_t = \begin{array}{|l|llll|}
\hline
+\mathrm{AR}_t & +\mathrm{AR}_{t-1} & +S_t & -F_t^c \\
+\mathrm{Inv}_t & +\mathrm{Inv}_{t-1} & +\Delta\mathrm{Inv}_t & \\
-\mathrm{AP}_t & -\mathrm{AP}_{t-1} & -\mathrm{COP}_t & +F_t^{cop} \\
-\mathrm{SWP}_t & -\mathrm{SWP}_{t-1} & -\mathrm{LC}_t & +F_t^{lc} \\
+\mathrm{NFA}_t & +\mathrm{NFA}_{t-1} & -\mathrm{Dep}_t & +F_t^{capex} \\
-\mathrm{TP}_t & -\mathrm{TP}_{t-1} & -\mathrm{T}_t & +F_t^T \\
+\mathrm{EC}_t & +\mathrm{EC}_{t-1} & +I_t^{ec} & -F_t^{ec} \\
+\mathrm{FA}_t & +\mathrm{FA}_{t-1} & +I_t^{fa} & -F_t^{fa} \\
\hline
+\mathrm{LO}_t & +\mathrm{LO}_{t-1} & +I_t^{lo} & -F_t^{lo} \\
+\mathrm{BO}_t & +\mathrm{BO}_{t-1} & +I_t^{bo} & -F_t^{bo} \\
+\mathrm{NP}_t & +\mathrm{NP}_{t-1} & +I_t^{np} & -F_t^{np} \\
+C_t^e & +C_{t-1}^e & +I_t^e & -F_t^e \\
\hline
\end{array} \quad (3.40)
$$

Equation (3.40) is the expanded version of Eq. (3.39), which is the expanded version of Eq. (3.1), which in turn is the expanded version of (2.1) which in turn is the expanded version of Eq. (1.1):

(1.1)

$+C_t$	$+C_{t-1}$	$+I_t$	$-F_t$

expands to →

(2.1)

$+C_t^{\text{inv}}$	$+C_{t-1}^{\text{inv}}$	$+I_t^{\text{inv}}$	$-F_t^{\text{inv}}$
$+C_t^{\text{fin}}$	$+C_{t-1}^{\text{fin}}$	$+I_t^{\text{fin}}$	$-F_t^{\text{fin}}$

expands to →

(3.1)

$+C_t^o$	$+C_{t-1}^o$	$+I_t^o$	$-F_t^o$
$+C_t^l$	$+C_{t-1}^l$	$+I_t^l$	$-F_t^l$
$+C_t^d$	$+C_{t-1}^d$	$+I_t^d$	$-F_t^d$
$+C_t^e$	$+C_{t-1}^e$	$+I_t^e$	$-F_t^e$

expands to →

(3.39)

$+\text{AR}_t$	$+\text{AR}_{t-1}$	$+S_t$	$-F_t^c$
$-\text{NOL}_t$	$-\text{NOL}_{t-1}$	$-\text{OpC}_t$	$+F_t^{opc}$
$+\text{NFA}_t$	$+\text{NFA}_{t-1}$	$-\text{Dep}_t$	$+F_t^{capex}$
$-\text{TP}_t$	$-\text{TP}_{t-1}$	$-\text{T}_t$	$+F_t^T$
$+\text{EC}_t$	$+\text{EC}_{t-1}$	$+I_t^{ec}$	$-F_t^{ec}$
$+\text{FA}_t$	$+\text{FA}_{t-1}$	$+I_t^{fa}$	$-F_t^{fa}$
$+\text{LO}_t$	$+\text{LO}_{t-1}$	$+I_t^{lo}$	$-F_t^{lo}$
$+\text{BO}_t$	$+\text{BO}_{t-1}$	$-I_t^{bo}$	$-F_t^{bo}$
$+\text{NP}_t$	$+\text{NP}_{t-1}$	$-I_t^{np}$	$-F_t^{np}$
$+C_t^e$	$+C_{t-1}^e$	$-I_t^e$	$-F_t^e$

expands to →

(3.40)

$+\text{AR}_t$	$+\text{AR}_{t-1}$	$+S_t$	$-F_t^c$
$+\text{Inv}_t$	$+\text{Inv}_{t-1}$	$+\Delta\text{Inv}_t$	
$-\text{AP}_t$	$-\text{AP}_{t-1}$	$-\text{COP}_t$	$+F_t^{cop}$
$-\text{SWP}_t$	$-\text{SWP}_{t-1}$	$-\text{LC}_t$	$+F_t^{lc}$
$+\text{NFA}_t$	$+\text{NFA}_{t-1}$	$-\text{Dep}_t$	$+F_t^{capex}$
$-\text{TP}_t$	$-\text{TP}_{t-1}$	$-\text{T}_t$	$+F_t^T$
$+\text{EC}_t$	$+\text{EC}_{t-1}$	$+I_t^{ec}$	$-F_t^{ec}$
$+\text{FA}_t$	$+\text{FA}_{t-1}$	$+I_t^{fa}$	$-F_t^{fa}$
$+\text{LO}_t$	$+\text{LO}_{t-1}$	$+I_t^{lo}$	$-F_t^{lo}$
$+\text{BO}_t$	$+\text{BO}_{t-1}$	$+I_t^{bo}$	$-F_t^{bo}$
$+\text{NP}_t$	$+\text{NP}_{t-1}$	$+I_t^{np}$	$-F_t^{np}$
$+C_t^e$	$+C_{t-1}^e$	$+I_t^e$	$-F_t^e$

We call this algebraic structure the ***split-screen Matrix***, owing to its four partitions, or simply the *Matrix*. Matrix (3.40) displays the project (or firm) as a portfolio of 12 economic activities (or, mathematically speaking, as a first-order 12-dimensional discrete system); Matrix (3.39) describes it as a portfolio of 10 activities; Matrix (3.1) describes it as a portfolio of four activities; Matrix (2.1) describes it as a portfolio of two activities; Matrix (1.1) is the general law of motion for the capital, the nucleus Matrix.

A plus (minus) sign in a capital amount of the upper region indicates that the firm is investing in (borrowing from) that class. A plus (minus) sign in a capital amount in the lower region indicates that the firm is borrowing from (investing in) that class.

The split-screen Matrix condenses, in a simple and agile format, the breakdown of capital, income, and cash flow of (a firm or) a project. This breakdown generates the financial statements, namely, the balance sheet, income statement, and cash-flow statement.

A project has two dimensions: Diachronic (dynamics) and synchronic (statics). The split-screen Matrix represents a radiography of a project system, which is disaggregated into three basic elements: Capital, income, and cash flow. Three (static) equilibrium statements correspond to these elements. Respectively,

- **balance sheet** (time $t-1$: column 1, time t: column 2)
- **income statement** (time t: column 3)
- **cash-flow statement** (time t: column 4).

The breakdown may also be visualised in terms of flow charts (see Figs. 3.4, 3.5, and 3.6 at the end of this chapter). The next section shows that the Matrix is also a powerful analytical tool, capable of easily conveying significant pieces of information and guaranteeing logical consistency of the valuation model.

3.6 The *Split-Screen* Technique

We now show that the split-screen Matrix is a most powerful descriptive and working tool. As a quadruple-entry system, it condenses every relevant economic information about a project in four different ways, thereby making it easier to grasp all the relationships among the variables, help the analyst maintain consistency throughout the whole appraisal process, and, as we will see, enable extracting information in a simple and intuitive way.

3.6.1 The Working Rules

The split-screen Matrix is identified, horizontally, by a left region and a right region, and vertically, by an upper region and a lower region. Each region is in turn partitioned into two blocks (partitions), so that the Matrix is decomposed into four submatrices:

- Upper left (northwest) partition
- Upper right (northeast) partition
- Lower left (southwest) partition
- Lower right (southeast) partition

(see Eq. (3.41)).

	Left region	Right region
Upper region	Upper left partition	Upper right partition
Lower region	Lower left partition	Lower right partition

(3.41)

As anticipated, the horizontal and vertical bars which partition the Matrix express equilibrium; that is, mathematically, they express equality of adjacent submatrices. This means that the sum of the elements of the four partitions coincide. This in turn enables stating the following working rules.

The *split-screen* technique: Working rules

1. A row in the upper (lower) region may be shifted downward (upward) to the lower (upper) region, as long as the signs of its elements are changed
2. A column in the left (right) region may be shifted to the right (left) region, as long as the signs of its elements are changed
3. For any fixed row of the Matrix, the sum of its elements in the left region coincides with the sum of its elements in the right region. This implies that the elements of the opposite partitions may be swapped with no change in sign
4. For any fixed column of the Matrix, the sum of its elements in the upper region coincides with the sum of its elements in the lower region. This implies that the elements of the opposite partitions may be swapped with no change in sign
5. Any group of rows (columns) pertaining to the same region may be consolidated by algebraically summing the respective elements, so reducing the size of the Matrix. Symmetrically, each row may be broken down into further rows, so expanding the size of the Matrix
6. The rows (columns) belonging to the upper or lower (left or right) region may be permuted as long as permutation is made within the same region
7. Any number may be added to any cell c_1, provided that one of the four following procedures are followed:

   7.1. the same number is
   - subtracted in any cell c_2 in the same column as cell c_1 but in a different row of the same partition, and
   - subtracted in any cell c_3 in the same row as cell c_2 but in the opposite partition, and
   - added in the cell c_4 in the same column as c_3 and in the same row as c_1

7.2. the same number is
- added in any cell c_2 in the same row as cell c_1 but in the opposite partition, and
- added in any cell c_3 in the same column as the cell c_2 but in the opposite partition, and
- added in the cell c_4 in the same row as cell c_3 and in the same column as c_1

7.3. the same number is
- added in any cell c_2 in the same column as cell c_1 but in a different row of the opposite partition, and
- subtracted in any cell c_3 in the same row as cell c_2 and in the same partition, and
- subtracted in the cell c_4 in the same column as c_3 and in the same row as c_1

7.4. the same number is
- subtracted in any cell c_2 in the same column as cell c_1 but in a different row of the same partition, and
- added in any cell c_3 in the same row as cell c_2 and in the same partition, and
- subtracted in the cell c_4 in the same column as c_3 and in the same row as c_1.

Following is a graphical representation of Rules 7.1–7.4:

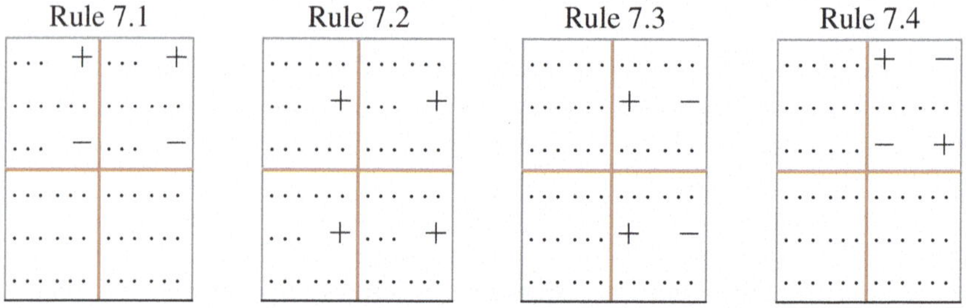

Graphically, Rule 7 may be stated in terms of an ideal quadrilateral, to be formed such that the four vertices are placed in two (left and right) partitions (Rule 7.1), four partitions (Rule 7.2), two (upper and lower) partitions (Rule 7.3), and same partition (Rule 7.4). It might then be called the *quadrilateral rule*.

The Matrix is permanently split up into four screens (the partitions) but it may be split up into different partitions by repeated moves of rows and columns. The Matrix may expand to an arbitrary size of $p \times 3$, $p \in \mathbb{N}$, by disaggregating rows, or shrinked by consolidating rows (and columns). The level of detail of the analysis (and therefore, the size of the Matrix) depends on the purpose of the analysis.

Therefore, the Matrix displays, so to say, a floating environment, in the sense that the working rules may be used for creating various split screens which convey different economic information. We will call the set of the working rules the split-screen technique.

Equation (3.40) is but one of the possible versions of the Matrix where the upper and lower region gives voice, respectively, to an investment perspective and a financing perspective. In particular, the upper left region informs on the composition of the various types of investments, net of the operating liabilities. We may then write its synchronic dimension as

$$C_t = C_t^{\text{inv}} = \boxed{\begin{array}{l} +\text{AR}_t \\ +\text{Inv}_t \\ -\text{AP}_t \\ -\text{SWP}_t \\ +\text{NFA}_t \\ -\text{TP}_t \\ +\text{EC}_t \\ +\text{FA}_t \end{array}}$$

The upper right region shows the invested capital diachronically, highlighting the sources of variation in terms of income components and cash-flow components:

$$C_t = C_t^{\text{inv}} = \boxed{\begin{array}{lll} +\text{AR}_{t-1} & +S_t & -F_t^c \\ +\text{Inv}_{t-1} & +\Delta\text{Inv}_t & \\ -\text{AP}_{t-1} & -\text{COP}_t & +F_t^{cop} \\ -\text{SWP}_{t-1} & -\text{LC}_t & +F_t^{lc} \\ +\text{NFA}_{t-1} & -\text{Dep}_t & +F_t^{capex} \\ -\text{TP}_{t-1} & -\text{T}_t & +F_t^T \\ +\text{EC}_{t-1} & +I_t^{ec} & -F_t^{ec} \\ +\text{FA}_{t-1} & +I_t^{fa} & -F_t^{fa} \end{array}}$$

The lower left region shows the financing structure of the capital raised:

$$C_t = C_t^{\text{fin}} = \boxed{\begin{array}{l} +\text{LO}_t \\ +\text{BO}_t \\ +\text{NP}_t \\ +C_t^e \end{array}}$$

The lower right partition shows the capital raised diachronically, highlighting the sources of variation from the BOP capital, in terms of income components and cash-flow components:

$$C_t = C_t^{\text{fin}} = \boxed{\begin{array}{lll} +\text{LO}_{t-1} & +I_t^{lo} & -F_t^{lo} \\ +\text{BO}_{t-1} & +I_t^{bo} & -F_t^{bo} \\ +\text{NP}_{t-1} & +I_t^{np} & -F_t^{np} \\ +C_{t-1}^e & +I_t^e & +F_t^e \end{array}}$$

Therefore, we have four conceptualizations of the same phenomenon.

A useful format of the Matrix is one where changes in capital are highlighted. This is done by shifting column 2 of (3.40) to the left region (while changing the signs of its elements) and then consolidating the first two columns. The first step is accomplished with Rule 2 and leads to

$$
\begin{array}{|ll|ll|}
\hline
+\mathrm{AR}_t & -\mathrm{AR}_{t-1} & +S_t & -F_t^c \\
+\mathrm{Inv}_t & -\mathrm{Inv}_{t-1} & +\Delta\mathrm{Inv}_t & \\
-\mathrm{AP}_t & +\mathrm{AP}_{t-1} & -\mathrm{COP}_t & +F_t^{cop} \\
-\mathrm{SWP}_t & +\mathrm{SWP}_{t-1} & -\mathrm{LC}_t & +F_t^{lc} \\
+\mathrm{NFA}_t & -\mathrm{NFA}_{t-1} & -\mathrm{Dep}_t & +F_t^{capex} \\
-\mathrm{TP}_t & +\mathrm{TP}_{t-1} & -\mathrm{T}_t & +F_t^T \\
+\mathrm{EC}_t & -\mathrm{EC}_{t-1} & +I_t^{ec} & -F_t^{ec} \\
+\mathrm{FA}_t & -\mathrm{FA}_{t-1} & +I_t^{fa} & -F_t^{fa} \\
\hline
+\mathrm{LO}_t & -\mathrm{LO}_{t-1} & +I_t^{lo} & -F_t^{lo} \\
+\mathrm{BO}_t & -\mathrm{BO}_{t-1} & +I_t^{bo} & -F_t^{bo} \\
+\mathrm{NP}_t & -\mathrm{NP}_{t-1} & +I_t^{np} & -F_t^{np} \\
+C_t^e & -C_{t-1}^e & +I_t^e & -F_t^e \\
\hline
\end{array}
$$

The second step is accomplished with Rule 5 leads to

$$
\begin{array}{|l|ll|}
\hline
+\Delta\mathrm{AR}_t & +S_t & -F_t^c \\
+\Delta\mathrm{Inv}_t & +\Delta\mathrm{Inv}_t & \\
-\Delta\mathrm{AP}_t & -\mathrm{COP}_t & +F_t^{cop} \\
-\Delta\mathrm{SWP}_t & -\mathrm{LC}_t & +F_t^{lc} \\
+\Delta\mathrm{NFA}_t & -\mathrm{Dep}_t & +F_t^{capex} \\
-\Delta\mathrm{TP}_t & -\mathrm{T}_t & +F_t^T \\
+\Delta\mathrm{EC}_t & +I_t^{ec} & -F_t^{ec} \\
+\Delta\mathrm{FA}_t & +I_t^{fa} & -F_t^{fa} \\
\hline
+\Delta\mathrm{LO}_t & +I_t^{lo} & -F_t^{lo} \\
+\Delta\mathrm{BO}_t & +I_t^{bo} & -F_t^{bo} \\
+\Delta\mathrm{NP}_t & +I_t^{np} & -F_t^{np} \\
+\Delta C_t^e & +I_t^e & -F_t^e \\
\hline
\end{array}
\tag{3.42}
$$

The above format is particularly significant, for it displays three statements: The first column is a statement of changes in capital, the second column is a statement of incomes, the third column is a statement of cash flows.

Example 3.2 ♦

Consider Matrix (3.39) and, using Rule 5, consolidate EC and FA on the investment side (upper region), and LO, BO, and NP on the financing side (lower region):

$+\mathrm{AR}_t$	$+\mathrm{AR}_{t-1}$	$+S_t$	$-F_t^c$
$-\mathrm{NOL}_t$	$-\mathrm{NOL}_{t-1}$	$-\mathrm{OpC}_t$	$+F_t^{opc}$
$+\mathrm{NFA}_t$	$+\mathrm{NFA}_{t-1}$	$-\mathrm{Dep}_t$	$+F_t^{capex}$
$-\mathrm{TP}_t$	$-\mathrm{TP}_{t-1}$	$-\mathrm{T}_t$	$+F_t^T$
$+C_t^l$	$+C_{t-1}^l$	$+I_t^l$	$-F_t^l$
$+C_t^d$	$+C_{t-1}^d$	$+I_t^d$	$-F_t^d$
$+C_t^e$	$+C_{t-1}^e$	$+I_t^e$	$-F_t^e$

Let us fill up the Matrix with the following numbers:

X	+90	+40	−30
−30	−25	−10	Y
+40	+60	−20	0
−4	−2	−1	−1
+4	W	+2	−1
+50	+46	+3	+1
+60	+80	Z	−28

X, Y, W, Z are four unknowns. They are independent and the derivation of them is straightforward by taking into account that the elements of rows and columns in opposite regions have the same sum: In this case, we only use the columns (law of conservation):

$$
\begin{aligned}
X &= +\mathrm{AR}_t &&= 50 + 60 - (-30 + 40 - 4 + 4) = 100\\
Y &= +F_t^{opc} &&= -30 - (-25 - 10) = +5\\
W &= +C_{t-1}^l &&= 46 + 80 - (90 - 25 + 60 - 2) = 3\\
Z &= +I_t^e &&= 40 - 10 - 20 - 1 + 2 - 3 - 8
\end{aligned}
$$

Example 3.3 ♦
Consider the same format above but filled with the following numbers:

120	180	90	−150
X	−40	−15	Y
420	520	−100	0
10	15	5	−10
202	220	2	−20
246	260	13	−27
W	Z	−31	−103

The four unknowns are not independent one another. Specifically, exploiting the law of conservation and the law of motion, and starting from column 4, one finds

$$
\begin{array}{llll}
& Y = F_t^{opc} & = -27 - 103 - (-150 - 10 - 20) = 50 & \text{[law of conservation]} \\
\text{whence} & X = -\text{NOL}_t & = -40 - 15 + 50 = -5 & \text{[law of motion]} \\
\text{whence} & W = C_t^e & = 120 - 5 + 420 + 10 + 202 - 246 = 501 & \text{[law of conservation]} \\
\text{whence} & Z = C_{t-1}^e & = 501 - (-31 - 103) = 635 & \text{[law of motion]}
\end{array}
$$

Example 3.4 ♦

Suppose one aims at deriving the change in the net financial position. One must then isolate the relevant items from the other available data. Let us start from (3.42) and, after consolidating the NOL items, let us focus on the accounts that form the net financial position (cells shaded in grey):

$+\Delta\text{AR}_t$	$+S_t$	$-F_t^c$
$-\Delta\text{NOL}_t$	$-\text{OpC}_t$	$+F_t^{opc}$
$+\Delta\text{NFA}_t$	$-\text{Dep}_t$	$+F_t^{capex}$
$-\Delta\text{TP}_t$	$-\text{T}_t$	$+F_t^T$
$+\Delta\text{EC}_t$	$+I_t^{ec}$	$-F_t^{ec}$
$+\Delta\text{FA}_t$	$+I_t^{fa}$	$-F_t^{fa}$
$+\Delta\text{LO}_t$	$+I_t^{lo}$	$-F_t^{lo}$
$+\Delta\text{BO}_t$	$+I_t^{bo}$	$-F_t^{bo}$
$+\Delta\text{NP}_t$	$+I_t^{np}$	$-F_t^{np}$
$+\Delta C_t^e$	$+I_t^e$	$-F_t^e$

One needs to isolate the area highlighted in grey so as to make it a partition of the Matrix. This implies shifting rows 5 and 6 to the lower region (and change signs):

$+\Delta\text{AR}_t$	$+S_t$	$-F_t^c$
$-\Delta\text{NOL}_t$	$-\text{OpC}_t$	$+F_t^{opc}$
$+\Delta\text{NFA}_t$	$-\text{Dep}_t$	$+F_t^{capex}$
$-\Delta\text{TP}_t$	$-\text{T}_t$	$+F_t^T$
$-\Delta\text{EC}_t$	$-I_t^{ec}$	$+F_t^{ec}$
$-\Delta\text{FA}_t$	$-I_t^{fa}$	$+F_t^{fa}$
$+\Delta\text{LO}_t$	$+I_t^{lo}$	$-F_t^{lo}$
$+\Delta\text{BO}_t$	$+I_t^{bo}$	$-F_t^{bo}$
$+\Delta\text{NP}_t$	$+I_t^{np}$	$-F_t^{np}$
$+\Delta C_t^e$	$+I_t^e$	$-F_t^e$

Finally, moving upward the equity row to the upper region,

Change in net financial position

$$\begin{array}{|l|ll|}
\hline
+\Delta \mathrm{AR}_t & +S_t & -F_t^c \\
-\Delta \mathrm{NOL}_t & -\mathrm{OpC}_t & +F_t^{opc} \\
+\Delta \mathrm{NFA}_t & -\mathrm{Dep}_t & +F_t^{capex} \\
-\Delta \mathrm{TP}_t & -\mathrm{T}_t & +F_t^T \\
-\Delta C_t^e & -I_t^e & +F_t^e \\
\hline
-\Delta \mathrm{EC}_t & -I_t^{ec} & +F_t^{ec} \\
-\Delta \mathrm{FA}_t & -I_t^{fa} & +F_t^{fa} \\
+\Delta \mathrm{LO}_t & +I_t^{lo} & -F_t^{lo} \\
+\Delta \mathrm{BO}_t & +I_t^{bo} & -F_t^{bo} \\
+\Delta \mathrm{NP}_t & +I_t^{np} & -F_t^{np} \\
\hline
\end{array} \tag{3.43}$$

The Matrix has then provided us, in an easy way, with four ways of conceptualizing (and computing) the change in the capital structure. One can

- use the change in the individual components of net debt (lower left submatrix)
- use the change in the individual components of the operating assets and the change in equity (upper left submatrix)
- use the interest incomes and interest expenses, as well as the cash generated by the respective accounts (lower right submatrix)
- use the items of the operating income and the net income, along with the respective cash flows (upper right submatrix).

Example 3.5 ♦
Consider (3.43). It may be useful to consolidate further the net debt by merging rows 6–7 and, on the other hand, merging rows 8–10:

$$\begin{array}{|l|ll|}
\hline
+\Delta \mathrm{AR}_t & +S_t & -F_t^c \\
-\Delta \mathrm{NOL}_t & -\mathrm{OpC}_t & +F_t^{opc} \\
+\Delta \mathrm{NFA}_t & -\mathrm{Dep}_t & +F_t^{capex} \\
-\Delta \mathrm{TP}_t & -\mathrm{T}_t & +F_t^T \\
-\Delta C_t^e & -I_t^e & +F_t^e \\
\hline
-\Delta C_t^l & -I_t^l & +F_t^l \\
+\Delta C_t^d & +I_t^d & -F_t^d \\
\hline
\end{array}$$

We may permute the last two rows and get

$$
\begin{array}{|l|ll|}
\hline
+\Delta \text{AR}_t & +S_t & -F_t^c \\
-\Delta \text{NOL}_t & -\text{OpC}_t & +F_t^{opc} \\
+\Delta \text{NFA}_t & -\text{Dep}_t & +F_t^{capex} \\
-\Delta \text{TP}_t & -\text{T}_t & +F_t^T \\
-\Delta C_t^e & -I_t^e & +F_t^e \\
\hline
+\Delta C_t^d & +I_t^d & -F_t^d \\
-\Delta C_t^l & -I_t^l & +F_t^l \\
\hline
\end{array} \tag{3.44}
$$

We may then avail ourselves of four different ways of calculating the net financial position, in terms of invested or financed capital (left region) and in terms of income form investments and income to net claimholders (right region).

If one consolidates further the last rows, one gets

$$
\begin{array}{|l|ll|}
\hline
+\Delta \text{AR}_t & +S_t & -F_t^c \\
-\Delta \text{NOL}_t & -\text{OpC}_t & +F_t^{opc} \\
+\Delta \text{NFA}_t & -\text{Dep}_t & +F_t^{capex} \\
-\Delta \text{TP}_t & -\text{T}_t & +F_t^T \\
-\Delta C_t^e & -I_t^e & +F_t^e \\
\hline
+\Delta C_t^{\mathcal{D}} & +I_t^{\mathcal{D}} & -F_t^{\mathcal{D}} \\
\hline
\end{array} \tag{3.45}
$$

Example 3.6 ♦

Consider the following format, where the NOL class is disaggregated in its constituent subclasses:

$$
\begin{array}{|l|ll|}
\hline
+\Delta \text{AR}_t & +S_t & -F_t^c \\
+\Delta \text{Inv}_t & +\Delta \text{Inv}_t & \\
-\Delta \text{AP}_t & -\text{COP}_t & +F_t^{cop} \\
-\Delta \text{SWP}_t & -\text{LC}_t & +F_t^{lc} \\
+\Delta \text{NFA}_t & -\text{Dep}_t & +F_t^{capex} \\
-\Delta \text{TP}_t & -\text{T}_t & +F_t^T \\
+\Delta C_t^l & +I_t^l & -F_t^l \\
\hline
+\Delta C_t^d & +I_t^d & -F_t^d \\
+\Delta C_t^e & +I_t^e & -F_t^e \\
\hline
\end{array} \tag{3.46}
$$

and suppose one aims at calculating the change in SWP: Moving all rows of the north region under the horizontal bar, one isolates the SWP row:

$-\Delta\text{SWP}_t$	$-\text{LC}_t$	$+F_t^{lc}$
$+\Delta C_t^d$	$+I_t^d$	$-F_t^d$
$+\Delta C_t^e$	$+I_t^e$	$-F_t^e$
$-\Delta\text{AR}_t$	$-S_t$	$+F_t^c$
$-\Delta\text{Inv}_t$	$-\Delta\text{Inv}_t$	
$+\Delta\text{AP}_t$	$+\text{COP}_t$	$-F_t^{cop}$
$-\Delta\text{NFA}_t$	$+\text{Dep}_t$	$-F_t^{capex}$
$+\Delta\text{TP}_t$	$+\text{T}_t$	$-F_t^T$
$-\Delta C_t^l$	$-I_t^l$	$+F_t^l$

Swapping the elements of the first row, one gets the change in the SWP:

$+\text{LC}_t - F_t^{lc}$	$+\Delta\text{SWP}_t$	
$+\Delta C_t^d$	$+I_t^d$	$-F_t^d$
$+\Delta C_t^e$	$+I_t^e$	$-F_t^e$
$-\Delta\text{AR}_t$	$-S_t$	$+F_t^c$
$-\Delta\text{Inv}_t$	$-\Delta\text{Inv}_t$	
$+\Delta\text{AP}_t$	$+\text{COP}_t$	$-F_t^{cop}$
$-\Delta\text{NFA}_t$	$+\text{Dep}_t$	$-F_t^{capex}$
$+\Delta\text{TP}_t$	$+\text{T}_t$	$-F_t^T$
$-\Delta C_t^l$	$-I_t^l$	$+F_t^l$

The change in SWP implies that the Matrix is now to be read as a two-column matrix. That is, the change in SWP is not equal to the sum of the elements of column 2 nor the sum of the elements of column 3 in the south region; rather, it is equal to the sum of all the elements of columns 2 and 3 in the south region.

Example 3.7 ♦
Suppose the analyst wants to derive the amount of principal repayments made by the firm to reduce the financial exposure. Attention must then be drawn to the loan balances. Starting, for example, from (3.39),

$+\text{AR}_t$	$+\text{AR}_{t-1}$	$+S_t$	$-F_t^c$
$-\text{NOL}_t$	$-\text{NOL}_{t-1}$	$-\text{OpC}_t$	$+F_t^{opc}$
$+\text{NFA}_t$	$+\text{NFA}_{t-1}$	$-\text{Dep}_t$	$+F_t^{capex}$
$-\text{TP}_t$	$-\text{TP}_{t-1}$	$-\text{T}_t$	$+F_t^T$
$+\text{EC}_t$	$+\text{EC}_{t-1}$	$+I_t^{ec}$	$-F_t^{ec}$
$+\text{FA}_t$	$+\text{FA}_{t-1}$	$+I_t^{fa}$	$-F_t^{fa}$
$+\text{LO}_t$	$+\text{LO}_{t-1}$	$+I_t^{lo}$	$-F_t^{lo}$
$+\text{BO}_t$	$+\text{BO}_{t-1}$	$+I_t^{bo}$	$-F_t^{bo}$
$+\text{NP}_t$	$+\text{NP}_{t-1}$	$+I_t^{np}$	$-F_t^{np}$
$+C_t^e$	$+C_{t-1}^e$	$+I_t^e$	$-F_t^e$

Principal repayments are just equal to $LO_{t-1} - LO_t$. Thus, we first isolate the row where LO_t and LO_{t-1} appears. To this end, we shift the last three rows of the lower region upward to the upper region). We find

$+AR_t$	$+AR_{t-1}$	$+S_t$	$-F_t^c$
$-NOL_t$	$-NOL_{t-1}$	$-OpC_t$	$+F_t^{opc}$
$+NFA_t$	$+NFA_{t-1}$	$-Dep_t$	$+F_t^{capex}$
$-TP_t$	$-TP_{t-1}$	$-T_t$	$+F_t^{\mathrm{T}}$
$+EC_t$	$+EC_{t-1}$	$+I_t^{ec}$	$-F_t^{ec}$
$+FA_t$	$+FA_{t-1}$	$+I_t^{fa}$	$-F_t^{fa}$
$-BO_t$	$-BO_{t-1}$	$-I_t^{bo}$	$+F_t^{bo}$
$-NP_t$	$-NP_{t-1}$	$-I_t^{np}$	$+F_t^{np}$
$-C_t^e$	$-C_{t-1}^e$	$-I_t^e$	$+F_t^e$
$+LO_t$	$+LO_{t-1}$	$+I_t^{lo}$	$-F_t^{lo}$

Then we shift columns 3–4 leftward to the west region:

$-S_t$	$+F_t^c$	$+AR_t$	$+AR_{t-1}$
$+OpC_t$	$-F_t^{opc}$	$-NOL_t$	$-NOL_{t-1}$
$+Dep_t$	$-F_t^{capex}$	$+NFA_t$	$+NFA_{t-1}$
$+T_t$	$-F_t^T$	$-TP_t$	$-TP_{t-1}$
$-I_t^{ec}$	$+F_t^{ec}$	$+EC_t$	$+EC_{t-1}$
$-I_t^{fa}$	$+F_t^{fa}$	$+FA_t$	$+FA_{t-1}$
$+I_t^{bo}$	$-F_t^{bo}$	$-BO_t$	$-BO_{t-1}$
$+I_t^{np}$	$-F_t^{np}$	$-NP_t$	$-NP_{t-1}$
$+I_t^e$	$-F_t^e$	$-C_t^e$	$-C_{t-1}^e$
$-I_t^{lo}$	$+F_t^{lo}$	$+LO_t$	$+LO_{t-1}$

Finally, we move column 3 rightward and get the principal repayments framed in four different ways:

Principal repayments

$-S_t$	$+F_t^c$	$+AR_{t-1}$	$-AR_t$
$+OpC_t$	$-F_t^{opc}$	$-NOL_{t-1}$	$+NOL_t$
$+Dep_t$	$-F_t^{capex}$	$+NFA_{t-1}$	$-NFA_t$
$+\mathrm{T}_t$	$-F_t^{\mathrm{T}}$	$-TP_{t-1}$	$+TP_t$
$-I_t^{ec}$	$+F_t^{ec}$	$+EC_{t-1}$	$-EC_t$
$-I_t^{fa}$	$+F_t^{fa}$	$+FA_{t-1}$	$-FA_t$
$+I_t^{bo}$	$-F_t^{bo}$	$-BO_{t-1}$	$+BO_t$
$+I_t^{np}$	$-F_t^{np}$	$-NP_{t-1}$	$+NP_t$
$+I_t^e$	$-F_t^e$	$-C_{t-1}^e$	$+C_t^e$
$-I_t^{lo}$	$+F_t^{lo}$	$+LO_{t-1}$	$-LO_t$

or, using the change-in-capital format,

Principal repayments

$-S_t$	$+F_t^c$	$-\Delta \text{AR}_t$
$+\text{Dep}_t$	$-F_t^{capex}$	$-\Delta \text{NFA}_t$
$+\text{OpC}_t$	$-F_t^{opc}$	$+\Delta \text{NOL}_t$
$+\text{T}_t$	$-F_t^{\text{T}}$	$+\Delta \text{TP}_t$
$-I_t^{ec}$	$+F_t^{ec}$	$-\Delta \text{EC}_t$
$-I_t^{fa}$	$+F_t^{fa}$	$-\Delta \text{FA}_t$
$+I_t^{bo}$	$-F_t^{bo}$	$+\Delta \text{BO}_t$
$+I_t^{np}$	$-F_t^{np}$	$+\Delta \text{NP}_t$
$+I_t^{e}$	$-F_t^{e}$	$+\Delta C_t^e$
$-I_t^{lo}$	$+F_t^{lo}$	$-\Delta \text{LO}_t$

Principal repayments may then be derived (and conceptualized) in four different ways starting from four groups of parameters. The upper left partition computes principal repayments in terms of income components and cash-flow components unrelated to loans. The lower left partition computes principal repayments as the difference between the instalments and the interest expenses. The lower right partition frames principal repayments as the additive inverse of the change in loans. The upper right partition expresses principal repayments in terms of algebraic sum of the capital changes of all those assets and liabilities other than loans.

Example 3.8 ♦
Owing to Rule 4, swapping of partitions is possible with no change in sign. For example, consider the following format of the Matrix, highlighting the debt items:

$+\Delta \text{AR}_t$	$+S_t$	$-F_t^c$
$-\Delta \text{NOL}_t$	$-\text{OpC}_t$	$+F_t^{opc}$
$+\Delta \text{NFA}_t$	$-\text{Dep}_t$	$+F_t^{capex}$
$-\Delta \text{TP}_t$	$-\text{T}_t$	$+F_t^T$
$-\Delta C_t^e$	$-I_t^e$	$+F_t^e$
$+\Delta \text{EC}_t$	$+I_t^{ec}$	$-F_t^{ec}$
$+\Delta \text{FA}_t$	$+I_t^{fa}$	$-F_t^{fa}$
$+\Delta \text{LO}_t$	$+I_t^{lo}$	$-F_t^{lo}$
$+\Delta \text{BO}_t$	$+I_t^{bo}$	$-F_t^{bo}$
$+\Delta \text{NP}_t$	$+I_t^{np}$	$-F_t^{np}$

One may swap the upper left partition and the lower left partition, getting the following format:

$+\Delta\text{LO}_t + \Delta\text{BO}_t + \Delta\text{NP}_t$	$+S_t - F_t^c$ $-\text{OpC}_t + F_t^{opc}$ $-\text{Dep}_t + F_t^{capex}$ $-\text{T}_t + F_t^T$ $-I_t^e + F_t^e$ $+I_t^{ec} - F_t^{ec}$ $+I_t^{fa} - F_t^{fa}$
$+\Delta\text{AR}_t$ $-\Delta\text{NOL}_t$ $+\Delta\text{NFA}_t$ $-\Delta\text{TP}_t$ $-\Delta C_t^e$ $+\Delta\text{EC}_t$ $+\Delta\text{FA}_t$	$+I_t^{lo} - F_t^{lo} + I_t^{bo} - F_t^{bo} + I_t^{np} - F_t^{np}$

or, equivalently,

$+\Delta C_t^d$	$+S_t - F_t^c$ $-\text{OpC}_t + F_t^{opc}$ $-\text{Dep}_t + F_t^{capex}$ $-\text{T}_t + F_t^T$ $-I_t^e + F_t^e$ $+I_t^{ec} - F_t^{ec}$ $+I_t^{fa} - F_t^{fa}$
$+\Delta C_t^o$ $-\Delta C_t^e$	$+I_t^{lo} - F_t^{lo} + I_t^{bo} - F_t^{bo} + I_t^{np} - F_t^{np}$

Note that, working on the partitions, the rows and columns of the Matrix have been aggregated so that the Matrix is now a 2×2 matrix.

Example 3.9 ♦
Consolidation will turn out to be extremely useful. For the sake of illustrative purposes, let us start from

$+\Delta\text{AR}_t$	$+S_t$	$-F_t^c$
$-\Delta\text{NOL}_t$	$-\text{OpC}_t$	$+F_t^{opc}$
$+\Delta\text{NFA}_t$	$-\text{Dep}_t$	$+F_t^{capex}$
$-\Delta\text{TP}_t$	$-\text{T}_t$	$+F_t^T$
$+\Delta C_t^l$	$+I_t^l$	$-F_t^l$
$+\Delta C_t^d$	$+I_t^d$	$-F_t^d$
$+\Delta C_t^e$	$+I_t^e$	$-F_t^e$

We may merge rows 1, 2 and 4 into one row, to highlight the NOWC and its income and cash-flow component:

$$\begin{array}{|l|lr|}\hline +\Delta \text{NOWC}_t & +S_t - \text{OpC}_t - \text{T}_t & -F_t^{nowc} \\ +\Delta \text{NFA}_t & -\text{Dep}_t & +F_t^{capex} \\ +\Delta C_t^l & +I_t^l & -F_t^l \\ \hline +\Delta C_t^d & +I_t^d & -F_t^d \\ +\Delta C_t^e & +I_t^e & -F_t^e \\ \hline \end{array} \tag{3.47}$$

where we let $F_t^{nowc} = F_t^c - F_t^{opc} - F_t^T$ be the cash flow from NOWC. Then, we may shift row 3 onto the lower region and merge with row 4 to get a capital-structure perspective:

$$\begin{array}{|l|lr|}\hline +\Delta \text{NOWC}_t & +S_t - \text{OpC}_t - \text{T}_t & -F_t^{nowc} \\ +\Delta \text{NFA}_t & -\text{Dep}_t & +F_t^{capex} \\ \hline +\Delta C_t^{\mathcal{D}} & +I_t^{\mathcal{D}} & -F_t^{\mathcal{D}} \\ +\Delta C_t^e & +I_t^e & -F_t^e \\ \hline \end{array} \tag{3.48}$$

The latter framing emphasizes the NOWC and the NFA as the main asset classes of a project or firm. This framing is especially deep-rooted in finance.

Example 3.10 ♦
Particular assumptions may nullify some entries of the Matrix. For example, consider the latest version above. Suppose that

- taxes are paid within the year in which they accrue, so that $\text{TP}_t = 0$ and $F^T = \text{T}_t$
- purchases are made on cash, so that $\text{AP}_t = 0$ and $\text{COP}_t = F_t^{cop}$
- salaries and wages are paid in the year where they accrue, so that $\text{SWP}_t = 0$ and $\text{LC}_t = F_t^{lc}$
- no cash outflow for capital expenditure, that is, $F_t^{capex} = 0$
- the project is unlevered, which implies $C_t^{\mathcal{D}} = 0$.

Therefore,

$$\begin{array}{|l|lr|}\hline +\Delta \text{AR}_t + \Delta \text{Inv}_t & +S_t - \text{OpC}_t - \text{T}_t & -F_t^{nowc} \\ +\Delta \text{NFA}_t & -\text{Dep}_t & +0 \\ \hline +0 & +0 & -0 \\ +\Delta C_t^e & +I_t^e & -F_t^e \\ \hline \end{array}$$

or, isolating the tax row and deleting the zero row,

$$\begin{array}{|l|lr|}\hline +\Delta \text{AR}_t + \Delta \text{Inv}_t & +S_t - \text{OpC}_t & -(F_t^c - F_t^{opc}) \\ +\Delta \text{NFA}_t & -\text{Dep}_t & +0 \\ +0 & -\text{T}_t & +\text{T}_t \\ \hline +\Delta C_t^e & +I_t^e & -F_t^e \\ \hline \end{array}$$

From the first row, one gets the equality $S_t - \text{OpC}_t - F^c + F_t^{opc} = \Delta\text{AR}_t + \Delta\text{Inv}_t$. Also, considering that suppliers and employees are paid within the period, then $F_t^{sga} = \text{SGA}_t$, whence

$$\Delta\text{AR}_t + \Delta\text{Inv}_t = S_t - \text{OpC}_t - F_t^c + F_t^{opc} \tag{3.49}$$

$$= S_t - (\text{COGS}_t + \text{SGA}_t) - F^c + (F_t^{cogs} + \text{SGA}_t) \tag{3.50}$$

$$= (S_t - F_t^c) - (\text{COGS}_t - F_t^{cogs}) \tag{3.51}$$

or, in matrix format,

$+\Delta\text{AR}_t + \Delta\text{Inv}_t$	$+S_t - \text{COGS}_t - \text{SGA}_t$	$-F_t^c + F_t^{cogs} + \text{SGA}_t$
$+\Delta\text{NFA}_t$	$-\text{Dep}_t$	$+0$
$+0$	$-\text{T}_t$	$+\text{T}_t$
$+\Delta C_t^e$	$+I_t^e$	$-F_t^e$

If, in addition, no inventory exists (e.g., manufacturing of the product is made on demand or the firm is a service firm), so that $\text{Inv}_t = 0$, then $\text{COGS}_t = F_t^{cogs}$ and the Matrix turns to

$+\Delta\text{AR}_t$	$+S_t - \text{COGS}_t - \text{SGA}_t$	$-F_t^c + \text{COGS}_t + \text{SGA}_t$
$+\Delta\text{NFA}_t$	$-\text{Dep}_t$	$+0$
$+0$	$-\text{T}_t$	$+\text{T}_t$
$+\Delta C_t^e$	$+I_t^e$	$-F_t^e$

Suppose sales are not affected by the undertaking of the project.[10] This is described as follows:

$+0$	$-\text{COGS}_t - \text{SGA}_t$	$+\text{COGS}_t + \text{SGA}_t$
$+\Delta\text{NFA}_t$	$-\text{Dep}_t$	$+0$
$+0$	$-\text{T}_t$	$+\text{T}_t$
$+\Delta C_t^e$	$+I_t^e$	$-F_t^e$

The change in equity is then equal to the change in NFA, which is in turn equal to the depreciation charge (changed in sign). (Note that it is not possible to cancel out COGS and SGA, otherwise equality of lower and upper columns in opposite regions does not hold. For the same reason, one must not cancel out the two occurrences of T_t.)

---

[10]This occurs in costs-only project, such as the decision regarding replacement of existing assets or technologies or substitution of suppliers (replacement projects). In these cases, $S_t = F_t^c = \Delta\text{AR}_t = 0$ for every t and the project generates cost savings which increase the income (see Examples 1.12 and 1.13).

Example 3.11 ♦
We now show how to apply Rule 7.1. Let us consider the following format:

$+AR_t$	$-AR_{t-1}$	$+S_t$	$-F_t^c$
$-NOL_t$	$+NOL_{t-1}$	$-OpC_t$	$+F_t^{opc}$
$+NFA_t$	$-NFA_{t-1}$	$-Dep_t$	$+F_t^{capex}$
$-TP_t$	$+TP_{t-1}$	$-T_t$	$+F_t^T$
$+EC_t$	$-EC_{t-1}$	$+I_t^{ec}$	$-F_t^{ec}$
$+FA_t$	$-FA_{t-1}$	$+I_t^{fa}$	$-F_t^{fa}$
$+LO_t$	$-LO_{t-1}$	$+I_t^{lo}$	$-F_t^{lo}$
$+BO_t$	$-BO_{t-1}$	$+I_t^{bo}$	$-F_t^{bo}$
$+NP_t$	$-NP_{t-1}$	$+I_t^{np}$	$-F_t^{np}$
$+C_t^e$	$-C_{t-1}^e$	$+I_t^e$	$-F_t^e$

Let (i, j) be the cell of row i and column j, $i \in \mathbb{N}_{10}^1$, $j \in \mathbb{N}_4^1$ and suppose X is added to cell (4, 4) (F_t^T). Then, by the law of conservation, we need to subtract X (add $-X$) to a cell in the same column but in a different row of the same region, for example, cell (2, 4). Then, by the law of motion, we need to subtract X to an element of the same row as cell (2, 4) but in the opposite partition, for example, (2, 1). Finally, by the law of conservation, one needs to add X to cell (4, 1).

Therefore, the cells that are involved are (4, 4), (2, 4), (2, 1), (4, 1). The result is

$+AR_t$	$-AR_{t-1}$	$+S_t$	$-F_t^c$
$-NOL_t - \boxed{X}$	$+NOL_{t-1}$	$-OpC_t$	$+F_t^{opc} - \boxed{X}$
$+NFA_t$	$-NFA_{t-1}$	$-Dep_t$	$+F_t^{capex}$
$-TP_t + \boxed{X}$	$+TP_{t-1}$	$-T_t$	$+F_t^T + \boxed{X}$
$+EC_t$	$-EC_{t-1}$	$+I_t^{ec}$	$-F_t^{ec}$
$+FA_t$	$-FA_{t-1}$	$+I_t^{fa}$	$-F_t^{fa}$
$+LO_t$	$-LO_{t-1}$	$+I_t^{lo}$	$-F_t^{lo}$
$+BO_t$	$-BO_{t-1}$	$+I_t^{bo}$	$-F_t^{bo}$
$+NP_t$	$-NP_{t-1}$	$+I_t^{np}$	$-F_t^{np}$
$+C_t^e$	$-C_{t-1}^e$	$+I_t^e$	$-F_t^e$

Example 3.12 ♦
We now show how to apply Rule 7.2. Let us start from the same format as in the previous example:

$+\text{AR}_t$	$-\text{AR}_{t-1}$	$+S_t$	$-F_t^c$
$-\text{NOL}_t$	$+\text{NOL}_{t-1}$	$-\text{OpC}_t$	$+F_t^{opc}$
$+\text{NFA}_t$	$-\text{NFA}_{t-1}$	$-\text{Dep}_t$	$+F_t^{capex}$
$-\text{TP}_t$	$+\text{TP}_{t-1}$	$-\text{T}_t$	$+F_t^T$
$+\text{EC}_t$	$-\text{EC}_{t-1}$	$+I_t^{ec}$	$-F_t^{ec}$
$+\text{FA}_t$	$-\text{FA}_{t-1}$	$+I_t^{fa}$	$-F_t^{fa}$
$+\text{LO}_t$	$-\text{LO}_{t-1}$	$+I_t^{lo}$	$-F_t^{lo}$
$+\text{BO}_t$	$-\text{BO}_{t-1}$	$+I_t^{bo}$	$-F_t^{bo}$
$+\text{NP}_t$	$-\text{NP}_{t-1}$	$+I_t^{np}$	$-F_t^{np}$
$+C_t^e$	$-C_{t-1}^e$	$+I_t^e$	$-F_t^e$

Let (i, j) be the cell of row i and column j, with $i \in \mathbb{N}_{10}^1$, $j \in \mathbb{N}_4^1$ and suppose X is added to cell $(4, 4)$ (F_t^T), just as in the previous example. Then, by the law of motion, we may add X to a cell in the same row but in the opposite partition, for example, cell $(4, 2)$. Then, by the law of conservation, we need to add X to any one of the four cells of column 2 which are placed in the opposite partition, for example, cell $(8, 2)$. Finally, by the law of motion, one needs to add X to cell $(8, 4)$.

Then, the cells to which X is added are $(4, 4)$, $(4, 2)$, $(8, 2)$, $(8, 4)$. The result is

$+\text{AR}_t$	$-\text{AR}_{t-1}$	$+S_t$	$-F_t^c$
$-\text{NOL}_t$	$+\text{NOL}_{t-1}$	$-\text{OpC}_t$	$+F_t^{opc}$
$+\text{NFA}_t$	$-\text{NFA}_{t-1}$	$-\text{Dep}_t$	$+F_t^{capex}$
$-\text{TP}_t$	$+\text{TP}_{t-1} + \boxed{X}$	$-\text{T}_t$	$+F_t^T + \boxed{X}$
$+\text{EC}_t$	$-\text{EC}_{t-1}$	$+I_t^{ec}$	$-F_t^{ec}$
$+\text{FA}_t$	$-\text{FA}_{t-1}$	$+I_t^{fa}$	$-F_t^{fa}$
$+\text{LO}_t$	$-\text{LO}_{t-1}$	$+I_t^{lo}$	$-F_t^{lo}$
$+\text{BO}_t$	$-\text{BO}_{t-1} + \boxed{X}$	$+I_t^{bo}$	$-F_t^{bo} + \boxed{X}$
$+\text{NP}_t$	$-\text{NP}_{t-1}$	$+I_t^{np}$	$-F_t^{np}$
$+C_t^e$	$-C_{t-1}^e$	$+I_t^e$	$-F_t^e$

Note that, joining the four X's with segment lines, a rectangle is formed whose four vertices are placed in the four partitions. As anticipated, Rule 7.2 may be stated in terms of an ideal quadrilateral, to be formed such that the four vertices are placed in the four partitions. (Analogous quadrilaterals are created via Rules 7.3–7.4, but the corners are not of the same sign.)

Now, as an additional example, let us add Y to cell $(1, 1)$ (AR_t) in the above Matrix. One just has to place the other three vertices of the rectangle so that each partition contains one vertex. For example, cells $(1, 1)$, $(1, 3)$, $(10, 3)$, $(10, 1)$.

$+\text{AR}_t + \textcircled{Y}$	$-\text{AR}_{t-1}$	$+S_t + \textcircled{Y}$	$-F_t^c$
$-\text{NOL}_t$	$+\text{NOL}_{t-1}$	$-\text{OpC}_t$	$+F_t^{opc}$
$+\text{NFA}_t$	$-\text{NFA}_{t-1} + \boxed{X}$	$-\text{Dep}_t$	$+F_t^{capex} + \boxed{X}$
$-\text{TP}_t$	$+\text{TP}_{t-1}$	$-\text{T}_t$	$+F_t^T$
$+\text{EC}_t$	$-\text{EC}_{t-1}$	$+I_t^{ec}$	$-F_t^{ec}$
$+\text{FA}_t$	$-\text{FA}_{t-1}$	$+I_t^{fa}$	$-F_t^{fa}$
$+\text{LO}_t$	$-\text{LO}_{t-1}$	$+I_t^{lo}$	$-F_t^{lo}$
$+\text{BO}_t$	$-\text{BO}_{t-1} + \boxed{X}$	$+I_t^{bo}$	$-F_t^{bo} + \boxed{X}$
$+\text{NP}_t$	$-\text{NP}_{t-1}$	$+I_t^{np}$	$-F_t^{np}$
$+C_t^e + \textcircled{Y}$	$-C_{t-1}^e$	$+I_t^e + \textcircled{Y}$	$-F_t^e$

Example 3.13 ♦
Rule 7.2 may be useful to analyze relations between variables or groups of variables belonging to different rows and different columns. For illustrative purposes, suppose the analyst wants to compute (for whatever reason) the difference between the change in debt and the sales revenues. Let us start from (3.44):

$+\Delta\text{AR}_t$	$+S_t$	$-F_t^c$
$-\Delta\text{NOL}_t$	$-\text{OpC}_t$	$+F_t^{opc}$
$+\Delta\text{NFA}_t$	$-\text{Dep}_t$	$+F_t^{capex}$
$-\Delta\text{TP}_t$	$-\text{T}_t$	$+F_t^T$
$-\Delta C_t^e$	$-I_t^e$	$+F_t^e$
$+\Delta C_t^d$	$+I_t^d$	$-F_t^d$
$-\Delta C_t^l$	$-I_t^l$	$+F_t^l$

As the elements of each partition coincide, one may swap the lower left partition and the lower right partition, then consolidate the columns of the upper right submatrix as well as the columns of the lower left submatrix

$+\Delta\text{AR}_t$	$+S_t - F_t^c$
$-\Delta\text{NOL}_t$	$-\text{OpC}_t + F_t^{opc}$
$+\Delta\text{NFA}_t$	$-\text{Dep}_t + F_t^{capex}$
$-\Delta\text{TP}_t$	$-\text{T}_t + F_t^T$
$-\Delta C_t^e$	$-I_t^e + F_t^e$
$+I_t^d - F_t^d$	$+\Delta C_t^d$
$-I_t^l + F_t^l$	$-\Delta C_t^l$

The Matrix has now only 2 columns. Then, one shifts the first row and the last row to the opposite regions:[11]

$+I_t^l - F_t^l$	$+\Delta C_t^l$
$-\Delta\text{NOL}_t$	$-\text{OpC}_t + F_t^{opc}$
$+\Delta\text{NFA}_t$	$-\text{Dep}_t + F_t^{capex}$
$-\Delta\text{TP}_t$	$-\text{T}_t + F_t^T$
$-\Delta C_t^e$	$-I_t^e + F_t^e$
$+I_t^d - F_t^d$	$+\Delta C_t^d$
$-\Delta\text{AR}_t$	$-S_t + F_t^c$

Finally, using Rule 7.2, one creates a rectangle by subtracting F_t^c from cells $(7, 2)$, $(7, 1)$, $(1, 5)$, $(5, 2)$:

$-I_t^l + F_t^l$	$-\Delta C_t^l$
$-\Delta\text{NOL}_t$	$-\text{OpC}_t + F_t^{opc}$
$-\Delta\text{NFA}_t$	$-\text{Dep}_t + F_t^{capex}$
$-\Delta\text{TP}_t$	$-T_t + F_t^T$
$-\Delta C_t^e - F_t^c$	$-I_t^e + F_t^e - F_t^c$
$-I_t^d + F_t^d$	$+\Delta C_t^d$
$-\Delta\text{AR}_t - F_t^c$	$-S_t + F_t^c - F_t^c$

Considering that $F_t^c - F_t^c = 0$, one gets

$-I_t^l + F_t^l$	$-\Delta C_t^l$
$-\Delta\text{NOL}_t$	$-\text{OpC}_t + F_t^{opc}$
$-\Delta\text{NFA}_t$	$-\text{Dep}_t + F_t^{capex}$
$-\Delta\text{TP}_t$	$-T_t + F_t^T$
$-\Delta C_t^e - F_t^c$	$-I_t^e + F_t^e - F_t^c$
$-I_t^d + F_t^d$	$+\Delta C_t^d$
$-\Delta\text{AR}_t - F_t^c$	$-S_t$

We have found three ways (in addition to the direct method) of computing the difference between change in debt and sales revenues.

Example 3.14 ♦

The split-screen technique may be fruitfully applied to find single items: One only needs to shift rows and columns in such a way that the single element under con-

[11] Note that the signs of the elements of the rows are changed.

sideration exhausts a region. For example, suppose one wants to represent OpC and start from the following format:

$+\Delta \mathrm{AR}_t$	$+S_t$	$-F_t^c$
$-\Delta \mathrm{NOL}_t$	$-\mathrm{OpC}_t$	$+F_t^{opc}$
$+\Delta \mathrm{NFA}_t$	$-\mathrm{Dep}_t$	$+F_t^{capex}$
$-\Delta \mathrm{TP}_t$	$-\mathrm{T}_t$	$+F_t^T$
$+\Delta C_t^l$	$+I_t^l$	$-F_t^l$
$+\Delta C_t^d$	$+I_t^d$	$-F_t^d$
$+\Delta C_t^e$	$+I_t^e$	$-F_t^e$

Shifting downward all the rows in the upper region (except the second one), one gets

$-\Delta \mathrm{NOL}_t$	$-\mathrm{OpC}_t$	$-F_t^{opc}$
$-\Delta \mathrm{AR}_t$	$-S_t$	$+F_t^c$
$-\Delta \mathrm{NFA}_t$	$+\mathrm{Dep}_t$	$-F_t^{capex}$
$+\Delta \mathrm{TP}_t$	$+\mathrm{T}_t$	$-F_t^T$
$-\Delta C_t^l$	$-I_t^l$	$+F_t^l$
$+\Delta C_t^d$	$+I_t^d$	$-F_t^d$
$+\Delta C_t^e$	$+I_t^e$	$-F_t^e$

Swapping first and second column (with due changes in sign), the operating costs are found:

(Pre-tax) operating costs

$+\mathrm{OpC}_t$	$+\Delta \mathrm{NOL}_t$	$+F_t^{opc}$
$+S_t$	$+\Delta \mathrm{AR}_t$	$+F_t^c$
$-\mathrm{Dep}_t$	$+\Delta \mathrm{NFA}_t$	$-F_t^{capex}$
$-\mathrm{T}_t$	$-\Delta \mathrm{TP}_t$	$-F_t^T$
$+I_t^l$	$+\Delta C_t^l$	$+F_t^l$
$-I_t^d$	$-\Delta C_t^d$	$-F_t^d$
$-I_t^e$	$-\Delta C_t^e$	$-F_t^e$

or, equivalently,

(Pre-tax) operating costs

$+\mathrm{COGS}_t + \mathrm{SGA}_t$	$+\Delta \mathrm{NOL}_t$	$+F_t^{opc}$
$+S_t$	$+\Delta \mathrm{AR}_t$	$+F_t^c$
$-\mathrm{Dep}_t$	$+\Delta \mathrm{NFA}_t$	$-F_t^{capex}$
$-\mathrm{T}_t$	$-\Delta \mathrm{TP}_t$	$-F_t^T$
$+I_t^l$	$+\Delta C_t^l$	$+F_t^l$
$-I_t^d$	$-\Delta C_t^d$	$-F_t^d$
$-I_t^e$	$-\Delta C_t^e$	$-F_t^e$

This result supplies four ways of computing OpC_t. If one wants to additionally derive the COGS, it suffices to subtract SGA_t from the cells (1, 1), (1, 2), (3, 1), (3, 2) (Rule 7.2):

Cost of goods sold

$+\text{COGS}_t$	$+\Delta\text{NOL}_t - \text{SGA}_t + F_t^{opc}$
$+S_t$	$+\Delta\text{AR}_t \quad +F_t^c$
$-\text{Dep}_t - \text{SGA}_t$	$+\Delta\text{NFA}_t - \text{SGA}_t \quad +F_t^{capex}$
$-\text{T}_t$	$-\Delta\text{TP}_t \quad -F_t^{\text{T}}$
$+I_t^l$	$+\Delta C_t^l \quad +F_t^l$
$-I_t^d$	$-\Delta C_t^d \quad -F_t^d$
$-I_t^e$	$-\Delta C_t^e \quad -F_t^e$

We have found three ways of interpreting the COGS in addition to its standard definition (see (3.19)).

3.6.2 *Rule 7: Recording Transactions*

Among other things, Rule 7 of the split-screen technique is extremely powerful in that it enables recording any business transaction made by the firm minimizing the risk of modelling error. It suffices to remember that, for any transaction, both the law of motion and the law of conservation must be fulfilled, which is graphically described by a quadrilateral in the split-screen Matrix.

For simplicity, we analyze the following transactions: Purchase of fixed assets made on credit, purchase of fixed assets made on cash, disposal of fixed assets on credit, disposal of fixed assets on cash, disposal of fixed assets on cash with gain or loss on sale. Disposal of assets is particularly important in capital budgeting decisions because

- at time n, the project/firm is liquidated and the salvage value of the residual assets must be included in the appraising process (see also Sect. 4.4)
- at time 0, assets must be sold if the project is an abandonment project or a replacement project (e.g., see Bumps example in Sect. 12.1).

Purchase of Fixed Assets Made on Credit

Let y be the price of a piece of equipment (or other depreciable asset) and let the purchase be made on credit. Let us start from Matrix (3.40) and consolidate it to get a more economical format:

$$
\mathbf{M}_t = \begin{array}{|l|llll|}
\hline
+\mathrm{AR}_t & +\mathrm{AR}_{t-1} & +S_t & -F_t^c \\
+\mathrm{Inv}_t & +\mathrm{Inv}_{t-1} & +\Delta\mathrm{Inv}_t & \\
-\mathrm{AP}_t & -\mathrm{AP}_{t-1} & -\mathrm{COP}_t & +F_t^{cop} \\
-\mathrm{SWP}_t & -\mathrm{SWP}_{t-1} & -\mathrm{LC}_t & +F_t^{lc} \\
+\mathrm{NFA}_t & +\mathrm{NFA}_{t-1} & -\mathrm{Dep}_t & +F_t^{capex} \\
-\mathrm{TP}_t & -\mathrm{TP}_{t-1} & -\mathrm{T}_t & +F_t^T \\
+C_t^l & +C_{t-1}^l & +I_t^l & -F_t^l \\
\hline
+C_t^d & +C_{t-1}^d & +I_t^d & -F_t^d \\
+C_t^e & +C_{t-1}^e & +I_t^e & -F_t^e \\
\hline
\end{array}
$$

Let $\mathrm{NFA}_t^{\#}$ and $\mathrm{AP}_t^{\#}$ be the book value of NFA and AP, respectively, gross of the purchase price. The purchase of the fixed asset makes NFA increase by y while (by law of conservation) the accounts payable increase by the same amount: $\mathrm{NFA}_t = \mathrm{NFA}_t^{\#} + y$ and $\mathrm{AP}_t = \mathrm{AP}_t^{\#} + y$. However, to fulfill the law of motion, elements in the left region and elements in the right regions must balance, so the same amount must be added in the NFA row (in the opposite region) and subtracted in the AP row (in the opposite region). Thus, the Matrix may be framed as

$$
\mathbf{M}_t = \begin{array}{|l|lll|}
\hline
+\mathrm{AR}_t & +\mathrm{AR}_{t-1} & +S_t & -F_t^c \\
+\mathrm{Inv}_t & +\mathrm{Inv}_{t-1} & +\Delta\mathrm{Inv}_t & \\
-\mathrm{AP}_t^{\#} - \boxed{y} & -\mathrm{AP}_{t-1} - \boxed{y} & -\mathrm{COP}_t & +F_t^{cop} \\
-\mathrm{SWP}_t & -\mathrm{SWP}_{t-1} & -\mathrm{LC}_t & +F_t^{lc} \\
+\mathrm{NFA}_t^{\#} + \boxed{y} & +\mathrm{NFA}_{t-1} + \boxed{y} & -\mathrm{Dep}_t & +F_t^{capex} \\
-\mathrm{TP}_t & -\mathrm{TP}_{t-1} & -\mathrm{T}_t & +F_t^T \\
+C_t^l & +C_{t-1}^l & +I_t^l & -F_t^l \\
\hline
+C_t^d & +C_{t-1}^d & +I_t^d & -F_t^d \\
+C_t^e & +C_{t-1}^e & +I_t^e & -F_t^e \\
\hline
\end{array}
\tag{3.52}
$$

(Rule 7.1 is used). Observe that the capital of the firm has not increased nor decreased, nor any additional cash flow has flown in or out of the system. Only, the structure of the operating assets has been altered.

Purchase of Fixed Assets Made on Cash

Suppose the purchase of the same asset is made on cash and the cash is withdrawn from the liquid accounts. Let $C_t^{l\#}$ be the amount of liquid assets at time t, before the transaction is recorded (i.e., excluding the amount paid for the asset purchase); let $F_t^{capex\#}$ and $F_t^{l\#}$ be, respectively, the amount of cash outflow for capital expenditures and the amount of cash flow from liquid assets, excluding the amount paid for the asset purchase. The purchase of the asset makes the NFA be greater than $\mathrm{NFA}_t^{\#}$ by y and, by law of conservation, liquid assets are smaller than $C_t^{l\#}$ by y: $C_t^l = C_t^{l\#} - y$. Likewise, capital expenditures increase by y, an amount which is subtracted from liquid assets, which means that $-y$ is added to $-F_t^{l\#}$: $F_t^{capex} = F_t^{capex\#} + y$ and $-F_t^l = -F_t^{l\#} - y$.

$$\mathbf{M}_t = \left|\begin{array}{l|lll}
+\text{AR}_t & +\text{AR}_{t-1} & +S_t & -F_t^c \\
+\text{Inv}_t & +\text{Inv}_{t-1} & +\Delta\text{Inv}_t & \\
-\text{AP}_t & -\text{AP}_{t-1} & -\text{COP}_t & +F_t^{cop} \\
-\text{SWP}_t & -\text{SWP}_{t-1} & -\text{LC}_t & +F_t^{lc} \\
+\text{NFA}_t^{\#} + \boxed{y} & +\text{NFA}_{t-1} & -\text{Dep}_t & +F_t^{capex\#} + \boxed{y} \\
-\text{TP}_t & -\text{TP}_{t-1} & -\text{T}_t & +F_t^T \\
+C_t^{l\#} - \boxed{y} & +C_{t-1}^l & +I_t^l & -F_t^{l\#} - \boxed{y} \\
\hline
+C_t^d & +C_{t-1}^d & +I_t^d & -F_t^d \\
+C_t^e & +C_{t-1}^e & +I_t^e & -F_t^e
\end{array}\right|$$

(Rule 7.1 is used). The firm's capital has not increased nor decreased; only its structure has been altered, exchanging liquid assets for fixed assets.

One might also express the same transactions in terms of net debt, by shifting downward the liquid asset row, thereby forming a different quadrilateral:

$$\mathbf{M}_t = \left|\begin{array}{l|lll}
+\text{AR}_t & +\text{AR}_{t-1} & +S_t & -F_t^c \\
+\text{Inv}_t & +\text{Inv}_{t-1} & +\Delta\text{Inv}_t & \\
-\text{AP}_t & -\text{AP}_{t-1} & -\text{COP}_t & +F_t^{cop} \\
-\text{SWP}_t & -\text{SWP}_{t-1} & -\text{LC}_t & +F_t^{lc} \\
+\text{NFA}_t^{\#} + \boxed{y} & +\text{NFA}_{t-1} & -\text{Dep}_t & +F_t^{capex\#} + \boxed{y} \\
-\text{TP}_t & -\text{TP}_{t-1} & -\text{T}_t & +F_t^T \\
\hline
+C_t^{\mathcal{D}\#} + \boxed{y} & +C_{t-1}^{\mathcal{D}} & +I_t^{\mathcal{D}} & -F_t^{\mathcal{D}\#} + \boxed{y} \\
+C_t^e & +C_{t-1}^e & +I_t^e & -F_t^e
\end{array}\right|$$

where $C_t^{\mathcal{D}\#}$ and $F_t^{\mathcal{D}\#}$ are, respectively, the net debt and the CF$\mathcal{D}$, without taking into account the asset price. Observe that the reduction in cash makes the net debt increase (Rule 7.2 is applied). Under this framing, it appears that cash flow has flown into the system, thereby increasing the capital. However, the capital-structure disaggregates only the operating capital, not the total firm's (or project's) capital. It is the *operating* capital which has been increased, not the firm's or project's capital. As seen above, the latter is unvaried and no net investment has been made by the investors (equityholders and debtholders).

Alternatively, suppose that the asset is purchased partly on cash and partly on credit and the cash outflow is financed by debholders, by equityholders, and by liquid accounts. Precisely, suppose that y_1 is raised from debtholders, y_2 is raised by issuing shares, y_3 is withdrawn from the liquid account, while y_4 is granted on credit by suppliers, such that

$$\overbrace{y_1 + y_2 + y_3}^{\text{purchase made on cash}} \quad + \quad \overbrace{y_4}^{\text{purchase made on credit}} \quad = y.$$

Then, the transactions are easily summarized in the Matrix with Rule 7 as follows (with obvious meaning of the symbols):

$$\mathbf{M}_t = \begin{array}{|l|llll|}
\hline
+\mathrm{AR}_t & +\mathrm{AR}_{t-1} & +S_t & -F_t^c & \\
+\mathrm{Inv}_t & +\mathrm{Inv}_{t-1} & +\Delta\mathrm{Inv}_t & & \\
-\mathrm{AP}_t^{\#} - \boxed{y_4} & -\mathrm{AP}_{t-1} - \boxed{y_4} & -\mathrm{COP}_t & +F_t^{cop} & \\
-\mathrm{SWP}_t & -\mathrm{SWP}_{t-1} & -\mathrm{LC}_t & +F_t^{lc} & \\
+\mathrm{NFA}_t^{\#} + \boxed{y} & +\mathrm{NFA}_{t-1} + \boxed{y_4} & -\mathrm{Dep}_t & +F_t^{capex\#} & + \boxed{y_1 + y_2 + y_3} \\
-\mathrm{TP}_t & -\mathrm{TP}_{t-1} & -\mathrm{T}_t & +F_t^T & \\
+C_t^{l\#} - \boxed{y_3} & +C_{t-1}^l & +I_t^l & -F_t^{l\#} - \boxed{y_3} & \\
\hline
+C_t^{d\#} + \boxed{y_1} & +C_{t-1}^d & +I_t^d & -F_t^{d\#} + \boxed{y_1} & \\
+C_t^{e\#} + \boxed{y_2} & +C_{t-1}^e & +I_t^e & -F_t^{e\#} + \boxed{y_2} & \\
\hline
\end{array}$$

In this case, we have used both Rule 7.1 (for y_3 and y_4) and Rule 7.2 (for y_1 and y_2). It should be clear that the firm's capital has increased, after the purchase, by $y_1 + y_2$, and cash flow has flown into the system for the same amount. The amount $y_1 + y_2$ is the firm's net investment, while the amount $y_3 + y_4$ has only determined a change in the structure of the assets. In other words, the transactions with suppliers (purchase on credit by y_4) and financial markets (sales of marketable securities or withdrawal from cash accounts by y_3) have only changed the assets' structure, whereas the transactions with debtholders (by y_1) and equityholders (by y_2) have determined an additional net investment in the firm.

Disposal of Fixed Assets Made on Credit

Let us now consider the case of disposals of fixed assets. For simplicity, let us use a more economical format with the NOL class replacing its single constituents:

$$\mathbf{M}_t = \begin{array}{|l|lll|}
\hline
+\mathrm{AR}_t & +\mathrm{AR}_{t-1} & +S_t & -F_t^c \\
-\mathrm{NOL}_t & -\mathrm{NOL}_{t-1} & -\mathrm{OpC}_t & +F_t^{opc} \\
+\mathrm{NFA}_t & +\mathrm{NFA}_{t-1} & -\mathrm{Dep}_t & +F_t^{capex} \\
-\mathrm{TP}_t & -\mathrm{TP}_{t-1} & -\mathrm{T}_t & +F_t^T \\
+C_t^l & +C_{t-1}^l & +I_t^l & -F_t^l \\
\hline
+C_t^d & +C_{t-1}^d & +I_t^d & -F_t^d \\
+C_t^e & +C_{t-1}^e & +I_t^e & -F_t^e \\
\hline
\end{array}$$

Suppose the selling price of the disposed assets is equal to its book value, x (no gains nor losses on disposal of assets). Let $\mathrm{NFA}_t^{\#}$ and $\mathrm{AR}_t^{\#}$ be, respectively, the NFA and the AR excluding any consideration about the disposal. If one takes account of the disposal of assets, NFA_t is smaller than $\mathrm{NFA}_t^{\#}$ by x while AR_t is greater than $\mathrm{AR}_t^{\#}$

by the same amount: $\text{NFA}_t = \text{NFA}_t^{\#} - x$ and $\text{AR}_t = \text{AR}_t^{\#} + x$. However, to fulfill the law of motion, opposite partitions must balance, so the same amount must be subtracted in the NFA row (in the opposite partition) and must be added in the AR row (in the opposite partition). Thus, the Matrix becomes

$$\mathbf{M}_t = \begin{array}{|l|ll|}\hline +\text{AR}_t^{\#} + \boxed{x} & +\text{AR}_{t-1} + \boxed{x} & +S_t \quad -F_t^c \\ -\text{NOL}_t & -\text{NOL}_{t-1} & -\text{OpC}_t \quad +F_t^{opc} \\ +\text{NFA}_t^{\#} - \boxed{x} & +\text{NFA}_{t-1} - \boxed{x} & -\text{Dep}_t \quad +F_t^{capex} \\ -\text{TP}_t & -\text{TP}_{t-1} & -\text{T}_t \quad +F_t^T \\ +C_t^l & +C_{t-1}^l & +I_t^l \quad -F_t^l \\ \hline +C_t^d & +C_{t-1}^d & +I_t^d \quad -F_t^d \\ +C_t^e & +C_{t-1}^e & +I_t^e \quad -F_t^e \\ \hline \end{array} \tag{3.53}$$

(Rule 7.1 is used). There is no cash effect and the capital effect consists of a change in the structure of the operating assets (fixed assets are exchanged for accounts receivable), with total amount unvaried.

Remark 3.1 (*Decant*) In Part I, we have shown that the level of the capital C_t is affected by two sources: Income and cash flow. Breaking the project system down into subsystems, we have seen that each subsystem is perturbated by its own income component and cash flow component. However, this section pictorializes another way the level of a subsystem's capital may be affected: Decant of some of one subsystem's capital into another subsystem. For example, Matrix (3.53) pictorializes a decant of some NFA into the AR system. Likewise, Matrix (3.52) presents a decant of some NFA into the AP system (NFA increases, −AP decreases). The income components and cash-flow components are unaffected, and the net effect on the level of the operating-asset system is zero; only the distribution of operating assets is changed. This implies that, all considered, a subsystem is perturbated by three sources:

- a flow of income
- a flow of cash
- a decant from one subsystem to another.

All considered, one may state that Rule 7 shows that the Matrix and the split-screen technique represent a *quadruple-entry book-keeping system*, for both statics and dynamics are characterized by a double-entry system such that the regions always balance out.

Disposal of Fixed Assets Made on Cash

We apply Rule 7.1 to the case where a fixed asset is disposed and sold on a cash basis and the cash inflow is deposited in the cash account. Let $C_t^{l\#}$, $F_t^{capex\#}$, $F_t^{l\#}$ be, respectively, the liquid assets, the capital expenditures, the cash flow from liquid assets, before any consideration of the book value of disposed assets. The amount $\text{NFA}_t^{\#}$ decreases by x, and the same occurs to the net capital expenditures $F_t^{capex\#}$. Owing to the cash inflow x, the liquid assets increase by x. The above Matrix becomes

$$\mathbf{M}_t = \left|\begin{array}{l|llll}
+\mathrm{AR}_t & +\mathrm{AR}_{t-1} & +S_t & -F_t^c & \\
-\mathrm{NOL}_t & -\mathrm{NOL}_{t-1} & -\mathrm{OpC}_t & +F_t^{opc} & \\
+\mathrm{NFA}_t^{\#} - \boxed{x} & +\mathrm{NFA}_{t-1} & -\mathrm{Dep}_t & +F_t^{capex\#} & -\boxed{x} \\
-\mathrm{TP}_t & -\mathrm{TP}_{t-1} & -\mathrm{T}_t & +F_t^T & \\
+C_t^{l\#} + \boxed{x} & +C_{t-1}^l & +I_t^l & -F_t^{l\#} + & \boxed{x} \\
\hline
+C_t^d & +C_{t-1}^d & +I_t^d & -F_t^d & \\
+C_t^e & +C_{t-1}^e & +I_t^e & -F_t^e &
\end{array}\right| \tag{3.54}$$

(No net cash effect occurs, only a redistribution of the operating structure.)

Suppose, alternatively, that the cash inflow generated by the disposal is distributed to equityholders, rather than retained in the firm. Then, the equity decreases by x and the CFE increases by x (i.e., $-F_t^{e\#}$ decreases by x). Rule 7.2 is used, which reveals a net distribution to equityholders.

$$\mathbf{M}_t = \left|\begin{array}{l|llll}
+\mathrm{AR}_t & +\mathrm{AR}_{t-1} & +S_t & -F_t^c & \\
-\mathrm{NOL}_t & -\mathrm{NOL}_{t-1} & -\mathrm{OpC}_t & +F_t^{opc} & \\
+\mathrm{NFA}_t^{\#} - \boxed{x} & +\mathrm{NFA}_{t-1} & -\mathrm{Dep}_t & +F_t^{capex\#} & -\boxed{x} \\
-\mathrm{TP}_t & -\mathrm{TP}_{t-1} & -\mathrm{T}_t & +F_t^T & \\
+C_t^l & +C_{t-1}^l & +I_t^l & -F_t^l & \\
\hline
+C_t^d & +C_{t-1}^d & +I_t^d & -F_t^d & \\
+C_t^{e\#} - \boxed{x} & +C_{t-1}^e & +I_t^e & -F_t^{e\#} & -\boxed{x}
\end{array}\right| \tag{3.55}$$

Disposal of Fixed Assets with Gain or Loss on Sale

Suppose x is the book value of a plant asset and let z be a gain (or loss, if negative) on the disposal of asset at time t, such that $x + z$ is the selling price. As usual, we use the superscript # to denote the amount of an item before the transaction is recorded. After the sale, revenues increases as well as taxes (revenues are taxed at a tax rate of τ): $S_t = S_t^{\#} + z$ and $\mathrm{T}_t = \mathrm{T}_t^{\#} + \tau z$, implying that the net income increases by z (if positive) but decreases by the taxes paid on z: $I_t^e = I_t^{e\#} + z - \tau z$. Suppose that

i. the extra tax component, τz, is paid in the year where it occurs (so that taxes payable are not affected by the transaction)
ii. the asset is sold on a cash basis
iii. the cash collected (net of the extra tax payment) is distributed to equityholders.

We rest on the law of conservation and the law of motion (and, in particular, on Rule 7) to describe the transaction:

$$
\mathbf{M}_t = \left[\begin{array}{l|lll}
+\mathrm{AR}_t & +\mathrm{AR}_{t-1} & +S_t^{\#} + \textcircled{z} & -F_t^{c\#} - \textcircled{z} \\
-\mathrm{NOL}_t & -\mathrm{NOL}_{t-1} & -\mathrm{OpC}_t & +F_t^{opc} \\
+\mathrm{NFA}_t^{\#} - \boxed{x} & +\mathrm{NFA}_{t-1} & -\mathrm{Dep}_t & +F_t^{capex\#} - \boxed{x} \\
-\mathrm{TP}_t & -\mathrm{TP}_{t-1} & -\mathrm{T}_t^{\#} - \boxed{\tau z} & +F_t^{T\#} + \boxed{\tau z} \\
+C_t^l & +C_{t-1}^l & +I_t^l & -F_t^l \\
\hline
+C_t^d & +C_{t-1}^d & +I_t^d & +F_t^d \\
+C_t^{e\#} - \boxed{x} & +C_{t-1}^e & +I_t^{e\#} + \textcircled{z} - \boxed{\tau z} & -F_t^{e\#} - \textcircled{z} - \boxed{x} + \boxed{\tau z}
\end{array}\right] \tag{3.56}
$$

As explained earlier, Rule 7 turns the transaction into three quadrilaterals: One is described with boxed corners (book value, x, Rule 7.2), one has circled corners (tax effect, τz, Rule 7.3), and the third quadrilateral has rounded corners (gain/loss on sale, z, Rule 7.3). A subtraction of resources from the system occurs, as a consequence of the disposal of the asset. Specifically, the capital decreases by x. This is the effect of the (after-tax) gain, $z(1-\tau)$ (income-component), and the distribution of additional CFE to equityholders, $z(1-\tau)+x$ (cash-flow component).

Matrix (3.55) is equal to Matrix (3.56) if $z = 0$.

Suppose, alternatively, that the sale is made on credit. Then, the Matrix becomes

$$
\mathbf{M}_t = \left[\begin{array}{l|lll}
+\mathrm{AR}_t + \boxed{x} + \textcircled{z} & +\mathrm{AR}_{t-1} + \boxed{x} & +S_t^{\#} + \textcircled{z} & -F_t^{c\#} \\
-\mathrm{NOL}_t & -\mathrm{NOL}_{t-1} & -\mathrm{OpC}_t & +F_t^{opc} \\
+\mathrm{NFA}_t^{\#} - \boxed{x} & +\mathrm{NFA}_{t-1} - \boxed{x} & -\mathrm{Dep}_t & +F_t^{capex\#} \\
-\mathrm{TP}_t & -\mathrm{TP}_{t-1} & -\mathrm{T}_t^{\#} - \boxed{\tau z} & +F_t^{T\#} + \boxed{\tau z} \\
+C_t^l & +C_{t-1}^l & +I_t^l & -F_t^l \\
\hline
+C_t^d & +C_{t-1}^d & +I_t^d & +F_t^d \\
+C_t^{e\#} + \textcircled{z} & +C_{t-1}^e & +I_t^{e\#} + \textcircled{z} - \boxed{\tau z} & -F_t^{e\#} + \boxed{\tau z}
\end{array}\right] \tag{3.57}
$$

In this case, Rule 7.1 is applied for x, Rule 7.2 is applied for z, Rule 7.3 is applied for τz. The net result is a change in the structure of the operating capital and a net increase in the assets, owing to the gain on the sale, z, which increases the operating assets (accounts receivable) and the equity capital by the same amount (this is the net effect of an increase in income and a decrease in equity distribution, amounting to $z(1-\tau)$ and τz, respectively).

Matrix (3.53) is equal to Matrix (3.57) if $z = 0$ (no gain/loss on sale).

Both the statics and the dynamics of an economic system are expressed by two fundamental equations: The law of motion and the law of conservation. Each equation subsumes a double-entry system, for addition of a monetary amount on one side of each equation must be matched by an addition of the same amount on the other side of the same equation. As a result, the Matrix and the split-screen technique give rise to a **quadruple-entry book-keeping system**.

3.7 Reconciling the Matrix with the Standard Financial Statements

The split-screen Matrix simultaneously represents a detailed snapshot of the project's or firm's capital at a given date (statics) and the sources of variation of the project's capital in a given period (dynamics), broken down in income component and cash-flow component. As such, it contains, in a single visual representation, the *balance sheet* (BS), the *income statement* (IS), and the *cash-flow statement* (CFS).

In traditional accounting and finance textbooks, these financial statements are presented as separate documents, where information is congealed and economic analysis is only possible off-statement with a high degree of cognitive effort upon the learner/reader. In contrast, the Matrix is a floating environment and the associated split-screen technique represent a powerful and simple method of displaying and reframing information, accomplishing sophisticated economic analysis while minimizing the risk of errors and the cognitive effort required on the part of the analyst. It therefore embeds the three statements into a unified system of statics and dynamics, which may be managed, without any mathematical ability, for appraising projects and making decisions.

Below we explicitly show that the Matrix incorporates the traditional standard financial statements as special cases of possible configurations of the Matrix.

Balance Sheet

Table 3.11 represents a standard account-form BS as may be found in any finance or accounting textbook. It collects the book values (i.e., carrying amounts) of assets and liabilities, which express the capital invested (assets) or borrowed (liabilities) by a firm. Consider the expanded Matrix in (3.40), which we report below:

$$\mathbf{M}_t = \begin{array}{|l|llll|}
\hline
+\mathrm{AR}_t & +\mathrm{AR}_{t-1} & +S_t & -F_t^r \\
+\mathrm{Inv}_t & +\mathrm{Inv}_{t-1} & +\Delta\mathrm{Inv}_t & \\
-\mathrm{AP}_t & -\mathrm{AP}_{t-1} & -\mathrm{COP}_t & -F_t^{cop} \\
-\mathrm{SWP}_t & -\mathrm{SWP}_{t-1} & -\mathrm{LC}_t & -F_t^{lc} \\
+\mathrm{NFA}_t & +\mathrm{NFA}_{t-1} & -\mathrm{Dep}_t & +F_t^{capex} \\
-\mathrm{TP}_t & -\mathrm{TP}_{t-1} & -\mathrm{T}_t & +F_t^{T} \\
+\mathrm{EC}_t & +\mathrm{EC}_{t-1} & +I_t^{ec} & -F_t^{ec} \\
+\mathrm{FA}_t & +\mathrm{FA}_{t-1} & +I_t^{fa} & -F_t^{fa} \\
\hline
+\mathrm{LO}_t & +\mathrm{LO}_{t-1} & +I_t^{lo} & -F_t^{lo} \\
+\mathrm{BO}_t & +\mathrm{BO}_{t-1} & +I_t^{bo} & -F_t^{bo} \\
+\mathrm{NP}_t & +\mathrm{NP}_{t-1} & +I_t^{np} & -F_t^{np} \\
+C_t^e & +C_{t-1}^e & +I_t^e & -F_t^e \\
\hline
\end{array}$$

Table 3.11 The account form balance sheet

Investments	Financings
Operating assets	**Debt**
NET OPERATING WORKING CAPITAL	Loans
Accounts receivable	Bonds
Inventory	Notes payable
Less: Accounts payable	
Less: Salaries and wages payable	
Less: Taxes payable	**Equity**
	Contributed capital and accumulated retained earnings
NET FIXED ASSETS	
Property, plant & equipment (net of accumulated depreciation)	
Liquid assets	
Excess cash	
Financial assets	

Columns 1 and 2 of the Matrix now showcase all 12 items contained in Table 3.11, displayed in a tabular format, and with the proper sign, referred at time $t-1$ and t, respectively.

Income Statement

Table 3.12 Multiple-step income statement

+ Sales	$+ S_t$
− Cost of Goods Sold	$-\text{COGS}_t$
= **Gross profit**	$= \mathbf{GP}_t$
− Selling, general and administrative expenses	$-\text{SGA}_t$
= **EBITDA**	$= \mathbf{EBITDA}_t$
− Depreciation and amortization	$-\text{Dep}_t$
= **EBIT**	$= \mathbf{EBIT}_t$
+ Interest income	$+ I_t^l$
− Interest expenses	$- I_t^d$
= **EBT**	$= \mathbf{EBT}_t$
− Income taxes	$-\text{T}_t$
= **Net income**	$= \boldsymbol{I_t^e}$

Table 3.12 shows a standard multiple-step IS:

Step 1. The amount of sales is written down in the first line[12]
Step 2. Subtracting COGS from sales one finds the *gross profit* (GP)
Step 3. Subtracting SGA from GP one gets the *earnings before interest, taxes, depreciation and amortization* (EBITDA). This is a pre-tax and pre-depreciation operating income
Step 4. Subtracting depreciation from EBITDA one gets the *earnings before interest and taxes* (EBIT). This is a pre-tax operating income
Step 5. Subtracting net interest ($I_t^{\mathcal{D}}$) from EBIT one gets the *earnings before taxes* (EBT)
Step 6. Subtracting income taxes (T_t) from EBT one gets the net income (I_t^e), which is the income pertaining to equityholders.

We now show that the standard IS in Table 3.12 is embedded in the Matrix, which, in addition, visually and logically associates each component of IS to each component of BS and to each component of CFS, thereby providing, for each type of income, both the corresponding capital component and also the corresponding cash flow component.

Further, it provides four different ways of computing the same income. The four perspectives may be classified into two subclasses (top-down approach and bottom-up approach), each of which is split up into two subgroups (income approach, capital-cash approach). Thus, for any income we will have a *fourfold* split screen providing four different-but-equivalent representations.

Consider the operating matrix in (3.32), which portraits the manufacturing and nonmanufacturing liabilities. Using this framing, the Matrix is pictured as in Eq. (3.58).

$$\begin{array}{|l|llr|}
\hline
+\text{AR}_t & +\text{AR}_{t-1} & +S_t & -F_t^c \\
-\text{NOL}_t^m & -\text{NOL}_{t-1}^m & -\text{COGS}_t & +F_t^{cogs} \\
-\text{NOL}_t^{nm} & -\text{NOL}_{t-1}^{nm} & -\text{SGA}_t & +F_t^{sga} \\
+\text{NFA}_t & +\text{NFA}_{t-1} & -\text{Dep}_t & +F_t^{capex} \\
-\text{TP}_t & -\text{TP}_{t-1} & -\text{T}_t & +F_t^T \\
+C_t^l & +C_{t-1}^l & +I_t^l & -F_t^l \\
\hline
+C_t^d & +C_{t-1}^d & +I_t^d & -F_t^d \\
+C_t^e & +C_{t-1}^e & +I_t^e & -F_t^e \\
\hline
\end{array} \qquad (3.58)$$

Moving column 2 leftward to the west region and consolidating, one gets

[12]If gains/losses on disposal of fixed assets occur, they must be added/subtracted at this step. See also Sect. 4.4.

$$
\begin{array}{|l|ll|}
\hline
+\Delta \mathrm{AR}_t & +S_t & -F_t^c \\
-\Delta \mathrm{NOL}_t^m & -\mathrm{COGS}_t & +F_t^{cogs} \\
-\Delta \mathrm{NOL}_t^{nm} & -\mathrm{SGA}_t & +F_t^{sga} \\
+\Delta \mathrm{NFA}_t & -\mathrm{Dep}_t & +F_t^{capex} \\
-\Delta \mathrm{TP}_t & -\mathrm{T}_t & +F_t^T \\
+\Delta C_t^l & +I_t^l & -F_t^l \\
\hline
+\Delta C_t^d & +I_t^d & -F_t^d \\
+\Delta C_t^e & +I_t^e & -F_t^e \\
\hline
\end{array}
\tag{3.59}
$$

Moving rows 3–6 downward to the south region and shifting column 3 leftward to the west region, the *fourfold* GP is created:

Gross Profit

$$
\begin{array}{|ll|l|}
\hline
+F_t^c & +\Delta \mathrm{AR}_t & +S_t \\
-F_t^{cogs} & -\Delta \mathrm{NOL}_t^m & -\mathrm{COGS}_t \\
\hline
+F_t^{sga} & +\Delta \mathrm{NOL}_t^{nm} & +\mathrm{SGA}_t \\
+F_t^{capex} & -\Delta \mathrm{NFA}_t & +\mathrm{Dep}_t \\
+F_t^T & +\Delta \mathrm{TP}_t & +\mathrm{T}_t \\
-F_t^l & -\Delta C_t^l & -I_t^l \\
+F_t^d & +\Delta C_t^d & +I_t^d \\
+F_t^e & +\Delta C_t^e & +I_t^e \\
\hline
\end{array}
\tag{3.60}
$$

Using the income approach, one may adopt a top-down approach starting from sales and proceeding downward, or a bottom-up approach starting from net income and proceeding upward (right region):

$$
\mathrm{GP}_t = \overbrace{S_t - \mathrm{COGS}_t}^{\text{top-down income approach}}
$$

$$
\mathrm{GP}_t = \overbrace{I_t^e + I_t^d - I_t^l + \mathrm{T}_t + \mathrm{Dep}_t + \mathrm{SGA}_t}^{\text{bottom-up income approach}}\,.
$$

Alternatively, one may adopt a top-down capital-cash approach starting from $F_t^c + \Delta\mathrm{AR}_t$ and proceeding downward; or one may employ a bottom-up capital-cash approach, starting from $F_t^e + \Delta C_t^e$ and proceeding upward (left region):

$$
\mathrm{GP}_t = \overbrace{F_t^c - F_t^{cogs} + \Delta\mathrm{AR}_t - \Delta\mathrm{NOL}_t^m}^{\text{top-down capital-cash approach}}
$$

$$
\mathrm{GP}_t = \overbrace{F_t^e + \Delta C_t^e + F_t^d + \Delta C_t^d - F_t^l - \Delta C_t^l + F_t^T + \Delta\mathrm{TP}_t + F_t^{capex} - \Delta\mathrm{NFA}_t + F_t^{sga} + \Delta\mathrm{NOL}_t^{nm}}^{\text{bottom-up capital-cash approach}}\,.
$$

Shifting up row 3, one gets the fourfold EBITDA:

$$
\begin{array}{|ll|l|}
\multicolumn{3}{c}{\textbf{EBITDA}} \\
\hline
+F_t^{c} & +\Delta\mathrm{AR}_t & +S_t \\
-F_t^{cogs} & -\Delta\mathrm{NOL}_t^{m} & -\mathrm{COGS}_t \\
-F_t^{sga} & -\Delta\mathrm{NOL}_t^{nm} & -\mathrm{SGA}_t \\
\hline
-F_t^{capex} & -\Delta\mathrm{NFA}_t & +\mathrm{Dep}_t \\
+F_t^{T} & +\Delta\mathrm{TP}_t & +\mathrm{T}_t \\
-F_t^{l} & -\Delta C_t^{l} & -I_t^{l} \\
+F_t^{d} & +\Delta C_t^{d} & +I_t^{d} \\
+F_t^{e} & +\Delta C_t^{e} & +I_t^{e} \\
\hline
\end{array}
\tag{3.61}
$$

Like the GP, the EBITDA may be conceptualized in a top-down approach or a bottom-up approach, either using income components or using capital-and-cash components (we leave it to the reader). Shifting up the depreciation row, the fourfold EBIT is found:

$$
\begin{array}{|ll|l|}
\multicolumn{3}{c}{\textbf{EBIT}} \\
\hline
+F_t^{c} & +\Delta\mathrm{AR}_t & +S_t \\
-F_t^{cogs} & -\Delta\mathrm{NOL}_t^{m} & -\mathrm{COGS}_t \\
-F_t^{sga} & -\Delta\mathrm{NOL}_t^{nm} & -\mathrm{SGA}_t \\
-F_t^{capex} & +\Delta\mathrm{NFA}_t & -\mathrm{Dep}_t \\
\hline
+F_t^{T} & +\Delta\mathrm{TP}_t & +\mathrm{T}_t \\
-F_t^{l} & -\Delta C_t^{l} & -I_t^{l} \\
+F_t^{d} & +\Delta C_t^{d} & +I_t^{d} \\
+F_t^{e} & +\Delta C_t^{e} & +I_t^{e} \\
\hline
\end{array}
\tag{3.62}
$$

Shifting upward the rows pertaining to debt and liquid assets, one finds the fourfold EBT:

$$
\begin{array}{|ll|l|}
\multicolumn{3}{c}{\textbf{EBT}} \\
\hline
+F_t^{c} & +\Delta\mathrm{AR}_t & +S_t \\
-F_t^{cogs} & -\Delta\mathrm{NOL}_t^{m} & -\mathrm{COGS}_t \\
-F_t^{sga} & -\Delta\mathrm{NOL}_t^{nm} & -\mathrm{SGA}_t \\
-F_t^{capex} & +\Delta\mathrm{NFA}_t & -\mathrm{Dep}_t \\
+F_t^{l} & +\Delta C_t^{l} & +I_t^{l} \\
-F_t^{d} & -\Delta C_t^{d} & -I_t^{d} \\
\hline
+F_t^{T} & +\Delta\mathrm{TP}_t & +\mathrm{T}_t \\
+F_t^{e} & +\Delta C_t^{e} & +I_t^{e} \\
\hline
\end{array}
\tag{3.63}
$$

Finally, moving upward the tax row, the fourfold Net Income is found:

Net Income

$+F_t^c$	$+\Delta AR_t$	$+S_t$
$-F_t^{cogs}$	$-\Delta NOL_t^m$	$-COGS_t$
$-F_t^{sga}$	$-\Delta NOL_t^{nm}$	$-SGA_t$
$-F_t^{capex}$	$+\Delta NFA_t$	$-Dep_t$
$+F_t^l$	$+\Delta C_t^l$	$+I_t^l$
$-F_t^d$	$-\Delta C_t^d$	$-I_t^d$
$-F_t^T$	$-\Delta TP_t$	$-T_t$
$+F_t^e$	$+\Delta C_t^e$	$+I_t^e$

(3.64)

Consolidating the manufacturing and non-manufacturing items,

Net Income

$+F_t^c$	$+\Delta AR_t$	$+S_t$
$-F_t^{opc}$	$-\Delta NOL_t$	$-OpC_t$
$-F_t^{capex}$	$+\Delta NFA_t$	$-Dep_t$
$+F_t^l$	$+\Delta C_t^l$	$+I_t^l$
$-F_t^d$	$-\Delta C_t^d$	$-I_t^d$
$-F_t^T$	$-\Delta TP_t$	$-T_t$
$+F_t^e$	$+\Delta C_t^e$	$+I_t^e$

(3.65)

Alternatively, highlighting the constituent elements of NOWC,

Net Income

$+F_t^c$	$+\Delta AR_t$	$+S_t$
$+0$	$+\Delta Inv_t$	$+\Delta Inv_t$
$-F_t^{cop}$	$-\Delta AP_t$	$-COP_t$
$-F_t^{lc}$	$-\Delta SWP_t$	$-LC_t$
$-F_t^{capex}$	$+\Delta NFA_t$	$-Dep_t$
$+F_t^l$	$+\Delta C_t^l$	$+I_t^l$
$-F_t^d$	$-\Delta C_t^d$	$-I_t^d$
$-F_t^T$	$-\Delta TP_t$	$-T_t$
$+F_t^e$	$+\Delta C_t^e$	$+I_t^e$

(3.66)

Example 3.15 ♦
Consolidation may bring about several (fourfold) information. For example, consider (3.64). Consolidating

(i) rows 1, 2, 3, and 7
(ii) rows 5 and 6,

net income may be framed as

Net Income

$$\begin{array}{|ll|l|}
\hline
+F^{nowc} & +\Delta\text{NOWC}_t & +\text{EBITDA}_t - \text{T}_t \\
-F_t^{capex} & +\Delta\text{NFA}_t & -\text{Dep}_t \\
-F_t^{\mathcal{D}} & -\Delta C_t^{\mathcal{D}} & -I_t^{\mathcal{D}} \\
\hline
+F_t^e & +\Delta C_t^e & +I_t^e \\
\hline
\end{array} \qquad (3.67)$$

where, as we remind, $F_t^{nowc} = F_t^c - F_t^{cogs} - F_t^{sga} - F_t^T$ is the cash flow generated by NOWC. Shifting the net debt row downward to the lower region and consolidating, one gets the fourfold OI:

Operating Income (EBIT$_t$ − T$_t$)

$$\begin{array}{|ll|l|}
\hline
+F^{nowc} & +\Delta\text{NOWC}_t & +\text{EBITDA}_t - \text{T}_t \\
-F_t^{capex} & +\Delta\text{NFA}_t & -\text{Dep}_t \\
\hline
+F_t^o & +\Delta C_t^o & +I_t^o \\
\hline
\end{array}$$

(compare with (3.12)).

Example 3.16 ♦
Suppose the analyst has the following data for getting the gross profit: $I_t^d = 7$, $I_t^e = 4$, $\Delta\text{NOL}_t^{nm} = -2$, $\Delta C_t^l = -3$, $F_t^l = 5$ $F_t^{sga} = 4.5$, $T_t = 1$, $\text{Dep}_t = 4$. Filling in the cells in the gross profit Matrix (3.60),

Gross Profit

$$\begin{array}{|ll|l|}
\hline
+F_t^c & +\Delta\text{AR}_t & +S_t \\
-F_t^{cogs} & -\Delta\text{NOL}_t^m & -\text{COGS}_t \\
\hline
+4.5 & -2 & +\text{SGA}_t \\
+F_t^{capex} & -\Delta\text{NFA}_t & +4 \\
+F_t^T & +\Delta\text{TP}_t & +1 \\
-5 & +3 & -I_t^l \\
+F_t^d & +\Delta C_t^d & +7 \\
+F_t^e & +\Delta C_t^e & +4 \\
\hline
\end{array}$$

Swapping the elements of row 3 and the elements of row 6,

Gross Profit

$+F_t^c$	$+\Delta\text{AR}_t$	$+S_t$
$-F_t^{cogs}$	$-\Delta\text{NOL}_t^m$	$-\text{COGS}_t$
$+\text{SGA}_t$		$+4.5-2$
$+F_t^{capex}$	$-\Delta\text{NFA}_t$	$+4$
$+F_t^T$	$+\Delta\text{TP}_t$	$+1$
$-I_t^l$		$-5+3$
$+F_t^d$	$+\Delta C_t^d$	$+7$
$+F_t^e$	$+\Delta C_t^e$	$+4$

The GP is then equal to $16.5 = +4.5 - 2 + 4 + 1 - 5 + 3 + 7 + 4$.

Cash Flow Statement

The cash flow statement or statement of cash flows collects the various types of cash inflows and outflows that characterize a project (or firm). In the Matrix, each type of cash flow is clearly associated with its income component and capital component. In particular, consider (3.46) which we report below:

$+\Delta\text{AR}_t$	$+S_t$	$-F_t^c$
$+\Delta\text{Inv}_t$	$+\Delta\text{Inv}_t$	
$-\Delta\text{AP}_t$	$-\text{COP}_t$	$+F_t^{cop}$
$-\Delta\text{SWP}_t$	$-\text{LC}_t$	$+F_t^{lc}$
$+\Delta\text{NFA}_t$	$-\text{Dep}_t$	$+F_t^{capex}$
$-\Delta\text{TP}_t$	$-\text{T}_t$	$+F_t^T$
$+\Delta C_t^l$	$+I_t^l$	$-F_t^l$
$+\Delta C_t^d$	$+I_t^d$	$-F_t^d$
$+\Delta C_t^e$	$+I_t^e$	$-F_t^e$

(3.68)

Breaking down liquid assets and debt into their constituent classes, one gets

$$
\begin{array}{|l|ll|}
\hline
+\Delta \mathrm{AR}_t & +S_t & -F_t^c \\
+\Delta \mathrm{Inv}_t & +\Delta \mathrm{Inv}_t & \\
-\Delta \mathrm{AP}_t & -\mathrm{COP}_t & +F_t^{cop} \\
-\Delta \mathrm{SWP}_t & -\mathrm{LC}_t & +F_t^{lc} \\
+\Delta \mathrm{NFA}_t & -\mathrm{Dep}_t & +F_t^{capex} \\
-\Delta \mathrm{TP}_t & -\mathrm{T}_t & +F_t^{T} \\
+\Delta \mathrm{EC}_t & +I_t^{ec} & -F_t^{ec} \\
+\Delta \mathrm{FA}_t & +I_t^{fa} & -F_t^{fa} \\
\hline
+\Delta \mathrm{LO}_t & +I_t^{lo} & -F_t^{lo} \\
+\Delta \mathrm{BO}_t & +I_t^{bo} & -F_t^{bo} \\
+\Delta \mathrm{NP}_t & +I_t^{np} & -F_t^{np} \\
+\Delta C_t^e & +I_t^{e} & -F_t^{e} \\
\hline
\end{array}
\tag{3.69}
$$

The cash-flow statement is highlighted by isolating the third column. To this end, one shifts column 1 to the right and column 3 to the left, so getting

$$
\begin{array}{|l|ll|}
\multicolumn{3}{c}{\text{Cash Flow Statement}} \\
\hline
+F_t^c & +S_t & -\Delta \mathrm{AR}_t \\
 & +\Delta \mathrm{Inv}_t & -\Delta \mathrm{Inv}_t \\
-F_t^{cop} & -\mathrm{COP}_t & +\Delta \mathrm{AP}_t \\
-F_t^{lc} & -\mathrm{LC}_t & +\Delta \mathrm{SWP}_t \\
-F_t^{capex} & -\mathrm{Dep}_t & -\Delta \mathrm{NFA}_t \\
-F_t^{T} & -\mathrm{T}_t & +\Delta \mathrm{TP}_t \\
+F_t^{ec} & +I_t^{ec} & -\Delta \mathrm{EC}_t \\
+F_t^{fa} & +I_t^{fa} & -\Delta \mathrm{FA}_t \\
\hline
+F_t^{lo} & +I_t^{lo} & -\Delta \mathrm{LO}_t \\
+F_t^{bo} & +I_t^{bo} & -\Delta \mathrm{BO}_t \\
+F_t^{np} & +I_t^{np} & -\Delta \mathrm{NP}_t \\
+F_t^{e} & +I_t^{e} & -\Delta C_t^e \\
\hline
\end{array}
\tag{3.70}
$$

The CFS highlights each cash flow in a fourfold perspective. In particular, a top-down approach and a bottom-up approach may be employed to derive a given notion of cash flow; each of these approaches may be framed in terms of cash (cash approach) or in terms of difference between income components and capital components (income-capital approach).

Table 3.13 Statement of cash flows (traditional format)

Operating activities	
1 Net Income	$+I_t^e$
2 + Depreciation and amortization	$+\text{Dep}_t$
3 − Changes in accounts receivables	$-\Delta\text{AR}_t$
4 − Changes in inventory	$-\Delta\text{Inv}_t$
5 + Changes in salaries and wages payable	$+\Delta\text{SWP}_t$
6 + Changes in accounts payable	$+\Delta\text{AP}_t$
7 + Changes in taxes payable	$+\Delta\text{TP}_t$
8 **Cash from operating activities**	
Investment activities	
8 − Change in financial assets	$-\Delta\text{FA}_t$
9 − Capital expenditures	$-F_t^{capex}$
10 **Cash from investing activities**	
Financing activities	
11 + Change in notes payable	$+\Delta\text{NP}_t$
12 + Changes in bonds	$+\Delta\text{BO}_t$
13 + Changes in loans	$+\Delta\text{LO}_t$
14 + Issuance of new stock	$-F_t^e$
15 − Distributions to equityholders	
16 **Cash from financing activities**	
17 **Change in cash** (8+10+16)	$= \Delta\text{EC}_t$

The standard CFS described in finance and accounting textbooks is represented in a report form as illustrated in Table 3.13. It is divided into three parts:

- cash from operating activities (line 8)
- cash from investing activities (line 10)
- cash from financing activities (line 16).

The sum of these three types of cash flows supplies the change in excess cash, ΔEC_t (line 17 in Table 3.13).

What is the relation between (3.70) and the standard (report form) statement of cash flows in Table 3.13? We show that the latter is but a reframing of the former, which is easily carved out via the split-screen technique. To see it, we only need to isolate the change-in-excess-cash cell. To this end, let us take Matrix (3.69) and move all the rows of the north region (except EC) down to the south region. This isolates the excess cash item, which exhausts the upper left partition:

$+\Delta \text{EC}_t$	$+I_t^{ec}$	$-F_t^{ec}$
$-\Delta \text{AR}_t$	$-S_t$	$+F_t^c$
$-\Delta \text{Inv}_t$	$-\Delta \text{Inv}_t$	
$+\Delta \text{AP}_t$	$+\text{COP}_t$	$-F_t^{cop}$
$+\Delta \text{SWP}_t$	$+\text{LC}_t$	$-F_t^{lc}$
$-\Delta \text{NFA}_t$	$+\text{Dep}_t$	$-F_t^{capex}$
$+\Delta \text{TP}_t$	$+\text{T}_t$	$-F_t^T$
$-\Delta \text{FA}_t$	$-I_t^{fa}$	$+F_t^{fa}$
$+\Delta \text{LO}_t$	$+I_t^{lo}$	$-F_t^{lo}$
$+\Delta \text{BO}_t$	$+I_t^{bo}$	$-F_t^{bo}$
$+\Delta \text{NP}_t$	$+I_t^{np}$	$-F^{np}$
$+\Delta C_t^e$	$+I_t^e$	$-F_t^e$

Now, let us swap (i) the elements of the equity row that are placed in opposite partitions and (ii) the elements of the NFA row that are placed in opposite partitions (no change in sign is needed here. See Rule 3 of the split-screen technique). The result is

Change in excess cash

$+\Delta \text{EC}_t$	$+I_t^{ec}$	$-F_t^{ec}$
$-\Delta \text{AR}_t$	$-S_t$	$+F_t^c$
$-\Delta \text{Inv}_t$	$-\Delta \text{Inv}_t$	
$+\Delta \text{AP}_t$	$+\text{COP}_t$	$-F_t^{cop}$
$+\Delta \text{SWP}_t$	$+\text{LC}_t$	$-F_t^{lc}$
$+\Delta \text{TP}_t$	$+T_t$	$-F_t^T$
$+\text{Dep}_t - F_t^{capex}$	$-\Delta \text{NFA}_t$	
$-\Delta \text{FA}_t$	$-I_t^{fa}$	$+F_t^{fa}$
$+\Delta \text{LO}_t$	$+I_t^{lo}$	$-F_t^{lo}$
$+\Delta \text{BO}_t$	$+I_t^{bo}$	$-F_t^{bo}$
$+\Delta \text{NP}_t$	$+I_t^{np}$	$-F_t^{np}$
$+I_t^e - F_t^e$	$+\Delta C_t^e$	

(3.71)

The 13 items contained in the lower left submatrix are exactly the same as the ones appearing in Table 3.13.

This means that textbooks adopt a bottom-up change-in-capital approach, while replacing ΔNFA_t and ΔC_t^e with the respective differences of income and cashflow components. (Note that, if no interest is earned on excess cash, then the change in excess cash is a cash flow: $\Delta\text{EC}_t = -F_t^{ec}$.) Therefore, the traditional statement of cash flows represents but one single column of a special configuration of the Matrix.

Remark 3.2 (*Terminology*) One should not confuse the notion of *cash from operating activities* as defined in the traditional statement of cash flows with the notion of operating cash flow (CFO) we use in the book. The CFO is an after-tax cash flow generated by the operations, which is equal to $F_t^o = I_t^o - \Delta C_t^o$. The latter includes the capital expenditures (via ΔC_t^o), which represent a capital injection in the firm's operations, and abstracts from interest incomes and interest expenses (but includes their tax effect via T_t). In contrast, the cash from operating activities of a traditional CFS excludes capital expenditures (which are included in the cash from investing activities) and includes interest income and interest expenses (incorporated in the IS). Indeed, the reader may check that the difference between the cash from operating activities as calculated in Table 3.13 and the CFO is equal to $F_t^{capex} - I_t^d + I_t^l$. ▲

3.8 A Managerial Perspective on COGS

The cost of goods sold is an essential notion in managerial accounting, corporate finance, and engineering economics. If the project is undertaken by a manufacturing firm (one which makes use of manufacturing fixed assets), the COGS may be defined either by including depreciation in the manufacturing overhead or excluding depreciation from it, depending on the purpose of the analysis. The alternative does not affect the appraising process, it only affects the gross profit (GP). Financial models prefer not to include noncash charges such as depreciation, while accounting models privilege a comprehensive approach, inclusive of depreciation.[13] To avoid ambiguities and errors, we here reconcile the two perspectives.

Let Dep_t^m be the depreciation charge for manufacturing assets (plant and machinery depreciation, factory depreciation, etc.) and let

$$\text{COGS}_t^\circ = \text{COGS}_t + \text{Dep}_t^m \tag{3.72}$$

denote the cost of goods sold inclusive of depreciation of manufacturing assets. Also, let Dep_t^{nm} be the depreciation charges for nonmanufacturing assets (e.g., office equipment) and let

$$\text{SGA}_t^\circ = \text{SGA}_t + \text{Dep}_t^{nm} \tag{3.73}$$

be the so-called ***operating expenses***.[14] All (pre-tax) operational costs are then divided into the two above categories:

---

[13] Berk and DeMarzo (2014) use the expression "clean expenses" to refer to costs that do not include depreciation charges. Clean expenses are implicitly used in finance textbooks (a separate line for depreciation is a signal of it). Berk and DeMarzo (2014) explicitly use this approach (see p. 235, footnote 2, see also glossary, p. 1049, and Example 23, p. 266, where COGS is referred to as "Manufacturing expenses (*other than depreciation*)" (italics added). Titman and Martin (2011) also use the expressions "operating expenses (before depreciation)" (p. 32) or "operating expenses (excluding depreciation)" (p. 343) to refer to SGA.

[14] Not to be confused with operating costs, $\text{OpC}_t = \text{COGS}_t + \text{SGA}_t$.

Table 3.14 EBIT—financial and accounting perspective

Financial	Accounting	Difference
$+S_t$	$+S_t$	
$-\text{COGS}_t$	$-(\underbrace{\text{COGS}_t + \text{Dep}_t^m}_{\text{COGS}_t^\circ})$	$+\text{Dep}_t^m$
= **Gross Profit** (GP_t)	= **Gross Profit**(GP_t°)	= Dep_t^m
$-\text{SGA}_t$	$-(\underbrace{\text{SGA}_t + \text{Dep}_t^{nm}}_{\text{SGA}_t^\circ})$	$+\text{Dep}_t^{nm}$
= **EBITDA**		
$-\text{Dep}_t$		$-\text{Dep}_t$
= **EBIT**$_t$	= **EBIT**$_t$	= 0

$$\text{OpC}_t^\circ = \text{COGS}_t^\circ + \text{SGA}_t^\circ.$$

The gross profit (GP_t°) is then redefined as $\text{GP}_t^\circ = S_t - \text{COGS}_t^\circ$. Hence,

$$\begin{aligned} \text{EBIT}_t &= \text{GP}_t^\circ - \text{SGA}_t^\circ \\ &= S_t - \text{COGS}_t^\circ - \text{SGA}^\circ \\ &= S_t - \underbrace{\left(\text{COGS}_t + \text{Dep}_t^m + \text{SGA}_t + \text{Dep}_t^{nm}\right)}_{\text{OpC}_t^\circ}. \end{aligned} \tag{3.74}$$

The difference between financial and accounting perspective lies in the treatment of depreciation (see Table 3.14). In particular, the difference between OpC_t and OpC_t° is equal to Dep_t, while the difference between GP_t and GP_t° is equal to Dep_t^m.

In terms of the Matrix, if the analyst computes a comprehensive COGS instead of a clean one, the NFA row is split up into manufacturing and nonmanufacturing assets, here denoted as NFA_t^m and NFA_t^{nm}. Consistently, capital expenditures must be classified as relating to manufacturing activities (plant and machinery) and nonmanufacturing activities (buildings, office equipment, etc.): $F_t^{capex} = F_t^{cem} + F_t^{cenm}$. Thus,

$$\text{NFA}_t = \begin{cases} \text{NFA}_t^m &= \text{NFA}_{t-1}^m - \text{Dep}_t^m + F_t^{cem} \\ \text{NFA}_t^{nm} &= \text{NFA}_{t-1}^{nm} - \text{Dep}_t^{nm} + F_t^{cenm}. \end{cases}$$

The GP Matrix presented in Eq. (3.60) may be reframed as

Gross profit (GP°) – accounting perspective (3.75)

$+F_t^c$	$+\Delta\mathrm{AR}_t$	$+S_t$
$-F_t^{cogs}$	$-\Delta\mathrm{NOL}_t^m$	$-\mathrm{COGS}_t$
$+F_t^{cem}$	$-\Delta\mathrm{NFA}_t^m$	$+\mathrm{Dep}_t^m$
$+F_t^{sga}$	$+\Delta\mathrm{NOL}_t^{nm}$	$+\mathrm{SGA}_t$
$+F_t^{cenm}$	$-\Delta\mathrm{NFA}_t^{nm}$	$+\mathrm{Dep}_t^{nm}$
$+F_t^T$	$+\Delta\mathrm{TP}_t$	$+\mathrm{T}_t$
$-F_t^l$	$-\Delta C_t^l$	$-I_t^l$
$+F_t^d$	$+\Delta C_t^d$	$+I_t^d$
$+F_t^e$	$+\Delta C_t^e$	$+I_t^e$

or, consolidating rows 2–3 and rows 4–5,

Gross profit (GP°) – accounting perspective (3.76)

$+F_t^c$	$+\Delta\mathrm{AR}_t$	$+S_t$
$-F_t^{cogs} - F_t^{cem}$	$-\Delta\mathrm{NOL}_t^m + \Delta\mathrm{NFA}_t^m$	$-\mathrm{COGS}_t - \mathrm{Dep}_t^m$
$+F_t^{sga} + F_t^{cenm}$	$+\Delta\mathrm{NOL}_t^{nm} - \Delta\mathrm{NFA}_t^{nm}$	$+\mathrm{SGA}_t + \mathrm{Dep}_t^{nm}$
$+F_t^T$	$+\Delta\mathrm{TP}_t$	$+\mathrm{T}_t$
$-F_t^l$	$-\Delta C_t^l$	$-I_t^l$
$+F_t^d$	$+\Delta C_t^d$	$+I_t^d$
$+F_t^e$	$+\Delta C_t^e$	$+I_t^e$

or, equivalently,

Gross profit (GP°) – accounting perspective

$+F_t^c$	$+\Delta\mathrm{AR}_t$	$+S_t$
$-F_t^{cogs} - F_t^{cem}$	$-\Delta\mathrm{NOL}_t^m + \Delta\mathrm{NFA}_t^m$	$-\mathrm{COGS}_t^\circ$
$+F_t^{sga} + F_t^{cenm}$	$+\Delta\mathrm{NOL}_t^{nm} - \Delta\mathrm{NFA}_t^{nm}$	$+\mathrm{SGA}_t^\circ$
$+F_t^T$	$+\Delta\mathrm{TP}_t$	$+\mathrm{T}_t$
$-F_t^l$	$-\Delta C_t^l$	$-I_t^l$
$+F_t^d$	$+\Delta C_t^d$	$+I_t^d$
$+F_t^e$	$+\Delta C_t^e$	$+I_t^e$

3.9 Cost-Volume-Profit Analysis: An Engineering Perspective

A third perspective on costs, commonly employed in engineering projects, is one where all the pre-tax operational costs OpC_t are classified as either fixed costs or variable costs. Fixed costs are independent of the activity level, whereas variable costs vary under changes in the level of activity. In these cases, COGS and SGA are considered in terms of their variable and fixed components. In general, COGS are variable costs (direct labor, direct materials, variable manufacturing overhead) and SGAs are fixed costs (employees' and executives' salaries, rent, insurance, office supplies, etc.). However, there are also fixed COGS (fixed manufacturing overhead, factory insurance, plant managers' salaries, etc.) as well as variable SGAs (sales commissions, shipping expenses, etc.):

$$\text{OpC}_t = \overbrace{(\text{variable COGS}_t + \text{variable SGA}_t)}^{\text{Variable Costs}} + \overbrace{(\text{fixed COGS}_t + \text{fixed SGA}_t)}^{\text{Fixed Costs}}. \tag{3.77}$$

Whenever costs are classified as variable costs or fixed costs, the resulting IS is said to be a *cost-volume-profit* IS. The cost-volume-profit IS appears as follows:[15]

Cost-volume-profit IS
+ Sales
− Variable costs
Variable COGS
Variable SGA
= **Contribution margin**
− Fixed costs
Fixed COGS
Fixed SGA
= **EBITDA**
− Depreciation
= **EBIT**
− Income Taxes
= **Operating income**

Practically, variable costs are often estimated starting from the known variable unit cost, so that the operating costs are calculated as

$$\text{OpC}_t = \text{Variable unit cost} \times \text{Annual production (in units)} + \text{Fixed costs}. \tag{3.78}$$

The variable unit cost may itself be constant or, rather, a function of the production output. Letting x be the production units, one may then write

$$\text{OpC}_t(x_t) = \text{c}(x_t) \cdot x_t + \text{FC}_t \tag{3.79}$$

where $\text{c}(x_t)$ is the variable unit cost and FC_t denote the fixed costs.[16]

[15]See also Rainbow project in Sect. 12.3.

[16]In some cases, the variable unit cost may be constant under changes in the production output: $\text{c}(x_t) = \text{c}$.

Example 3.17 ♦
An engineering project consists of launching a product at a price of \$1.5. Fixed costs for production in a given year are estimated at $\mathrm{FC}_t = \$100{,}000$ while the estimated variable unit cost is constant and equal to $\mathrm{c}(x) = \$1.2$. Depreciation charges amount to \$3,500. Therefore, letting x_t be the production in the tth year, the EBIT is

$$\begin{aligned}\mathrm{EBIT}_t(x_t) &= S_t(x_t) - \mathrm{OpC}_t(x_t) - \mathrm{Dep}_t = 1.5 \cdot x_t - (100{,}000 + 1.2 \cdot x_t) - 3{,}500\\ &= 0.3x_t - 103{,}500\end{aligned}$$

(we are assuming that the units produced, x_t, equate the units sold). If the production in the year is $x_t = 400{,}000$, then

$$\mathrm{EBIT}_t(400{,}000) = 120{,}000 - 103{,}500 = 16{,}500.$$

What is the minimum amount necessary to break even the EBIT? From $\mathrm{EBIT}_t = 0.3x_t - 103{,}500 = 0$ one gets $x_t = 345{,}000$.

Example 3.18 ♦
Let us consider the project in the example above and suppose $\Delta\mathrm{NOWC}_t = 10{,}000$, $F_t^{capex} = 0$. How many units of production will be needed to generate an operating cash flow equal to \$4,000?

$$\begin{aligned}F_t^o &= S_t(x_t) - \mathrm{OpC}_t(x_t) - \mathrm{Dep}_t - \Delta\mathrm{NFA}_t - \Delta\mathrm{NOWC}_t\\ &= S_t(x_t) - \mathrm{OpC}_t(x_t) - \mathrm{Dep}_t + \mathrm{Dep}_t - F_t^{capex} - \Delta\mathrm{NOWC}_t\\ &= 1.5 \cdot x_t - (100{,}000 + 1.2 \cdot x_t) - 10{,}000 =\\ &= 0.3x_t - 110{,}000 = 4{,}000\end{aligned}$$

whence $x_t = 380{,}000$. How many units of production are necessary to break even the CFO? From $0.3 \cdot x_t - 110{,}000 = 0$, one gets $x_t = 366{,}666.67$.

3.10 Key Points

- A project system is classified into four subsystems: Operating assets, non-operating assets, debt, and equity. Operating assets are in turn divided into net fixed assets (property, plant and equipment, net of depreciation) and net operating working capital (accounts receivable and inventories, net of accounts payable, taxes payable, and salaries and wages payable). Non-operating assets consist of excess cash, marketable securities and other financial assets. Debt consists of loans, bonds, and notes payable. Equity consists of paid-in capital and retained earnings
- Depreciation is an income component of net fixed assets and represents a decrease in invested capital; capital expenditures are a cash-flow component and represent an increase in invested capital

- Accounts payable, salaries and wages payable, net of inventories, form the class of net operating liabilities. The cost of goods sold is the income component of this class; payments to suppliers and employees for purchases of material, direct labor, and manufacturing overhead represent its cash-flow component
- Accounts receivable represent the amount that customers owe to the firm. Sales revenues represent the income component of the class while payments from customers is the cash-flow component
- Income taxes payable represent the amount owed by the firm to the government. It increases by the income component, the taxes, and decreases by the payments made by the firm in the period. In project analysis, this class is usually at zero level in every period (taxes are assumed to be paid when they are due)
- All information of a project or firm is fruitfully collected in the (diachronic-synchronic) split-screen *Matrix*. Following are three equivalent configurations of the Matrix:

The Matrix

$+\Delta \text{NOWC}_t$	$\text{EBITDA}_t - \text{T}_t$	$-F_t^{nowc}$
$+\Delta \text{NFA}_t$	$-\text{Dep}_t$	$+F_t^{capex}$
$+\Delta C_t^l$	$+I_t^l$	$-F_t^l$
$+\Delta C_t^d$	$+I_t^d$	$-F_t^d$
$+\Delta C_t^e$	$+I_t^e$	$-F_t^e$

The Matrix

$+\Delta \text{AR}_t$	$+S_t$	$-F_t^c$
$-\Delta \text{NOL}_t$	$-\text{OpC}_t$	$+F_t^{opc}$
$-\Delta \text{TP}_t$	$-\text{T}_t$	$+F_t^{\text{T}}$
$+\Delta \text{NFA}_t$	$-\text{Dep}_t$	$+F_t^{capex}$
$+\Delta \text{EC}_t$	$+I_t^{ec}$	$-F_t^{ec}$
$+\Delta \text{FA}_t$	$+I_t^{fa}$	$-F_t^{fa}$
$+\Delta \text{LO}_t$	$+I_t^{lo}$	$-F_t^{lo}$
$+\Delta \text{BO}_t$	$+I_t^{bo}$	$-F_t^{bo}$
$+\Delta \text{NP}_t$	$+I_t^{np}$	$-F_t^{np}$
$+\Delta C_t^e$	$+I_t^e$	$-F_t^e$

The Matrix

$+\Delta \text{AR}_t$	$+S_t$	$-F_t^c$
$+\Delta \text{Inv}_t$	$+\Delta \text{Inv}_t$	
$-\text{AP}_t$	$-\text{COP}_t$	$+F_t^{cop}$
$-\text{SWP}_t$	$-\text{LC}_t$	$+F_t^{lc}$
$-\Delta \text{TP}_t$	$-\text{T}_t$	$+F_t^{\text{T}}$
$+\Delta \text{NFA}_t$	$-\text{Dep}_t$	$+F_t^{capex}$
$+\Delta \text{EC}_t$	$+I_t^{ec}$	$-F_t^{ec}$
$+\Delta \text{FA}_t$	$+I_t^{fa}$	$-F_t^{fa}$
$+\Delta \text{LO}_t$	$+I_t^{lo}$	$-F_t^{lo}$
$+\Delta \text{BO}_t$	$+I_t^{bo}$	$-F_t^{bo}$
$+\Delta \text{NP}_t$	$+I_t^{np}$	$-F^{np}$
$+\Delta C_t^e$	$+I_t^e$	$-F_t^e$

  Any format of the Matrix is partitioned into four screens; the inner (horizontal and vertical) bars express equality of adjacent partitions, which then collect quantities of equal amount. The rows express the law of motion for any subsystem; the columns express the law of conservation for capital, income, and cash flow
- The three columns of the Matrix represent three statements: The statement of financial position (balance sheet), the income statement, and the cash-flow statement
- The Matrix enjoys invariance properties such that a set of quadruple-entry working rules may be derived, called *split-screen technique*. The latter enables framing the Matrix into different configurations which favor understanding of the relations among variables and help the analyst make complex analysis with no need of mathematical skills
- The gross profit, EBITDA, EBIT, EBT, and net income are types of income, and are obtained from the sales by subtracting, respectively, the operating costs (cost of goods sold and selling, general, and administrative expenses), the depreciation charges, the net interest, the income tax. For each type of income, the resulting partition highlights the relations with the respective capital and cash-flow components

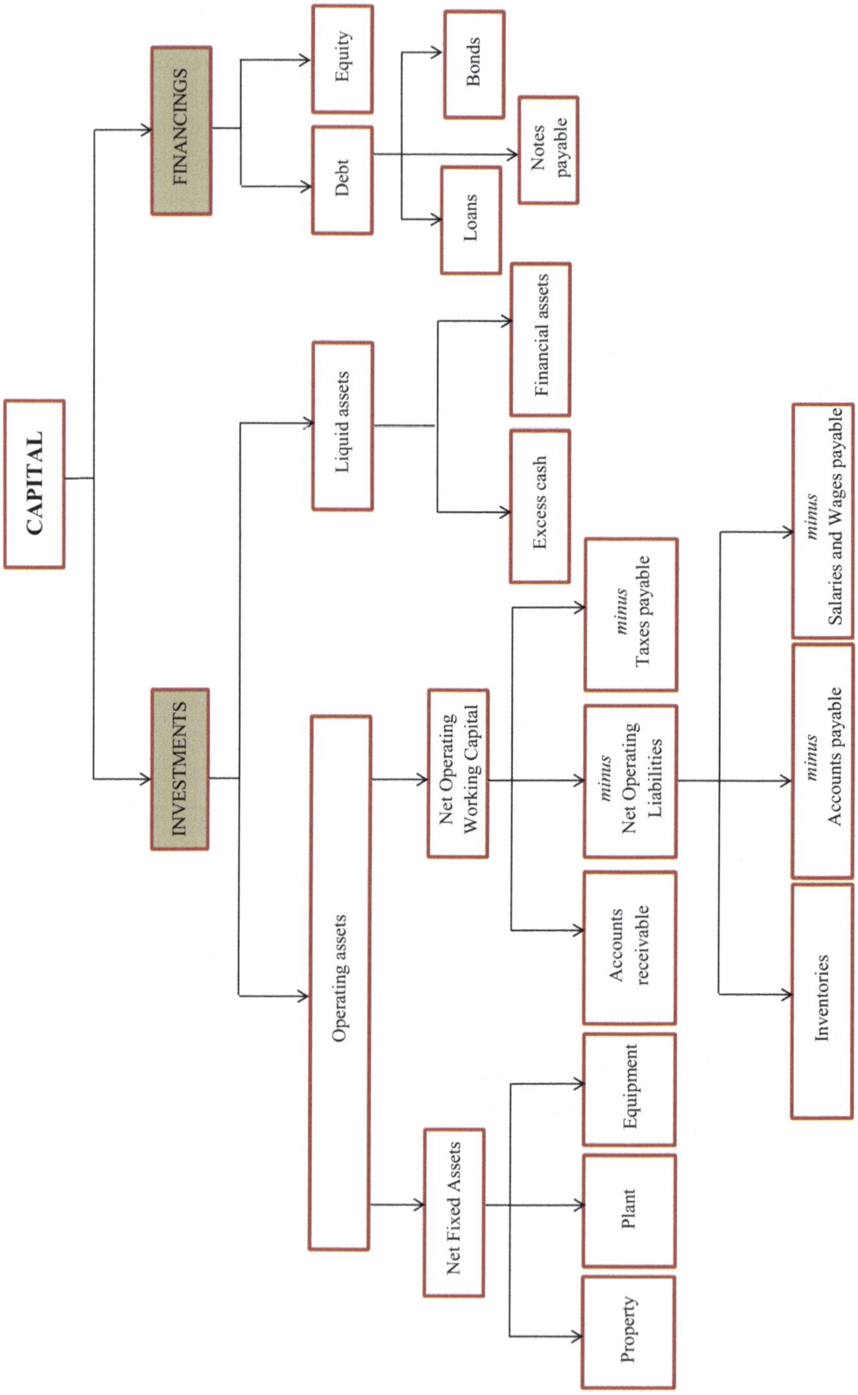

Fig. 3.4 Breakdown chart of capital

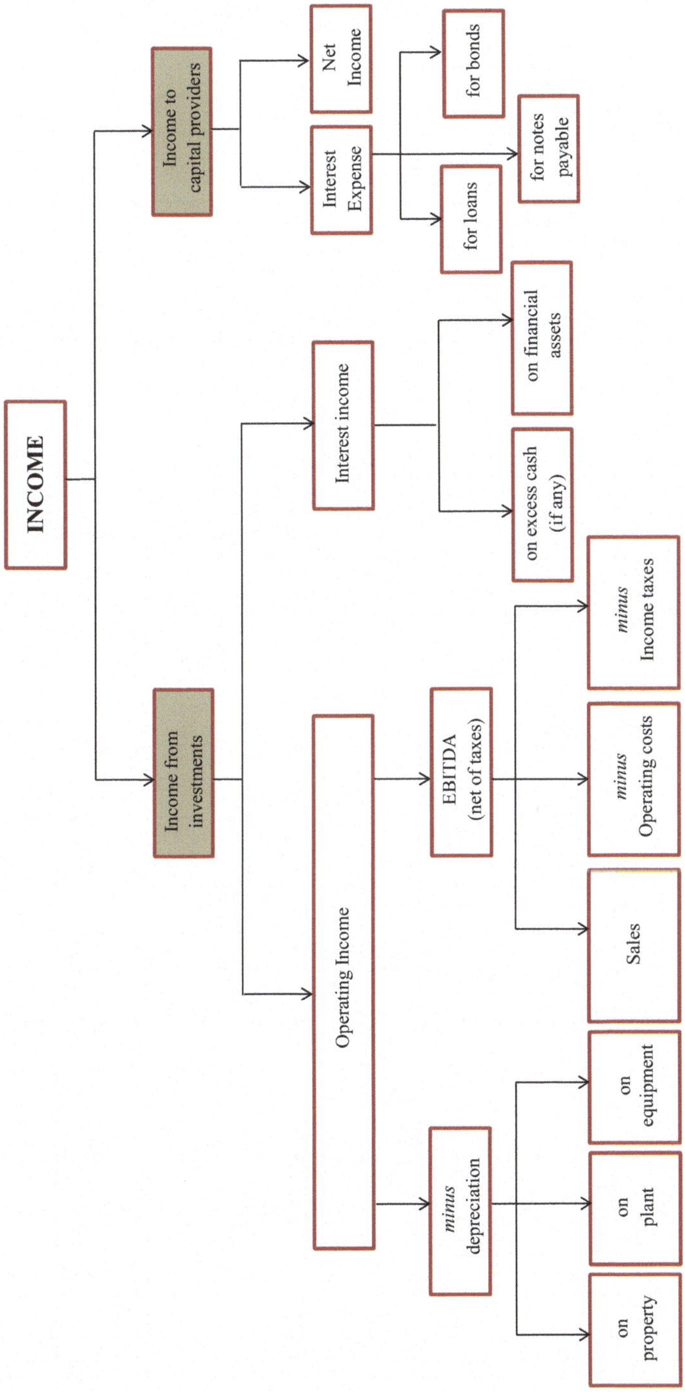

Fig. 3.5 Breakdown chart of income

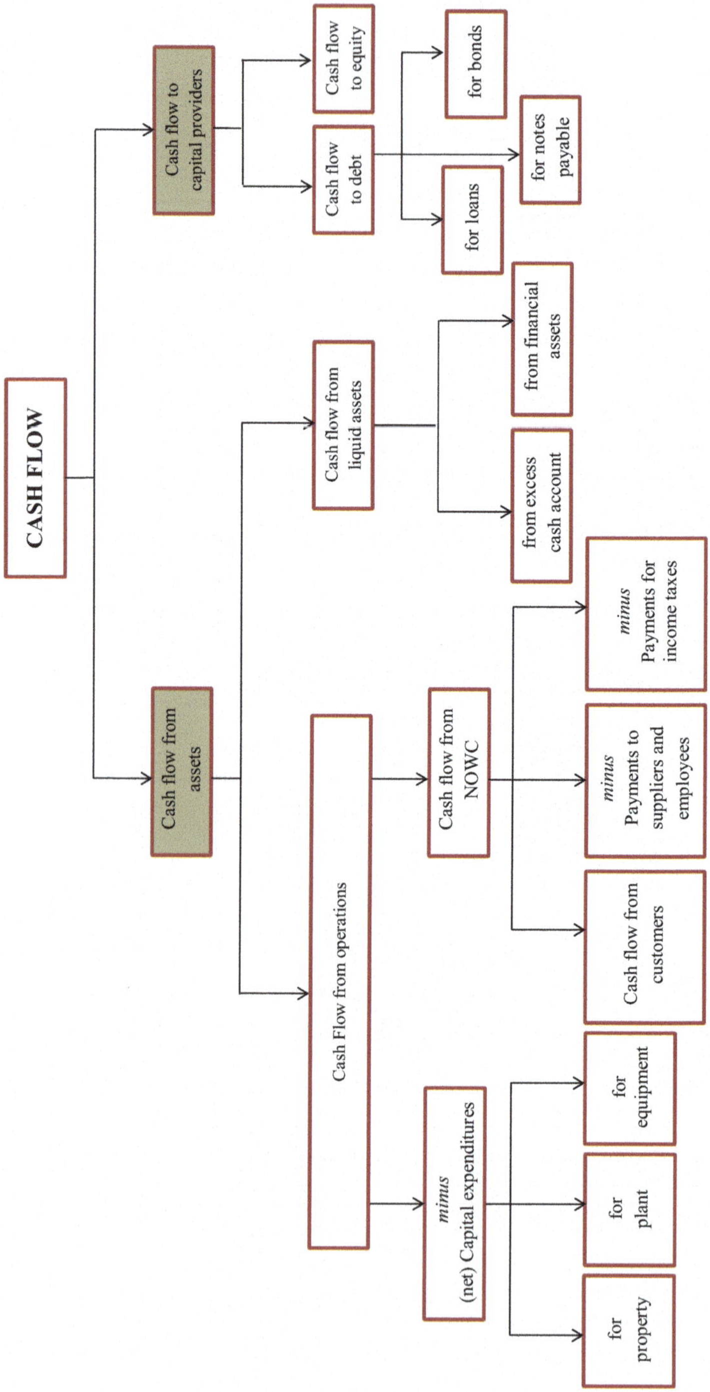

Fig. 3.6 Breakdown chart of cash flow

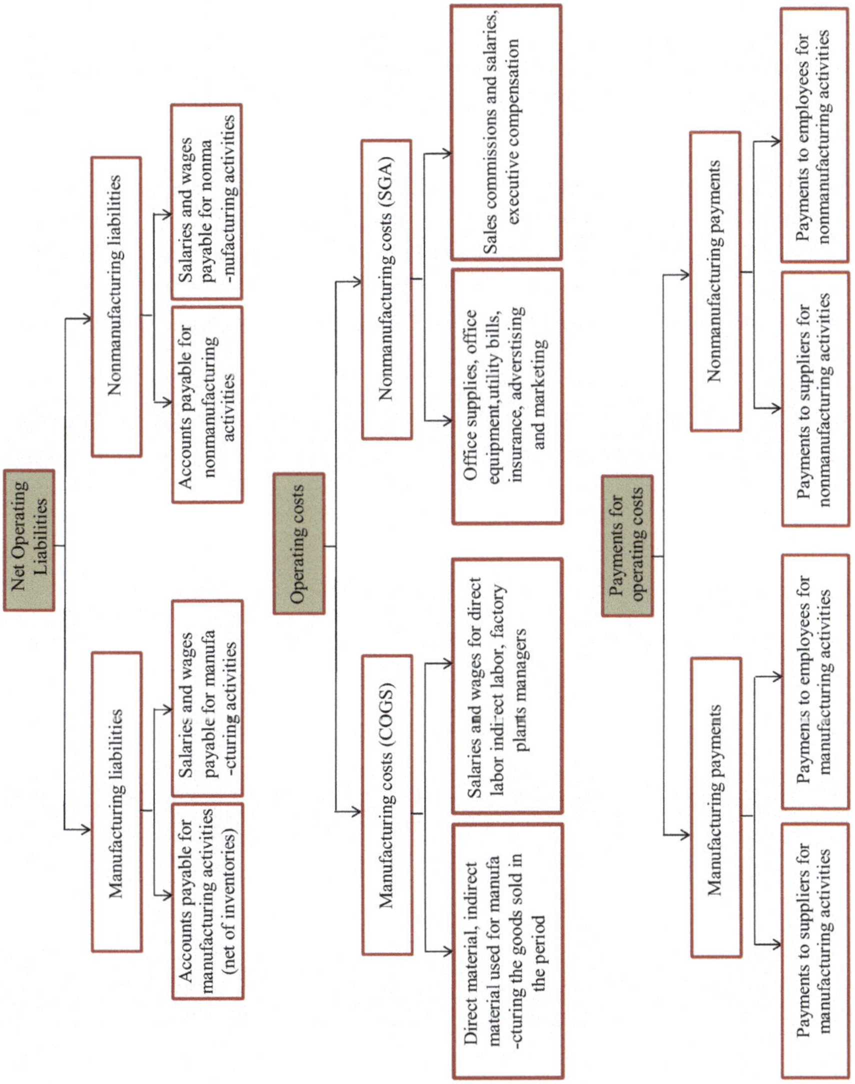

Fig. 3.7 Breakdown chart of net operating liabilities: capital component, income component, and cash flow component

- In a financial perspective, the cost of goods sold does not include depreciation of manufacturing assets; in a managerial accounting perspective, the cost of goods sold does include depreciation of manufacturing assets. Likewise, in financial models, selling, general, and administrative expenses do not include depreciation of non-manufacturing assets, whereas in accounting models they do include depreciation of non-manufacturing assets and are called *operating expenses*
- Under an engineering or managerial accounting perspective, operating costs are divided into fixed costs and variable costs. Subtracting variable costs from the sales revenues, the contribution margin is obtained. Hence, subtracting the fixed costs, the EBITDA is reached.

Figures 3.4, 3.5, and 3.6 provide a breakdown chart for each basic component: Capital, income, and cash flow. Note that the flow chart is structurally a single one. Each rectangle in the capital chart (Fig. 3.4) has an equivalent rectangle in the income chart (Fig. 3.5) and an equivalent rectangle in the cash flow chart (Fig. 3.6).

A breakdown table is also provided in Fig. 3.7 for the NOL's capital, income, and cash flow.

Chapter 4
Estimating the Cash Flows

> Accounting variables are the 'independent' variables and net dividends the 'dependent' variable, not the other way around.
>
> Brief (1996, p. 28).

C. A. Magni, *Investment Decisions and the Logic of Valuation*,
https://doi.org/10.1007/978-3-030-27662-1_4

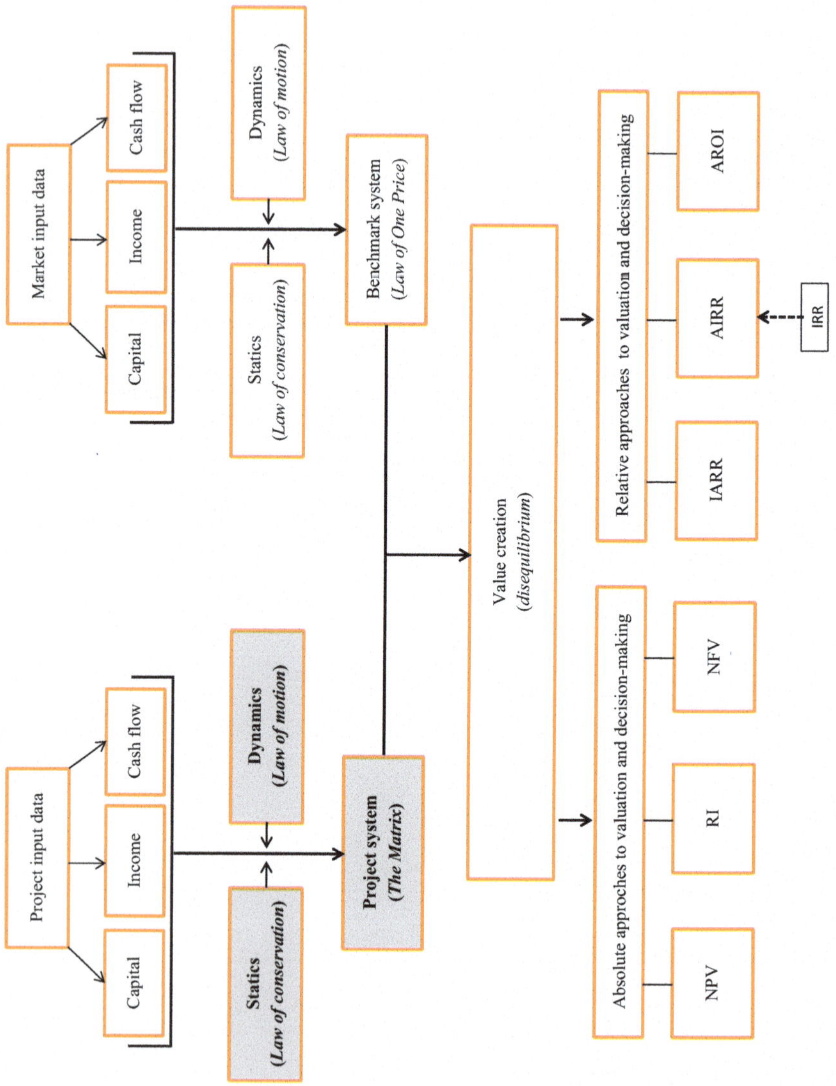

Fig. 4.1 Chapter 4: The Matrix (the cash flows)

This chapter focuses on cash flows and how they are unraveled from the forecasts of incomes and capitals.

The starting point is the determination of the **project input data**, which are the value drivers of the project. Typical value drivers include prices, operating costs (fixed, variable, manufacturing, selling, administrative, etc.), units of production, inventory levels, accounts receivable and payable, depreciation charges, sources of financing, loan rates, interest rates on liquid assets, tax rate, etc.

Combining the input data and processing them, the analyst draws the project's **pro forma income statements** (where operating income, interest expense, non-operating income, and net income are recorded) and **pro forma balance sheets** (where operating and non-operating capitals, debt capital, and equity capital are displayed) for each period of the planned interval. At this point, by use of the split-screen technique, the various types of the **project's cash flows** are derived: Cash flow from assets, cash flow from operations, cash flow from non-operating assets, free cash flow, cash flow to debt, cash flow to equity.

We also consider the notion of **potential dividends** and the impact of **payout policy** on the project system and subsystems.

After completing this chapter, the reader should be able to:

- compute any type of cash flow (in particular, CFA, CFO, CFL, CFD and CFE) in four different ways starting from the input data
- conceptualize the project as the unfolding of accounting snapshots describing capital, income, and cash flow in each year and navigate with confidence through the subtle relations they bear one another and the relations they bear to the various income components and capital components
- understand the notion of depreciation tax shield and interest tax shield, and quantify the effect of depreciation and interest on taxes and cash flows
- understand the difference between the after-tax operating income (OI) and the net operating profit after taxes (NOPAT), as well as the related difference between operating cash flow (CFO) and free cash flow (FCF)
- compute the return on invested capital (ROI) and express the return on equity (ROE) as a function of debt/equity ratios
- distinguish the *operating* FCF from the *asset* FCF
- understand the notions of potential dividends (PD), free cash flow to equity (FCFE) and undistributed/retained cash (RC), thus differentiating the cash *available* for distribution from the cash *actually* distributed to shareholders (CFE)
- master the subtle relations (among incomes and) among cash flows with the help of a mosaic structure framing the static equilibrium in different-but-equivalent ways
- use the Matrix as a diagnostic tool for checking internal consistency of the model as well as for squeezing every bit of economic information out of a project
- conceptualize a project as a strip of consecutive snapshots which may be summarized in a transposed Matrix, suited for spreadsheet analysis.

4.1 Cash Flow from Operations

Consider the CF statement in (3.70). Shifting downward the liquid accounts (and consolidating the NOL classes for reducing the size), one gets the CFO/CCF:

$$
\begin{array}{c}
\textbf{CFO} \\
\begin{array}{|l|ll|}
\hline
+F_t^c & +S_t & -\Delta\text{AR}_t \\
-F_t^{opc} & -\text{OpC}_t & +\Delta\text{NOL}_t \\
-F_t^{capex} & -\text{Dep}_t & -\Delta\text{NFA}_t \\
-F_t^T & -\text{T}_t & +\Delta\text{TP}_t \\
\hline
-F_t^{ec} & -I_t^{ec} & +\Delta\text{EC}_t \\
-F_t^{fa} & -I_t^{fa} & +\Delta\text{FA}_t \\
+F_t^{lo} & +I_t^{lo} & -\Delta\text{LO}_t \\
+F_t^{bo} & +I_t^{bo} & -\Delta\text{BO}_t \\
+F_t^{np} & +I_t^{np} & -\Delta\text{NP}_t \\
+F_t^e & +I_t^e & -\Delta C_t^e \\
\hline
\end{array}
\end{array}
\tag{4.1}
$$

Consolidating the items of the south region, an even more compact form of CFO is found:

$$
\begin{array}{c}
\textbf{CFO} \\
\begin{array}{|l|ll|}
\hline
+F_t^c & +S_t & -\Delta\text{AR}_t \\
-F_t^{opc} & -\text{OpC}_t & +\Delta\text{NOL}_t \\
-F_t^{capex} & -\text{Dep}_t & -\Delta\text{NFA}_t \\
-F_t^T & -\text{T}_t & +\Delta\text{TP}_t \\
\hline
+F_t^{\mathcal{D}} & +I_t^{\mathcal{D}} & -\Delta C_t^{\mathcal{D}} \\
+F_t^e & +I_t^e & -\Delta C_t^e \\
\hline
\end{array}
\end{array}
\tag{4.2}
$$

We find four conceptualizations (and formalizations) for CFO; two of them employ an investment perspective, the other two employ a financing perspective:

$$
\begin{aligned}
F_t^o &= F_t^c - F_t^{opc} - F_t^{capex} - F_t^T \\
F_t^o &= \left(S_t - \text{OpC}_t - \text{Dep}_t - \text{T}_t\right) + \left(-\Delta\text{AR}_t + \Delta\text{NOL}_t - \Delta\text{NFA}_t + \Delta\text{TP}_t\right) \\
F_t^o &= F_t^{\mathcal{D}} + F_t^e \\
F_t^o &= \left(I_t^{\mathcal{D}} + I_t^e\right) + \left(-\Delta C_t^{\mathcal{D}} - \Delta C_t^e\right)
\end{aligned}
\tag{4.3}
$$

The second expression is particularly important:

$$
F_t^o = \overbrace{\underbrace{\left(S_t - \text{OpC}_t - \text{Dep}_t - \text{T}_t\right)}_{\text{EBIT}_t}}^{\text{income components}} + \overbrace{\underbrace{\left(-\Delta\text{AR}_t + \Delta\text{NOL}_t + \Delta\text{TP}_t - \Delta\text{NFA}_t\right)}_{-\Delta\text{NOWC}_t}}^{\text{capital components}} \tag{4.4}
$$

for $t \in \mathbb{N}_n^0$ with $-\Delta\text{NFA}_0 - \Delta\text{NOWC}_0 = -\text{NFA}_0 - \text{NOWC}_0 = -C_0^o$. This equality represents the inferential process of the estimation of CFO/CCF by specifying it as being derived from the ex ante accounting estimates.

In other words, the analyst must estimate several classes of income components and capital components to find the CFO.[1]

This implies that cash flow is not *objectively* determinable; it must be inferred from the accounting magnitudes. The CFO should then be more properly understood as a function of income and (change in) capital: $F_t^o = F_t^o(\Delta C_t^o, I_t^o)$ where ΔC_t^o and I_t^o are themselves functions of the estimations of income components and capital components. Which just means that, in capital asset investments, the operating cash flows are not primitive variables, but derived ones.[2]

One may consolidate rows 1, 2, and 4 of Matrix (4.2), getting

CFO

$+F_t^{nowc}$	$\text{EBITDA}_t - \text{T}_t$	$-\Delta\text{NOWC}_t$
$-F_t^{capex}$	$-\text{Dep}_t$	$-\Delta\text{NFA}_t$
$+F_t^{\mathcal{D}}$	$+I_t^{\mathcal{D}}$	$-\Delta C_t^{\mathcal{D}}$
$+F_t^e$	$+I_t^e$	$-\Delta C_t^e$

(4.5)

The first row of the above Matrix is nothing but a reframing of the law of motion for NOWC:

$$\text{NOWC}_t = \text{NOWC}_{t-1} + (\text{EBITDA}_t - \text{T}_t) - F_t^{nowc}. \tag{4.6}$$

Finally, one may consolidate further and write

CFO

$+F_t^o$	$\text{EBIT}_t - \text{T}_t$	$-\Delta\text{NOWC}_t - \Delta\text{NFA}_t$
$+F_t^{\mathcal{D}}$	$+I_t^{\mathcal{D}}$	$-\Delta C_t^{\mathcal{D}}$
$+F_t^e$	$+I_t^e$	$-\Delta C_t^e$

(4.7)

The second column of Eq. (4.7) is the operating income (OI), viewed in an investment perspective (upper region) or in a financing perspective (lower region):

$$I_t^{\mathcal{D}} + I_t^e = I_t^o = \text{EBIT}_t - \text{T}_t. \tag{4.8}$$

The OI is the income component of ΔC_t^o, the CFO is the cash-flow component of ΔC_t^o. The triplet $(I_t^o, \Delta C_t^o, F_t^o)$ summarizes information on the operating activities. Table 4.1 showcases a detailed breakdown of this triplet.

---

[1]Specifically, the analyst must estimate: Sales revenues, COGS and SGA, depreciation, income taxes, changes in accounts receivable, changes in inventory, changes in accounts payable, changes in salaries and wages payable, changes in taxes payable, and net capital expenditures. For most projects, taxes and labor costs are paid when they are due, so that $\text{TP}_t = \text{SWP}_t = 0$. The estimation of these income and capital components depends in turn on the estimation of key parameters that affect them, such as prices, unit costs, unit of production, operating and cash cycles, interest rates, tax rate, etc. The primitive variables may amount to tens or hundreds, depending on the complexity of the project and the estimation methodology used by the analyst.

[2]This is particularly important when assessing the economic profitability of a project. While it is true that value depends on cash flows, cash flows in turn depends on the accounting estimates of incomes and capitals.

Table 4.1 The basic elements of operating assets: a breakdown

Operating Income	=	$\overbrace{\underbrace{S_t - \overbrace{(\underbrace{\mathrm{COM}_t + \mathrm{DL}_t + \mathrm{MOH}_t - \Delta\mathrm{Inv}_t}_{\mathrm{COGS}_t} + \mathrm{SGA}_t)}^{\mathrm{OpC}_t} - \mathrm{T}_t}_{\mathrm{EBITDA}_t - \mathrm{T}_t} - \mathrm{Dep}_t}^{\mathrm{EBIT}_t - T_t}$
Change in operating capital	=	$\overbrace{\underbrace{\Delta\mathrm{AR}_t - \overbrace{(\Delta\mathrm{AP}_t + \Delta\mathrm{SWP}_t - \Delta\mathrm{Inv}_t)}^{\Delta\mathrm{NOL}_t}}_{\Delta\mathrm{NOWC}_t} + \Delta\mathrm{TP}_t + \Delta\mathrm{NFA}_t}^{\Delta C_t^o}$
Operating cash flow	=	$\overbrace{\underbrace{F_t^c - \overbrace{(F_t^{cogs} + F_t^{sga})}^{F_t^{opc}}}_{F_t^{nowc}} - F_t^T + F_t^{capex}}^{F_t^o}$

Remark 4.1 (*Simplified form of the Matrix*) The CFO Matrix may be simplified under suitable assumptions. For example, if salaries and wages are paid within the year (i.e., $\mathrm{SWP}_t = 0$ for all t), and assuming that sales are made on a cash-only basis (i.e., $\mathrm{AR}_t = 0$ for all t) and suppliers are paid on cash (i.e., $\mathrm{AP}_t = 0$ for all t), then $F_t^c = S_t$ (i.e. $\Delta\mathrm{AR}_t = 0$) and

$$\mathrm{OpC}_t = F_t^{opc} + \Delta\mathrm{NOL}_t = F_t^{opc} - \Delta\mathrm{Inv}_t.$$

In particular, splitting up OpC_t into COGS_t and SGA_t, one finds $F_t^{sga} = \mathrm{SGA}_t$ and $F_t^{cogs} = \mathrm{COGS}_t + \Delta\mathrm{Inv}_t$ or, from (3.23), $F_t^{cogs} = \mathrm{COP}_t + \mathrm{DL}_t + \mathrm{MOH}_t$. If, in addition, no net capital expenditures occur after the initial investment (i.e., $F_t^{capex} = 0$ for all $t \in \mathbb{N}_n^1$) and taxes are paid when they are due (i.e., $F_t^T = \mathrm{T}_t$ for all t), then the Matrix may be framed as in (4.9).

CFO

$+S_t$	$+S_t$	0
$-\mathrm{COP}_t - \mathrm{DL}_t - \mathrm{MOH}_t$	$-\mathrm{COGS}_t$	$-\Delta\mathrm{Inv}_t$
$-\mathrm{SGA}_t$	$-\mathrm{SGA}_t$	0
-0	$-\mathrm{Dep}_t$	$+\mathrm{Dep}_t$
$-\mathrm{T}_t$	$-\mathrm{T}_t$	0
$+F_t^{\mathcal{D}}$	$+I_t^{\mathcal{D}}$	$-\Delta C_t^{\mathcal{D}}$
$+F_t^e$	$+I_t^e$	$-\Delta C_t^e$

(4.9)

(note the rather interesting relation stemming from column 3: $\Delta C_t^{\mathcal{D}} + \Delta C_t^e = \Delta\mathrm{Inv}_t - \mathrm{Dep}_t$). This format is a rather usual one. It is typically found in many finance textbooks and engineering economy textbooks, albeit the assumptions are usually implicit.

4.2 Cash Flow from Assets

Consider the CFO Matrix in (4.2). To find the CFA (which is equal to the CFP), we need to disaggregate the net-debt row into debt and liquid assets:

$$
\textbf{CFA} \quad
\begin{array}{|l|ll|}
\hline
+F_t^{c} & +S_t & -\Delta \mathrm{AR}_t \\
-F_t^{opc} & -\mathrm{OpC}_t & +\Delta \mathrm{NOL}_t \\
-F_t^{capex} & -\mathrm{Dep}_t & -\Delta \mathrm{NFA}_t \\
-F_t^{T} & -\mathrm{T}_t & +\Delta \mathrm{TP}_t \\
\hline
+F_t^{d} & +I_t^{d} & -\Delta C_t^{d} \\
-F_t^{l} & -I_t^{l} & +\Delta C_t^{l} \\
+F_t^{e} & +I_t^{e} & -\Delta C_t^{e} \\
\hline
\end{array}
\tag{4.10}
$$

Shifting up the liquid asset row, the CFA is found:

$$
\textbf{CFA} \quad
\begin{array}{|l|ll|}
\hline
+F_t^{c} & +S_t & -\Delta \mathrm{AR}_t \\
-F_t^{opc} & -\mathrm{OpC}_t & +\Delta \mathrm{NOL}_t \\
-F_t^{capex} & -\mathrm{Dep}_t & -\Delta \mathrm{NFA}_t \\
-F_t^{T} & -\mathrm{T}_t & +\Delta \mathrm{TP}_t \\
+F_t^{l} & +I_t^{l} & -\Delta C_t^{l} \\
\hline
+F_t^{d} & +I_t^{d} & -\Delta C_t^{d} \\
+F_t^{e} & +I_t^{e} & -\Delta C_t^{e} \\
\hline
\end{array}
\tag{4.11}
$$

Like any other item, CFA presents four conceptualizations (two investment perspectives, two financing perspectives):

$$
\begin{aligned}
F_t &= F_t^{c} - F_t^{opc} - F_t^{capex} - F_t^{T} + F_t^{l} \\
F_t &= \left(I_t^{d} + I_t^{e}\right) + \left(-\Delta C_t^{d} - \Delta C_t^{e}\right) \\
F_t &= F_t^{d} + F_t^{e} \\
F_t &= \Big(S_t - \underbrace{(\mathrm{COGS}_t + \mathrm{SGA}_t)}_{\mathrm{OpC}_t} - \mathrm{Dep}_t - \mathrm{T}_t + I_t^{l}\Big) \\
&\quad + \left(-\Delta \mathrm{AR}_t + \Delta \mathrm{NOL}_t - \Delta \mathrm{NFA}_t + \Delta \mathrm{TP}_t - \Delta C_t^{l}\right).
\end{aligned}
\tag{4.12}
$$

Consolidating the first 4 entries of the Matrix,

$$
\textbf{CFA} \quad
\begin{array}{|l|ll|}
\hline
+F_t^{o} & +\mathrm{EBIT}_t - \mathrm{T}_t & -\Delta \mathrm{NOWC}_t - \Delta \mathrm{NFA}_t \\
+F_t^{l} & +I_t^{l} & -\Delta C_t^{l} \\
\hline
+F_t^{d} & +I_t^{d} & -\Delta C_t^{d} \\
+F_t^{e} & +I_t^{e} & -\Delta C_t^{e} \\
\hline
\end{array}
\tag{4.13}
$$

Also, looking at the lower right partition and considering that the lower and the upper part of the second column coincide, that is,

$$\text{EBIT}_t - \text{T}_t + I_t^l = I_t = I_t^d + I_t^e,$$

one may write

$$F_t = F_t^d + F_t^e = \text{EBIT}_t - \text{T}_t + I_t^l - (\Delta C_t^e + \Delta C_t^d). \tag{4.14}$$

Other perspectives on the CFA Matrix are possible. For example, disaggregating the NOL components in order to highlight the constituent assets of NOWC, one gets

$$\begin{array}{c}
\textbf{CFA} \\
\begin{array}{|l|ll|}
\hline
+F_t^c & +S_t & -\Delta\text{AR}_t \\
+0 & +\Delta\text{Inv}_t & -\Delta\text{Inv}_t \\
-F_t^{cop} & -\text{COP}_t & +\Delta\text{AP}_t \\
-F_t^{lc} & -\text{LC}_t & +\Delta\text{SWP}_t \\
-F_t^{capex} & -\text{Dep}_t & -\Delta\text{NFA}_t \\
-F_t^T & -\text{T}_t & +\Delta\text{TP}_t \\
+F_t^l & +I_t^l & -\Delta C_t^l \\
\hline
+F_t^d & +I_t^d & -\Delta C_t^d \\
+F_t^e & +I_t^e & -\Delta C_t^e \\
\hline
\end{array}
\end{array} \tag{4.15}$$

4.3 Cash Flow to Equity

One may start from (4.11) and move the debt row upward to the north region. This results in

$$\begin{array}{c}
\textbf{CFE} \\
\begin{array}{|l|ll|}
\hline
+F_t^c & +S_t & -\Delta\text{AR}_t \\
-F_t^{opc} & -\text{OpC}_t & +\Delta\text{NOL}_t \\
-F_t^{capex} & -\text{Dep}_t & -\Delta\text{NFA}_t \\
-F_t^T & -\text{T}_t & +\Delta\text{TP}_t \\
+F_t^l & +I_t^l & -\Delta C_t^l \\
-F_t^d & -I_t^d & +\Delta C_t^d \\
\hline
+F_t^e & +I_t^e & -\Delta C_t^e \\
\hline
\end{array}
\end{array} \tag{4.16}$$

Consolidating, one may write

$$\begin{array}{c}
\textbf{CFE} \\
\begin{array}{|l|ll|}
\hline
+F_t^o & \text{EBIT}_t - \text{T}_t & -\Delta\text{NOWC}_t - \Delta\text{NFA}_t \\
-F_t^{\mathcal{D}} & -I_t^{\mathcal{D}} & +\Delta C_t^{\mathcal{D}} \\
\hline
+F_t^e & +I_t^e & -\Delta C_t^e \\
\hline
\end{array}
\end{array} \tag{4.17}$$

The CFE is the difference between net income and change in equity and is also equal to the difference of operating cash flow and CF$\mathcal{D}$. One may also break down CFE into components of the operating income and net financing costs and the related changes of capital:

$$F_t^e = \text{EBIT}_t - \text{T}_t - I_t^{\mathcal{D}} - \Delta\text{NOWC}_t - \Delta\text{NFA}_t + \Delta C_t^{\mathcal{D}}.$$

4.4 The Last Cash Flows

At time n, the project ends. As we know, this means that the BOP capital values at time n are zero: $C_n^o = C_n^l = C_n^d = C_n^e = 0$ and, in particular, $\text{NFA}_n = \text{NOWC}_n = 0$. This implies that $\Delta C_n^o = -C_{n-1}^o$, $\Delta C_n^l = -C_{n-1}^l$, $\Delta C_n^d = -C_{n-1}^d$, $\Delta C_n^e = -C_{n-1}^e$, $\Delta\text{NFA}_n = -\text{NFA}_{n-1}$, $\Delta\text{NOWC}_n = -\text{NOWC}_{n-1}$. Therefore, the last cash flows are equal to the EOP capitals:

$$\begin{aligned} E_n^o &= F_n^o = I_n^o + C_{n-1}^o \\ E_n^l &= F_n^l = I_n^l + C_{n-1}^l \\ E_n^d &= F_n^d = I_n^d + C_{n-1}^d \\ E_n^e &= F_n^e = I_n^e + C_{n-1}^e. \end{aligned}$$

One should not forget to include gain/loss on disposal of assets in the sales revenues. More precisely, we define the ***salvage value*** as the residual book value of the fixed assets at the end of period n, before the final distribution of cash.[3] We denote it as SV_n:

$$\text{SV}_n = \sum_{t=0}^{n} F_t^{capex} - \sum_{t=1}^{n} \text{Dep}_t.$$

If some assets are disposed at time $t < n$ and sold on credit, their residual book values must be excluded:

$$\text{SV}_n = \sum_{t=0}^{n} F_t^{capex} - \sum_{t=1}^{n} \text{Dep}_t - \text{residual book value of assets sold on credit before } n$$

(residual book value of assets sold on cash before n is included in F_t^{capex}). If assets are fully depreciated, then $\text{SV}_n = 0$.

[3]More generally, we define salvage value of an asset (or a set of assets) as being the residual book value of the asset (or the set of assets) whenever it is liquidated.

When the project's operations cease, plant, property, and equipment are disposed of, and the disposal value may be greater, equal, or smaller than SV_n. If the disposal value is greater than the salvage value, a gain on disposal of assets is recorded in the income statement as an extra revenue; if the disposal value is smaller, a loss on disposal value is recorded in the income statement as a negative income component. Therefore,

$$\mathrm{EBITDA}_n = \overbrace{S_n^{\#} + z}^{S_n} - \mathrm{OpC}_n$$

where $S_n^{\#}$ represents sales revenues, excluding the gain/loss on disposal of asset, and z is the gain/loss on sale. For example, consider a 5-year project with initial (and unique) capital expenditure equal to \$1,000 and depreciation charges equal to $\mathbf{Dep} = (100, 100, 100, 60, 40)$, constant inventory equal to \$50 (fully recovered at the end of the project), and tax rate equal to 25%. The project is assumed to be unlevered; purchases and sales are made on a cash-basis only (i.e., $\mathrm{AP}_t = \mathrm{AR}_t = 0$) and salaries and wages are paid when they are due, which implies $-\mathrm{NOL}_t = \mathrm{Inv}_t = 50$. Taxes are paid as soon as they occur, so that $\mathrm{TP}_t = 0$. Suppose the operating costs are \$300 in each period, whereas the sales revenues are equal to \$500 in each period. Suppose the assets are sold, after 5 years, at \$680 (sale is made on cash).

The data imply that the salvage value is $\mathrm{SV}_5 = 1000 - 400 = 600$, so there is a gain on disposal of assets equal to $\$80 = 680 - 600$, which is recorded as an extra revenue. Therefore, the EBITDAs are

$$\begin{aligned}
\mathrm{EBITDA}_1 &= 500 - 300 = 200\\
\mathrm{EBITDA}_2 &= 500 - 300 = 200\\
\mathrm{EBITDA}_3 &= 500 - 300 = 200\\
\mathrm{EBITDA}_4 &= 500 - 300 = 200\\
\mathrm{EBITDA}_5 &= 580 - 300 = 280.
\end{aligned}$$

Hence, subtracting Dep_t,

$$\begin{aligned}
\mathrm{EBIT}_1 &= 200 - 100 = 100\\
\mathrm{EBIT}_2 &= 200 - 100 = 100\\
\mathrm{EBIT}_3 &= 200 - 100 = 100\\
\mathrm{EBIT}_4 &= 200 - 60 \;\; = 140\\
\mathrm{EBIT}_5 &= 280 - 40 \;\; = 240.
\end{aligned}$$

Consider that $\mathrm{NFA}_4 = 1000 - 360 = 640$ is the book value of the fixed asset in $t = 4$. Therefore, $-\Delta\mathrm{NFA}_5 = 640$. Thus, the last CFO is

$$F_5^o = \overbrace{240 \cdot (1 - 0.25)}^{\mathrm{EBIT}_5 - \mathrm{T}_5} + \overbrace{(50 - 0)}^{-\Delta\mathrm{NOWC}_5} + \overbrace{(640 - 0)}^{-\Delta\mathrm{NFA}_5} = 870.$$

It is possible to break down the CFO into its constituents as follows:

$$\begin{aligned}
F_5^c &= +S_5 - \Delta\text{AR}_5 &&= +580 - 0 &&= +580\\
-F_5^{opc} &= -\text{OpC}_5 + \Delta\text{NOL}_5 &&= -300 + 50 &&= -250\\
-F_5^{capex} &= -\text{Dep}_5 - \Delta\text{NFA}_5 &&= -40 + 640 &&= +600\\
-F_5^T &= -\text{T}_5 + \Delta\text{TP}_5 &&= -0.25 \cdot 240 + 0 &&= -60.
\end{aligned}$$

Hence,

$$\begin{aligned}
F_5^o &= F_5^c - F_5^{opc} - F_5^{capex} - F_5^T\\
&= 580 - 250 + 600 - 60\\
&= 870.
\end{aligned}$$

Note that this implies $-F_5^{capex} = \text{SV}_5 = 600 \neq 680$, the latter being the disposal value. The gain on disposal asset, equal to 80, is included in F_5^c.

This transaction may be fruitfully described by making use of the Matrix and employing Rule 7 (see Sect. 3.6.2). Specifically, let us start from Matrix (3.56). As the project is unlevered, we may suppress the liquid asset row and the debt row. Also, owing to the assumptions, $\text{AR}_4 = \text{TP}_4 = 0$ and $-\text{NOL}_4 = \text{Inv}_4 = 50$. Further, at time $n = 5$, all the capital amounts are equal to zero, so the first column of Matrix (3.56) is the null column. Hence, using the symbol SV_5 in place of x, Matrix (3.56) becomes

$$\mathbf{M}_5 = \left[\begin{array}{c|cll}
+0 & +0 & +S_5^{\#} + \textcircled{z} & -F_5^{c\#} - \textcircled{z}\\
-0 & -\text{NOL}_4 & -\text{OpC}_5 & +F_5^{opc}\\
+0 & +\text{NFA}_4 & -\text{Dep}_5 & +F_5^{capex\#} - \boxed{\text{SV}_5}\\
-0 & -0 & -\text{T}_5^{\#} - \fbox{τz} & +F_5^{T\#} + \fbox{τz}\\
\hline
+0 & +C_4^e & +I_5^{e\#} + \textcircled{z} - \fbox{τz} & -F_5^{e\#} - \textcircled{z} - \boxed{\text{SV}_5} + \fbox{τz}
\end{array}\right]$$

Substituting numbers,

$$\mathbf{M}_5 = \left[\begin{array}{c|cll}
+0 & +0 & +500 + \textcircled{80} & -500 - \textcircled{80}\\
-0 & +50 & -300 & +250\\
+0 & +640 & -40 & +0 - \boxed{600}\\
-0 & -0 & -40 - \fbox{$0.25 \cdot 80$} & +40 + \fbox{$0.25 \cdot 80$}\\
\hline
+0 & +690 & +120 + \textcircled{80} - \fbox{$0.25 \cdot 80$} & -210 - \textcircled{80} - \boxed{600} + \fbox{$0.25 \cdot 80$}
\end{array}\right]$$

(the $-\$600$ on the right region are included in the left region, in the NFA row and in the equity row, resulting in a zero cell). Hence,

$$\mathbf{M}_5 = \begin{array}{|c|rrr|} \hline +0 & +0 & +580 & -580 \\ -0 & +50 & -300 & +250 \\ +0 & +640 & -40 & -600 \\ -0 & -0 & -60 & +60 \\ \hline +0 & +690 & +180 & -870 \\ \hline \end{array}$$

Column 2 is the BS at time 4, column 3 is the IS at time 5, column 4 is the CFS at time 5.

4.5 The Project as a Split-Screen Film

In the light of what we have learned, a project may be viewed as a set of accounting pictures ordered diachronically. Mathematically, the triplet $(\boldsymbol{C}, \boldsymbol{I}, \boldsymbol{F})$ is broken down into its basic constituents via the Matrix.

It should now be natural to accept the following definition of project, which is consistent with Definition (1.2).

Definition 4.1 (*Project*) An n-period project is wholly described by a set of $n+1$ Matrices, $\mathbf{M}_t$, $t \in \mathbb{N}_n^0$, which incorporate the three basic statements for capitals, incomes, and cash flows at every date. Grouping the Matrices of each period, the project Matrix (4.18) is found. The invariance properties of the Matrix do not depend on the date t, so the split-screen technique may be employed in each Matrix. We will also use the symbol $\mathbf{M} = \{\mathbf{M}_t\}_{t \in \mathbb{N}_n^0}$ to denote a project.[4]

The project Matrix $\mathbf{M}$, expressed by (4.18), consists of the diachronic unfolding of the $n+1$ snapshots provided by the $n+1$ Matrices $\mathbf{M}_t$, $t \in \mathbb{N}_n^0$. It is a strip which conveys economic information to the analyst.[5] Such an information may be as rich as one prefers, owing to its flexibility.

The strip of the project Matrix may also be displayed in terms of capital amounts so as to display the BSs instead of the statements of change in capital (which are but differences of consecutive BSs). In addition, it may also be transposed and rearranged so that columns refer to the dates $t = 1, 2, \ldots, n$, as shown in Eq. (4.19). The first block contains the BSs, the second block contains the ISs, the third block contains the CFSs.

[4] A usual convention is to define a project as a sequence of cash flows $\boldsymbol{F} = (F_0, F_1, \ldots, F_n)$. The reader should now realize that this definition is unsatisfactory: Cash flows represent but one aspect of a project (derived from accounting magnitudes).

[5] The project is, so to say, a "split-screen movie"; the strip provides the "plot".

The latter format lends itself to easy use for spreadsheets. Abusing terminology (and symbology), we call the new format the ***transposed Matrix*** and denote it as $\mathbf{M}^T$ or $(\boldsymbol{C}, \boldsymbol{I}, \boldsymbol{F})$.[6] A shrinked transposed Matrix is presented in (4.20). We will sometimes use this format to refer to a project and whenever we need to summarize a project in terms of the three basic components.

To recap, the Matrix is a diachronic-synchronic, variable-size framework associated with floating *split screens* obtained by performing elementary operations on the Matrix (shifting rows and columns as well as their elements and possibly giving rise to quadrilaterals for coping with additional amounts). The association of the Matrix and the split-screen technique brings about several benefits:

- it visually translates, in a simple way, the static equilibrium and the dynamical system of a project
- it highlights the fact that projects and firms are, essentially, a set of relationships among the three basic notions: Capital, income, and cash flows
- it condenses into one single statement the three financial statements that in textbooks are presented as isolated statements (balance sheets, income statements, cash flow statement), thereby making it easier to grasp their interrelations with a single snapshot
- it enables squeezing every last drop of accounting information from the project, finding any possible relations among the accounting data and constructing new relationships among variables which are considered relevant for financial and accounting analysis
- it enables visualizing relationships in a more intuitive way than traditional arrays of equations and identities
- it always supplies four different-but-equivalent interpretation of a given set of items, so enhancing comprehension of accounting and financial relationships
- it is a diagnostic tool that reveals self-consistency (or inconsistency) in the model; thus, it facilitates fulfilling logical consistency
- it is a tool for recording accounting transactions
- it lends itself to an expansion or a reduction of the number of items depending on the level of detail the analyst is willing to deal with. The (non)complexity of the relations remains unvaried because the structure remains the same and the rules for retrieving information do not change
- it may be applied to any kind of economic assets as well as projects (e.g., loans or portfolio of loans, financial assets and portfolios of financial assets, funds, firms, etc.).

[6] The transposed Matrix is not workable, so the split-screen technique may not be employed.

$$
\mathbf{M} = \{\mathbf{M}_t\}_{t \in \mathbb{N}_n^0} =
\left\|
\begin{array}{c}
\begin{array}{|l|l|}
\hline
\boxed{t=0} & \\
+F_0^o & +I_0^o \; -\Delta C_0^o \\
+F_0^l & +I_0^l \; -\Delta C_0^l \\
\hline
+F_0^d & +I_0^d \; -\Delta C_0^d \\
+F_0^e & +I_0^e \; -\Delta C_1^e \\
\hline
\end{array} \\
\downarrow \\
\begin{array}{|l|l|}
\hline
\boxed{t=1} & \\
+F_1^o & +I_1^o \; -\Delta C_1^o \\
+F_1^l & +I_1^l \; -\Delta C_1^l \\
\hline
+F_1^d & +I_1^d \; -\Delta C_1^d \\
+F_1^e & +I_1^e \; -\Delta C_1^e \\
\hline
\end{array} \\
\downarrow \\
\begin{array}{|l|l|}
\hline
\boxed{t=2} & \\
+F_2^o & +I_2^o \; -\Delta C_2^o \\
+F_2^l & +I_2^l \; -\Delta C_2^l \\
\hline
+F_2^d & +I_2^d \; -\Delta C_2^d \\
+F_2^e & +I_2^e \; -\Delta C_2^e \\
\hline
\end{array} \\
\downarrow \\
\vdots \quad \vdots \quad \vdots \\
\vdots \quad \vdots \quad \vdots \\
\downarrow \\
\begin{array}{|l|l|}
\hline
\boxed{t=n} & \\
+F_n^o & +I_n^o \; -\Delta C_n^o \\
+F_n^l & +I_n^l \; -\Delta C_n^l \\
\hline
+F_n^d & +I_n^d \; -\Delta C_n^d \\
+F_n^e & +I_n^e \; -\Delta C_n^e \\
\hline
\end{array}
\end{array}
\right\|
\tag{4.18}
$$

$$
\begin{array}{c}
(\boldsymbol{C}, \boldsymbol{I}, \boldsymbol{F}) \\
\text{Transposed Matrix} \\
\begin{array}{|lllll|}
\hline
t=0 & t=1 & t=2 & \ldots & t=n \\
\hline
\multicolumn{5}{|l|}{\textbf{Capital}} \\
C_0^o & C_1^o & C_2^o & \ldots & C_n^o = 0 \\
C_0^l & C_1^l & C_2^l & \ldots & C_n^l = 0 \\
\hline
C_0^d & C_1^d & C_2^d & \ldots & C_n^d = 0 \\
C_0^e & C_1^e & C_2^e & \ldots & C_n^e = 0 \\
\hline
\multicolumn{5}{|l|}{\textbf{Income}} \\
I_0^o & I_1^o & I_2^o & \ldots & I_n^o \\
I_0^l & I_1^l & I_2^l & \ldots & I_n^l \\
\hline
I_0^d & I_1^d & I_2^d & \ldots & I_n^d \\
I_0^e & I_1^e & I_2^e & \ldots & I_n^e \\
\hline
\multicolumn{5}{|l|}{\textbf{Cash flow}} \\
F_0^o & F_1^o & F_2^o & \ldots & F_n^o \\
F_0^l & F_1^l & F_2^l & \ldots & F_n^l \\
\hline
F_0^d & F_1^d & F_2^d & \ldots & F_n^d \\
F_0^e & F_1^e & F_2^e & \ldots & F_n^e \\
\hline
\end{array}
\end{array}
\tag{4.19}
$$

$$
\begin{array}{c}
(\boldsymbol{C}, \boldsymbol{I}, \boldsymbol{F}) \\
\text{Transposed Matrix (concise form)} \\
\begin{array}{|llllllll|}
\hline
\text{Time} & & 0 & 1 & 2 \ldots & & \ldots & n \\
\textbf{Capital} & \boldsymbol{C} & C_0 & C_1 & C_2 & \ldots & \ldots & 0 \\
\hline
\textbf{Income} & \boldsymbol{I} & I_0 & I_1 & I_2 & \ldots & \ldots & I_n \\
\hline
\textbf{Cash Flow} & \boldsymbol{F} & F_0 & F_1 & F_2 & \ldots & \ldots & F_n \\
\hline
\end{array}
\end{array}
\tag{4.20}
$$

From a mathematical point of view, the Matrix and the split-screen technique represent a unique powerful tool for solving systems of equations which would otherwise be solved with considerable effort, especially when the number of entries (and, therefore, the number of equations) is high.[7] It can be used even by those practitioners who have no knowledge whatsoever of systems of equations.

In the numerical examples we will make use of thumbnails to provide information on the structure of investments and financings of a given project. Specifically, consider that a project may consist of investments in operations and/or financial assets and the capital may be raised by either debt or equity or both. This brings about 9 different structures, each of which may be graphically signalled via a 2×2 grid where absence of a given type of capital for all t is displayed by an empty grey area:

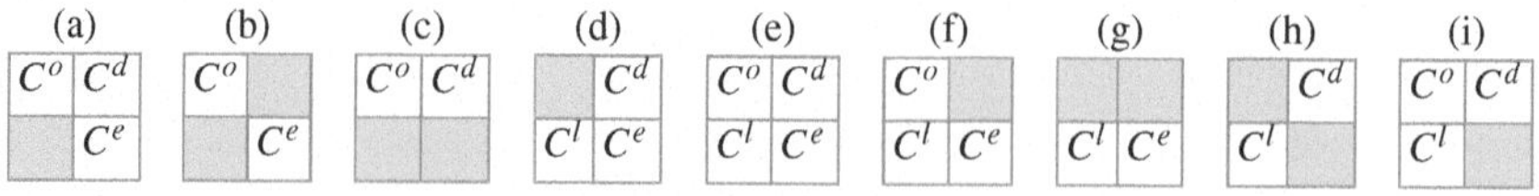

[7]Consider that any row in the Matrix may in principle be broken down into tens of rows detailing the composition of the items.

A project falls within one among classes (a), (b), (c) if it consists of investments in operations and non-operating assets are zero at every date; a project falls within one among classes (e), (f), or (i) if non-operating assets are nonzero at some date t. The analysis of a financial investment (portfolio, fund, etc.) has to do with framings (d), (g), (h) (typically, these framings are considered in investment performance measurement and attribution).

We focus on projects where equity capital is nonzero at least in one period, so we exclude cases (c), (h), and (i) from the analysis, which assume $C_t^e = 0$ for all t. Hence, the following 6 thumbnails remain:

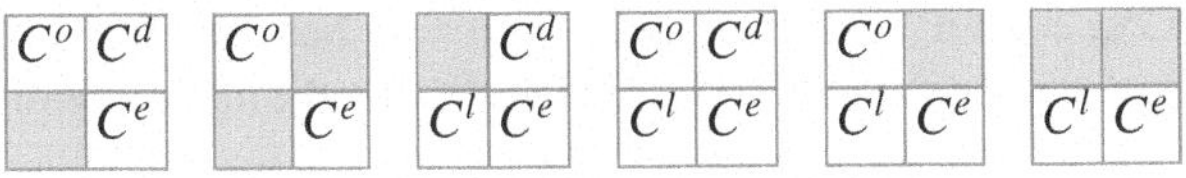

We will use these thumbnails at the beginning of many examples to signal to the reader which kind of structure the project illustrated enjoys.

Example 4.1 ♦

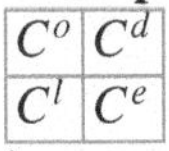

ML Inc. faces the opportunity of undertaking a three-year project with an investment in NFA equal to \$200 and an investment in raw materials equal to \$40. While the investment in NFA is made on cash, raising funds from either debtholders or equityholders, the investment in NOWC is made on credit, so $C_0^l = 0$ (no cash is withdrawn from the firm's cash account) and $\text{NFA}_0 = 200$ while $\text{Inv}_0 = \text{AP}_0 = 40$. Hence, $C_0 = \text{NFA}_0 - \text{NOL}_0 = \text{NFA}_0 + \text{Inv}_0 - \text{AP}_0 = 200$.[8] The manufacturing costs incurred in the various periods are as follows:

Time	1	2	3
COM	3	50	15
DL	10	25	5
MOH	5	5	6
Total	18	80	26

All the purchases of raw material will be cashed out at the end of the project, whence $\text{AP}_1 = 40 + 3 = 43$, $\text{AP}_2 = 43 + 50 = 93$, $\text{AP}_3 = 0$ (at the beginning of the fourth period, all suppliers will have been paid). The DL and MOH are paid within the year in which they are incurred, as well as taxes. This implies that there will be no accrued liabilities for direct labor and manufacturing overhead nor taxes payable nor any other manufacturing liability than AP. Therefore, $\text{NOL}_t = \text{AP}_t - \text{Inv}_t$ for every t. The analyst estimates that inventories will increase by a total \$10 in both the first

---

[8]If one assumed that the initial investment in NOWC is made on cash, raising funds from either equityholders or debtholders, then one would have $\text{NFA}_0 = 200$, $\text{Inv}_0 = 40$, $\text{AP}_0 = 0$, whence $C_0 = 240$. If one assumed that both the investment in NFA and in NOWC is made on cash, and the cash is drawn from the firm's liquid assets, then the net investment would be zero: $C_0 = \text{NFA}_0 + C_0^l + \text{Inv}_0 = 200 - 240 + 40 = 0$.

and second period. Then, $Inv_1 = 50$, $Inv_2 = 60$, $Inv_3 = 0$. This enables computing the COGS via the following equation:[9]

$$\text{COGS}_t = (\text{COM}_t + \text{DL}_t + \text{MOH}_t) - \Delta\text{Inv}_t.$$

In our case,

$$\begin{aligned}\text{COGS}_1 &= 18 - 10 = 8\\ \text{COGS}_2 &= 80 - 10 = 70\\ \text{COGS}_3 &= 26 + 60 = 86.\end{aligned}$$

(Note that $\text{COGS}_0 = \text{COP}_0 - \Delta\text{Inv}_0 = 40 - 40 = 0$. See also Table 3.6.) Manufacturing cash payments are carved out from the COGS with the following equation:

$$F_t^{cogs} = \text{COGS}_t - \Delta\text{NOL}_t^m.$$

In our case, $\text{NOL}_t^m = \text{NOL}_t$, and

$$\begin{aligned}\text{NOL}_t &= \text{AP}_t - \text{Inv}_t\\ \hline \text{NOL}_0 &= 40 - 40 = 0\\ \text{NOL}_1 &= 43 - 50 = -7\\ \text{NOL}_2 &= 93 - 60 = 33\\ \text{NOL}_3 &= 0\end{aligned}$$

Observe that the NOL is negative at the beginning of the second period. As we know from Chap. 1, this means that, in that period, NOL represents an investment, not a liability (inventories exceeds the suppliers' financings), so it adds to the other assets of the firm. Hence,

$$\begin{aligned}F_t^{cogs} &= \text{COGS}_t - \Delta\text{NOL}_t\\ \hline F_1^{cogs} &= 8 - (-7 - 0) = 15\\ F_2^{cogs} &= 70 - (33 + 7) = 30\\ F_3^{cogs} &= 86 - (0 - 33) = 119\end{aligned}$$

Let us assume that sales projections are $S_1 = 250$, $S_2 = 200$, $S_3 = 300$, made on a cash-basis only (i.e., $\text{AR}_t = 0$ for every t). The NOWC is equal to the difference between AR and NOL, for taxes payable is zero in each period:

$$\begin{aligned}\text{NOWC}_t &= \text{AR}_t - \text{NOL}_t\\ \hline \text{NOWC}_0 &= 0 - 0 = 0\\ \text{NOWC}_1 &= 0 - (-7) = 7\\ \text{NOWC}_2 &= 0 - 33 = -33\end{aligned}$$

[9]We use a financial perspective on COGS (see Sect. 3.8).

(see Table 4.2). Note that the NOWC is negative at the beginning of the third period. This means that the firm is absorbing resources from the project. Specifically, the firm benefits from the fact that customers pays in cash each year, while suppliers will only be paid later. Customers and suppliers are then "financing" the firm in the third period. Or, in other terms, *the project* is financing the firm.

Table 4.2 Net operating WC of ML Inc

Time	**0**	**1**	**2**	**3**
Accounts receivables	0	0	0	0
– *Accounts payable*	−40	−43	−93	– 0
+ *Inventory*	40	50	60	0
– Net Operating Liabilities	0	7	−33	0
Net operating WC	0	7	−33	0

Assume SGAs are forecasted to be equal to 30% of sales and are paid on cash: $\text{SGA}_1 = F_1^{sga} = 75$, $\text{SGA}_2 = F_2^{sga} = 60$, $\text{SGA}_3 = F_3^{sga} = 90$, which implies

$$\begin{aligned}\text{OpC}_1 &= \text{COGS}_1 + \text{SGA}_1 = 8 + 75 = 83\\ \text{OpC}_2 &= \text{COGS}_2 + \text{SGA}_2 = 70 + 60 = 130\\ \text{OpC}_3 &= \text{COGS}_3 + \text{SGA}_3 = 86 + 90 = 176.\end{aligned}$$

The investment in fixed assets consists of a single cash flow equal to $|F_0^{capex}| = 200$, depreciated according to the following depreciation plan: $\mathbf{Dep} = (60, 42, 98)$. The tax rate is $\tau = 33\%$. The project is sustained by a loan of \$60, to be reimbursed at $i^d = 3\%$ with a single instalment at the end of the third period. Focusing on column 3 of Matrix (3.64), the income statement is drawn in Table 4.3. The BS of the project is depicted in Table 4.4.

As we know, the operating profit is the EBIT net of actual taxes:

$$\begin{aligned}I_1^o &= \text{EBIT}_1 - T_1 = 107 - 34.7 = 72.3\\ I_2^o &= \text{EBIT}_2 - T_2 = 28 - 8.6 = 19.4\\ I_3^o &= \text{EBIT}_3 - T_3 = 26 - 7.9 = 18.1.\end{aligned}$$

The operating cash flow may be obtained in several different ways (see Matrix (4.2)). Using an investment perspective and focusing on cash flows,

$$F_t^o = \underbrace{F_t^c - \overbrace{(F_t^{cogs} + F_t^{sga})}^{F_t^{opc}} - F_t^T}_{F_t^{nowc}} - F_t^{capex}.$$

Table 4.3 Income statement of ML Inc

Time	1	2	3
Sales	250	200	300
– COGS	–8	–70	–86
– SGA	–75	–60	–90
EBITDA	167	70	124
– Depreciation	–60	–42	–98
EBIT	107	28	26
– Interest expenses	–1.80	–1.85	–1.91
EBT	105.2	26.1	24.1
– Taxes	–34.7	–8.6	–7.9
Net Income	70.5	17.5	16.1

Table 4.4 Balance sheet of ML Inc

Time	0	1	2	3
NET OPERATING WC	0	7	–33	0
Accounts receivables	0	0	0	0
+ *Inventory*	40	50	60	0
– *Accounts payable*	–40	–43	–93	0
– Net Operating Liabilities	0	7	–33	0
NET FIXED ASSETS	200	140	98	0
INVESTMENTS	200	147	65	0
Debt	60	61.8	63.65	0
Equity	140	85.2	1.35	0
FINANCINGS	200	147	65	0

In our case, $F_t^T = \mathrm{T}_t$, $F^c = S_t$, $F_t^{sga} = \mathrm{SGA}_t$ and $F_t^{capex} = 0$ for $t > 0$, so that

$$\underline{F_t^o = S_t - (F_t^{cogs} + \mathrm{SGA}_t) - \mathrm{T}_t - F_t^{capex}}$$

$$\begin{aligned}
F_0^o &= 0 - (0 + 0) - 0 - 200 = -200 \\
F_1^o &= 250 - (15 + 75) - 34.7 = 125.3 \\
F_2^o &= 200 - (30 + 60) - 8.6 = 101.4 \\
F_3^o &= 300 - (119 + 90) - 7.9 = 83.1.
\end{aligned}$$

Or, as a second alternative, one may use an investment perspective while focusing on incomes and capitals (e.g., see Matrix (4.7)):

$$F_t^o = \overbrace{\text{EBIT}_t - \text{T}_t}^{I_t^o} - \overbrace{(\Delta\text{NFA}_t + \Delta\text{NOWC}_t)}^{\Delta C_t^o}.$$

Hence,

$$\underline{F_t^o = I_t^o - \Delta C_t^o}$$

$$\begin{aligned} F_0^o &= 0 - 200 = -200 \\ F_1^o &= 72.3 + 53 = 125.3 \\ F_2^o &= 19.4 + 82 = 101.4 \\ F_3^o &= 18.1 + 65 = 83.1. \end{aligned}$$

(Using the lower partitions of Matrix (4.7), one may also use a financing perspective; we leave it to the reader.)

The CFD is zero at time 1 and 2. At time 3, the principal is reimbursed and the accumulated interest is paid: $F_3^d = 60(1.03)^3 = 65.56$. Or, alternatively, with the usual law of motion,

$$\underline{F_t^d = I_t^d - \Delta C_t^d}$$

$$\begin{aligned} F_0^d &= 0 - (60 - 0) = -60 \\ F_1^d &= 1.80 - (61.8 - 60) = 0 \\ F_2^d &= 1.85 - (63.65 - 61.8) = 0 \\ F_3^d &= 1.91 - (0 - 63.65) = 65.56. \end{aligned}$$

The CFE is straightforward as well, with any of the four relations provided by the Matrix (see (4.16) or (4.17)). For example, using the top left partition of Matrix (4.17): $F_t^e = F_t^o - F_t^{\mathcal{D}}$. Considering that $F_t^d = F_t^{\mathcal{D}}$ (no liquid assets),

$$\underline{F_t^e = F_t^o - F_t^d}$$

$$\begin{aligned} F_0^e &= -200 + 60 = -140 \\ F_1^e &= 125.3 - 0 = 125.3 \\ F_2^e &= 101.4 - 0 = 101.4 \\ F_3^e &= 83.1 - 65.56 = 17.54. \end{aligned}$$

We now use the project Matrix to check whether the model fulfills internal consistency and no error occurs: Summing by rows, by columns, by partitions must provide the same result at each and every date (see the "film" of ML Inc. in Eq. (4.21) below). The invariance properties hold at each date, so no error has been made.

$\mathbf{M} =$ (4.21)

$t = 0$	Cash flow	Income	$-\Delta$Capital
Operating assets	−200	+0	−200
Liquid assets	+0	+0	−0
Debt	+60	−0	+60
Equity	−140	+0	−140

↓

$t = 1$	Cash flow	Income	$-\Delta$Capital
Operating assets	+125.3	+72.3	+53.0
Liquid assets	+0.0	+0.0	−0.0
Debt	+0.0	−1.8	+1.8
Equity	+125.3	+70.5	+54.8

↓

$t = 2$	Cash flow	Income	$-\Delta$Capital
Operating assets	+101.4	+19.4	+82.0
Liquid assets	+0.0	+0.0	−0.0
Debt	+0.0	−1.9	+1.9
Equity	+101.4	+17.5	+83.9

↓

$t = 3$	Cash flow	Income	$-\Delta$Capital
Operating assets	+83.1	+18.0	+65.0
Liquid assets	+0.0	+0.0	−0.0
Debt	−65.6	−1.9	−63.7
Equity	+17.5	+16.1	+1.3

Following is a summary table which collects all the four main classes of capital, income, and cash flow (based on the transposed Matrix $(\boldsymbol{C}, \boldsymbol{I}, \boldsymbol{F})$ presented in (4.19)). It is a further way of confirming that the model fulfills its laws of conservation.[10]

[10]Numbers are rounded to one decimal place for illustrative purposes, so small rounding errors appear. The use of spreadsheet makes this point immaterial.

ML Inc. – $(\boldsymbol{C}, \boldsymbol{I}, \boldsymbol{F})$

Time		0	1	2	3
Capital					
Operating assets	C^o	200	147	65	0
Non-operating assets	C^l	0	0	0	0
Debt	C^d	60	61.8	63.65	0
Equity	C^e	140	85.2	1.35	0
Income					
Operating income	I^o	0	72.3	19.4	18.1
Non-operating income	I^l	0	0	0	0
Interest expense	I^d	0	1.8	1.9	1.9
Net Income	I^e	0	70.5	17.5	16.1
Cash flow					
Cash flow from operations	F^o	−200	125.3	101.4	83.1
Cash flow from non-operating assets	F^l	0	0	0	0
Cash flow to debt	F^d	−60	0	0	65.6
Cash flow to equity	F^e	−140	125.3	101.4	17.5

Example 4.2 ♦

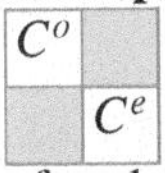

(*Opening project*) Consider a firm, named Opening, facing the opportunity of undertaking a new project, a five-year investment whereby a new product will be produced and sold. The investment cost of Opening is \$400,000, consisting of \$50,000 working capital requirements and \$350,000 fixed assets (property plant and equipment). The fixed assets are depreciated evenly ($\text{Dep}_t = 350.000/5 = 70.000$) and the salvage value is zero ($\text{SV}_5 = 0$). No cash is retained in the business (full distribution policy) and the project is equity financed (the project is then unlevered). The managers of firm Opening will manage the operating cycle in such a way that the NOWC will decrease by an annual \$10,000. In the first year, the units produced and sold will amount to 2,500 and will grow by 2% a year. The unit price is constant and equal to \$150. The operating costs (i.e., COGS and SGA) are divided into variable costs and fixed costs (see Eq. (3.78)). In particular, the variable unit cost is \$60; fixed costs are \$60,000 (see estimated data in Table 4.5). We also assume that operating costs are paid in the year when they are incurred (this implies $\text{AP}_t = \text{SWP}_t = 0$ for every t) and that the tax rate is zero: $\tau = 0$, so that $\text{TP}_t = \text{T}_t = 0$.

These assumptions imply that the EBIT represents, at one time, the project income, the operating income, and the net income: This implies that income at time t is

Table 4.5 Opening – project input data

Parameter	Estimate
Investment cost:	
Fixed assets	$350,000
Working capital	$50,000 (linear decrease)
Salvage value	$0
Unit sales	2,500
growth rate	2%
Unit price	$150
Variable unit cost	$60
Fixed costs	$60,000
NFA depreciation	straight-line
Horizon	5 years

$$I_t = I_t^e = I_t^o = \text{EBIT}_t = \overbrace{150 \cdot 2,500 \cdot (1.02)^{t-1}}^{S_t} - \overbrace{\big(60 \cdot 2,500 \cdot (1.02)^{t-1} + 60,000\big)}^{\text{OpC}_t} - \overbrace{70,000}^{\text{Dep}_t}$$

for $t \in \mathbb{N}_5^1$. (We display the pro forma financial statements in Table 4.6). The project's cash-flow stream

$$\boldsymbol{F} = \boldsymbol{F}^e = (-400, 175, 179.5, 184.1, 188.8, 193.5)$$

(in thousands) is derived from the capital stream $\boldsymbol{C} = (400, 320, 240, 160, 80)$ by using Eq. (4.11) or Eq. (4.16).

A different capital sequence would lead to a different cash-flow stream. In particular, suppose that the marketing managers select a different credit policy, such that the NOWC will remain constant at $50,000. Other things equal, the result is a different invested capital: $\boldsymbol{C} = (400, 330, 260, 190, 120)$. This reverberates on the profile of cash flows:

$$\boldsymbol{F} = \boldsymbol{F}^e = (-400, 165, 169.5, 174.1, 178.8, 233.5)$$

(see Table 4.7). The differences in the alternative cash flow streams reflect the difference in the alternative NOWCs.

For clarity, the diagram in Fig. 4.2 describes the functional relations of this project, from project input data (value drivers) to CFE.

C^o	C^d
	C^e

Consider now the alternative case where the Opening project is financed by a loan which covers 15% of the total invested capital. The loan is repaid with even principal repayments, which means that debt changes by $\Delta C_t^d = -\$60,000/5 = -\$12,000$ a year. The interest rate is assumed to be equal to a constant $i^d = 3\%$.

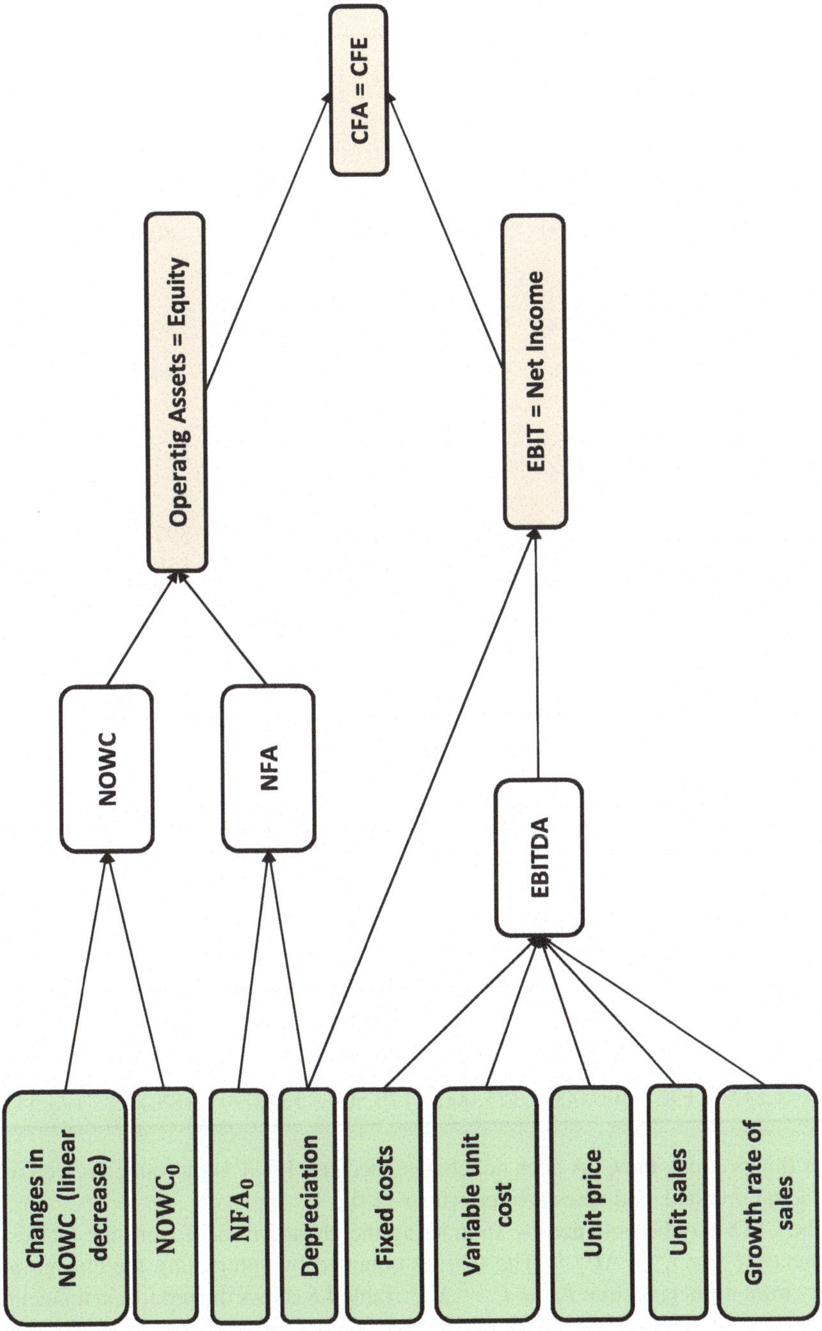

Fig. 4.2 Influence diagram indicating the functional relationships of Opening project

Table 4.6 Project opening unlevered: pro forma financial statements

Time	0	1	2	3	4	5
BALANCE SHEET						
Net Fixed Assets	350,000	280,000	210,000	140,000	70,000	0
Net Operating WC	50,000	40,000	30,000	20,000	10,000	0
INVESTMENTS	400,000	320,000	240,000	160,000	80,000	0
Debt	0	0	0	0	0	0
Equity	400,000	320,000	240,000	160,000	80,000	0
FINANCINGS	400,000	320,000	240,000	160,000	80,000	0
INCOME STATEMENT						
Sales		375,000	382,500	390,150	397,953	405,912
– Variable costs		–150,000	–153,000	–156,060	–159,181	–162,365
– Fixed costs		–60,000	–60,000	–60,000	–60,000	–60,000
EBITDA		165,000	169,500	174,090	178,772	183,547
– Depreciation		–70,000	–70,000	–70,000	–70,000	–70,000
EBIT		95,000	99,500	104,090	108,772	113,547
–Interest		0	0	0	0	0
–Taxes		0	0	0	0	0
NET INCOME		95,000	99,500	104,090	108,772	113,547
CASH FLOW						
EBIT		95,000	99,500	104,090	108,772	113,547
$-\Delta$NOWC	–50,000	10,000	10,000	10,000	10,000	10,000
$-\Delta$NFA	–350,000	70,000	70,000	70,000	70,000	70,000
CFA=CFO=CFE	–400,000	175,000	179,500	184,090	188,772	193,547

Other things equal, the CFA does not change, because EBIT is the same, as well as the capital invested, and taxes are zero (for $\tau = 0$).

The CFD may be obtained by subtracting the change in debt from the interest payments: $F_t^d = I_t^d - \Delta C_t^d$.[11] The CFE is obtained by subtracting the change in equity from the net income: $F_t^e = I_t^e - \Delta C_t^e$. Table 4.8 shows the pro forma financial statements and the resulting profile of the three types of cash flows. (We leave to the

[11]In this example, $C_t^l = 0$ so $C_t^{\mathcal{D}} = C_t^d$ and $I_t^{\mathcal{D}} = I_t^d$. The project is levered (see Sect. 2.2).

reader the construction of the 5 Matrices, which confirm that the model is correct and self-consistent.) Table 4.9 provides the transposed Matrix for the levered case and the (first) unlevered case.

Table 4.7 Project opening unlevered: pro forma financial statements changing NOWC projections

Time	**0**	**1**	**2**	**3**	**4**	**5**
BALANCE SHEET						
Net Fixed Assets	350,000	280,000	210,000	140,000	70,000	0
Net Operating WC	50,000	50,000	50,000	50,000	50,000	0
INVESTMENTS	400,000	330,000	260,000	190,000	120,000	0
Debt	0	0	0	0	0	0
Equity	400,000	330,000	260,000	190,000	120,000	0
FINANCINGS	400,000	330,000	260,000	190,000	120,000	0
INCOME STATEMENT						
Sales		375,000	382,500	390,150	397,953	405,912
– Variable costs		−150,000	−153,000	−156,060	−159,181	−162,365
– Fixed costs		−60,000	−60,000	−60,000	−60,000	−60,000
EBITDA		165,000	169,500	174,090	178,772	183,547
– Depreciation		−70,000	−70,000	−70,000	−70,000	−70,000
EBIT		95,000	99,500	104,090	108,772	113,547
−Interest		0	0	0	0	0
−Taxes		0	0	0	0	0
NET INCOME		95,000	99,500	104,090	108,772	113,547
CASH FLOW						
EBIT		95,000	99,500	104,090	108,772	113,547
−ΔNOWC	−50,000	0	0	0	0	50,000
−ΔNFA	−350,000	70,000	70,000	70,000	70,000	70,000
CFA=CFO=CFE	−400,000	165,000	169,500	174,090	178,772	233,547

Table 4.8 Project opening levered: pro forma financial statements (without taxes)

Time	**0**	**1**	**2**	**3**	**4**	**5**
BALANCE SHEET						
Net Fixed Assets	350,000	280,000	210,000	140,000	70,000	0
Net Operating WC	50,000	40,000	30,000	20,000	10,000	0
INVESTMENTS	400,000	320,000	240,000	160,000	80,000	0
Debt	60,000	48,000	36,000	24,000	12,000	0
Equity	340,000	272,000	204,000	136,000	68,000	0
FINANCINGS	400,000	320,000	240,000	160,000	80,000	0
INCOME STATEMENT						
Sales		375,000	382,500	390,150	397,953	405,912
– Variable costs		–150,000	–153,000	–156,060	–159,181	–162,365
– Fixed costs		–60,000	–60,000	–60,000	–60,000	–60,000
EBITDA		165,000	169,500	174,090	178,772	183,547
– Depreciation		–70,000	–70,000	–70,000	–70,000	–70,000
EBIT		95,000	99,500	104,090	108,772	113,547
– Interest		–1,800	–1,440	–1,080	–720	–360
– Taxes		0	0	0	0	0
NET INCOME		93,200	98,060	103,010	108,052	113,187
CASH FLOW						
EBIT		95,000	99,500	104,090	108,772	113,547
$-\Delta$NOWC	–50,000	10,000	10,000	10,000	10,000	10,000
$-\Delta$NFA	–350,000	70,000	70,000	70,000	70,000	-70,000
CFA=CFO	–400,000	175,000	179,500	184,090	188,772	193,547
Net Income		93,200	98,060	103,010	108,052	113,187
$-\Delta C^e$	–340,000	68,000	68,000	68,000	68,000	68,000
CFE	–340,000	161,200	166,060	171,010	176,052	181,187
Interest expenses		1,800	1,440	1,080	720	360
$-\Delta C^d$	–60,000	12,000	12,000	12,000	12,000	12,000
CFD	–60,000	13,800	13,440	13,080	12,720	12,360

Table 4.9 Transposed matrix of opening project (unlevered and levered scenario)

(C,I,F) Unlevered scenario – see Table 4.6

Time		0	1	2	3	4	5
Capital							
Operating assets	C^o	400,000	320,000	240,000	160,000	80,000	0
Non-operating assets	C^l	0	0	0	0	0	0
Debt	C^d	0	0	0	0	0	0
Equity	C^e	400,000	320,000	240,000	160,000	80,000	0
Income							
Operating income	I^o		95,000	99,500	104,090	108.772	113,547
Non-operating income	I^l		0	0	0	0	0
Interest expense	I^d		0	0	0	0	0
Net Income	I^e		95,000	99,500	104,090	108,772	113,547
Cash flow							
Cash flow from operations	F^o	−400,000	175,000	179,500	184,090	188,772	193,547
Cash flow from non-operating assets	F^l	0	0	0	0		
Cash flow to debt	F^d	0	0	0	0		
Cash flow to equity	F^e	−400,000	175,000	179,500	184,090	188,772	193,547

(C,I,F) Levered scenario – see Table 4.8

Time		0	1	2	3	4	5
Capital							
Operating assets	C^o	400,000	320,000	240,000	160,000	80,000	0
Non-operating assets	C^l	0	0	0	0	0	0
Debt	C^d	60,000	48,000	36,000	24,000	12,000	0
Equity	C^e	340,000	272,000	204,000	136,000	68,000	0
Income							
Operating income	I^o		95,000	99,500	104,090	108.772	113,547
Non-operating income	I^l		0	0	0	0	0
Interest expense	I^d		0	0	0	0	0
Net Income	I^e		93,200	98,060	103,010	108,052	113,187
Cash flow							
Cash flow from operations	F^o	−400,000	175,000	179,500	184,090	188,772	193,547
Cash flow from non-operating assets	F^l	0	0	0	0		
Cash flow to debt	F^d	−60,000	13,800	13,440	13,080	12,720	12,360
Cash flow to equity	F^e	−340,000	161,200	166,060	171,010	176,052	181,187

4.6 Taxes and Depreciation Tax Shield

The Opening project in Example 4.2 illustrates a before-tax analysis (it assumes no taxes). How do taxes affect incomes and cash flows?

Because $\text{EBIT}_t = \text{EBITDA}_t - \text{Dep}_t$ and $\text{T}_t = \tau\text{EBT}_t = \tau(\text{EBIT}_t - I_t^{\mathcal{D}})$, then

$$I_t^o = \text{EBIT}_t - \text{T}_t = \text{EBITDA}_t - \text{Dep}_t - \tau(\text{EBIT}_t - I_t^{\mathcal{D}}),$$

and remembering that $\Delta\text{NFA}_t = F_t^{capex} - \text{Dep}_t$, one may write the CFO/CCF as a function of tax rate (other things unvaried):[12]

$$F_t^o = (\text{EBITDA}_t - I_t^{\mathcal{D}})(1-\tau) + I_t^{\mathcal{D}} - \Delta\text{NOWC}_t - F_t^{capex} + \tau \cdot \text{Dep}_t. \quad (4.22)$$

In terms of CFA:

$$\begin{aligned} F_t &= (\text{EBITDA}_t - I_t^{\mathcal{D}})(1-\tau) - \Delta\text{NOWC}_t - F_t^{capex} + \tau \cdot \text{Dep}_t + I_t^{\mathcal{D}} + I_t^l - \Delta C_t^l \\ &= (\text{EBITDA}_t - I_t^{\mathcal{D}})(1-\tau) - \Delta\text{NOWC}_t - F_t^{capex} + \tau \cdot \text{Dep}_t + I_t^d - \Delta C_t^l. \end{aligned} \quad (4.23)$$

As for the CFE stream,

$$F_t^e = \overbrace{(\text{EBITDA}_t - \text{Dep}_t - I_t^{\mathcal{D}})(1-\tau)}^{I_t^e} - \overbrace{\left(\Delta\text{NFA}_t + \Delta\text{NOWC}_t - \Delta C_t^{\mathcal{D}}\right)}^{\Delta C_t^e} \quad (4.24)$$

$$= (\text{EBITDA}_t - I_t^{\mathcal{D}})(1-\tau) - \Delta\text{NOWC}_t - F_t^{capex} + \tau\text{Dep}_t + \Delta C_t^{\mathcal{D}}. \quad (4.25)$$

The amount $\tau \cdot \text{Dep}_t$ is called *depreciation tax shield*; this tax shield shelters the income whenever the tax rate is nonzero and is the key to understanding the role of depreciation in affecting cash flows. Specifically, other things unvaried, F_t, F_t^o, and F_t^e are linear functions of the depreciation charge: The higher the depreciation charge, the higher the cash flow. A change of Dep_t affects the cash flow by an amount equal to the product of τ and the change in Dep_t.

In other words, considering after-tax cash flows, $\boldsymbol{F^o}$, $\boldsymbol{F}$, and $\boldsymbol{F^e}$ are functions of the depreciation schedule $\textbf{Dep} = (\text{Dep}_1, \text{Dep}_2, \ldots, \text{Dep}_n)$.

Depreciation is the income component of fixed assets. It does not affect cash flow in a before-tax analysis; however, it does affect cash flow in an after-tax analysis via the depreciation tax shield. (The other assets affect cash flows regardless of whether the analysis is before- or after-tax.)

By constructing the sequence of pro forma income statements and the sequence of pro forma balance sheets for every period of the project's life, the analyst should

[12]Manipulating algebraically,

$$\begin{aligned} F_t^o &= \overbrace{\text{EBIT}_t - \text{T}_t}^{I_t^o} - \overbrace{(\Delta\text{NFA}_t + \Delta\text{NOWC}_t)}^{\Delta C_t^o} \\ &= \text{EBITDA}_t - \text{Dep}_t - \tau(\text{EBIT}_t - I_t^{\mathcal{D}}) - F_t^{capex} + \text{Dep}_t - \Delta\text{NOWC}_t \\ &= \text{EBITDA}_t - \tau(\text{EBITDA}_t - \text{Dep}_t - I_t^{\mathcal{D}}) - F^{capex} - \Delta\text{NOWC}_t \\ &= (\text{EBITDA}_t - I_t^{\mathcal{D}}) + I_t^{\mathcal{D}} - \tau(\text{EBITDA}_t - I_t^{\mathcal{D}}) + \tau\text{Dep}_t - F^{capex} - \Delta\text{NOWC}_t \\ &= (\text{EBITDA}_t - I_t^{\mathcal{D}})(1-\tau) + I_t^{\mathcal{D}} + \tau\text{Dep}_t - F^{capex} - \Delta\text{NOWC}_t. \end{aligned}$$

try to correctly describe the future *actual* reported statements of the project, so as to capture the appropriate change in capital and therefore, correctly estimate the amount of cash flows that will be received by shareholders.

Example 4.3 ♦

C^o	
	C^e

(*Opening project – changing depreciation*) Consider the Opening Example (Example 4.2) and suppose, again, that capital invested in fixed assets decreases on a straight-line basis, such that **Dep** = (70, 70, 70, 70, 70), and assume that the tax rate is $\tau = 30\%$, other things equal (in particular, the project is unlevered). The net income coincides with OI, for $I_t^d = I_t^l = 0$. The CFA coincides with CFE. In particular,

$$\boldsymbol{F} = \boldsymbol{F}^o = \boldsymbol{F}^e = (-400, 146.5, 149.6, 152.9, 156.1, 159.5).$$

We may summarize the main features of the project in the transposed Matrix (Table 4.10. See also Table 4.11).

Let us now see what happens if a different depreciation plan is employed. In particular, suppose that the depreciation plan is accelerated, such that **Dep** = (150, 150, 50, 0, 0); it is easy to see that the after-tax cash-flow stream would be

Table 4.10 Transposed matrix of opening project (unlevered with taxes) (C, I, F) (see Table 4.11)

Time		0	1	2	3	4	5
Capital							
Operating assets	C^o	400,000	320,000	240,000	160,000	80,000	0
Non-operating assets	C^l	0	0	0	0	0	0
Debt	C^d	0	0	0	0	0	0
Equity	C^e	400,000	320,000	240,000	160,000	80,000	0
Income							
Operating income	I^o		66,500	69,650	72,863	76,140	79,483
Non-operating income	I^l		0	0	0	0	0
Interest expense	I^d		0	0	0	0	0
Net Income	I^e		66,500	69,650	72,863	76,140	79,483
Cash flow							
Cash flow from operations	F^o	–400,000	146,500	149,650	152,863	156,140	159,483
Cash flow from non-operating assets	F^l	0	0	0	0	0	0
Cash flow to debt	F^d	0	0	0	0		
Cash flow to equity	F^e	–400,000	146,500	149,650	152,863	156,140	159,483

Table 4.11 Project opening unlevered: pro forma financial statements and cash flows (with taxes)

Time	**0**	**1**	**2**	**3**	**4**	**5**
BALANCE SHEET						
Net Fixed Assets	350,000	280,000	210,000	140,000	70,000	0
Net Operating WC	50,000	40,000	30,000	20,000	10,000	0
INVESTMENTS	400,000	320,000	240,000	160,000	80,000	0
Debt	0	0	0	0	0	0
Equity	400,000	320,000	240,000	160,000	80,000	0
FINANCINGS	400,000	320,000	240,000	160,000	80,000	0
INCOME STATEMENT						
Sales		375,000	382,500	390,150	397,953	405,912
– Variable costs		−150,000	−153,000	−156,060	−159,181	−162,365
– Fixed costs		−60,000	−60,000	−60,000	−60,000	−60,000
EBITDA		165,000	169,500	174,090	178,772	183,547
– Depreciation		−70,000	−70,000	−70,000	−70,000	−70,000
EBIT		95,000	99,500	104,090	108,772	113,547
– Taxes		−28,500	−29,850	−31,227	−32,632	−34,064
Net income= OI		66,500	69,650	72,863	76,140	79,483
CASH FLOW						
= EBIT – T		66,500	69,650	72,863	76,140	79,483
−ΔNOWC	−50,000	10,000	10,000	10,000	10,000	10,000
−ΔNFA	−350,000	70,000	70,000	70,000	70,000	70,000
CFA=CFE=CFO	−400,000	146,500	149,650	152,863	156,140	159,483

$$\boldsymbol{F} = \boldsymbol{F}^o = \boldsymbol{F}^e = (-400, 170.5, 173.6, 146.9, 135.1, 138.5).$$

The difference between the two vectors is $(0, 24, 24, -6, -21, -21)$ and is just equal to the product of the tax rate and the change in depreciation:

$$(0, 0.3 \cdot (150 - 70), 0.3 \cdot (150 - 70), 0.3 \cdot (50 - 70), 0.3 \cdot (0 - 70), 0.3(0 - 70)).$$

Switching to accelerated depreciation, one gets greater cash flows in the early years owing to the higher shelter provided by the tax shield. The relation is reversed in the later years: Smaller depreciation charges bring about smaller cash flows owing to the greater shelter offered by straight-line depreciation. (Other depreciation patterns will lead to different cash-flow profiles.)

Example 4.4 ♦

C^o	C^d
	C^e

(*BHP Billiton*) The following example builds upon a real-life case illustrated in Hartman (2007, p. 260). BHP Billiton, an Australian mining and energy company, analyzes a \$550 million investment in a liquified natural gas (LNG) import facility off the coast of Southern California. The company will supply 800 million cubic feet of natural gas per day beginning at time 1 for a total 20 years. NFA will reduce on a straight-line basis and no working capital will be used in the operations. We assume that the firm borrows \$200 million to finance the project. Interest payments will be paid periodically at the interest rate of 8%. The principal will be due at the end of the 10th year. We also assume that cash flows generated by the project will be entirely distributed to equityholders and debtholders, which implies that no cash will be retained in the business, so that $C_t^l = 0$ for all t and $C_t^{\mathcal{D}} = C_t^d$ (i.e., 100% payout ratio).

Table 4.12 depicts the order-of-magnitude prospective data. The projected revenues are

$$S_t = \frac{3.5}{1000 \text{ ft}^3} \cdot 800\text{M ft}^3 \cdot 300 \text{ days/year} = 840\text{M}$$

while the projected operating costs are

$$\text{OpC}_t = \overbrace{\$15\text{M}}^{\text{fixed costs}} + \overbrace{\frac{2}{1000\text{ft}^3} \cdot 800\text{M ft}^3 \cdot 300 \text{ days/year}}^{\text{variable costs}} = 495\text{M}$$

for $t \in \mathbb{N}_{19}^1$. At the end of the twentieth year, disposal costs are incurred, so that $\text{OpC}_{20} = 495 + 100 = 595$. Also, $I_t^d = 16$ for $t \in \mathbb{N}_{10}^1$ and $\text{Dep}_t = 550/20 = 27.5$ for $t \in \mathbb{N}_{20}^1$. Table 4.13 reports the resulting income statements, period by period (to be read from left to right), while Table 4.14 shows the BSs with the relative capital amounts and the cash flows.

Table 4.12 BHP Billiton – project input data

Parameter	Estimate
Investment cost	$550 million
Salvage value	$0
Annual Operating Costs (fixed)	$15 million
Daily Production Level	800 million ft^3
LNG Material Costs	$2 per 1000 ft^3
LNG Price	$3.5 per 1000 ft^3
Disposal Cost	$100 million
NFA depreciation	straight-line
Borrowing	$200 million
Type of borrowing	Interest payments
Loan term	10 years
Interest rate	8%
Horizon	20 years
Tax rate	30%

Table 4.13 BHP Billiton: income statements

Time	S_t	$-\text{OpC}_t$	EBITDA	$-\text{Dep}_t$	EBIT	$-I_t^d$	EBT	$-\text{T}_t$	I_t^e
1	840.0	−495.0	345.0	−27.5	317.5	−16.0	301.5	−90.5	211.1
2	840.0	−495.0	345.0	−27.5	317.5	−16.0	301.5	−90.5	211.1
3	840.0	−495.0	345.0	−27.5	317.5	−16.0	301.5	−90.5	211.1
4	840.0	−495.0	345.0	−27.5	317.5	−16.0	301.5	−90.5	211.1
5	840.0	−495.0	345.0	−27.5	317.5	−16.0	301.5	−90.5	211.1
6	840.0	−495.0	345.0	−27.5	317.5	−16.0	301.5	−90.5	211.1
7	840.0	−495.0	345.0	−27.5	317.5	−16.0	301.5	−90.5	211.1
8	840.0	−495.0	345.0	−27.5	317.5	−16.0	301.5	−90.5	211.1
9	840.0	−495.0	345.0	−27.5	317.5	−16.0	301.5	−90.5	211.1
10	840.0	−495.0	345.0	−27.5	317.5	−16.0	301.5	−90.5	211.1
11	840.0	−495.0	345.0	−27.5	317.5	0.0	317.5	−95.3	222.3
12	840.0	−495.0	345.0	−27.5	317.5	0.0	317.5	−95.3	222.3
13	840.0	−495.0	345.0	−27.5	317.5	0.0	317.5	−95.3	222.3
14	840.0	−495.0	345.0	−27.5	317.5	0.0	317.5	−95.3	222.3
15	840.0	−495.0	345.0	−27.5	317.5	0.0	317.5	−95.3	222.3
16	840.0	−495.0	345.0	−27.5	317.5	0.0	317.5	−95.3	222.3
17	840.0	−495.0	345.0	−27.5	317.5	0.0	317.5	−95.3	222.3
18	840.0	−495.0	345.0	−27.5	317.5	0.0	317.5	−95.3	222.3
19	840.0	−495.0	345.0	−27.5	317.5	0.0	317.5	−95.3	222.3
20	840.0	−595.0	245.0	−27.5	217.5	0.0	217.5	−65.3	152.3

Table 4.14 BHP Billiton: balance sheets and cash flows (without NOWC)

Time	NFA	C_t^d	C_t^e	F_t	F_t^d	F_t^e
0	550.0	200	350.0	−550.0	−200.0	−350.0
1	522.5	200	322.5	254.6	16.0	238.6
2	495.0	200	295.0	254.6	16.0	238.6
3	467.5	200	267.5	254.6	16.0	238.6
4	440.0	200	240.0	254.6	16.0	238.6
5	412.5	200	212.5	254.6	16.0	238.6
6	385.0	200	185.0	254.6	16.0	238.6
7	357.5	200	157.5	254.6	16.0	238.6
8	330.0	200	130.0	254.6	16.0	238.6
9	302.5	200	102.5	254.6	16.0	238.6
10	275.0	0	275.0	254.6	216.0	38.6
11	247.5	0	247.5	249.8	0	249.8
12	220.0	0	220.0	249.8	0	249.8
13	192.5	0	192.5	249.8	0	249.8
14	165.0	0	165.0	249.8	0	249.8
15	137.5	0	137.5	249.8	0	249.8
16	110.0	0	110.0	249.8	0	249.8
17	82.5	0	82.5	249.8	0	249.8
18	55.0	0	55.0	249.8	0	249.8
19	27.5	0	27.5	249.8	0	249.8
20	0	0	0.0	179.8	0	179.8

Table 4.15 BHP Billiton project input data (with NOWC)

Parameter	Estimate
Investment cost (NFA_0)	$550 million
Initial NOWC ($NOWC_0$)	$50 million
NOWC-to-revenues ratios	5%
Salvage value	$0
Annual Operating Costs (fixed)	$15 million
Daily Production Level	800 million ft^3
LNG Material Costs	$2 per 1000 ft^3
LNG Price	$3.5 per 1000 ft^3
Disposal Cost	$100 million
NFA depreciation	straight-line
Borrowing	$200 million
Type of borrowing	Periodic interest payments
Loan term	10 years
Interest rate	8%
Horizon	20 years
Tax rate	30%

Now, consider the same project but assume that it requires some amount of NOWC. In particular, suppose the initial working capital is \$50 million and prospective NOWCs are estimated at 5% of revenues in each period: $\text{NOWC}_t = 0.05 \cdot S_t = 0.05 \cdot 840 = 42$ for $t > 0$. This implies that the capital stream $\boldsymbol{C}$ changes (with respect to the previous assumptions) and, hence, the profiles of CFAs and CFEs change. We assume that the project is financed with the same amount of debt as before, which implies that the CFDs do not change. We report the project input data with the new additions in Table 4.15. Table 4.16 depicts the BSs and the associated cash flows for this case.[13] The diagram in Fig. 4.3 describes the functional relations of the project, from input data to CFE.

Table 4.16 BHP Billiton: balance sheets and cash flows (with NOWC)

Time	NFA	NOWC	C_t^d	C_t^e	F_t	F_t^d	F_t^e
0	550.0	50.0	200	400.0	−600.0	−200	−400.0
1	522.5	42.0	200	364.5	262.6	16	246.6
2	495.0	42.0	200	337.0	254.6	16	238.6
3	467.5	42.0	200	309.5	254.6	16	238.6
4	440.0	42.0	200	282.0	254.6	16	238.6
5	412.5	42.0	200	254.5	254.6	16	238.6
6	385.0	42.0	200	227.0	254.6	16	238.6
7	357.5	42.0	200	199.5	254.6	16	238.6
8	330.0	42.0	200	172.0	254.6	16	238.6
9	302.5	42.0	200	144.5	254.6	16	238.6
10	275.0	42.0	0	317.0	254.6	216	38.6
11	247.5	42.0	0	289.5	249.8	0	249.8
12	220.0	42.0	0	262.0	249.8	0	249.8
13	192.5	42.0	0	234.5	249.8	0	249.8
14	165.0	42.0	0	207.0	249.8	0	249.8
15	137.5	42.0	0	179.5	249.8	0	249.8
16	110.0	42.0	0	152.0	249.8	0	249.8
17	82.5	42.0	0	124.5	249.8	0	249.8
18	55.0	42.0	0	97.0	249.8	0	249.8
19	27.5	42.0	0	69.5	249.8	0	249.8
20	0.0	0.0	0	0.0	221.8	0	221.8

[13]For the sake of illustration, we have implicitly assumed that the project's COGS do not vary if NOWC is added (which in turn implies that OpC_t do not change). In general, assuming COM_t, DL_t, MOH_t are fixed, if NOWC changes owing to a change in Inv_t, then COGS changes as well (for $\text{COGS}_t = \text{COM}_t + \text{DL}_t + \text{MOH}_t - \Delta\text{Inv}_t$); if NOWC changes owing to a change in AP_t and/or AR_t, then COGS is not affected.

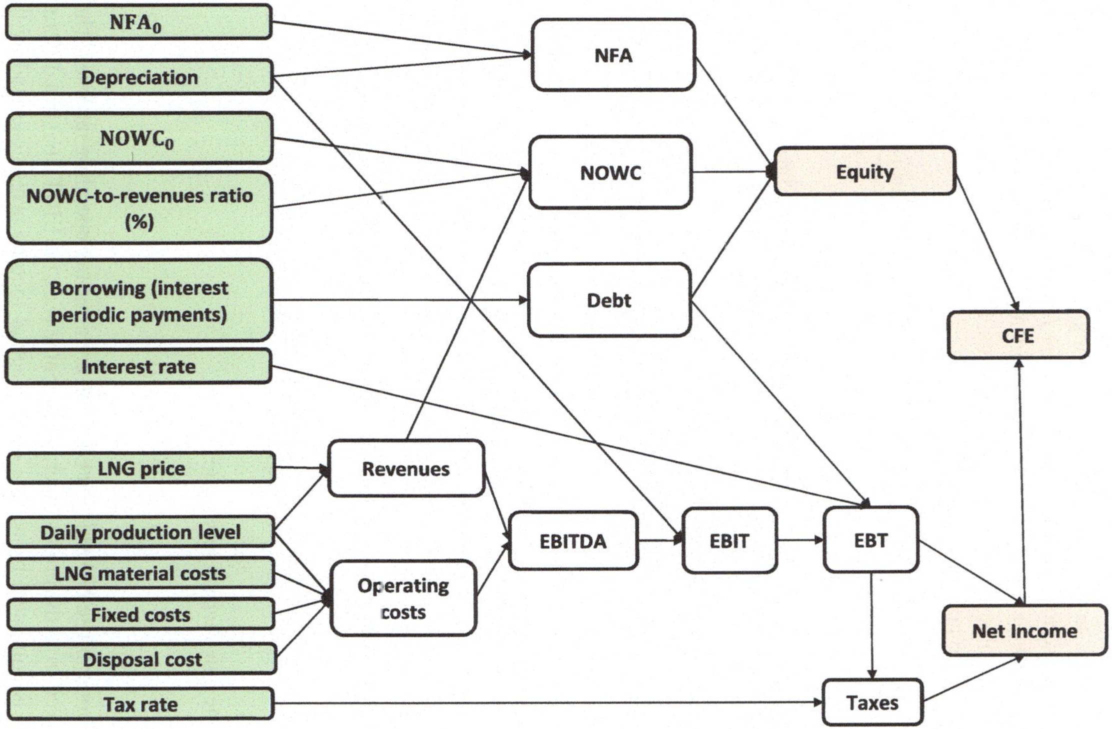

Fig. 4.3 Influence diagram indicating the functional relationships for BHP Billiton project (with NOWC), from project input data to CFE. In general, interest rate affects debt. However, in this case, the interest rate effect on the debt amount is offset by the CFD, which coincides with the interest payments ($F_t^d = 0.03 \cdot C_{t-1}^d = I_t^d$). Therefore, the debt is constant: $C_t^d = 200$ for $t \in \mathbb{N}_9^0$

4.7 NOPAT and Free Cash Flow

Operating income (OI) and cash flow from operations (CFO/CCF) are the sources of variation of the operating capital: $\Delta C_t^o = I_t^o - F_t^o$. As seen, OI is an operating income net of taxes on EBT. In corporate and managerial finance as well as engineering economics, an alternative pair of operating income and operating cash flow is popular among scholars and practitioners: The *Net Operating Profit After Taxes* (NOPAT) and the *Free Cash Flow* (FCF). The idea of NOPAT stems from the idea of splitting the income taxes into two components. Specifically, letting τ denote the tax rate, the income taxes amount to

$$\mathrm{T}_t = \tau \cdot \mathrm{EBT}_t = \tau \cdot (\mathrm{EBIT}_t - I_t^{\mathcal{D}}) = \overbrace{\tau \cdot \mathrm{EBIT}_t}^{\text{Taxes on EBIT}} - \overbrace{\tau I_t^{\mathcal{D}}}^{\text{Interest tax shield}}.$$

The first summand in the last expression quantifies the taxes on EBIT, which is reduced by the taxes on net interest, commonly known as *interest tax shield.* Hence,

$$\begin{aligned} I_t^e &= \mathrm{EBT}_t(1-\tau) \\ &= \mathrm{EBIT}_t(1-\tau) - I_t^{\mathcal{D}}(1-\tau). \end{aligned} \tag{4.26}$$

The first component, $\mathrm{EBIT}_t(1-\tau)$, is the *Net Operating Profit After Taxes* (NOPAT), while the second component, $I_t^{\mathcal{D}}(1-\tau)$, is the after-tax net interest. The NOPAT may also be interpreted as the after-tax operating income that the project would generate if it were unlevered. For this reason, we will also refer to NOPAT as the *unlevered operating income*. However, if the project were unlevered, the OI would coincide with the net income. Thus, one might also conceptualize NOPAT as the *unlevered net income*.

We will henceforth often use the symbol I_t^{uo} to denote the unlevered operating income:

$$I_t^{uo} = \overbrace{\mathrm{EBIT}_t(1-\tau)}^{\mathrm{NOPAT}_t}. \tag{4.27}$$

Therefore, one can alternatively write:

$$I_t^e = I_t^{uo} - I_t^{\mathcal{D}}(1-\tau) \tag{4.28}$$

One should not confuse NOPAT and OI. Both are after-tax operating incomes. They are both derived from the pre-tax operating income (i.e., the EBIT), but in a different way. The OI is computed by netting out the actual income taxes generated by the project, whereas the NOPAT is computed by netting out the taxes on EBIT.

This has relevant implications: The OI is affected by $I_t^{\mathcal{D}}$ and, therefore, by the capital structure; specifically, given a net interest rate equal to $i_t^{\mathcal{D}}$, the higher $I_t^{\mathcal{D}}$, the smaller T_t. In particular, remembering that $\text{EBT}_t = \text{EBIT}_t - I_t^{\mathcal{D}}$,

$$\begin{aligned} I_t^o &= \text{EBIT}_t - \tau(\text{EBIT}_t - I_t^{\mathcal{D}}) \\ &= \text{EBIT}_t(1-\tau) + \tau I_t^{\mathcal{D}} \\ &= \text{NOPAT}_t + \tau I_t^{\mathcal{D}} \\ &= I_t^{uo} + \tau I_t^{\mathcal{D}} \end{aligned} \tag{4.29}$$

As anticipated, the amount $\tau I_t^{\mathcal{D}}$ is called *interest tax shield*. It shelters the income whenever $I_t^{\mathcal{D}} > 0$ and the tax rate is nonzero: For any given NOPAT, the higher the tax shield, the higher the operating income. In a nutshell: I_t^o is the actual after-tax operating income earned by investors, whereas I_t^{uo} is the part of it that is not affected by the project's capital structure.

The cash flow associated with NOPAT is the Free Cash Flow, which is an unlevered operating cash flow.

Definition 4.2 (*Free Cash Flow*) Assume $\text{TP}_t = 0$. The Free Cash Flow (FCF), denoted as F_t^{uo}, is the difference between NOPAT and the change in operating capital:

$$F_t^{uo} = \text{NOPAT}_t - \Delta C_t^o = I_t^{uo} - \Delta C_t^o \tag{4.30}$$

where $F_0^{uo} = F_0^o$.[14]

Like the CFO/CCF, the FCF is an after-tax operating cash-flow, but the taxes are computed on EBIT. The FCF may then be interpreted as the hypothetical after-tax CFO that the project would generate if net interest were zero[15] (or if the tax rate were zero). For this reason, we will also call the FCF the *unlevered CFO*. More simply, the FCF is the after-tax cash flow generated by the operations, *neglecting the effects of liquid assets and debt on taxes* (see also below). It is worth noting that, if the project is unlevered, the CFO coincides with the CFE. Thus, the FCF is, at the same time, the *unlevered CFE*.

The above definition excludes the taxes generated by non-operating assets. Therefore, the pre-tax CFO is equal to $F_t^{\bullet} = \text{EBIT}_t - \Delta C_t^o$. This implies (assuming $\text{TP}_t = 0$)

[14]The definition is appropriate only insofar as the company is not able to defer some taxes (i.e., $\text{TP}_t = 0$). Otherwise, it is necessary to adjust the definition so as to take account of the actual tax payments attributable to the operations as opposed to the tax payments attributable to the income components of debt and liquid assets. That is, ΔTP_t should be split up into operating component and financing component; the latter should be excluded from the definition of FCF. The FCF would then be defined as $F_t^{uo} = \text{NOPAT}_t - (\Delta\text{NFA}_t + \Delta\text{NOWC}_t - \Delta \text{ non-operating TP}_t)$. See also Eq. (6.121). We henceforth stick with the assumption $\text{TP}_t = 0$ for every t, which means $\text{T}_t = F_t^T$.

[15]A particular case of this situation is when the project has no liquid assets outstanding nor debt.

$$\begin{aligned} F_t^o &= F^\bullet - F_t^T = F^\bullet - \tau \cdot (\text{EBIT}_t - I_t^{\mathcal{D}}) \\ F_t^{uo} &= F^\bullet - \tau \cdot \text{EBIT}_t \end{aligned} \tag{4.31}$$

The difference between the pair (I_t^o, F_t^o) and the pair (I_t^{uo}, F_t^{uo}) may then be best appreciated by making use of the Matrix tool. Let us start from the CFO Matrix (4.2), after performing consolidation of rows 1–3 (and assuming $\text{TP}_t = 0$ for all t):

$$\textbf{CFO}$$
$$\begin{array}{|l|ll|} \hline +F_t^c - F_t^{opc} - F_t^{capex} & +\text{EBIT}_t & -(\Delta\text{AR}_t - \Delta\text{AP}_t + \Delta\text{Inv}_t + \Delta\text{NFA}_t) \\ -\text{T}_t & -\text{T}_t & \\ \hline +F_t^{\mathcal{D}} & +I_t^{\mathcal{D}} & -\Delta C_t^{\mathcal{D}} \\ +F_t^e & +I_t^e & -\Delta C_t^e \\ \hline \end{array} \tag{4.32}$$

The upper left element (i.e., first row, first column) is the pre-tax CFO: $F_t^\bullet = F_t^c - F_t^{opc} - F_t^{capex}$. One may then write the CFO as $F_t^o = F_t^\bullet - \text{T}_t$:

$$\textbf{CFO}$$
$$\begin{array}{|l|ll|} \hline +F^\bullet & +\text{EBIT}_t & -\Delta C_t^o \\ -\text{T}_t & -\text{T}_t & \\ \hline +F_t^{\mathcal{D}} & +I_t^{\mathcal{D}} & -\Delta C_t^{\mathcal{D}} \\ +F_t^e & +I_t^e & -\Delta C_t^e \\ \hline \end{array} \tag{4.33}$$

We now split up row 2 into two rows associated with the taxes on EBIT (negative component) and the taxes on net interest (positive component), respectively:

$$\textbf{CFO}$$
$$\begin{array}{|l|ll|} \hline +F_t^\bullet & +\text{EBIT}_t & -\Delta C_t^o \\ -\tau\ \text{EBIT}_t & -\tau\ \text{EBIT}_t & \\ +\tau I_t^{\mathcal{D}} & +\tau I_t^{\mathcal{D}} & \\ \hline +F_t^{\mathcal{D}} & +I_t^{\mathcal{D}} & -\Delta C_t^{\mathcal{D}} \\ +F_t^e & +I_t^e & -\Delta C_t^e \\ \hline \end{array} \tag{4.34}$$

Shifting the interest tax shield to the lower part, and consolidating with net debt, one gets the FCF:

$$\textbf{FCF}$$
$$\begin{array}{|l|ll|} \hline +F_t^\bullet & +\text{EBIT}_t & -\Delta C_t^o \\ -\tau\ \text{EBIT}_t & -\tau\ \text{EBIT}_t & \\ \hline +F_t^{\mathcal{D}} - \tau I_t^{\mathcal{D}} & +I_t^{\mathcal{D}}(1-\tau) & -\Delta C_t^{\mathcal{D}} \\ +F_t^e & +I_t^e & -\Delta C_t^e \\ \hline \end{array} \tag{4.35}$$

We pair the two Matrices below, so that the difference between CFO and FCF may be clearly identified in the interest tax shield, $\tau I_t^{\mathcal{D}}$ (it suffices to contrast any partition in the CFO Matrix with the corresponding partition in the FCF Matrix):

CFO

$+F^{\bullet}_t$	$+\text{EBIT}_t$	$-\Delta C^o_t$
$-\tau\,(\text{EBIT}_t - I^{\mathcal{D}}_t)$	$-\tau\,(\text{EBIT}_t - I^{\mathcal{D}}_t)$	
$+F^{\mathcal{D}}_t$	$+I^{\mathcal{D}}_t$	$-\Delta C^{\mathcal{D}}_t$
$+F^e_t$	$+I^e_t$	$-\Delta C^e_t$

FCF

$+F^{\bullet}_t$	$+\text{EBIT}_t$	$-\Delta C^o_t$
$-\tau\text{EBIT}_t$	$-\tau\text{EBIT}_t$	
$+F^{\mathcal{D}}_t - \tau I^{\mathcal{D}}_t$	$+I^{\mathcal{D}}_t(1-\tau)$	$-\Delta C^{\mathcal{D}}_t$
$+F^e_t$	$+I^e_t$	$-\Delta C^e_t$

(4.36)

One may then write

$$F^o_t - F^{uo}_t = \tau I^{\mathcal{D}}_t \iff F^o_t = F^{uo}_t + \tau I^{\mathcal{D}}_t. \tag{4.37}$$

CFO = FCF + tax shield

CFO = operating cash flow/capital cash flow = CCF

tax shield = tax rate × net interest

The interest tax shield shelters the net income in the sense that the higher the net interest, the higher the CFO (other things unvaried). This means that FCF notion abstracts from the effect of financing decisions on the project's cash flow. The Matrices in (4.36) and Eq. (4.37) tell us that FCF is that part of CFO/CCF that is not affected by the project's capital structure. Capital structure affects the interest tax shield, $\tau I^{\mathcal{D}}_t$, and in turn, the CFO. Hence, the CFO may be decomposed into two components, the FCF and the (interest) tax shield, the former reflecting the investment policy, the latter reflecting the financing policy (borrowing and payout policy).[16] The CFO Matrix and FCF Matrix depicted above make it clear that the difference between CFO and FCF is exactly equal to the difference between OI and NOPAT (see column 2 in each Matrix of (4.36)).

Also, contrasting the upper and lower parts of the first two columns of the FCF Matrix,

$$\begin{aligned} I^{uo}_t &= \overbrace{I^{\mathcal{D}}_t(1-\tau) + I^e_t}^{\text{EBIT}_t(1-\tau)} \\ F^{uo}_t &= \overbrace{F^{\mathcal{D}}_t - \tau I^{\mathcal{D}}_t + F^e_t}^{F^{\bullet}_t - \tau\text{EBIT}_t} \end{aligned} \tag{4.38}$$

[16]We are assuming that the financing policy has no impact on the investment policy. There may well be situations where the latter is affected by the former (e.g., restrictions to investment via debt covenants). In this case, FCF will necessarily be the result of both investment and financing policies.

In Matrix form,

FCF

$$\begin{array}{|l|l|}\hline +F_t^{uo} & +\text{NOPAT}_t \quad -\Delta C_t^o \\ \hline +F_t^{\mathcal{D}} - \tau I_t^{\mathcal{D}} & +I_t^{\mathcal{D}}(1-\tau) \; -\Delta C_t^{\mathcal{D}} \\ +F_t^e & +I_t^e \quad -\Delta C_t^e \\ \hline \end{array} \tag{4.39}$$

Equivalently,

$$\begin{aligned} \overbrace{I_t^{uo} + \tau I_t^{\mathcal{D}}}^{I_t^o} &= I_t^{\mathcal{D}} + I_t^e \\ \overbrace{F_t^{uo} + \tau I_t^{\mathcal{D}}}^{F_t^o} &= F_t^{\mathcal{D}} + F_t^e. \end{aligned} \tag{4.40}$$

The NOPAT (I_t^{uo}) is then equal to the after-tax net interest ($I_t^{\mathcal{D}}(1-\tau)$) plus the net income ($I_t^e$), while the FCF can be viewed as an adjusted CF$\mathcal{D}$ net of the tax shield ($F_t^{\mathcal{D}} - \tau I_t^{\mathcal{D}}$) plus the CFE ($F_t^e$).

The NOPAT may be computed either by netting the EBIT out of the taxes on EBIT or by subtracting the tax shield from the operating income.

The FCF may be computed either by deducting the change in operating assets from the NOPAT or by subtracting the tax shield from the CFO/CCF.

If interest expenses exceed interest income ($I_t^{\mathcal{D}} > 0$), the CFO/CCF is greater than the FCF ($F_t^o > F_t^{uo}$) because the interest expenses are deductible. However, it may occur that $I_t^{\mathcal{D}} < 0$ (interest income exceeds interest expenses) so that CFO/CCF is smaller than FCF. In this case, the tax shield does not shelter income; rather, it exposes income to an additional tax amount (since interest income is taxable).

FCF coincides with CFO/CCF only in three cases:

- no tax ($\tau = 0$)
- no liquid assets nor debt ($C_{t-1}^d = C_{t-1}^l = 0$)
- no interest ($I_t^{\mathcal{D}} = 0$).[17]

A special case of particular importance occurs whenever no excess cash is retained in the business (i.e., $C_t^l = 0$ for all t). In this case, the CFO/CCF notion coincides with the CFA notion ($F_t^o = F_t$) so that

$$F_t = F_t^{uo} + \tau I_t^d \iff F_t^{uo} = F_t - \tau I_t^d. \tag{4.41}$$

We summarize below the three different top-down income approaches to find the net income, starting from EBIT.

[17] This occurs whenever $i_t^d C_{t-1}^d = i_t^l C_{t-1}^l$.

(a)	(b)	(c)		(a)	(b)	(c)
+ EBIT	+ EBIT	+ EBIT	or	$+\text{EBIT}_t$	$+\text{EBIT}_t$	$+\text{EBIT}_t$
− Net Interest	− Taxes on EBT	− Taxes on EBIT	in	$-I_t^{\mathcal{D}}$	$-\text{T}_t$	$-\tau\cdot\text{EBIT}_t$
= EBT	= OI	= NOPAT	symbols	$=\text{EBT}_t$	$=I_t^o$	$=I_t^{uo}$
− Taxes on EBT	− Net Interest	− After-tax net interest		$-\text{T}_t$	$-I_t^{\mathcal{D}}$	$-I_t^{\mathcal{D}}(1-\tau)$
= **Net Income**	= **Net Income**	= **Net Income**		$=\boldsymbol{I_t^e}$	$=\boldsymbol{I_t^e}$	$=\boldsymbol{I_t^e}$

(4.42)

Remark 4.2 (***Financing policy***) The expression “financing policy” may be ambiguous. It sometimes refers to the debt policy (choice of the amount and the kind of borrowings which will sustain the project). Some other times, it refers to both debt policy and payout policy (decisions on cash flow distribution and withdrawals). However, even when it refers to debt policy, the payout policy is involved. More precisely, given a fixed investment policy (i.e. given a triplet $(\boldsymbol{C}^o, \boldsymbol{I}^o, \boldsymbol{F}^o)$), the choice of increasing/reducing debt is a choice of increasing/reducing equity, which is in turn a choice of increasing/reducing liquid assets. For example, for a given investment outlay, the firm must decide how much of it must be raised from debtholders, how much of it must be raised from equityholders, and how much of it must be raised from cash withdrawals (internal financing). The concepts of financing policy, borrowing policy, distribution policy, payout policy are intertwined and reciprocally dependent (see Sect. 4.10). Even the investment policy may be correlated with the financing policy. For example, debt covenants tend to set restrictions on the investment policy (to avoid excessively risky investment activities), on dividend payments and use of liquid assets (to avoid erosion of the firm’s cash flow to the detriment of the lenders), and on subsequent debt policy (to avoid claim dilution).[18] Besides, a given investment policy might be impossible to pursue without debt; in other words, the very act of raising debt may be made in order to undertake a desired investment policy which could not be possible with internal financing or equity capital (this is especially true for small-sized firms). In all these cases, raising new debt affects the investment decisions which in turn affect EBITDAs and FCFs as well.

In conclusion, decisions on investments, financing, and distributions are reciprocally dependent and subtly intertwined with the estimates of the accounting magnitudes. ▲

Example 4.5 ♦

C^o	C^d
	C^e

(*Opening project* – *Free cash flow*) Consider the Opening project again. In particular, let us go back to Example 4.3, where a nonzero tax rate is assumed: $\tau = 0.3$. Table 4.17 depicts the case where the project is levered (with $i^d = 3\%$), using the same data as in Table 4.8, with the only addition of the (nonzero) tax row. As noted above, the NOPAT is nothing but the unlevered OI or, equivalently, the unlevered net income. Therefore, it is already portrayed in Table 4.11 as net income

[18]See Smith and Warner (1979), DeAngelo et al. (2002), Bathala et al. (2006).

Table 4.17 Project opening levered: pro forma financial statements and cash flows (with taxes, $\tau = 30\%$)

Time	**0**	**1**	**2**	**3**	**4**	**5**
BALANCE SHEET						
Net Fixed Assets	350,000	280,000	210,000	140,000	70,000	0
Net Operating WC	50,000	40,000	30,000	20,000	10,000	0
INVESTMENTS	400,000	320,000	240,000	160,000	80,000	0
Debt	60,000	48,000	36,000	24,000	12,000	0
Equity	340,000	272,000	204,000	136,000	68,000	0
FINANCINGS	400,000	320,000	240,000	160,000	80,000	0
INCOME STATEMENT						
Sales		375,000	382,500	390,150	397,953	405,912
– Variable costs		–150,000	–153,000	–156,060	–159,181	–162,365
– Fixed costs		–60,000	–60,000	–60,000	–60,000	–60,000
EBITDA		165,000	169,500	174,090	178,772	183,547
– Depreciation expense		–70,000	–70,000	–70,000	–70,000	–70,000
EBIT		95,000	99,500	104,090	108,772	113,547
– Interest expenses		–1,800	–1,440	–1,080	–720	–360
EBT		93,200	98,060	103,010	108,052	113,187
– Taxes		–27,960	–29,418	–30,903	–32,416	–33,956
Net Income		65,240	68,642	72,107	75,636	79,231
CASH FLOW						
EBIT–T		67,040	70,082	73,187	76,356	79,591
$-\Delta$ NOWC	–50,000	10,000	10,000	10,000	10,000	10,000
$-\Delta$ NFA	–350,000	70,000	70,000	70,000	70,000	70,000
CFA=CFO	–400,000	147,040	150,082	153,187	156,356	159,591
Net Income		65,240	68,642	72,107	75,636	79,231
$-\Delta C^e$	–340,000	68,000	68,000	68,000	68,000	68,000
CFE	–340,000	133,240	136,642	140,107	143,636	147,231
Interest expense		1,800	1,440	1,080	720	360
$-\Delta C^d$	–60,000	12,000	12,000	12,000	12,000	12,000
CFD		13,800	13,440	13,080	12,720	12,360
NOPAT(see net income in Table 4.11)		66,500	69,650	72,863	76,140	79,483
$-\Delta C_t^o$	–400,000	80,000	80,000	80,000	80,000	80,000
FCF (see CFA in Table 4.11)	–400,000	146,500	149,650	152,863	156,140	159,483

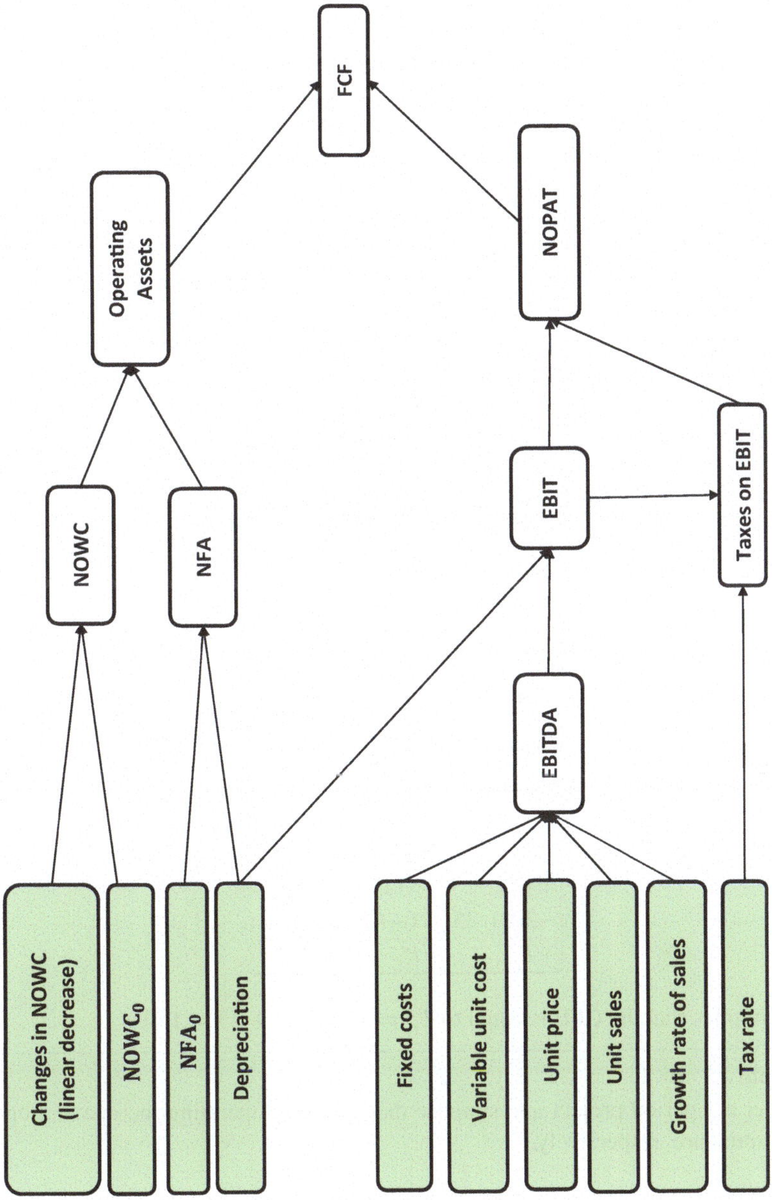

Fig. 4.4 Influence diagram indicating the functional relationships of Opening project (levered, with taxes). The project is levered, but the FCF is calculated as the CFE that would be distributed to investors if the project were not levered

(or OI). The FCFs are also already portrayed in the same table as (unlevered) CFA and CFE. The difference between NOPAT and OI and between FCF and CFO is measured by the interest tax shield.

The FCF line represents the cash flow that would be distributed to investors (equityholders) if the Opening project were unlevered. It is then that part of the (levered) CFO that is unaffected by the net interest expenses.

Note also that the CFLs are zero for all t, because no excess cash is involved: $C_t^l = 0$ for all t. Therefore, the CFA and the CFO concide: $F_t = F_t^o$.

Figure 4.4 describes the functional relations of this levered case, from project input data to FCF.

Example 4.6 ♦

Suppose that the FCF at a given date t is equal to \$145, the after-tax net interest is \$30 and the change in net debt is \$25. Suppose the analyst wants to compute the CFE. Starting from (4.39), one can substitute the figures and write

FCF

$+145$	$+\text{NOPAT}_t$	$-\Delta C_t^o$
$+F_t^{\mathcal{D}} - \tau I_t^{\mathcal{D}}$	$+30$	-25
$+F_t^e$	$+I_t^e$	$-\Delta C_t^e$

Swapping the elements of the second row,

FCF

$+145$	$+\text{NOPAT}_t$	$-\Delta C_t^o$
$+30 - 25$	$+(F_t^{\mathcal{D}} - \tau I_t^{\mathcal{D}})$	
$+F_t^e$	$+I_t^e$	$-\Delta C_t^e$

Shifting the second row above the horizontal bar, we isolate the CFE:

CFE

$+145$	$+\text{NOPAT}_t$	$-\Delta C_t^o$
$-30 + 25$	$(-F_t^{\mathcal{D}} + \tau I_t^{\mathcal{D}})$	
$+F_t^e$	$+I_t^e$	$-\Delta C_t^e$

which implies that the CFE is equal to $F_t^e = 145 - 30 + 25 = 140$.

Example 4.7 ♦

Consider a 4-period project and suppose that (pre-tax) operating income and operating capital are, respectively,

$$\textbf{EBIT} = (0, 100, 200, -100, 300) \qquad \boldsymbol{C^o} = (800, 600, 300, 200, 0).$$

Assume a tax rate equal to $\tau = 25\%$. Using (4.30) and remembering that $\text{NOPAT}_t = \text{EBIT}_t(1 - \tau)$, the FCF may be computed as

$$F_1^{uo} = 100(1 - 0.25) - (600 - 800) = 275$$
$$F_2^{uo} = 200(1 - 0.25) - (300 - 600) = 450$$
$$F_3^{uo} = -100(1 - 0.25) - (200 - 300) = 25$$
$$F_4^{uo} = 300(1 - 0.25) - (0 - 200) = 125.$$

Example 4.8 ♦
Let $\text{EBITDA}_t = 100$ and $\tau = 20\%$ and assume $I_t^{\mathcal{D}} = 20$, $\Delta\text{NFA}_t = -40$, $\Delta\text{NOWC}_t = -90$, $F_t^{capex} = 0$. The data imply $\text{Dep}_t = 40$ and $\text{EBIT}_t = \text{EBITDA}_t - \text{Dep}_t = 60$. Hence, the FCF is $F_t^{uo} = 60(1 - 0.2) + 40 + 90 = 178$. Knowledge of the net interest expenses is pleonastic for computing the FCF. However, adding the tax shield generated by the net interest, one may compute the CFO: $F_t^o = 178 + 0.2 \cdot 20 = 182$. This means that the operating cash flow is greater than the FCF by \$4. We are not given any information about liquid assets; however, we know that, if the balance of liquid assets were zero, then the CFO would equal CFA: $F_t^o = F_t = 182$.

Example 4.9 ♦

C^o	C^d
	C^e

(BHP Billiton – Free cash flow) Consider BHP Billiton's project again and consider the case where the capital is entirely invested in NFA (no NOWC employed). See, in particular, Tables 4.13 and 4.14. In the first 19 years, the FCFs are constant and equal to \$249.8 because NOPAT and depreciation are constant and no working capital nor capital expenditures exist after $t = 0$:

$$F_t^{uo} = \overbrace{(345 - 27.5)(1 - 0.3)}^{\text{NOPAT}_t} + 27.5 = 249.8 \qquad t \in \mathbb{N}_{19}^1.$$

In the last year, the NOPAT changes (owing to the different EBITDA),[19] so FCF changes as well:

$$F_{20}^{uo} = \overbrace{(245 - 27.5)(1 - 0.3)}^{\text{NOPAT}_{20}} + 27.5 = 179.8.$$

One can also compute the FCFs directly from the CFAs (which equate the CFOs) by subtracting the interest tax shield. In the first 10 years,

$$F_t^{uo} = F_t - \tau \cdot I_t^d = 254.6 - 0.3 \cdot 16 = 249.8 \qquad t \in \mathbb{N}_{10}^1.$$

In the remaining years, the interest tax shield is zero because the debt has been repaid.

---

[19]The last EBITDA is smaller than the previous EBITDAs by \$100, owing to the disposal costs.

Table 4.18 FCFs, CFAs, and CFEs of BHP Billiton with or without NOWC (100% payout)

Time	with NOWC			without NOWC		
	FCF	**CFA**	**CFE**	**FCF**	**CFA**	**CFE**
0	−600.0	−600.0	−400.0	−550.0	−550.0	−350.0
1	257.8	262.6	246.6	249.8	254.6	238.6
2	249.8	254.6	238.6	249.8	254.6	238.6
3	249.8	254.6	238.6	249.8	254.6	238.6
4	249.8	254.6	238.6	249.8	254.6	238.6
5	249.8	254.6	238.6	249.8	254.6	238.6
6	249.8	254.6	238.6	249.8	254.6	238.6
7	249.8	254.6	238.6	249.8	254.6	238.6
8	249.8	254.6	238.6	249.8	254.6	238.6
9	249.8	254.6	238.6	249.8	254.6	238.6
10	249.8	254.6	38.6	249.8	254.6	38.6
11	249.8	249.8	249.8	249.8	249.8	249.8
12	249.8	249.8	249.8	249.8	249.8	249.8
13	249.8	249.8	249.8	249.8	249.8	249.8
14	249.8	249.8	249.8	249.8	249.8	249.8
15	249.8	249.8	249.8	249.8	249.8	249.8
16	249.8	249.8	249.8	249.8	249.8	249.8
17	249.8	249.8	249.8	249.8	249.8	249.8
18	249.8	249.8	249.8	249.8	249.8	249.8
19	249.8	249.8	249.8	249.8	249.8	249.8
20	221.8	221.8	221.8	179.8	179.8	179.8

Therefore, in the last 10 years, FCFs and CFAs are equal:

$$F_t^{uo} = F_t - \overbrace{\tau \cdot I_t^d}^{=0} = F_t = 249.8 \qquad t \in \mathbb{N}_{19}^{11}$$

$$F_{20}^{uo} = F_{20} - \overbrace{\tau \cdot I_{20}^d}^{=0} = F_{20} = 179.8.$$

The presence of a loan does not impact FCFs: They are only affected by estimates on revenues, costs, tax rate, and by the depreciation schedule. The case where NOWC is present is analogous. (See Table 4.18)

EBITDA and Dep_t do not change, so the only difference lies in F_1^{uo} and F_{20}^{uo}. Specifically, from time $t = 0$ to time $t = 1$, the NOWC decreases by \$8, which makes F_1^{uo} increase by \$8 with respect to the no-NOWC case (\$257.8 as opposed to \$249.8). From time $t = 19$ to time $t = 20$, the NOWC decreases by \$42, which makes the FCF increase by \$42 with respect to the no-NOWC case (\$221.8 as opposed to \$179.8).

4.8 Unlevered Operating ROI

We can express (4.30) in relative terms by building the notion of *unlevered* (operating) ROC or ROI, which expresses the operating ROC that the project would generate if it were unlevered:

$$i_t^{uo} = \frac{\text{NOPAT}_t}{\text{NOWC}_{t-1} + \text{NFA}_{t-1}} = \frac{I_t^{uo}}{C_{t-1}^o}. \tag{4.43}$$

It is then an operating ROC, which abstracts from the effect of debt and liquid assets, as opposed to i_t^o which is affected by debt and liquid assets via the interest tax shield:

$$i_t^o = \frac{\text{EBIT}_t - \text{T}_t}{\text{NOWC}_{t-1} + \text{NFA}_{t-1}} = \frac{\text{NOPAT}_t + \tau I_t^{\mathcal{D}}}{\text{NOWC}_{t-1} + \text{NFA}_{t-1}}.$$

The relation with the ROE and RO$\mathcal{D}$ is easily established by reframing the CFO Matrix in terms of income rates:

$$\begin{array}{c}\textbf{CFO}\\ \begin{array}{|c|cc|}\hline +F_t^o & +i_t^o C_{t-1}^o & -\Delta C_t^o \\ \hline +F_t^{\mathcal{D}} & +i_t^{\mathcal{D}} C_{t-1}^{\mathcal{D}} & -\Delta C_t^{\mathcal{D}} \\ +F_t^e & +i_t^e C_{t-1}^e & -\Delta C_t^e \\ \hline \end{array}\end{array} \tag{4.44}$$

whence, considering that the upper part and the lower part of column 2 coincide:

$$i_t^o = \frac{i_t^e C_{t-1}^e + i_t^{\mathcal{D}} C_{t-1}^{\mathcal{D}}}{C_{t-1}^e + C_{t-1}^{\mathcal{D}}} \tag{4.45}$$

which is nothing but (2.27). Likewise, from the FCF Matrix,

$$\begin{array}{c}\textbf{FCF}\\ \begin{array}{|l|lr|}\hline +F_t^{uo} & +i_t^{uo} C_{t-1}^o & -\Delta C_t^o \\ \hline +F_t^{\mathcal{D}} - \tau I_t^{\mathcal{D}} & +i_t^{\mathcal{D}}(1-\tau) C_{t-1}^{\mathcal{D}} & -\Delta C_t^{\mathcal{D}} \\ +F_t^e & +i_t^e C_{t-1}^e & -\Delta C_t^e \\ \hline \end{array}\end{array} \tag{4.46}$$

one gets

$$i_t^{uo} = \frac{i_t^e \cdot C_{t-1}^e + i_t^{\mathcal{D}}(1-\tau) \cdot C_{t-1}^{\mathcal{D}}}{C_{t-1}^e + C_{t-1}^{\mathcal{D}}}. \tag{4.47}$$

By reverse-engineering (4.45), (4.47), and from the well-known relation

$$i_t = \frac{i_t^e C_{t-1}^e + i_t^d C_{t-1}^d}{C_{t-1}^e + C_{t-1}^d},$$

one gets the ROE as a function of debt/equity ratios:

$$
\begin{aligned}
i^e_t &= i^o_t + \frac{C^{\mathcal{D}}_{t-1}}{C^e_{t-1}} \cdot (i^o_t - i^{\mathcal{D}}_t) \\
i^e_t &= i^{uo}_t + \frac{C^{\mathcal{D}}_{t-1}}{C^e_{t-1}} \cdot \left(i^{uo}_t - i^{\mathcal{D}}_t(1-\tau)\right) \\
i^e_t &= i_t + \frac{C^d_{t-1}}{C^e_{t-1}} \cdot (i_t - i^d_t).
\end{aligned} \tag{4.48}
$$

It is observed that two different ratios C^d_{t-1}/C^e_{t-1} and $C^{\mathcal{D}}_{t-1}/C^e_{t-1}$ are involved, depending on whether the financial structure or the capital structure is considered.

Also, from (4.29), $I^o_t - I^{uo}_t = \tau I^{\mathcal{D}}_t$, which means

$$
C^o_{t-1}\left(i^o_t - i^{uo}_t\right) = \tau I^{\mathcal{D}}_t \tag{4.49}
$$

whence

$$
i^o_t = i^{uo}_t + \frac{\tau I^{\mathcal{D}}_t}{C^o_{t-1}} \tag{4.50}
$$

(the same relation may be found directly from (4.45)–(4.47)).

Finally, we may summarize the differences among the operating ROI, the unlevered ROI, and the asset ROI, starting from the EBIT estimates:

$$
\begin{aligned}
i^o_t &= \frac{I^o_t}{C^o_{t-1}} = \frac{\text{EBIT}_t - \text{T}_t}{C^o_{t-1}} \\
i^{uo}_t &= \frac{I^{uo}_t}{C^o_{t-1}} = \frac{\text{EBIT}_t(1-\tau)}{C^o_{t-1}} \\
i_t &= \frac{I_t}{C_{t-1}} = \frac{\text{EBIT}_t - \text{T}_t + I^l_t}{C_{t-1}}
\end{aligned} \tag{4.51}
$$

4.9 The Asset FCF

The FCF represents the CFO that would be generated if the project were unlevered. The notion of *asset* FCF refers to the CFA that would be generated by the assets if the project were equity financed (see Definition 2.2. See also Tham and Vélez-Pareja 2004 on this notion). Therefore, asset FCF is a financially unlevered CFA ($C^d_t = 0$ for all t), in the same sense FCF is an operationally unlevered CFA ($C^d_t = 0$ and C^l_t=0 for all t). To better appreciate the difference between the (operating) FCF analyzed in the previous section and the asset FCF, one may avail oneself of the FCF Matrix, which we report below:

FCF

$+F_t^{\bullet}$	$+\text{EBIT}_t$	$-\Delta C_t^o$
$-\tau\cdot\text{EBIT}_t$	$-\tau\cdot\text{EBIT}_t$	
$+F_t^{\mathcal{D}}-\tau I_t^{\mathcal{D}}$	$+I_t^{\mathcal{D}}(1-\tau)$	$-\Delta C_t^{\mathcal{D}}$
$+F_t^e$	$+I_t^e$	$-\Delta C_t^e$

Let us disaggregate the net debt and consolidate the first two rows:

FCF

$+F_t^{uo}$	$+\text{EBIT}_t(1-\tau)$	$-\Delta C_t^o$
$+F_t^d-\tau I_t^d$	$+I_t^d-\tau I_t^d$	$-\Delta C_t^d$
$-F_t^l+\tau I_t^l$	$-I_t^l+\tau I_t^l$	$+\Delta C_t^l$
$+F_t^e$	$+I_t^e$	$-\Delta C_t^e$

Then, let us move the liquid assets upward above the line. The result is the asset FCF:

asset FCF

$+F_t^{uo}$	$+\text{EBIT}_t(1-\tau)$	$-\Delta C_t^o$
$+F_t^l-\tau I_t^l$	$+I_t^l-\tau I_t^l$	$-\Delta C_t^l$
$+F_t^d-\tau I_t^d$	$+I_t^d-\tau I_t^d$	$-\Delta C_t^d$
$+F_t^e$	$+I_t^e$	$-\Delta C_t^e$

or, equivalently,

asset FCF (4.52)

$+F_t^{uo}$	$+\text{NOPAT}_t$	$-\Delta C_t^o$
$+F_t^l-\tau I_t^l$	$+I_t^l(1-\tau)$	$-\Delta C_t^l$
$+F_t^d-\tau I_t^d$	$+I_t^d(1-\tau)$	$-\Delta C_t^d$
$+F_t^e$	$+I_t^e$	$-\Delta C_t^e$

Let F_t^u denote this asset FCF. As usual, we have four perspectives:

$$\begin{aligned}F_t^u &= \text{NOPAT}_t + I_t^l(1-\tau) - \Delta C_t\\ &= F_t^{uo} + F_t^l - \tau I_t^l\\ &= F_t^d - \tau I_t^d + F_t^e\\ &= I_t^d(1-\tau) + I_t^e - \Delta C_t^d - \Delta C_t^e.\end{aligned} \tag{4.53}$$

The FCF is to CFO what asset FCF is to CFA. This relation can be easily grasped by placing the corresponding Matrices side by side:

$$
\begin{array}{c|c|c}
\multicolumn{3}{c}{\textbf{CFO}} \\
\hline
+F_t^o & +\text{EBIT}_t - \text{T}_t & -\Delta C_t^o \\
\hline
+F_t^d & +I_t^d & -\Delta C_t^d \\
-F_t^l & -I_t^l & +\Delta C_t^l \\
+F_t^e & +I_t^e & -\Delta C_t^e \\
\hline
\end{array}
\qquad
\begin{array}{c|c|c}
\multicolumn{3}{c}{\textbf{(operating) FCF}} \\
\hline
+F_t^{uo} & +\text{EBIT}_t(1-\tau) & -\Delta C_t^o \\
\hline
+F_t^d - \tau I_t^d & +I_t^d - \tau I_t^d & -\Delta C_t^d \\
-F_t^l + \tau I_t^l & -I_t^l + \tau I_t^l & +\Delta C_t^l \\
+F_t^e & +I_t^e & -\Delta C_t^e \\
\hline
\end{array}
\tag{4.54}
$$

$$
\begin{array}{c|c|c}
\multicolumn{3}{c}{\textbf{CFA}} \\
\hline
+F_t^o & +\text{EBIT}_t - \text{T}_t & -\Delta C_t^o \\
+F_t^l & +I_t^l & -\Delta C_t^l \\
\hline
+F_t^d & +I_t^d & -\Delta C_t^d \\
+F_t^e & +I_t^e & -\Delta C_t^e \\
\hline
\end{array}
\qquad
\begin{array}{c|c|c}
\multicolumn{3}{c}{\textbf{asset FCF}} \\
\hline
+F_t^{uo} & +\text{EBIT}_t(1-\tau) & -\Delta C_t^o \\
+F_t^l - \tau I_t^l & +I_t^l(1-\tau) & -\Delta C_t^l \\
\hline
+F_t^d - \tau I_t^d & +I_t^d(1-\tau) & -\Delta C_t^d \\
+F_t^e & +I_t^e & -\Delta C_t^e \\
\hline
\end{array}
\tag{4.55}
$$

Contrasting the lower left submatrices, it is clear that

$$
\begin{aligned}
F_t^o - F_t^{uo} &= \tau I_t^{\mathcal{D}} \text{ or } F_t^o = F_t^{uo} + \tau I_t^{\mathcal{D}} \\
F_t - F_t^u &= \tau I_t^d \text{ or } F_t = F_t^u + \tau I_t^d.
\end{aligned}
\tag{4.56}
$$

The asset FCF may then be interpreted as that part of CFA which is unaffected by the amount of debt borrowed.

The difference between the asset FCF (F_t^u) and the *operating* FCF (F_t^{uo}) is the *non-operating* FCF, which is the after-tax CFL (i.e., after-tax interest income minus the change in liquid assets. See upper right submatrix of the asset FCF Matrix). Therefore, one may decompose the asset FCF into two components, just in the same way the CFA is decomposed into an operating component (i.e., the CFO/CCF) and a non-operating component (i.e., the CFL):

$$
\begin{array}{|l|}
\overbrace{F_t^u}^{\text{Asset FCF}} = \overbrace{F_t^{uo}}^{\text{Operating FCF}} + \overbrace{F_t^{ul}}^{\text{Non-operating FCF}} \\
+ \overbrace{\tau I_t^d}^{\text{Gross tax shield}} = + \overbrace{\tau I_t^{\mathcal{D}}}^{\text{Tax shield}} + \overbrace{\tau I_t^l}^{\text{Taxes on interest income}} \\
\hline
\overbrace{F_t}^{\text{CFA}} = \overbrace{F_t^o}^{\text{Operating CFA}} + \overbrace{F_t^l}^{\text{Non-operating CFA}}
\end{array}
\tag{4.57}
$$

Since $F_t^{ul} = F_t^l - \tau I_t^l$, the non-operating FCF may be derived as after-tax interest income net of change in liquid assets:

$$
F_t^{ul} = \overbrace{I_t^l(1-\tau)}^{\text{after-tax interest income}} - \overbrace{\Delta C_t^l}^{\text{change in liquid assets}}. = F_t^l - \tau I_t^l.
\tag{4.58}
$$

Note that the (operating) FCF is unaffected by the financing policy (i.e., unaffected by both debt and liquid assets), whereas the asset FCF is unaffected by debt, but is affected by the non-operating investments.[20]

In an investment perspective, the income associated with the asset FCF is the sum of the NOPAT and the after-tax interest. In a financing perspective, it is the sum of the after-tax interest expenses and the net income (see second column of the asset FCF Matrix). We use the symbol I_t^u to denote this unlevered income:

$$I_t^{uo} + I_t^l(1-\tau) = I_t^u = I_t^d(1-\tau) + I_t^e. \tag{4.59}$$

The income (cash flow) I_t^u (F_t^u) is the income (cash flow) that would be generated if the project were financially unlevered (i.e., equity financed); the income (cash flow) I_t^{uo} (F_t^{uo}) is the income (cash flow) that would be generated if the project were operationally unlevered (i.e., with no debt nor liquid assets). (See Definition of unlevered project on Sect. 2.2).

The capital evolves according to the following law of motion:

$$C_t = C_{t-1} + I_t^u - F_t^u$$

or, in relative terms, $C_t = C_{t-1}(1 + i_t^u) - F_t^u$ where

$$i_t^u = \frac{I_t^u}{C_{t-1}} = \frac{I_t^{uo} + I_t^l(1-\tau)}{C_{t-1}} = \frac{(\text{EBIT}_t + I_t^l)(1-\tau)}{C_{t-1}}$$

so that

$$C_0 = \sum_{t=1}^{n} \frac{F_t^u}{(1+i_1^u)\cdot(1+i_2^u)\cdot\ldots\cdot(1+i_t^u)}.$$

We call the rate i_t^u the *unlevered asset ROI (ROC)* or *financially unlevered ROI (ROC)*. It is straightforward that this ROI is equal to

$$i_t^u = \frac{i_t^e C_{t-1}^e + i_t^d(1-\tau)C_{t-1}^d}{C_{t-1}^e + C_{t-1}^d} \tag{4.60}$$

as well as

$$i_t^e = i_t^u + \frac{C_{t-1}^d}{C_{t-1}^e}(i_t^u - i_t^d(1-\tau)) \tag{4.61}$$

or, equivalently,

$$i_t^u = i_t - \frac{\tau I_t^d}{C_{t-1}}. \tag{4.62}$$

[20] In academia and in practice, the overwhelming convention is to use the term "FCF" (with no other qualification) to refer to the operating component of the asset FCF. We comply with this convention.

To recap,
−the asset FCF is the CFA that would be generated by the operations if the project were financially unlevered (i.e., equity financed: $C_t^d = 0$);

−the operating FCF is the CFO that would be generated if the project were (operationally) unlevered (i.e., $C_t^d = C_t^l = 0$);

−the non-operating FCF is the CFL net of taxes on interest income.

If the operating assets exhaust the assets, $C_t^o = C_t$ for all t, then the operating FCF and asset FCF are the same notion: $F_t^{uo} = F_t^u$.

Remark 4.3 (*The laws of motion revisited*) The Matrices in (4.54) provide us with a reframing of the law of motion for operating assets. More explicitly, disaggregating the change-in-capital column and swapping the columns as appropriate, one finds

Operating assets		**OI**	**CFO**
$+C_t^o$	$+C_{t-1}^o$	$+\text{EBIT}_t - \text{T}_t$	$-F_t^o$
$-C_t^l$	$-C_{t-1}^l$	$-I_t^l$	$+F_t^l$
$+C_t^d$	$+C_{t-1}^d$	$+I_t^d$	$-F_t^d$
$+C_t^e$	$+C_{t-1}^e$	$+I_t^e$	$-F_t^e$

Operating assets		**NOPAT**	**FCF**
$+C_t^o$	$+C_{t-1}^o$	$+\text{EBIT}_t(1-\tau)$	$-(F_t^o - \tau I_t^{\mathcal{D}})$
$-C_t^l$	$-C_{t-1}^l$	$-I_t^l(1-\tau)$	$+F_t^l - \tau I_t^l$
$+C_t^d$	$+C_{t-1}^d$	$+I_t^d(1-\tau)$	$-(F_t^d - \tau I_t^d)$
$+C_t^e$	$+C_{t-1}^e$	$+I_t^e$	$-F_t^e$

(4.63)

Reading the left Matrix by row, one finds the laws of motion for each class of capital, as we have presented them in Chap. 1 and used thus far. On the right-hand Matrix, one finds the laws of motion for the same classes using an operationally unlevered perspective. Alternatively, one can merge debt and liquid assets:

Operating assets		**OI**	**CFO**
$+C_t^o$	$+C_{t-1}^o$	$+\text{EBIT}_t - \text{T}_t$	$-F_t^o$
$+C_t^{\mathcal{D}}$	$+C_{t-1}^{\mathcal{D}}$	$+I_t^{\mathcal{D}}$	$-F_t^{\mathcal{D}}$
$+C_t^e$	$+C_{t-1}^e$	$+I_t^e$	$-F_t^e$

Operating assets		**NOPAT**	**FCF**
$+C_t^o$	$+C_{t-1}^o$	$+\text{EBIT}_t(1-\tau)$	$-(F_t^o - \tau I_t^{\mathcal{D}})$
$+C_t^{\mathcal{D}}$	$+C_{t-1}^{\mathcal{D}}$	$+I_t^{\mathcal{D}}(1-\tau)$	$-(F_t^{\mathcal{D}} - \tau I_t^{\mathcal{D}})$
$+C_t^e$	$+C_{t-1}^e$	$+I_t^e$	$-F_t^e$

(4.64)

In terms of total assets,

Total assets		I_t	**CFA**
$+C_t^o$	$+C_{t-1}^o$	$+\text{EBIT}_t - \text{T}_t$	$-F_t^o$
$+C_t^l$	$+C_{t-1}^l$	$+I_t^l$	$-F_t^l$
$+C_t^d$	$+C_{t-1}^d$	$+I_t^d$	$-F_t^d$
$+C_t^e$	$+C_{t-1}^e$	$+I_t^e$	$-F_t^e$

Total assets		I_t^u	**Asset FCF**
$+C_t^o$	$+C_{t-1}^o$	$+\text{EBIT}_t(1-\tau)$	$-(F_t^o - \tau I_t^{\mathcal{D}})$
$+C_t^l$	$+C_{t-1}^l$	$+I_t^l(1-\tau)$	$-(F_t^l - \tau I_t^l)$
$+C_t^d$	$+C_{t-1}^d$	$+I_t^d(1-\tau)$	$-(F_t^d - \tau I_t^d)$
$+C_t^e$	$+C_{t-1}^e$	$+I_t^e$	$-F_t^e$

(4.65)

The two perspectives are logically equivalent.

Using income rates instead of incomes,

$$\begin{array}{c|cc c}
\multicolumn{2}{c}{\textbf{Operating assets}} & \textbf{OI} & \textbf{CFO} \\
\hline
+C_t^o & +C_{t-1}^o & +i_t^o C_{t-1}^o & -F_t^o \\
\hline
+C_t^{\mathcal{D}} & +C_{t-1}^{\mathcal{D}} & +i_t^{\mathcal{D}} C_{t-1}^{\mathcal{D}} & -F_t^{\mathcal{D}} \\
+C_t^e & +C_{t-1}^e & +i_t^e C_{t-1}^e & -F_t^e \\
\hline
\end{array}
\quad
\begin{array}{c|cc c}
\multicolumn{2}{c}{\textbf{Operating assets}} & \textbf{NOPAT} & \textbf{FCF} \\
\hline
+C_t^o & +C_{t-1}^o & +i_t^{uo} C_{t-1}^o & -(F_t^o - \tau i_t^{\mathcal{D}} C_{t-1}^{\mathcal{D}}) \\
\hline
+C_t^{\mathcal{D}} & +C_{t-1}^{\mathcal{D}} & +i_t^{\mathcal{D}} C_t^{\mathcal{D}}(1-\tau) & -(F_t^{\mathcal{D}} - \tau i_t^{\mathcal{D}} C_{t-1}^{\mathcal{D}}) \\
+C_t^e & +C_{t-1}^e & +i_t^e C_{t-1}^e & -F_t^e \\
\hline
\end{array} \tag{4.66}$$

$$\begin{array}{c|cc c}
\textbf{Total assets} & I_t & & \textbf{CFA} \\
\hline
+C_t^o & +C_{t-1}^o & +i_t^o C_{t-1}^o & -F_t^o \\
+C_t^l & +C_{t-1}^l & +i_t^l C_{t-1}^l & -F_t^l \\
\hline
+C_t^d & +C_{t-1}^d & +i_t^d C_{t-1}^d & -F_t^d \\
+C_t^e & +C_{t-1}^e & +i_t^e C_{t-1}^e & -F_t^e \\
\hline
\end{array}
\quad
\begin{array}{c|cc c}
\textbf{Total assets} & & I_t^u & \textbf{Asset FCF} \\
\hline
+C_t^o & +C_{t-1}^o & +i_t^{uo} C_{t-1}^o & -(F_t^o - \tau i_t^{\mathcal{D}} C_{t-1}^{\mathcal{D}}) \\
+C_t^l & +C_{t-1}^l & +i_t^l (1-\tau) C_{t-1}^l & -(F_t^l - \tau i_t^l C_{t-1}^l) \\
\hline
+C_t^d & +C_{t-1}^d & +i_t^d (1-\tau) C_{t-1}^d & -(F_t^d - \tau i_t^d C_{t-1}^d) \\
+C_t^e & +C_{t-1}^e & +i_t^e C_{t-1}^e & -F_t^e \\
\hline
\end{array} \tag{4.67}$$

▲

F^u ⟶ Asset FCF	= financially unlevered cash flow	= CFO net of taxes on interest expenses
F^{uo} ⟶ Operating FCF	= operationally unlevered cash flow	= CFA net of taxes on *net* interest expenses
F^{ul} ⟶ Non-operating FCF		= CFL net of taxes on interest income

4.10 Potential Dividends

The operating assets generate the after-tax operating cash flow (CFO). This cash flow may be interpreted as the (after-tax) cash flow generated in a given period that is available for distribution to the capital providers (which is the reason why the CFO is often called capital cash flow, CCF). Subtracting the cash flow paid to debtholders from it one gets the *cash flow available for distribution* to shareholders, also called ***potential dividends*** (PD) or ***free cash flow to equity*** (FCFE):

$$F_t^{pd} = F_t^o - F_t^d. \tag{4.68}$$

Hence, one deducts the CFE, that is, the cash flow that is actually distributed to shareholders; (assuming the result is positive) one gets the excess retained cash (or undistributed cash):

$$\text{RC}_t = F_t^{pd} - F_t^e. \tag{4.69}$$

One may represent the situation with the Matrix. Starting from the CFO Matrix in (4.54), let us move the debt row to the upper part. The result is the (fourfold) potential dividends (left Matrix in Eq. (4.70)) Additionally moving the equity row to the upper region, one gets (fourfold) retained cash (right Matrix in Eq. (4.70)):

$$
\begin{array}{c}
\textbf{PD}\\
\begin{array}{|c|lr|}
\hline
+F_t^o & +\text{EBIT}_t - \text{T}_t & -\Delta C_t^o \\
-F_t^d & -I_t^d & +\Delta C_t^d \\
\hline
-F_t^l & -I_t^l & +\Delta C_t^l \\
+F_t^e & +I_t^e & -\Delta C_t^e \\
\hline
\end{array}
\end{array}
\quad
\begin{array}{c}
\textbf{RC}\\
\begin{array}{|c|lr|}
\hline
+F_t^o & +\text{EBIT}_t - \text{T}_t & -\Delta C_t^o \\
-F_t^d & -I_t^d & +\Delta C_t^d \\
-F_t^e & -I_t^e & +\Delta C_t^e \\
\hline
-F_t^l & -I_t^l & +\Delta C_t^l \\
\hline
\end{array}
\end{array}
\tag{4.70}
$$

It is then clear that the retained cash is nothing but the CFL changed in sign:

$$\text{RC}_t = -F_t^l. \tag{4.71}$$

In other words, assuming $\text{RC}_t > 0$ (i.e., $F_t^l < 0$), the undistributed cash is deposited in the liquid assets. If, instead, $\text{RC}_t < 0$ (i.e., $F_t^l > 0$), the CFO is not sufficient to cover the payments to claimholders: $F_t^o - F_t^e - F_t^d < 0$ (see northwest partition of RC); there is no retained cash but, rather, a net withdrawal from the liquid assets.[21]

If a full payout policy is employed, the liquid assets are zero, and the CFL is zero as well. If excess cash is retained in the firm at some date, the cash will be, sooner or later, distributed to investors. This means that, sooner or later, a negative CFL is followed by a positive CFL.[22]

Shifting the debt row down to the south part of (4.70), one gets

$$
\begin{array}{c}
\text{Splitting up CFO}\\
\begin{array}{|c|lr|}
\hline
+F_t^o & +\text{EBIT}_t - \text{T}_t & -\Delta C_t^o \\
\hline
+F_t^d & +I_t^d & -\Delta C_t^d \\
+F_t^e & +I_t^e & -\Delta C_t^e \\
-F_t^l & -I_t^l & +\Delta C_t^l \\
\hline
\end{array}
\end{array}
\quad
\begin{array}{c}
\text{Splitting up PD}\\
\begin{array}{|c|lr|}
\hline
+F_t^o & +\text{EBIT}_t - \text{T}_t & -\Delta C_t^o \\
-F_t^d & -I_t^d & +\Delta C_t^d \\
\hline
+F_t^e & +I_t^e & -\Delta C_t^e \\
-F_t^l & -I_t^l & +\Delta C_t^l \\
\hline
\end{array}
\end{array}
\tag{4.72}
$$

which tells that CFO may be decomposed into three components:

- cash flow to debtholders
- cash flow to equityholders
- cash flow to liquid assets (undistributed cash)

while the PD is split up into two components:

- cash flow to equityholders
- cash flow to liquid assets (undistributed cash)

(see also Fig. 4.5).

Keeping an eye on the upper right submatrix of the PD Matrix (Eq. (4.70)), one gets the PD from the OI, the change in operating capital, and the cash flow to debt:

[21] Given a fixed CFO and a fixed CFD, a negative undistributed cash represents an excess distribution to equityholders over and above the cash available for distributions: Cash flows are withdrawn from liquid assets.

[22] For example, if $F_t^l < 0$ for all $0 \leq t < n$, then $F_n^l \geq 0$ (if $i_t^l = -1$ for all t, then $F_n = 0$): When the project is liquidated at time n, the EOP balance of liquid assets must be entirely distributed: $E_n^l = C_{n-1}^l + I_n^l = F_n^l$.

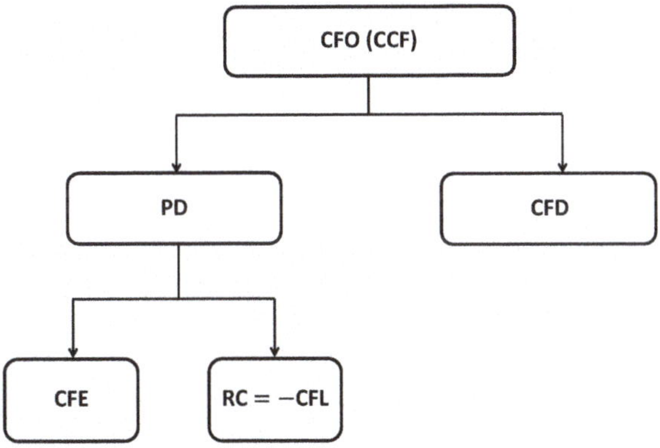

Fig. 4.5 Splitting up the CFO

$$F_t^{pd} = \overbrace{\text{EBIT}_t - \text{T}_t}^{I_t^o} - \overbrace{(\Delta\text{NFA}_t + \Delta\text{NOWC}_t)}^{\Delta C_t^o} - \overbrace{(I_t^d - \Delta C_t^d)}^{F_t^d}. \tag{4.73}$$

One may reframe it as

$$F_t^{pd} = I_t^o - I_t^d - (\Delta C_t^o - \Delta C_t^d). \tag{4.74}$$

Let $C^{pd} = C_t^o - C_t^d$ and $I_t^{pd} = I_t^o - I_t^d$. Then,

$$C_t^{pd} = C_{t-1}^{pd} + I_t^{pd} - F_t^{pd} \tag{4.75}$$

or

$$C_t^{pd} = C_{t-1}^{pd}(1 + i_t^{pd}) - F_t^{pd} \tag{4.76}$$

where $i_t^{pd} = (i_t^o C_{t-1}^o - i_t^d C_{t-1}^d)/(C_{t-1}^o - C_{t-1}^d)$. The above equation represents the law of motion for what one might call the *potential capital* (this is the amount of equity capital that would be owed to shareholders if liquid assets were zero). In Matrix format,

Potential (equity) capital

$$\begin{array}{|c|ccc|}
\hline
+C_t^{pd} & +C_{t-1}^{pd} & +I_t^{pd} & -F_t^{pd} \\
\hline
+C_t^e & +C_{t-1}^e & +I_t^e & -F_t^e \\
-C_t^l & -C_{t-1}^l & -I_t^l & -\text{RC}_t \\
\hline
\end{array} \tag{4.77}$$

Obviously, subtracting the potential capital from the actual equity capital, one finds the balance of liquid assets (it suffices to swap PD row and liquid-asset row):

Liquid assets

$$\begin{array}{|c|ccc|}
\hline
+C_t^l & +C_{t-1}^l & +I_t^l & +\text{RC}_t \\
\hline
+C_t^e & +C_{t-1}^e & +I_t^e & -F_t^e \\
-C_t^{pd} & -C_{t-1}^{pd} & -I_t^{pd} & +F_t^{pd} \\
\hline
\end{array} \tag{4.78}$$

This Matrix supplies a practical method for computing the BOP balance of liquid asset at time t: Add the interest income and the undistributed cash to the BOP balance at time $t-1$:

$$C_t^l = C_{t-1}^l + I_t^l + \overbrace{(F_t^{pd} - F_t^e)}^{\text{RC}_t} \tag{4.79}$$

The above equation says that, to compute C_t^l, one must first compute the CFE, which is a decision variable: The company may select to distribute all available cash (full payout policy) or part of the available cash (or even more than the available cash).

We express the amount of CFE that is actually distributed to shareholders as a portion of potential dividends: $F_t^e = \alpha_t \cdot F_t^{pd}$ for $t \in \mathbb{N}_{n-1}^0$, where $\alpha_t \geq 0$ is the ***payout ratio***, expressing the fraction of the available cash that is paid out to shareholders. This is a decision variable.[23]

From (4.69) and (4.71) one gets, for every $t \in \mathbb{N}_{n-1}^0$,

$$-F_t^l = \text{RC}_t = F_t^{pd} - F_t^e = (1 - \alpha_t)F_t^{pd}, \tag{4.80}$$

so the BOP balance of non-operating assets may be written

$$C_t^l = C_{t-1}^l + I_t^l + \overbrace{F_t^{pd}(1 - \alpha_t)}^{-F_t^l}.$$

This means that, out of any dollar available for distribution, only α_t of it will be distributed to shareholders; the remaining amount, $(1 - \alpha_t)$, is deposited in the excess cash account or invested in marketable securities. Since α_t is a *payout* ratio, then $(1 - \alpha_t)$ is a *retention* rate (or an extra-distribution rate if $\alpha_t > 1$); the payout ratio will then be applied to PD to get CFE and the retention rate will be applied to PD to get retained cash, RC$= -$CFL. If the available cash is negative ($F_t^{pd} < 0$), the debt may be serviced by withdrawing cash from the liquid assets ($F_t^l > 0$) and/or by issuing new equity shares ($F_t^e < 0$) and/or by raising additional debt from new sources ($F_t^d < 0$).[24]

In any case, the computation of CFE is a matter of financing policy; the latter will affect the undistributed cash and, therefore, the balance, C_t^l, of the liquid assets. Assuming $F_t^{pd} \geq 0$, there are many ways to select the amount of cash flow distributed to equityholders (legal constraints may also play a role in distribution). For example, a company may distribute nothing if net income is zero or negative and, if positive, no more than the net income. In this case,

[23] At time $t = 0$, if $C_0^e > 0$, α_0 is a *pay-in* ratio, rather than a payout ratio, expressing the share of investment cost (net of debt borrowing) which will be covered by equityholders. The remaining part, $1 - \alpha_0$, will be covered by internal financing. At time $t = n$, every residual capital is distributed to shareholders; therefore, no decision is made on CFE. The last CFE is necessarily equal to $F_n^e = E_n^e = C_{n-1}^{pd} + I_n^{pd} + C_{n-1}^l + I_n^l$.

[24] The available cash may be negative for several reasons (e.g., a negative CFO due to additional capital expenditures or a positive CFO that is smaller than the CFD).

$$F_t^e = \begin{cases} 0, & \text{if } I_t^e \le 0; \\ \beta_t I_t^e, & \text{otherwise} \end{cases} \tag{4.81}$$

with $0 \le \beta_t \le 1$, which implies

$$\alpha_t = \begin{cases} 0, & \text{if } I_t^e \le 0; \\ \beta_t I_t^e / F_t^{pd}, & \text{otherwise.} \end{cases}$$

Alternatively, suppose the firm distributes no more than the smaller between net income and potential dividends:

$$F_t^e = \beta_t \cdot \max\{\min\{I_t^e,\ F_t^{pd}\},\ 0\}; \tag{4.82}$$

then

$$\alpha_t = \beta_t \cdot \max\{\min\{I_t^e / F_t^{pd},\ 1\},\ 0\}.$$

In the latter case, if $\max\{\min\{I_t^e,\ F_t^{pd}\},\ 0\} = F_t^{pd}$, α_t and β_t coincide. Distribution may also be linked to previous-year net income; for example, $F_t^e = \beta_t \cdot \max\{\min\{I_{t-1}^e,\ F_t^{pd}\},\ 0\}$. Many other alternatives are possible.[25] Unless otherwise stated, we assume that the payout ratio is constant from time 1 to time $n-1$: $\alpha_t = \alpha$ for all $t \in \mathbb{N}_{n-1}^1$.

Remark 4.4 (*FCF and PD*) It is worth stressing that FCF and PD (=FCFE) are different. Notwithstanding the adjective "free", the FCF may not be interpreted as the cash which is freely available for distribution to shareholders. We may establish the formal relation between the cash available for distribution to shareholders and FCF by remembering that $F_t^o = F_t^{uo} + \tau I_t^{\mathcal{D}}$. Hence, $F_t^{pd} + F_t^d = F_t^o = F_t^{uo} + \tau I_t^{\mathcal{D}}$. This implies $F_t^{uo} = F_t^{pd} + F_t^d - \tau I_t^{\mathcal{D}}$. Neither is FCF the cash flow which is freely available for distribution to *all* investors. The latter concept is quantified by the CFO/CCF notion. ▲

Remark 4.5 (*Terminology*) Paradoxically, *cash flow from operations* and *free cash flow* seem to be reciprocally misnamed: The cash flow from operations might in principle be called free cash flow, since it is the cash flow that is *freely* available for distribution to all investors after tax payments, whereas the free cash flow should be called cash flow from operations, since it is a cash flow generated by the operations (before tax savings), $\tau I_t^{\mathcal{D}}$; in other words, it is that part of the cash available for distribution to investors which is not affected by the financing policy (assuming that the investment policy is fixed and the financing policy has no impact on investments). Adding the remaining part (the tax savings), one indeed gets the cash which is freely available for distribution. Unfortunately, we may not change a well-rooted habit and

[25] If additional cash is withdrawn from liquid assets and distributed to shareholders over and above the PD, then we say that PD has not changed but the payout ratio is greater than 1 (see Remark 4.6 below).

call "free cash flow" the cash flow from operations (which, by the way, is best known in finance "capital cash flow"). ▲

Remark 4.6 (*Extra distribution*) Extra distribution of cash flows to (current) equity-holders may be accomplished in two ways:

- by increasing the potential dividends
- by keeping the potential dividends unvaried.

In the first case, new debt is raised from debtholders, in the second case, either extra cash is withdrawn from the firm's liquid assets or new equity shares are raised from new shareholders.

If new debt is raised, the CFD is reduced, which brings about an increase in the distributable cash. Let $F_t^{d\#}$ and $F_t^{pd\#}$ be, respectively, the CFD and the potential dividends at time t before raising new debt: $F_t^{pd\#} = F_t^o - F_t^{d\#}$. Now, if the firm borrows an extra h_t from debtholders, potential dividends increase to $F_t^{pd} = F_t^{pd\#} + h_t = F_t^o - F_t^{d\#} + h_t$, so that

$$F_t^e = F_t^{pd\#} + h_t = F_t^o - \overbrace{(F_t^{d\#} - h_t)}^{F_t^d} = F_t^{pd}.$$

(The payout ratio is $\alpha = 1$.)

If extra cash is withdrawn from liquid assets, then CFEs exceed F_t^{pd}:

$$F_t^e = \overbrace{F_t^o - F_t^d}^{F_t^{pd}} + h_t > F_t^{pd},$$

where h_t denotes the additional amount withdrawn from the liquid account and distributed to shareholders. (The payout ratio is $\alpha > 1$).

If new equity shares are issued, the firm raises $h_t > 0$ from *new* shareholders and distribute it to *existing* shareholders. Overall, the CFEs do not change: What is raised from new shareholders is distributed to old shareholders:

$$F_t^e = \overbrace{F_t^{pd} + h_t}^{\text{cash distributed to current shareholders}} + \overbrace{-h_t.}^{\text{cash raised from new shareholders}} = F_t^{pd}$$

(The payout ratio is equal to $\alpha = 1$).

The latter case is the case analyzed by Miller and Modigliani (1961). To prove their famous dividend irrelevance theorem, they assume a full payout policy ($\alpha = 1$) and investigate what happens if new stocks are issued for extra distribution to existing shareholders, that is, if existing shareholders receive more than 100% of the cash available for distribution (see DeAngelo and DeAngelo 2006, 2007).[26]

[26] Miller and Modigliani (1961) prove that, in perfect markets, if the new shares are fairly priced (i.e., they are raised at the cost of equity), appraising firms with discounting either potential dividends

The differences among the three cases may be fruitfully represented by the Matrix, where the superscript # denotes the amount of an item before the extra distribution of h_t (see Rule 7 in Sect. 3.6.2).

Extra distribution: New debt

$$\begin{array}{|l|l|}\hline +C_t^o & +C_{t-1}^o \ +I_t^o \ -F_t^o \\ +C_t^l & +C_{t-1}^l \ +I_t^l \ -F_t^l \\ \hline +C_t^{d\#} + \boxed{h_t} & +C_{t-1}^d \ +I_t^d \ -F_t^{d\#} + \boxed{h_t} \\ +C_t^{e\#} - \boxed{h_t} & +C_{t-1}^e \ +I_t^e \ -F_t^{e\#} - \boxed{h_t} \\ \hline \end{array}$$

Extra distribution: Cash withdrawals

$$\begin{array}{|l|l|}\hline +C_t^o & +C_{t-1}^o \ +I_t^o \ -F_t^o \\ +C_t^{l\#} - \boxed{h_t} & +C_{t-1}^l \ +I_t^l \ -F_t^{l\#} - \boxed{h_t} \\ \hline +C_t^d & +C_{t-1}^d \ +I_t^d \ -F_t^d \\ +C_t^{e\#} - \boxed{h_t} & +C_{t-1}^e \ +I_t^e \ -F_t^{e\#} - \boxed{h_t} \\ \hline \end{array}$$

Extra distribution: New shares

$$\begin{array}{|l|l|}\hline +C_t^o & +C_{t-1}^o \ +I_t^o \ -F_t^o \\ +C_t^l & +C_{t-1}^l \ +I_t^l \ -F_t^l \\ \hline +C_t^d & +C_{t-1}^d \ +I_t^d \ -F_t^d \\ +C_t^e & +C_{t-1}^e \ +I_t^e \ -F_t^e + \boxed{h_t} - \boxed{h_t} \\ \hline \end{array}$$

The first case and the second case represent an extra distribution made by modifying the net financial position. They may be encompassed in the following format:

Extra distribution from net debt

$$\mathbf{M}_t = \begin{array}{|l|l|}\hline +C_t^o & +C_{t-1}^o \ +I_t^o \ -F_t^o \\ \hline +C_t^{\mathcal{D}\#} + \boxed{h_t} & +C_{t-1}^{\mathcal{D}} \ +I_t^{\mathcal{D}} \ -F_t^{\mathcal{D}\#} + \boxed{h_t} \\ +C_t^{e\#} - \boxed{h_t} & +C_{t-1}^e \ +I_t^e \ -F_t^{e\#} - \boxed{h_t} \\ \hline \end{array}$$

This framework means that either new debt is raised or cash is withdrawn from the liquid assets, without specifying which one of two alternatives is actually employed. ▲

Remark 4.7 (*Excess cash and potential dividends*) The relation of the change in excess cash with the notion of potential dividends is straightforward:

$$F_t^{pd} = F_t^e - F_t^l = F_t^e + \Delta C_t^l - I_t^l = F_t^e + \Delta \text{FA}_t + \Delta \text{EC}_t - I_t^l.$$

or CFEs provides the same result. The same result holds if, instead of distributing more than 100% of the available cash, the firm distributes only part of it, thereby retaining cash in the liquid assets (see Sect. 6.4. See also Magni 2010c).

Hence,

$$\Delta \mathrm{EC}_t = F_t^{pd} - F_t^e + I_t^l - \Delta \mathrm{FA}_t \tag{4.83}$$

is another way of computing the change in excess cash.

If $\mathrm{FA}_t = 0$, then $\mathrm{EC}_t = C_t^l$ and the change in cash, $\Delta \mathrm{EC}_t$, is equal to the change in liquid assets, $\Delta \mathrm{EC}_t = \Delta C_t^l = F_t^{pd} - F_t^e + I_t^l$. If return on cash is zero, as is usual, then the change in cash is the retained cash: $\Delta \mathrm{EC}_t = F_t^{pd} - F_t^e = \mathrm{RC}_t$. ▲

Remark 4.8 (*Other non-operating assets and retained cash*) If other non-operating assets are included (e.g., rental property, real estate assets, land, etc.), F_t^l is made up of two types of cash flow: The cash from liquid assets (F_t^{liquid}) and the cash from other non-operating assets ($F_t^{\text{other non-op}}$). That is, $F_t^l = F_t^{\text{liquid}} + F_t^{\text{other non-op}}$.[27] This implies that

$$\begin{array}{llll} C_t^{\text{liquid}} & = C_{t-1}^{\text{liquid}} & + I_t^{\text{liquid}} & - F_t^{\text{liquid}} \\ C_t^{\text{other non-op}} & = C_{t-1}^{\text{other non-op}} & + I_t^{\text{other non-op}} & - F_t^{\text{other non-op}} \\ \hline C_t^l & = C_{t-1}^l & + I_t^l & - F_t^l \end{array}$$

The amount $-F_t^{\text{liquid}}$ represents cash and cash equivalents which are retained in the business. The amount F_t^l is calculated as $-(1-\alpha_t)F_t^{pd}$ (see Eq. (4.80)). Assuming $F_t^{\text{other non-op}}$ has already been estimated (e.g., some rental payment from a real estate asset), the cash retained in liquid assets is derived as

$$-F_t^{\text{liquid}} = -F_t^l + F_t^{\text{other non-op}} = (1-\alpha_t)F_t^{pd} + F_t^{\text{other non-op}} \tag{4.84}$$

(e.g., see HomeNet example in Chap. 12). ▲

Example 4.10 ♦

Consider the following data for a company at a given point in time:

- the interest payments to debtholders are \$2,700 (no principal repayments)
- operating assets have decreased by \$87,000
- equity capital have decreased by \$63,463
- operating income is \$33,386
- the interest rate on liquid assets is zero.

Let us find the remaining values: Net Income, CFO, retained cash, change in liquid assets, CFD.

One may use (4.72). Either version of the Matrix works. For example, considering the left Matrix and filling in the cells with the given data, one gets

---

[27] Note that $F_t^{\text{liquid}} = F_t^{ec} + F_t^{fa}$, so that $F_t^l = F_t^{ec} + F_t^{fa} + F_t^{\text{other non-op}}$. In other words, non-operating assets usually consist of liquid assets such as cash, marketable securities, and other financial assets. However, other non-operating assets may be present which have a degree of liquidity smaller than F_t^{liquid}.

	Cash flow	Income	$-\Delta$Capital
Operating assets	$+F_t^o$	+33, 386	+87, 000
Debt	$+F_t^d$	+2, 700	
Equity	$+F_t^e$	$+I_t^e$	+63, 463
Liquid assets	$-F_t^l$		$+\Delta C_t^l$

Columns 2 and 3 enable one to compute the implicit net income and change in liquid assets, respectively: $I_t^e = 33,386 - 2,700 = 30,686$ and $\Delta C_t^l = 87,000 - 63,463 = 23,537$. The latter is also the undistributed cash. Remembering that $F_t^l = -\text{RC}_t$,

	Cash flow	Income	$-\Delta$Capital
Operating assets	$+F_t^o$	+33, 386	+87, 000
Debt	$+F_t^d$	+2, 700	
Equity	$+F_t^e$	+30, 686	+63, 463
Liquid assets	$+\text{RC}_t$		+23, 537

The cash flows can then be derived from incomes and capitals:[28]

$$
\begin{aligned}
F_t^o &= 33,386 + 87,000 = 120,386\\
F_t^d &= 2,700 + 0 \qquad\quad\;\; = 2,700\\
F_t^e &= 30,686 + 63,463 = 94,149\\
\text{RC}_t &= 0 + 23,537 \qquad\; = 23,537.
\end{aligned}
$$

The Matrix is then completed:

	Cash flow	Income	Δ Capital
Operating assets	+120, 386	+33, 386	+87, 000
Debt	+2, 700	+2, 700	
Equity	+94, 149	+30, 686	+63, 463
Liquid assets	+23, 537		+23, 537

Example 4.11 ♦

C^o	C^d
C^l	C^e

(*Mockup project – 80% payout*) Mockup Inc. faces the opportunity of undertaking a 5-period new project with an investment cost equal to \$500,000, including \$65,000 initial NOWC. The firm forecasts incremental sales amounting to 2,950 units in the first year, growing by an annual 2.5%. The price of the product will be \$120. Operating costs are divided into variable costs and fixed costs. The former are estimated at \$50 per unit, the latter are estimated at \$71,000. The project has no salvage value, and the tax rate is 33%. The project is levered, financed with a bullet bond with nominal value equal to \$90,000, reimbursed with coupon payments at 3% of the face value and reimbursement of the principal after 5 years. Initial NOWC is equal to \$65,000 and will remain constant until the liquidation date. The input data are reported in Table 4.19. Table 4.20 depicts the pro forma financial statements.

[28]Mathematically: The cash-flow column is linearly dependent on the income column and change-in-capital column.

Table 4.19 Mockup – Project input data

Parameter	Estimate	Parameter	Estimate
Investment cost:		Borrowing	$90,000 (bullet bond)
Fixed assets	$435,000 (straight-line depreciation)		
Working capital	$65,000 (constant)	Coupon rate	3%
Salvage value	$0	Loan term	5 years
Unit sales	2,950	Horizon	5 years
growth rate	2.5%	Tax rate	33%
Unit price	$120	payout ratio	80%
Variable unit cost	$50	Interest rate	
Fixed costs	$71,000	on liquid assets	2%

Table 4.20 Mockup Project – Pro forma BSs and ISs (80% payout)

Time	**0**	**1**	**2**	**3**	**4**	**5**
BALANCE SHEET						
Net Fixed Assets	435,000	348,000	261,000	174,000	87,000	0
Net Operating WC	65,000	65,000	65,000	65,000	65,000	0
Liquid Assets	0	23,537	48,206	74,044	101,092	0
INVESTMENTS	500,000	436,537	374,206	313,044	253,092	0
Debt	90,000	90,000	90,000	90,000	90,000	0
Equity	410,000	346,537	284,206	223,044	163,092	0
FINANCINGS	500,000	436,537	374,206	313,044	253,092	0
INCOME STATEMENT						
Sales		354,000	362,850	371,921	381,219	390,750
– Variable costs		−147,500	−151,188	−154,967	−158,841	−162,812
– Fixed costs		−71,000	−71,000	−71,000	−71,000	−71,000
EBITDA		135,500	140,663	145,954	151,378	156,937
−Depreciation		−87,000	−87,000	−87,000	−87,000	−87,000
EBIT		48,500	53,663	58,954	64,378	69,937
– Interest expense		−2,700	−2,700	−2,700	−2,700	−2,700
+ Interest income			+471	+964	+1,481	+2,022
EBT		45,800	51,433	57,218	63,159	69,259
– Taxes		−15,114	−16,973	−18,882	−20,842	−22,856
Net Income		30,686	34,460	38,336	42,316	46,404

Table 4.21 Mockup project – Calculation of CFO (80% payout)

Time	0	1	2	3	4	5
EBIT−T	0	33,386	36,690	40,072	43,536	47,082
$-\Delta C^o$	−500,000	87,000	87,000	87,000	87,000	152,000
CFO	−500,000	120,386	123,690	127,072	130,536	199,082
NOPAT	0	32,495	35,954	39,499	43,133	46,858
$-\Delta C^o$	−500,000	87,000	87,000	87,000	87,000	152,000
FCF	−500,000	119,495	122,954	126,499	130,133	198,858
+ tax shield		+891	+736	+573	+402	+224
CFO	−500,000	120,386	123,690	127,072	130,536	199,082

Table 4.22 Mockup project – Potential dividends, CFE, and undistributed cash (80% payout)

Time	0	1	2	3	4	5
CFO	−500,000	120,386	123,690	127,072	130,536	199,082
- CFD	90,000	−2,700	−2,700	−2,700	−2,700	−92,700
PD	−410,000	117,686	120,990	124,372	127,836	106,382
−CFE	410,000	−94,149	−96,792	−99,498	−102,268	−209,496
RC− −CFL	0	23,537	24,198	24,874	25,567	−103,114

To draw the balance sheets, one needs to compute the balance of non-operating assets, C_t^l, for every t. To this end, we assume a payout ratio equal to $\alpha = 80\%$, which means that 20% of the potential dividends is retained as excess cash and/or invested in marketable securities. We assume that the interest rate on liquid assets is $i_t^l = 0.02$ for every t. Table 4.21 illustrates the computation of the CFO (in two ways, from OI and from NOPAT) while Table 4.22 shows the undistributed cash. The latter is added to the BOP balance of liquid assets along with interest income (see Eq. (4.79)), so the asset side is completed.

For example, focusing on $t = 4$, the balance is $C_4^l = 101,092$. This is equal to the previous balance, $C_3^l = 74,044$ plus the interest income ($I_4^l = 0.02 \cdot 74,044$) and the retained cash ($\text{RC}_4 = -F_4^l = 25,567$):

$$101,092 = 74,044 + 1,481 + 25,567.$$

Table 4.23 Mockup project – Prospective cash flows (80% payout)

Time	**0**	**1**	**2**	**3**	**4**	**5**
EBIT–T		33,386	36,690	40,072	43,536	47,082
+ Interest income			+471	+964	+1,481	+2,022
$-\Delta C^o$	–500,000	87,000	87,000	87,000	87,000	152,000
$-\Delta C^l$	0	23,537	24,669	25,839	27,048	–101,092
CFA	–500,000	96,849	99,492	102,198	104,968	302,196
Net Income		30,686	34,460	38,336	42,316	46,404
$-\Delta C^e$	–410,000	63,463	62,331	61,161	59,952	163,092
CFE	–410,000	94,149	96,792	99,498	102,268	209,496
Interest expense		2,700	2,700	2,700	2,700	2,700
$-\Delta C^d$	–90,000	0	0	0	0	90,000
CFD	–90,000	2,700	2,700	2,700	2,700	92,700
Interest income		0	471	964	1,481	2,022
$-\Delta C^l$		0	–23,537	–24,669	–25,839	–27,048	101,092
CFL (–RC)		0	–23,537	–24,198	–24,874	–25,567	103,114
CFD	–90,000	2,700	2,700	2,700	2,700	92,700
–CFL (+RC)	0	23,537	24,198	24,874	25,567	–103,114
CF$\mathcal{D}$	–90,000	26,237	26,898	27,574	28,267	–10,414
+ CFE	–410,000	94,149	96,792	99,498	102,268	209,496
CFO	–500,000	120,386	123,690	127,072	130,536	199,082

The final CFL is positive (retained cash is negative): $F_5^l = -\text{RC}_5 = 103,114$. This is because everything that has been invested in the liquid assets in the previous periods must necessarily be paid out to equityholders (see also footnote 22). Therefore, in the last period, the CFE is greater than the potential dividends generated by the operating activity: 209,496 > 106,382. The CFA and CFE may be computed in several ways (e.g., see Eqs. (4.11) and (4.16)). Table 4.23 reports the CFA, CFE, CFL, CFD for every t. In the table, we also compute back the CFO from the CFD, by subtracting the CFL (which supplies the CF$\mathcal{D}$) and, hence, adding the CFE.

We summarize the project using the transposed project Matrix in Table 4.24. The laws of conservation hold, which confirms the logical consistency of the model.

Figure 4.6 is an influence diagram illustrating the functional relations of this case.

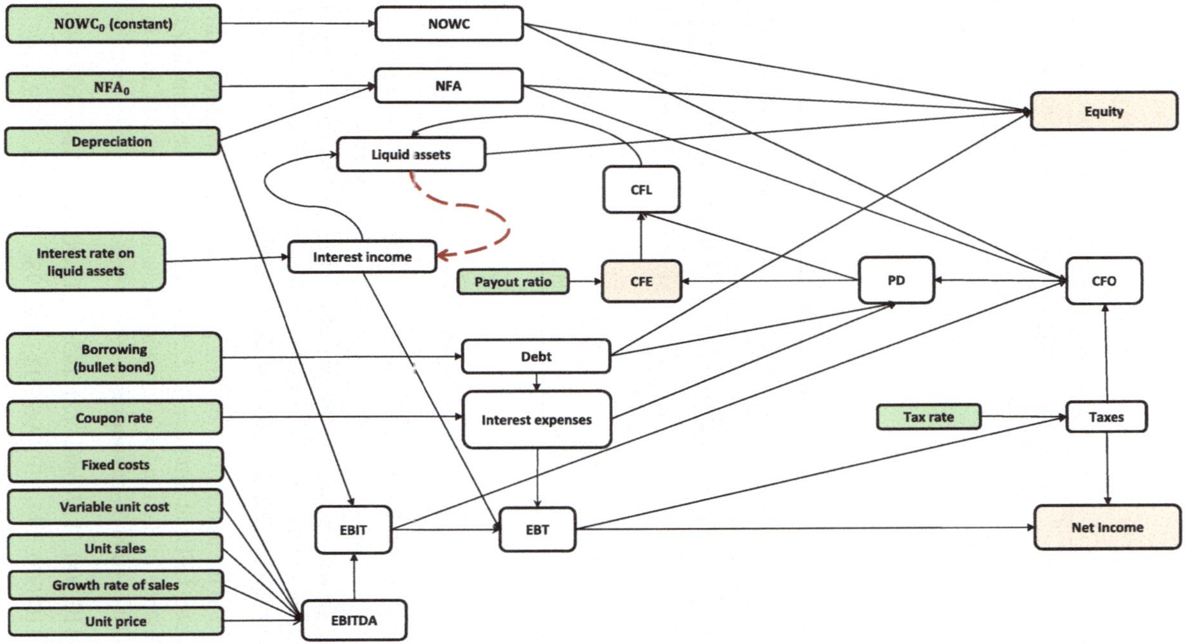

Fig. 4.6 Influence diagram indicating the functional relationships of Mockup project (80% payout). The dashed line refers to time $t-1$: the balance of liquid assets at time $t-1$ affects, along with i_t^l, the interest income as of time t, $I_t^l = i_t^l C_{t-1}^l$. The latter affects in turn the balance of liquid assets as of time t, alongside F_t^l: $C_t^l = C_{t-1}^l + I_t^l - F_t^l$. Since the payout ratio is applied to potential dividends, net income does not affect CFE. If, conversely, the CFE is made to depend on net income, an arrow from it to the CFE box should be drawn and the arrow from PD to CFE should be removed. If CFE is made to depend on both PD and NI (e.g., Eq. (4.82)), then an arrow must be drawn from both PD and NI to CFE. The CFD coincides with interest expense from time 1 to time 4. At time 5, CFD is the sum of interest expense and the outstanding debt

Table 4.24 Mockup project – Summary (80% payout)

(C, I, F)							
Time		0	1	2	3	4	5
Capital	C	500,000	436,537	374,206	313,044	253,092	0
Operating assets	C^o	500,000	413,000	326,000	239,000	152,000	0
Non-operating assets	C^l	0	23,537	48,206	74,044	101,092	0
Debt	C^d	90,000	90,000	90,000	90,000	90,000	0
Equity	C^e	410,000	346,537	284,206	223,044	163,092	0
Income	I		33,386	37,160	41,036	45,016	49,104
Operating income	I^o		33,386	36,690	40,072	43,536	47,082
Non-operating income	I^l		0	471	964	1,481	2,022
Interest expense	I^d		2,700	2,700	2,700	2,700	2,700
Net Income	I^e		30,686	34,460	38,336	42,316	46,404
Cash flow	F	-500,000	96,849	99,492	102,198	104,968	302,196
Cash flow from operations	F^o	-500,000	120,386	123,690	127,072	130,536	199,082
Cash flow from non-operating assets	F^l	0	-23,537	-24,198	-24,874	-25,567	103,114
Cash flow to debt	F^d	-90,000	2,700	2,700	2,700	2,700	92,700
Cash flow to equity	F^e	-410,000	94,149	96,792	99,498	102,268	209,496

Example 4.12 ♦

C^o	C^d
	C^e

(Mockup project – 100% payout) Consider the Mockup project (Example 4.11) but assume now that the payout ratio is $\alpha = 100\%$ (all the available cash after servicing the debt is distributed to equityholders). This implies $F_t^l = 0$ and $F_t^{pd} = F_t^e$ for all t. Table 4.25 reports the pro forma BSs and ISs under the new assumption, while Table 4.26 reports the cash flows. In this case, owing to zero CFL for every t, the CFO and the CFA coincide, as well as CFD and CF$\mathcal{D}$.

Note that, as opposed to the assumption of 80% payout ratio, equityholders pull out from the project greater CFEs at time $t = 1, 2, 3, 4$, but will receive a smaller CFE at $t = 5$. For convenience of the reader, we report below the vectors of CFEs in the alternative cases of 100 and 80% payout ratio:

$$\boldsymbol{F}^e(100\%) = (-410{,}000,\ 117{,}686,\ 121{,}145,\ 124{,}690,\ 128,324,\ 107,049)$$
$$\boldsymbol{F}^e(80\%) = (-410{,}000,\ 94{,}149,\ 96{,}792,\ 99{,}498,\ 102{,}268,\ 209{,}496)$$

The difference between the two streams of cash flows is

$$\boldsymbol{F}^e(100\%) - \boldsymbol{F}^e(80\%) = (0,\ 23{,}537,\ 24{,}353,\ 25{,}193,\ 26{,}056,\ -102{,}447).$$

We summarize the project in Table 4.27.

Table 4.25 Mockup project – Pro forma BSs and ISs (100% payout)

Time	**0**	**1**	**2**	**3**	**4**	**5**
BALANCE SHEET						
Net Fixed Assets	435,000	348,000	261,000	174,000	87,000	0
Net Operating WC	65,000	65,000	65,000	65,000	65,000	0
Liquid Assets	0	0	0	0	0	0
INVESTMENTS	500,000	413,000	326,000	239,000	152,000	0
Debt	90,000	90,000	90,000	90,000	90,000	0
Equity	410,000	323,000	236,000	149,000	62,000	0
FINANCINGS	500,000	413,000	326,000	239,000	152,000	0
INCOME STATEMENT						
Sales		354,000	362,850	371,921	381,219	390,750
– Variable costs		–147,500	–151,188	–154,967	–158,841	–162,812
– Fixed costs		–71,000	–71,000	–71,000	–71,000	–71,000
EBITDA		135,500	140,663	145,954	151,378	156,937
–Depreciation		–87,000	–87,000	–87,000	–87,000	–87,000
EBIT		48,500	53,663	58,954	64,378	69,937
– Interest expense		–2,700	–2,700	–2,700	–2,700	–2,700
+ Interest income		0	0	0	0	0
EBT		45,800	50,963	56,254	61,678	67,237
– Taxes		–15,114	–16,818	–18,564	–20,354	–22,188
Net Income		30,686	34,145	37,690	41,324	45,049

Example 4.13 ♦

C^o	C^d
C^l	C^e

(*BHP Billiton – 90% payout*) Let us go back to BHP Billiton (Example 4.4). We now assume that the payout ratio is $\alpha = 90\%$ (i.e., $F_t^e = 0.9 \cdot F_t^{pd}$) and the retained cash is invested in liquid assets at an interest rate equal to $i_t^l = 0.03$ for all t. Tables 4.28 and 4.29 show the project's pro forma balance sheets, the incomes and the cash flows.

To compute the balance of liquid assets, we first compute the available cash and then subtract the cash flow that will be distributed to equityholders. For example, let us focus on the third period. The BOP balance at time 2 is $C_2^l = 49.2$. The interest income in the third period is then $I_3^l = 0.03 \cdot 49.2 = 1.5$. Therefore, the EOP balance is $49.2 + 1.5 = 50.7 = C_2^l + I_3^l = E_3^l$. The balance at the beginning of the fourth period (time 3) is obtained by adding the retained cash, and to compute the latter the analyst must first compute the potential dividends. To accomplish this task, one first

Table 4.26 Mockup project – Prospective cash flows (100% payout)

Time	**0**	**1**	**2**	**3**	**4**	**5**
EBIT–T		33,386	36,845	40,390	44,024	47,749
$-\Delta C^o$	−500,000	87,000	87,000	87,000	87,000	152,000
CFO=CFA	−500,000	120,386	123,845	127,390	131,024	199,749
– tax shield		−891	−891	−891	−891	−891
FCF	−500,000	119,495	122,954	126,499	130,133	198,858
Net Income		30,686	34,145	37,690	41,324	45,049
$-\Delta C^e$	−410,000	87,000	87,000	87,000	87,000	62,000
CFE	−410,000	117,686	121,145	124,690	128,324	107,049
Interest expense		2,700	2,700	2,700	2,700	2,700
$-\Delta C^d$	−90,000	0	0	0	0	90,000
CFD	−90,000	2,700	2,700	2,700	2,700	92,700

Table 4.27 Mockup project – Summary (100% payout)

(C, I, F)

Time		0	1	2	3	4	5
Capital	C	500,000	413,000	326,000	239,000	152,000	0
Operating assets	C^o	500,000	413,000	326,000	239,000	152,000	0
Non-operating assets	C^l	0	0	0	0	0	0
Debt	C^d	90,000	90,000	90,000	90,000	90,000	0
Equity	C^e	410,000	323,000	236,000	149,000	62,000	0
Income	I		33,386	36,845	40,390	44,024	47,749
Operating income	I^o		33,386	36,845	40,390	44,024	47,749
Non-operating income	I^l		0	0	0	0	0
Interest expense	I^d		2,700	2,700	2,700	2,700	2,700
Net Income	I^e		30,686	34,145	37,690	41,324	45,049
Cash flow	F	−500,000	120,386	123,845	127,390	131,024	199,749
Cash flow from operations	F^o	−500,000	120,386	123,845	127,390	131,024	199,749
Cash flow from non-operating assets	F^l	0	0	0	0		0
Cash flow to debt	F^d	−90,000	2,700	2,700	2,700	2,700	92,700
Cash flow to equity	F^e	−410,000	117,686	121,145	124,690	128,324	107,049

Table 4.28 BHP Billiton, 90% payout balance sheets and cash flows

Time	NFA$_t$	NOWC$_t$	C_t^l	C_t^d	C_t^e	F_t	F_t^d	F_t^e	F_t^{uo}
0	550.0	50.0	0.0	200	400.0	−600.0	−200.0	−400.0	−600.0
1	522.5	42.0	24.7	200	389.2	237.9	16.0	221.9	257.8
2	495.0	42.0	49.2	200	386.2	230.5	16.0	214.5	249.8
3	467.5	42.0	74.5	200	384.0	230.3	16.0	214.3	249.8
4	440.0	42.0	100.5	200	382.5	230.1	16.0	214.1	249.8
5	412.5	42.0	127.3	200	381.8	229.9	16.0	213.9	249.8
6	385.0	42.0	154.9	200	381.9	229.7	16.0	213.7	249.8
7	357.5	42.0	183.2	200	382.7	229.4	16.0	213.4	249.8
8	330.0	42.0	212.4	200	384.4	229.2	16.0	213.2	249.8
9	302.5	42.0	242.5	200	387.0	229.0	16.0	213.0	249.8
10	275.0	42.0	253.4	0.0	570.4	248.7	216.0	32.7	249.8
11	247.5	42.0	285.7	0.0	575.2	222.7	0.0	222.7	249.8
12	220.0	42.0	319.0	0.0	581.0	222.5	0.0	222.5	249.8
13	192.5	42.0	353.3	0.0	587.8	222.2	0.0	222.2	249.8
14	165.0	42.0	388.5	0.0	595.5	221.9	0.0	221.9	249.8
15	137.5	42.0	424.8	0.0	604.3	221.6	0.0	221.6	249.8
16	110.0	42.0	462.1	0.0	614.1	221.3	0.0	221.3	249.8
17	82.5	42.0	500.6	0.0	625.1	221.0	0.0	221.0	249.8
18	55.0	42.0	540.1	0.0	637.1	220.7	0.0	220.7	249.8
19	27.5	42.0	580.8	0.0	650.3	220.4	0.0	220.4	249.8
20	0.0	0.0	0.0	0.0	0.0	814.7	0.0	814.7	221.8

Table 4.29 BHP Billiton, 90% payout retained cash, CFEs, and liquid assets

Time	EBIT$_t$	−T$_t$	$-\Delta C_t^o$	F_t^o	$-F_t^d$	F_t^{pd}	$-F_t^e$	$-F_t^l$	E_t^l	C_t^l
0			−600	−600.0	200.0	−400	400	0.0	0.0	0.0
1	317.5	−90.5	35.5	262.6	−16.0	246.6	−221.9	24.7	0.0	24.7
2	317.5	−90.7	27.5	254.3	−16.0	238.3	−214.5	23.8	25.4	49.2
3	317.5	−90.9	27.5	254.1	−16.0	238.1	−214.3	23.8	50.7	74.5
4	317.5	−91.1	27.5	253.9	−16.0	237.9	−214.1	23.8	76.8	100.5
5	317.5	−91.4	27.5	253.6	−16.0	237.6	−213.9	23.8	103.6	127.3
6	317.5	−91.6	27.5	253.4	−16.0	237.4	−213.7	23.7	131.1	154.9
7	317.5	−91.8	27.5	253.2	−16.0	237.2	−213.4	23.7	159.5	183.2
8	317.5	−92.1	27.5	252.9	−16.0	236.9	−213.2	23.7	188.7	212.4
9	317.5	−92.4	27.5	252.6	−16.0	236.6	−213.0	23.7	218.8	242.5
10	317.5	−92.6	27.5	252.4	−216.0	36.4	−32.7	3.6	249.7	253.4
11	317.5	−97.5	27.5	247.5	0.0	247.5	−222.7	24.7	261.0	285.7
12	317.5	−97.8	27.5	247.2	0.0	247.2	−222.5	24.7	294.3	319.0
13	317.5	−98.1	27.5	246.9	0.0	246.9	−222.2	24.7	328.6	353.3
14	317.5	−98.4	27.5	246.6	0.0	246.6	−221.9	24.7	363.9	388.5
15	317.5	−98.7	27.5	246.3	0.0	246.3	−221.6	24.6	400.2	424.8
16	317.5	−99.1	27.5	245.9	0.0	245.9	−221.3	24.6	437.6	462.1
17	317.5	−99.4	27.5	245.6	0.0	245.6	−221.0	24.6	476.0	500.6
18	317.5	−99.8	27.5	245.2	0.0	245.2	−220.7	24.5	515.6	540.1
19	317.5	-100.1	27.5	244.9	0.0	244.9	−220.4	24.5	556.3	580.8
20	217.5	−70.5	69.5	216.5	0.0	216.5	−814.7	−598.2	598.2	0.0

subtracts from the EBIT the taxes T so as to get the OI: $I_3^o = 317.5 - 90.9 = 226.6$. From it, the change in the operating capital is subtracted to get the operating cash flow: $F_3^o = 226.6 - (-27.5) = 254.1$. Subtracting the CFD, the potential dividends are found: $F_3^{pd} = 254.1 - 16 = 238.1$. This enables calculating the CFE: $F_3^e = 0.9 \cdot 238.1 = 214.3$. Subtracting the latter from the potential dividends, the retained cash is found: $\text{RC}_3 = -F_3^l = 238.1 - 214.3 = 23.8$. Finally, this enables computing the BOP balance of liquid assets: $C_3^l = 23.8 + 50.7 = 74.5$ (see Table 4.29).

Example 4.14 ♦

C^o	C^d
C^l	C^e

(*BHP Billiton – 90% payout – change in excess cash*) Consider the BHP Billiton example above (Example 4.13) and assume that 93% of liquid assets consists of cash (i.e., $\text{FA}_t = 0.07C_t^l$). Table 4.30 illustrates how to compute the change in cash, based upon Eq. (4.83). Column 2 collects the value of the FA account (7% of liquid assets), on the basis of which the change in FA is calculated (column 6). Note also that, summing the change in EC and the change in FA, one gets back the change in liquid assets: $\Delta\text{FA}_t + \Delta\text{EC}_t = \Delta C_t^l$.

Table 4.30 BHP Billiton, calculation of change in excess cash

Time	FA_t	F_t^{pd}	$-F_t^e$	$+I_t^l$	$-\Delta\text{FA}_t$	ΔEC_t	ΔC_t^l
0	0.0						
1	1.7	246.6	-221.9	0.0	-1.7	22.9	24.7
2	3.4	238.3	-214.5	0.7	-1.7	22.9	24.6
3	5.2	238.1	-214.3	1.5	-1.8	23.5	25.3
4	7.0	237.9	-214.1	2.2	-1.8	24.2	26.0
5	8.9	237.6	-213.9	3.0	-1.9	24.9	26.8
6	10.8	237.4	-213.7	3.8	-1.9	25.6	27.6
7	12.8	237.2	-213.4	4.6	-2.0	26.4	28.4
8	14.9	236.9	-213.2	5.5	-2.0	27.1	29.2
9	17.0	236.6	-213.0	6.4	-2.1	27.9	30.0
10	17.7	36.4	-32.7	7.3	-0.8	10.1	10.9
11	20.0	247.5	-222.7	7.6	-2.3	30.1	32.3
12	22.3	247.2	-222.5	8.6	-2.3	31.0	33.3
13	24.7	246.9	-222.2	9.6	-2.4	31.9	34.3
14	27.2	246.6	-221.9	10.6	-2.5	32.8	35.3
15	29.7	246.3	-221.6	11.7	-2.5	33.7	36.3
16	32.4	245.9	-221.3	12.7	-2.6	34.7	37.3
17	35.0	245.6	-221.0	13.9	-2.7	35.7	38.4
18	37.8	245.2	-220.7	15.0	-2.8	36.8	39.5
19	40.7	244.9	-220.4	16.2	-2.8	37.8	40.7
20	0.0	216.5	-814.7	17.4	40.7	-540.1	-580.8

4.11 The Mosaics of Incomes and Cash Flows

We now summarize, in a handy and user friendly format, the static equilibrium of a project's (or firm's) cash flows as well as the static equilibrium of related incomes. Figure 4.7 portrays a mosaic of relations among cash flows and among incomes, under the assumption of positive CFL (extra distribution), while Fig. 4.8 describes a mosaic under the assumption of zero CFL (full payout policy and no extra distribution). Precisely,

- the upper part of Fig. 4.7 articulates the equilibrium among cash flows of a project (firm) into 9 equivalent conceptualizations. It makes use of the notions of cash flow from assets (CFA), cash flow to capital providers (CFP), free cash flow (FCF) and asset FCF, operating cash flow (CFO), cash flow from non-operating assets (CFL), cash flow to debt (CFD), cash flow to net debt (CF$\mathcal{D}$), cash flow to equity (CFE). Columns 1–2–3 clarify the relationships among FCF, asset FCF, and CFA, as well as the role of the tax shield, $\tau I_t^{\mathcal{D}} = \tau(I_t^d - I_t^l)$, as opposed to the role of

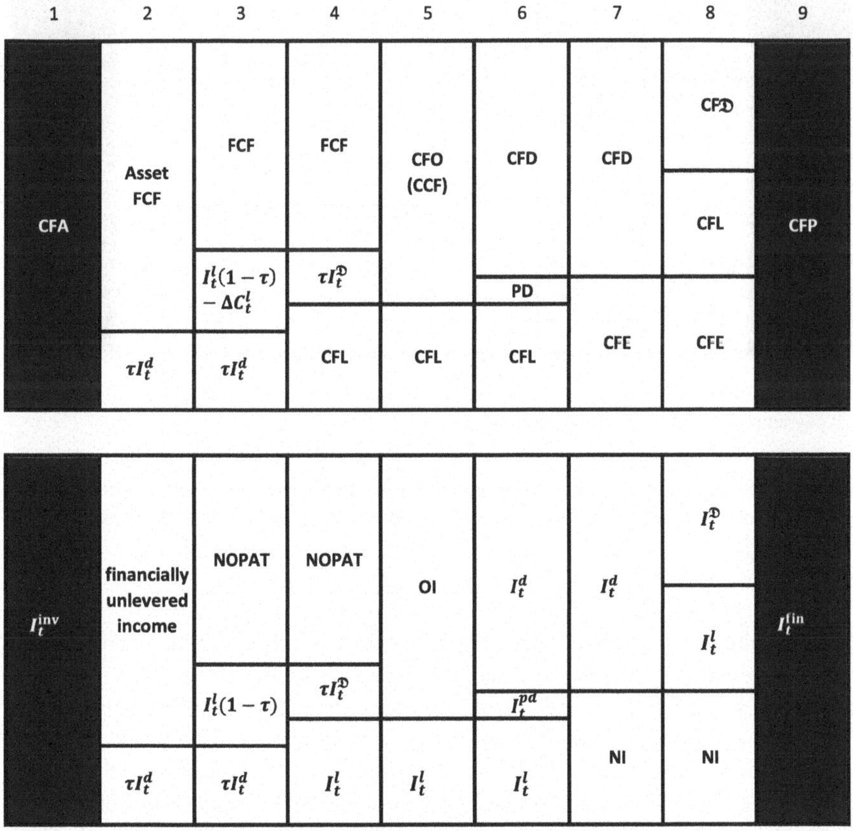

Fig. 4.7 The Mosaics of a project's (or firm's) cash flows and incomes, under the assumption of positive CFL (extra distribution of cash to equityholders)

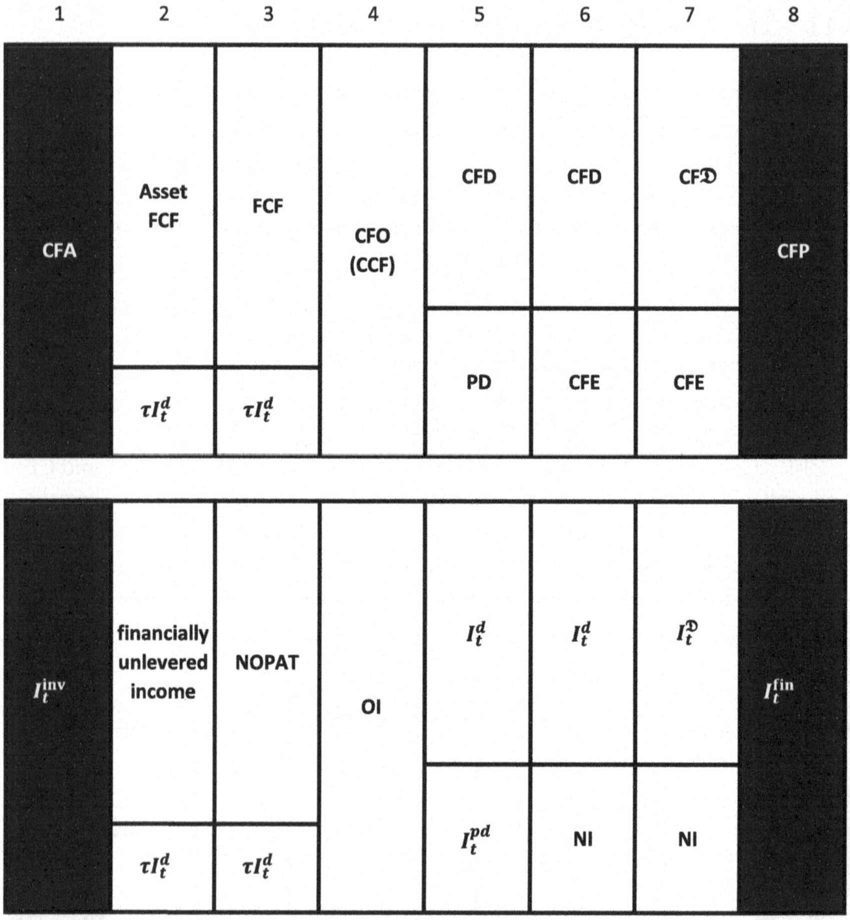

Fig. 4.8 The Mosaics of a project's (or firm's) cash flows and incomes, under the assumption of full payout policy and no extra distribution

the gross tax shield, τI_t^d. Note that column 2 contains the after-tax interest income net of change in liquid assets, which coincides with the CFL net of taxes on liquid assets: $I_t^l(1-\tau) - \Delta C_t^l = F_t^l - \tau I_t^l$. This is what we have called *non-operating* FCF (F_t^{ul}). Columns 4–5 clarify the relationship between CFO/CCF and FCF. Column 5-6 make it clear that PD is equal the difference between CFO and CFD, while columns 6–7 make it clear that PD is also the difference between CFE and CFL. Columns 7–8–9 illustrate the relationships of CFP, CFD, CF$\mathcal{D}$, and CFL. All indicated cash flows are positive and, in particular, CFE is greater than PD, that is, CFL is positive (additional cash flow is withdrawn from liquid assets for distribution to equityholders). In other words, the payout ratio is $\alpha > 1$

- the lower part of Fig. 4.7 expresses the equivalent static equilibrium in terms of incomes. It shows equivalent conceptualizations of a firm's or project's income, making use of income from investments (I_t^{inv}), income to capital providers (I_t^{fin}), financially unlevered income (I_t^u), NOPAT (I_t^{uo}), operating income (OI), non-operating income (I_t^l), interest expense (I_t^d), net interest ($I_t^{\mathcal{D}}$), net income (NI)

- the upper part of Fig. 4.8 presents the static equilibrium of cash flows of a project (firm) under the assumption of full payout policy ($\alpha = 1$, no retained cash flow and no extra distribution). In this case, relationships are simplified, owing to absence of liquid assets: The CFL is zero, the FCF coincides with the asset FCF, for no interest income arises, the PD coincides with the CFE, and the CFD concides with the CF$\mathcal{D}$
- the lower part of Fig. 4.8 presents the equilibrium among incomes under the same assumption of 100% payout. The financially unlevered income equates the NOPAT; the potential income coincides with the net income; interest coincides with net interest (columns 2 and 3 become equally partitioned as well as the same columns for the cash flow mosaic).

We leave to the reader the description of the case where CFL is negative, that is, excess cash is retained and deposited in liquid assets ($\alpha < 1$).

4.12 Key Points

- Once the project input data (income and capital components) are estimated, the Matrix is put at work to find the cash flows at each date. For example, the (fourfold) CFA may be obtained as

CFA

$+F_t^c$	$+S_t$	$-\Delta\text{AR}_t$
$-F_t^{opc}$	$-\text{OpC}_t$	$+\Delta\text{NOL}_t$
$-F_t^{capex}$	$-\text{Dep}_t$	$-\Delta\text{NFA}_t$
$-F_t^T$	$-\text{T}_t$	$+\Delta\text{TP}_t$
$+F_t^l$	$+I_t^l$	$-\Delta C_t^l$
$+F_t^d$	$+I_t^d$	$-\Delta C_t^d$
$+F_t^e$	$+I_t^e$	$-\Delta C_t^e$

or, consolidating rows 1–3 (and assuming $F_t^T = \text{T}_t$ and $\Delta\text{TP}_t = 0$, as usual),

CFA

$+F_t^\bullet$	$+\text{EBIT}_t$	$-\Delta C_t^o$
$-\text{T}_t$	$-\text{T}_t$	
$+F_t^l$	$+I_t^l$	$-\Delta C_t^l$
$+F_t^d$	$+I_t^d$	$-\Delta C_t^d$
$+F_t^e$	$+I_t^e$	$-\Delta C_t^e$

where $F_t^\bullet$ is the pre-tax CFO. Any other cash flow may be found by performing invariant operations on the Matrix via the split-screen technique

- Depreciation shelters the project income by the amount $\tau\,\text{Dep}_t$; net interest shelters the project income by $\tau I_t^{\mathcal{D}}$
- Both the pair (NOPAT, FCF) and the pair (OI, CFO) represent the profit and the cash flow generated by the operations. However, (NOPAT, FCF) only takes account of the taxes on EBIT, whereas (OI, CFO) also takes account of the taxes on net interest. In other words, (NOPAT, FCF) results from an unlevered perspective, whereas (OI, CFO) refers to a levered perspective. Put it differently: (NOPAT, FCF) is that part of (OI, CFO) which only depends on investment policy. The interest tax shield, $\tau I_t^{\mathcal{D}}$ expresses the effect on OI and on CFO of the financing policy (i.e., payout policy and borrowing policy)
- The unlevered (operating) ROC/ROI, here denoted as i_t^{uo}, is the amount of NOPAT per unit of operating capital invested. It measures the growth rate of the operating assets in an unlevered perspective
- The asset FCF is the CFA that would be obtained if the project were equity financed (financially unlevered). The asset FCF is to CFA what FCF is to CFO
- Potential dividends represent the cash available for distribution. It is the cash flow that remains from the CFO after servicing the debt and differs from the cash flow actually distributed to equityholders, which is the CFE. Other things equal, if the potential dividends are greater (smaller) than CFE, it means that some cash is retained in (withdrawn from) the project. The CFE is then, to some extent, a decision variable. If the payout ratio is 100%, then potential dividends and CFE coincide
- Potential dividends and change in excess cash (or change in liquid assets) are different notions. The change in liquid assets is equal to the sum of the interest income and the cash flow to liquid assets (retained cash) or, if cash is subtracted from liquid assets, the difference between the interest income and the cash flow from the liquid assets (see also Eq. (4.79))
- Any project can be defined and represented by $n+1$ triplets of capital, income, and cash flow, denoted as $(\boldsymbol{C}, \boldsymbol{I}, \boldsymbol{F})$ which we call the *Transposed Matrix*:

Time		0	1	2	3 ...	n
Capital	C	C_0	C_1	C_2	C_3 ...	C_n
Operating assets	C^o	C_0^o	C_1^o	C_2^o	C_3^o ...	C_n^o
Non-operating assets	C^l	C_0^l	C_1^l	C_2^l	C_3^l ...	C_n^l
Debt	C^d	C_0^d	C_1^d	C_2^d	C_3^d ...	C_n^d
Equity	C^e	C_0^e	C_1^e	C_2^e	C_3^e ...	C_n^e
Income	I	I_0	I_1	I_2	I_3 ...	I_n
Operating income	I^o	I_0^o	I_1^o	I_2^o	I_3^o ...	I_n^o
Non-operating income	I^l	I_0^l	I_1^l	I_2^l	I_3^l ...	I_n^l
Interest expense	I^d	I_0^d	I_1^d	I_2^d	I_3^d ...	I_n^d
Net Income	I^e	I_0^e	I_1^e	I_2^e	I_3^e ...	I_n^e
Cash flow	F	F_0	F_1	F_2	F_3 ...	F_n
Cash flow from operations	F^o	F_0^o	F_1^o	F_2^o	F_3^o ...	F_n^o
Cash flow from non-operating assets	F^l	F_0^l	F_1^l	F_2^l	F_3^l ...	F_n^l
Cash flow to debt	F^d	F_0^d	F_1^d	F_2^d	F_3^d ...	F_n^d
Cash flow to equity	F^e	F_0^e	F_1^e	F_2^e	F_3^e ...	F_n^e

- After the model is realized and transferred to a spreadsheet, the Matrix may be fruitfully employed as a diagnostic tool for checking that the model does not contain error and is internally consistent.

Table 4.31 summarizes all the types of cash flows introduced so far with the respective incomes and capital values, as well as the respective laws of motion.

Table 4.32 derives each and every capital as the discounted sum of the respective cash flows, calculated at the respective income rates.

Table 4.33 is a top-down synchronic representation that connects all the cash flows at a given date starting from the accounting estimates, under the assumption that some available cash flow is not distributed (undistributed cash is nonzero).

Table 4.34 reports the same schedule assuming a 100% payout ratio. In this case, the relations are highly simplified, for

- the CFO/CCF coincides with the CFA
- the gross debt coincides with the net debt
- the tax shield coincides with the gross tax shield: $\tau I_t^{\mathcal{D}} = \tau I_t^d$
- the FCF concides with the asset FCF
- the PDs coincide with the CFE.

Table 4.35 presents a summary of some fundamental (synchronic and diachronic) relations in terms of capital, income, and cash flows that can be drawn from the Matrix.

Table 4.36 summarizes the income rates as ratios of income to capital.

Table 4.37 presents reciprocal relations among income rates.

Figure 4.9 is one among many possible diagrams illustrating a network of relations among some cash flows, starting from the EBIT.

Figure 4.10 represents the mosaic of a project's cash flows, assuming positive CFL (extra distribution of cash to equityholders via withdrawals from liquid assets). Figure 4.11 represents the mosaic of a project's cash flows, assuming zero CFL (full payout). This structure will be preserved when passing from cash flows to market values (see Fig. 6.2).

Equation (4.85) below makes use of the thumbnail Matrices and provides a handy guide to helpful implications in terms of capitals and rates of return related to different financial structures.

$$\begin{array}{|c|c|} \hline C^o & C^d \\ \hline \; & C^e \\ \hline \end{array} \Longrightarrow C^o_t = C_t \qquad \begin{array}{l} \text{and } F^o_t = F_t \\ \text{and } F^{uo}_t = F^u_t \end{array} \qquad \begin{array}{l} \text{and } i^o_t = i_t \\ \text{and } i^{uo}_t = i^u_t \end{array}$$

$$\begin{array}{|c|c|} \hline C^o & \; \\ \hline \; & C^e \\ \hline \end{array} \Longrightarrow C_t = C^o_t = C^e_t \qquad \begin{array}{l} \text{and } F_t = F^o_t = F^e_t \\ \text{and } F_t = F^{uo}_t = F^u_t \end{array} \qquad \begin{array}{l} \text{and } i_t = i^o_t = i^e_t \\ \text{and } i_t = i^{uo}_t = i^u_t \end{array}$$

$$\begin{array}{|c|c|} \hline C^o & \; \\ \hline C^l & C^e \\ \hline \end{array} \Longrightarrow C_t = C^e_t \qquad \text{and } F_t = F^u_t = F^e_t \qquad \text{and } i_t = i^u_t = i^e_t$$

$$\begin{array}{|c|c|} \hline C^o & C^d \\ \hline C^l & C^e \\ \hline \end{array} \Longrightarrow \text{no general implications}$$

$$\begin{array}{|c|c|} \hline \; & C^d \\ \hline C^l & C^e \\ \hline \end{array} \Longrightarrow C^l_t = C_t \qquad \text{and } F^l_t = F_t \qquad \text{and } i^l_t = i_t$$

$$\begin{array}{|c|c|} \hline \; & \; \\ \hline C^l & C^e \\ \hline \end{array} \Longrightarrow C^l_t = C_t = C^e_t \qquad \text{and } F^l_t = F_t = F^e_t \qquad \text{and } i^l_t = i_t = i^e_t$$

(4.85)

Table 4.31 Cash flows – Acronyms, symbols, and relations

TYPES OF CASH FLOW[(a)]			INCOME[(b)]	CAPITAL[(c)]	RELATION
Cash flow from assets	CFA	F_t^{inv}	I_t^{inv} (income from assets)	C_t^{inv}	$F_t = I_t - \Delta C_t$
Cash flow from operations Capital cash flow	CFO CCF	F_t^o	I_t^o (operating income)	C_t^o	$F_t^o = I_t^o - \Delta C_t^o$
Cash flow from liquid assets	CFL	F_t^l	I_t^l (interest income)	C_t^l	$F_t^l = I_t^l - \Delta C_t^l$
Cash flow to equity	CFE	F_t^e	I_t^e (net income)	C_t^e	$F_t^e = I_t^e - \Delta C_t^e$
Cash flow to debt	CFD	F_t^d	I_t^d (interest expense)	C_t^d	$F_t^d = I_t^d - \Delta C_t^d$
Cash flow to net debt	CF$\mathcal{D}$	$F_t^{\mathcal{D}}$	$I_t^{\mathcal{D}}$ (net interest)	$C_t^{\mathcal{D}}$	$F_t^{\mathcal{D}} = I_t^{\mathcal{D}} - \Delta C_t^{\mathcal{D}}$
Cash flow to capital providers	CFP	F_t^{fin}	I_t^{fin} (income to capital providers)	C_t^{fin}	$F_t = I_t - \Delta C_t$
Potential dividends Free cash flow to equity	PD FCFE	F_t^{pd}	I_t^{pd} (potential income)	C_t^{pd}	$F_t^{pd} = I_t^{pd} - \Delta C_t^{pd}$
UNLEVERED					
Free cash flow	FCF	F_t^{uo}	I_t^{uo} (NOPAT)	C_t^o	$F_t^{uo} = I_t^{uo} - \Delta C_t^o$
Asset free cash flow		F_t^u	I_t^u (financially unlevered income)	C_t	$F_t^u = I_t^u - \Delta C_t$

(a) $F_t^{\text{inv}} = F_t^{\text{fin}} = F_t$
(b) $I_t^{\text{inv}} = I_t^{\text{fin}} = I_t$
(c) $C_t^{\text{inv}} = C_t^{\text{fin}} = C_t$

Table 4.32 Summary of discounted values at income rates

TYPES OF CASH FLOW[(a)]			DISCOUNTED SUM
Cash flow from assets	CFA	F_t^{inv}	$C_t = \sum_{j=t+1}^{n} \frac{F_j}{\prod_{h=t+1}^{j}(1+i_h)}$
Cash flow from operations Capital cash flow	CFO CCF	F_t^o	$C_t^o = \sum_{j=t+1}^{n} \frac{F_j^o}{\prod_{h=t+1}^{j}(1+i_h^o)}$
Cash flow from liquid assets	CFL	F_t^l	$C_t^l = \sum_{j=t+1}^{n} \frac{F_j^l}{\prod_{h=t+1}^{j}(1+i_h^l)}$
Cash flow to equity	CFE	F_t^e	$C_t^e = \sum_{j=t+1}^{n} \frac{F_j^e}{\prod_{h=t+1}^{j}(1+i_h^e)}$
Cash flow to debt	CFD	F_t^d	$C_t^d = \sum_{j=t+1}^{n} \frac{F_j^d}{\prod_{h=t+1}^{j}(1+i_h^d)}$
Cash flow to net debt	CF$\mathcal{D}$	$F_t^{\mathcal{D}}$	$C_t^{\mathcal{D}} = \sum_{j=t+1}^{n} \frac{F_j^{\mathcal{D}}}{\prod_{h=t+1}^{j}(1+i_h^{\mathcal{D}})}$
Cash flow to capital providers	CFP	F_t^{fin}	$C_t = \sum_{j=t+1}^{n} \frac{F_j}{\prod_{h=t+1}^{j}(1+i_h)}$
Potential dividends Free cash flow to equity	PD FCFE	F_t^{pd}	$C_t^{pd} = \sum_{j=t+1}^{n} \frac{F_j^{pd}}{\prod_{h=t+1}^{j}(1+i_h^{pd})}$
UNLEVERED			
Free cash flow	FCF	F_t^{uo}	$C_t^o = \sum_{j=t+1}^{n} \frac{F_j^{uo}}{\prod_{h=t+1}^{j}(1+i_h^{uo})}$
Asset free cash flow		F_t^u	$C_t = \sum_{j=t+1}^{n} \frac{F_j^u}{\prod_{h=t+1}^{j}(1+i_h^u)}$

[(a)] $F_t^{\text{inv}} = F_t^{\text{fin}} = F_t$

Table 4.33 Top-down schedule of cash flows (payout < 100%)

Sales revenues	S_t
– Operating costs	$-\text{OpC}_t$
EBITDA	$\textbf{EBITDA}_t$
– Depreciation	$-\text{Dep}_t$
EBIT	$\textbf{EBIT}_t$
– Taxes on EBIT	$-\tau \cdot \text{EBIT}_t$
NOPAT	$\boldsymbol{I_t^{uo}}$
– Change in net fixed assets	$-\Delta\text{NFA}_t$
– Change in net operating working capital	$-\Delta\text{NOWC}_t$
Free cash flow	$\boldsymbol{F_t^{uo}}$
+ tax shield	$+\tau I_t^{\mathcal{D}}$
Cash flow from operations = Capital cash flow	$\boldsymbol{F_t^{o}}$
– Cash flow to debt	$\boldsymbol{-F_t^{d}}$
Cash flow available for distribution	$\boldsymbol{F_t^{pd}}$
+ Cash flow from liquid assets (– Retained Cash)	$+F_t^{l}$
Cash flow to equity	$\boldsymbol{F_t^{e}}$
+ Cash flow to debt	$+F_t^{d}$
Cash flow from assets	$\boldsymbol{F_t}$
– (gross) tax shield	$-\tau I_t^{d}$
Asset free cash flow	$\boldsymbol{F_t^{u}}$

Table 4.34 Top-down schedule of cash flows (100% payout)

Sales revenues	S_t
– Operating costs	$-\text{OpC}_t$
EBITDA	$\textbf{EBITDA}_t$
– Depreciation	$-\text{Dep}_t$
EBIT	$\textbf{EBIT}_t$
– Taxes on EBIT	$-\tau \cdot \text{EBIT}_t$
NOPAT	$\boldsymbol{I_t^{uo} = I_t^{u}}$
– Change in net fixed assets	$-\Delta\text{NFA}_t$
– Change in net operating working capital	$-\Delta\text{NOWC}_t$
Free cash flow	$\boldsymbol{F_t^{uo} = F_t^{u}}$
+ gross tax shield	$+\tau I_t^{d} = +\tau I_t^{\mathcal{D}}$
Cash flow from assets = Cash flow from operations	$\boldsymbol{F_t = F_t^{o}}$
– Cash flow to debt	$\boldsymbol{-F_t^{d}}$
Cash flow available for distribution= Cash flow to equity	$\boldsymbol{F_t^{pd} = F_t^{e}}$

Table 4.35 Summary of diachronic and synchronic relations for a project

Diachronic relations (law of motion)	Synchronic relations (law of conservation)
$F_t = \text{EBIT}_t - \text{T}_t + I^l_t - (\Delta\text{NFA}_t + \Delta\text{NOWC}_t) - \Delta C^l_t$	$F_t = F^o_t + F^l_t$
$F_t = I^e_t + I^d_t - (\Delta C^e_t + \Delta C^d_t)$	$F_t = F^e_t + F^d_t$
$F^e_t = I^e_t - (\Delta\text{NFA}_t + \Delta\text{NOWC}_t - \Delta C^{\mathcal{D}}_t)$	$F^e_t = F^{pd}_t + F^l_t$
$F^{uo}_t = \text{EBIT}_t(1-\tau) - (\Delta\text{NFA}_t + \Delta\text{NOWC}_t)$	$F^{uo}_t = F^o_t - \tau I^{\mathcal{D}}_t$
$F^o_t = \text{EBIT}_t - \text{T}_t - (\Delta\text{NFA}_t + \Delta\text{NOWC}_t)$	$F^o_t = F^{pd}_t + F^d_t$
$F^{\mathcal{D}}_t = I^d_t - I^l_t - (\Delta C^d_t - \Delta C^l_t)$	$F^{\mathcal{D}}_t = F^o_t - F^e_t$
$F^u_t = (\text{EBIT}_t + I^l_t)(1-\tau) - (\Delta\text{NFA}_t + \Delta\text{NOWC}_t + \Delta C^l_t)$	$F^u_t = F_t - \tau I^d_t$
$C^o_t = (\text{NFA}_{t-1} + \text{NOWC}_{t-1})(1 + i^o_t) - F^o_t$	$C_t = \text{NFA}_t + \text{NOWC}_t + C^l_t$
$C^o_t = (\text{NFA}_{t-1} + \text{NOWC}_{t-1})(1 + i^{uo}_t) - F^{uo}_t$	$C_t = C^o_t + C^l_t$
$C^o_t = (C^e_{t-1} + C^{\mathcal{D}}_{t-1})(1 + i^{uo}_t) - F^{uo}_t$	$C^o_t = C^{\mathcal{D}}_t + C^e_t$
$C^o_t = C^o_{t-1}(1 + i^o_t) - F^o_t$	$C_t = C^d_t + C^e_t$
$C_t = (C^o_{t-1} + C^l_{t-1})(1 + i_t) - F_t$	$C^{pd}_t = C^e_t - C^l_t$
$C_t = (C^e_{t-1} + C^d_{t-1})(1 + i^u_t) - F^u_t$	$I^e_t + I^{\mathcal{D}}_t = \text{EBIT}_t - \text{T}_t$
$C^d_t = C^d_{t-1}(1 + i^d_t) - F^d_t$	$I^e_t + I^{\mathcal{D}}_t = I^{uo}_t + \tau I^{\mathcal{D}}_t$
$C^e_t = C^e_{t-1}(1 + i^e_t) - F^e_t$	$I^e_t + I^d_t = I^u_t + \tau I^d_t$
$C_t = (C^e_{t-1} + C^d_{t-1})(1 + i_t) - F_t$	$I^o_t = I^{uo}_t + \tau I^{\mathcal{D}}_t$
$C^l_t = C^l_{t-1}(1 + i^l_t) - F^l_t$	$I^o_t = I^{\mathcal{D}}_t + I^e_t$
$C^{\mathcal{D}}_t = C^{\mathcal{D}}_{t-1}(1 + i^{\mathcal{D}}_t) - F^{\mathcal{D}}_t$	$I_t = I^o_t + I^l_t$

Table 4.36 Project's income rates

operating ROI	
$i_t^o = \dfrac{\text{EBIT}_t - \text{T}_t}{C_{t-1}^o}$	$i_t^o = \dfrac{I_t^{\mathcal{D}} + I_t^e}{C_{t-1}^{\mathcal{D}} + C_{t-1}^e}$
asset ROI	
$i_t = \dfrac{\text{EBIT}_t + I_t^l - \text{T}_t}{C_{t-1}}$	$i_t = \dfrac{I_t^d + I_t^e}{C_{t-1}^d + C_{t-1}^e}$
unlevered (operating) ROI	
$i_t^{uo} = \dfrac{\text{EBIT}_t(1-\tau)}{C_{t-1}^o}$	$i_t^{uo} = \dfrac{I_t^{\mathcal{D}}(1-\tau) + I_t^e}{C_{t-1}^{\mathcal{D}} + C_{t-1}^e}$
unlevered asset ROI (financially unlevered ROI)	
$i_t^u = \dfrac{(\text{EBIT}_t + I_t^l)(1-\tau)}{C_{t-1}}$	$i_t^u = \dfrac{I_t^d(1-\tau) + I_t^e}{C_{t-1}^d + C_{t-1}^e}$
ROE	
$i_t^e = \dfrac{I_t^e}{C_{t-1}^e}$	$i_t^e = \dfrac{(\text{EBIT}_t - I_t^{\mathcal{D}})(1-\tau)}{C_{t-1}^e}$
RO$\mathcal{D}$	
$i_t^{\mathcal{D}} = \dfrac{I_t^{\mathcal{D}}}{C_{t-1}^{\mathcal{D}}}$	$i_t^{\mathcal{D}} = \dfrac{\text{EBIT}_t - \text{T}_t - I_t^e}{C_{t-1}^o - C_{t-1}^e}$

Table 4.37 Summary of relations for a project's income rates

operating ROI	$i^o_t = \dfrac{C^o_t + F^o_t}{C^o_{t-1}} - 1$
$i^o_t = \dfrac{i_t C_{t-1} - i^l_t C^l_{t-1}}{C_{t-1} - C^l_{t-1}}$	$i^o_t = \dfrac{i^e_t C^e_{t-1} + i^{\mathcal{D}}_t C^{\mathcal{D}}_{t-1}}{C^e_{t-1} + C^{\mathcal{D}}_{t-1}}$
asset ROI	$i_t = \dfrac{C_t + F_t}{C_{t-1}} - 1$
$i_t = \dfrac{i^o_t C^o_{t-1} + i^l_t C^l_{t-1}}{C^o_{t-1} + C^l_{t-1}}$	$i_t = \dfrac{i^e_t C^e_{t-1} + i^d_t C^d_{t-1}}{C^e_{t-1} + C^d_{t-1}}$
unlevered operating ROI	$i^{uo}_t = \dfrac{C^o_t + F^{uo}_t}{C^o_{t-1}} - 1$
$i^{uo}_t = i^o_t - \dfrac{\tau I^{\mathcal{D}}_t}{C^o_{t-1}}$	$i^{uo}_t = \dfrac{i^e_t C^e_{t-1} + i^{\mathcal{D}}_t (1-\tau) C^{\mathcal{D}}_{t-1}}{C^e_{t-1} + C^{\mathcal{D}}_{t-1}}$
unlevered asset ROI (financially unlevered ROI)	$i^u_t = \dfrac{C_t + F^u_t}{C_{t-1}} - 1$
$i^u_t = i_t - \dfrac{\tau I^d_t}{C_{t-1}}$	$i^u_t = \dfrac{i^e_t C^e_{t-1} + i^d_t (1-\tau) C^d_{t-1}}{C^e_{t-1} + C^d_{t-1}}$
ROE	$i^e_t = \dfrac{C^e_t + F^e_t}{C^e_{t-1}} - 1$
$i^e_t = i^{uo}_t + \dfrac{C^{\mathcal{D}}_{t-1}}{C^e_{t-1}}(i^{uo}_t - i^{\mathcal{D}}_t) + \dfrac{\tau I^{\mathcal{D}}_t}{C^e_{t-1}}$	$i^e_t = i^{uo}_t + \dfrac{C^{\mathcal{D}}_{t-1}}{C^e_{t-1}}(i^{uo}_t - i^{\mathcal{D}}_t(1-\tau))$
$i^e_t = i^u_t + \dfrac{C^d_{t-1}}{C^e_{t-1}}(i^u_t - i^d_t) + \dfrac{\tau I^d_t}{C^e_{t-1}}$	$i^e_t = i^u_t + \dfrac{C^d_{t-1}}{C^e_{t-1}}(i^u_t - i^d_t(1-\tau))$
$i^e_t = i^o_t + \dfrac{C^{\mathcal{D}}_{t-1}}{C^e_{t-1}}(i^o_t - i^{\mathcal{D}}_t)$	$i^e_t = i_t + \dfrac{C^d_{t-1}}{C^e_{t-1}}(i_t - i^d_t)$
RO$\mathcal{D}$	$i^{\mathcal{D}}_t = \dfrac{i^d_t C^d_{t-1} - i^l_t C^l_{t-1}}{C^d_{t-1} - C^l_{t-1}}$

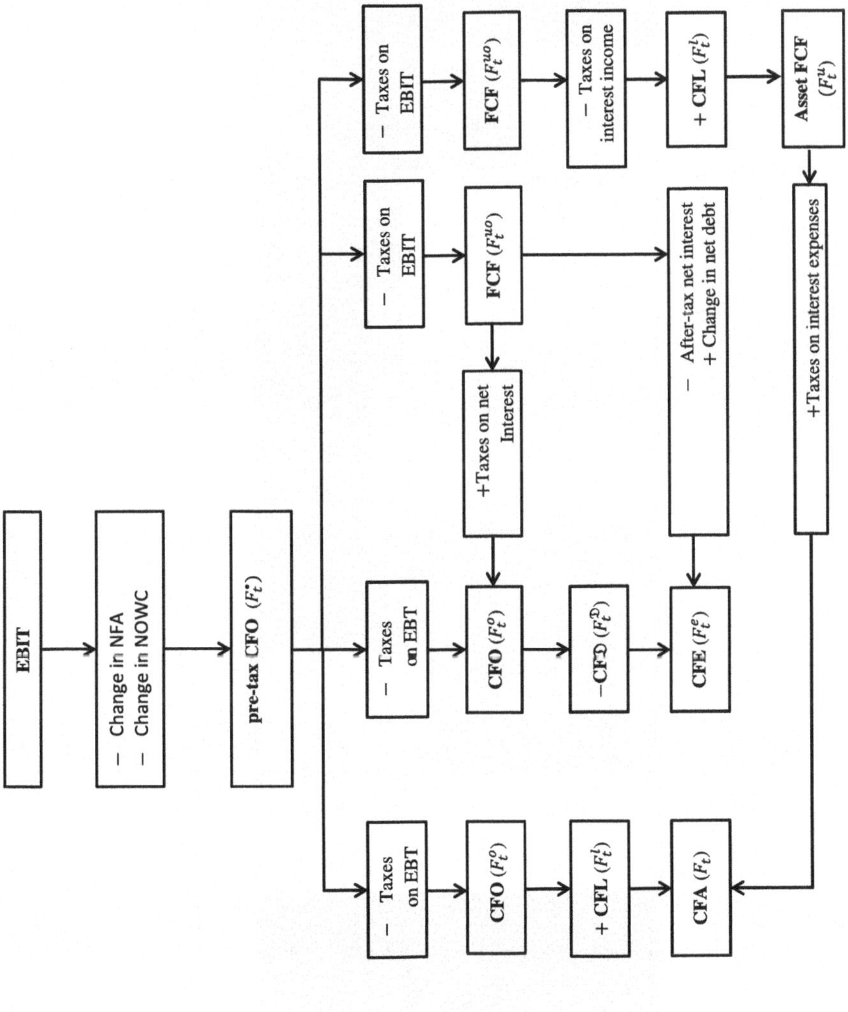

Fig. 4.9 A network of cash flows, starting from EBIT

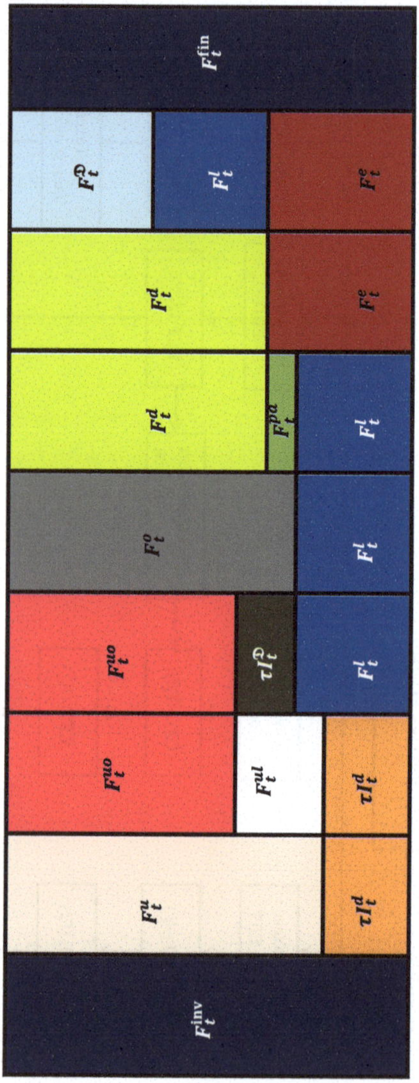

Fig. 4.10 The Mosaic of a project's (firm's) cash flows with $\alpha > 1$

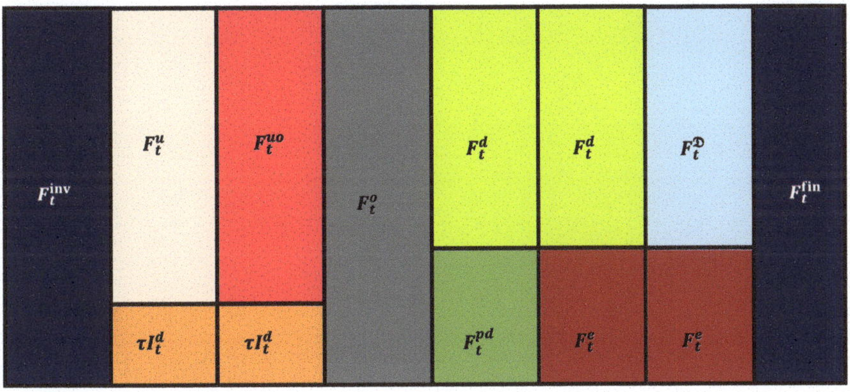

Fig. 4.11 The Mosaic of a projects's (firm's) cash flow with $\alpha = 1$

Part III
Valuation and Decision-making: The Absolute Approaches

In this part we put to full use the accounting-and-finance engineering system presented in the previous parts.

We introduce the notions of **value creation** and (economically) **rational decision-making**. Value creation is reflected in the reciprocal equilibrium/disequilibrium between the project system and a **benchmark** system, which represents the capital market.

We present three benchmark systems, associated with the basic notions of cash flow, income, and capital, respectively. The use of these three benchmark systems gives rise to three approaches to valuation and decision-making:

Net Present Value (NPV) method
Residual Income (RI) method
Net Future Value (NFV) or **Value Added** (VA) method.

As the three measures of worth are expressed in monetary units, they are **absolute measures** of worth.

Each of the approaches is presented in different framings:

Investment perspective vs. financing perspective
Levered perspective vs. unlevered perspective
Financial structure vs. capital structure
Standard discounting vs. modified discounting
Actual cash flows vs. potential cash flows.

We show that each framing of each method provides **equivalent** values and leads to the same, rational decision.

Chapter 5
Valuation and Value Creation

> This law of one price is the logic upon which virtually all valuation is based. It is important that you realize that this means that value in finance is defined in relative terms.
>
> Welch (2009, p. 2)

> Like most good ideas, the net present value rule is "obvious when you think about it." But notice what an important idea it is. The NPV rule allows thousands of shareholders, who may have vastly different levels of wealth and attitudes toward risk, to participate in the same enterprise and to delegate its operation to a professional manager. They give the manager one simple instruction: "Maximize net present value."
>
> Brealey, Myers and Allen (2011, p. 900)

> Irving Fisher, in *The Rate of Interest: Its Nature, Determination and Relation to Economic Phenomena* (1907), may have been the first to propose that any capital project should be evaluated in terms of its present value. Using an arbitrage argument, he compared the stream of cash flows from the project to the cash flows from a portfolio of securities constructed to match the project
>
> Rubinstein (2003, p. 48)

C. A. Magni, *Investment Decisions and the Logic of Valuation*,
https://doi.org/10.1007/978-3-030-27662-1_5

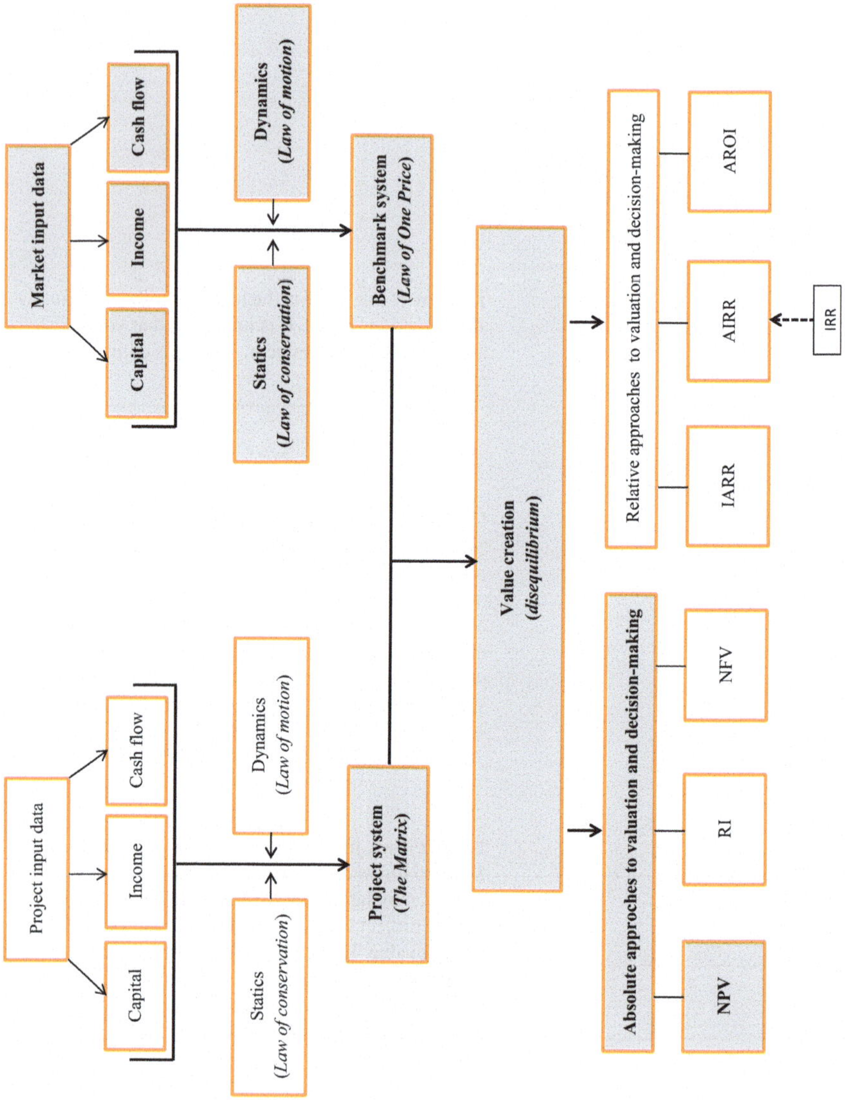

Fig. 5.1 Chapter 5: Value creation, NPV, and the benchmark system

This chapter supplies the basis of **valuation theory** and the principle of **economic rationality**. Specifically, it introduces the **Law of One Price**. It is also called the *no-arbitrage principle* and is a law of conservation of value which is guaranteed by a normal market where financial assets are traded.

A **benchmark system** is introduced, based on a replicating portfolio which is representative of the market equilibrium. The project system is compared with the benchmark system for assessing the **reciprocal equilibrium or disequilibrium**. The **net present value** (NPV) measures the degree of disequilibrium and its direction (positive or negative) and is shown to lead to economically rational decisions.

The NPV is strictly linked with the notion of **cost of capital** (COC), which is the income rate of the benchmark system. As such, it is a **market input** (as opposed to the project inputs) for the accounting-and-finance engineering system. It acts as a benchmark return, a **minimum attractive (or acceptable) rate of return** (MARR).

Risk is coped with in terms of risk-adjusted cash flows or in terms of **risk-adjusted COC/MARR**, which includes a risk premium. The latter is by far and large the preferred method in real-life applications. In practice, the risk premium may well be based on informed judgment as well as on a "normative" risk-adjusted rate of return drawn from some pricing models.

After completing this chapter, the reader should be able to:

- understand the *law of one price*, which is a principle of (social) economic rationality, derived from a fundamental law of economics: The law of demand and supply
- understand the related notions of
  - arbitrage
  - equilibrium price and equilibrium asset
  - value additivity
- understand the notion of disequilibrium and value creation
- measure the economic value created by a project with the Net Present Value (NPV)
- understand the basic principles of (individual) economic rationality and the relation between the shareholders' wealth increase and the NPV
- consider the NPV as a rational criterion for decision-making
- familiarize with the notions of normal profit and *cost of capital* (COC), also called *minimum attractive rate of return* (MARR)
- use the accounting-and-finance engineering system developed in the book for valuing both certain and uncertain projects, the latter being assessed in terms of an *expected* NPV, which incorporates risk either by adjusting the cash flows or adjusting the cost of capital to include risk (the latter being a much more common approach)
- regard the estimation of the COC/MARR of capital as a delicate issue which involves managerial skills and informed judgment. From this point of view, normative theoretical models may provide guidance but may not substitute "soft skills and expertise".

5.1 The Law of One Price

In this section, we introduce the benchmark system which will be contrasted with the project system.

Consider a capital market where securities are bought and sold (including firms' stocks and governments bonds and bills). We assume the market is *perfect* and *efficient*. A perfect market is a competitive market (i.e., a large number of investors exist who cannot affect prices and rates) with no constraints on transactions, no information costs, no transaction costs, and no costs of financial intermediation. Also, the financial assets are infinitely divisible. An efficient market is one where prices always fully reflect available information and quickly adjust to fully reflect new information.[1] We will use the expression *normal market* to refer to a perfect and efficient market.

Definition 5.1 (*Normal market*) A *normal market* is perfect and efficient.

For the time being, we assume that cash flows are certain and let r be the (assumed constant) interest rate for investment and financing. That is, investors can invest in the market at a lending rate equal to r or borrow at a financing rate equal to r. The market value of a financial asset (or portfolio of assets) expresses the capital invested (if positive) or borrowed (if negative) by the trading agents. In financial markets, a positive capital committed in a portfolio is also referred to as a *long* position, whereas a negative capital is referred to as a *short* position. A long position boils down to investing money, a short position boils down to borrowing money (see footnote 4 in this chapter). Let V_t be the market value of a financial asset (or portfolio of assets) traded in the market. Let F_t be its prospective cash flow at time $t \in \mathbb{N}_n^1$. The dynamics of such an asset may be described using (1.16) with V_t replacing C_t and r replacing i:

$$V_t = V_{t-1}(1+r) - F_t \quad \text{or, proceeding backward, } V_{t-1} = \frac{V_t + F_t}{1+r} \tag{5.1}$$

for $t \in \mathbb{N}_n^1$.[2] Remembering that $V_n = 0$ (the system reaches zero level at time n) and iterating, one finds

$$V_t = \sum_{k=t+1}^{n} \frac{F_k}{(1+r)^{k-t}} \tag{5.2}$$

which is just (1.18) with V_t and r replacing C_t and i, respectively.

---

[1]The notion of efficient market has been studied extensively. Three forms of efficiency have been defined: Weak, semi-strong, and strong (see Malkiel 1962; Fama 1965, 1970; Shiller 1981; Malkiel 2003 on this topic). A discussion of this concept is beyond the scope of this book.

[2]Time n is to be intended as the time when the security expires.

Example 5.1 ♦
Suppose the market rate is 5% and an investor purchases a security which generates cash flows equal to \$10 for the following six periods. Applying (1.18), and reminding the well-known result

$$(1+i)^{-1} + (1+i)^{-2} + \cdots + (1+i)^{-n} = \frac{1-(1+i)^{-n}}{i}, \tag{5.3}$$

the current price is

$$V_0 = 10 \cdot \frac{1-(1.05)^{-6}}{0.05} = 50.76.$$

Using the usual law of motion, the market value at each date is

$$\begin{aligned}
V_0 &= 50.76 \\
V_1 &= 50.76 \cdot (1.05) - 10 = 43.29 \\
V_2 &= 43.29 \cdot (1.05) - 10 = 35.46 \\
V_3 &= 35.46 \cdot (1.05) - 10 = 27.23 \\
V_4 &= 27.23 \cdot (1.05) - 10 = 18.59 \\
V_5 &= 18.59 \cdot (1.05) - 10 = 9.52 \\
V_6 &= 9.52 \cdot (1.05) - 10 = 0
\end{aligned}$$

or, proceeding backwards,

$$\begin{aligned}
V_6 &= 0 \\
V_5 &= (0+10)/(1.05) = 9.52 \\
V_4 &= (9.52+10)/(1.05) = 18.59 \\
V_3 &= (18.59+10)/(1.05) = 27.23 \\
V_2 &= (27.23+10)/(1.05) = 35.46 \\
V_1 &= (35.46+10)/(1.05) = 43.29 \\
V_0 &= (43.29+10)/(1.05) = 50.76
\end{aligned}$$

The market value represents the capital invested in the financial security in each period.

In a normal market prices are *unique*, in the sense that two assets may not be traded at a different price at a given point in time if their prospective cash flows are equal; if this situation occurred, market forces would quickly arbitrage away the mispricing. This is the so-called *Law of One Price*. This law is a natural derivation of a fundamental law of economics, the law of demand and supply.

Law of demand and supply. As the price of a good increases/decreases
- the quantity demanded decreases/increases
- the quantity supplied increases/decreases

Suppose that the law of one price does not hold. In particular, suppose that two assets or portfolios, A and B, generating the same cash-flow stream $\boldsymbol{F}$, are traded in the financial market at different prices. Let us assume, with no loss of generality, that $p_0^A > p_0^B$. This means that the cash-flow stream for an investor purchasing A is $\boldsymbol{F}^A = (-p_0^A, F_1, F_2, \dots, F_n)$ whereas the cash-flow stream for an investor purchasing B is $\boldsymbol{F}^B = (-p_0^B, F_1, F_2, \dots, F_n)$. Assuming investors aim at maximizing their wealth, and given that the market is normal, no investor will purchase A, because B is preferable (B dominates A). The demand for A will then quickly fall and the price of A will shrink to that point where A and B have the same price, $p_0^A = p_0^B = V_0 = \sum_{t=1}^n F_t(1+r)^{-t}$ (the same holds at every date). We may then state the following proposition.

Proposition 5.1 (Law of One Price) *Consider a normal market where the market rate is r and two assets (or portfolios) A and B, traded at time $t \geq 0$, generate the same cash flow F_k for $k \in \mathbb{N}_n^{t+1}$. If A and B are traded at prices p_t^A and p_t^B, then*

$$p_t^A = p_t^B = V_t = \sum_{k=t+1}^{n} \frac{F_k}{(1+r)^{k-t}}.$$

The law of one price is subtly linked to the notion of *arbitrage*. An arbitrage, also known as *Dutch Book*, is a sure-win strategy with no cash outlay.[3] A more precise definition is the following one.

Definition 5.2 (*Arbitrage*) An arbitrage is a trading strategy resulting in a vector $\boldsymbol{F} = (F_0, F_1, \dots, F_n)$ of cash flows such that $F_t \geq 0$ and there exists some $t \in \mathbb{N}_n^0$ such that $F_t > 0$. We call $\boldsymbol{F}$ an *arbitrage vector* or *arbitrage stream*.

If $p_0^A > p_0^B$, a quick-witted investor will soon realize that the mispricing of A and B offers one the opportunity of earning a net gain with no outlay. It suffices to short portfolio A and purchase portfolio B. Shorting portfolio A means that the investor "borrows" p_0^A and uses the cash flow received to invest in B incurring an outlay of p_0^B. The cash flows F_t, $t \in \mathbb{N}_n^1$ generated by B are used to close off the short position.[4]

The net cash-flow stream of this trading strategy (A short, B long) is

[3] de Finetti (1931) was the first to give a precise formal dress to arbitrage. Working under uncertainty, he defined subjective probabilities in terms of the rates at which individuals are willing to bet money on events. Its fundamental theory of subjective probability is also known as the principle of coherence or the *Dutch Book Theorem*. It is a principle of rationality that links arbitrage and risk-neutral probabilities (see Nau 1999, 2001 for details).

[4] In practice, in a short sale, the investor borrows the security from someone who owns it and sells it in the market. The short position is later closed off (i.e., the debt is repaid) by buying back the

$$\begin{aligned}\boldsymbol{F}^B - \boldsymbol{F}^A &= (-p_0^B, F_1, F_2, \ldots, F_n) - (-p_0^A, +F_1, +F_2, \ldots, +F_n) \\ &= (p_0^A - p_0^B, 0, 0, \ldots, 0).\end{aligned}$$

As $p_0^A > p_0^B$, this is a vector where all cash flows are nonnegative and the first one is positive. Therefore, portfolio $\boldsymbol{F}^B - \boldsymbol{F}^A$ represents an arbitrage vector. As soon as this trading strategy becomes apparent, more and more investors will exploit this arbitrage opportunity; hence, the supply of A will increase alongside the demand of asset B. Owing to the law of demand and supply, this will cause an increase in the price of B and a decrease in the price of A. The two prices adjust to a point where the two prices coincide: $p_0^A = p_0^B = V_0$. (The price of A will not fall below the price of B, otherwise a symmetric arbitrage opportunity would arise, consisting in shorting B and purchasing A.)

A market is said to be in **equilibrium** at time t if, at that time, no arbitrage opportunities arise. A market where the law of one price holds is a market where prices are or tend to be in equilibrium, owing to investors who seek arbitrage opportunities. In other words, it is just by seeking arbitrage opportunities that investors sweep away arbitrage opportunities.

The market value V_t of an asset or portfolio is also called the *no-arbitrage price* or *equilibrium price*.

We say that an asset is in equilibrium if it is traded in the market at a price p_0 which is equal to its equilibrium price V_0, that is, the discounted value, at the market rate r, of its future cash flows. Suppose that the market rate is $r = 10\%$ and asset A generates cash flows equal to 40, 30, 20 at time 1, 2, 3, respectively. Suppose also that its current price is $p_0^A = 76.18$. This means that the asset is in equilibrium, for its equilibrium price is

$$V_0^A = \frac{40}{1.1} + \frac{30}{1.1^2} + \frac{20}{1.1^3} = 76.18.$$

In other words, the asset is in line with the prices of all other assets in the market, which guarantees an income rate equal to 10%. If an asset is not in equilibrium, then it is either overpriced (price is higher than the equilibrium price) or underpriced (price is smaller than the equilibrium price). In this case, market forces move to restore the equilibrium.

We have thus far assumed that the market rate is constant. For a more realistic framework, we now generalize and allow for time-varying market rates. Let $\boldsymbol{r} = (r_1, r_2, \ldots, r_n)$ be the vector of equilibrium forward rates holding in the various periods $[t-1, t]$. The fundamental law of motion (5.1) is generalized to

$$V_t = V_{t-1}(1 + r_t) - F_t \quad \text{or, proceeding backward,} \quad V_{t-1} = \frac{V_t + F_t}{1 + r_t}, \tag{5.4}$$

security and giving it back to the lender (as well as paying out any interim cash flows to the lender) (see Broverman 2008; Kellison 2009 for details).

and (5.2) becomes

$$V_t = \sum_{k=t+1}^{n} \frac{F_k}{(1+r_{t+1})(1+r_{t+2}) \cdot \ldots \cdot (1+r_k)}. \tag{5.5}$$

The vector $\boldsymbol{r} = (r_1, r_2, \ldots, r_n)$ describes the so-called *term structure of interest rates*. If $r_t = r$ for all t, the term structure is said to be flat. We will henceforth use the discount factor $\mathrm{d}_{j,h}$, $j \geq h$ to denote the value as of time h of a monetary unit available at time j, with $\mathrm{d}_{j,j} = 1$:

$$\mathrm{d}_{j,h} = \frac{1}{(1+r_{h+1})(1+r_{h+2}) \cdot \ldots \cdot (1+r_j)}.$$

The symbol $\mathrm{u}_{h,j}$ for $h \leq j$ will be used for denoting the value at time j of a monetary unit invested or borrowed at time h, with $\mathrm{u}_{h,h} = 1$:

$$\mathrm{u}_{h,j} = (1+r_{h+1})(1+r_{h+2}) \cdot \ldots \cdot (1+r_j).$$

The discount factor $\mathrm{d}_{j,h}$ discounts cash flows from j to h, while the compound factor $\mathrm{u}_{h,j}$ compounds cash flows from h to j. Obviously, $\mathrm{u}_{h,j} = 1/\mathrm{d}_{j,h}$, so one can write

$$V_t = \sum_{k=t+1}^{n} F_k \mathrm{d}_{k,t} = \sum_{k=t+1}^{n} F_k/\mathrm{u}_{t,k} \tag{5.6}$$

Remark 5.1 (***Spot rates and forward rates***) The equilibrium rates r_t, $t \in \mathbb{N}_n^1$, are single-period *forward* rates expressing the interest on any investment or borrowing starting at time $t-1$ and ending at time t. More generally, a forward rate, $r_{h,j}$, expresses the interest on a \$1 lent or borrowed at time h for $j-h$ periods. The law of one price guarantees that the following equality holds for every $h < j$, $h, j \in \mathbb{N}_n^1$:

$$(1+r_{h,j})^{j-h} = (1+r_{h+1})(1+r_{h+2}) \cdot \ldots \cdot (1+r_j) = \mathrm{u}_{h,j} = 1/\mathrm{d}_{j,h}$$

(if $h = j-1$, the rates are single-period rates: $r_{j-1,j} = r_j$).

The term structure of interest rates may also be described in terms of *spot* rates. A spot rate, $s_t = r_{0,t}$, is the interest on a \$1 investment or financing that starts at time 0 and ends up at time t. The law of one price guarantees that, for every $t \in \mathbb{N}_n^1$, the following equality holds:

$$(1+s_t)^t = (1+r_1)(1+r_2) \cdot \ldots \cdot (1+r_t) = \mathrm{u}_{0,t} = 1/\mathrm{d}_{t,0} \tag{5.7}$$

whence

$$s_t = \left(\frac{1}{\mathrm{d}_{t,0}}\right)^{1/t} - 1$$

(note that $s_1 = r_{0,1} = r_1$). In general, the law of one price implies the following relation between the structure of spot rates and the structure of forward rates:

$$(1+s_h)^h \cdot (1+r_{h,t})^{t-h} = (1+s_t)^t \tag{5.8}$$

for every $h < t$, whence

$$r_{h,t} = \left[\frac{(1+s_t)^t}{(1+s_h)^h}\right]^{1/(t-h)} - 1.$$

For simplicity, we will use single-period forward rates throughout the book: $r_{t-1,t} = r_t$. ▲

Example 5.2 ♦

The term structure of interest rates of a normal market is $\boldsymbol{r} = (0.01, 0.04, 0.05, 0.07)$. A financial asset is introduced in the market which will release cash flows equal to \$10 from time 1 to time 4, as well as a terminal cash flows of \$100 at the end of the fourth period. Using (5.6) for $t = 0$, and assuming equilibrium, the asset will then be necessarily priced at

$$121.7 = \frac{10}{1.01} + \frac{10}{(1.01)(1.04)} + \frac{10}{(1.01)(1.04)(1.05)} + \frac{110}{(1.01)(1.04)(1.05)(1.07)}.$$

The law of one price implies the principle of **value additivity**. Consider asset A and assume that the prospective cash flows of A can be replicated with a portfolio of n_1 units of assets A_1, n_2 units of assets A_2, ..., n_m units of asset A_m. We write $A = \sum_{j=1}^m n_j \cdot A_j$ to denote this portfolio. The no-arbitrage price of A is the linear combination of the no-arbitrage prices of the constituent assets: $V_t^A = \sum_{j=1}^m n_j \cdot V_t^{A_j}$. As a particular case, if $n_j = 1$ for all j, then

$$V_t^A = \sum_{j=1}^m V_t^{A_j} \tag{5.9}$$

which formalizes the principle of value additivity.

Indeed, if the law of one price holds, then $V_0^A = p_0^A$ and $\sum_{j=1}^m n_j \cdot V_0^{A_j} = \sum_{j=1}^m n_j \cdot p_0^{A_j}$. If p_0^A were different from $\sum_{j=1}^m n_j \cdot p_0^{A_j}$, a state of temporary disequilibrium would arise and rational investors would take advantage of the arbitrage opportunity, combining $A, A_1, A_2, \ldots, A_m$ into an arbitrage trading strategy such that the resulting cash-flow stream $\boldsymbol{F} = (F_0, F_1, \ldots, F_n)$ is an arbitrage stream, with $F_t = F_t^A - \sum_{j=1}^m n_j F_t^{A_j}$ if $p_0^A < \sum_{j=1}^m n_j \cdot p_0^{A_j}$ or $F_t = -F_t^A + \sum_{j=1}^m n_j F_t^{A_j}$ if $p_0^A > \sum_{j=1}^m n_j \cdot V_0^{A_j}$.

Example 5.3 ♦
Consider asset A, traded in a normal market, with prospective cash flows equal to \$100, \$120, and \$75 at time 1, 2, and 3. Consider three zero-coupon bonds (zcb) A_1, A_2, A_3. The first zcb generates a cash flow of \$50 after one period, the second zcb generates a cash flow equal to \$30 after two periods, the third zcb generates a cash flow equal to \$15 after three periods. Let $V_0^{A_1} = 47.85$, $V_0^{A_2} = 27.47$, $V_0^{A_3} = 13.14$ be the respective equilibrium prices of the zcb's. This means that the portfolio $2A_1 + 4A_2 + 5A_3$ replicates asset A's prospective cash flows. Hence, its equilibrium price is

$$V_0^A = 2V_0^{A_1} + 4V_0^{A_2} + 5V_0^{A_3} = 271.3.$$

If the price of asset A were higher or lower (overpricing or underpricing), then investors would exploit the arbitrage opportunities and would quickly sweep away the state of disequilibrium, so that the price of asset A would equal the price of the replicating portfolio.

Example 5.4 ♦
Consider an asset, A_1, and suppose it will generate cash flows equal to \$100, \$80, and \$50 at time 1, 2, and 3 respectively. The term structure of interest rates in the (assumed normal) market is flat; in particular, the market rate is 10%. The current price of A_1 is \$194.6. This means that A_1 is in equilibrium, because

$$194.6 = \frac{100}{1.1} + \frac{80}{1.1^2} + \frac{50}{1.1^3}.$$

Now, consider assets A_2, A_3, A_4, such that A_2 generates a cash flow equal to \$20 in one period, A_3 generates a cash flow equal to \$10 in two periods, and A_4 generates a cash flow equal to \$50 in three periods. Suppose A_2 is currently traded at $p_0^{A_2} = 19$, A_3 is traded at $p_0^{A_3} = 9$, and A_4 is traded at $p_0^{A_4} = 40$. These assets are overpriced, their equilibrium prices being $V_0^{A_2} = 20(1.1)^{-1} = 18.18$, $V_0^{A_3} = 10(1.1)^{-2} = 8.26$, $V_0^{A_4} = 50(1.1)^{-3} = 37.57$, respectively. Rational investors soon realize that this state of disequilibrium creates an arbitrage opportunity. Therefore, they short 5 units of A_2, 8 units of A_3, and 1 unit of A_4 and, at the same time, purchase asset A_1.[5] The resulting portfolio is $1A_1 - 5A_2 - 8A_3 - 1A_4$ which generates an arbitrage vector equal to

$$\boldsymbol{F} = (12.4, 0, 0, \dots, 0)$$

(see Table 5.1).

[5] We recall that shorting is financially equivalent to borrowing.

Table 5.1 An example of arbitrage trading strategy

Time	purchase 1 unit of A_1	sell short 5 units of A_2	sell short 8 units of A_3	sell short 1 unit of A_4	Net cash flow
0	−194.6	$5 \cdot 19$	$8 \cdot 9$	$1 \cdot 40$	12.4
1	100	$-5 \cdot 20$	0	0	0
2	80	0	$-8 \cdot 10$	0	0
3	50	0	0	$-1 \cdot 50$	0

The prices of A_2, A_3, A_4 will soon decrease, as an effect of greater supply, up to that point where they are traded at their equilibrium prices, and disequilibrium is swept away.

5.2 A Benchmark System

The law of one price and the related principle of value additivity represent static equilibria determined by the forces that act on the market where securities are traded. However, the law of conservation presented in Part I refers to the equilibrium of the two sides of one system (investment and financing); in contrast, the one-price law refers to the reciprocal equilibrium of any two different economic entities traded in the market. Thus, if the law of one price holds, *all* assets traded in the market are in (reciprocal) equilibrium. For this reason, one may say that the replicating portfolio represents the capital market as a whole. One may use the following concise transposed Matrix to describe the replicating portfolio:

$(\boldsymbol{V}, \boldsymbol{I}^V, \boldsymbol{F}^V)$

Time		0	1	2	...	$n-1$	n
Capital	$\boldsymbol{V}$	V_0	V_1	V_2	...	V_{n-1}	0
Income	$\boldsymbol{I}^V$	0	I_1^V	I_2^V	...	I_{n-1}^V	I_n^V
Cash Flow	$\boldsymbol{F}^V$	$-V_0$	F_1	F_2	...	F_{n-1}	F_n

(where $I_t^V = r_t V_{t-1}$). This system acts as a benchmark system, as opposed to the project system, characterized by the following concise transposed Matrix:

$(\boldsymbol{C}, \boldsymbol{I}, \boldsymbol{F})$

Time		0	1	2	...	$n-1$	n
Capital	C	C_0	C_1	C_2	...	C_{n-1}	0
Income	I	I_0	I_1	I_2	...	I_{n-1}	I_n
Cash Flow	F	F_0	F_1	F_2	...	F_{n-1}	F_n

The replicating system is then a benchmark system, which acts as a yardstick, whereby the market is contrasted with the project. We have written "a" benchmark system, not "the" benchmark system, for there are other ways of benchmarking a project (and, therefore, other ways of representing the market) which give rise to other measures of worth. We will analyze them in the next chapters. In the next section, we show that shareholders' wealth is increased if and only if the project system and the replicating system are not in reciprocal equilibrium, that is, if and only if a state of disequilibrium occurs such that the outflow $|F_0|$ implied by the project is smaller than the outflow V_0 implied by the replicating system.

Remark 5.2 (***MM*** 1958, ***Proposition I***) The law of conservation and the law of one price imply that, in the market, monetary values at a particular date can be split up or combined together in several different ways with no effect: The sum of the parts will always be equal to the whole and the same amount will be shared by any asset. This is the core of Modigliani and Miller's (1958) Proposition I: Assuming a normal market, the authors show that (i) given a fixed investment policy, the value of a firm does not depend on the equity/debt ratio and (ii) any equal-risk firm has the same value, provided that the investment policy is the same (i.e., provided that the prospective cash flows of all equally risky firms are the same) (see also Proposition 6.1 later in the book). ▲

5.3 Rational Decision-Making and Net Present Value

Economic rationality is one where investors prefer more money to less money (wealth maximization). In a rational market, arbitrage opportunities are not possible or, if they occur, they vanish away very soon, in few minutes or even seconds, for rational investors pursue wealth maximization and abide by the law of demand and supply. This is also known as the *arbitrage principle* or *no-arbitrage principle*. This principle may be considered the primal characterization of economic rationality with respect to which the law of one price represents a dual characterization (see Ross 1978; Varian 1987; Nau and McCardle 1991; Nau 1999).

At the market level, the no-arbitrage principle represents the unifying principle of rational choice theory in economics (see Nau and McCardle 1991): A normal market is rational, in the sense that the group of individuals that invest in it does not give rise to arbitrage opportunities. At the individual level, decision makers are rational if and only if they exploit any riskless profit opportunity whenever it is available.

In other words, groups in financial markets rationally tend to ensure no-arbitrage (i.e, tend to equilibrium), whereas individual investors search for arbitrage gains (i.e., tend to disequilibrium). One might even say that no-arbitrage is a principle of social rationality, while arbitrage-seeking is a principle of individual rationality, equivalent to the principle of wealth maximization.[6]

Therefore, a firm's management should rationally tend to create the conditions (via new products, new technologies, etc.) that trigger a temporary disequilibrium such that the wealth of the firm's shareholders is maximized. If this occurs, the firm's management is said to *create value* (or *create wealth* or *add value*) for shareholders. Assuming a firm's shares are traded in the financial market, a rational decision made by the firm's managers is to pursue any action that leads to an increase in the firm's share price.

Definition 5.3 (*Rational decision-making*) A decision made by the firm's managers is rational if and only if the firm's share price increases as a consequence of it. A decision criterion is rational if it leads to decisions that are rational.

Let n_0 be the number of outstanding shares of firm f, currently traded in the market at an equilibrium price of p_0. The current market value of the firm's equity is then $p_0 \cdot n_0$. Assume that the firm's (fixed) investment policy is expected to generate a prospective equity cash-flow vector equal to $(f_1^e, f_2^e, \dots f_N^e)$, $N \leq +\infty$. Suppose the firm has the opportunity of undertaking a project P whose cost is C_0 and the prospective stream of CFAs is estimated to be $(F_1, F_2, \dots F_n)$.

For the time being, we assume that the project is unlevered and will be financed with issuance of new shares. This implies that the CFAs coincide with the CFEs: $F_t = F_t^e$.[7]

Let us analyze what happens when the firm's management announces the undertaking of the project (we assume, for illustrative purposes, that $F_0 < 0$, i.e., the project starts with an initial investment):

[6] As anticipated in the previous section, it is precisely the rational behavior of the individuals seeking disequilibrium that guarantees that the market as a whole tends to equilibrium.

[7] We will remove this assumption in Chap. 6, showing that the same result holds for levered projects.

- the market reacts to the announcement and the share price changes by Δp_0
- the firm issues $\Delta n_0 = C_0/(p_0 + \Delta p_0)$ shares at the new price $p_0 + \Delta p_0$ to finance the project
- the new equity value becomes $(n_0 + \Delta n_0) \cdot (p_0 + \Delta p_0)$

Consider a portfolio consisting of (i) firm f without the project and (ii) a security which replicates the project's cash flows; the portfolio's market value is

$$\overbrace{n_0 \cdot p_0}^{\text{value of the firm without the project}} \qquad + \qquad \overbrace{V_0}^{\text{value of the replicating security}}$$

where $V_0 = \sum_{t=1}^{n} F_t \mathrm{d}_{t,0}$ and $\mathrm{d}_{t,0}$ is the present value of a monetary unit available at time t (see Sect. 5.1). The law of one price implies that this portfolio and the equity of the firm with the project are traded at the same price (for they generate the same cash-flows):

$$n_0 \cdot p_0 + V_0 = \overbrace{(n_0 + \Delta n_0) \cdot (p_0 + \Delta p_0)}^{\text{value of the firm with the project}} \tag{5.10}$$

whence, recalling that $\Delta n_0 \cdot (p_0 + \Delta p_0) = -F_0$,

$$n_0 \cdot p_0 + V_0 = n_0 \cdot (p_0 + \Delta p_0) - F_0$$

which implies

$$V_0 + F_0 = n_0 \cdot \Delta p_0. \tag{5.11}$$

Owing to Eq. (5.11), $\Delta p_0 > 0$ if and only if $V_0 + F_0 > 0$.

In Eq. (5.11), V_0 is the price of the security replicating the project's cash-flow stream and represents the so-called *market value of project P*:

price of the replicating security = market value of project P.

V_0 is then the amount of money that has to be invested in the security market for replicating the stream of the project's cash flows. Put another way, the market value of a project is "the price the project would have if it were traded" (Mason and Merton 1985, pp. 38–39). The left-hand side of (5.11) measures the difference between the project's market value and its cost: The firm pays $|F_0|$ to get an asset whose value is V_0. Using the equality $V_0 = \sum_{t=1}^{n} F_t \mathrm{d}_{t,0}$, we may then state the following definition.

Definition 5.4 (*Project value and NPV*) Given a cash-flow stream $\boldsymbol{F} = (F_0, F_1, \ldots, F_n) \in \mathbb{R}^{n+1}$, the *market value* of a project at time 0 is the discounted sum of its prospective cash flows:

$$V_0 = \sum_{t=1}^{n} F_t \, \mathrm{d}_{t,0}. \tag{5.12}$$

The net present value (NPV), also known as net present worth (NPW), is defined as the discounted sum of all cash flows:

$$\text{NPV} = V_0 + F_0 = \sum_{t=0}^{n} F_t \, \mathrm{d}_{t,0}. \tag{5.13}$$

The latter may be reframed as

$$\text{NPV} = V_0 + I_0 - C_0 = \sum_{t=0}^{n} F_t \, \mathrm{d}_{t,0} \tag{5.14}$$

or, if $I_0 = 0$,

$$\text{NPV} = V_0 - C_0 = \sum_{t=0}^{n} F_t \, \mathrm{d}_{t,0}. \tag{5.15}$$

Assuming a flat structure of interest rates (i.e., $r_t = r$ for all t), the NPV is equal to $\text{NPV} = \sum_{t=0}^{n} F_t \, (1+r)^{-t}$, where the market rate r represents the growth rate of capital required by investors for assets traded in the financial market. From the point of view of the firm facing the opportunity of investing in project P, the market rate expresses a foregone income rate, in the sense that, if the firm undertakes the project, it will forego the opportunity of earning incomes at a rate of r. For this reason, it is called (opportunity) ***cost of capital*** (see Sect. 5.4). The NPV is then the algebraic sum of the project's CFAs discounted at the cost of capital. (If the term structure of interest rates is not flat, then the project's cash flows are discounted at a time-varying cost of capital r_t: $\text{NPV} = \sum_{t=0}^{n} F_t (1+r_1)^{-1}(1+r_2)^{-1} \cdot \ldots \cdot (1+r_t)^{-1}$.) Investors' wealth increases if and only if the project is worth more than it costs, that is, if the NPV is positive. As $\text{NPV} = n_0 \Delta p_0$, a positive NPV signals an arbitrage gain: The share price increases by Δp_0 so shareholders will benefit from an arbitrage vector equal to $(\Delta p_0, 0, 0, \ldots, 0)$ for each share. Therefore, a rational decision criterion is to accept all projects with positive NPV and reject all projects with negative NPV.

Definition 5.5 (*NPV criterion*) A project generating a cash-flow vector equal to $\boldsymbol{F}$ is worth undertaking (i.e., creates value) if and only if its NPV is positive

$$\text{NPV} = \sum_{t=0}^{n} F_t \mathrm{d}_{t,0} > 0.$$

Owing to Definition 5.3, we can state the following proposition.

Proposition 5.2 (Rationality of NPV criterion) *The NPV criterion is rational. Managers create value for shareholders if and only if they undertake positive-NPV projects.*

In a normal market, the law of one price holds; then, the NPV of a security transaction is necessarily zero. We may then state the following proposition.

Proposition 5.3 (Law of One Price and NPV) *In a normal market, any (investment or financing) transaction is value-neutral, that is, NPV = 0. In words: What one pays is what one gets.*

In general, we call ***equilibrium asset*** any asset which is in equilibrium, that is, it is value-neutral (see Sect. 5.1). An equilibrium asset neither creates nor destroys value, because its NPV is zero. In contrast, we call ***disequilibrium asset*** any activity whose NPV is nonzero. The NPV criterion states that managers should invest in disequilibrium projects with positive NPV (investing in disequilibrium projects with negative NPV destroys value).

As anticipated, the notion of economic value created (and, therefore, net present value) is the result of the comparison of two alternative economic systems: The project system and a benchmark (replicating) system. In this case, the replicating system shares the same prospective inflows and outflows as the project's, but the dynamics is different, owing to the difference in the income rates (i_t as opposed to r_t). As seen, the two equations of motion are, respectively,[8]

$$C_t = C_{t-1} + I_t - F_t = [\text{if } C_{t-1} \neq 0] = C_{t-1}(1 + i_t) - F_t$$
$$V_t = V_{t-1} + I_t^V - F_t = V_{t-1}(1 + r_t) - F_t.$$

[8]Notice that the market income rate, r_t, is always defined, and then it is applied to some capital, whereas the project income rate may not be defined, since it derives from capital, which may well be zero.

Both systems enjoy static equilibrium at any given point in time, in the sense that each possesses two sides, investment and financing (purchasers of the replicating portfolios will raise funds by equity and/or debt as well as the firm undertaking the project).

The mechanics of the two economic systems is affected by different forces. The forces acting on the project system have to do with transactions with customers, suppliers, employees, etc. The forces acting on the benchmark system are the same forces that act on the capital market, which have to do with demand and offer of securities by investors who trade in the market.

The comparison of the project system and the capital market (via the replicating system) is a comparison of alternative forces. The two systems are in reciprocal equilibrium if they share the same equilibrium in the sense that NPV $= 0$; they are in **reciprocal disequilibrium** if NPV $\neq 0$, which means that the forces acting on the project system do not match the forces acting on the market.

NPV is a measure of the degree of reciprocal disequilibrium. Wealth is increased if and only if disequilibrium arises in a positive direction (NPV > 0).

Remark 5.3 (*Old and new shares*) The shareholders who purchase the new shares issued to finance the project undertake a value-neutral activity, for they purchase at a price C_0 an asset whose value is just C_0. To see this, consider that the shareholders who purchase the new shares expect to get a cash flow equal to $(\Delta n_0/(n_0+\Delta n_0)) \cdot (f_t^e + F_t)$ at time $t \in \mathbb{N}_N^1$. The market value of this cash-flow stream is

$$\sum_{t=1}^{N} \frac{\Delta n_0}{n_0+\Delta n_0} \cdot \frac{f_t^e + F_t}{(1+r)^t} = \frac{\Delta n_0}{n_0+\Delta n_0}(n_0 p_0 + V_0) \tag{5.16}$$

(with $F_t = 0$ for $t > n$).[9] The NPV generated by investing in the new shares is equal to the difference between the value reported in (5.16) and the price paid for buying the shares (which is equal to $\Delta n_0(p_0+\Delta p_0)$). This means

$$\text{NPV of new shares} = \overbrace{\frac{\Delta n_0}{n_0+\Delta n_0}(n_0 p_0 + V_0)}^{\text{value of investing in new shares}} - \overbrace{\Delta n_0(p_0+\Delta p_0)}^{\text{cost of investing in new shares}}.$$

From (5.10), the above difference is zero. Therefore, the new shares are value-neutral; the purchasers of the new shares just get what they pay for. Put it equivalently,

[9]Equation (5.16) uses the equality $\sum_{t=1}^{N} f_t^e/(1+r)^t = n_0 p_0$ which holds because, before undertaking the project, the equity of firm f is an equilibrium asset.

$$\sum_{t=1}^{N} \frac{\Delta n_0}{n_0 + \Delta n_0} \cdot \frac{f_t^e + F_t}{(1+r)^t} = C_0 \tag{5.17}$$

As for the "old" (i.e., already existing) shares, consider that shareholders owning the stock before the project announcement expect to receive f_t at time t; after the project announcement and the issuance of Δn_0 shares, they expect to receive $n_0/(n_0 + \Delta n_0) \cdot (f_t + F_t)$ at time t. The cash-flow difference between the two scenarios is $n_0/(n_0 + \Delta n_0) \cdot (f_t + F_t) - f_t$ for $t \in \mathbb{N}_N^1$. The market value of this difference is the project's value for the shareholders:

$$\sum_{t=1}^{N} \frac{n_0}{n_0 + \Delta n_0} \frac{f_t^e + F_t}{(1+r)^t} - \sum_{t=1}^{N} \frac{f_t^e}{(1+r)^t} = \sum_{t=1}^{N} \frac{n_0}{n_0 + \Delta n_0} \frac{f_t^e + F_t}{(1+r)^t} - n_0 p_0.$$

Since $\sum_{t=1}^{N}(f_t^e + F_t)/(1+r)^t = n_0 p_0 + V_0$, and using (5.10), the above amount is equal to

$$\begin{aligned}\frac{n_0}{n_0 + \Delta n_0} \cdot (n_0 p_0 + V_0) - n_0 p_0 &= \frac{n_0 \cdot (n_0 + \Delta n_0) \cdot (p_0 + \Delta p_0)}{n_0 + \Delta n_0} - n_0 p_0 \\ &= n_0 \cdot (p_0 + \Delta p_0) - n_0 p_0 \\ &= n_0 \Delta p_0 = \text{NPV}.\end{aligned} \tag{5.18}$$

The value of undertaking the project to the existing shareholders is equal to the project's NPV. Therefore, value is created only for the current shareholders, not for the new shareholders:

> The definition of "old" or "current" stocks and bonds is useful when considering financing-investment decisions at a point in time, e.g., at time 0. The objective is to maximize the wealth of shareholders in the firm at the time the decisions are made. If new shares are issued at time 0, it is in order to benefit holders at time 0 of shares other than the new shares, i.e., to benefit the "old" shareholders at time 0. (Haley and Schall 1979, p. 212, footnote 4 in Chap. 2)

(See also Remark 6.3.) ▲

> The NPV is equal to the increase in value of the old shares: NPV $= n_0 \Delta p_0$.

Remark 5.4 (*The paradox of investment policy irrelevance*) The assumption that a market is in equilibrium should not be intended in the sense that it remains in a state of perpetual equilibrium, with no disequilibrium shifts. If one were to assume perpetual equilibrium, one would incur the following paradox:

> Whenever any project undertaking is announced, the firm's share does not change for, by assumption, no disequilibrium occurs. This means that any project is value-neutral (zero

NPV). This implies that the firm's investment policy is irrelevant and, therefore, that the decisions made by the firm's management on investments are irrelevant: Whatever the decision made, the same (neutral) effect on shareholder wealth will be produced. Hence, if the management dissipates money, shareholders' wealth will be unaffected.

This paradox, which we call the paradox of investment policy irrelevance, demonstrates that the assumption of a perpetual equilibrium is theoretically unacceptable (even more so from an applicative point of view). One should never think of a normal market as one where equilibrium persists in perpetuity. Rather, one should conceive the state of equilibrium as a temporary one, followed by a state of disequilibrium, which is arbitraged away by rational investors, followed by another state of equilibrium, which is followed in turn by another disequilibrium shift, which peters out via arbitrage, and so on. From this point of view, a financial market may be thought of as a pricing mechanism which perpetually yo-yos between states of equilibrium and states of disequilibrium. The tendency to equilibrium derives from rational investors willing to exploit arbitrage opportunities by purchasing and selling financial assets in the market; the tendency to disequilibrium is caused by several factors, including the firms' decisions about investment policy (new products, new technologies, expansion of existing projects, etc.).

Indeed, the role of managers is precisely to actively find and exploit, in the "product" markets, the conditions for positive-NPV investments and, hence, to perturbate the financial market (where the firms' shares are traded) so as to cause a state of disequilibrium, which is soon pumped away by the investors operating in the financial market, until other positive-NPV investments are undertaken (which cause new perturbations in the financial market, and so on in an endless buzz). ▲

Example 5.5 ♦

Let $\boldsymbol{F} = (-100, 40, 50, -30, 120)$ be the vector of an unlevered project's CFAs and let $r = 6\%$ be the assumed constant cost of capital. The market value of the project is

$$V_0 = \frac{40}{1.06} + \frac{50}{1.06^2} - \frac{30}{1.06^3} + \frac{120}{1.06^4} = 152.1.$$

Therefore, the NPV is NPV $= V_0 + F_0 = 152.1 - 100 = 52.1$. The project creates value and is worth undertaking.

Example 5.6 ♦

Consider a firm facing an unlevered project and let $\boldsymbol{F} = (-40, 20, 15, -5, 25)$ be the sequence of the project's CFAs. Suppose the term structure of equilibrium rates is $\boldsymbol{r} = (10\%, 15\%, 25\%, 35\%)$. The vector of discount factors is $\mathbf{d} = (1, 0.90909, 0.79051, 0.63241, 0.468453)$. The market value of P is

$$V_0 = +20 \cdot 0.90909 + 15 \cdot 0.79051 - 5 \cdot 0.63241 + 25 \cdot 0.46845 = 38.589.$$

The NPV is negative because NPV $= 38.589 - 40 = -1.411$, so the project destroys value. Therefore, it should not be undertaken.

Example 5.7 ♦
Let $\boldsymbol{F} = (-88, 40, 50.1, 19.75)$ be the vector of CFAs and $\boldsymbol{r} = (0.09, 0.15, 0.39)$ be the vector of costs of capital of an unlevered project. The vector of discount factors is $\mathbf{d} = (1, 0.91743, 0.79777, 0.57393)$. This project is value-neutral, since $\text{NPV} = -88 + 40 \cdot 0.91743 + 50.1 \cdot 0.79777 + 19.75 \cdot 0.57393 = 0$. The project is an equilibrium asset: Undertaking the project is equivalent to investing the same amount of money in the financial market.

Remark 5.5 (Historical account on NPV) The pivotal role of NPV maximization in finance, accounting, and engineering economics is unanimously accepted (e.g., Finnerty 1986; Copeland and Weston 1988; Rao 1992; Hartman 2007; Brealey et al. 2011; Park 2011; Blank and Tarquin 2014; Berk and DeMarzo 2014) and has been given a solid theoretical status by Fisher (1930). However, discounting and compounding for valuing a stream of cash flows were already known since Medieval times. Across centuries, mathematicians, actuaries, engineers, and economists played a major role in the development of the so-called discounted-cash-flow methods. The first to theoretically envisage the method of present values for comparing alternative courses of action was the Italian mathematician Leonardo [da Pisa] Fibonacci in his *Liber Abaci* [1202] (Goetzmann 2004; Goetzmann and Rouwenhorst 2005). Luca Pacioli's *Summa de Arithmetica, Geometria, Proportioni et Proportionalità* [1494] treated problems on simple and compound interest and systematically presented, for the first time, the double-entry book-keeping system used by merchants of Venice, Genoa, Pisa, and Florence.[10] Problems of financial mathematics were treated by the Dutch mathematician and engineer Simon Stevin (1548–1620), who apparently was the first to describe the NPV rule for decision-making, although limiting its application to loans (Parker 1968). Arthur M. Wellington, an American civil engineer, was the first one to have recommended the use of NPV in industry for selecting among different projects (Wellington 1887. See also Scorgie 1965; Dulman 1989). Eugene L. Grant's (1930) *Principles of Engineering Economy* had a primary role in the development of engineering economics by "emphasizing management's role and the need for a thorough investigation of investment proposals" (Scorgie 1965, p. 87. See also Parker 1968 for a detailed account of the roles of actuaries, political economists, engineers, and accounting scholars in the birth and dissemination of discounted-cash-flow methods).

In microeconomics, the NPV criterion is sanctioned as the solution of a constrained optimization process, where the investor maximizes his utility function with respect to the consumption plan (see MacMinn 2005. See therein for the Fisher model under uncertainty. See also Copeland and Weston 1988). Net present value maximization is then commonplace: "The firm attempts to maximize the present value of its net cash flow over an infinite horizon" (Abel 1990, p. 755); "the net present value rule is also the basis for the neoclassical theory of investment …Much of the theoretical and empirical literature on the economics of investment deals with issues

[10] Curiously, Luca Pacioli might have been taught mathematics by the Italian Renaissance painter Piero della Francesca and, in turn, taught mathematics to Leonardo da Vinci (Gleeson-White 2012, p. 8).

of this sort" (Dixit and Pindyck 1994, p. 5); "The NPV analysis is equivalent to that of Jorgensen (1963) and to the q theory of investment [Tobin 1969]" (Magni 2009b, p. 124. See also Yoshigawa 1980); "In all of this, the underlying principle is the basic net present value rule" (Dixit and Pindyck 1994, p. 5) and "the net present value rule is appropriate for decision making" (MacMinn 2005, p. 1):

> The neoclassical theory of corporate investment is based on the assumption that the management seeks to maximize the present net worth of the company, the market value of the outstanding common shares. An investment project should be undertaken if and only if it increased the value of the shares. The securities markets appraise the project, its expected contributions to the future earnings of the company and its risks. If the value of the project as appraised by investors exceeds the cost, then the company's shares will appreciate to the benefit of existing stockholders. That is, the market will value the project more than the cash used to pay for it. If new debt or equity securities are issued to raise the cash, the prospectus leads to an increase of share prices . . . Clearly, it is the q ratio on the margin that matters for investment: the ratio of the increment of market valuation to the cost of associated investment. (Tobin and Brainard 1977, p. 242, as quoted in Yoshigawa 1980, p. 739)

▲

5.4 Cost of Capital

To better understand the meaning and the role of the cost of capital, we need to introduce the notion of *opportunity cost*. Opportunity cost is the income of the benchmark system. Historically, economists have conceptualized this income in terms of *counterfactual conditionals*. We will briefly focus on this notion, because it is strictly connected with the notion of value creation.

Counterfactual conditionals are defined by undoing a given *factual* scenario and constructing a suitable *counterfactual* scenario which describes what might occur (or what might have occurred) if the factual scenario did not occur (had not occurred).[11] A typical counterfactual conditional has the form

'If it were the case that A, then it would be the case that B'.

Counterfactual conditionals are used in almost every domain and are ubiquitous in daily life (Kahneman and Tversky 1982; Wells et al. 1987; Roese and Olson 1995; Harris et al. 1996; Roese 1997; Byrne and Tasso 1999; Byrne 2002; Byrne 2016a, b). The role of counterfactuals in causation has been investigated for a long time by philosophers of science (Chisholm 1946; Will 1947; Goodman 1947, 1983; Popper 1949; Kneale 1950; Lewis 1973; Sosa 1975) and they are pervasive in economic thinking as well: They are tools economists often adopt to explore the world and construct their concepts and models (Sugden 2000; Hülsmann 2003).

The notion of opportunity cost has been developed by Austrian economists (in particular Ludwig von Mises) and economists of the London School of Economics (e.g., Coase, Thirlby, Shackle):

[11] The term counterfactual was coined by Goodman (1947).

> The person is faced with the possibility of taking one or other of (at least) two courses of action, but not both. He considers the relative significance to him of the two courses of action, and finds that one course is of higher significance than the other. He prefers one course to the other. His prospective opportunity of taking the less preferred course of action becomes the prospective cost of his taking the more preferred course. (Thirlby 1946, p. 33)

The connection of the opportunity-cost notion with counterfactual conditionals was explicitly recognized, as well as the link with profit or wealth maximization:

> The cost of doing anything consists of the receipts which *could have been obtained if that particular decision had not been taken*. When someone says that a particular course of action 'is not worth the cost', this merely means that he prefers some other courses . . . This particular concept of cost would seem to be the only one which is of use in the solution of business problems, since it concentrates attention to the alternative courses of action which are open to the businessman. Costs will only be covered if he chooses, out of the various courses of action which seem open to him, that one which maximizes profit. (Coase 1938, p. 123, as quoted in Buchanan 1969, italics supplied)

In this light, opportunity cost (does not measure an outflow, but) quantifies the outcome of a course of action which is alternative to the factual one; it is the outcome the decision maker would receive (have received) if she undertook (had undertaken) a different course of action (see also Magni 2009b). In terms of our accounting-and-finance engineering system, the factual scenario faced by the firm consists in the project system (investing in the project), while the counterfactual scenario consists in the benchmark system (investing in a replicating portfolio[12]). A benchmark system is then associated with a counterfactual conditional:

> 'If the firm invested X in the security market instead of investing in the project, then the return would be Y'

or, as a variant of it,

> 'If any investor invested in the security market the same amount X that the firm invests in the project, the investor's return would be Y.'

Specifically, the return Y is the opportunity cost of investing in the project. Opportunity cost is then the profit of the foregone opportunity; it is the counterfactual income associated with the investment in the market, as opposed to the factual income received (or to be received) in actual facts from the project. This foregone profit acts as a benchmark, a norm in the sense of Kahneman and Miller (1986), and is also called '*normal profit*' (e.g. Edey 1957; Bodenhorn 1964; Carsberg 1966; Archer and D'Ambrosio 1972; Begg et al. 1984).

The market rate r_t expresses the normal profit per unit of invested capital: It is called (opportunity) *cost of capital* (henceforth, often COC).

It should be stressed that the notions of opportunity cost and (opportunity) cost of capital are not synonymous. The former denotes the normal profit, the latter denotes the *ratio* of the normal profit to the BOP capital.

---

[12] Or, equivalently, letting the capital providers invest themselves in the normal market. If the market is efficient, transactions costs are negligible and it is irrelevant whether the firm or the firm's shareholders invest in the market.

Definition 5.6 (*Normal or benchmark profit*) The normal or benchmark profit (or opportunity cost) is the counterfactual profit that would be earned if funds were invested in a benchmark system. (If capital is negative, it is the interest that would be paid by borrowing from the market as opposed to borrowing from the project). (See also Definition 6.6)

The normal profit may also be named "market return". The expressions "opportunity cost", "market return", "counterfactual return", "normal profit", "benchmark profit" represent synonymous expressions. This notion acts as a benchmark in the sense that, assuming the normal profit earned in a normal market is the highest profit available to investors, any agent aims at undertaking economic activities which, for a given amount invested (borrowed), generate an income above (below) the normal profit.

Remark 5.6 The expression "counterfactual" literally means "contrary to fact". As a matter of fact, if the project's NPV is negative (i.e., the project's income is smaller than the normal profit), then the firm will not pursue the counterfactual alternative (i.e., invest in the market at the cost of capital). Rather, it will seek other value-creating (i.e., positive NPV) opportunities. The market is only used as a benchmark system. If no investment opportunities are found that generate income over and above the normal profit, then, from an economic point of view, the firm should cease operations and distribute back the funds to shareholders, who may invest themselves in the market with passive investments earning the cost of capital. ▲

Definition 5.7 (*Cost of capital*) The cost of capital (COC), denoted as r or r_t if it is time-varying, is the *normal* rate of return. It is a benchmark (rate of) return at which investors may alternatively invest their funds. It is the income rate of *any* benchmark system. The COC is also known as *required rate of return* or *minimum attractive rate of return* (MARR).

Remark 5.7 It is worthwhile reminding that an income rate can be an investment rate or a financing rate (see Table 1.4); thus, the COC (and the normal profit) may be interpreted accordingly. For example, if $C_{t-1} < 0$ for every t, then r_t measures the financing rate foregone by the economic agent. In other words, it represents the interest rate that might be paid if one borrowed in the market instead of borrowing from the project. We will also use the expression *maximum attractive financing rate* (MAFR) to denote this financing COC. ▲

Example 5.8 ♦

(*Normal profit*) Formally, since we are using the benchmark system illustrated in Sect. 5.2, the normal profit is $I_t^V = r_t \cdot V_{t-1}$. For example, consider a one-period investment of \$150 producing income at a rate of $i = 10\%$, and let $r = 8\%$ be the market rate. The normal profit is $0.08 \cdot V_0 = 0.08 \cdot (150 \cdot 1.1)/1.08 = 0.08 \cdot 152.778 = \$12.222$, whereas the actual income is $0.1 \cdot C_0 = 0.1 \cdot 150 = \$15$. The difference between actual income and normal income gives the above-normal profit: $I_1 - I_1^V = 15 - 12.222 = 2.778$. This is exactly equal to the NPV: $150 \cdot 1.1/1.08 - 150 = 2.778 = \text{NPV}$.

However, this is not the only way of defining a normal profit. In the next chapters, we will see two other ways of benchmarking a project and, therefore, two other ways of defining a normal profit. To anticipate, just consider that, by investing in the project, an investor earns \$15. However, he might invest the same amount at 8% market rate. This makes an income of $8\% \cdot 150 = \$12$. The difference between the two alternative incomes is $\$3 = \$15 - \$12$. Note that, discounting this amount at 8% one finds back the NPV: $3/1.08 = 2.778 = \text{NPV}$. In this case, the implicit definition of normal profit is not $I_t^V = r_t V_{t-1}$ but $I_t^* = r_t C_{t-1}$. This alternative definition presupposes a different benchmark system, which replicates capitals (not cash flows) (see Sect. 6.8).

However, regardless of the way normal profit is computed, the normal rate of return does not change, for any benchmark system represents the same equilibrium, which is condensed in the market rate (see Definition 6.6).

Example 5.9 ♦
(*Financing project*) Consider firm A, which receives an upfront payment of \$100 at time 0 from a customer for the production of a good whose price is \$130. At times 1, 2, and 3 the firm spends \$60, \$40, and \$35 for the operations necessary to produce the good. At time 3, the customer settles the account by paying the remaining \$30. Let us assume that the incomes are equal to the CFAs (i.e., $I_t = F_t$), which means $\Delta C_t = 0$ for $t = 1, 2$ and $\Delta C_3 = 0 - C_0 = -C_0$. The stream of CFAs is then $\boldsymbol{F} = (100, -60, -40, -5)$ and the capital stream is $\boldsymbol{C} = (-100, -100, -100, 0)$.[13] The firm is in a financing position in each period (it raises funds from the customer). Assume the market rate is $r = 5\%$. This is the MAFR (firm A might alternatively borrow \$100 at 5%). The NPV criterion informs that this financing project creates value: $\text{NPV} = 100 - 60(1.05)^{-1} - 40(1.05)^{-2} - 5(1.05)^{-3} = 2.26$. Thus, the firm makes money out of a borrowing. Such a borrowing is obtained from the firm's assets (a negative cash cycle and a sufficiently small amount of fixed assets, so that $C_t = \text{NOWC}_t + \text{NFA}_t + C_t^l < 0$ in each period).

5.5 Considering Risk

Under certainty, all assets are easily comparable: The NPV is the sum of the cash flows discounted at the COCs, which represent the equilibrium rates r_t that form the term structure of interest rates in a normal market.

Under uncertainty, comparisons are not straightforward. For example, an expected cash flow of \$150 available at time t is not necessarily preferable to an expected cash

[13] This project has been introduced in Example 1.10.

flow of \$100 available at the same time, if the former is riskier than the latter: The greater amount of the former may be more than offset by the smaller risk of the latter (risk-aversion is assumed). So, if alternatives are not equivalent in risk, the law of one price is limited to assets equivalent in risk: Riskless arbitrage is possible only for assets that, at any date t, generate the same cash flows in the same states of nature.

Under uncertainty, the notion of equilibrium refers to equal-risk assets: All equal-risk assets traded in the market have the same (stochastic) rate of return and the law of one price guarantees that two assets generating the same (stochastic) cash flows are traded in a normal market at the same price (possible arbitrage opportunities are soon wiped out).

If the cash flow is a random variable, investors will determine the price of it taking into account the associated risk. Usually, two alternative ways are employed to account for risk: The *certainty-equivalent* approach and the *risk-adjusted approach.* The former consists of computing the certain cash flow CE_t that is considered to be worth the same as the random cash flow $\tilde{F}_t$ and discounting it at the interest rate for riskless financial assets, so-called *risk-free rate*.

A time-invariant risk-free rate will be denoted as r_f, while $\boldsymbol{r}_f = (r_{f1}, r_{f2}, \ldots, r_{fn})$ will denote the term structure of time-varying risk-free rates.

While the certainty-equivalent approach is considered a theoretical correct one (e.g., Smith and Nau 1995), it is hardly ever used in practice, as it is most difficult to single out the certainty equivalent of a given cash flow. The most common method adopted in practical applications is the risk-adjusted rate-of-return method, which consists of discounting the estimated value of $\tilde{F}_t$, at a risk-adjusted COC, r. Typically, the estimated value is an expected value, here denoted as $F_t = \mathbb{E}(\tilde{F}_t)$. The risk-adjusted COC is equal to the risk-free rate, r_f, plus a risk premium, π, which takes account of the project risk: $r = r_f + \pi$.

In general, for a given expected value of a risky cash flow, the higher its riskiness, the higher the premium, the higher the risk-adjusted discount rate, the smaller the price:

$$\text{riskiness} \uparrow \Longrightarrow \pi \uparrow \Longrightarrow r \uparrow \Longrightarrow V_0 = \frac{F_t}{(1+r)^t} \downarrow .$$

Thus, if A is riskier than B, but the expected cash flow is the same, a necessary condition for A to be bought by investors is that its price is smaller than the price of B.

Example 5.10 ♦

Consider a given portfolio A and suppose that it consists of a single random cash flow $\tilde{F}_1^A$ available in one period. Assume that only two scenarios are possible, with equal probabilities; in the favorable scenario, the cash flow will be equal to \$120; in the unfavorable one, the cash flow will be \$80. Then, the expected amount is $F_1^A = 0.5 \cdot 120 + 0.5 \cdot 80 = 100$. Assume the term structure of interest rates is flat and, specifically, let $r_f = 10\%$ be the risk-free rate. Let $\pi^A = 3\%$ be the risk premium for A, so that the risk-adjusted COC is $r^A = 13\%$. Then, the equilibrium price of A in the market is $V_0^A = F_1^A/(1 + r^A) = 100/(1.13) = 88.5$.

Consider now asset B, which will generate a random cash flow equal to $\tilde{F}_1^B$ at time 1. Suppose that, in the favorable scenario, the cash flow will be \$140; in the unfavorable one, the cash flow will be \$60. Therefore, asset B's expected amount is the same as A: $F_1^B = 100 = F_1^A$. However, B is riskier,[14] so the risk premium is higher, say $\pi^B = 5\%$, whence the risk-adjusted COC is $r^B = 15\%$. The current price of B will then be smaller: $V_0^B = 100/(1.15) = 86.96$. The market evaluates B less than A, because, given the same expected amount, B is riskier.

Note that a riskless zero-coupon bond (zcb) with a face value equal to \$100 has a higher price than both A and B (precisely, \$90.91 $= 100/1.1 = 100/(1 + r_f)$), because it has the same expected payoff as portfolios A and B but is less risky (indeed, riskless). Vice versa, a riskless zcb or portfolio of riskless zcb's whose current price is equal to \$88.5 ($= V_0^A$) will generate an end-of-period cash flow smaller than \$100 (precisely, $97.35 = 88.5 \cdot 1.1$); analogously, a zcb whose current price is \$86.96 ($= V_0^B$) will generate an even smaller end-of-period cash flow (precisely, $95.66 = 86.95 \cdot 1.1$).

The price of a risky asset with a single prospective cash flow $\tilde{F}_t$ may be alternatively stated in terms of certainty equivalents or in terms of risk premiums (and, therefore, risk-adjusted income rates); the law of one price guarantees that the price is unique:[15]

$$V_0 = \frac{F_t}{(1 + r_f + \pi)^t} \tag{5.19}$$

$$V_0 = \frac{CE_t}{(1 + r_f)^t}. \tag{5.20}$$

For single-period assets, this implies $V_0 = CE_1/(1 + r_f)$ and $\pi = (F_1 - CE_1)/(CE_1/(1 + r_f))$. These two methods are *theoretically* equivalent. They are only two different ways of computing the same value.

The setting is easily generalized for assets with multiple cash flows and for non-flat structures of term interest rates $\boldsymbol{r}_f = (r_{f1}, r_{f2}, \ldots, r_{fn})$ and risk premiums $\boldsymbol{\pi} = (\pi_1, \pi_2, \ldots, \pi_n)$.

Remark 5.8 The relation between certainty equivalents and risk premiums can be easily derived as follows. Let $(\tilde{F}_1, \tilde{F}_2, \ldots, \tilde{F}_n)$ be the stream of cash flows of the asset under consideration. The dynamical relation (1.10) may be written in terms of certainty equivalents or in terms of risk premiums as

---

[14] The risk is higher in the sense that B's cash flow has a higher variance than A's cash flow. However, variance is not the only measure of risk. The notion of risk itself is not a simple one and no agreed upon model of risk exists in academia or in practice. Hurdle rates are often used which are aimed to account for various sources of risk (see Sect. 5.6).

[15] The same asset cannot be traded at different prices in a competitive, normal market, otherwise arbitrage opportunities would arise that would immediately sweep away the state of disequilibrium (see Sect. 5.1). See also below.

$$V_t = V_{t-1}(1 + r_{ft}) - CE_t \tag{5.21}$$
$$V_t = V_{t-1}(1 + r_{ft} + \pi_t) - F_t. \tag{5.22}$$

Equations (5.21)–(5.22) imply

$$\pi_t = \frac{F_t - CE_t}{V_{t-1}} \qquad t \in \mathbb{N}_n^1 \tag{5.23}$$

or, equivalently,

$$CE_t = F_t - \pi_t \cdot V_{t-1}. \tag{5.24}$$

These two equations can be solved iteratively. One finds, respectively,

$$\pi_t = \frac{F_t - CE_t}{\sum_{k=t}^{n} CE_k \cdot \mathrm{d}_{k,t-1}} \qquad t = 1, 2, \ldots, n \tag{5.25}$$

and

$$CE_t = F_t - \pi_t \cdot \sum_{k=t}^{n} \frac{F_k}{\prod_{j=t}^{n}(1 + r_{fj} + \pi_j)}. \tag{5.26}$$

▲

While theoretically equivalent, there are pros and cons in using either method in practice. As anticipated, the preferred method in real-life applications is, by far and large, the risk-adjusted method or a variant of it where several variables other than risk are taken into account (see Sect. 5.6).[16]

As anticipated, two equivalent-risk cash-flow streams with the same future cash flows in the same states of nature are equally priced in a normal market. Whenever arbitrage opportunities arise, they are soon swept away by the market driving forces of economic rationality. As an example, let

$$\tilde{\boldsymbol{F}}^A = (-p_0^A, \tilde{F}_1, \tilde{F}_2, \ldots, \tilde{F}_n) \qquad \tilde{\boldsymbol{F}}^B = (-p_0^B, \tilde{F}_1, \tilde{F}_2, \ldots, \tilde{F}_n)$$

be the cash-flow vectors of assets A and B and assume they are equal in risk. Then, the two assets are equally priced:

$$p_0^A = p_0^B = V_0 = \sum_{t=1}^{n} \frac{F_t}{\prod_{k=1}^{t}(1 + r_k + \pi_k)},$$

so that $\tilde{\boldsymbol{F}}^A - \tilde{\boldsymbol{F}}^B = 0$. Suppose a mispricing occurs; a temporary state of disequilibrium arises, such that $p_0^A \neq p_0^B$. Rational investors immediately take advantage of

[16] See Robichek and Myers (1965) and Haley and Schall (1979, Chap. 9) for a detailed treatment of the two approaches.

the arbitrage opportunity, selling the overpriced asset and purchasing the underpriced asset. This process tends to adjust the prices and restore the equilibrium condition $p_0^A = p_0^B = V_0$ (analogously for a bundle of m equivalent-risk assets).

An implication of this is that value additivity holds under uncertainty as well. Consider m risky assets A_k, $k \in \mathbb{N}_m^1$. Let $\tilde{\boldsymbol{F}}^{A_k} = (\tilde{F}_1^{A_k}, \tilde{F}_2^{A_k}, \ldots, \tilde{F}_n^{A_k})$ be the cash-flow vector of A_k and r^{A_k} be the (assumed constant) risk-adjusted discount rate for A_k (i.e., inclusive of the risk premiums). Then, the equilibrium price of A_k is

$$V_0^{A_k} = \sum_{t=1}^{n} \frac{F_t^{A_k}}{(1+r^{A_k})^t}. \tag{5.27}$$

Assume $\tilde{F}_t^A = \tilde{F}_t^{A_1} + \tilde{F}_t^{A_2} + \ldots + \tilde{F}_t^{A_m}$ is the cash flow of asset A for every $t > 0$ and let V_0^A be the equilibrium price of A. Then, value additivity ensures that $V_0^A = V_0^{A_1} + V_0^{A_2} + \ldots + V_0^{A_m}$ or, explicitly,

$$V_0^A = \sum_{t=1}^{n} \frac{F_t^{A_1}}{(1+r^{A_1})^t} + \sum_{t=1}^{n} \frac{F_t^{A_2}}{(1+r^{A_2})^t} + \ldots + \sum_{t=1}^{n} \frac{F_t^{A_m}}{(1+r^{A_m})^t}. \tag{5.28}$$

The result keeps on holding if the risk-adjusted COC is time-varying with $\mathrm{d}_{t,k} = 1/[(1+r_{k+1})(1+r_{k+2})\cdots(1+r_t)]$ replacing $1/(1+r)^{(t-k)}$ as a discount factor:

$$V_0^A = \sum_{t=1}^{n} \frac{F_t^{A_1}}{(1+r_1^{A_1})(1+r_2^{A_1})\cdots(1+r_t^{A_1})} + \sum_{t=1}^{n} \frac{F_t^{A_2}}{(1+r_1^{A_2})(1+r_2^{A_2})\cdots(1+r_t^{A_2})} + \ldots$$
$$\ldots + \sum_{t=1}^{n} \frac{F_t^{A_m}}{(1+r_1^{A_m})(1+r_2^{A_m})\cdots(1+r_t^{A_m})}. \tag{5.29}$$

This also means that the risk-adjusted stream of COCs for A is $(r_1^A, r_2^A, \ldots, r_n^A)$ such that

$$V_0^A = \sum_{t=1}^{n} \frac{F_t^A}{(1+r_1^A)(1+r_2^A)\cdots(1+r_t^A)}.$$

As previously seen, the market value of a project P is

- the price that the project would have if it were traded in the market or, equivalently
- the current price of a portfolio actually traded in the market that generates the same cash flows as the project's (from 1 to n).

Therefore, given a project with a vector of cash flows $\tilde{F}_t$ for $t = 1, 2, \ldots n$ and a portfolio generating the same cash flows, the firm might either undertake the project by investing $C_0 = -F_0$ or invest $V_0 = \sum_{t=1}^{n} F_t(1+r)^{-t}$ in an equivalent-risk asset traded in the market generating the same cash flows from time 1 to time n. The risk-adjusted rate r is then the project COC. In case of time-varying COCs, $\boldsymbol{r} = (r_1, r_2, \ldots, r_n)$ and $V_0 = \sum_{t=1}^{n} F_t \mathrm{d}_{t,0}$, where F_t denotes the expected values of $\tilde{F}_t$. The NPV may then be interpreted as the difference between the price (value) of a replicating portfolio and the investment cost or, equivalently, as the difference between the price (value) the project would have if it were traded in a normal market and the investment cost.

Example 5.11 ♦
Consider three one-period risky assets: Asset A pays off, after one period, a cash flow equal to $\tilde{F}_1^A$ whose expected value is $F_1^A = 1,500$. The payoffs of assets B and C are $\tilde{F}_1^B$ and $\tilde{F}_1^C$ such that $\tilde{F}_1^A = \tilde{F}_1^B + \tilde{F}_1^C$ and $F_1^B = 906$, $F_1^C = 594$, respectively. Let 5% be the risk-free rate and 9% and 6% be the risk-adjusted discount rates for B and C.[17] Then $V_0^B = 906/1.09 = 831.2$ and $V_0^C = 594/1.06 = 560.4$. Then, $V_0^A = V_0^B + V_0^C = 831.2 + 560.4 = 1{,}391.57$. This also implies that the risk-adjusted income rate for A is 7.79%, since $V_0^A = 1{,}500/1.0779 = 1{,}391.57$. It also means that the implied risk premium is $\pi^A = 2.79\% = 7.79\% - 5\%$.

Example 5.12 ♦
Consider a project P whose expected CFA stream is $\boldsymbol{F} = (-100, 80, -10, 40, 60)$. Suppose the term structure of (risk-free) equilibrium rates is flat, such that $r_{ft} = r_f = 0.08$. Let $\pi = 2\%$ be the risk premium. Then, the (risk-adjusted) COC is $r = 10\%$ and the market value of the project is

$$V_0 = 80(1.1)^{-1} - 10(1.1)^{-2} + 40(1.1)^{-3} + 60(1.1)^{-4} = 135.5.$$

The cost is \$100 so the NPV is NPV $= V_0 - C_0 = 135.5 - 100 = 35.5$. The project is expected to create value and is worth undertaking, because the price of an equivalent-risk asset traded in the market producing the same cash flows from time 1 to time n is higher than the cost of the project by \$35.5. The value created is precisely \$35.5.

Example 5.13 ♦
Consider project P again with $\boldsymbol{F} = (-100, 80, -10, 40, 60)$ and suppose that the term structure of interest rates for risk-free assets changes to $\boldsymbol{r}_f = (0.36, 0.28, 0.18, 0.03)$. Assuming the risk premium is still constant and equal to $\pi = 2\%$, the sequence of risk-adjusted COCs is $\boldsymbol{r} = (0.38, 0.3, 0.2, 0.05)$ and the profile of (risk-adjusted) discount factors is $\mathbf{d} = (1, 0.7246, 0.5574, 0.4645, 0.4424)$. Thus, $V_0 = 80 \cdot 0.7246 - 10 \cdot 0.5574 + 40 \cdot 0.4645 + 60 \cdot 0.4424 = 97.5$, whence NPV $= 97.5 - 100 = -2.5$. Thus, the project is expected to destroy value and is not worth undertaking (it would be preferable to build a replicating portfolio, whose current price would be 97.5).

---

[17] This implies risk premiums are equal to $\pi^B = 4\%$ and $\pi^C = 1\%$, respectively.

5.6 Cost of Capital in Practice[18]

As seen, under uncertainty, the COC is a risk-adjusted, market-determined income rate which is interpretable as the minimum attractive rate of return (MARR) (or the maximum attractive financing rate, MAFR, if the transaction is a financing project). Assuming the financial market is a competitive, normal market, it expresses the expected rate of return of an equilibrium asset traded in the market which is equivalent in risk to the project under consideration. Several models are available in the academic literature to practically compute a rational, normative COC.[19] However, empirical evidence shows that, while actual decision makers do use the NPV criterion for making decisions, they employ a subjectively determined *hurdle rate* instead of using a rational COC (i.e., derived from some market model).

The hurdle rate is a cutoff rate that identifies a MARR that decision makers fix using subjective evaluations, which may or may not involve the calculation of market-determined risk premiums. This means that many investors are not maximizers, they are *satisficers*; that is, they use ***satisficing*** strategies (Simon 1955). Given a project, they subjectively determine a threshold which is regarded as a satisfying one, and invest in the project if its NPV, computed with that threshold as discount rate, is positive. This means that they use ***heuristics*** rather than economically rational methods for selecting the MARR. Thus, they depart from economic rationality as defined in Definition 5.3.[20]

The problem arises whether such a *heuristic* can be considered rational from some point of view. The answer is positive. Several authors have shown that the use

---

[18]The first part of this section is adapted from Magni (2009d). Investment decisions, net present value and bounded rationality. *Quantitative Finance*, 9(8) (December), 967–979.

[19]Among the various models developed in the financial literature, the Capital Asset Pricing Model (CAPM) (Sharpe 1964; Mossin 1966) is by far and large the most widely used. In essence, the risk premium should be computed as the product of the systematic risk of the project (so-called beta) and the market premium (difference between expected market return and risk-free rate). The rationality of the use of the CAPM for investment decisions has been shown by many scholars since 1960s (e.g. Tuttle and Litzenberger 1968; Hamada 1969; Bierman and Hass 1973; Bogue and Roll 1974; Mossin 1969; Rubinstein 1973. See also Senbet and Thompson 1978 and Magni 2007a, b). However, some problems of inconsistency with no-arbitrage pricing exist for CAPM-based decision-making; albeit unearthed long since (Dybvig and Ingersoll 1982), they are hardly ever mentioned (see Magni 2002, 2008a, b, 2009a).

[20]The satisficing concept is particularly relevant in the engineering practice, to such an extent that engineering textbooks make a distinction between COC and MARR, where COC is a normative, market-determined rate of return, while MARR is a hurdle rate subjectively determined, which may well differ from the former (see also Rogers and Duffy 2012, pp. 10–11). More generally, small-sized, family-owned, companies often set MARRs subjectively when valuing projects. The use of a rigorous decision-making procedure alongside the use of a subjective hurdle rate was practiced since 13th century in the evaluations of properties, trees, lands, collieries, coppices, buildings, leases, shops etc. made by the legal, banking, business communities (see, for example, Wing 1965; Edwards and Warman 1981; Miller and Napier 1993; Scorgie 1965; Brackenborough et al. 2001).

of subjectively determined COCs is not necessarily unreliable for decision-making. For example, McDonald (2000), focusing on real options,[21]

> for a wide range of project characteristics, fixed hurdle rate and profitability index rule can provide a good approximation to optimal investment timing decisions, in the sense that the ex ante loss from following the suboptimal rule is small; it is possible to follow the wrong investment rule without losing much of the ex ante value of the investment timing option. In fact, as the investment timing option becomes worth more and it becomes optimal to wait longer to invest, the option value becomes less sensitive to errors in investment rules. (McDonald 2000, p. 15)

The author underlines that "for a variety of parameters, particular hurdle-rate ...rules can provide close-to-optimal investment decisions. Thus, it may be that firms using seemingly arbitrary rules of thumb are approximating optimal decisions" (McDonald 2000, p. 13). Jagannathan and Meier (2002) maintain that the use of hurdle rates should not be deemed less reliable than the use of a normative cost of capital: "managers ...[take] the right decisions ...because they use a hurdle rate that is higher than the cost of capital to capture the option to wait" (p. 12). They concede that there are several advantages in using a hurdle-rate rule as a MARR for a project: "the advantage of using a hurdle rate is that modelling all possible future options is not necessary" (Jagannathan and Meier 2002, p. 4) and decision makers may "find it useful to use a rule that best justifies making intuitively plausible investment decisions" (McDonald 2000, p. 26); "managers adjust these rules ...when an investment is strategic and expiring." (McDonald 2000, p. 30); "use [of a hurdle-rate rule] in practice might stem from the success of apparently arbitrary rules that are revealed over time to be close to optimal. Managers likely observe the capital budgeting practices, in their own and other companies, and in many cases probably mimic what seems to work" (ibidem).[22]

[21] Real options are investments that allow for some kind of flexibility; for example, investment may be deferred to next year or may be expanded in following years or abandoned.

[22] Essentially, these three reasons fit the premises on which Gigerenzer (2001) bases the notion of adaptive toolbox: psychological plausibility (i.e., regard for constraints in time, knowledge, and computational abilities), domain specificity (a rule may be appropriate in one domain but not in another one, so the rule is changed or combined with other rules), ecological rationality (imitation and social learning may be ecologically rational due to success over time). See Magni (2009d) for details.

Brigham (1975) *surveyed 33 large, relatively sophisticated firms. Although 94% of them used DCF methodology, only 61% of the firms using DCF [Discounted Cash Flow] adopted the cost of capital as the discount rate.* ... Dixit (1992) *recognizes that "firms invest in projects that they expect to yield a return in excess of a required or hurdle rate" (p. 107). "Finance scholars have always been puzzled by the durability of ...the hurdle rate rule"* (Ross 1995) *and, in actual fact, "we know that hurdle rates ...are used in practice"* (McDonald 2000, p. 30); *"it appears common for firms to use investment criteria that do not strictly implement the NPV criterion" (ibidem, p. 13), so that their "actions do not reflect the application of current financial theory"* (Gitman and Mercurio 1982, *p. 29*). Graham and Harvey (2002) *surveyed 392 companies, and despite the fact that many companies claimed they do employ the NPV technique and use CAPM for estimating the cost of capital, they point out that the use of hurdle rates is predominant. In particular, "small firms are significantly less likely to use the NPV criterion or the capital asset pricing model and its variants" (p. 22). They find that sometimes the use of hurdle rates is explicitly acknowledged: "small firms were inclined to use a cost of equity determined by "what investors tell us to require" [and a] majority (in fact, nearly 60%) of the companies said that they would use a single-company wide discount rate to evaluate a new investment project, even though different projects are likely to have different risk characteristics" (p. 12).* Jagannathan and Meier (2002) *observe that "managers use a ...hurdle rate" (p. 3) instead of employing a CAPM-derived cost of capital. Relying on* Poterba and Summers (1995) *analysis, they find that "hurdle rates are not ...linked to the cost of capital" (p. 22).*

Magni (2009d). Investment decisions, net present value and bounded rationality. *Quantitative Finance*, 9(8) (December), 967–979 (pp. 968–969), https://doi.org/10.1080/14697680902849338

Furthermore, limited supply of managerial resources may be a reason to add a premium to a normative COC. This is consistent with the resource-based theory and the literature on Top Management Teams, according to which managerial and firm-specific skills play a crucial role in value creation (Barney 1986, 1991, 2001; Grant 1991; Grant and Robert 1995; Levinthal 1995; Bromiley 2005). Furthermore, if a project is strategic, it may be worth undertaking even if its NPV (computed with the normative, market-determined COC) is negative, because it brings about future growth opportunities. In this case, the use of a MARR which is smaller than the market-derived COC (thereby increasing the project's value) is an appropriate choice.

The use of hurdle rates as MARRs for projects is also consistent with the management strategic literature, which largely acknowledges that strategic decisions are key factors for competitive advantage (Collis and Montgomery 1995; Quinn and Mintzberg 1996). Strategic decisions are ones where pricing models are of little help for determining the appropriate COC and an allegedly rational decision.

Agency theory (Jensen and Meckling 1976; Jensen 1986) may also play a role in determining a hurdle rate different from a normative, market-driven COC.

The flexibility of hurdle rates is evidenced by empirical research, which suggests that companies may adjust the MARR depending on the size of the project, on the product and market conditions, which are determined by some fundamental driving forces such as rivalry, supplier power, buyer power, threat of substitutes, and entry barriers, whose impact on value creation is substantial in several circumstances.

Berg and Lien (2003), in an expected-utility framework, show that tracking-error decision-makers who, in addition to mean and variance, care about a subjectively determined benchmark, gain greater shares of accumulated wealth than those decision makers who base their decisions on pure mean-variance analysis.

Miller and Shapira (2004) admit that actual choice behaviors take account of key variables neglected by normative models: "normative models for pricing options overlook key aspects of the behavioral and organizational contexts in which investment decisions occur" (p. 282).

As for the risk, it should be stressed that the notion of risk which is helpful for actual decision makers is not necessarily equivalent to a normative one. "Threshold levels will be revalued ... In this revising of threshold levels ... projects which would have a minor loss if they failed would face a less stringent screening than those that could bankrupt the organization" (Carter 1971, p. 426). That is, the distinction between nonruinous losses versus ruinous losses is relevant for real-life decision makers. For example, bankruptcy costs are not easily quantifiable[23] and analysts may be willing to increase the MARR to cope with this problem.

The perceived risk may also take the shape of a ***payback period***. The payback period is the first period when the project cost is recovered in cash. As such, it supplies information about a break-even point (when the cash balance of the project's cash flow turns from negative to positive), an information that the normative COC is not aimed to provide.[24] Given any two projects, A and B, based on a normative, market-determined COC, decision makers will nonetheless be more prone to undertake the project with the smaller payback (or discounted payback) period.

This type of concern about risk is "consistent with the bounded rationality approach of Simon (1955)" (Carter 1971, p. 426) and may often conflict with usual capital budgeting textbook prescriptions: "the available models do not capture the essence of risk as defined by decision makers" (Laughhunn et al. 1980, p. 1248).

Moreover, the analyst/decision maker may regard the market as not being efficient and, therefore, not completely reliable. Value creation might not be recognized instantaneously and investors may take some time to recognize the project's value so that the disequilibrium-to-equilibrium process described in Sect. 5.3 may be grad-

[23]Bankruptcy costs include direct costs (legal and administrative expenses) and indirect cots (decrease in sales due to customers's perception of the increased default risk).

[24]For example, consider the cash-flow sequence $(-100, 40, 30, 20, 25, 40, 10)$. The payback period is 4, because time 4 is the first date where the cumulative cash flow is positive: $-100 + 40 + 30 + 20 + 25 = 15 > 0$. A variant of the payback period is the discounted payback period, where present value of cash flows is considered. For example, if the discount rate is 10%, the discounted payback period is 5, for time 5 is the first date where the cumulative discounted cash flow is positive: $-100 + 40/1.1 + 30/1.1^2 + 20/1.1^3 + 25/1.1^4 + 40/1.1^5 = 18.1 > 0$.

ual rather than instantaneous.[25] The analyst may consider adjusting the normative market-determined COC to take account of this market's imperfection.

In addition, the shares of the firm undertaking the project may not be traded in the security market. In this case, one may refer to comparable firms in the market and estimate the value creation that an efficient market would guarantee if the firm's shares were traded in the market. This is a useful metaphor, a good starting point, but by no means perfect, also because the firm's shareholders may be satisficers rather than maximizers (which is a common case when the firm is small and family-owned). In these cases, a comparison with the market is irrelevant. More relevant is the aspiration level (i.e., threshold) of the decision makers.

Overall, these findings suggest that decision makers condense into a base aspiration level several considerations related to

- uncertainty
- decision flexibility
- market conditions
- future opportunities
- limited managerial skills
- agency costs
- strategic considerations.

This heuristic is eco-logically rational and is exploited in a non-rigid way so that the aspiration level may change as the above mentioned elements change.

Thus, academic models may provide a basis for determining the appropriate COC but not replace the analyst's judgment:

> Any decision-making framework, however, can only improve a manager's understanding of the problem at hand and help him/her to make a more informed and consistent decision. No decision-making framework can guarantee a "good" outcome and there is no substitute for managerial effort, creativity, experience, knowledge, and critical thinking" (Lander and Pinches 1998, p. 542).

The "environmental indifference that is often assumed by many financial theories and models is not a reasonable assumption for the future" (Crum and Derkinderen 1981, p. 237).

Finally, there is an additional issue that calls for the use of **informed judgment** in setting the appropriate MARR: The normative COC is not unambiguously defined, to such an extent that subjectivity and informed judgment is inescapable. Indeed, even assuming a 'perfect' world of unbounded rationality, the actual derivation of the COC from a rational asset pricing model is unclear and less simple than one may think. More precisely, assuming that capital markets are perfect and efficient and that any financial and nonfinancial feature of a project can be embraced in the expected rate of return of a traded portfolio replicating the project's random cash flows (the COC), there is no univocal way of determining practically such an expected rate of return, for three main reasons:

---

[25]This implies that the notion of residual income becomes compelling, as it triggers the idea that value is created gradually over time (see Sect. 6.6.1 for details on this notion).

- more than one normative model exist in the literature, which are not always consistent one another. For example, mean-variance pricing based on CAPM contradicts arbitrage pricing theory under certain circumstances, and ambiguities arise on the application of such models (see Dybvig and Ingersoll 1982; Magni 2002, 2008a, b, 2009a, 2010d)
- there is no agreement in academia on which model should be used for determining the COC (CAPM, arbitrage theory, multifactor models, etc.). This implies that "no general analytic solution to the full-blown financing and investment problem of the firm is currently available. The only recommendation that can be made at present is that management must evaluate all options and *do the best it can*" (Pinches 1982, p. 12, italics supplied)
- even if one solved the previous problem, that is, if one assumed that a well-defined normative pricing model takes into proper account the managerial, strategic, behavioral, organizational features specific to a given project/firm and if agreement were established among scholars on the model, the following question would arise: How should one apply the agreed-upon model in practice? For example, suppose that the CAPM is such a model: How should the risk of the project (so-called beta) be determined? Should it be accounting-based or market-based? And how should one determine the forward-looking covariance which affects the beta? Would a historic approach (i.e., based on historic data as proxies for the beta) be adequate for estimating the beta and, therefore, the COC? Would monthly data or weekly data or daily data about rates of return be adequate? Would a survey approach (interviews with experts or with a firm's shareholders or with a suitable sample of market investors) be reliable or even possible? Should a mix of criteria be used? Should adjustments be made and how should they be done? To answer these and other questions we would need to rest on a yet-to-be-conceived *meta-theory* of cost of capital, that is, a theory which explains how to correctly apply the agreed-upon model of cost of capital.

Practically, the analyst may well start from the term structure of interest rates and then add an incremental (positive or negative) premium which, in addition to risk, takes account of the domain-specific factors that affect the value drivers (and, therefore, the wealth increase) of the project under consideration. This task is a difficult task and necessarily requires informed judgment on the part of the analyst/decision maker.

Overall, the role of informed judgment is essential in the appropriate determination of the MARR. However, regardless of the approach followed, the analyst should be careful in estimating MARRs in such a way that the laws of motion and conservation are fulfilled, to avoid ambiguity and multiple, conflicting evaluations.

Particularly important is then the role of the analyst. As we have seen, managers of a firm should follow an investment policy maximizing the shareholders' wealth. We have shown that maximization of the NPV leads to maximization of shareholders' wealth. Therefore, managers and analysts should determine the COC in such a way that the decision made leaves shareholders better off than they would be if they invested their funds in the securities market. In this case, there is a delicate issue of trade-off between the requirement of acting in shareholders' interests and the requirement of using informed judgment for taking into account all possible aspects of the project under consideration. Also, if the firm is a small, possibly nontraded firm, and the number of shareholders is low, managers might be able to elicit shareholders' premiums over and above the risk-free rate to account for their preferences. The reference to expected rate of return on comparable firms or comparable projects (as often recommended by authors) may be of help, with the understanding that informed judgment is necessary in any case (it is not an easy task to ascertain the expected rate of return of a comparable firm nor is the notion of comparable firm an unambiguous one; and, in addition, there is no unique way of aggregating the expected rates of return from a sample of comparable firms).[26]

Now, if the determination of COC is then a delicate issue, the estimation of earnings and capital is no less delicate. And the estimation of cash flows and COC is interrelated: Some non-measurable aspects may be considered by adding a premium to COCs or subtracting a premium from cash flows, and care should be taken to avoid double counting. Managerial judgement and use of 'soft' skills play then a major role. How estimation of value drivers impacts on value creation is a serious issue that may be handled via sensitivity analysis.

Given the complexity of the estimation process, the very notion of *market value* becomes a slippery one, to such an extent that some authors distinguish the market value as derived by resting upon a normative model from the *economic* or *intrinsic* value obtained by discounting monetary values at a hurdle rate (MARR) which incorporates the above mentioned aspects of the project. However, as we have seen, even the alleged adoption of a normative model contains several elements of subjectivity in treating the available data, a meta-theory being currently unavailable.

We argue that normative models may provide guidance as to how an adequate COC/MARR should be determined but there is no alternative to using managerial skills and informed judgment.

As a result, we do not commit ourselves on the problem relating to how input data (including COCs) are selected. We will use data assumptions as exogenously given. This book being about logical relations, we aim to show how to keep consistency in the financial modeling and derive a consistent appraisal and decision-making approach on the basis of the selected inputs.

> We will use the expressions ***market value***, ***intrinsic value***, ***economic value*** as synonymous with discounted sum of the prospective cash flows at the given MARR. And we will also continue to use COC and MARR as synonyms, no matter how the analyst estimates them.

[26]"Unfortunately, given the data, there is no completely satisfactory way to specify exactly how much higher or lower we should go in setting risk-adjusted costs of capital." (Ehrhardt and Brigham 2016, p. 401)

Following are two simple examples where same incremental cash flows occur but the MARR is different, due to the different perspectives of the investors: Satisficers in the first case, maximizers in the second case.

Example 5.14 ♦
(*Satisficers*) Mr. and Mrs. Smith have the opportunity of expanding their family-owned business. They estimate that an incremental $C_0 = \$68{,}000$ investment will lead to incremental CFAs equal to $F_t = \$5{,}000$ for the following 25 years. They consider that they might invest the same amount of money in risk-free assets earning an annual 3%. Considering the risk of the project and their fervent commitment to the business, they believe they will be adequately remunerated if the expansion will bring a MARR of $r = 5\%$, two percentage points higher than the risk-free rate. The NPV is

$$\text{NPV} = 5{,}000 \cdot \frac{1 - 1.05^{-25}}{0.05} - 68{,}000 = 2{,}469.7.$$

With a 5% hurdle rate, subjectively determined as an appropriate aspiration level, they regard the project as worth undertaking.

Example 5.15 ♦
(*Maximizers*) A firm's management team faces the opportunity of expanding the business by investing \$68,000 equity and hence increase the expected CFAs by \$5,000 for the next 25 years. Managers aim at maximizing shareholder's wealth and, to this end, they take into careful consideration the risk of the project, the strategic importance, the possible competitors' reactions. They also are aware that there is some possibility in the future of developing the project if things go well, or abandoning the project in case of unfavorable circumstances. Upon careful inspection of the various aspects of the capital asset investment, they believe that, overall, 6% is an adequate risk premium to compensate shareholders for risk. In other words, they believe that, considering every aspect, measurable and non-measurable, tangible and intangible, 9% (= 6% + 3%) is the expected rate of return of a portfolio traded in the market equivalent in risk to the project under consideration. Based on these data, the shareholder created value associated with the project is

$$\text{NPV} = 5{,}000 \frac{1 - 1.09^{-25}}{0.09} - 68{,}000 = -18{,}887.$$

Therefore, the firm's managers reject the project. Note that, in the previous example, the Smith family has accepted the project, although they might earn an expected 9% by investing in an equivalent-risk portfolio (assuming the firm's management has correctly valued the premium).

5.7 Key Points

- The law of one price states that, in a normal capital market, two assets generating the same cash-flow stream must be equally priced. If they are not, arbitrage opportunities arise (trading strategies where cash flows are nonnegative and at least one is positive) which disappear as soon as investors exploit them
- The law of one price derives from a fundamental law of demand and supply, according to which the higher the price of a good, the lower the quantity demanded of that good (and the higher the quantity supplied)
- Equilibrium among assets traded in the capital market occurs whenever no arbitrage opportunities arise. Owing to the one-price law, only temporary states of disequilibrium are possible in a normal (i.e., perfect and efficient) market. However, equilibrium is temporary as well, due to investment decisions of the firms' managers who seek value-creating opportunities (disequilibrium is also triggered by new exogenous information): The market fluctuates from states of equilibrium to states of disequilibrium and vice versa
- The value of a project is the price that the project would have if it were traded in the market or, equivalently, the price of a portfolio of financial assets that replicates the project's cash flows from time 1 to time n. The price of such a portfolio is the sum of its prospective cash flows, discounted at the equilibrium market rates, r_t: $V_0 = \sum_{t=1}^{n} F_t / \prod_{k=1}^{t}(1 + r_k)$
- The replicating portfolio generates a benchmark system, with its own mechanics. The equilibrium rates are the income rates of this benchmark system and represent the growth rates of the benchmark capital. The benchmark system is representative of the market equilibrium
- The equilibrium rate is known as the cost of capital (COC) and is the rate at which investors might alternatively invest their funds in each period if they did not undertake the project (counterfactual alternative). Therefore, it is the minimum attractive rate of return (MARR) or required rate of return
- A project is undertaken by a firm with the intent of creating a state of disequilibrium in the market such that the firm's share price increases, which is also expressed by saying that shareholders' wealth is increased or shareholders' value is created. The degree of the disequilibrium is measured by the net present value (NPV). The NPV is equal to the difference between the value of the project and its cost: NPV $=$ $V_0 - C_0$. More generally, in those cases where the initial income is nonzero, $I_0 \neq 0$ (e.g., gain/loss on disposed assets), the NPV is defined as NPV $= V_0 + F_0 =$ $V_0 + I_0 - C_0$
- Rational decision-making means that only positive-NPV projects should be undertaken. A positive NPV signals that value is created
- Under uncertainty, the COC/MARR includes a risk premium which accounts for the risk of the project (the higher the risk, the higher the premium, the higher the risk-adjusted COC/MARR). In this context, the law of one price operates only for assets equivalent in risk: If two assets with the same expected cash flows and are equivalent in risk, then they must be priced equally. A risky project is then worth

undertaking if and only if a replicating portfolio (with the same cash flows in the same state of nature) has a higher price than the project's cost. Value creation is measured by the (expected) NPV, which is calculated by discounting cash flows at the risk-adjusted COC/MARR. If the NPV is positive, the project creates value and is worth undertaking. If the NPV is negative, the project should be rejected (if the NPV is zero, the project is equivalent to investing in the replicating portfolio)

- COCs/MARRs are market input data. In practice, the estimation of a COC/MARR (as well as the estimation of incomes and capital amounts) is an extremely delicate issue. Decision makers often set the MARR by making recourse to a hurdle rate (*aspiration level*), which takes account of additional aspects (quantitative and qualitative) that normative pricing models may not be able to capture: Size of the project, size of the firm, strategic value of the project, decision flexibility, managerial and firm-specific skills, possibility of ruinous losses, competitors' reactions, behavioral and organizational aspects, risk factors not included in the normative models, inefficiency of the security market. Theoretically, there is no agreement among scholars upon which model should be used (although Capital Asset Pricing Model seems to be the preferred method). Practically, the application of the model from which the normative COC should be drawn (even if agreed upon) is not straightforward: Informed judgment and managerial skills are essential in fixing the appropriate MARR.

Chapter 6
Project Appraisal

> The fundamental economic and financial decision of the firm is that of determining the level and composition of its investment. Firms can invest in either physical assets or financial assets (securities). Such investment decisions determine the firm's future income . . . A firm must make two additional and related financial decisions: first, how much of its current income to pay out to the owners of the firm (the dividend decision), and second, the amount and type of financial securities to issue (the financial structure decision). These constitute the financing decisions of the firm. Therefore, firms are considered to make two general types of financial decisions: *investment decisions* and *financing decisions*.
>
> Haley and Schall (1979, p. 3)

> We agree that managers with reliable intuition (that is, a discriminating gut) and the experience to make sound judgements will – and should – ultimately make the major corporate investment choices. However, analytical tools . . .can help see through the complexity . . .inherent in the evaluation of a major investment. It is our belief that, by using these tools and taking a more disciplined approach to valuation, managers' judgment will be enhanced and they will make better investment choices.
>
> Titman and Martin (2011, p. 3)

C. A. Magni, *Investment Decisions and the Logic of Valuation*,
https://doi.org/10.1007/978-3-030-27662-1_6

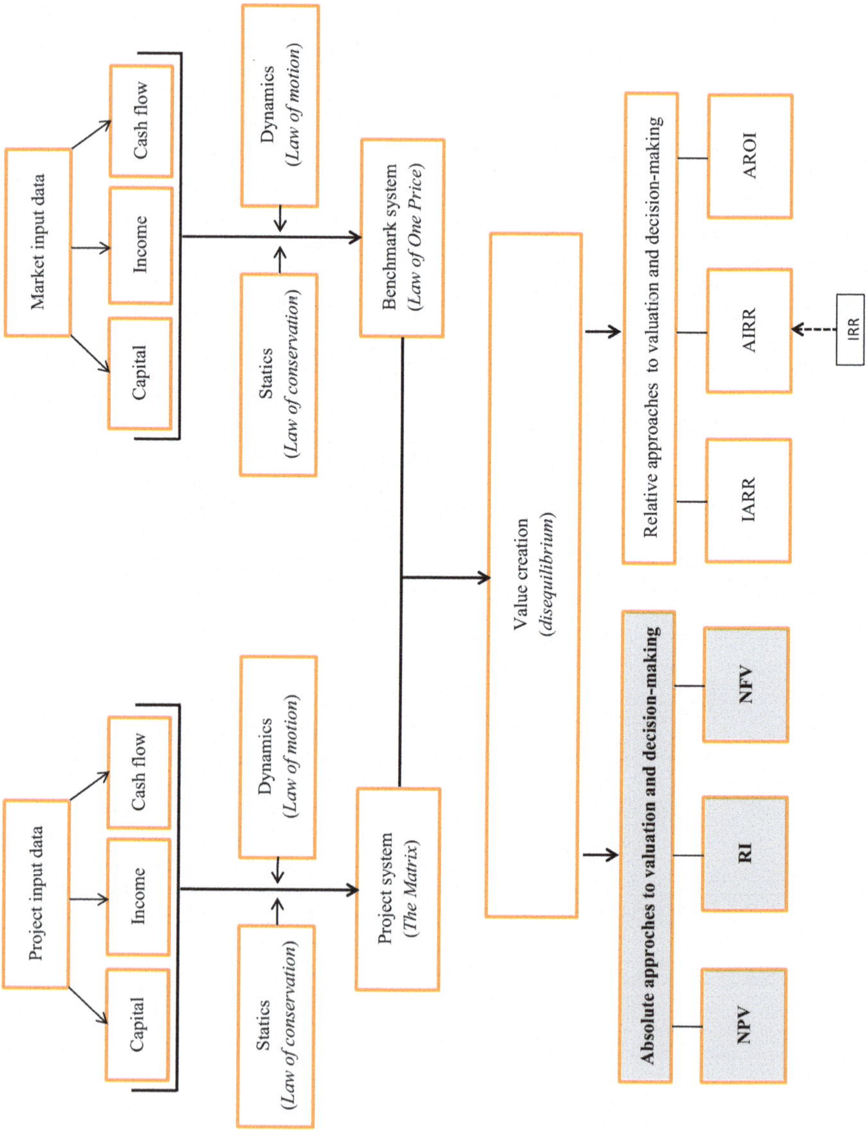

Fig. 6.1 Chapter 6: Cash-flow-based valuation, income-based valuation, capital-based valuation

This chapter makes full use of the notions, relations and techniques presented in the previous chapters for evaluating a project, assuming the (project and market) input data have been previously selected and the pro forma financial statements have been drawn. We show that, depending on the perspective and on the data available, several different approaches may be used to determine the project value and make a rational decision. Specifically, consistently with the basic triplet (Capital, Income, Cash Flow), the approaches may be conveniently classified into three groups:

- **cash-flow** methods, linked to the **Net Present Value** (NPV) criterion
- **income** methods, linked to the **Residual Income** (RI) criterion
- **capital** methods, linked to the **Net Future Value** (NFV) criterion.

The first one is suitably interpreted in terms of the project's prospective cash flows (including CFEs, CFOs/CCFs, CFAs, FCFs); the second one is conceptualized in terms of (residual) income; the third one focuses on the valuation of the terminal capital. Formally, each approach is associated with a well-defined benchmark system.

Exploiting the static equilibrium of any economic entity and, in particular, the accounting-and-finance engineering system, each valuation approach may be framed in terms of symmetric perspectives:

investment versus **financing**
levered versus **unlevered**
financial structure versus **capital structure**
standard discounting versus **modified discounting**
actual cash flows versus **potential dividends**

After completing this chapter, the reader should be able to:

- evaluate a project via an investment perspective, discounting CFOs and CFLs, or a financing perspective, discounting CFEs and CFDs, and make consistent decisions
- appreciate the rationality of the equity NPV for making decisions and appraise the relationships between equity NPV and project NPV and, therefore, understand the reason why a negative-NPV (positive-NPV) project may increase (decrease) shareholders' wealth
- dominate the subtle relations among the plethora of costs of capital implied by a project, including cost of equity, cost of (net) debt, cost of assets, cost of operating assets, cost of non-operating assets, cost of unlevered assets
- decompose shareholders' value creation into operating NPV, debt NPV and non-operating NPV, and distinguish the economic value created by the financing and distribution policy as opposed to the economic value created by the investment policy
- have a deeper understanding of the relation between the notion of cost of capital and the expected rate of return of an equivalent-risk asset traded in the security market
- apprehend the notion of weighted average cost of capital (WACC) and have a clear idea of the various kinds of WACC and their interrelations
- evaluate a project in an unlevered perspective with the WACC method
- catch the logic of the Adjusted Present Value (APV) method and its relation with the Capital Cash Flow (CCF) method
- appreciate the modified CFE discounting approach for valuing the project's equity and the issue of *circularity*
- understand how to correctly use potential dividends for appraising a project (or a firm) with the appropriate discount rates (PD method)
- grasp the reason why PDs may not be discounted at the cost of equity unless CFLs are raised/invested at the cost of equity (FCFE method)
- appreciate the equivalence among all the 12 discounted-cash-flow methods presented, irrespective of whether a levered or unlevered perspective is employed, a financial structure or a capital structure is adopted, an investment perspective or a financing perspective is followed, a modified discounting approach or a standard discounting approach is used, a potential-dividend setting or an actual-cash-flow setting is put to use
- understand the notion of Residual Income (RI), manage various types of RIs and use one or more of them for evaluating a project and making rational decisions

- fathom out the notion of Value Added (VA), based on terminal capitals, and appreciate the equivalence with the notion of Net Future Value (NFV)
- comprehend that the logical consistency of all valuation methods presented directly derives from the law of motion, the law of conservation, and the law of one price, and, therefore, from the accounting-and-finance engineering system. They should then be considered different-but-equivalent framings
- appreciate the resultant logical equivalence of the three accept-reject decision criteria based on NPV, RI, and VA/NFV. They are economically rational and are suitably associated with the three basic elements: Cash flow, income, and capital, respectively.

6.1 Valuing Projects with Cash Flows: Levered Perspective

Valuing projects with cash flows means to make use of a benchmark system which mimics the project's prospective cash flows: $(\boldsymbol{V}, \boldsymbol{I}^V, \boldsymbol{F}^V)$ (see Chap. 5). For each kind of cash flows, there corresponds a benchmark system. By comparing the project system and the benchmark system one may derive value creation or destruction. This section deals with valuing projects in a levered perspective, from both the investment side and the financing side. Section 6.2 deals with valuing projects in an unlevered perspective (via FCF), Sect. 6.3 presents the method of modified cash-flow streams, a method for valuing projects via modified discounting, Sect. 6.4 analyzes project appraisal with potential dividends, and Sect. 6.5 summarizes the 12 discounted-cash-flow methods illustrated, relating them to the market input data available to the analyst.

6.1.1 Valuation in a Financing Perspective

As seen in Chap. 5, V_t is the market value of the project, that is, the price of a portfolio that replicates the project's cash flows from time 1 to time n. It is then the capital invested (and borrowed) of a benchmark system representing the market equilibrium: By law of conservation, $V_0^{\text{inv}} = V_0 = V_0^{\text{fin}}$.

Likewise, let V_t^e and V_t^d be the price of two assets that replicate, respectively, the project's CFEs and the project's CFDs, from time 1 to time n.[1] The cash-flow streams associated with the three replicating portfolios are, respectively,

$$
\begin{aligned}
\boldsymbol{F}^V &= \left(-V_0, F_1, F_2, \ldots, F_{n-1}, F_n\right)\\
\boldsymbol{F}^{Ve} &= \left(-V_0^e, F_1^e, F_2^e, \ldots, F_{n-1}^e, F_n^e\right)\\
\boldsymbol{F}^{Vd} &= \left(-V_0^d, F_1^d, F_2^d, \ldots, F_{n-1}^d, F_n^d\right).
\end{aligned}
\tag{6.1}
$$

Let r_t be the expected return on V_{t-1}, r_t^e the expected return on V_{t-1}^e, and r_t^d be the expected return on V_{t-1}^d, as determined by the market. Then, the fundamental equation of motion (1.1) may be applied to the three replicating portfolios, taking the form of

$$V_t = V_{t-1}(1 + r_t) - F_t \tag{6.2}$$

$$V_t^e = V_{t-1}^e(1 + r_t^e) - F_t^e \tag{6.3}$$

$$V_t^d = V_{t-1}^d(1 + r_t^d) - F_t^d. \tag{6.4}$$

---

[1]The associated benchmark systems are then described by the respective transposed Matrices $(\boldsymbol{V}^e, \boldsymbol{I}^{Ve}, \boldsymbol{F}^{Ve})$ and $(\boldsymbol{V}^d, \boldsymbol{I}^{Vd}, \boldsymbol{F}^{Vd})$.

This leads to the prospective relations

$$V_t = \sum_{k=t+1}^{n} F_k \cdot \mathrm{d}_{k,t} \tag{6.5}$$

$$V_t^e = \sum_{k=t+1}^{n} F_k^e \cdot \mathrm{d}_{k,t}^e \tag{6.6}$$

$$V_t^d = \sum_{k=t+1}^{n} F_k^d \cdot \mathrm{d}_{k,t}^d \tag{6.7}$$

where $\mathrm{d}_{k,t} = \prod_{j=t+1}^{k}(1+r_j)^{-1}, \mathrm{d}_{k,t}^e = \prod_{j=t+1}^{k}(1+r_j^e)^{-1}, \mathrm{d}_{k,t}^d = \prod_{j=t+1}^{k}(1+r_j^d)^{-1}$ are the respective discount factors (see Eq. (5.6)).

The three different rates, r_t, r_t^e, and r_t^d, express the COCs for the streams of CFAs, CFEs, and CFDs, respectively. The rate r_t^e acts as a MARR for shareholders, and r_t^d acts as a MARR for debtholders, while r_t acts as a MARR for capital providers as a whole (equityholders + debtholders).[2]

The values V_t^e and V_t^d are the market value of the equity component of the project and the market value of the debt component of the project, respectively. The law of one price implies that, for every $t \geq 0$, the project's market value must equal the sum of market value of equity and market value of debt: $V_t = V_t^e + V_t^d$. In other words, it is indifferent for an investor to hold an asset which releases the same stream of the project's prospective CFAs or a portfolio of two assets that release the stream of the prospective CFEs and the stream of prospective CFDs, respectively. This is just the principle of value additivity described in Eqs. (5.28)–(5.29), with $m = 2$ and $\boldsymbol{F}^{A_1}$ and $\boldsymbol{F}^{A_2}$ now representing the CFE and CFD streams.

The three values V_0, V_0^e, V_0^d give rise to three NPVs, namely, the ***project*** NPV, the ***equity*** NPV, and the ***debt*** NPV:[3]

$$\text{NPV} = V_0 - C_0 \tag{6.11}$$

$$\text{NPV}^e = V_0^e - C_0^e \tag{6.12}$$

$$\text{NPV}^d = V_0^d - C_0^d. \tag{6.13}$$

---

[2]If $V_{t-1}^e < 0$, the rate r_t^e acts as a MAFR for shareholders; if $V_{t-1}^d < 0$ (admittedly, a rather-peculiar-but-not-impossible case), r_t^d acts as a MAFR for debtholders; if $V_{t-1} < 0$, r_t acts as a MAFR for capital "borrowers" (equityholders + debtholders).

[3]More generally, one gets

$$\text{NPV} = V_0 + F_0 = V_0 + I_0 - C_0 \tag{6.8}$$

$$\text{NPV}^e = V_0^e + F_0^e = V_0^e + I_0^e - C_0^e \tag{6.9}$$

$$\text{NPV}^d = V_0^d + F_0^d = V_0^d + I_0^d - C_0^d. \tag{6.10}$$

If $I_0 = I_0^e = I_0^d = 0$, then $F_0 = -C_0$, $F_0^e = -C_0^e$, and $F_0^d = -C_0^d$.

Given that $C_0 = C_0^e + C_0^d$, the NPV notion is additive: $\text{NPV} = \text{NPV}^e + \text{NPV}^d$. The project NPV measures the value created (or destroyed) by the project and may be split into equity NPV and debt NPV; the former measures the share of project NPV that is captured by equityholders, the latter measures the share of project NPV that is captured by debtholders.

Note that V_t is the market value of the capital *invested* in the project, while $V_t^e + V_t^d$ is the market value of the capital *raised* from the capital providers. The one-price law states that they are equal. This is nothing but the static principle for capital values (see Sect. 2.1.1), with V_t, V_t^e, V_t^d replacing C_t, C_t^e, C_t^d (see Eq. (2.5)). We can formalize the result as follows.

Proposition 6.1 (NPV as sum of equity NPV and debt NPV) *The law of one price guarantees that the static equilibrium of an asset traded in a normal market is fulfilled: The capital raised by the firm from the capital providers has the same market value as the capital V_t generated by the firm's assets:*

$$V_t = V_t^e + V_t^d = V_t^{\text{fin}} \tag{6.14}$$

(see also Remark 4.6). This implies that the NPV *generated by the project is entirely distributed to equityholders and debtholders:*

$$\text{NPV} = \text{NPV}^e + \text{NPV}^d = \text{NPV}^{\text{fin}}. \tag{6.15}$$

This proposition is equivalent to Modigliani and Miller's (MM) (1958) Proposition I, according to which the unlevered firm/project value is equal to the levered firm/project value (logically, evaluation of projects and firms or any other asset is equivalent because the law of one price applies to any kind of assets). If the project is equity financed (i.e., $C_t^d = 0$) then $F_t = F_t^e$ and $\text{NPV} = \text{NPV}^e$. We have used this assumption in Sect. 5.3 where we have proved that the project is worth undertaking if and only if $\text{NPV} = \text{NPV}^e > 0$. However, in general, projects are partially financed with debt. How does the NPV criterion behave in these cases? We analyze the only two possible cases:

a. $C_0^d = V_0^d$
b. $C_0^d \neq V_0^d$.

If case a. holds, $\text{NPV}^d = 0$ and, therefore, $\text{NPV} = \text{NPV}^e$. This implies that maximization of the project NPV is equivalent to maximization of the equity NPV (and, therefore, to maximization of shareholders' wealth).

It is most usual in finance and engineering to implicitly assume $i_t^d = r_t^d$ for every t, which just implies $C_0^d = V_0^d$. The assumption $i_t^d = r_t^d$ is particularly appropriate whenever the firm borrows funds by issuing bonds whose prices are equilibrium prices. In this case, debtholders make a value-neutral investment (i.e., bonds are equilibrium assets. See Chap. 5).[4]

If case b. holds, maximization of NPV is not necessarily equivalent to maximization of NPV^e. Since maximization of shareholders' wealth is the goal of investment policy, and since shareholders' wealth is increased if and only if the equity NPV is positive, then the project is worth undertaking if and only if NPV^e (not NPV) is positive. Equation (5.10) holds in this case as well. However, in this case the firm issues a number of shares which does not entirely cover the initial investment: $\Delta n_0 < C_0/(p_0 + \Delta p_0)$, because part of the initial investment C_0 is financed with debt. As a result, CFAs and CFEs do not coincide any more: $F_t \neq F_t^e$ and Eq. (5.11) becomes

$$\overbrace{V_0 - C_0}^{\text{project NPV}} = \overbrace{n_0 \Delta p_0}^{\text{equity NPV}} + \overbrace{(V_0^d - C_0^d)}^{\text{debt NPV}}. \tag{6.16}$$

The first summand of the right-hand side is the equity NPV: $n_0 \Delta p_0 = V_0^e - C_0^e$. It differs from the project NPV ($= V_0 - C_0$) as long as $C_0^d \neq V_0^d$. In principle, it may even occur that the project NPV is negative and the equity NPV is positive (this occurs whenever NPV^d < NPV < 0), which implies that the value destroyed by the investment policy is borne by debtholders, whereas shareholders benefit from a low-cost financing, to such an extent that the shareholder value created by the financing policy outweighs the value destroyed by the investment policy. Symmetrically, a positive-NPV project may cause a loss in value to shareholders if the financing is too costly, that is, if NPV^d > NPV > 0.[5] Since shareholders' wealth increase is reflected by the equity NPV, a project is worth undertaking if and only if the equity NPV (not the project NPV) is positive. We need then to reformulate the NPV criterion in its most general form, allowing for debt financing.

---

[4]Symmetrically, the firm makes a value-neutral financing.

[5]The sign and magnitude of the debt NPV depends on the ability of the company of borrowing at a higher or lower cost than the cost of financing in a normal market as opposed to the ability of debtholders to lend money at an above-normal interest rate (this also depends on several types of imperfections that may arise in the market). A positive project NPV associated with a negative equity NPV may also reveal a debt overhang problem, which results in under-investment (see Berk and DeMarzo 2014, pp. 554–555; Myers 1977). Evidently, if the firm borrows from a lender at a rate i^d such that $r^d < i^d$ (implying $C_t^d < V_t^d$), then the firm's managers should preferably borrow from the market, not from that lender; however, as long as the equity NPV is positive, the undertaking of a project with positive debt NPV increases equityholders' wealth. Vice versa, if the firm's managers happen to find a lender available to lend money at $i^d < r^d$, then the firm is able to create value out of that borrowing. The ROD may also differ from r_t^d (implying nonzero debt NPV) whenever the analysis of the project is made not when the firm raises the debt capital but, rather, at a later stage. In that case, the firm may be using debt on which it pays an interest rate i_t^d different from r_t^d.

Definition 6.1 (*NPV criterion—general form*) Any economic activity generating a cash-flow vector equal to $\boldsymbol{F}$ is worth undertaking (i.e., creates value for shareholders) if and only if the equity NPV is positive:

$$\text{NPV}^e = F_0^e + V_0^e = \sum_{t=0}^{n} F_t^e \text{d}_{t,0}^e > 0.$$

Definition 5.5 is a special case of Definition 6.1 under the assumption that $C_t^d = 0$ for all t. Hence, if the objective of the firm is to maximize shareholders' wealth, the equity NPV (not the project NPV) will be of greatest importance to managers.

Proposition 6.2 (Rationality of NPV criterion—general form) *The NPV criterion is a rational criterion. Managers create value for shareholders if and only if they undertake projects with positive equity NPVs.*

To recap, shareholders aim to maximize their wealth. Therefore, in general, the analyst must draw attention on the equity NPV, not the project NPV. However, the project NPV and the equity NPV provide the same value (and, therefore, the same decision) if

- the project is not financed with debt ($C_t^d = 0$ for all t)
- the market value of debt coincides with the book value of debt (e.g., because the return on debt is equal to the required return on debt).

Whenever these assumptions are not fulfilled, some of the project NPV is grasped by debtholders (and lost by shareholders) or, vice versa, is lost by debtholders (and grasped by shareholders). More precisely, Eq. (6.15) may be viewed as a decomposition of the value created by the project into two parts: The value accrued to the equityholders and the value accrued to debtholders. Taking the point of view of shareholders, one may reframe the same identity as

$$\text{NPV}^e = \text{NPV} + (-\text{NPV}^d). \tag{6.17}$$

Assuming NPV > 0,

- if $\text{NPV}^d = 0$, shareholders entirely grasp the value created by the project
- if $\text{NPV}^d > 0$, debtholders capture part of the value created by the project at the expense of shareholders (who lose part of the value created by the project)
- if $\text{NPV}^d < 0$, shareholders snap up additional value at the expense of the firm's creditors (who give up value).

In general, the idea that both equity NPV and debt NPV may be positive is not awkward, if one remembers that an economic entity generates profits and cash flows which are shared out by two kinds of investors, debtholders and equityholders, both aiming at investing money at an above-normal profit. A positive-NPV project may well be one where the value created is splitted up between the two classes of investors (see also footnote 5).

The benchmark system brings about a *benchmark* (replicating) Matrix describing the replicating portfolio. In it, the capital values and incomes are determined by the market. Highlighting the financing components and expressing incomes in terms of return rates ($I_t = i_t C_{t-1}$, $I_t^V = r_t V_{t-1}$), consider the following pair of Matrices:

$$\begin{array}{cc} \textbf{Project} & \textbf{Benchmark} \\ \begin{array}{|c|c|c|} \hline +\Delta C_t & +i_t C_{t-1} & -F_t \\ \hline +\Delta C_t^d & +i_t^d C_{t-1}^d & -F_t^d \\ +\Delta C_t^e & +i_t^e C_{t-1}^e & -F_t^e \\ \hline \end{array} & \begin{array}{|c|c|c|} \hline +\Delta V_t & +r_t V_{t-1} & -F_t \\ \hline +\Delta V_t^d & +r_t^d V_{t-1}^d & -F_t^d \\ +\Delta V_t^e & +r_t^e V_{t-1}^e & -F_t^e \\ \hline \end{array} \end{array} \tag{6.18}$$

with $t \in \mathbb{N}_n^1$.[6] The two Matrices share the same cash flows, while the profits and the capitals are different.

From the second column of the benchmark Matrix, $r_t V_{t-1} = r_t^d V_{t-1}^d + r_t^e V_{t-1}^e$ whence

$$r_t = \frac{r_t^e \cdot V_{t-1}^e + r_t^d \cdot V_{t-1}^d}{V_{t-1}^e + V_{t-1}^d}. \tag{6.19}$$

The above equation says that the project COC, r_t, is a weighted average of the cost of equity and the cost of debt, just in the same way as the project ROC is a weighted average of the ROE and the ROD:

$$i_t = \frac{i_t^e \cdot C_{t-1}^e + i_t^d \cdot C_{t-1}^d}{C_{t-1}^e + C_{t-1}^d} \tag{6.20}$$

(see second column of the project Matrix. See also Sect. 2.1.4). Several different expressions may be used for these COCs:

- r_t—*asset COC, cost of assets, required return on assets,*
- r_t^e—*equity COC, cost of equity, required return on equity,*
- r_t^d—*debt COC, cost of debt, required return on debt.*

The cost of assets, r_t, is, to all purposes and effects, a return on capital, where *capital* is evaluated at market price. Precisely, it is the counterfactual ROC that the investors would earn if they invested V_{t-1} into a portfolio that replicates the project's CFA from time 1 to time n. Note that the asset COC is a weighted average of the equity COC and the debt COC, with weights equal to the market values of equity and debt. The asset COC is, so to say, the market-based companion of the asset ROC: The former is the growth rate of the project system's capital, the latter is the growth rate of the benchmark system's capital. Analogously, the cost of equity (cost of debt) is the counterfactual ROC that equityholders (debtholders) would earn if they invested

[6] For $t = 0$, the benchmark Matrix as such is not valid. For it to be valid, one should define $r_0 V_{-1}$ as being equal to $F_0 + V_0$.

V^e_{t-1} (V^d_{t-1}) in a portfolio replicating the CFEs (CFDs) from time 1 to time n. It is, so to say, the market-based companions of the ROE (ROD).

The equity-debt ratio affects the ROE and the cost of equity in the same way:

$$i^e_t = i_t + \frac{C^d_{t-1}}{C^e_{t-1}}(i_t - i^d_t) \tag{6.21}$$

$$r^e_t = r_t + \frac{V^d_{t-1}}{V^e_{t-1}}(r_t - r^d_t). \tag{6.22}$$

One should not confound the two debt/equity ratios. They refer to two alternative courses of action, the factual one (undertaking the project) and the counterfactual one (investing in a replicating portfolio). The former are then book values, the latter are market values. Therefore, i_t is a book rate, whereas r_t is a market rate. The confusion between the two may cause errors and misunderstandings.

Remark 6.1 (*Required return on equity*) It should be underlined that, as illustrated in Chap. 5, the costs of capital incorporate a risk premium: The higher the risk, the higher the risk premium and the COC. The analyst should also bear in mind that the risk of the equityholders is higher than the risk of the debtholders. Equityholders are residual claimholders, so they will be paid out only after the CFD has been paid out in full. Therefore, the equity COC is necessarily greater than the debt COC: $r^e_t > r^d_t$. The analyst should always check that this inequality is fulfilled for every t. ▲

Remark 6.2 (*Required return on debt*) The cost of debt, r^d_t, reflects the capability of the firm to repay the debt. If the debt is non-recourse (as is the case for project financing transactions), the debtholders may not seek out the firm if the project's CFAs are not sufficient to cover the CFDs. In contrast, if the debt is recourse, the debtholders have recourse to the firm's other assets in case the project's CFA are not sufficient. Therefore, in the first case, the risk for debtholders will be higher than in the latter case and, consequently, the cost of debt r^d_t will be higher than it would be in case of a recourse debt. In other words, the risk of debtholders may be tied either to the project's risk or, rather, to the firm's risk, depending on the type of debt. ▲

Remark 6.3 (*ROE as an "average" of cost of equity and rate of return of existing shareholders*) A frequent error is to believe that the equity COC, r^e_t, is the cost the firm bears for raising incremental funds from the shareholders. In contrast, the equity COC is the expected rate of return of the shareholders purchasing the new shares. The financing rate for the firm is the project's ROE, i^e_t, not the equity COC. We now clarify this issue, assuming the project is partially financed with issuance of new shares.[7]

[7]If no shares are issued at time 0 (e.g., debt financing or internal financing is used), the ROE in the first period is not defined, so that (6.23) may not be written as such. It may be written as

$$(C^e_0 + I_1)(1 + i^e_2)(1 + i^e_3) \cdot \ldots \cdot (1 + i^e_n) = \sum_{t=1}^{n} F^e_t (1 + i^e_{t+1})(1 + i^e_{t+2}) \cdot \ldots \cdot (1 + i^e_n).$$

When the project is undertaken, the firm issues Δn_0 shares and receives, at time 0, an equity capital injection of C_0^e from the purchasers of the new shares. Later, it will distribute an expected amount equal to F_t^e at time $t \in \mathbb{N}_n^1$ to *all* shareholders (existing shareholders and new shareholders) (see also Remark 5.3). However, we know from Chap. 1 that

$$C_0^e = \sum_{t=1}^{n} \frac{F_t^e}{\prod_{k=1}^{t}(1+i_k^e)} \tag{6.23}$$

(see also Table 4.32) which means that the cost of raising C_0^e is the (expected) ROE. From the law of conservation (Chap. 2), this is both an investment rate (for the equityholders, which owns the equity) and a financing rate (for the firm, which owes the equity).[8]

The equity COC is different. On one hand, it is a foregone rate of return: Instead of injecting an additional C_0^e in the firm, investors might inject V_0^e in the capital market and purchase a portfolio replicating the equity cash flows of the project. This investment is such that

$$V_0^e = \sum_{t=1}^{n} \frac{F_t^e}{\prod_{k=1}^{t}(1+r_k^e)}. \tag{6.24}$$

On the other hand, the equity COC is the rate of return the new shareholders expect to receive as a consequence of purchasing the shares, for

$$C_0^e = \sum_{t=1}^{n} \frac{(1-\theta)(f_t^e + F_t^e)}{\prod_{k=1}^{t}(1+r_k^e)} \tag{6.25}$$

where $\theta = n_0/(n_0 + \Delta n_0)$ denotes the relative number of old shares over the total shares existing after the issuance.[9]

One may decompose the sequence $(-C_0^e, F_1^e, F_2^e, \ldots, F_n^e)$ into two subsequences referred to old shares, owned by the existing shareholders before the issuance, and new shares, owned by the purchasers of the new shares. New shareholders expected to receive part of the CFEs of the firm-with-the-project, namely, $(1-\theta)(f_t^e + F_t^e)$. Owing to (6.25), the new shareholders' NPV is zero:

$$-C_0^e + \sum_{t=1}^{n} \frac{(1-\theta)(f_t^e + F_t^e)}{\prod_{k=1}^{t}(1+r_k^e)} = 0.$$

In contrast, the old shareholders give up part of the expected CFEs generated by the firm-without-the-project, namely, $(1-\theta)f_t^e$, while grasping part of the CFEs generated by the firm-with-the-project, namely, θF_t^e. Put another way, they exchange $(1-\theta)f_t^e$ for θF_t^e at every date:

[8] See also Sect. 1.3.

[9] Equation (6.25) is just Eq. (5.17) (we remind that in Sect. 5.3 we assumed the project to be unlevered, so that $C_0 = C_0^e$).

Cash flows to existing equityholders

Time	Firm without the project	Firm with the project	Project
1	f_1^e	$\theta(f_1^e + F_1^e)$	$\theta F_1^e - (1-\theta) f_1^e$
2	f_2^e	$\theta(f_2^e + F_2^e)$	$\theta F_2^e - (1-\theta) f_2^e$
$\vdots$	$\vdots$	$\vdots$	$\vdots$
n	f_n^e	$\theta(f_n^e + F_n^e)$	$\theta F_n^e - (1-\theta) f_n^e$

Their gain is the project's equity NPV:

$$\begin{aligned}\sum_{t=1}^{n} \frac{\theta F_t^e - (1-\theta) f_t^e}{\prod_{k=1}^{t}(1+r_k^e)} &= \sum_{t=1}^{n} \frac{F_t^e}{\prod_{k=1}^{t}(1+r_k^e)} - \sum_{t=1}^{n} \frac{(1-\theta)(f_t^e + F_t^e)}{\prod_{k=1}^{t}(1+r_k^e)} \\ &= V_0^e - C_0^e \\ &= \text{NPV}^e. \end{aligned} \tag{6.26}$$

The whole equity NPV is then captured by the current shareholders. The new shareholders invest C_0^e in a value-neutral asset whose expected rate of return is r_t^e. Table 6.1 couples the two subsequences of cash flows for new and existing shares.

The ROE is then an "average" of the new shareholders' rate of return and the old shareholders's rate of return, denoted as i_t^{old} and i_t^{new}. We can then make the following statement:

The cost of financing of the firm when raising new shares is the (forecast) ROE, i_t^e. The ROE is a function of

- the new shareholders' rate of return, i_t^{new}, which is equal to the cost of equity, r_t^e, and
- the old shareholders' rate of return, i_t^{old}.

Table 6.1 Decomposition of the project CFEs between old shares and new shares[a]

Time	CFE (old shares)	CFE (new shares)	Total
0	0	$-C_0^e$	$-C_0^e$
1	$\theta F_1^e - (1-\theta) f_1^e$	$(1-\theta)(f_1^e + F_1^e)$	F_1^e
2	$\theta F_2^e - (1-\theta) f_2^e$	$(1-\theta)(f_2^e + F_2^e)$	F_2^e
$\vdots$	$\vdots$	$\vdots$	$\vdots$
n	$\theta F_n^e - (1-\theta) f_n^e$	$(1-\theta)(f_n^e + F_n^e)$	F_n^e
Gain	NPV^e	0	NPV^e
Rate of return	i_t^{old}	$i_t^{\text{new}} = r_t^e$	i_t^e

[a]See also Example 9.12

(Existing shareholders may well purchase the new shares as well, which means that they will own a portfolio of old shares and new shares. Thus, they will expect to earn a rate of return on this portfolio which is a weighted average of r_t^e and i_t^{old}.) ▲

Example 6.1 ♦
(*A positive-NPV project which increases both shareholders' wealth and debtholders' wealth*) A firm, whose shares are traded in a normal market, borrows $C_0^d = 60$ from a bank and raises $C_0^e = 40$ from equitholders in order to undertake a one-year project whose cash-flow stream is $\boldsymbol{F} = (-100, 150)$. The ROD is $i^d = 33.33\%$, the required return on assets is $r = 20\%$, the required return on debt (cost of debt) is $r^d = 8\%$. The project value is $V_0 = 150/1.2 = 125$, the debt value is $V_0^d = 60 \cdot 1.3333/1.08 = 74.07$. This implies that the equity value is $V_0^e = 125 - 74.07 = 50.93$. Given these values, both equityholders and debtholders take advantage of the project. Equityholders contribute \$40 to the firm and get an asset whose worth in a normal market is \$50.93. The difference, equal to \$10.93, represents the equity NPV, captured by the firm's shareholders: $\text{NPV}^e = 10.93$. The bank contributes \$60 and, assuming the investment is equity financed, its equityholders get an asset whose worth in a normal market is \$74.07. The difference, equal to \$14.07, represents the debt NPV, captured by the firm's debtholders (i.e., the bank's shareholders). Suppose $n_0^e = 100$ and $n_0^d = 15$ are, respectively, the number of firm's shares and bank's shares. Then, the increase in the firm's share price is $\Delta p_0^e = 10.93/10 = 1.093$; the increase in the bank's share price is $\Delta p_0^d = 14.07/15 = 0.938$.[10]

Why should the bank ask for an interest rate which is greater than the required rate of return? Because, in this way, the debt NPV is positive and, therefore, the bank's shareholders increase their wealth.

Why should the firm accept to pay interest at a rate which is greater than the required return on debt? Because, notwithstanding this fact, the equity NPV remains positive and, therefore, the firm's shareholders increase their wealth.

The firm's investment policy is favorable. But is the financing policy optimal? The answer is no: Shareholders would maximize their wealth if the firm's managers raised \$60 from the market, not from the bank. In this way, all the project NPV would be grasped by the firm's shareholders. Making recourse to the bank, a part of the project NPV is lost.

Example 6.2 ♦
(*A negative-NPV project which increases shareholders' wealth*) Tom undertakes a 2-year project whose cost is \$150 with prospective CFAs equal to \$120 and \$70, respectively. To undertake the project, Tom borrows \$65 from Jerry. The debt is repayable with two instalments equal to \$35 and \$41, respectively.[11] Suppose the cost of assets is $r = 22.76\%$ and the cost of debt is $r^d = 20\%$. The project NPV is negative:

---

[10]Strictly speaking, we must assume that either the bank's investment of \$60 is equity financed or that it is financed with debt and the debt is value-neutral; otherwise, \$0.938 measures the bank's NPV but not the bank's equity NPV.

[11]Assuming the debt rate, i^d, is constant, then $i^d = 10.78\%$, for $65 = 35/1.1078 + 41/1.1078^2$.

$$\text{NPV} = -150 + \frac{120}{1.2276} + \frac{70}{1.2276^2} = -5.8.$$

However, Tom will benefit from undertaking it, since the debt NPV is negative and smaller than the project NPV (i.e., greater in absolute value):

$$\text{NPV}^d = -65 + \frac{35}{1.2} + \frac{41}{1.2^2} = -7.36.$$

Hence, the equity NPV is $\text{NPV}^e = \text{NPV} - \text{NPV}^d = -5.8 - (-7.36) = 1.56$.[12]

In essence, Tom takes advantage from the fact that the negative-project NPV is more than compensated by a borrowing cost, $i^d = 10.78\%$, which is smaller than the required return on debt, $r^d = 20\%$. In this case, value creation is not obtained from an optimal investment policy, but from the ability of Tom of borrowing at favorable conditions.[13]

Example 6.3 ♦

(*A positive-NPV project which decreases shareholders' wealth*) A customer deposits \$1,000 in a bank account. The bank sets the interest rate at 7.24%, with a contract clause according to which the customer is prevented to withdraw the money before two years.

The bank immediately uses those funds and adds an equal amount of \$1,000 to grant a loan to a company, which will repay the debt with two instalments equal to \$450 and \$2,350 respectively. From the point of view of the bank, the transaction with the company is a levered project, partially financed with debt (the customer's injection of \$1,000). Suppose the cost of assets is $r = 20\%$, while the cost of debt is $r^d = 5\%$. This means that the project NPV is

$$\text{NPV} = -2000 + \frac{450}{1.2} + \frac{2{,}350}{1.2^2} = 6.94.$$

The project is value-creating, because the instalments are sufficiently high to cover the required return. Yet, the bank does not benefit from this course of action, because the debt NPV is positive as well and greater than the project NPV:

$$\text{NPV}^d = -1000 + \frac{1{,}000 \cdot 1.0724^2}{1.05^2} = 43.12.$$

As a result, the equity NPV for the bank is negative: $\text{NPV}^e = 6.94 - 43.12 = -36.18$.

---

[12] Assuming the equity COC is constant, this implies $r^e = 25\%$. To see it, consider that $V_0 = \text{NPV} + C_0 = -5.8 + 150 = 144.2$ and $V_0^d = \text{NPV}^d + C_0 = -7.36 + 65 = 57.64$ whence $V_0^e = 144.2 - 57.64 = 86.56$. Also, $F_1^e = 120 - 35 = 85$ and $F_2^e = 70 - 41 = 29$. Hence, $86.56 = 85/1.25 + 29/1.25^2$.

[13] The reason why Jerry lends money to Tom at conditions which are better (for Tom) than the security market may have to do with some kind of market imperfections or with bounded rationality or with other side effects on the part of Jerry's economic activities. Whatever the reasons, these situations do occur in real life.

This situation is symmetrical to the one illustrated in Example 6.2: A project exists with positive NPV which is not worth undertaking for the equityholder (i.e., the bank). The rationale is the same, albeit symmetrical: The bank creates value by lending \$2,000 to the company, but destroys it by borrowing from the customer. The favorable investment policy is frustrated by an unfavorable financing policy: Given the assumption of a market rate equal to 5%, the bank should not grant 7.24% to the customer. This transaction destroys shareholder value and the value destruction is only partially compensated by the favorable transaction with the company. The net effect is value destruction for the bank's shareholders. Here, shareholder value destruction is determined by the interactions of investment and borrowing policy.

Note that it is not necessary that the bank borrows from the customer at an interest rate smaller than 5%. For example, suppose the bank sets the account rate equal to 5.2%, other things unvaried. It is easy to check that this implies that the debt NPV reduces to $\text{NPV}^d = 3.81$ and, given a project NPV of \$6.94, the shareholder value created is now $\text{NPV}^e = 6.94 - 3.81 = 3.13$.[14]

The above examples illustrate in a simple way why the project NPV is not the relevant driver for decisions.

The **equity NPV** is the 'gold standard' for decisions. Shareholder value may be created/destroyed even though the project NPV is negative/positive (If $r_t^d = i_t^d$, the project NPV correctly measures shareholder value creation).

To compute the equity NPV one must compute the equity value, V_0^e. It may be computed in two different ways, depending on whether r_t^e or r_t is known (along with r_t^d):

- **CFE method**, embedded in (6.6) for $t = 0$, consists of discounting the CFEs at r_t^e.
- **CFA method**, epitomized by (6.5) and (6.7) for $t = 0$, consists of discounting the CFAs at r_t and then subtracting the debt value, obtained as discounted sum of CFDs at r_t^d.

Once calculated the equity value, the nitial investment in equity, C_0^e, is substracted (or F_0^e is added, if $I_0^e \neq 0$), and the equity NPV is obtained.

Example 6.4 ♦

C^o	C^d
	C^e

Consider a levered project P whose forecasted expected CFEs and CFDs are reported in Table 6.2. The table also reports the assumed ROE and ROD. This enables calculating the capital amounts for both equity and debt, via the usual fundamental relation (1.10). As for the equity capital, one finds

---

[14]The optimal financing policy for the bank would be to set the account rate as small as possible, so as to snap additional value from the customer (obviously, this conflicts with the optimal investment policy of the customer, who aims at maximizing the return on the capital invested. If the account rate is 5%, the value is neither created nor destroyed for both the bank and the customer).

$$\begin{aligned}
C_0^e &= 0 - (-145.94) &&= 145.9\\
C_1^e &= 145.94 \cdot 1.06 - 20 &&= 134.7\\
C_2^e &= 134.7 \cdot 1.09 - 40 &&= 106.8\\
C_3^e &= 106.8 \cdot 1.1 - 55 &&= 62.5\\
C_4^e &= 62.5 \cdot 1.12 - 70 &&= 0
\end{aligned}$$

while, for the debt capital, one finds

$$\begin{aligned}
C_0^d &= 0 - (-57.3) &&= 57.3\\
C_1^d &= 57.3 \cdot 1.05 - 15 &&= 45.2\\
C_2^d &= 45.2 \cdot 1.05 - 15 &&= 32.4\\
C_3^d &= 32.4 \cdot 1.05 - 15 &&= 19\\
C_4^d &= 19 \cdot 1.05 - 20 &&= 0.
\end{aligned}$$

Symmetrically, Table 6.3 reports the exogenously estimated cost of equity and cost of debt (cash flows are repeated). This enables computing the market values of equity and debt. We use the law of motion relation by proceeding backwards (see Eq. (5.1)):

$$\begin{aligned}
V_4^e &= 0\\
V_3^e &= (0 + 70)/1.09 &&= 64.2\\
V_2^e &= (64.2 + 55)/1.08 &&= 110.4\\
V_1^e &= (110.4 + 40)/1.07 &&= 140.6\\
V_0^e &= (140.6 + 20)/1.06 &&= 151.5
\end{aligned}$$

while for the debt capital one finds

$$\begin{aligned}
V_4^d &= 0\\
V_3^d &= (0 + 20)/(1.05) &&= 19\\
V_2^d &= (19 + 15)/(1.05) &&= 32.4\\
V_1^d &= (32.4 + 15)/(1.05) &&= 45.2\\
V_0^d &= (45.2 + 15)/(1.05) &&= 57.3
\end{aligned}$$

(note that $C_t^d = V_t^d$ because we assume $i_t^d = r_t^d$ for all t).

Table 6.2 Project P—cash flows and income rates

Time	CFA	CFE	CFD	i_t^e	i_t^d
0	−203.2	−145.9	−57.3		
1	35.0	20.0	15.0	6%	5%
2	55.0	40.0	15.0	9%	5%
3	70.0	55.0	15.0	10%	5%
4	90.0	70.0	20.0	12%	5%

Table 6.3 Project P—cash flows and costs of capital

Time	CFA	CFE	CFD	r_t^e	r_t^d
0	−203.2	−145.9	−57.3		
1	35.0	20.0	15.0	6%	5%
2	55.0	40.0	15.0	7%	5%
3	70.0	55.0	15.0	8%	5%
4	90.0	70.0	20.0	9%	5%

Table 6.4 Project P—ROI and COC

Time	i_t	r_t	C_t	V_t
0			203.2	208.8
1	5.72%	5.73%	179.9	185.7
2	8.00%	6.51%	139.2	142.8
3	8.84%	7.32%	81.5	83.3
4	10.36%	8.08%	0.0	0.0

Table 6.4 reports the asset ROIs and the asset COCs. They can be determined from (6.19) and (6.20) and the respective capital amounts are determined either by additivity (i.e., by summing the equity capital and the debt capital) or, directly, by using the usual laws of motion. Hence, the NPVs are

$$\begin{aligned} \text{NPV} &= V_0 - C_0 = 208.8 - 203.2 = 5.6 \\ \text{NPV}^e &= V_0^e - C_0^e = 151.5 - 145.9 = 5.6 \\ \text{NPV}^d &= V_0^d - C_0^d = 57.3 - 57.3 \quad = 0 \end{aligned}$$

Example 6.5 ♦

C^o	C^d
C^l	C^e

(*Equity NPV of BHP Billiton, 90% payout*) Consider Billiton project assuming the presence of NOWC and a 90% payout ratio (see Example 4.3). Assume that the cost of equity is time-invariant and equal to $r^e = 0.15$. We have already computed the CFEs (see Table 4.29). Discounting the CFEs from time 1 to time 20 at 15%, one gets the equity value: $V_0^e = 1{,}347.2$. The equity NPV is $V_0^e - C_0^e = 1{,}347.2 - 400 = 947.2$.

As for the project NPV, it depends on the debt NPV, and the latter depends on the cost of debt. For example, if one assumes that the required return on debt is equal to the ROD ($i^d = r^d = 0.08$), then $V_0^d = C_0^d = 200$ and $\text{NPV}^d = 0$, which implies that the project NPV coincides with the equity NPV: $\text{NPV} = \text{NPV}^e = 947.2$.

Example 6.6 ♦

C^o	C^d
C^l	C^e

(*NPV of Mockup, 80% payout*) Consider the Mockup project (Example 4.11), where a payout ratio of 80% is assumed. We compute the asset ROCs by dividing the project income, $EBIT_t + I_t^l - T_t$, to the invested capital, C_{t-1} (see (4.51) and Table 4.37), resulting in the ROC vector

$$\boldsymbol{i} = (6.68\%, 8.51\%, 10.97\%, 14.38\%, 19.4\%).$$

(It is easy to see that discounting the CFAs at the ROCs the invested capital is found back, as one expects.) Suppose the cost of assets is time-invariant: $\boldsymbol{r} = (9\%, 9\%, 9\%, 9\%, 9\%)$. Discounting CFAs at the cost of assets, the market value of the project is obtained:

$$V_0 = 522{,}277 = \frac{96{,}849}{1.09} + \frac{99{,}492}{1.09^2} + \frac{102{,}198}{1.09^3} + \frac{104{,}968}{1.09^4} + \frac{302{,}196}{1.09^5}.$$

The project NPV is $NPV = V_0 - C_0 = 522{,}277 - 500{,}000 = 22{,}277$. Whether this is equal or not to the equity NPV depends on whether $i_t^d = r_t^d$ or not. Let us assume that $i^d = r^d = 0.03$. Then, the book value of debt is equal to the market value of debt:

$$V_0^d = \frac{2{,}700}{1.03} + \frac{2{,}700}{1.03^2} + \frac{2{,}700}{1.03^3} + \frac{2{,}700}{1.03^4} + \frac{92{,}700}{1.03^5} = 90{,}000$$

and the debt NPV is zero: $NPV^d = V_0^d - C_0^d = 90{,}000 - 90{,}000 = 0$. The market value of equity is found by additivity: $V_0^e = V_0 - V_0^d = 522{,}277 - 90{,}000 = 432{,}277$. The equity NPV is $NPV^e = V_0^e - C_0^e = 432{,}277 - 410{,}000 = 22{,}277$, equal to the project NPV, as expected.

Using (6.21)–(6.22) one can find the ROEs and the equity COCs, respectively:

$$\boldsymbol{i}^e = (7.5\%, 9.9\%, 13.5\%, 19.0\%, 28.5\%)$$
$$\boldsymbol{r}^e = (10.2\%, 10.4\%, 10.7\%, 11.1\%, 11.9\%).$$

(It should also be clear that the book value of equity and the market value of equity may be found back by discounting CFEs at the ROEs and at the costs of equity, respectively.)

Table 6.5 collects, for every date t, the market values of the project as well as the equity values and debt values.

Table 6.5 Mockup project—market values (financing perspective, 80% payout)

Time	0	1	2	3	4	5
V	522,277	472,433	415,460	350,654	277,244	0
V^d	90,000	90,000	90,000	90,000	90,000	0
V^e	432,277	382,433	325,460	260,654	187,244	0.0

Example 6.7 ♦
Consider Example 2.8 where firm XYZ faces the opportunity of undertaking a project, named EngPro, either with corporate financing or with project financing. The risk faced by equityholders as well as debtholders is different in the two cases. Corporate financing is less risky for debtholders and riskier for equityholders, because, should things go wrong with the project, debtholders have recourse to the firm's assets, whereas in project financing, they only have recourse to the SPV's assets. In terms of COC, it means that the debt COC, for EngPro, r_t^d, will be higher if EngPro is incorporated as a firm than if EngPro is financed on-balance sheet. In contrast, the equity COC for EngPro, r_t^e, will be smaller if project financing is used.

6.1.2 *Valuation in an Investment Perspective*

In Sect. 6.1.1 we have evaluated the project by interpreting it as a portfolio of equity and debt. However, taking into account the law of conservation for capital, the project is also interpretable as a portfolio of the operating assets and the non-operating assets. Let V_t^o and V_t^l be the market value of two assets that replicate, respectively, the project's CFOs and the project's CFLs from time 1 to time n.[15] The associated cash-flow streams are

$$\begin{aligned} \boldsymbol{F}^{Vo} &= \left(-V_0^o, F_1^o, F_2^o, \ldots, F_{n-1}^o, F_n^o\right) \\ \boldsymbol{F}^{Vl} &= \left(-V_0^l, F_1^l, F_2^l, \ldots, F_{n-1}^l, F_n^l\right) \end{aligned} \tag{6.27}$$

Let r_t^o be the expected return rate on V_{t-1}^o, and r_t^l be the expected return rate on V_{t-1}^l, as determined by the market. Then, the equation of motion (1.1) boils down to

$$V_t^o = V_{t-1}^o(1 + r_t^o) - F_t^o \tag{6.28}$$

$$V_t^l = V_{t-1}^l(1 + r_t^l) - F_t^l \tag{6.29}$$

which entail, respectively, the prospective relations

$$V_t^o = \sum_{k=t+1}^{n} F_k^o \cdot \mathrm{d}_{k,t}^o \tag{6.30}$$

$$V_t^l = \sum_{k=t+1}^{n} F_k^l \cdot \mathrm{d}_{k,t}^l \tag{6.31}$$

---

[15]The associated benchmark systems are then described by the respective transposed Matrices $(\boldsymbol{V}^o, \boldsymbol{I}^{Vo}, \boldsymbol{F}^{Vo})$ and $(\boldsymbol{V}^l, \boldsymbol{I}^{Vl}, \boldsymbol{F}^{Vl})$.

where $\mathrm{d}^o_{k,t} = \prod_{j=t+1}^{k}(1+r^o_j)^{-1}$, $\mathrm{d}^l_{k,t} = \prod_{j=t+1}^{k}(1+r^l_j)^{-1}$ are the respective discount factors.[16] The two rates, r^o_t and r^l_t, express the respective COCs for CFOs and CFLs. Several different expressions may be used for these COCs:

- r^o_t—*operating COC, cost of operating assets, required return on operating assets*
- r^l_t—*non-operating COC, cost of non-operating (or liquid) assets, required return on non-operating (or liquid) assets.*

The law of one price implies that, for every $t \geq 0$, the project's market value must equal the sum of market value of operating assets and market value of liquid assets: $V_t = V^o_t + V^l_t$. This means that it is indifferent for an investor to hold an asset which releases the same stream of the prospective project's CFAs or a portfolio of two assets that release the stream of the prospective CFOs and the stream of prospective CFLs. And this is just the principle of value additivity described in Eqs. (5.28)–(5.29), with $m = 2$ and $\boldsymbol{F}^{A_1}$ and $\boldsymbol{F}^{A_2}$ now representing the CFO and CFL streams.

Hence, one finds what we call the ***operating*** NPV and the ***non-operating*** (or *liquid-asset*) NPV:

$$\mathrm{NPV}^o = V^o_0 - C^o_0 \tag{6.32}$$

$$\mathrm{NPV}^l = V^l_0 - C^l_0 \tag{6.33}$$

(more generally, $\mathrm{NPV}^o = V^o_0 + F^o_0$ and $\mathrm{NPV}^l = V^l_0 + F^l_0$). Given that $C_0 = C^o_0 + C^l_0$ (the investment cost is fully distributed to operating assets and financial assets), the project NPV is such that $\mathrm{NPV} = \mathrm{NPV}^o + \mathrm{NPV}^l$. Therefore, the project NPV may now be split into operating NPV and non-operating NPV; the former measures the share of project NPV that is generated by the operations, the latter measures the share of project NPV that is generated by financial investments. Note that V_t is the market value of the capital raised by the firm from the capital providers, while $V^o_t + V^l_t$ is the market value of the capital invested in the project. The one-price law states that they are equal. This is nothing but the static principle for capital values introduced in Sect. 2.1.1 with V_t, V^o_t, V^l_t replacing C_t, C^o_t, C^l_t (see Eq. (2.5)).

We can then formalize the result in the following proposition, which is the mirror image of Proposition 6.1.

Proposition 6.3 (NPV as sum of operating NPV and non-operating NPV) *The law of one price guarantees that the static equilibrium of an asset traded in a normal market is fulfilled: The capital V_t raised by the firm from the capital providers has the same market value as the capital generated by the firm's assets:*

$$V_t = V^o_t + V^l_t = V^{\mathrm{inv}}_0. \tag{6.34}$$

[16]The value of a firm's operating assets is also known as the *enterprise* value (see Berk and DeMarzo 2014, p. 28). In terms of the project, V^o_t is the increase of the enterprise value caused by the project undertaking.

This implies that the NPV distributed to equityholders and debtholders is entirely generated by the project's assets:

$$\text{NPV} = \text{NPV}^o + \text{NPV}^l = \text{NPV}^{\text{inv}}. \tag{6.35}$$

Propositions 6.1 and 6.3 express the law of conservation of value creation.

Law of conservation for value creation. The intrinsic value of the assets, V^{inv}, coincides with the intrinsic value of the financings, V^{fin}. Hence, the economic value created by the investments, NPV^{inv}, is equal to the economic value created by the financings, NPV^{fin}. This boils down to saying that the economic value created by the operating and non-operating assets is entirely distributed to claimholders:

$$\overbrace{V_0^o + V_0^l}^{V_0^{\text{inv}}} = V_0 = \overbrace{V_0^e + V_0^d}^{V_0^{\text{fin}}}.$$

$$\overbrace{\text{NPV}^o + \text{NPV}^l}^{\text{NPV}^{\text{inv}}} = \text{NPV} = \overbrace{\text{NPV}^e + \text{NPV}^d}^{\text{NPV}^{\text{fin}}}.$$

Assuming $\text{NPV} > 0$,

- if $\text{NPV}^l = 0$, the value created by the project is entirely generated by the operations
- if $\text{NPV}^l > 0$, some of the value created by the project's assets is generated by the non-operating assets
- if $\text{NPV}^l < 0$, some of the value created by the project's asset is destroyed by the investment in liquid assets.

Like the financing perspective, the investment perspective may be visually grasped by coupling the project Matrix and the benchmark Matrix, with the caveat that replication is now made upon the CFOs and the CFLs, rather than the CFEs and the CFDs, as follows:

$$\begin{array}{c}\textbf{Project}\\ \begin{array}{|c|c|c|}\hline +\Delta C_t^o & +i_t^o C_{t-1}^o & -F_t^o \\ +\Delta C_t^l & +i_t^l C_{t-1}^l & -F_t^l \\ \hline +\Delta C_t & +i_t C_{t-1} & -F_t \\ \hline \end{array}\end{array} \quad \begin{array}{c}\textbf{Benchmark}\\ \begin{array}{|c|c|c|}\hline +\Delta V_t^o & +r_t^o V_{t-1}^o & -F_t^o \\ +\Delta V_t^l & +r_t^l V_{t-1}^l & -F_t^l \\ \hline +\Delta V_t & +r_t V_{t-1} & -F_t \\ \hline \end{array}\end{array} \tag{6.36}$$

From the second column of the benchmark Matrix, $r_t^o V_{t-1}^o + r_t^l V_{t-1}^l = r_t V_{t-1}$ whence

$$r_t = \frac{r_t^o \cdot V_{t-1}^o + r_t^l \cdot V_{t-1}^l}{V_{t-1}^o + V_{t-1}^l}. \tag{6.37}$$

The above equality says that the cost of assets is a weighted average of the operating COC and the non-operating COC, just in the same way the asset ROC is a weighted average of the operating ROC and the non-operating ROC:[17]

$$i_t = \frac{i_t^o \cdot C_{t-1}^o + i_t^l \cdot C_{t-1}^l}{C_{t-1}^o + C_{t-1}^l}. \tag{6.38}$$

We may join the investment perspective and the financing perspective in the following pair of Matrices, one for the project, the other one for the replicating portfolios:

$$\begin{array}{c} \textbf{Project} \\ \begin{array}{|c|cc|} \hline +\Delta C_t^o & +i_t^o C_{t-1}^o & -F_t^o \\ +\Delta C_t^l & +i_t^l C_{t-1}^l & -F_t^l \\ \hline +\Delta C_t^d & +i_t^d C_{t-1}^d & -F_t^d \\ +\Delta C_t^e & +i_t^e C_{t-1}^e & -F_t^e \\ \hline \end{array} \end{array} \quad \begin{array}{c} \textbf{Benchmark} \\ \begin{array}{|c|cc|} \hline +\Delta V_t^o & +r_t^o V_{t-1}^o & -F_t^o \\ +\Delta V_t^l & +r_t^l V_{t-1}^l & -F_t^l \\ \hline +\Delta V_t^d & +r_t^d V_{t-1}^d & -F_t^d \\ +\Delta V_t^e & +r_t^e V_{t-1}^e & -F_t^e \\ \hline \end{array} \end{array} \tag{6.39}$$

Shifting the second row to the lower part for both Matrices,

$$\begin{array}{c} \textbf{Project} \\ \begin{array}{|c|cc|} \hline +\Delta C_t^o & +i_t^o C_{t-1}^o & -F_t^o \\ \hline +\Delta C_t^d & +i_t^d C_{t-1}^d & -F_t^d \\ +\Delta C_t^e & +i_t^e C_{t-1}^e & -F_t^e \\ -\Delta C_t^l & -i_t^l C_{t-1}^l & +F_t^l \\ \hline \end{array} \end{array} \quad \begin{array}{c} \textbf{Benchmark} \\ \begin{array}{|c|cc|} \hline +\Delta V_t^o & +r_t^o V_{t-1}^o & -F_t^o \\ \hline +\Delta V_t^d & +r_t^d V_{t-1}^d & -F_t^d \\ +\Delta V_t^e & +r_t^e V_{t-1}^e & -F_t^e \\ -\Delta V_t^l & -r_t^l V_{t-1}^l & +F_t^l \\ \hline \end{array} \end{array} \tag{6.40}$$

From the income column of the benchmark Matrix, $r_t^o V_{t-1}^o = r_t^d V_{t-1}^d + r_t^e V_{t-1}^e - r_t^l V_{t-1}^l$, whence, remembering that $V_t^{\mathcal{D}} = V_t^d - V_t^l$.

$$r_t^o = \frac{r_t^e V_{t-1}^e + r_t^{\mathcal{D}} V_{t-1}^{\mathcal{D}}}{V_{t-1}^e + V_{t-1}^{\mathcal{D}}} \tag{6.41}$$

where

$$r_t^{\mathcal{D}} = \frac{r_t^d V_{t-1}^d - r_t^l V_{t-1}^l}{V_{t-1}^d - V_{t-1}^l} \tag{6.42}$$

[17]In principle, one might define a mirror equation of (6.21):

$$i_t^l = i_t + \frac{C_{t-1}^o}{C_{t-1}^l}(i_t - i_t^o)$$

$$r_t^l = r_t + \frac{V_{t-1}^o}{V_{t-1}^l}(r_t - r_t^o).$$

Again, we stress that the two rates refer to project system and benchmark system, respectively. That is, they refer to two alternative courses of action, the factual one (undertaking the project) and the counterfactual one (replicating the project by investing in the market). In the first formula, pro forma book values are considered; in the latter formula, market values are taken into account.

is the *net-debt COC*, *cost of net debt*, or *required return on net debt*. Equation (6.41) shows that the operating COC is a weighted average of the equity COC and the net-debt COC. It is often called (operating) *pre-tax weighted average cost of capital* (**pre-tax WACC**). The pre-tax WACC and the net-debt COC are the market-based companions of the operating ROC and the return on net debt, respectively:

$$i_t^o = \frac{i_t^e C_{t-1}^e + i_t^{\mathcal{D}} C_{t-1}^{\mathcal{D}}}{C_{t-1}^e + C_{t-1}^{\mathcal{D}}} \qquad r_t^o = \frac{r_t^e V_{t-1}^e + r_t^{\mathcal{D}} V_{t-1}^{\mathcal{D}}}{V_{t-1}^e + V_{t-1}^{\mathcal{D}}}$$

$$i_t^{\mathcal{D}} = \frac{i_t^d C_{t-1}^d - i_t^l C_{t-1}^l}{C_{t-1}^d - C_{t-1}^l} \qquad r_t^{\mathcal{D}} = \frac{r_t^d V_{t-1}^d - r_t^l V_{t-1}^l}{V_{t-1}^d - V_{t-1}^l}$$

(compare the second columns of the project Matrix and the benchmark Matrix. Compare also Table 4.37 with Tables 6.30 and 6.31).

The net-debt COC may be used for computing the market value of net debt:

$$V_t^{\mathcal{D}} = V_{t-1}^{\mathcal{D}}(1 + r_t^{\mathcal{D}}) - F_t^{\mathcal{D}} \tag{6.43}$$

so that

$$V_0^{\mathcal{D}} = \sum_{t=1}^{n} F_t^{\mathcal{D}}\, \mathrm{d}_{t,0}^{\mathcal{D}} \tag{6.44}$$

where $\mathrm{d}_{t,0}^{\mathcal{D}} = \prod_{k=1}^{t}(1 + r_k^{\mathcal{D}})^{-1}$ and $\mathrm{d}_{0,0}^{\mathcal{D}}=1$. The corresponding NPV is then $\text{NPV}^{\mathcal{D}} = V_0^{\mathcal{D}} - C_0^{\mathcal{D}}$. It should not surprise the reader the fact that, owing to value additivity, $\text{NPV}^{\mathcal{D}}$ may be also found as the difference between the debt NPV and the non-operating NPV, that is, $\text{NPV}^{\mathcal{D}} = \text{NPV}^d - \text{NPV}^l$.

Assuming one has exogenous information on the pre-tax WACC, r_t^o, as well as the net-debt COC, $r_t^{\mathcal{D}}$, an equivalent method to value the equity of a project consists of discounting the CFOs at pre-tax WACC and subtracting the market value of net debt.[18] In symbols,

$$V_0^e = V_0^o - V_0^{\mathcal{D}}. \tag{6.45}$$

Equations (6.30), (6.44), and (6.45) summarize the **CFO method** (or CCF method).

As a consequence, the equity NPV can be framed in terms of financial structure (debt and equity) or in terms of capital structure (net debt and equity):

$$\text{NPV}^e = \text{NPV} - \text{NPV}^d \tag{6.46}$$

$$\text{NPV}^e = \text{NPV}^o - \text{NPV}^{\mathcal{D}}. \tag{6.47}$$

[18] The net-debt COC is known whenever both the debt COC, r_t^d, and the non-operating COC, r_t^l, are known or, alternatively, whenever the debt COC is known and the financial assets have the same risk as the debt, which implies that the non-operating COC is equal to the debt COC.

Correspondingly, the asset COC is the weighted mean of equity COC and debt COC (financial-structure perspective), while the operating COC (the pre-tax WACC) is a weighted mean of equity COC and net-debt COC (capital-structure perspective):

$$r_t = \frac{r_t^e V_{t-1}^e + r_t^d V_{t-1}^d}{V_{t-1}^e + V_{t-1}^d}$$

$$r_t^o = \frac{r_t^e V_{t-1}^d + r_t^{\mathcal{D}} V_{t-1}^{\mathcal{D}}}{V_{t-1}^e + V_{t-1}^{\mathcal{D}}}.$$

We will also use the expression **asset pre-tax WACC** to refer to r_t.

While the financing perspective and the investment perspective lead to the same result in terms of decision-making, they jointly provide an enriched economic analysis of the project. On one hand, the financing perspective decomposes the economic value created (i.e., the project NPV) into the equity component and the debt component; on the other hand, the investment perspective decomposes the economic value created into the value created by the operations and the value created by the financial investments.

To sum up, the operating assets and the non-operating assets generate two NPVs (investment perspective). The overall amount is distributed to equityholders and debtholders (financing perspective).

Table 6.6 illustrates the twofold decomposition. Summing by row, the NPV (with due sign) is obtained as the total; summing by column, the net-debt NPV (with due sign) is obtained as the total. The algebraic sums of the total (both by row and by column) are zero, which just reflects the laws of conservation of value creation. The equity NPV is boxed in order to remind the reader that NPV^e is the objective function that must be maximized in order to maximize shareholders' wealth. The table may also be read as providing a decomposition of the shareholder value created into three parts:

- the shareholder value created as a result of the investment policy NPV^o
- the shareholder value created as a result of the debt policy, $-\text{NPV}^d$
- the shareholder value created as a result of the distribution policy, NPV^l.

That is, $\text{NPV}^e = \text{NPV}^o + \text{NPV}^l - \text{NPV}^d$. If $i_t^d = r_t^d$, the debt policy is value-neutral ($\text{NPV}^d = 0$); if $i_t^l = r_t^l$, the distribution policy is value-neutral ($\text{NPV}^l = 0$). In general, the financing policy (debt+distribution policy) is value-neutral if $\text{NPV}^d - \text{NPV}^l = \text{NPV}^{\mathcal{D}} = 0$. In this case, $\text{NPV} = \text{NPV}^o = \text{NPV}^e$.

Example 6.8 ♦

C^o	C^d
C^l	C^e

(*NPV of Mockup, 80% payout—continued*) Let us pick up Example 6.6 with 80% payout ratio. We now repeat the analysis by finding the equity NPV as the difference between operating NPV and net-debt NPV.

We first remind that we have assumed $i^d = 0.03 = r^d$, which means that the market value of debt is equal to the book value of debt, namely, a constant \$90,000

Table 6.6 Decomposition of NPV investment side and financing side

		Total
investment side	$\text{NPV}^l + \text{NPV}^o$	NPV
− financing side	$-\text{NPV}^d - \boxed{\text{NPV}^e}$	−NPV
Total	$-\text{NPV}^{\mathcal{D}} + \text{NPV}^{\mathcal{D}}$	0

(see also Table 6.5). Assume that the required return on liquid assets is $r^l = 0.03$ as well (different from the return on liquid assets, $i^l = 0.02$). Since $r^d = r^l = 0.03$, then $r^{\mathcal{D}} = 0.03$. Therefore, the value of net debt can equivalently be computed by (i) discounting CF$\mathcal{D}$s at 3% or by (ii) subtracting the market value of non-operating assets from the value of debt. If (i) is followed, then, remembering that

$$\boldsymbol{F}^{\mathcal{D}} = (-90{,}000,\ 26{,}237,\ 26{,}898,\ 27{,}574,\ 28{,}267,\ -10{,}414)$$

(see Table 4.23), the value of the net financial position is computed as

$$V_0^{\mathcal{D}} = \frac{26{,}237}{1.03} + \frac{26{,}898}{1.03^2} + \frac{27{,}574}{1.03^3} + \frac{28{,}267}{1.03^4} + \frac{-10{,}414}{1.03^5} = 92{,}193.$$

If (ii) is followed, then $V_t^{\mathcal{D}} = V_t^d - V_t^l = 90{,}000 - V_t^l$ and the value of liquid assets may be computed by discounting the CFLs at the cost of non-operating assets, $r^l = 3\%$, where the CFLs are found in Table 4.22. This results in

$$V_0^l = \frac{-23{,}537}{1.03} + \frac{-24{,}198}{1.03^2} + \frac{-24{,}874}{1.03^3} + \frac{-25{,}567}{1.03^4} + \frac{103{,}114}{1.03^5} = -2{,}193.$$

Subtracting this (negative) value from the debt value, we get back to $V_0^{\mathcal{D}} = 90{,}000 + 2{,}193 = 92{,}193$.[19]

Now we need the operating value. We first compute the pre-tax WACC as a weighted mean of the cost of net debt ($r^{\mathcal{D}} = 3\%$) and the cost of equity, with the latter being represented by the vector

$$\boldsymbol{r}^e = (10.2\%, 10.4\%, 10.7\%, 11.1\%, 11.9\%)$$

[19]The negative value of non-operating assets signals that, undertaking this project, the firm absorbs additional resources from non-operating assets. In particular, consider that, at time 1, the CFL is negative and equal to $F_1^l = -23{,}537$. If one replicates the CFL vector in the market, one receives, at time 1, an asset which is worth $V_1^l = 21{,}278$ (see Table 6.7) but will have to disburse $|F_1^l| = 23{,}537$; the difference is $-2{,}259 = 21{,}278 - 23{,}537 = V_1^e + F_1^l$. Discounting back by one period at 3% one gets $V_0^l = -2{,}259/1.03 = -2{,}193$. Therefore, to invest in the replicating portfolio boils down to taking on a short position. The (negative) price of the replicating portfolio is the market value of the non-operating assets.

Table 6.7 Mockup project—market values (investment perspective, 80% payout)

Time	0	1	2	3	4	5
V	522,277	472,432	415,460	350,654	277,244	0
V^o	524,469	451,154	369,345	278,281	177,133	0
V^l	−2,193	21,278	46,115	72,373	100,111	0

(see Example 6.6). Hence, the sequence of pre-tax WACCs is, from (6.41),

$$\boldsymbol{r^o} = (8.97\%, 9.28\%, 9.75\%, 10.56\%, 12.39\%).$$

Reminding that $\boldsymbol{F^o}$=(−500,000, 120,386, 123,690, 127,072, 130,536, 199,082), one can find the value of operations in each period with the usual relation $V^o_{t-1} = (V^o_t + F^o_t)/(1 + r^o_t)$, which brings

$$\boldsymbol{V^o} = (524{,}469,\ 451{,}154,\ 369{,}345,\ 278{,}281,\ 177{,}133,\ 0).$$

Table 6.7 summarizes the result by decomposing the value of the assets into operating value and non-operating value: Note that $\text{NPV}^l = V^l_0 = -2{,}193 < 0$ since $C^l_0 = 0$. This means that, given the assumed data, the investments in financial assets destroy value (the firm pays nothing at time 0 but gets an asset whose value is negative). This occurs because the required return on liquid assets is 3% whereas the actual expected rate of return of financial assets is only 2%: Discounting CFLs at 3% makes the market value of liquid assets be smaller than the book value of liquid assets, which is the discounted value of the CFLs at 2%.

Note that the equity value (already computed in Example 6.6) is found back by subtracting the net-debt NPV from the operating NPV. The operating NPV is $\text{NPV}^o = V^o_0 - C^o_0 = 524{,}469 - 500{,}000 = 24{,}469$; the net-debt NPV is $\text{NPV}^{\mathcal{D}} = V^{\mathcal{D}}_0 - C^{\mathcal{D}}_0 = 92{,}193 - 90{,}000 = 2{,}193$. Hence, $\text{NPV}^e = \text{NPV}^o - \text{NPV}^{\mathcal{D}} = 24{,}469 - 2{,}193 = 22{,}277$ (the rounding error is due to approximation). This is obviously the same value we have obtained in Example 6.6, which confirms the logical consistency of the approach. Table 6.8 illustrates the various NPVs. Mockup project is value-creating and the value created is entirely grasped by equityholders, as the debt NPV is zero (the borrowing does not have any effect on shareholders' value as the 5-year loan's interest rate is equal to the required return on debt). However, part of the project NPV is lost owing to value-destroying financial activities. Specifically, the financial assets are invested at an expected rate of return which is smaller than their required rate of return, which precisely implies that part of the value generated by the operations is eroded by the liquid assets.

Table 6.8 Decomposition of Mockup NPV (80% payout ratio) investment side and financing side

investment side	$\overbrace{-2,193}^{\text{NPV}^l} + \overbrace{24,469}^{\text{NPV}^o} = \overbrace{27,277}^{\text{NPV}}$
financing side	$\overbrace{0}^{\text{NPV}^d} + \overbrace{27,277}^{\text{NPV}^e} = \overbrace{27,277}^{\text{NPV}}$

Example 6.9 ♦

C^o	C^d
	C^e

(*NPV of Mockup, 100% payout*) Let us consider Mockup project again and illustrate the same computations illustrated in Example 6.6 while assuming a 100% payout ratio instead of 80%. Keeping an eye on Table 4.25, one can compute the asset ROIs as income generated by the assets divided by the book value of assets: $i_t = (\text{EBIT}_t + I^l_t - \text{T}_t)/C_{t-1}$. One finds $\boldsymbol{i} = (6.68\%, 8.92\%, 12.39\%, 18.42\%, 31.41\%)$, higher than the respective rates in the 80% case (except the first-period ROIs, which are equal).[20] The reason of the increase is the result of two opposite effects:

- the invested capital in the 100% case is smaller in every period because all potential dividends are distributed, whereas in the 80% case some of the cash is retained in the business, becoming an asset
- the income generated by the asset in the 100% case is smaller in every period because interest income is zero.

The former effect tends to increase the ROIs while the latter effect tends to decrease them. The former effect prevails over the latter, so the net result is an increase of the ROIs.

Also, with a 100% payout, the CFOs are higher as opposed to the 80% case, because the taxes paid are smaller (interest income is zero)[21] and now the CFOs coincide with the CFAs:[22]

$$\boldsymbol{F}^o = \boldsymbol{F} = (-500{,}000,\ 120{,}386,\ 123{,}845,\ 127{,}390,\ 131{,}024,\ 199{,}749).$$

---

[20]In the 80% case, the vector of asset ROIs was $\boldsymbol{i} = (6.68\%, 8.51\%, 10.97\%, 14.38\%, 19.4\%)$.

[21]The increase in CFO is just equal to the product of the tax rate and the interest income generated by the 80% payout, τI^l_t. Precisely, the change in the CFO stream is $(0, 0, 155, 318, 489, 667)$.

[22]In the 80% case, the CFAs are smaller than the respective CFOs because some cash is retained.

Table 6.9 Mockup project—market values (100% payout)

Time	0	1	2	3	4	5
V	525,592	452,378	370,527	279,260	177,727	0
V^d	90,000	90,000	90,000	90,000	90,000	0
V^e	435,592	362,378	280,527	189,260	87,727	0

The operating pre-tax WACC coincides with asset pre-tax WACC:

$$\boldsymbol{r} = \boldsymbol{r}^o = (8.97\%, 9.28\%, 9.75\%, 10.56\%, 12.39\%).$$

Discounting these CFOs/CFAs, one gets $V_0^o = V_0 = 525{,}592$. The debt value is unvaried and equal to \$90,000. Hence, the equity value is $V_0^e = 525{,}592 - 90{,}000 = 435{,}592$. Subtracting the equity initial investment, one gets the equity NPV: $\text{NPV}^e = 435{,}592 - 410{,}000 = 25{,}592$, which also coincides with the project NPV.

A related point to consider is that, changing from 80 to 100% payout, the asset COC has changed from $\boldsymbol{r} = (9\%, 9\%, 9\%, 9\%, 9\%)$ to $\boldsymbol{r} = (8.97\%, 9.28\%, 9.75\%, 10.56\%, 12.39\%)$ (remember that, with 100%, $\boldsymbol{r} = \boldsymbol{r}^o$). As we assume that the cost of debt is unvaried ($r^d = 3\% = i^d$), this implies that the equity COC has changed. To calculate it, first subtract the market value of the debt from the market value of the assets. The former is equal to the book value of debt (\$90,000 for every t), the latter is found by discounting the prospective CFOs/CFAs at the pre-tax WACC. Table 6.9 collects the resulting equity values.

Using (6.22) one can finally find the equity costs of capital, which are higher than in the previous case:

$$\boldsymbol{r}^e(80\%) = (10.2\%, 10.4\%, 10.7\%, 11.1\%, 11.9\%)$$
$$\boldsymbol{r}^e(100\%) = (10.2\%, 10.8\%, 11.9\%, 14.2\%, 22.0\%).$$

The higher equity COCs are caused by the increased riskiness: In the 80% payout case, Mockup project was a portfolio of operating assets and (less risky) liquid assets, whereas in the 100% payout case all assets are operating assets. However, the profile of CFEs is different as well:

$$\boldsymbol{F}^e(80\%) = (-410{,}000,\ 94{,}149,\ 96{,}792,\ 99{,}498,\ 102{,}268,\ 209{,}496)$$
$$\boldsymbol{F}^e(100\%) = (-410{,}000,\ 117{,}686,\ 121{,}145,\ 124{,}690,\ 128{,}324,\ 107{,}049).$$

The net effect is an increase of the equity NPV, notwithstanding the higher risk for shareholders: The equity NPV increases from \$22,277 (80% payout) to \$25,592 (100% payout).

Table 6.10 decomposes the NPVs. The value created by the operating assets is entirely grasped by equityholders.

Table 6.10 Decomposition of Mockup NPV (100% payout) investment side and financing side

investment side	$\overbrace{0}^{\text{NPV}^l}$	+	$\overbrace{25{,}592}^{\text{NPV}^o}$	=	$\overbrace{25{,}592}^{\text{NPV}}$
financing side	$\overbrace{0}^{\text{NPV}^d}$	+	$\overbrace{25{,}592}^{\text{NPV}^e}$	=	$\overbrace{25{,}592}^{\text{NPV}}$

6.2 Valuing Projects with Cash Flows: Unlevered Perspective

The unlevered perspective to valuation may follow two alternative paths for calculating the levered value of the operating assets. Section 6.2.1 illustrates a one-step valuation: The unlevered cash flows (FCFs) are discounted at an after-tax WACC in order to directly find the *levered* value of operating assets. Section 6.2.2 describes a two-step valuation: The unlevered cash flows (FCFs) are discounted at an unlevered cost of capital in order to find the *unlevered* value of operating assets; the value of the tax savings is then added to finally get the *levered* value of operating assets. The first one is often called the ***WACC method*** while the second one is called the ***Adjusted Present Value (APV) method***.

6.2.1 FCF and the WACC Method

In Sect. 6.1.2, we have calculated the value of the operating assets by discounting the CFOs/CCFs at the pre-tax WACC, r_t^o. In this section we show that one may compute the same value by discounting FCFs at an appropriate **after-tax WACC**, a method known as the *WACC method* or *FCF-WACC method.*

As seen, FCF is that part of CFO which is unaffected by the interest tax shield: $F_t^o = F_t^{uo} + \tau I_t^{\mathcal{D}}$. We have also seen that $C_t^o = C_{t-1}^o(1 + i_t^{uo}) - F_t^{uo}$ (see Table 4.35). Similarly, one can define the COCs for the FCFs, denoted as r_t^{uo}, such that the fundamental dynamical relation $V_t^o = V_{t-1}^o(1 + r_t^{uo}) - F_t^{uo}$ is fulfilled. The value V_t^o is the equilibrium price of a portfolio traded in the market which replicates the FCFs of the project from time 1 to time n. The cash-flow stream generated by such a portfolio is $(-V_0^o, F_1^{uo}, F_2^{uo}, \ldots, F_n^{uo})$. In terms of Matrices, one may describe the situation by coupling the following framings of the project Matrix and benchmark Matrix:

$$\begin{array}{c c}
\textbf{Project} & \textbf{Benchmark} \\
\begin{array}{|l|l l|}
\hline
+F_t^{uo} & +\text{NOPAT}_t & -\Delta C_t^o \\
\hline
+F_t^{\mathcal{D}} - \tau I_t^{\mathcal{D}} & +I_t^{\mathcal{D}}(1-\tau) & -\Delta C_t^{\mathcal{D}} \\
+F_t^e & +I_t^e & -\Delta C_t^e \\
\hline
\end{array}
&
\begin{array}{|l|l l|}
\hline
+F_t^{uo} & +r_t^{uo} V_{t-1}^o & -\Delta V_t^o \\
\hline
+F_t^{\mathcal{D}} - \tau I_t^{\mathcal{D}} & +r_t^{\mathcal{D}} V_{t-1}^{\mathcal{D}} - \tau I_t^{\mathcal{D}} & -\Delta V_t^{\mathcal{D}} \\
+F_t^e & +r_t^e V_{t-1}^e & -\Delta V_t^e \\
\hline
\end{array}
\end{array} \tag{6.48}$$

From the second column of the benchmark Matrix, $r_t^{uo} \cdot V_{t-1}^o = r_t^{\mathcal{D}} \cdot V_{t-1}^{\mathcal{D}} - \tau I_t^{\mathcal{D}} + r_t^e \cdot V_{t-1}^e$, whence

$$r_t^{uo} = \frac{r_t^e \cdot V_{t-1}^e + r_t^{\mathcal{D}} \cdot V_{t-1}^{\mathcal{D}} - \tau I_t^{\mathcal{D}}}{V_{t-1}^e + V_{t-1}^{\mathcal{D}}}. \tag{6.49}$$

If one assumes $i_t^{\mathcal{D}} = r_t^{\mathcal{D}}$ for every t,[23] then $C_t^{\mathcal{D}} = V_t^{\mathcal{D}}$ and, therefore, $I_t^{\mathcal{D}} = r_t^{\mathcal{D}} V_{t-1}^{\mathcal{D}}$ for every t. Hence, one may reframe (6.48) as

Project

$+F_t^{uo}$	$+\text{NOPAT}_t$	$-\Delta C_t^o$
$+F_t^{\mathcal{D}} - \tau I_t^{\mathcal{D}}$	$+r_t^{\mathcal{D}} V_{t-1}^{\mathcal{D}}(1-\tau)$	$-\Delta C_t^{\mathcal{D}}$
$+F_t^e$	$+I_t^e$	$-\Delta C_t^e$

Benchmark

$+F_t^{uo}$	$+r_t^{uo} V_{t-1}^o$	$-\Delta V_t^o$
$+F_t^{\mathcal{D}} - \tau I_t^{\mathcal{D}}$	$+r_t^{\mathcal{D}} V_{t-1}^{\mathcal{D}}(1-\tau)$	$-\Delta V_t^{\mathcal{D}}$
$+F_t^e$	$+r_t^e V_{t-1}^e$	$-\Delta V_t^e$

(6.50)

Therefore, Eq. (6.49) becomes

$$r_t^{uo} = \frac{r_t^e \cdot V_{t-1}^e + r_t^{\mathcal{D}}(1-\tau) \cdot V_{t-1}^{\mathcal{D}}}{V_{t-1}^e + V_{t-1}^{\mathcal{D}}}. \tag{6.51}$$

A special case of (6.49) occurs whenever $C_t^l = 0$ (full payout policy) for every t. This implies

$$r_t^{uo} = \frac{r_t^e \cdot V_{t-1}^e + r_t^d \cdot V_{t-1}^d - \tau I_t^d}{V_{t-1}^e + V_{t-1}^d}. \tag{6.52}$$

If, in addition to the above assumption, $i_t^d = r_t^d$ for every t, then $I_t^d = r_t^d V_{t-1}^d$, so one may write

$$r_t^{uo} = \frac{r_t^e \cdot V_{t-1}^e + r_t^d(1-\tau) \cdot V_{t-1}^d}{V_{t-1}^e + V_{t-1}^d}. \tag{6.53}$$

The relationships between r_t^o and r_t^{uo} may be easily found by placing, side by side, two framings of the benchmark Matrix, where each partition supplies the CFE:

$+F_t^o$	$+r_t^o V_{t-1}^o$	$-\Delta V_t^o$
$-F_t^{\mathcal{D}}$	$-r_t^{\mathcal{D}} V_{t-1}^{\mathcal{D}}$	$+\Delta V_t^{\mathcal{D}}$
$+F_t^e$	$+r_t^e V_{t-1}^e$	$-\Delta V_t^e$

$+F_t^{uo}$	$+r_t^{uo} V_{t-1}^o$	$-\Delta V_t^o$
$-F_t^{\mathcal{D}} + \tau I_t^{\mathcal{D}}$	$-r_t^{\mathcal{D}} V_{t-1}^{\mathcal{D}} + \tau I_t^{\mathcal{D}}$	$+\Delta V_t^{\mathcal{D}}$
$+F_t^e$	$+r_t^e V_{t-1}^e$	$-\Delta V_t^e$

(6.54)

The second columns of the two framing coincide, which implies

$$r_t^o V_{t-1}^o - r_t^{\mathcal{D}} V_{t-1}^{\mathcal{D}} = r_t^{uo} V_{t-1}^o - r_t^{\mathcal{D}} V_{t-1}^{\mathcal{D}} + \tau I_t^{\mathcal{D}}$$

[23] This assumption means that the return on net debt is equal to the required return on net debt. It may occur for various reasons, including (but not limited to) the case where both the debt and the liquid assets are value-neutral operations (i.e. $i_t^d = r_t^d$ and $i_t^l = r_t^l$ for every t).

whence

$$r_t^{uo} = r_t^o - \frac{\tau I_t^{\mathcal{D}}}{V_{t-1}^o}. \tag{6.55}$$

If $C_t^l = 0$ for every t, then $r_t^o = r_t$ and

$$r_t^{uo} = r_t^o - \frac{\tau I_t^d}{V_{t-1}^o} = r_t - \frac{\tau I_t^d}{V_{t-1}}. \tag{6.56}$$

The rate r_t^{uo} is an *operating after-tax COC*. It is widely known in finance, accounting and engineering economics, as the ***weighted average cost of capital*** (WACC), with no other specification. It is often used alongside the assumption $i_t^{\mathcal{D}} = r_t^{\mathcal{D}}$.

As $V_{t-1}^o = (V_t^o + F_t^{uo})/(1 + r_t^{uo})$, the operating value can be determined by discounting FCFs at the respective WACCs:

$$V_0^o = \sum_{t=1}^{n} \frac{F_t^{uo}}{(1 + r_1^{uo}) \cdot (1 + r_2^{uo}) \cdot \ldots \cdot (1 + r_t^{uo})}. \tag{6.57}$$

In the light of what we have seen, the analyst may appraise the operating assets by using either unlevered cash flows (WACC method) or levered cash flows (CFO method). The two methods rest on two different-but-equivalent ways of considering the interest tax shield. The CFO method includes tax shield in the cash flow whereas the WACC method takes account of the tax shield in the COC (see Table 6.11).

Table 6.11 Market value of operating assets difference between WACC method and CFO/CCF method

Method	WACC method	CFO/CCF method
Cash flow	Free cash flow $F_t^{uo} = \text{NOPAT}_t - \Delta C_t^o$	Operating cash flow $F_t^o = \text{NOPAT}_t + \tau I_t^{\mathcal{D}} - \Delta C_t^o$
Value	Value of operations $\sum_{t=1}^{n} \frac{F_t^{uo}}{\prod_{k=1}^{t}(1 + r_k^{uo})}$	Value of operations $\sum_{t=1}^{n} \frac{F_t^o}{\prod_{k=1}^{t}(1 + r_k^o)}$
Cost of capital	after-tax WACC $r_t^{uo} = \frac{r_t^e V_{t-1}^e + r_t^{\mathcal{D}} V_{t-1}^{\mathcal{D}} - \tau I_t^{\mathcal{D}}}{V_{t-1}^e + V_{t-1}^{\mathcal{D}}}$	pre-tax WACC $r_t^o = \frac{r_t^e V_{t-1}^e + r_t^{\mathcal{D}} V_{t-1}^{\mathcal{D}}}{V_{t-1}^e + V_{t-1}^{\mathcal{D}}}$

Table 6.12 The four WACCs

WACC	Asset perspective (financially unlevered perspective)	Operating perspective (operationally unlevered perspective)
pre-tax	$r_t = \dfrac{r_t^e V_{t-1}^e + r_t^d V_{t-1}^d}{V_{t-1}^e + V_{t-1}^d}$	$r_t^o = \dfrac{r_t^e V_{t-1}^e + r_t^{\mathcal{D}} V_{t-1}^{\mathcal{D}}}{V_{t-1}^e + V_{t-1}^{\mathcal{D}}}$
after-tax	$r_t^u = \dfrac{r_t^e V_{t-1}^e + r_{t-1}^d V_{t-1}^d - \tau I_t^d}{V_{t-1}^e + V_{t-1}^d}$	$r_t^{uo} = \dfrac{r_t^e V_{t-1}^e + r_t^{\mathcal{D}} V_{t-1}^{\mathcal{D}} - \tau I_t^{\mathcal{D}}}{V_{t-1}^e + V_t^{\mathcal{D}}}$

Remark 6.4 (*The various kinds of WACC*) We have presented three kinds of weighted mean playing the role of COCs: An asset COC (r_t), an operating COC (r_t^o), an unlevered operating COC (r_t^{uo}). The first one is an *asset pre-tax* WACC, associated with CFAs; the second one is an *operating pre-tax* WACC, associated with CFOs/CCFs; the third one is an *operating after-tax* WACC, associated with FCFs. For completeness, one can build the notion of *asset after-tax* WACC, denoted as r_t^u associated with the asset FCF. We report the four kinds of WACCs in Eqs. (6.58)–(6.61) below, alongside their corresponding types of cash flow.

$$r_t^o = \frac{r_t^e V_{t-1}^e + r_t^{\mathcal{D}} V_{t-1}^{\mathcal{D}}}{V_{t-1}^e + V_{t-1}^{\mathcal{D}}} \qquad \to F_t^o \tag{6.58}$$

$$r_t^{uo} = \frac{r_t^e V_{t-1}^e + r_t^{\mathcal{D}} V_{t-1}^{\mathcal{D}} - \tau I_t^{\mathcal{D}}}{V_{t-1}^e + V_{t-1}^{\mathcal{D}}} \qquad \to F_t^{uo} \tag{6.59}$$

$$r_t = \frac{r_t^e V_{t-1}^e + r_t^d V_{t-1}^d}{V_{t-1}^e + V_{t-1}^d} \qquad \to F_t \tag{6.60}$$

$$r_t^u = \frac{r_t^e V_{t-1}^e + r_t^d V_{t-1}^d - \tau I_t^d}{V_{t-1}^e + V_{t-1}^d} \qquad \to F_t^u. \tag{6.61}$$

If liquid assets are not present, then the asset pre-tax WACC and the operating pre-tax WACC coincide: $r_t = r_t^o$, as well as the asset (after-tax) WACC and the operating (after-tax) WACC: $r_t^u = r_t^{uo}$. (See also Table 6.12.)[24] ▲

Remark 6.5 (*WACC method with asset FCFs*) The rate r_t^u may be used for discounting asset FCFs to get the (levered) value of assets, V_0. First, let us define $r_t^{ul} = (V_t^l + F_t^{ul} - V_{t-1}^l)/V_{t-1}^l$ so that $V_t^l = V_{t-1}^l(1 + r_t^{ul}) - F_t^{ul}$, whence $V_0^l = \sum_{t=1}^n F_t^{ul} \cdot \prod_{k=1}^t (1 + r_k^{ul})^{-1}$ where $F_t^{ul} = F_t^l - \tau I_t^l$ is the non-operating FCF (see Sect. 4.9). This implies $r_t^{ul} = r_t^l - \tau I_t^l / V_{t-1}^l$ and

[24] The operating after-tax WACC is, by far and large, the most widely known COC in finance, which explains the reasons why it is often called "WACC" with no other specifications.

$$V_t = \overbrace{V^o_{t-1}(1+r^{uo}_t) - F^{uo}_t}^{V^o_t} + \overbrace{V^l_{t-1}(1+r^{ul}_t) - F^{ul}_t}^{V^l_t}$$

or, looking forward,

$$V_{t-1} = \frac{V^o_t + F^{uo}_t}{1+r^{uo}_t} + \frac{V^l_t + F^{ul}_t}{1+r^{ul}_t}.$$

This also enables computing r^u as a weighted average of r^{uo}_t and r^{ul}_t:

$$r^u_t = \frac{r^{uo}_t \cdot V^o_{t-1} + r^{ul}_t V^l_{t-1}}{V^o_{t-1} + V^l_{t-1}} \tag{6.62}$$

(contrast with (6.61)). Discounting asset FCFs, F^u_t ($= F^{uo}_t + F^{ul}_t$) with the asset (after-tax) WACC, the market value of the assets is easily found:

$$V_0 = \sum_{t=1}^{n} \frac{F^u_t}{(1+r^u_1)\cdot(1+r^u_2)\cdot\dots\cdot(1+r^u_t)}. \tag{6.63}$$

This valuation method, here labeled as **WACC method with asset FCFs**, is a WACC method performed under a financially unlevered perspective. Specifically, it replaces operating FCFs with assets FCFs and operating after-tax WACC, r^{uo}_t, with asset after-tax WACC, r^u_t, to find the value of the assets (rather than the value of the operations). The equity value is determined by subtracting the debt value, V^d_0.

However, we have previously introduced r^u_t in a different way, so we must reconcile the two expressions and show that, indeed,

$$\frac{r^{uo}_t V^o_{t-1} + r^{ul}_t V^l_{t-1}}{V^o_{t-1} + V^l_{t-1}} = r^u_t = \frac{r^e V^e_{t-1} + r^d V^d_{t-1} - \tau I^d_t}{V^e_{t-1} + V^d_{t-1}}. \tag{6.64}$$

It suffices to show that the numerators are equal. In order to do so, we disaggregate the net debt in Matrix (6.54).

$$\begin{array}{|l|l|l|}\hline +F^o_t & +r^o_t V^o_{t-1} & -\Delta V^o_t \\ +F^l_t & +r^l_t V^l_{t-1} & -\Delta V^l_t \\ -F^d_t & -r^d_t V^d_{t-1} & +\Delta V^d_t \\ \hline +F^e_t & +r^e_t V^e_{t-1} & -\Delta V^e_t \\ \hline \end{array} \quad \begin{array}{|l|l|r|}\hline +F^{uo}_t & +r^{uo}_t V^o_{t-1} & -\Delta V^o_t \\ +F^l_t - \tau I^l_t & +r^l_t V^l_{t-1} - \tau I^l_t & -\Delta V^l_t \\ -F^d_t + \tau I^d_t & -r^d_t V^d_{t-1} + \tau I^d_t & +\Delta V^d_t \\ \hline +F^e_t & +r^e_t V^e_{t-1} & -\Delta V^e_t \\ \hline \end{array} \tag{6.65}$$

or, equivalently,

$$\begin{array}{|l|l|l|}\hline +F^o_t & +r^o_t V^o_{t-1} & -\Delta V^o_t \\ +F^l_t & +r^l_t V^l_{t-1} & -\Delta V^l_t \\ -F^d_t & -r^d_t V^d_{t-1} & +\Delta V^d_t \\ \hline +F^e_t & +r^e_t V^e_{t-1} & -\Delta V^e_t \\ \hline \end{array} \quad \begin{array}{|l|l|r|}\hline +F^{uo}_t & +r^{uo}_t V^o_{t-1} & -\Delta V^o_t \\ +F^{ul}_t & +r^{ul}_t V^l_{t-1} & -\Delta V^l_t \\ -F^d_t + \tau I^d_t & -r^d_t V^d_{t-1} + \tau I^d_t & +\Delta V^d_t \\ \hline +F^e_t & +r^e_t V^e_{t-1} & -\Delta V^e_t \\ \hline \end{array} \tag{6.66}$$

The upper right partitions coincide, so that

$$r_t^o V_{t-1}^o + r_t^l V_{t-1}^l - r_t^d V_{t-1}^d = r_t^{uo} V_{t-1}^o + r_t^{ul} V_{t-1}^l - r_t^d V_{t-1}^d + \tau I_t^d$$

whence

$$r_t^o V_{t-1}^o + r_t^l V_{t-1}^l - \tau I_t^d = r_t^{uo} V_{t-1}^o + r_t^{ul} V_{t-1}^l.$$

Since $r_t^o V_{t-1}^o + r_t^l V_{t-1}^l = r_t^e V_{t-1}^e + r_t^d V_{t-1}^d$, (6.64) is proved. ▲

Remark 6.6 (*WACCs and book values*) The WACCs, either pre-tax or after-tax, are COCs weighted by market values, not book values. Therefore, rates such as

$$\frac{r_t^e C_{t-1}^e + r_t^d C_{t-1}^d}{C_{t-1}^e + C_{t-1}^d} \quad \text{or} \quad \frac{r_t^e C_{t-1}^e + r_t^d C_{t-1}^d (1-\tau)}{C_{t-1}^e + C_{t-1}^d},$$

are not WACCs. They express, respectively, the pre-tax return and after-tax return on a portfolio consisting of C_{t-1}^e invested at r_t^e and C_{t-1}^d invested at r_t^d. Which is not the same as the return on a portfolio consisting of V_{t-1}^e invested at r_t^e and V_{t-1}^d invested at r_t^d. The two portfolios have different rates of return because the invested amounts are different:

$$\frac{r_t^e C_{t-1}^e + r_t^d C_{t-1}^d}{C_{t-1}^e + C_{t-1}^d} \neq \frac{r_t^e V_{t-1}^e + r_t^d V_{t-1}^d}{V_{t-1}^e + V_{t-1}^d} = r_t.$$ ▲

Remark 6.7 (*Terminology*) As we have seen, the rates i_t^{uo} and r_t^{uo} are after-tax rates. However, care should be taken in interpreting the expression "after-tax". In this case, the expression means "after tax shield is deducted"; that is, these rates are net of the tax-shield component, which is equal to the amount of tax savings on net interest, $\tau I_t^{\mathcal{D}}$:

$$i_t^{uo} = i_t^o - \frac{\tau I_t^{\mathcal{D}}}{C_{t-1}^o}, \qquad r_t^{uo} = r_t^o - \frac{\tau I_t^{\mathcal{D}}}{V_{t-1}^o}.$$

Therefore, i_t^{uo} may well be said to be an (operating) *after-tax* ROI in the same sense r_t^{uo} is said to be an (operating) *after-tax* WACC. Unlike i_t^{uo} and r_t^{uo}, the rates i_t^o and r_t^o are gross of the tax shield, so that i_t^o is an (operating) *pre-tax* ROI in the same sense r_t^o is an (operating) *pre-tax* WACC. Analogously, i_t^u and r_t^u are after-tax rates. Specifically, the former is an asset after-tax ROI and the latter is an asset after-tax WACC, respectively. Accordingly, i_t and r_t are pre-tax rates: Asset pre-tax ROI and asset pre-tax WACC, respectively. ▲

Example 6.10 ♦

C^o	C^d
C^l	C^e

(*Mockup—80% payout—WACC*) In Example 6.8 we have evaluated the Mockup project with 80% payout policy by discounting the CFOs at the pre-tax WACCs. We now re-evaluate the operating assets assuming that the after-tax WACC is time-varying and equal to $\boldsymbol{r^{uo}} = (8.805\%, 9.12\%, 9.594\%, 10.416\%, 12.265\%)$.

The corresponding discount vector is $\mathbf{d}^{uo} = (1, 0.91907, 0.84226, 0.76853, 0.69603, 0.61999)$. We remind that the FCFs are $\boldsymbol{F}^{uo}$=(−500,000, 119,495, 122,954, 126,499, 130,133, 198,858). Hence,

$$V_t^o = 119{,}495 \cdot 0.91907 + 122{,}954 \cdot 0.84226 + 126{,}499 \cdot 0.76853$$
$$+ 130{,}133 \cdot 0.69603 + 198{,}858 \cdot 0.61999 = 524{,}469.$$

This is the same value we have obtained by discounting CFOs at the operating pre-tax WACCs. The reason is that the vectors of (operating) pre-tax WACCs and (operating) after-tax WACCs, namely,

$$\boldsymbol{r}^o = (8.97\%, 9.28\%, 9.75\%, 10.56\%, 12.39\%),$$
$$\boldsymbol{r}^{uo} = (8.805\%, 9.12\%, 9.594\%, 10.416\%, 12.265\%),$$

are logically equivalent (i.e., $r_t^o = r_t^{uo} + \tau I_t^{\mathcal{D}} / V_{t-1}^o$ for every t).

Example 6.11 ♦

C^o	C^d
	C^e

(*Mockup—100% payout—WACC*) Consider valuation of Mockup project with 100% payout, presented in Example 6.9 and keep the assumption that the operating pre-tax WACC is

$$\boldsymbol{r}^o = \boldsymbol{r} = (8.97\%, 9.28\%, 9.75\%, 10.56\%, 12.39\%).$$

We can unravel the (time-varying) after-tax WACC from Eq. (6.56):

$$\boldsymbol{r}^{uo} = \boldsymbol{r}^u = (8.805\%, 9.086\%, 9.509\%, 10.241\%, 11.89\%)$$

whence

$$\mathbf{d}^{uo} = \mathbf{d}^u = (0.91907, 0.84252, 0.76936, 0.69789, 0.62373).$$

The FCFs are the same as in the case of 80% payout (as we know, the financing policy does not affect FCF):

$$\boldsymbol{F}^{uo} = \boldsymbol{F}^u = (-500{,}000,\ 119{,}495,\ 122{,}954,\ 126{,}499,\ 130{,}133,\ 198{,}858).$$

We can recalculate the value of the operations (now identical to the value of assets) by discounting the FCFs at the WACCs (Eq. (6.57)):

$$V_0^o = V_0 = 119{,}495 \cdot 0.91907 + 122{,}954 \cdot 0.84252 + 126{,}499 \cdot 0.76936$$
$$+ 130{,}133 \cdot 0.69789 + 198{,}858 \cdot 0.62373 = 525{,}592$$

as expected.

Example 6.12 ♦

C^o	C^d
	C^e

(*Mockup—100% payout—WACC with* $i^d \neq r^d$) Consider again Mockup Inc. with 100% payout ratio, whose NPV has been computed in Example 6.9. While we have so far assumed that the ROD is equal to 3%, equal to the cost of debt, let us now assume that the ROD and the cost of debt differ, other things unvaried. In particular, while we keep the assumption $r^d = 3\%$, we also assume $i^d = 4\% \neq r^d$. This means that interest expenses are greater than $2,700 (see the interest line in Table 4.25); namely, $I_t^d = 0.04 \cdot 90{,}000 = 3{,}600$ for all t. The FCFs do not increase (they do not depend on I_t^d), but the WACCs (slightly) decrease in each period:

$$\boldsymbol{r}^{uo} = (8.75\%, 9.02\%, 9.43\%, 10.14\%, 11.72\%).$$

Discounting the unvaried FCFs with smaller WACCs makes the project value increase ($V_0 = 526{,}730$), as well as the project NPV, which rises to NPV = 26,730. The market value of the debt increases as well, because the CFDs are higher (due to higher interest expenses): $V_0^d = 94{,}122$, so the debt's NPV is nonzero and equal to $\text{NPV}^d = 94{,}122 - 90{,}000 = 4{,}122$. The equity market value decreases ($V_0^e = 432{,}608$) as well as the equity NPV, which becomes $\text{NPV}^e = 432{,}608 - 410{,}000 = 22{,}608$. Part of the project NPV is then lost by equityholders to the benefit of the debtholders (see Table 6.13).

Table 6.13 Decomposition of Mockup NPV (100% payout), $i^d \neq r^d$ investment side and financing side

investment side	$\overbrace{0}^{\text{NPV}^l} + \overbrace{26{,}730}^{\text{NPV}^o} = \overbrace{26{,}730}^{\text{NPV}}$
financing side	$\overbrace{4{,}122}^{\text{NPV}^d} + \overbrace{22{,}608}^{\text{NPV}^e} = \overbrace{26{,}730}^{\text{NPV}}$

6.2.2 *FCF and the APV Method*

As opposed to the CFO, the FCF is an unlevered cash flow; it is the cash flow that would be distributed to investors if the project were unlevered. Therefore, it seems to be economically meaningful to link FCF with the *unlevered* value of the operating assets rather than the *levered* value of the operating assets. Under this perspective, let V_t^{uo} denote the unlevered operating value of the project and let k_t^{uo} be the expected return on it. The fundamental law of motion may be written as

$$V_t^{uo} = V_{t-1}^{uo}(1 + k_t^{uo}) - F_t^{uo} \quad \Longleftrightarrow \quad V_{t-1}^{uo} = \frac{V_t^{uo} + F_t^{uo}}{1 + k_t^{uo}} \tag{6.67}$$

whence

$$V_0^{uo} = \sum_{t=1}^{n} \frac{F_t^{uo}}{(1+k_1^{uo}) \cdot (1+k_2^{uo}) \cdot \ldots \cdot (1+k_t^{uo})}. \tag{6.68}$$

The rate k_t^{uo} is the expected rate of return on an asset which generates cash flows equal to the project's FCFs from time 1 to time n and is the *cost of unlevered (operating) assets* or, simply, *unlevered cost of assets*, or even *unlevered cost of equity* (if the project is unlevered, assets and equity coincide). Likewise, let V_t^{TS} be the value of the interest tax shield, $\tau I_t^{\mathcal{D}}$:

$$V_t^{\text{TS}} = V_{t-1}^{\text{TS}}(1+r_t^{\text{TS}}) - \tau I_t^{\mathcal{D}} \quad \Longleftrightarrow \quad V_{t-1}^{\text{TS}} = \frac{V_t^{\text{TS}} + \tau I_t^{\mathcal{D}}}{(1+r_t^{\text{TS}})}.$$

whence

$$V_0^{\text{TS}} = \sum_{t=1}^{n} \frac{\tau I_t^{\mathcal{D}}}{(1+r_1^{\text{TS}}) \cdot (1+r_2^{\text{TS}}) \cdot \ldots \cdot (1+r_t^{\text{TS}})}. \tag{6.69}$$

The rate r_t^{TS} is the expected rate of return on an asset generating a stream of cash flows equal to the interest tax savings from time 1 to time n. Now, given that $F_t^o = F_t^{uo} + \tau I_t^{\mathcal{D}}$ and $I_t^o = I_t^{uo} + \tau I_t^{\mathcal{D}}$, the law of one price implies that the value of operations is equal to the sum of the unlevered value plus the value of tax shield:

$$V_t^o = V_t^{uo} + V_t^{\text{TS}}. \tag{6.70}$$

In this way, the project's operating value is split up into two components, one which is unaffected by the financing policy, the other one measuring the tax savings generated by the use of (net) debt. Using (6.68) and (6.69),

$$V_t^o = \overbrace{\sum_{j=t+1}^{n} \frac{F_j^{uo}}{(1+k_{t+1}^{uo}) \cdot (1+k_{t+2}^{uo}) \cdot \ldots \cdot (1+k_j^{uo})}}^{V_t^{uo}} + \overbrace{\sum_{j=t+1}^{n} \frac{\tau I_j^{\mathcal{D}}}{(1+r_{t+1}^{\text{TS}}) \cdot (1+r_{t+2}^{\text{TS}}) \cdot \ldots \cdot (1+r_j^{\text{TS}})}}^{V_t^{\text{TS}}} \quad t \in \mathbb{N}_n^0. \tag{6.71}$$

Equation (6.71) represents the so-called **Adjusted Present Value** (APV) method. It is straightforward to find the relations between (operating) pre-tax WACC and k_t^{uo}, as well as between WACC and k_t^{uo}. From the laws of motion one deduces

$$r_t^o V_{t-1}^o = k_t^{uo} \cdot V_{t-1}^{uo} + r_t^{\text{TS}} \cdot V_{t-1}^{\text{TS}}$$

whence

$$r_t^o = k_t^{uo} \frac{V_{t-1}^{uo}}{V_{t-1}^o} + r_t^{\text{TS}} \cdot \frac{V_{t-1}^{\text{TS}}}{V_{t-1}^o}. \tag{6.72}$$

Alternatively,

$$\begin{aligned} r_t^o V_{t-1}^o &= k_t^{uo} \cdot V_{t-1}^{uo} + r_t^{\text{TS}} \cdot V_{t-1}^{\text{TS}} \\ &= k_t^{uo} \cdot V_{t-1}^{uo} + r_t^{\text{TS}} \cdot V_{t-1}^{\text{TS}} + k_t^{uo} V_{t-1}^{\text{TS}} - k_t^{uo} V_{t-1}^{\text{TS}} \\ &= k_t^{uo} \cdot V_{t-1}^o + (r_t^{\text{TS}} - k_t^{uo}) V_{t-1}^{\text{TS}}. \end{aligned}$$

Dividing by the operating value,

$$r_t^o = k_t^{uo} + (r_t^{\text{TS}} - k_t^{uo}) \frac{V_{t-1}^{\text{TS}}}{V_{t-1}^o} \iff k_t^{uo} = \frac{r_t^o V_{t-1}^o - r_t^{\text{TS}} V_{t-1}^{\text{TS}}}{V_{t-1}^o - V_{t-1}^{\text{TS}}}. \tag{6.73}$$

That is, the unlevered cost of operating assets is a (generalized) weighted mean of the pre-tax WACC and the discount rate for the tax shield. Since $F_t^{uo} + \tau I_t^{\mathcal{D}} = F_t^o$, if the expected return on the tax shield, r_t^{TS}, is equal to the unlevered cost of assets, k_t^{uo}, then the latter coincides with the (operating) pre-tax WACC:

$$r_t^{\text{TS}} = k_t^{uo} \implies k_t^{uo} = r_t^{\text{TS}} = r_t^o. \tag{6.74}$$

This holds whenever the risk of CFOs and the risk of FCFs is the same. Considering that the unlevered cost of assets is the discount rate for the FCF, which is unaffected by the tax shield (and, therefore, by the net financial position), one may state the following proposition.

Proposition 6.4 *If the discount rate for tax shield is the same as the unlevered cost of (operating) assets, then the latter coincides with the (operating) pre-tax WACC: $k_t^{uo} = r_t^o$. In this case, the pre-tax WACC does not depend on the financing policy.*[25]

Under the assumption of Proposition 6.4, the APV method is equivalent to discounting the CFOs/CCFs at k_t^{uo}. This is the **conventional CCF method** or **compressed**

[25] What is, in general, the proper discount rate for tax shield, r^{TS}? The problem of finding the correct discount rate has been investigated extensively in the literature but, to date, no agreed upon solution have been advanced. The determination of r^{TS} is a subtle issue because it depends on the risk of $\tau I_t^{\mathcal{D}}$, which in turn depends on the set of assumptions made. In particular, if the risk of the tax shield is equal to the risk of debt, then $r_t^{\text{TS}} = r_t^{\mathcal{D}}$; if the risk of the tax shield is equal to the risk of the unlevered operating assets, then $r_t^{\text{TS}} = k_t^{uo}$. Other assumptions are possible and the determination of r_t^{TS} will depend on the financing policy and on the level of the prospective operating income as opposed to net interest (if the latter is greater than the former, no tax saving arises) (see Myers 1974; Miles and Ezzell 1980, 1985; Inselbag and Kaufold 1997; Harris and Pringle 1985; Lewellen and

APV (see Kaplan and Ruback 1995; Ruback 2002; Fernández 2002; Tham and Vélez-Pareja 2004; Booth 2007; Cooper and Nyborg 2007; Jagannathan et al. 2017) and is mainly presented under the additional assumption of constant rates:

$$V_0^o = \overbrace{\sum_{t=1}^{n} \frac{F_t^o}{(1+k^{uo})^t}}^{\text{CCF method}} = \overbrace{\sum_{t=1}^{n} \frac{F_t^{uo}}{(1+k^{uo})^t} + \sum_{t=1}^{n} \frac{\tau I_t^{\mathcal{D}}}{(1+k^{uo})^t}}^{\text{APV method assuming } r^{\text{TS}}=k^{uo}} . \tag{6.75}$$

As such, this method may be considered a special case of the more general CFO method presented in Sect. 6.1.2, which allows for time-varying rates and for any possible pre-tax WACC, $r_t^o \neq k_t^{uo}$. Indeed, the left-hand side of (6.75) is equal to Eq. (6.30) under the assumption $r_t^o = k_t^{uo} = k^{uo}$.

Remark 6.8 (*Modigliani and Miller* 1963) Remembering that $r_t^{uo} = r_t^o - \tau I_t^{\mathcal{D}} / V_{t-1}^o$ (see Eq. (6.55)) and using (6.73), the relation between the unlevered cost of assets and the WACC is easily established:

$$r_t^{uo} = k_t^{uo} + (r_t^{\text{TS}} - k_t^{uo}) \frac{V_{t-1}^{\text{TS}}}{V_{t-1}^o} - \frac{\tau I_t^{\mathcal{D}}}{V_{t-1}^o} \iff k_t^{uo} = \frac{r^{uo} V_{t-1}^o + \tau I_t^{\mathcal{D}} - r^{\text{TS}} V_{t-1}^{\text{TS}}}{V_{t-1}^{uo}} . \tag{6.76}$$

If $r_t^{\text{TS}} = k_t^{uo}$, then the above relation becomes $r_t^{uo} = k_t^{uo} - \tau I_t^{\mathcal{D}} / V_{t-1}^o$.

From an equity perspective, considering that $r_t^o V_{t-1}^o = r_t^e V_{t-1}^e + r_t^{\mathcal{D}} V_{t-1}^{\mathcal{D}}$ and $r_t^o = r_t^{uo} + \tau I_t^{\mathcal{D}} / V_{t-1}^o$, one finds, after simple algebraic manipulations upon (6.76), the relation between r_t^e and k_t^{uo}:

$$r_t^e = k_t^{uo} + (k_t^{uo} - r_t^{\mathcal{D}}) \frac{V_{t-1}^{\mathcal{D}}}{V_{t-1}^e} + (r_t^{\text{TS}} - k_t^{uo}) \frac{V_{t-1}^{\text{TS}}}{V_{t-1}^e} . \tag{6.77}$$

If the discount rate for tax shield is equal to the unlevered cost of assets, that is, if $r_t^{\text{TS}} = k_t^{uo}$, then

$$r_t^e = k_t^{uo} + (k_t^{uo} - r_t^{\mathcal{D}}) \frac{V_{t-1}^{\mathcal{D}}}{V_{t-1}^e} \tag{6.78}$$

and, considering that $k_t^{uo} = r_t^o$ (by (6.74)), one gets

$$r_t^e = r_t^o + (r_t^o - r_t^{\mathcal{D}}) \frac{V_{t-1}^{\mathcal{D}}}{V_{t-1}^e} . \tag{6.79}$$

One may also reframe the equity COC in (6.77) as a simple weighted average of k_t^{uo}, $r_t^{\mathcal{D}}$, and r_t^{TS}:

Emery 1986; Taggart 1991; Fernández 2002, 2004; Arzac and Glosten 2005; Cooper and Nyborg 2006, 2007, 2008; Grinblatt and Liu 2008; Massari et al. 2008; Dempsey 2013).

$$
\begin{aligned}
r_t^e &= k_t^{uo}\left(1+\frac{V_{t-1}^{\mathcal{D}}-V_{t-1}^{\mathrm{TS}}}{V_{t-1}^e}\right)-r_t^{\mathcal{D}}\cdot\frac{V_{t-1}^{\mathcal{D}}}{V_{t-1}^e}+r_t^{\mathrm{TS}}\cdot\frac{V_{t-1}^{\mathrm{TS}}}{V_{t-1}^e} \\
&= k_t^{uo}+\frac{V_{t-1}^{uo}}{V_{t-1}^e}-r_t^{\mathcal{D}}\cdot\frac{V_{t-1}^{\mathcal{D}}}{V_{t-1}^e}+r_t^{\mathrm{TS}}\cdot\frac{V_{t-1}^{\mathrm{TS}}}{V_{t-1}^e}
\end{aligned}
\tag{6.80}
$$

If $r_t^{\mathrm{TS}} = k_t^{uo}$, then

$$
\begin{aligned}
r_t^e &= k_t^{uo}\left(1+\frac{V_{t-1}^{\mathcal{D}}}{V_{t-1}^e}\right)-r_t^{\mathcal{D}}\cdot\frac{V_{t-1}^{\mathcal{D}}}{V_{t-1}^e} \\
&= r_t^o\left(1+\frac{V_{t-1}^{\mathcal{D}}}{V_{t-1}^e}\right)-r_t^{\mathcal{D}}\cdot\frac{V_{t-1}^{\mathcal{D}}}{V_{t-1}^e}
\end{aligned}
\tag{6.81}
$$

which is just (6.78)–(6.79).[26]

We now show that (6.80) implies Modigliani and Miller's (1963) famous result, according to which the cost of equity increases, as the leverage increases, by a factor of $[k^{uo} - r^{\mathcal{D}}] \cdot (1 - \tau)$. To this end, we abide by Modigliani and Miller's assumptions:

- k^{uo} is constant
- r^{TS} and $r^{\mathcal{D}}$ are constant and equal[27]
- $n = \infty$, that is, the project/firm is a perpetuity generating a perpetual FCF stream ($F_t^{uo} = F^{uo}$ for all t)
- the debt level is fixed (and equal to its nominal value): $C_t^{\mathcal{D}} = V_t^{\mathcal{D}} = V_0^{\mathcal{D}}$ for every $t \in \mathbb{N}_\infty^0$.

Under these assumptions, the discounted value of the tax shield is

$$
V_0^{\mathrm{TS}} = \sum_{t=1}^{\infty}\frac{\tau r^{\mathcal{D}} V_0^{\mathcal{D}}}{(1+r^{\mathcal{D}})^t} = \frac{\tau r^{\mathcal{D}} V_0^{\mathcal{D}}}{r^{\mathcal{D}}} = \tau V_0^{\mathcal{D}},
$$

the equity value is constant ($V_t^e = V_0^e$), and (6.80) becomes

$$
r_t^e = r^e = k^{uo}\left(1+\frac{V_0^{\mathcal{D}}-\tau V_0^{\mathcal{D}}}{V_0^e}\right)-r^{\mathcal{D}}\cdot\frac{V_0^{\mathcal{D}}}{V_0^e}+r^{\mathcal{D}}\cdot\frac{\tau V_0^{\mathcal{D}}}{V_0^e}. \tag{6.82}
$$

Hence,

$$
r^e = k^{uo} + [k^{uo} - r^{\mathcal{D}}](1-\tau)\frac{V_0^{\mathcal{D}}}{V_0^e} \tag{6.83}
$$

which is exactly Modigliani and Miller's (1963), Eq. (12.c), as expected.

[26] Equation (6.80) is equal to Eq. (7) in Dempsey (2013). See also Cooper and Nyborg (2006).

[27] In particular, the authors assumed a riskless debt: $r^{\mathcal{D}} = r_f$.

Solving (6.83) for k^{uo}, one gets

$$k^{uo} = \frac{r^e V_0^e + r^{\mathcal{D}}(1-\tau)V_0^{\mathcal{D}}}{V_0^e + (1-\tau)V_0^{\mathcal{D}}} \tag{6.84}$$

(see also Fernández 2002, Eq. (17.12) on p. 379). Hence,

$$\begin{aligned} k^{uo} &= \frac{r^e V_0^e + r^{\mathcal{D}} V_0^{\mathcal{D}} - r^{\mathcal{D}}\tau V_0^{\mathcal{D}}}{V_0^e + V_0^{\mathcal{D}} - \tau V_0^{\mathcal{D}}} \\ &= \frac{r^o V_0^o - r^{\mathcal{D}}\tau V_0^{\mathcal{D}}}{V_0^o - \tau V_0^{\mathcal{D}}}. \end{aligned}$$

The latter equality may also be derived directly from (6.73), omitting subscript t, replacing subscript $t-1$ with subscript 0, replacing r^{TS} with $r^{\mathcal{D}}$, and remembering that $V_0^{\text{TS}} = \tau V_0^{\mathcal{D}}$. ▲

Remark 6.9 (*APV with asset FCFs*) The APV method may be framed in terms of asset FCFs, as long as the tax shield $\tau I_t^{\mathcal{D}}$ is replaced by the gross tax shield, τI_t^d:

$$V_t = \overbrace{\sum_{j=t+1}^{n} \frac{F_j^u}{(1+k_{t+1}^u)\cdot(1+k_{t+2}^u)\cdot\ldots\cdot(1+k_j^u)}}^{V_t^u} + \overbrace{\sum_{j=t+1}^{n} \frac{\tau I_j^d}{(1+r_{t+1}^{\text{GTS}})\cdot(1+r_{t+2}^{\text{GTS}})\cdot\ldots\cdot(1+r_j^{\text{GTS}})}}^{V_t^{\text{GTS}}} \quad t \in \mathbb{N}_n^0 \tag{6.85}$$

where V_t^u is the value of the financially unlevered assets and V_t^{GTS} is the value of the gross tax shield, while k_t^u and r_t^{GTS} are the corresponding expected growth rates, such that $V_t^u = V_{t-1}^u(1+k_t^u) - F_t^u$ and $V_t^{\text{GTS}} = V_{t-1}^{\text{GTS}}(1+r_t^{\text{GTS}}) - \tau I_t^d$. We call this method the **financially unlevered (APV) method**.

Equation (6.73) is replaced by its asset counterpart,

$$r_t = k_t^u + (r_t^{\text{GTS}} - k_t^u)\frac{V_{t-1}^{\text{GTS}}}{V_{t-1}} \iff k_t^u = \frac{r_t V_{t-1} - r_t^{\text{GTS}} V_{t-1}^{\text{GTS}}}{V_{t-1} - V_{t-1}^{\text{GTS}}}. \tag{6.86}$$

The rate k_t^u may be said to be the *financially unlevered cost of assets* or the *required return on financially unlevered assets*. We note again that, if no liquid assets exist, the operating perspective and the asset perspective coincide, so that $F_t^{uo} = F_t^u$, $k_t^{uo} = k_t^u$ and $r_t^{\text{TS}} = r_t^{\text{GTS}}$.

Let k_t^{ul} be the cost of capital for non-operating FCFs, such that $V_t^{ul} = V_{t-1}^{ul}(1 + k_t^{ul}) - F_t^{ul}$ is the intrinsic value of non-operating FCFs. Hence,

$$V_0^{ul} = \sum_{t=1}^{n} \frac{F_t^{ul}}{\prod_{k=1}^{t}(1 + k_t^{ul})}.$$

The financially unlevered cost of assets, k_t^u, may be viewed as a weighted mean of k_t^{uo} and k_t^{ul}:

$$k_t^u = \frac{k_t^{uo} V_{t-1}^{uo} + k_t^{ul} V_{t-1}^{ul}}{V_{t-1}^{uo} + V_{t-1}^{ul}}.$$

The financially unlevered value of assets may then be viewed as

$$V_0^u = V_0^{uo} + V_0^{ul} = \sum_{t=1}^{n} \frac{F_t^{uo}}{\prod_{k=1}^{t}(1 + k_t^{uo})} + \sum_{t=1}^{n} \frac{F_t^{ul}}{\prod_{k=1}^{t}(1 + k_t^{ul})}. \tag{6.87}$$

▲

Example 6.13 ♦
(*APV method with* $r^{\text{TS}} = r^{\mathcal{D}}$) A project is expected to generate an (assumed constant) FCF equal to $F_t^{uo} = \$100$ for $t \in \mathbb{N}_{10}^1$. The net debt is expected to be constant and equal to $C_t^{\mathcal{D}} = \$350$ for all $t < 10$. At time $t = 10$, it will be wholly reimbursed ($C_{10}^{\mathcal{D}} = 0$). Let $\tau = 40\%$ be the tax rate and $i^{\mathcal{D}} = 3\%$ be the (assumed constant) RO$\mathcal{D}$, equal to the cost of net debt (i.e., $i^{\mathcal{D}} = r^{\mathcal{D}} = 3\%$), which implies that the tax shield is $\tau \cdot r^{\mathcal{D}} \cdot C_{t-1}^{\mathcal{D}} = 0.4 \cdot 0.03 \cdot \$350 = \$4.2$. The unlevered cost of operating assets is estimated to be equal to $k^{uo} = 10\%$. The analyst believes that the tax saving has the same risk as the debt, so $r^{\mathcal{D}}$ is the appropriate discount rate for the tax shield: $r^{\text{TS}} = 3\%$. Then, the value of the operations is obtained from (6.71):

$$\begin{aligned} V_0^o &= V_0^{uo} + V_0^{\text{TS}} \\ &= \sum_{t=1}^{10} \frac{100}{(1.1)^t} + \sum_{t=1}^{10} \frac{4.2}{(1.03)^t} \\ &= 100 \cdot \frac{1-(1.1)^{-10}}{0.1} + 4.2 \cdot \frac{1-(1.03)^{-10}}{0.03} \\ &= 614.46 + 35.83 = 650.28. \end{aligned}$$

Note that one can derive the implicit pre-tax WACCs of this project by applying (6.73). In our case, one finds

$$\boldsymbol{r^o} = (9.61\%, 9.62\%, 9.63\%, 9.64\%, 9.65\%, 9.66\%, 9.67\%, 9.68\%, 9.69\%, 9.70\%)$$

(we leave the computations to the reader).[28] The CFOs are $F_t^o = \$100 + \$4.2 = \$104.2$ for every $t \in \mathbb{N}_{10}^1$. It may be checked that, discounting back the CFOs with the pre-tax WACCs, one gets back the operating value, $V_0^o = \$650.28$.

6.3 Valuing Projects with Cash Flows: Modified Discounting

In the following two subsections, we show that (i) equity may be valued by discounting CFEs at some discount rate which is different from the equity cost of capital, as long as CFEs are suitably modified, and (ii) this modified approach is, so to say, the solution of a related issue of circularity arising whenever the cost of equity is not a known parameter (i.e., it is not exogenously given and is derived from some WACC or some unlevered cost of asset).

6.3.1 Modifying CFEs

We have so far associated the CFE stream with the cost of equity, which is the expected rate of return of the equity benchmark system, such that

$$V_t^e = V_{t-1}^e(1 + r_t^e) - F_t^e \quad \text{or, equivalently,} \quad V_{t-1}^e = \frac{V_t^e + F_t^e}{1 + r_t^e},$$

whence

$$V_0^e = F_1^e \mathrm{d}_1^e + F_2^e \mathrm{d}_2^e + \ldots + F_n^e \mathrm{d}_n^e$$

where d_t^e is the usual discount factor for the CFEs. However, one may well adjust the CFEs so that they are discounted with some WACC (r_t^o, r_t, r_t^{uo}, or r_t^u) or an unlevered cost of assets (k_t^{uo} or k_t^u). In this case, the analyst may directly compute the equity value by discounting a suitably modified CFE stream at the given discount rate.

More precisely, consider Eqs. (6.58)–(6.61). By reverse-engineering each equality, one gets the equity COC in terms of debt/equity ratio and various market rates:

[28] The variations of the WACCs are very small, owing to the simple structure of this operation. In particular, (6.73) may be reframed in such a way that the pre-tax WACC is computed as a weighted mean of k^{uo} and r^{TS}, where the weights are V_{t-1}^{uo}/V_{t-1}^o and $V_{t-1}^{\mathrm{TS}}/V_{t-1}^o$:

$$r_t^o = \frac{k_t^{uo} V_{t-1}^{uo} + r_t^{\mathrm{TS}} V_{t-1}^{\mathrm{TS}}}{V_{t-1}^{uo} + V_{t-1}^{\mathrm{TS}}} = \frac{0.1 V_{t-1}^{uo} + 0.03 V_{t-1}^{\mathrm{TS}}}{V_{t-1}^o}.$$

It may be checked that the weight V_{t-1}^{uo}/V_{t-1}^o gradually increases as t increases, from 94.49% to 95.71%, which gradually pushes the pre-tax WACC toward $k^{uo} = 10\%$ (at a pace of about 0.01% a year).

$$
\begin{aligned}
r_t^e &= r_t^o + \frac{V_{t-1}^{\mathcal{D}}}{V_{t-1}^e} \cdot (r_t^o - r_t^{\mathcal{D}}) \\
r_t^e &= r_t^{uo} + \frac{V_{t-1}^{\mathcal{D}}}{V_{t-1}^e} \cdot \left(r_t^{uo} - r_t^{\mathcal{D}}\right) + \frac{\tau I_t^{\mathcal{D}}}{V_{t-1}^e} \\
r_t^e &= r_t^{uo} + \frac{V_{t-1}^{\mathcal{D}}}{V_{t-1}^e} \cdot \left(r_t^{uo} - r_t^{\mathcal{D}}(1-\tau)\right) \;\; [\text{if } I_t^{\mathcal{D}} = r_t^{\mathcal{D}} \cdot V_{t-1}^{\mathcal{D}}] \\
r_t^e &= r_t + \frac{V_{t-1}^d}{V_{t-1}^e} \cdot (r_t - r_t^d) \\
r_t^e &= r_t^u + \frac{V_{t-1}^d}{V_{t-1}^e} \cdot \left(r_t^u - r_t^d\right) + \frac{\tau I_t^d}{V_{t-1}^e} \\
r_t^e &= r_t^u + \frac{V_{t-1}^d}{V_{t-1}^e} \cdot \left(r_t^u - r_t^d(1-\tau)\right) \;\; [\text{if } I_t^d = r_t^d \cdot V_{t-1}^d]
\end{aligned}
\tag{6.88}
$$

(contrast with Eq. (4.48) where pro forma book values are used in place of market values).[29]

These relations enable calculating the intrinsic value of equity, V_0^e, in a direct way (i.e., with no calculation of the value of assets or operating assets) whenever the cost of equity is not exogenously given. For example, assume that the pre-tax WACC, r_t^o, and the cost of debt, r_t^d, are the known market input data. Using the first equation in (6.88) and considering that $r_t^e = (V_t^e + F_t^e)/V_{t-1}^e - 1$, the equation may be rewritten as

$$\frac{V_t^e + F_t^e}{V_{t-1}^e} - 1 = r_t^o + \frac{V_{t-1}^{\mathcal{D}}}{V_{t-1}^e}(r_t^o - r_t^{\mathcal{D}}). \tag{6.89}$$

Solving for V_{t-1}^e, one gets the following prospective relation:

$$V_{t-1}^e = \frac{V_t^e + F_t^e - (r_t^o - r_t^{\mathcal{D}}) \cdot V_{t-1}^{\mathcal{D}}}{1 + r_t^o}. \tag{6.90}$$

This means that the equity benchmark system fulfills the following law of motion:

$$V_t^e = V_{t-1}^e(1 + r_t^o) - \Phi_t^e \tag{6.91}$$

where $\Phi_t^e = F_t^e - (r_t^o - r_t^{\mathcal{D}})V_{t-1}^{\mathcal{D}}$. The above equation describes the equity benchmark system in terms of r_t^o instead of r_t^e. This is possible because the CFE is adjusted giving rise to a modified CFE, Φ_t^e, which is associated with the same expected return as the CFO stream. Hence, this stream may be properly discounted at the operating COC, r_t^o. Equation (6.91) has a direct solution in terms of discounted cash flows:

$$V_0^e = \Phi_1^e \mathrm{d}_1^o + \Phi_2^e \mathrm{d}_2^o + \ldots + \Phi_n^e \mathrm{d}_n^o \tag{6.92}$$

[29] See Table 6.31 for other relations involving the unlevered cost of assets.

where $\mathrm{d}_t^o = \prod_{j=1}^t (1+r_j^o)^{-1}$ is the discount factor for the CFOs/CCFs, as usual. This means that, if the market input data are given by the operating pre-tax WACC and the cost of debt, the equity value may be computed by directly discounting the modified CFE stream $(\Phi_1^e, \Phi_2^e, \ldots, \Phi_n^e)$ at the operating pre-tax WACC.

The modified CFE may be interpreted as the difference between the CFO/CCF and a modified CF$\mathcal{D}$. To see it, first consider that the CF$\mathcal{D}$ may be written as $F_t^{\mathcal{D}} = V_{t-1}^{\mathcal{D}}(1+r_t^{\mathcal{D}}) - V_t^{\mathcal{D}}$. Replacing $r_t^{\mathcal{D}}$ with r_t^o one may define a modified CF$\mathcal{D}$ equal to $\Phi_t^{\mathcal{D}} = V_{t-1}^{\mathcal{D}}(1+r_t^o) - V_t^{\mathcal{D}}$, which implies that the modified CFE is equal to the difference between the CFO and the modified CF$\mathcal{D}$:

$$\begin{aligned}
\Phi_t^e &= F_t^e - (r_t^o - r_t^{\mathcal{D}})V_{t-1}^{\mathcal{D}} \\
&= F_t^e - r_t^o V_{t-1}^{\mathcal{D}} + r_t^{\mathcal{D}} V_{t-1}^{\mathcal{D}} \\
&= F_t^e - r_t^o V_{t-1}^{\mathcal{D}} + F_t^{\mathcal{D}} + (V_t^{\mathcal{D}} - V_{t-1}^{\mathcal{D}}) \\
&= F_t^o - r_t^o V_{t-1}^{\mathcal{D}} + (V_t^{\mathcal{D}} - V_{t-1}^{\mathcal{D}}) \\
&= F_t^o - \big(V_{t-1}^{\mathcal{D}}(1+r_t^o) - V_t^{\mathcal{D}}\big) \\
&= F_t^o - \Phi_t^{\mathcal{D}}.
\end{aligned}$$

Equation (6.91) may then be reframed as

$$V_t^e = V_{t-1}^e(1+r_t^o) - F_t^o + \Phi_t^{\mathcal{D}} \tag{6.93}$$

or, equivalently,

$$V_t^e = \big(V_{t-1}^o(1+r_t^o) - F_t^o\big) - \big(V_{t-1}^{\mathcal{D}}(1+r_t^o) - \Phi_t^{\mathcal{D}}\big) \tag{6.94}$$

which implies

$$V_0^e = \sum_{t=1}^n F_t^o \mathrm{d}_t^o - \sum_{t=1}^n \Phi_t^{\mathcal{D}} \mathrm{d}_t^o.$$

Since $V_0^o = \sum_{t=1}^n F_t^o \mathrm{d}_t^o$ is the market value of operating assets and $V_0^e = V_0^o - V_0^{\mathcal{D}}$, then $\sum_{t=1}^n \Phi_t^{\mathcal{D}} \mathrm{d}_t^o = V_0^{\mathcal{D}}$ is the market value of the (net) debt. In other words, to discount the standard CF$\mathcal{D}$s at the cost of debt $(r_t^{\mathcal{D}})$ is equivalent to discounting the modified CF$\mathcal{D}$s at the operating pre-tax WACC (r_t^o).

Now, suppose that, instead of r_t^o, the input market rate is the after-tax WACC (r_t^{uo}). Even in this case, the CFEs may be suitably modified so that they may be discounted at the after-tax WACC. Specifically, assuming, for illustrative purposes, that the ROD and the cost of debt coincide (i.e., $i_t^{\mathcal{D}} = r_t^{\mathcal{D}}$), the third equation of (6.88) holds, which implies

$$V_{t-1}^e = \frac{V_t^e + F_t^e - \big(r_t^{uo} - r_t^{\mathcal{D}}(1-\tau)\big) \cdot V_{t-1}^{\mathcal{D}}}{1 + r_t^{uo}}. \tag{6.95}$$

This means that the equity benchmark system fulfills the following law of motion:

$$V_t^e = V_{t-1}^e(1 + r_t^{uo}) - \Phi_t^{e'} \tag{6.96}$$

where $\Phi_t^{e'}$ is defined as $\Phi_t^{e'} = F_t^e - (r_t^{uo} - r_t^{\mathcal{D}}(1-\tau))V_{t-1}^{\mathcal{D}}$. The above equation describes the equity benchmark system in terms of r_t^{uo} instead of r_t^e. This is possible because the CFE is adjusted giving rise to a modified CFE, $\Phi_t^{e'}$, which is associated with the same expected return as the FCF stream. Hence, this stream may be properly discounted at the after-tax WACC, r_t^{uo}. Equation (6.96) has a direct solution in terms of discounted cash flows:

$$V_0^e = \Phi_1^{e'}\mathrm{d}_1^{uo} + \Phi_2^{e'}\mathrm{d}_2^{uo} + \ldots + \Phi_n^{e'}\mathrm{d}_n^{uo} \tag{6.97}$$

where $\mathrm{d}_t^{uo} = \prod_{j=1}^t (1 + r_j^{uo})^{-1}$ is the discount factor for the FCFs, as usual.[30] Even in this case, the modified CFE may be interpreted as the difference between the FCF and a suitably modified CF$\mathcal{D}$. Specifically, let us define this modified CF$\mathcal{D}$ as $\Phi_t^{\mathcal{D}'} = V_{t-1}^{\mathcal{D}}(1 + r_t^{uo}) - V_t^{\mathcal{D}}$. Then, the modified CFE becomes

$$\begin{aligned}
\Phi_t^{e'} &= F_t^e - (r_t^{uo} - r_t^{\mathcal{D}}(1-\tau))V_{t-1}^{\mathcal{D}} \\
&= F_t^e - r_t^{uo}V_{t-1}^{\mathcal{D}} + r_t^{\mathcal{D}}V_{t-1}^{\mathcal{D}} - \tau r_t^{\mathcal{D}}V_{t-1}^{\mathcal{D}} \\
&= F_t^e + F_t^{\mathcal{D}} + (V_t^{\mathcal{D}} - V_{t-1}^{\mathcal{D}}) - r_t^{uo}V_{t-1}^{\mathcal{D}} - \tau r_t^{\mathcal{D}}V_{t-1}^{\mathcal{D}} \\
&= F_t^o - V_{t-1}^{\mathcal{D}}(1 + r_t^{uo} + \tau r_t^{\mathcal{D}}) + V_t^{\mathcal{D}} \\
&= F_t^{uo} - V_{t-1}^{\mathcal{D}}(1 + r_t^{uo} + \tau r_t^{\mathcal{D}}) + V_t^{\mathcal{D}} + \tau I_t^{\mathcal{D}} \\
&= F_t^{uo} - \left(V_{t-1}^{\mathcal{D}}(1 + r_t^{uo}) - V_t^{\mathcal{D}}\right) \\
&= F_t^{uo} - \Phi_t^{\mathcal{D}'}.
\end{aligned}$$

Equation (6.91) may be reframed as

$$V_t^e = V_{t-1}^e(1 + r_t^{uo}) - F_t^{uo} + \Phi_t^{\mathcal{D}'} \tag{6.99}$$

[30] Analogous relations may be set up for modified CFEs so that they are discounted at r_t or at r_t^u:

$$\begin{aligned}
V_{t-1}^e &= \frac{V_t^e + F_t^e - (r_t - r_t^d)\cdot V_{t-1}^d}{1 + r_t} \\
V_{t-1}^e &= \frac{V_t^e + F_t^e - \left(r_t^u - r_t^d(1-\tau)\right)\cdot V_{t-1}^d}{1 + r_t^u}.
\end{aligned} \tag{6.98}$$

or, equivalently,

$$V_t^e = \left(V_{t-1}^o(1+r_t^{uo}) - F_t^{uo}\right) - \left(V_{t-1}^{\mathcal{D}}(1+r_t^{uo}) - \Phi_t^{\mathcal{D}'}\right) \tag{6.100}$$

which implies

$$V_0^e = \sum_{t=1}^{n} F_t^{uo} \mathrm{d}_t^{uo} - \sum_{t=1}^{n} \Phi_t^{\mathcal{D}'} \mathrm{d}_t^{uo}.$$

Since $V_0^o = \sum_{t=1}^n F_t^{uo}\mathrm{d}_t^{uo}$ is the market value of operating assets and $V_0^e = V_0^o - V_0^{\mathcal{D}}$, then $\sum_{t=1}^n \Phi_t^{\mathcal{D}'}\mathrm{d}_t^{uo} = V_0^{\mathcal{D}}$ is the market value of the (net) debt. In other words, to discount the CF$\mathcal{D}$s at the cost of debt ($r_t^{\mathcal{D}}$) is equivalent to discounting these newly modified CF$\mathcal{D}$s at the operating after-tax WACC (r_t^{uo}).

Assume now that the known parameters are the unlevered cost of (operating) assets, k_t^{uo}, and the expected rate on tax shield r_t^{TS}. Then, one may use (6.77), which we rewrite as

$$\frac{V_t^e + F_t^e}{V_{t-1}^e} - 1 = k_t^{uo} + (k_t^{uo} - r_t^{\mathcal{D}})\frac{V_{t-1}^{\mathcal{D}}}{V_{t-1}^e} + (r_t^{\mathrm{TS}} - k_t^{uo})\frac{V_{t-1}^{\mathrm{TS}}}{V_{t-1}^e}. \tag{6.101}$$

Manipulating algebraically,

$$V_{t-1}^e = \frac{V_t^e + F_t^e - (k_t^{uo} - r_t^{\mathcal{D}})V_{t-1}^{\mathcal{D}} - (r_t^{\mathrm{TS}} - k_t^{uo})V_{t-1}^{\mathrm{TS}}}{1 + k_t^{uo}} \tag{6.102}$$

and

$$V_0^e = \sum_{t=1}^{n} \Phi_t^{e''}(1+k_1^{uo})^{-1} \cdot (1+k_2^{uo})^{-1} \cdot \ldots \cdot (1+k_t^{uo})^{-1} \tag{6.103}$$

where $\Phi_t^{e''}$ is defined as $\Phi_t^{e''} = F_t^e - (k_t^{uo} - r_t^{\mathcal{D}})V_{t-1}^{\mathcal{D}} - (r_t^{\mathrm{TS}} - k_t^{uo})V_{t-1}^{\mathrm{TS}}$.

If the discount rate for the tax savings is $r_t^{\mathrm{TS}} = k_t^{uo}$, then $r_t^o = k_t^{uo}$ and (6.102) boils down to (6.90), which may be rewritten as

$$V_{t-1}^e = \frac{V_{t-1}^e + F_t^e - (k_t^{uo} - r_t^{\mathcal{D}})V_{t-1}^{\mathcal{D}}}{1 + k_t^{uo}}$$

whence $V_0^e = \sum_{t=1}^n (F_t^e - (k_t^{uo} - r_t^{\mathcal{D}})V_{t-1}^{\mathcal{D}})/(1 + k_t^{uo})$.

If the discount rate for the tax savings is $r_t^{\mathrm{TS}} = r_t^{\mathcal{D}}$, Eq. (6.102) becomes

$$V_{t-1}^e = \frac{V_t^e + F_t^e - (k_t^{uo} - r_t^{\mathcal{D}})(V_{t-1}^{\mathcal{D}} - V_{t-1}^{\mathrm{TS}})}{1 + k_t^{uo}}.$$

(See related Examples 6.14 and 6.15.)

6.3.2 Adjustment Factors and Circularity

If the cost of equity is not exogenously derived, the standard discounting approach for CFEs creates an issue of circularity. Specifically,

$$V_{t-1}^e = \frac{V_t^e + F_t^e}{1 + r_t^e}.$$

To compute V_{t-1}^e one needs to know r_t^e. However, the latter depends in turn on V_{t-1}^e:

$$r_t^e = \frac{V_t^e + F_t^e}{V_{t-1}^e} - 1.$$

Therefore, to compute V_{t-1}^e one needs to know r_t^e and to compute r_t^e one needs to know V_{t-1}^e. This creates a loop, which has no analytical solution but can be solved numerically by a common spreadsheet (e.g., Excel).

This issue has attracted some interest by applied scholars and practitioners (see a literature survey in Mejía-Peláez and Vélez-Pareja 2011). From a theoretical perspective, it is interesting to note that the standard discounting approach for valuing equity and the corresponding modified discounting approach may be viewed as dual valuation methods, the former providing a circular valuation, the latter providing a non-circular analytical valuation. In a sense, the non-circular method is the solution of the circularity problem arisen with standard discounting.

We illustrate the point by analyzing six cases with different known parameters (market input data):

(1) Known parameters: r_t^o and $r_t^{\mathcal{D}}$.

The equity value may be alternatively computed as follows:

$$V_{t-1}^e = \frac{V_t^e + F_t^e}{1 + r_t^o + \underbrace{\frac{(r_t^o - r_t^{\mathcal{D}})V_{t-1}^{\mathcal{D}}}{V_{t-1}^e}}_{\text{adjustment factor}}} \tag{6.104}$$

$$V_{t-1}^e = \frac{V_t^e + F_t^e - \overbrace{(r_t^o - r_t^{\mathcal{D}})V_{t-1}^{\mathcal{D}}}^{\text{adjustment factor}}}{1 + r_t^o}. \tag{6.105}$$

In the first equation, the adjustment factor $(r_t^o - r_t^{\mathcal{D}})V_{t-1}^{\mathcal{D}}$ (suitably divided by V_{t-1}^e to get a relative measure) is added to r_t^o, whereas in the second equation the adjustment factor $(r_t^o - r_t^{\mathcal{D}})V_{t-1}^{\mathcal{D}}$ is subtracted from the CFE, resulting in the the modified CFE, Φ^e, which we have analyzed in Sect. 6.3.1. Therefore, the former method uses standard CFEs discounted at a modified pre-tax WACC, the latter uses modified CFEs discounted at the standard pre-tax WACC, r_t^o.

However, the first equation is circular (the equity value V^e_{t-1} appears in both the left-hand side and the right-hand side of the equation), whereas no circularity arises in the second equation (the equation may be iterated starting from $t = n$, when $V^e_n = 0$).

(2) Known parameters: r_t and r^d_t.

The equity value may be alternatively computed as follows:

$$V^e_{t-1} = \frac{V^e_t + F^e_t}{1 + r_t + \underbrace{\frac{(r_t - r^d_t)V^d_{t-1}}{V^e_{t-1}}}_{\text{adjustment factor}}} \tag{6.106}$$

$$V^e_{t-1} = \frac{V^e_t + F^e_t - \overbrace{(r_t - r^d_t)V^d_{t-1}}^{\text{adjustment factor}}}{1 + r_t} \tag{6.107}$$

The first equation is circular, the second equation is not. The former method uses standard CFEs discounted at a modified asset pre-tax WACC, the latter uses modified CFEs discounted at the standard asset pre-tax WACC.

(3) Known parameters: r^{uo}_t and $r^{\mathcal{D}}_t$ (assuming $i^{\mathcal{D}}_t = r^{\mathcal{D}}_t$).

The equity value may be alternatively computed as follows:

$$V^e_{t-1} = \frac{V^e_t + F^e_t}{1 + r^{uo}_t + \underbrace{\frac{(r^{uo}_t - r^{\mathcal{D}}_t(1-\tau))V^{\mathcal{D}}_{t-1}}{V^e_{t-1}}}_{\text{adjustment factor}}} \tag{6.108}$$

$$V^e_{t-1} = \frac{V^e_t + F^e_t - \overbrace{(r^{uo}_t - r^{\mathcal{D}}_t(1-\tau))V^{\mathcal{D}}_{t-1}}^{\text{adjustment factor}}}{1 + r^{uo}_t} \tag{6.109}$$

The first equation is circular, the second equation is not. The former method uses standard CFEs discounted at a modified operating after-tax WACC, the latter uses modified CFEs discounted at the standard operating after-tax WACC.

(4) Known parameters: r^u_t and r^d_t (assuming $i^d_t = r^d_t$).

The equity value may be alternatively computed as follows:

$$V^e_{t-1} = \frac{V^e_t + F^e_t}{1 + r^u_t + \underbrace{\frac{(r^u_t - r^d_t(1-\tau))V^d_{t-1}}{V^e_{t-1}}}_{\text{adjustment factor}}} \tag{6.110}$$

$$V_{t-1}^e = \frac{V_t^e + F_t^e - \overbrace{(r_t^u - r_t^d(1-\tau))V_{t-1}^d}^{\text{adjustment factor}}}{1 + r_t^u} \tag{6.111}$$

The first equation is circular, the second equation is not. The former method uses standard CFEs discounted at a modified asset after-tax WACC, the latter uses modified CFEs discounted at the standard asset after-tax WACC.

(5) Known parameters: k_t^{uo}, r_t^{TS}, and $r_t^{\mathcal{D}}$.

The equity value may be alternatively computed as follows:

$$V_{t-1}^e = \frac{V_t^e + F_t^e}{1 + k_t^{uo} + \underbrace{\frac{(k_t^{uo} - r_t^{\mathcal{D}})V_{t-1}^{\mathcal{D}} + (r_t^{\text{TS}} - k_t^{uo})V_{t-1}^{\text{TS}}}{V_{t-1}^e}}_{\text{adjustment factor}}} \tag{6.112}$$

$$V_{t-1}^e = \frac{V_t^e + F_t^e - \overbrace{\left((k_t^{uo} - r_t^{\mathcal{D}})V_{t-1}^{\mathcal{D}} + (r_t^{\text{TS}} - k_t^{uo})V_{t-1}^{\text{TS}}\right)}^{\text{adjustment factor}}}{1 + k_t^{uo}} \tag{6.113}$$

The first equation is circular, the second equation is not. The former method uses standard CFEs discounted at a modified unlevered cost of operating assets, the latter uses modified CFEs discounted at the standard unlevered cost of operating assets.

(6) Known parameters: k_t^u, r_t^{GTS}, and r_t^d.

The equity value may alternatively be computed as follows:

$$V_{t-1}^e = \frac{V_t^e + F_t^e}{1 + k_t^u + \underbrace{\frac{(k_t^u - r_t^d)V_{t-1}^d + (r_t^{\text{GTS}} - k_t^u)V_{t-1}^{\text{GTS}}}{V_{t-1}^e}}_{\text{adjustment factor}}} \tag{6.114}$$

$$V_{t-1}^e = \frac{V_t^e + F_t^e - \overbrace{\left((k_t^u - r_t^d)V_{t-1}^d + (r_t^{\text{GTS}} - k_t^u)V_{t-1}^{\text{GTS}}\right)}^{\text{adjustment factor}}}{1 + k_t^u} \tag{6.115}$$

The first equation is circular, the second equation is not. The former method uses standard CFEs discounted at a financially unlevered cost of assets suitably modified, the latter uses modified CFEs discounted at the standard financially unlevered cost of assets. This case has first been studied in Mejía-Peláez and Vélez-Pareja (2011. See Eq. (2) therein). In their informative paper, the authors illustrate the circularity issue and provide the analytical (i.e., non-circular) relation reported above (see also Mejía-

Table 6.14 Modified discounting approach with known r and r^d (input data)

Time	0	1	2	3	4	5
Market data						
$r^o = r$		0.1	0.1	0.1	0.1	0.1
$r^{\mathcal{D}} = r^d$		0.02	0.02	0.02	0.02	0.02
Project data						
CFO	−450	100	12	130	140	150
CFD	−188.5	40	40	40	40	40
CFL	0.0	0.0	0.0	0.0	0.0	0.0

Peláez and Vélez-Pareja 2010), also discussing the particular cases where $r_t^{\text{GTS}} = k_t^u$ or $r_t^{\text{GTS}} = r_t^d$.[31]

Table 6.35 in Sect. 6.9 reports all non-circular relations, associated with the various cases.

Example 6.14 ♦

C^o	C^d
	C^e

(*Modified discounting with known r and r^d*) A company has the opportunity of undertaking a levered project. We assume no liquid assets are involved and the firm' managers estimate that the appropriate pre-tax WACC is $r^o = r = 10\%$ while the required return on debt is $r^{\mathcal{D}} = r^d = 2\%$. Table 6.14 reports the market input data and the relevant project data. The application of the modified-discounting approach consists of the following steps (see Table 6.15):

(a) Calculate the debt values, V_{t-1}^d for every $t \in \mathbb{N}_5^1$ by discounting the CFDs at $r^d = 2\%$
(b) Calculate CFEs as $F_t^e = F_t^o + F_t^l - F_t^d$
(c) Find the modified CFEs, Φ_t^e, by subtracting $(r - r^d)V_{t-1}^d = (0.1 - 0.02) \cdot V_{t-1}^d$ from the CFEs (see Eq. (6.107))
(d) Discount all the modified CFEs, Φ_t^e, at $r = 10\%$ or, alternatively, use the backward formula $(V_t^e + \Phi_t^e)/(1 + r)$ recursively to find V_0^e.

The output is $V_0^e = 288$. Subtracting the equity investment, one gets the shareholder value created: $\text{NPV}^e = V_0^e - C_0^e = 288 - 261.5 = 26.5$ (which is also equal to NPV if $i^d = r^d = 2\%$).

Alternatively, one may compute the modified CFEs from the modified CFDs, which we may redenote as Φ_t^d (instead of $\Phi_t^{\mathcal{D}}$), since liquid assets are absent. In this case, steps (b) and (c) are replaced by the following ones:

(b) Calculate the modified CFDs as $\Phi_t^d = V_{t-1}^d(1 + 0.1) - V_t^d$

[31] We remind the reader that r^{TS} is the discount rate for $\tau I_t^{\mathcal{D}} = \tau(I_t^d - I_t^l)$, whereas r^{GTS} is the discount rate for τI_t^d. The first one follows a capital structure perspective (net tax savings), the second one follows a financial structure perspective (gross tax savings).

Table 6.15 Modified discounting approach with known r and r^d (first version)

Time	0	1	2	3	4	5
Equity valuation						
V^d	188.5	152.3	115.4	77.7	39.2	0.0
CFE	−261.5	60	80	90	100	110
Modified CFE, Φ_t^e		44.9	67.8	80.8	93.8	106.9
V_t^e	288.0	271.9	231.2	173.6	97.1	0.0
$\text{NPV}^e = V_0^e - C_0^e$	26.5					

Table 6.16 Modified discounting approach with known r and r^d (second version)

Time	0	1	2	3	4	5
Equity valuation						
V_t^d	188.5	152.3	115.4	77.7	39.2	0.0
Modified CFD, Φ_t^d		55.1	52.2	49.2	46.2	43.1
Modified CFE, Φ_t^e		44.9	67.8	80.8	93.8	106.9
V_t^e	288.0	271.9	231.2	173.6	97.1	0.0
$\text{NPV}^e = V_0^e - C_0^e$	26.5					

(c) Find the modified CFEs by subtracting the modified CFDs from the CFOs: $\Phi_t^e = F_t^o - \Phi_t^d$

(see Table 6.16). The two procedures are equivalent and none of them utilizes the equity COC: The CFEs are precisely modified so that they may be directly discounted at 10%, which is an input market rate (see Table 6.16):

Example 6.15 ♦

C^o	C^d
	C^e

(Modified discounting with known k^u, r^{GTS}, and r^d, and APV) Consider a 5-year investment and assume that the managers estimate the (constant) financially unlevered cost of assets, k^u, and the discount rate for (gross) tax shield, which is assumed to be (constant) and equal to the cost of debt: $r^{\text{GTS}} = r^d$. Equation (6.115) becomes

$$V_{t-1}^e = \frac{V_t^e + F_t^e - \overbrace{(k^u - r^d)(V_{t-1}^d - V_{t-1}^{\text{GTS}})}^{\text{adjustment factor}}}{1 + k^u}. \tag{6.116}$$

Let us also assume $r^d = i^d$ so that the tax saving is $\tau r^d V_{t-1}^d$. Table 6.17 reports the key assumptions on project data (including tax rate at 30%) and market data. The modified discounting approach consists of the following steps (described in Table 6.18):

(a) Calculate the debt values, V_{t-1}^d, for every $t \in \mathbb{N}_5^1$ by discounting the CFDs at $r^d = 2\%$

Table 6.17 Modified discounting approach with known k^u, r^{GTS} and r^d (input data)

Time	0	1	2	3	4	5
Market data						
k^u (= k^{uo})		0.09	0.09	0.09	0.09	0.09
$r^d = i^d$		0.02	0.02	0.02	0.02	0.02
Project data						
FCF	−450	98.87	119.09	129.31	139.53	149.76
CFD	−188.54	40	40	40	40	40
CFL	0.0	0.0	0.0	0.0	0.0	0.0
Tax rate	30%					

Table 6.18 Modified discounting approach with known k^u, r^{GTS} and r^d (valuation)

Time	0	1	2	3	4	5
Equityvaluation						
V^d	188.54	152.31	115.36	77.66	39.22	0.00
Tax saving		1.13	0.91	0.69	0.47	0.24
V^{GTS}	3.28	2.22	1.35	0.68	0.23	0.00
CFE	−261.46	60	80	90	100	110.00
Modified CFE		47.03	69.49	82.02	94.61	107.27
V_t^e	301.71	281.84	237.71	177.09	98.41	0.00
$\text{NPV}^e = V_0^e - C_0^e$	40.25					

(b) Calculate the tax savings as $r^{\text{GTS}} = 0.3 \cdot 0.02 \cdot V_{t-1}^d$
(c) Calculate the values of the tax savings, V_{t-1}^{GTS}, by discounting the tax savings at $r^{\text{GTS}} = 2\%$
(d) Calculate the CFEs as $F_t^e = F_t^{uo} + \tau r^d V_{t-1}^d + F_t^l - F_t^d$
(e) Calculate the modified CFEs by subtracting the adjustment factors $(k^u - r^d)(V_{t-1}^d - V_{t-1}^{\text{GTS}}) = (0.09 - 0.02)(V_{t-1}^d - V_{t-1}^{\text{GTS}})$ from the CFEs
(f) Find V_0^e via repeated application of (6.116) or, which is the same, by discounting all modified CFEs at $k^u = 9\%$.

The output is $V_0^e = 301.71$. Subtracting the equity investment, one gets $\text{NPV}^e = 40.26$.

The same valuation may be obtained by applying the APV method (see Sect. 6.2.2). Specifically, discounting these FCFs at $k^u = 9\%$, one gets the unlevered value of the project (i.e., the value that the project would have if it were unlevered), resulting in $V_0^u = 486.97$. Adding the value of the tax shield, $V^{\text{GTS}} = 3.28$, the project value is obtained: $V_0 = 486.97 + 3.28 = 490.25$. Subtracting the debt value, the equity value is found: $V_0^e = 490.25 - 188.54 = 301.71$ (whence, the equity NPV).

6.4 Valuing Projects with Cash Flows: Potential Dividends

We stress that the equity value of a project, V_0^e (and, therefore, value creation), depends on the CFEs actually distributed, not on the cash available for distribution to shareholders, F_t^{pd}. Therefore, the equity value may not be computed by discounting PDs. To see why, just remember that $F_t^{pd} = F_t^e - F_t^l$ (see column 1 of left Matrix in (4.70)) and consider the following equalities:

$$V_0^e = \sum_{t=1}^{n} F_t^e \cdot \mathrm{d}_{t,0}^e$$
$$= \sum_{t=1}^{n} (F_t^{pd} + F_t^l) \cdot \mathrm{d}_{t,0}^e.$$

The equity value is equal to the present value of the PDs plus the present value of the CFLs, calculated at the cost(s) of equity. Let us then focus on the CFLs for $t > 0$; they represent the excess cash retained (if negative) or distributed (if positive). We may note that, if $\sum_{t=1}^{n} F_t^l \mathrm{d}_{t,0}^e = 0$, it is irrelevant whether PDs or CFEs are discounted:

$$V_0^e = \sum_{t=1}^{n} F_t^{pd} \cdot \mathrm{d}_{t,0}^e. \tag{6.117}$$

This method is called the ***FCFE method***. A sufficient condition for the FCFE method to be a rational valuation method is that the undistributed cash ($\mathrm{RC}_t = -F_t^l > 0$) is invested at an expected interest rate equal to the cost of equity and the extra distributed cash ($\mathrm{RC}_t = -F_t^l < 0$) is raised by borrowing at an interest rate equal to the cost of equity. For example, assuming $n = 2$, the condition $\sum_{t=1}^{n} F_t^l \mathrm{d}_{t,0}^e = 0$ simplifies to $F_1^l = -F_2^l/(1 + r_2^e)$. Assuming $F_1^l < 0$ (i.e., $\mathrm{RC}_1 > 0$), this equality may be interpreted as follows. At time 1, shareholders receive less cash than what could be distributed; this reduction in dividends is compensated by the distribution of an extra dividend $F_2^l > 0$ ($\mathrm{RC}_2 < 0$) at time 2, in such a way that the time-1 reduction of CFEs is perfectly neutralized by the extra dividend. This means that, in terms of shareholder value creation, it is irrelevant whether excess cash is retained at time 1 or distributed.

Notably, the irrelevance implied by condition $\sum_{t=1}^{n} F_t^l \mathrm{d}_{t,0}^e = 0$ expresses (a generalization of) Miller and Modigliani's (1961) **dividend irrelevance theorem**, according to which the dividend policy of a firm is irrelevant.[32] Practically, this proposition means that one should not bother forecasting the CFEs for it suffices to discount FCFEs (potential dividends), given that the retained cash is invested at r^e. However, for capital asset projects (and for firms as well), the retained cash is not usually invested at the cost of equity. (For a thorough treatment of the relevance of the payout policy, see Magni 2010c. See also Vélez-Pareja and Magni 2009; Magni

[32] See also Remark 4.6.

and Vélez-Pareja 2009). The use of PDs for valuing equity may then result in over- or under-estimation, depending on whether $\sum_{t=1}^{n} F_t^l \mathrm{d}_{t,0}^e < 0$ or $\sum_{t=1}^{n} F_t^l \mathrm{d}_{t,0}^e > 0$. All in all, the FCFE method should be discouraged, unless there are reasons to believe that the above condition is fulfilled.

DeAngelo and DeAngelo (2007) *argue that a firm's equity is not given by potential dividends but by cash flows paid to shareholders. The former (they label them 'free cash flow') determines the investment value, the latter leads to the distribution value: "Investment value is the discounted value of the FCF to the firm generated by its investment policy, which determines the firm's capacity to make payouts. Distribution value is the discounted value of the cash payouts to currently outstanding shares, i.e., the cash flow paid to stockholders, which determines the market value of equity" (p. 16). They underline that "value is generated for investors only to the extent that this capacity is transformed into actual payouts"* (DeAngelo and DeAngelo 2006, *p. 309). Only if "the full PV of FCF is distributed to investors, variation in the timing of the stream of payouts and in their form (e.g., dividends versus stock repurchase) has no effect on stockholder wealth"* (DeAngelo and DeAngelo 2007, *p. 25).*

…

If a firm does not pay out cash flow, no value is created for the shareholders. Firm is an entity separated from the shareholders. If the firm generates large amounts of cash flows, this does not automatically make the shareholders richer. It depends both on the magnitude of the cash flows and on the fact that those cash flows are actually paid out to the shareholders. Inflows for capital providers are outflows for the firm and vice versa, so the value of the firm for shareholders does not lie in the funds retained by the firm, but in the funds that are withdrawn from the firm by the shareholders. To retain funds is a good choice only if those funds will provide, sooner or later, sufficient payouts to shareholders above the cost of capital. Shareholders wealth does not increase if cash is not actually pulled out from the firm and distributed to the shareholders. In other words, the investment value of a firm is different from the distribution value of a firm (DeAngelo and DeAngelo 2007*). Only distribution value counts for shareholders. To assume that potential dividends are actually pulled out of the firm is like trying to pull potential rabbits out of actual hats.*

Magni and Vélez-Pareja (2009). Potential dividends versus actual cash flows in firm valuation. *The ICFAI Journal of Applied Finance*, 15(7) (July), 51–66 (p. 58).

Even if the condition $\sum_{t=1}^{n} F_t^l \mathrm{d}_{t,0}^e = 0$ is not met, one may correctly use PDs for valuing the project's or firm's equity. It suffices to discount PDs at the correct discount rate, reflecting the risk of PDs. This may be done easily, if one bears in mind that the benchmark system fulfills its own statics and dynamics and that, owing to the law of one price, its equilibrium is representative of the market equilibrium. Specifically, in the benchmark system, the *potential capital* is valued as $V_t^{pd} = V_{t-1}^{pd}(1 + r_t^{pd}) - F_t^{pd}$ where r_t^{pd} is the income rate for the benchmark system (required return on potential capital, cost of potential capital). Since V_t^{pd} is equal the difference between the value of operations and the value of debt, one may write $V_t^{pd} = V_t^o - V_t^d$. Hence, the required return on potential capital is

$$r_t^{pd} = \frac{r_t^o V_{t-1}^o - r_t^d V_{t-1}^d}{V_{t-1}^o - V_{t-1}^d} \tag{6.118}$$

or $r_t^{pd} = (V_t^{pd} + F_t^{pd})/V_{t-1}^{pd} - 1$. If $r_t^o = k_t^{uo}$, then

$$r_t^{pd} = \frac{k_t^{uo} V_{t-1}^o - r_t^d V_{t-1}^d}{V_{t-1}^o - V_{t-1}^d}.$$

PDs may well be discounted at r_t^{pd} to get V_0^{pd}:

$$V_0^{pd} = \sum_{t=1}^{n} F_t^{pd} \cdot \mathrm{d}_{t,0}^{pd} \tag{6.119}$$

where $\mathrm{d}_{t,0}^{pd} = \prod_{j=1}^{t}(1 + r_j^{pd})^{-1}$ is the discount factor for PDs. Since $V_0^e = V_0^{pd} + V_0^l$, then

$$V_0^e = \sum_{t=1}^{n} F_t^{pd} \cdot \mathrm{d}_{t,0}^{pd} + \sum_{t=1}^{n} F_t^l \cdot \mathrm{d}_{t,0}^l$$

which describes what we call the **PD method**. If the payout ratio is 100% so that $F_t^l = 0$ (and $F_t^e = F_t^{pd}$) for all t, the market equity value is indeed the market value of the PDs:

$$V_0^e = \sum_{t=1}^{n} F_t^{pd} \cdot \mathrm{d}_{t,0}^{pd}.$$

If the project/firm is equity financed ($C_t^d = 0$ for all t), then $F_t^{pd} = F_t^o$ and $r_t^{pd} = r_t^o$. Therefore, PDs may be discounted at the pre-tax WACC to find the market value of PDs and, hence, the market value of equity:

$$V_0^e = \sum_{t=1}^{n} F_t^{pd} \cdot \mathrm{d}_{t,0}^o + \sum_{t=1}^{n} F_t^l \cdot \mathrm{d}_{t,0}^l. \tag{6.120}$$

The potential NPV is then defined as $\text{NPV}^{pd} = V_0^{pd} - C_0^{pd} = \sum_{t=1}^{n} F_t^{pd} \cdot \mathrm{d}_{t,0}^{pd} + F_0^{pd}$ where $F_0^{pd} = I_0^{pd} - C_0^{pd}$. This is the potential NPV that the firm project would generate if the payout ratio were 100% (i.e., if $C_0^l = 0$ for all t).

Example 6.16 ♦

C^o	C^d
C^l	C^e

(*BHP Billiton—90% payout—Potential dividends*) Let us consider the case where BHP Billiton has a 90% payout ratio. The PDs and CFLs are reported in Table 4.29. Suppose the (assumed constant) cost of equity is $r^e = 0.15$, as in Example 6.5, where we have computed the equity market value: $V_0^e = 1{,}347.2$. It is straightforward to check that the use of PDs for valuing equity overestimates the equity value, leading to an incorrect amount equal to 1,454.8. The difference, −107.6, measures the present value (calculated at the cost of equity) of the CFLs. The discrepancy between the two evaluations reflects the discrepancies between the interest rate on liquid assets and the cost of equity.

In order to use PDs correctly, one may calculate the correct COC for PDs as $r_t^{pd} = (V_t^o - V_t^d + F_t^{pd})/(V_{t-1}^o - V_{t-1}^d)$. This leads to the following rates:

- from 1 to 10: (15.0%,15.2%, 15.5%, 15.7%, 16.0%, 16.4%, 16.7%, 17.2%, 17.7%, 18.3%)
- from 11 to 20: (18.0%, 18.6%, 19.4%, 20.4%, 21.6%, 23.3%, 25.9%, 30.3%, 39.2%, 69.6%).

Discounting the PDs with such rates one gets $V_0^{pd} = 1{,}347.2$, which is equal to the market value of equity, as expected.

If the cost of equity were equal to the interest rate on liquid assets, then the present value of the CFLs would be zero and PDs would lead to the same equity value than CFEs. In particular, if i^l is set equal to the equity COC (i.e., $i^l = r^e = 0.15$), the equity value becomes $V_0^e = 1{,}386.5$, regardless of whether PDs or CFEs are discounted.[33] Alternatively, if one set the equity COC equal to the expected return on liquid assets, so that $r^e = i^l = 0.03$, then one would get $V_0^e = 3{,}432.1$, irrespective of whether PDs or CFEs are discounted.[34]

Figures 6.2 and 6.3 in Sect. 6.9 report the **Mosaic of values** of a project (or firm), under different assumptions of payout policy ($\alpha > 1$ and $\alpha = 1$). We point out that the structure of the mosaic mirrors the structures of the mosaics for incomes and cash flows (see Figs. 4.7 and 4.8). This derives from the statics of a project (firm), whereby capital, income, and cash flow fulfill the same kind of conservation laws.

[33] In this case, the interest rate on debt is smaller than the interest rate on liquid assets: $0.15 = i^l > i^d = 0.08$. However, for a capital asset investment, it might be unrealistic (albeit not impossible) to borrow funds at a given interest rate i^d and invest them in financial assets i^l with a higher expected return (i.e., the condition $i^l > i^d$ is usually not fulfilled).

[34] If this assumption is made, one may not make the additional usual assumption $i^d = r^d = 0.08$, because this would imply $r^d > r^e$ which is not legitimate (see Remark 6.1).

6.5 Summary of 12 Discounted-Cash-Flow Valuation Methods

In academia and in practice, the WACC method seems to be the preferred method. In general, the notions of NOPAT and FCF seem to be preferred to the respective notions of OI and CFO/CCF.

However, as we have seen, from a logical point of view, the way one frames investments and financings and defines cash flows is irrelevant, for the result in terms of valuation (and decision-making) is the same. The choice of a valuation scheme is an important issue from a *practical* point of view, one which has to do with the market input data: Depending on the known parameters, some method may be more appropriate than some other ones. As explained in the Introduction of this book, we take market input data as given: This book aims at explaining how to reach rational and consistent valuations and decisions *whatever the perspective adopted and whatever the input data.*

The WACC method, the APV method, and the CFO/CCF method are often compared one another. The former two are based on the notion of FCF, while the third method is based on the notion of CFO. Conceptually, the notion of CFO seems to have some conceptual advantages over the notion of FCF:

- the triplet (OI/CFO/operating assets) is more naturally originated from the statics and dynamics of an economic system. There are biunivocal correspondences among the components of operating assets, the components of operating income, the components of CFO:

Capital component	Income component	Cash-flow component
+ Accounts receivable	+ Sales	+ Cash flow from customers
− Accounts payable	− Cost of purchases	− Cash flow to suppliers
− Salaries & Wages payable	− Labor costs	− Cash flow to employees
+ Inventory	+ Inventory income	
− Taxes payable	− Taxes	− Tax payments
+ Net Fixed Assets	− Depreciation	− Capital expenditures
= Operating assets	= Operating income	= Cash flow from operations

  Note that taxes payable are part of the operating assets. So, income taxes are naturally part of the operating notion of after-tax profit
- in those cases where $\Delta \text{TP}_t \neq 0$ for some t, the introduction of the FCF notion raises the additional question of how tax payments should be split up into an operating component and a non-operating component such that

$$\begin{array}{c} \text{Changes in operating TP} = \text{Taxes on EBIT} - \text{payment for operating taxes} \\ \underline{\text{Changes in non-operating TP} = \text{Taxes on net interest} - \text{payment for non-operating taxes}} \\ \text{Changes in TP} = \text{Income taxes} - \text{payment for taxes} \end{array} \tag{6.121}$$

Beside the problem of how the breakdown of payments should be done, the non-operating component of taxes payable must be excluded from the operating assets, which implies that NOPAT and FCF are not matched by the operating assets; they are matched by the operating assets before (i.e., gross of) the non-operating component of TP:

Capital component	Income component	Cash-flow component
+ Accounts receivable	+ Sales	+ Cash flow from customers
− Accounts payable	− Cost of purchases	− Cash flow to suppliers
− Salaries & Wages payable	− Labor costs	− Cash flow to employees
+ Inventory	+ Inventory income	
− Operating Taxes payable	− Taxes on EBIT	− Payments for operating taxes
+ Net Fixed Assets	− Depreciation	− Capital expenditures
= Operating assets (before non-operating TP)	= NOPAT	= FCF

This means that the FCF should be redefined as

$$F_t^{uo,\text{redef}} = \text{NOPAT} - \text{change in operating assets before non-operating TP}.$$

Even with this redefinition of FCF, things tend to become messy, because the WACC should itself be redefined, such that

$$r_t^{uo,\text{redef}} = \frac{r_t^e V_{t-1}^e + r_t^{\mathcal{D}} V_{t-1}^{\mathcal{D}} - \tau I_t^{\mathcal{D}} - \Delta \text{ non-operating TP}_t}{V_{t-1}^e + V_{t-1}^d}.$$

This rate is not easily understandable. Further, we can say that the non-operating component of TP is thrown out the front door (i.e., excluded from the numerator) only to allow it to sneak in the back (i.e., included in the denominator, via the redefined WACC).[35]

- by discounting the CFOs/CCFs at the pre-tax WACC, r^o, one discounts levered cash flows to get a levered value, V_0^o. In contrast, by discounting FCFs at the WACC, r^{uo}, one discounts an unlevered cash flow to get a levered value. This is due to the presence of the interest tax shield in the denominator (with negative sign). Hence, the WACC turns an unlevered cash-flow stream into a levered value. We can say, again, that the levered perspective is thrown out the front door only to allow it to sneak in the back
- the notion of FCF seems to be more natural when used with the APV method. Discounting *unlevered* cash flows to get an *unlevered* value (APV method) may be viewed just as natural as discounting *levered* cash flows to get a *levered* value (CFO/CCF method); both methods are consistent with its own (unlevered and levered) perspective. In contrast, discounting *unlevered* cash flows to get a *levered* value is a hybrid scheme which mixes oranges and apples. Therefore, the CFO method and the APV method may be regarded as more intuitive than the WACC method.

[35] Admittedly, if one assumes, as usual in capital budgeting, that $\text{TP}_t = 0$ for every $t \in \mathbb{N}_n^0$, this problem does not arise.

A practical advantage of the use of WACC method as opposed to the CFO method, the CFA method, and the CFE method seems to be that the computation of CFO (CFA, CFE) requires estimation of the taxes on EBT, which in turn requires estimation of the interest expenses and interest incomes, which in turn depend on the financing policy. In contrast, the FCF does not depend on any particular financing policy. However, the advantage is illusory, for the FCF must be discounted at the (after-tax) WACC, which includes the tax shield, which in turn depends on the financing policy.

There is indeed a case where the use of FCFs is preferable to CFOs/CCFs. Specifically, suppose that the following assumptions are made:

1. $i_t^{\mathcal{D}} = r_t^{\mathcal{D}}$. This implies that the tax shield is $\tau \cdot r_t^{\mathcal{D}} V_{t-1}^{\mathcal{D}}$ and the normal profit to debt, net of tax shield, is $r_t^{\mathcal{D}}(1-\tau)V_{t-1}^{\mathcal{D}}$
2. the firm periodically adjusts its debt so as to maintain a target, prespecified leverage ratio ϑ_t (in market values): $\vartheta_t = V_{t-1}^{\mathcal{D}} / V_{t-1}^{o}$.

If the above assumptions are made, then the WACC, r_t^{uo}, can be determined without making any specific estimation about debt levels year by year or levels of liquid assets year by year. It suffices to estimate the cost of equity and the cost of net debt and apply the target leverage ratio so that $r_t^{uo} = r_t^e(1-\vartheta_t) + r_t^{\mathcal{D}}(1-\tau)\vartheta_t$.

When used with Assumptions 1. and 2. above, Eq. (6.57) is often called the ***textbook WACC method*** and is often presented with constant r^e, constant $r^{\mathcal{D}}$, and constant ϑ, so that $r^{uo} = r^e(1-\vartheta) + r^{\mathcal{D}}(1-\tau)\vartheta$ is time-invariant (see also Damodaran 2006a, b; Berk and DeMarzo 2011, Brealey et al. 2011; Titman and Martin 2016). This implies

$$V_0^o = \sum_{t=1}^{n} \frac{F_t^{uo}}{(1+r^{uo})^t}. \tag{6.122}$$

Assumption 2. is critical. In principle, a firm's leverage ratio may be, to some extent, rebalanced year by year so as to keep the leverage ratio at a prefixed level, but this is not common.[36] Focusing on capital budgeting projects, the project's leverage ratio is not usually predetermined and, rather, is the result of explicit forecast of the prospective debt policy, from which the tax savings are explicitly derived.

Further, it should be stressed that the textbook WACC method avoids *explicit* estimates of CF$\mathcal{D}$s and CFEs, not *implicit* estimates. Specifically,

(a) the value of operations is calculated as $V_{t-1}^o = (V_t^o + F_t^{uo})/(1+r^{uo})$ for every t
(b) the implicit debt value and the equity value are unambiguously determined by applying the leverage ratio to the value of operations:

$$V_{t-1}^{\mathcal{D}} = \vartheta_t V_{t-1}^o$$
$$V_{t-1}^e = (1-\vartheta_t) V_{t-1}^o$$

(c) using the laws of motion (6.3) and (6.43), the implicit CFEs and the CF$\mathcal{D}$s are univocally determined as

[36] Block (2011) surveyed 255 top-ranking financial officers. Only 20.3% of them use debt rebalancing.

$$F_t^e = V_{t-1}^e(1 + r_t^e) - V_t^e$$
$$F_t^{\mathcal{D}} = V_{t-1}^{\mathcal{D}}(1 + r_t^{\mathcal{D}}) - V_t^{\mathcal{D}}.$$

Therefore, the advantage of removing explicit modeling of debt and related cash flows CFDs is offset by an implicit estimation of them.[37]

The textbook WACC method is widely employed (presumably, the most widely employed valuation method). The reason may reside on the implication of the aforementioned assumptions, namely, that it is not necessary to make explicit estimation on the financing policy (debt policy and distribution policy): The levels of debt and liquid assets are assumed to be kept at a preselected (usually constant) proportion of the value of operating assets. However, as mentioned, the two aforementioned assumptions are usually violated and, even when they are fulfilled, removal of CFE and CF$\mathcal{D}$ estimates is illusory, for they are univocally determined implicitly (and they may well turn out to be unreasonable estimates, inconsistent with the financial requirements of the firm). We also add that, in many cases, even FCF itself is affected by the financing policy. This occurs if debt covenants exist which set restrictions on the firm's investment decisions; it also occurs if equity capital and/or excess cash are not sufficient to fund some investment so that raising debt becomes a sine qua non for investing and the amount of debt raised affects the investment choices (see also Remark 4.2). In these cases, the debt policy contributes to determine whether and how a given course of action may be undertaken and, therefore, explicit modeling of the financing policy and its associated cash flows will be necessary to determine EBITDA and FCF.

Analogous reasons induce many analysts to use the FCFE method to value a project's equity: It suffices to *pretend* that a full distribution policy will be followed, thereby eluding, in actual facts, the problem of forecasting the CFEs, which are made to artificially coincide with the PDs. However, if the retained cash is invested (or the extra distribution is raised) at a rate which is significantly different from the cost of equity, the equity may be substantially under-valued or over-valued (also depending on the amount of retained cash and/or extra distribution).

To wrap things up, we have presented a **unified paradigm** for valuing projects and making accept-reject decisions by discounting several types of cash flows at risk-adjusted costs of capital (discounted-cash-flow approach). However, this paradigm may take different shapes. One may list (no less than) twelve formulations of this paradigm: WACC method (with either FCFs or asset FCFs), textbook WACC method, APV method, financially unlevered method (APV with asset FCFs), CFO method, conventional CCF method, CFE method, modified CFE method, CFA method, PD method, FCFE method. The textbook WACC method and the FCFE method are special cases of the WACC method and the PD method, respectively. Since they are based on stringent assumptions, they are **critical schemes** and should be used with care and awareness.

[37] The relationships between the conventional CCF method and the textbook WACC method has been investigated in many academic contributions, especially in relation to firm valuation (e.g., see Arditti and Levy 1977; Ben-Horin 1979; Chambers et al. 1982; Ruback 2002; Fernández 2002; Tham and Vélez-Pareja 2004; Booth 2007; Cooper and Nyborg 2007; Jagannathan et al. 2017).

Theoretically, if the set of assumptions is the same, any one scheme provides correct values and rational recommendations for decision-making. This means that **the choice of a valuation scheme is a practical issue**, not a theoretical one: The best scheme depends on the key assumptions on input data. For example, if r^e is known, then the CFE method may be a natural choice; if k^{uo}, r^{TS}, and $r^{\mathcal{D}}$ are known, then the APV method or the modified CFE discounting approach may be selected; if r^o and $r^{\mathcal{D}}$ are known, then the CFO method would probably be preferred; if r^{uo} and $r^{\mathcal{D}}$ are known, the WACC method may be employed (or even the textbook WACC method, but only insofar as the firm will indeed be able to rebalance the debt year by year and follow the predetermined schedule of leverage ratios).

Whatever the formulations employed, one must always be aware of the assumptions and implications of the scheme one is following. One must be able to understand, work, and manage the model in any of the above mentioned perspectives (and any other conceivable one), with either constant or time-varying rates.

All perspectives are logically equivalent and must lead to the same result. However, the application of the schemes may cause some applicative problems. In particular, different sets of market input data and estimation techniques may bring different values.

A possible strategy for a reliable valuation and decision-making process is to rest on realistic assumptions and use several valuation schemes, starting from different market inputs, possibly estimated with different techniques. For example, suppose the analyst selects four different, independent techniques whereby she

- estimates k^{uo} and k^{TS}
- estimates r^e
- estimates WACC
- estimates r^o.

For each set of estimations, he might evaluate the project with several different schemes. For example, suppose she

- employs the APV method
- employs the CFE method
- employs the WACC method
- employs the CFO method
- employs the modified CFE method.

This brings about $5 \times 4 = 20$ evaluation runs. The multiple valuation process will serve as a system for spotting errors and inconsistencies. Specifically,

- for any given set of inputs, if different valuation schemes provide different values and NPVs, then some logical bugs are buried in the model and the analyst should then try to unearth them (in this respect, the split-screen Matrix may be helpful)
- for any given valuation scheme, if the four sets of inputs are not consistent (e.g., the estimated cost of equity is not consistent with the cost of equity implied by r^o

estimated independently, or the latter is not consistent with r^o as logically implied by the estimation of k^{uo} and k^{TS}), then some contradictions arise within the set of assumptions or in the estimation processes. The analyst should then spend some effort to fine-tune the estimation processes and reconcile the estimates, iterating the analysis until convergence to consistent values is provided by all schemes employed.

DISCOUNTED-CASH-FLOW APPROACH:

12 SCHEMES FOR VALUING PROJECTS AND MAKING DECISIONS

UNLEVERED PERSPECTIVE†

WACC method (with FCFs). Discount FCFs at the WACC, r^{uo}, to get the levered value of the operating assets, V_0^o. Subtract the value of net debt, $V_0^{\mathcal{D}}$, to get the value of equity, V_0^e. Subtract equity capital, C_0^e, to get the equity NPV.

WACC method (with asset FCFs). Discount asset FCFs at the asset after-tax WACC, r^u, to get the levered value of the assets, V_0. Subtract the value of debt, V_0^d, to get the value of equity, V_0^e. Subtract equity capital, C_0^e, to the get the equity NPV.

Textbook WACC method. Assume net debt is rebalanced year by year so as to be a predetermined proportion of the operating assets in market value; assume cost of net debt, $r^{\mathcal{D}}$, is equal to return on net debt, $i^{\mathcal{D}}$ (cost of equity and cost of net debt are usually assumed to be constant over time, as well as the leverage ratio). Discount FCFs at the WACC, r^{uo}, to get the levered value of the operating assets, V_0^o. Subtract operating capital, C_0^o, to get the operating NPV. This is equal to the equity NPV, by assumption.

APV method. Discount FCFs at the unlevered cost of (operating) assets, k_t^{uo}, to get the unlevered value of (operating) assets, V_0^{uo}. Add the value of tax shields, V_0^{TS}, to find the levered value of operating assets, V_0^o. Subtract the value of net debt, $V_0^{\mathcal{D}}$, to get the value of equity, V_0^e. Subtract equity capital, C_0^e, to get the equity NPV.

Financially unlevered (APV) method. Discount asset FCFs at the financially unlevered cost of assets, k^u, to get the financially unlevered value of assets, V_0^u. Add the value of gross tax shield, V_0^{GTS}, and subtract debt, V_0^d, to get the value of equity, V_0^e. Subtract equity capital, C_0^e, to get the equity NPV.

continued on next page

continued from previous page

LEVERED PERSPECTIVE

CFO/CCF method. Discount CFOs/CCFs at the pre-tax WACC, r^o, to get the levered value of the operating assets, V_0^o. Subtract the value of net debt, $V_0^{\mathcal{D}}$, to get the value of equity. Subtract equity capital, C_0^e, to get the equity NPV.

Conventional CCF method. Assume the discount rate for tax savings is equal to the pre-tax WACC and, therefore, equal to the unlevered cost of assets (usually assumed constant): $r^{\text{TS}} = r^o = k^{uo}$. Discount CFOs/CCFs at the unlevered cost of asset, k^{uo}, to get the levered value of operating assets, V_0^o. Subtract the value of net debt, $V_0^{\mathcal{D}}$, to get the value of equity. Subtract equity capital, C_0^e, to get the equity NPV.

CFE method. Discount CFEs at the equity cost of capital, r_t^e, to get the value of equity. Subtract equity capital, C_0^e, to get the equity NPV.

Modified CFE method. Modify CFEs and discount them at the appropriate (modified) cost of capital to get the value of equity (see Sect. 6.3). Subtract equity capital, C_0^e, to get the equity NPV.

CFA method. Discount CFA/CFP at the asset pre-tax WACC, r, to get the value of assets, V_0. Subtract debt, V_0^d, to find the equity value. Subtract equity capital, C_0^e, to get the equity NPV.

PD method. Discount PDs at the cost of potential capital, r^{pd}, to get the value of PD, V_0^{pd}. Add the value of the liquid assets, V_0^l, to get the equity value. Subtract equity capital, C_0^e, to get the equity NPV.

FCFE method. Assume retained cash (extra distribution) is reinvested (raised) at the cost of equity, r^e. Discount FCFEs (= PDs) at the cost of equity to get the value of equity. Subtract equity capital, C_0^e, to get the equity NPV.

---

†Subscripts of rates are omitted for simplicity.

Example 6.17 ♦

C^o	C^d
C^l	C^e

Consider the following key assumptions for a four-period levered project/firm:

Time	0	1	2	3	4
k^{uo}		14%	14%	14%	14%
$i^d = r^d$		5%	5%	5%	5%
$i^l = r^l$		4%	4%	4%	4%
Tax rate		25%	25%	25%	25%
Payout ratio		90%	90%	90%	
EBIT		10	10	20	25
Operating assets	100	70	30	10	0
Debt	70	40	20	10	0
r^{TS} is equal to $r^{\mathcal{D}}$					

Table 6.19 displays the balance sheets, the income statements, and the cash flow statements. As we know, the three pro forma financial statements are intertwined: To complete the BS one needs to know the balance of liquid assets in each period, which depends on the interest income and the CFL, the latter being calculated as $(1-\alpha)F_t^{pd} = 0.1F_t^{pd}$. In turn, the latter is the difference between the CFO and CFD, which in turn depends on the book value of debt and book value of operating assets as well as interest expenses and operating income (EBIT net of taxes), respectively.

Some of the available cash flow is retained in the liquid assets at time 1, 2, and 3. These interim cash flows are invested at $i^l = 4\%$, which is also assumed to be the required return on liquid assets, $r^l = 4\%$. This means that excess cash is invested in value-neutral assets ($\text{NPV}^l = 0$). At the end of the fourth and last period, the accumulated amount of these investments will be equal to the EOP balance of liquid assets, E_4^l, and will be distributed to equityholders, in addition to the last PD:

$$F_4^l = E_4^l = 5.88 = 0.49 \cdot 1.04^3 + 2.60 \cdot 1.04^2 + 2.42 \cdot 1.04.$$

There are several routes open for valuation; the shortest way is to use the modified CFE discounting approach. To compute $r_t^{\text{TS}} = r_t^{\mathcal{D}}$, one needs first compute the market values of debt and liquid assets by discounting the respective cash flows (in this case, book values and market values coincide, given the simplifying assumptions on rates of return). One may then derive $r_t^{\mathcal{D}}$ as $(r^d V_{t-1}^d - r^l V_{t-1}^l)/(V_{t-1}^d - V_{t-1}^l)$ or as $(V_{t+1}^{\mathcal{D}} + F_{t+1}^{\mathcal{D}})/V_t^{\mathcal{D}} - 1$.

With r_t^{TS} in hand, the equity values are straightforward from (6.113):

$$V^e = (41.62, 49.22, 36.20, 20.94, 0).$$

The project is worth undertaking, for $\text{NPV}^e = 41.62 - 30 = 11.62 > 0$.

With these market data, one may alternatively employ the APV method to find the value of operations and then subtract the value of net debt to find the equity value (see Table 6.20). As expected, consistency holds between the two methods.

Owing to the simplicity of the case, the various NPVs are all equal to 11.62, except the non-operating NPV and the debt NPV, which are equal to zero.

Time	0	1	2	3	4
V^d	70.00	40.00	20.00	10.00	0
V^l	0.00	0.49	3.11	5.65	0
$V^{\mathcal{D}}$	70.00	39.51	16.89	4.35	0
$r^{\mathcal{D}}$		5.00%	5.01%	5.18%	6.30%
r^{TS}		5.00%	5.01%	5.18%	6.30%
V^e	41.62	49.22	36.20	20.94	0

NPV^o	= 11.62
NPV	= 11.62
NPV^e	= 11.62
NPV^l	= 0.00
NPV^d	= 0.00
NPV	= 11.62

Table 6.19 IS, BS, and CFS (90% payout)

Time	0	1	2	3	4
Balance Sheet					
Operating assets	100.00	70.00	30.00	10.00	0.00
Liquid assets	0.00	0.49	3.11	5.65	0.00
Investments	100.00	70.49	33.11	15.65	0.00
Debt	70.00	40.00	20.00	10.00	0.00
Equity	30.00	30.49	13.11	5.65	0.00
Financings	100.00	70.49	33.11	15.65	0.00
Income Statement					
EBIT		10.00	10.00	20.00	25.00
− Interest expense		−3.50	−2.00	−1.00	−0.50
+ Interest Income		0.00	+0.02	+0.12	+0.23
=EBT		6.50	8.02	19.12	24.73
−T		−1.63	−2.00	−4.78	−6.18
= NI		4.88	6.01	14.34	18.54
Cash Flow Statement					
CFO	−100.00	38.38	48.00	35.22	28.82
CFD	−70.00	33.50	22.00	11.00	10.50
PD	−30.00	4.88	26.00	24.22	18.32
CFE	−30.00	4.39	23.40	21.80	24.20
CFL	0.00	−0.49	−2.60	−2.42	5.88
CFA	−100.00	37.89	45.40	32.80	34.70
FCF	−100.00	37.50	47.50	35.00	28.75
$\text{CF}\mathcal{D}$	−70.00	33.99	24.60	13.42	4.62

Table 6.20 APV method

APV eq. (6.71)					
Time	0	1	2	3	4
Tax savings ($\tau I^{\mathcal{D}}$)		0.875	0.495	0.219	0.068
V^{TS}	1.527	0.728	0.269	0.064	0
V^{uo}	110.09	88.00	52.82	25.22	
V^{o}	111.62	88.73	53.09	25.28	0
V^{d}	70.00	40.00	20.00	10.00	0
V^{e}	41.62	49.22	36.20	20.94	0
NPV^e	11.62				

If one assumed $r^{\mathrm{TS}} = k^{uo} = 14\%$, the conventional CCF method might be employed and the value of equity might be calculated by discounting the CFOs at 14% and then subtracting the value of net debt (in this case, the equity value would be $V_0^e = 41.43$ and the equity NPV would be $\mathrm{NPV}^e = 11.43$, as can be checked).

Example 6.18 ♦

C^o	C^d
C^l	C^e

(*Textbook WACC*) An analyst must appraise a 4-period project on the basis of the following data:

- the investment cost is \$1000, fully invested in property, plant and equipment ($\mathrm{NOWC}_t = 0$ for every t)
- the project's length is four years
- the estimated stream of incremental EBITs is **EBIT**= (0, 100, 120, 140, 110)
- the tax rate is $\tau = 25\%$
- the cost of equity is $r^e = 15\%$
- the cost of net debt is equal to the RO$\mathcal{D}$: $r^{\mathcal{D}} = i^{\mathcal{D}} = 3\%$
- the (net) debt-to-value ratio (in market values) is kept constant via rebalancing and equal to $\vartheta = 0.4$
- straight-line depreciation is used to depreciate the assets.

We calculate the project's (operating) value with the textbook WACC method.

Deducting the taxes on EBIT, the stream of NOPAT is **NOPAT**= (0, 75, 90, 105, 82.5). The depreciation charge for NFA is $\mathrm{Dep}_t = 1000/4 = 250$. As $\mathrm{NOWC}_t = 0$, then $F_t^{uo} = \mathrm{NOPAT}_t - \Delta\mathrm{NFA}_t = \mathrm{NOPAT}_t + 250 - F_t^{capex}$ with $F_0^{capex} = 1,000$ and $F_t^{capex} = 0$ for $t \in \mathbb{N}_4^1$, whence

$$\boldsymbol{F}^{uo} = (-1{,}000, 325, 340, 355, 332.5).$$

Using Eq. (6.53) with $0.4 = \vartheta = V_t^{\mathcal{D}} / V_t^o$ for $t \in \mathbb{N}_3^0$, one can compute the (constant) WACC:

$$r^{uo} = 0.15 \cdot 0.6 + 0.03(1 - 0.25) \cdot 0.4 = 0.099.$$

The discount factor for FCFs is then $\mathrm{d}^{uo}_{t,0} = 1/1.099^t$, whence

$$\mathbf{d}^{uo} = (1, 0.90992, 0.82795, 0.75337, 0.68550).$$

The discounted sum of the FCFs is equal to the project's (operating) NPV:

$$\begin{aligned}\text{NPV}^o &= -1{,}000 + \overbrace{325 \cdot 0.90992 + 340 \cdot 0.82795 + 355 \cdot 0.75337 + 332.5 \cdot 0.68550}^{V_0^o = 1{,}072.6} \\ &= 72.6.\end{aligned}$$

Since the RO$\mathcal{D}$ is equal to the required return on net debt (i.e., book value of net debt coincides with market value), the net debt NPV is zero and the operating NPV coincides with the equity NPV: Shareholder wealth increases by \$72.6.

Example 6.19 ♦

C^o	C^d
C^l	C^e

(*Textbook WACC—implicit financing policy*) Consider a 20-year project such that

$$F_0^{uo} = -600, \quad F_1^{uo} = 257.8, \quad F_2^{uo} = F_3^{uo} = \ldots = F_{19}^{uo} = 249.8, \quad F_{20}^{uo} = 221.8.$$

We assume that the project is levered, the cost of equity is constant and equal to $r^e = 15\%$, and the cost of net debt is constant as well and equal to $r^{\mathcal{D}} = 8\%$. We also assume that the RO$\mathcal{D}$ is equal to the cost of net debt, $i^{\mathcal{D}} = 8\%$, and the tax rate is $\tau = 0.3$. We assume that the (net) debt will be rebalanced each year so that the leverage remains constant at $\vartheta = 0.25$. This implies that the after-tax WACC is constant:

$$r^{uo} = 0.75 \cdot 0.15 + 0.25 \cdot 0.08(1 - 0.3) = 0.1265.$$

The project's operating value is obtained by discounting the FCFs at the constant WACC: $V_t^o = (V_{t+1}^o + F_{t+1}^{uo})/1.1265$ or, equivalently,

$$V_t^o = \frac{F_{t+1}^{uo}}{1.1265} + \frac{F_{t+2}^{uo}}{1.1265^2} + \ldots + \frac{F_{20}^{uo}}{1.1265^{20-t}}$$

for $t \in \mathbb{N}_{20}^0$ (see Table 6.21). In particular, at time 0, the project's operating value is

$$\begin{aligned}V_0^o &= \frac{257.8}{1.1265} + \frac{249.8}{1.1265^2} + \frac{249.8}{1.1265^3} + \ldots + \frac{249.8}{1.1265^{19}} + \frac{221.8}{1.1265^{20}} \\ &= \frac{257.8}{1.1265} + \frac{249.8}{1.1265} \cdot \frac{1 - (1.1265)^{-18}}{0.1265} + \frac{221.8}{1.1265^{20}} = 1{,}796.5.\end{aligned}$$

The economic value created by the operations is $\text{NPV}^o = 1{,}796.5 - 600 = 1{,}196.5$ which is equal to the equity NPV, since $i^{\mathcal{D}} = r^{\mathcal{D}}$.

The assumption of constant leverage (and constant r^e and $r^{\mathcal{D}}$) brings about an implicit assumption on the financing policy. Specifically, one can unravel the implicit

Table 6.21 Textbook WACC method—unraveling CF$\mathcal{D}$s, CFEs, and CCFs

Time	FCF	V^o	$V^{\mathcal{D}}$	V^e	CF$\mathcal{D}$	CFE	CCF
0	−600.0	1796.5	449.1	1347.4	−449.1	−1347.4	−600.0
1	257.8	1766.0	441.5	1324.5	43.6	225.0	268.5
2	249.8	1739.7	434.9	1304.8	41.9	218.4	260.3
3	249.8	1710.0	427.5	1282.5	42.2	218.0	260.2
4	249.8	1676.6	419.1	1257.4	42.6	217.5	260.0
5	249.8	1638.9	409.7	1229.2	42.9	216.9	259.8
6	249.8	1596.5	399.1	1197.4	43.4	216.2	259.6
7	249.8	1548.7	387.2	1161.5	43.9	215.5	259.3
8	249.8	1494.8	373.7	1121.1	44.4	214.6	259.0
9	249.8	1434.2	358.5	1075.6	45.1	213.7	258.7
10	249.8	1365.9	341.5	1024.4	45.8	212.6	258.4
11	249.8	1288.9	322.2	966.7	46.6	211.4	257.9
12	249.8	1202.2	300.5	901.6	47.5	210.0	257.5
13	249.8	1104.5	276.1	828.4	48.5	208.5	257.0
14	249.8	994.5	248.6	745.9	49.6	206.8	256.4
15	249.8	870.5	217.6	652.9	50.9	204.8	255.7
16	249.8	730.9	182.7	548.2	52.3	202.7	255.0
17	249.8	573.6	143.4	430.2	53.9	200.2	254.1
18	249.8	396.4	99.1	297.3	55.8	197.4	253.2
19	249.8	196.8	49.2	147.6	57.8	194.3	252.1
20	221.8	0.0	0	0.0	53.1	169.8	222.9

CF$\mathcal{D}$s by calculating the market value of net debt as $V_t^{\mathcal{D}} = \vartheta V_{t-1}^o = 0.25 V_{t-1}^o$ and then derive the CF$\mathcal{D}$s as $F_t^{\mathcal{D}} = V_{t-1}^{\mathcal{D}}(1+r^{\mathcal{D}}) - V_t^{\mathcal{D}}$. One may also compute the value of equity as $V_t^e = (1-\vartheta)V_{t-1}^o = 0.75 V_{t-1}^o$ and then derive the CFEs as $F_t^e = V_{t-1}^e(1+r^e) - V_t^e$. In Table 6.21, we report the market values and the cash flows.

6.6 Valuing Projects with Incomes

In the previous sections we have used cash flows for appraising a project and making decisions. In this section, we retrieve the notion of normal profit, encountered in Sect. 5.4, and build the notion of *residual income*. However, we now use a different benchmark system, and, therefore, a different conceptualization of a 'normal' profit. Specifically, the new benchmark system is based on replication of *capitals* (instead of prospective cash flows). The benchmark transposed Matrix will be $(\boldsymbol{C}^*, \boldsymbol{I}^*, \boldsymbol{F}^*)$ where $C_t^* = C_t$ for every t and $I_t^* = r_t C_{t-1}$ for $t \in \mathbb{N}_n^1$ ($I_0^* = 0$).

We show that a project can be appraised with residual incomes, which implies that the notions of income and capital suffice to compute value creation and make rational decisions.

6.6.1 Residual Income

The normal profit is the (counterfactual) profit that would be generated if investors invested in the market. In the cash-flow valuation methods, this means that the normal profit is $I_t^V = r_t V_{t-1}$, which is the profit that an investor would earn by investing V_{t-1} at the rate r_t (see Sect. 5.2). However, consider now an alternative interpretation. Suppose an investor invests the amount C_{t-1} in the market for $t \in \mathbb{N}_n^1$. This means that the benchmark system is different: The normal profit is now obtained by multiplying the market rate and the capital C_{t-1}. Let us denote this normal profit as $I_t^* = r_t C_{t-1}$. For example, consider a one-period project with initial investment equal to \$100 and suppose that the ROC is 10%; the project return is $0.1 \cdot \$100 = \$10$. Assuming a COC equal to 8%, the normal profit (opportunity cost) is $0.08 \cdot \$100 = \$8$ and the excess profit is $\$10 - \$8 = \$2$.[38]

Formally, consider a project P, described by the transposed Matrix $(\boldsymbol{C}, \boldsymbol{I}, \boldsymbol{F})$:

$(\boldsymbol{C}, \boldsymbol{I}, \boldsymbol{F})$

Time		0	1	2	...	$n-1$	n
Capital	$\boldsymbol{C}$	C_0	C_1	C_2	...	C_{n-1}	0
Income	$\boldsymbol{I}$	I_0	I_1	I_2	...	I_{n-1}	I_n
Cash Flow	$\boldsymbol{F}$	F_0	F_1	F_2	...	F_{n-1}	F_n

Consider a counterfactual scenario where investors invest in a benchmark portfolio, traded in the market, which replicates the capitals of the project. This portfolio acts as a benchmark system whose transposed Matrix is $(\boldsymbol{C}^*, \boldsymbol{I}^*, \boldsymbol{F}^*)$, where $\boldsymbol{C}^* = \boldsymbol{C}$ and $\boldsymbol{I}^*$ and $\boldsymbol{F}^*$ denote the corresponding vector of (normal) profits and cash flows. One may use the following concise transposed Matrix to describe the benchmark portfolio:

$(\boldsymbol{C}^*, \boldsymbol{I}^*, \boldsymbol{F}^*)$

Time		0	1	2	...	$n-1$	n
Capital	$\boldsymbol{C}^*$	C_0	C_1	C_2	...	C_{n-1}	0
Income	$\boldsymbol{I}^*$	0	I_1^*	I_2^*	...	I_{n-1}^*	I_n^*
Cash Flow	$\boldsymbol{F}^*$	$-C_0$	F_1^*	F_2^*	...	F_{n-1}^*	F_n

This system acts as a benchmark system, as opposed to the project system.

[38] In Example 5.8, the \$12 is just the project's normal profit under this perspective (and \$3 is the residual income).

This means that investors invest in the benchmark portfolio the same amounts C_{t-1} that are invested in project P. Consider, side by side, the two scenarios:

(a) for every $t \in \mathbb{N}_n^1$, firm f invests capital C_{t-1} in P (project scenario)
(b) for every $t \in \mathbb{N}_n^1$, firm f invests C_{t-1} in the benchmark portfolio (benchmark scenario).

The two scenarios give rise to two alternative dynamical systems, a factual one and a counterfactual one:

$$\begin{aligned} &\text{factual scenario (project system)} &(6.123)\\ &C_t = C_{t-1} + I_t - F_t = C_{t-1}(1+i_t) - F_t \\ &\text{counterfactual scenario (benchmark system)} &(6.124)\\ &C_t = C_{t-1} + I_t^* - F_t^* = C_{t-1}(1+r_t) - F_t^* \end{aligned}$$

where $F_t^* = C_{t-1}(1+r_t) - C_t$ denotes the cash flow withdrawn from (or deposited in) the benchmark portfolio (with $F_0^* = -C_0$).[39]

In this way, the counterfactual scenario consists of a portfolio that replicates the capital vector of the project, $\boldsymbol{C}^* = (C_0, C_1, \ldots, C_{n-1})$, while generating a vector of normal cash flows equal to $\boldsymbol{F}^* = (F_0^*, F_1^*, \ldots, F_n^*)$ and a vector of normal profits equal to $\boldsymbol{I}^* = (I_0^*, I_1^*, \ldots, I_n^*)$ with $I_0^* = 0$.

Definition 6.2 (*Residual income*) Residual income (RI) or excess profit is the difference between actual profit and normal profit: $I_t - I_t^* = I_t - r_t C_{t-1}$.

[39]The equality $F_t^* = C_{t-1}(1+r_t) - C_t$ logically derives from the assumption that the benchmark asset replicates the project's capital amounts: $C_t = C_t^*$. One may also write $F_t^* = I_t^* - (C_t - C_{t-1}) = [\text{using (6.123)}] = I_t^* - (I_t - F_t)$.

> *The term "residual income" was apparently coined by the General Electric Company, and adopted first in the literature by Solomons (*1965, *p. 63), although the same concept, differently labeled, was studied even earlier (e.g.* Preinreich 1936, 1938; Edwards and Bell 1961; Bodenhorn 1964*). The contributions of Peasnell (*1981, 1982a, b) *and* Ohlson (1989, 1995) *have caused a renewed interest in this notion among corporate finance and accounting scholars, with particular regard to firm valuation, performance measurement, incentive compensation (value-based management). A large number of theoretical and applied studies have appeared in both applied finance and accounting [e.g.,* Stewart 1991; Ohlson 1995; Stern et al. 1995; Feltham and Ohlson 1996; Rappaport 1998; Lundholm and O'Keefe 2001; Young and O'Byrne 2001; Martin et al. 2003; Weaver and Weston 2003; O'Byrne and Young 2006*; see also* Dunbar 2013*] and a large number of textbooks and professional publications in corporate finance, managerial finance and accounting directly deal with the topic [e.g.* Brealey and Myers 2000; Copeland et al. 2000; Palepu et al. 2000; Grinblatt and Titman 2002; Revsine et al. 2005; Arnold 2005; Brealey et al. 2011*].*
>
> Magni (2012). In search of the "lost capital". A theory for valuation, investment decisions, performance measurement. *Frontiers in Finance and Economics*, 9(1), 87–146 (p. 88)

We use the symbol RI_t to denote the residual income. It represents the profit over and above the normal profit. The conceptualization of RI with the Matrix is visually intuitive. Consider the project Matrix and the benchmark Matrix:

$$
\begin{array}{c}
\textbf{Project}\\
\begin{array}{|c|cc|}
\hline
+\Delta C_t^o & +I_t^o & -F_t^o\\
+\Delta C_t^l & +I_t^l & -F_t^l\\
\hline
+\Delta C_t^d & +I_t^d & -F_t^d\\
+\Delta C_t^e & +I_t^e & -F_t^e\\
\hline
\end{array}
\end{array}
\quad
\begin{array}{c}
\textbf{Benchmark}\\
\begin{array}{|c|cc|}
\hline
+\Delta C_t^o & +I_t^{o*} & -F_t^{o*}\\
+\Delta C_t^l & +I_t^{l*} & -F_t^{l*}\\
\hline
+\Delta C_t^d & +I_t^{d*} & -F_t^{d*}\\
+\Delta C_t^e & +I_t^{e*} & -F_t^{e*}\\
\hline
\end{array}
\end{array}
\tag{6.125}
$$

The first columns of the Matrices coincide; hence, the right partitions of the two Matrices coincide as well. Which implies that the difference between incomes is equal to the difference between cash flows:

$$I_t - F_t = I_t^* - F_t^* \implies I_t - I_t^* = F_t - F_t^* \tag{6.126}$$

or, equivalently,

$$I_t - I_t^* = \overbrace{(I_t - \Delta C_t)}^{F_t} - \overbrace{(I_t^* - \Delta C_t)}^{F_t^*}. \tag{6.127}$$

Therefore, residual income RI_t signals how larger/smaller is the income or, equivalently, how larger/smaller is the cash flow if capital is invested in the project rather than in the replicating portfolio.

Overall, cash-flow stream $\boldsymbol{F}$ can be viewed as a portfolio of $\boldsymbol{F}^*$ and the incremental vector $\mathbf{RI} = \Delta\boldsymbol{F}^*$ such that

$$\begin{aligned}\Delta\boldsymbol{F}^* &= \boldsymbol{F} - \boldsymbol{F}^* = (F_0 - F_0^*, F_1 - F_1^*, F_2 - F_2^*, \ldots, F_n - F_n^*)\\ &= (\text{RI}_0, \text{RI}_1, \text{RI}_2, \ldots, \text{RI}_n).\end{aligned} \tag{6.128}$$

(If $I_0 = 0$, then $\text{RI}_0 = 0$; otherwise, $\text{RI}_0 = I_0$ because $I_0^* = 0$.) Analogously, the income stream $\boldsymbol{I}$ can be viewed as a portfolio of $\boldsymbol{I}^*$ and the incremental vector $\mathbf{RI} = \Delta\boldsymbol{I}^*$ such that

$$\Delta\boldsymbol{I}^* = \boldsymbol{I} - \boldsymbol{I}^* = (I_0 - I_0^*, I_1 - I_1^*, I_2 - I_2^*, \ldots, I_n - I_n^*) = (\text{RI}_0, \text{RI}_1, \text{RI}_2, \ldots, \text{RI}_n). \tag{6.129}$$

Hence, the RI is, at the same time, an income measure (excess profit) and a cash-flow measure (excess cash flow).

> The notion of residual income is, intrinsically, a notion of residual cash flow as well. It is a cash-flow measure as much as it is an income measure.

In general, we have as many expressions for residual incomes as are the notions of income (and cash flows); to any notion there corresponds a normal profit and, therefore, a residual income. Specifically, we have analyzed several types of incomes (and cash flows), to each of which there corresponds a RI:

- project's RI, associated with project income (I_t) and CFA
- operating RI, associated with operating income (I_t^o) and CFO
- unlevered operating RI, associated with NOPAT (I_t^{uo}) and FCF
- unlevered RI, associated with financially unlevered income (I_t^u) and asset FCF
- non-operating RI, associated with interest income (I_t^l) and CFL
- equity RI, associated with net income (I_t^e) and CFE
- debt RI, associated with interest expenses (I_t^d) and CFD
- net-debt RI, associated with net interest ($I_t^{\mathcal{D}}$) and CF$\mathcal{D}$.[40]

Let $I_t^{j*} = r_t^j C_{t-1}^j$, denote the normal profits for category j, with $j = o, u, uo, l, e, d, \mathcal{D}$. One may formally define the following RIs:

[40] In principle, other RIs may be built. We will not pursue all the possibilities for constructing RIs and will confine ourselves to (a subset of) the above mentioned RIs.

$$\begin{aligned}
\mathrm{RI}_t &= I_t - I_t^* &&= I_t - r_t C_{t-1} && (6.130)\\
\mathrm{RI}_t^o &= I_t^o - I_t^{o*} &&= I_t^o - r_t^o C_{t-1}^o && (6.131)\\
\mathrm{RI}_t^u &= I_t^u - I_t^{u*} &&= I_t^u - r_t^u C_{t-1} && (6.132)\\
\mathrm{RI}_t^{uo} &= I_t^{uo} - I_t^{uo*} &&= I_t^{uo} - r_t^{uo} C_{t-1}^o && (6.133)\\
\mathrm{RI}_t^l &= I_t^l - I_t^{l*} &&= I_t^l - r_t^l C_{t-1}^l && (6.134)\\
\mathrm{RI}_t^e &= I_t^e - I_t^{e*} &&= I_t^e - r_t^e C_{t-1}^e && (6.135)\\
\mathrm{RI}_t^d &= I_t^d - I_t^{d*} &&= I_t^d - r_t^d C_{t-1}^d && (6.136)\\
\mathrm{RI}_t^{\mathcal{D}} &= I_t^{\mathcal{D}} - I_t^{\mathcal{D}*} &&= I_t^{\mathcal{D}} - r_t^{\mathcal{D}} C_{t-1}^{\mathcal{D}} && (6.137)
\end{aligned}$$

with obvious meaning of the symbols. For example, RI_t^e is the equity RI which measures the excess profit and the excess cash flow of equityholders over and above the normal equity profit, I^{e*}, and the normal equity cash flows, F_t^{e*}, respectively.

Using $I_t^{j*} = r_t^j C_{t-1}^j$, one may also write

$$\begin{aligned}
\mathrm{RI}_t &= C_{t-1}(i_t - r_t) && (6.138)\\
\mathrm{RI}_t^o &= C_{t-1}^o(i_t^o - r_t^o) && (6.139)\\
\mathrm{RI}_t^u &= C_{t-1}(i_t^u - r_t^u) && (6.140)\\
\mathrm{RI}_t^{uo} &= C_{t-1}^o(i_t^{uo} - r_t^{uo}) && (6.141)\\
\mathrm{RI}_t^l &= C_{t-1}^l(i_t^l - r_t^l) && (6.142)\\
\mathrm{RI}_t^e &= C_{t-1}^e(i_t^e - r_t^e) && (6.143)\\
\mathrm{RI}_t^d &= C_{t-1}^d(i_t^d - r_t^d) && (6.144)\\
\mathrm{RI}_t^{\mathcal{D}} &= C_{t-1}^{\mathcal{D}}(i_t^{\mathcal{D}} - r_t^{\mathcal{D}}) && (6.145)
\end{aligned}$$

Additivity holds for the RI notion. For example,

$$\begin{aligned}
\mathrm{RI}_t &= \mathrm{RI}_t^e + \mathrm{RI}_t^d\\
\mathrm{RI}_t &= \mathrm{RI}_t^o + \mathrm{RI}_t^l\\
\mathrm{RI}_t^o &= \mathrm{RI}_t^e + \mathrm{RI}_t^{\mathcal{D}},
\end{aligned}$$

as the reader can easily check.

Example 6.20 ♦

Consider a project and let the net income be equal to $I_t^e = 100$, and let $C_{t-1}^e = 60$ be the BOP equity capital. The project is levered and the debt capital is $C_{t-1}^d = 40$.

Assuming the ROD is $i_t^d = 7.5\%$, the cost of debt is $r_t^d = 5\%$, and the cost of equity is $r_t^e = 10\%$, the equity RI and the debt RI are, respectively,

$$\text{RI}_t^e = 100 - 0.1 \cdot 60 = 94$$
$$\text{RI}_t^d = 40 \cdot (0.075 - 0.05) = 1.$$

The equity RI is positive, meaning that, focusing on the t-th period, equityholders earn \$94 more than they would earn if they invested the same capital C_{t-1}^e in a portfolio (equivalent in risk to the project's equity) with expected return equal to 10%. The debt RI is positive, meaning that debtholders earn \$1 more than they would earn if they invested in the market at the equivalently-risky 5% instead of lending money to the firm at 7.5%. The project RI is

$$\text{RI}_t = \text{RI}_t^e + \text{RI}_t^d = 94 + 1 = 95$$

meaning that, overall, investors earn \$95 more than they would earn if they invested in the market in an equivalent-risk portfolio of assets.

6.6.2 *Residual Income Valuation*

The RI captures above-normal profits. Intuition seems to suggest that the sum of the excess profits could capture value creation. The corresponding decision criterion for accept-reject decisions is then as follows.

Definition 6.3 (*Residual income criterion*) Let $V_0(\text{RI}) = \sum_{j=0}^{n} \text{RI}_j \text{d}_{j,0}$ be the market value of $\mathbf{RI} = (\text{RI}_0, \text{RI}_1, \text{RI}_2, \ldots, \text{RI}_n)$ at time 0. The project is worth undertaking if and only if $V_0(\text{RI}) > 0$.

The proposition above implicitly assumes $C_0^d = V_0^d$ (which implies $V_0(\text{RI}) = V_0(\text{RI}^e) > 0$). If this does not hold, then the RI condition for acceptance becomes $V_0(\text{RI}^e) > 0$, where $V_0(\text{RI}^e) = \sum_{j=0}^{n} \text{RI}_j^e \text{d}_{j,0}^e$ is the market value of $\mathbf{RI}^e = (\text{RI}_0^e, \text{RI}_1^e, \text{RI}_2^e, \ldots, \text{RI}_n^e)$ at time 0.

To prove the rationality of the RI criterion, it suffices to prove consistency with the NPV. Manipulating algebraically and remembering that $C_{-1} = 0$,

$$
\begin{aligned}
V_0(\mathrm{RI}) &= \sum_{t=0}^{n} \mathrm{RI}_t \cdot \mathrm{d}_{t,0} \\
&= \sum_{t=0}^{n} (I_t - r_t \cdot C_{t-1}) \cdot \mathrm{d}_{t,0} \\
&= \sum_{t=0}^{n} (C_t + F_t - C_{t-1}(1 + r_t)) \cdot \mathrm{d}_{t,0} \\
&= \sum_{t=0}^{n} \left(-C_{t-1} + \frac{C_t + F_t}{1 + r_t} \right) \mathrm{d}_{t-1,0} \\
&= -\sum_{t=0}^{n} C_{t-1}\mathrm{d}_{t-1,0} + \sum_{t=0}^{n} C_t\mathrm{d}_{t,0} + \sum_{t=0}^{n} F_t\mathrm{d}_{t,0} \\
&= C_n\mathrm{d}_{n,0} + \sum_{t=0}^{n} F_t\mathrm{d}_{t,0} \\
&= \mathrm{NPV}.
\end{aligned}
\tag{6.146}
$$

(See also Sect. 10.6.) Hence, $V_0(\mathrm{RI}) > 0$ if and only if $\mathrm{NPV} > 0$. Therefore, remembering Definition 5.3 and Proposition 5.2, we may state the following proposition.

Proposition 6.5 (Rationality of RI criterion) *The RI criterion is economically rational.*

The above proposition shows that, in overall terms, the market value of RI supplies a precise measure of economic profitability and, compellingly, shows that the NPV notion is an overall RI. Additivity then holds in the following sense:

$$V_0^j(\mathrm{RI}) + V_0^k(\mathrm{RI}) = \mathrm{NPV}^j + \mathrm{NPV}^k = \mathrm{NPV}^{j+k} = V_0^{j+k}(\mathrm{RI}).$$

More generally, $V_t^j(\mathrm{RI}) + V_t^k(\mathrm{RI}) = V_t^{j+k}(\mathrm{RI})$ for every t.

It also confirms that, in a capital asset investment, **incomes and capitals are sufficient to value a project**. This should not be deemed surprising: The cash flows are second-order variables, they are derived from the estimates of incomes and capitals, so it is rather intuitive to expect that incomes and capitals provide a reliable valuation of a project.

Equation (6.146) holds for any type of RI. Therefore,

$$\text{NPV} = V_0(\text{RI}) = \sum_{t=0}^{n} \text{RI}_t \cdot \text{d}_{t,0} \tag{6.147}$$

$$\text{NPV}^o = V_0^o(\text{RI}) = \sum_{t=0}^{n} \text{RI}_t^o \cdot \text{d}_{t,0}^o \tag{6.148}$$

$$\text{NPV} = V_0^u(\text{RI}) = \sum_{t=0}^{n} \text{RI}_t^u \cdot \text{d}_{t,0}^u \tag{6.149}$$

$$\text{NPV}^o = V_0^{uo}(\text{RI}) = \sum_{t=0}^{n} \text{RI}_t^{uo} \cdot \text{d}_{t,0}^{uo} \tag{6.150}$$

$$\text{NPV}^l = V_0^l(\text{RI}) = \sum_{t=0}^{n} \text{RI}_t^l \cdot \text{d}_{t,0}^l \tag{6.151}$$

$$\text{NPV}^e = V_0^e(\text{RI}) = \sum_{t=0}^{n} \text{RI}_t^e \cdot \text{d}_{t,0}^e \tag{6.152}$$

$$\text{NPV}^d = V_0^d(\text{RI}) = \sum_{t=0}^{n} \text{RI}_t^d \cdot \text{d}_{t,0}^d \tag{6.153}$$

$$\text{NPV}^{\mathcal{D}} = V_0^{\mathcal{D}}(\text{RI}) = \sum_{t=0}^{n} \text{RI}_t^{\mathcal{D}} \cdot \text{d}_{t,0}^{\mathcal{D}}. \tag{6.154}$$

Note that (6.148) and (6.150) coincide, notwithstanding the fact that $\text{RI}_t^o \neq \text{RI}_t^{uo}$ (the same holds for (6.147) and (6.149)).[41]

If $I_0 = 0$, the NPV is a difference between a value and a cost. Therefore, we can now state the following corollary.

[41] The reader should not confuse $V_0^{uo}(\text{RI})$ with V_0^{uo}:

$$V_0^{uo} = \sum_{t=1}^{n} F_t^{uo}[(1+k_1^{uo})(1+k_2^{uo})\cdot \ldots \cdot(1+k_t^{uo})]^{-1}$$

$$V_0^{uo}(\text{RI}) = \sum_{t=0}^{n} \text{RI}_t^{uo}[(1+r_1^{uo})(1+r_2^{uo})\cdot \ldots \cdot(1+r_t^{uo})]^{-1}$$

The former is the discounted sum of FCFs, supplying the project's unlevered operating value; the latter is the discounted sum of RI^{uo}'s, supplying the project's levered operating NPV. Analogously, adopting an asset perspective, no confusion should be made between V_0^u and $V_0^u(\text{RI})$:

$$V_0^{u} = \sum_{t=1}^{n} F_t^{u}[(1+k_1^{u})(1+k_2^{u})\cdot \ldots \cdot(1+k_t^{u})]^{-1}$$

$$V_0^{u}(\text{RI}) = \sum_{t=0}^{n} \text{RI}_t^{u}[(1+r_1^{u})(1+r_2^{u})\cdot \ldots \cdot(1+r_t^{u})]^{-1}.$$

Corollary 6.1 *Consider an economic asset* $(\boldsymbol{C}, \boldsymbol{I}, \boldsymbol{F})$ *and assume* $I_0 = 0$. *Its market value can be found as the sum of initial book value and the market value at time 0 of the associated RI vector:*

$$\begin{aligned}
V_0 &= C_0 + V_0(\mathrm{RI}) \\
V_0^o &= C_0^o + V_0^o(\mathrm{RI}) \\
V_0 &= C_0 + V_0^u(\mathrm{RI}) \\
V_0^o &= C_0^o + V_0^{uo}(\mathrm{RI}) \\
V_0^l &= C_0^l + V_0^l(\mathrm{RI}) \\
V_0^e &= C_0^e + V_0^e(\mathrm{RI}) \\
V_0^d &= C_0^d + V_0^d(\mathrm{RI}) \\
V_0^{\mathcal{D}} &= C_0^{\mathcal{D}} + V_0^{\mathcal{D}}(\mathrm{RI}).
\end{aligned}$$

(If $I_0 \neq 0$, *the same relations hold, with* $-F_0 = C_0 - I_0$ *replacing* C_0.*)*

Example 6.21 ♦

C^o	C^d
	C^e

Suppose firm Res Inc. has the opportunity of investing \$80 in a capital asset investment. The project is summarized in its basic constituents below:[42]

$(\boldsymbol{C}, \boldsymbol{I}, \boldsymbol{F})$ Res Inc. - Project

Time		0	1	2	3	4	5
Capital							
Assets	$\boldsymbol{C}$	80	75	−5	65	35	0
Income							
Income from assets	$\boldsymbol{I}$	0	50	−10	40	60	30
Cash Flow							
Cash flow from assets	$\boldsymbol{F}$	−80	55	70	−30	90	65

We assume $C_t^l = 0$ for every t so that $I_t^o = I_t$ and $r_t^o = r_t$. From the forecasted capital amounts the vector I^* of normal profits is derived by assuming a constant pre-tax WACC equal to $r = 30\%$. As $I_t^* = 0.3 \cdot C_{t-1}$ for $t \in \mathbb{N}_5^1$,

[42] The BOP capital at time 2 is negative, while income in $t = 3$ is positive: The firm is subtracting funds from the project (borrowing position) and making money out of the borrowing (see Chaps. 1 and 2).

$$
\begin{aligned}
I_0^* &= 0 &&= 0\\
I_1^* &= 0.3 \cdot 80 &&= 24\\
I_2^* &= 0.3 \cdot 75 &&= 22.5\\
I_3^* &= 0.3 \cdot (-5) &&= -1.5\\
I_4^* &= 0.3 \cdot 65 &&= 19.5\\
I_5^* &= 0.3 \cdot 35 &&= 10.5.
\end{aligned}
$$

Hence, the normal cash flows are calculated as $F_t^* = I_t^* - \Delta C_t$:

$$
\begin{aligned}
F_0^* &= 0 - 80 &&= -80\\
F_1^* &= 24 - (-5) &&= 29\\
F_2^* &= 22.5 - (-80) &&= 102.5\\
F_3^* &= -1.5 - 70 &&= -71.5\\
F_4^* &= 19.5 - (-30) &&= 49.5\\
F_5^* &= 10.5 - (-35) &&= 45.5.
\end{aligned}
$$

We may then summarize the replicating course of action as follows:

(C^*, I^*, F^*) Res Inc. - Replicating Portfolio

Time		0	1	2	3	4	5
Capital							
Invested capital	C^*	80	75	−5	65	35	0
Income							
Normal profit	I^*	0	24.0	22.5	−1.5	19.5	10.5
Cash Flow							
Normal cash flow	F^*	−80	29.0	102.5	−71.5	49.5	45.5

The excess profits may be computed either as differential cash flows or as differential incomes. Using incomes,

$$
\begin{aligned}
\mathrm{RI}_0 &= 0 &&= 0\\
\mathrm{RI}_1 &= 50 - 24 &&= 26\\
\mathrm{RI}_2 &= -10 - 22.5 &&= -32.5\\
\mathrm{RI}_3 &= 40 - (-1.5) &&= 41.5\\
\mathrm{RI}_4 &= 60 - 19.5 &&= 40.5\\
\mathrm{RI}_5 &= 30 - 10.5 &&= 19.5.
\end{aligned}
$$

Discounting these excess profits at 30%, one gets the overall residual income:

$$V_0(\text{RI}) = \frac{26}{1.3} + \frac{-32.5}{(1.3)^2} + \frac{41.5}{(1.3)^3} + \frac{40.5}{(1.3)^4} + \frac{19.5}{(1.3)^5} = 39.1.$$

This is the project NPV. To verify it, let us discount the CFAs at 30%:

$$\text{NPV} = -80 + \frac{55}{1.3} + \frac{70}{(1.3)^2} + \frac{-30}{(1.3)^3} + \frac{90}{(1.3)^4} + \frac{65}{(1.3)^5} = 39.1.$$

Example 6.22 ♦

Suppose a firm is incorporated to undertake a five-year project. The estimated operating capital for each period is collected in the following vector: $\boldsymbol{C}^o = (150, 135, 90, 70, 60, 0)$. The vector of prospective unlevered ROCs is $\boldsymbol{i}^{uo} = (3\%, -2\%, 5\%, 10\%, -3\%)$. The structure of WACCs is $\boldsymbol{r}^{uo} = (2\%, 4\%, 5\%, 0\%, 2\%)$. This implies the following structure of residual incomes:

$$\begin{aligned}
\text{RI}_1^{uo} &= 150 \cdot (0.03 - 0.02) &&= 1.5\\
\text{RI}_2^{uo} &= 135 \cdot (-0.02 - 0.04) &&= -8.1\\
\text{RI}_3^{uo} &= 90 \cdot (0.05 - 0.05) &&= 0\\
\text{RI}_4^{uo} &= 70 \cdot (0.1 - 0) &&= 7\\
\text{RI}_5^{uo} &= 60 \cdot (-0.03 - 0.02) &&= -3
\end{aligned}$$

Using $\boldsymbol{r}^{uo}$ above, the profile of the discount factors is computed: $\text{d}_{1,0}^{uo} = 0.9804$, $\text{d}_{2,0}^{uo} = 0.9427$, $\text{d}_{3,0}^{uo} = 0.8978$, $\text{d}_{4,0}^{uo} = 0.8978$, $\text{d}_{5,0}^{uo} = 0.8802$, so the project NPV is

$$1.5 \cdot 0.9804 - 8.1 \cdot 0.9427 + 0 \cdot 0.8978 + 7 \cdot 0.8978 - 3 \cdot 0.8802 = -2.52.$$

Note that it is unnecessary to know the FCFs in order to compute the NPV: Incomes and capitals suffice. As anticipated, this should not come as a surprise, for a project is a triplet of incomes, capitals, and cash flows (and cash flows are derived from incomes and capitals). However, these can be easily unraveled via the usual law of motion, $F_t^{uo} = I_t^{uo} - \Delta C_t^o$:

$$\begin{aligned}
F_0^{uo} &= -150\\
F_1^{uo} &= 0.03 \cdot 150 - (135 - 150) = 19.5\\
F_2^{uo} &= -0.02 \cdot 135 - (90 - 135) = 42.3\\
F_3^{uo} &= 0.05 \cdot 90 - (70 - 90) = 24.5\\
F_4^{uo} &= 0.1 \cdot 70 - (60 - 70) = 17\\
F_5^{uo} &= -0.03 \cdot 60 - (0 - 60) = 58.2.
\end{aligned} \tag{6.155}$$

Example 6.23 ♦
Consider Example 6.21 above. To verify that the replicating portfolio is an equilibrium asset, let us discount the normal cash flows at the pre-tax WACC, $r = 30\%$:

$$-80 + \frac{29}{1.3} + \frac{102.5}{(1.3)^2} + \frac{-71.5}{(1.3)^3} + \frac{49.5}{(1.3)^4} + \frac{45.5}{(1.3)^5} = 0.$$

The NPV of $\boldsymbol{F}^*$ is zero, the portfolio is value-neutral and, therefore, it is an equilibrium asset. It then acts as a benchmark for the project. As seen, the investors earn \$39.1 more than they would if they invested in this benchmark portfolio.

Remark 6.10 (*Brief history of Residual Income*)[43] The early roots of the notion of RI date back to Hamilton (1777). He defined it as an "excess of gross profits above the interest of his stock . . . if the profit of his trade be less than his stock would have yielded at common interest, he may properly account it a losing one" (Hamilton 1777, vol. II, p. 246, as quoted in Arnold 2000, p. 14. Also quoted in Mepham 1980, p. 183). At the end of the 19th century, the concept gained some popularity, also due to Marshall (1890) (presumably inspired by Hamilton), and was used for valuation purposes. In particular, Carsberg (1966) reported that excess profits were sometimes discounted, rather than cash flows (e.g. Dicksee 1897) and stressed Leake's (1921) contribution to valuation of *Goodwill* (NPV), obtained by discounting the residual incomes, conceptualized as surplus of profit over a *normal* return on capital. In the early years of the 20th century the notion of normal profit was accepted in professional practice, and various labels were adopted including "reasonable", "fair", "equitable" return (Sloan 1929. See also Goetzmann and Garstka 1999). In later years, Preinreich (1936, 1937, 1938) made the link between residual incomes and NPV more explicit:

> Goodwill is commonly obtained by discounting 'excess earnings'. If the original investment (C) is added to the goodwill the same capital value results as from the discounting of 'services' [cash flows] (Preinreich 1936, p. 131).

> the discounted excess profits plus the recorded value will always give the true fair market value . . . This statement is a simple theorem of arithmetic (Preinreich 1937, p. 220)

> capital value equals the book value, plus the discounted excess profits (Preinreich 1938, p. 240).

Edey (1957) showed the link with some algebra. He assumed a constant perpetual cash flow $F_t = F$, which implies zero change in capital ($C_t = C_{t-1} = C$) so that $F_t = I_t = I$. Therefore, $\mathrm{RI}_t = \mathrm{RI}$ and

$$C + \sum_{t=1}^{\infty} \mathrm{RI}(1+r)^{-t} = C + \frac{\mathrm{RI}}{r} = C + \frac{I - rC}{r} = \frac{I}{r} = V_0.$$

A formal treatment in a finite-time setting was later carried out by Edwards and Bell (1961). In Appendix B of their book, the authors defined the *excess realizable profit* as $C_t + F_t - (1+r)C_{t-1}$, where the capital C_t is valued on the basis of

[43]Adapted from Magni CA (2009c), Splitting up value: A critical review of residual income theories, *European Journal of Operational Research*, 198(1) (October), 1–22, (p. 5).

Table 6.22 Expressions for 'residual income'[a]

Abnormal earnings	(Ohlson 1989, 1995; Feltham and Ohlson 1995; Francis, Olsson and Oswald 2000; Yee 2005; Revsine, Collins and Johnson 2005)
Abnormal accounting earnings	(Bao and Bao 1998)
Abnormal economic earnings	(Bao and Bao 1998)
Abnormal operating earnings	(Feltham and Ohlson 1995)
Abnormal gain	(Grant 1998)
Abnormal profit	(Bromwich and Walker 1998)
Accounting residual income	(Keef and Roush 2001a)
Adjusted income	(Peasnell 1981, 1982a, b)
Adjusted profit	(Carsberg 1966)
Economic income	(Grant 1998)
Economic profit	(Archer and D'Ambrosio 1972; Boadway and Bruce, 1984; Rao, 1992; Ehrbar 1998; Grant 1998; Kimball 1998; Martin and Petty, 2000; Fabozzi and Grant 2000; Arnold 2000, 2005, 2007; Shrieves and Wachowicz 2001; Weaver and Weston 2003; Magni 2008b)
Economic Rent	(Cnossen 1998; Ehrbar 1998)
Economic Value Added	(Stewart 1991; O'Byrne 1996; Rogerson, 1997; O'Hanlon and Peasnell 1998, 2002; Ehrbar 1998; Hartman 2000; Magni 2000a,b,c; 2001a,b; 2003, 2004, 2005, 2006; Ghiselli Ricci and Magni 2006; Stoughton and Zechner 2007)
Excess current income	(Edwards 1980)
Excess earnings	(Preinreich 1936; Pratt, Reilly and Schweihs 1996)
Excess economic income	(Pratt, Reilly and Schweihs 1996)
Excess income	(Peasnell 1982a)
Excess profit	(Preinreich 1937, 1938; Edwards and Bell 1961; Magni, 2001b, 2004, 2006)
Excess realizable profit	(Edwards and Bell 1961)
Excess return	(Rendleman 1978; Damodaran 2006a, b)
Financial margin	(Peccati 1987, 1989, 1991)
Marginal return	(Lohmann 1988)
Period contribution	(Peccati 1989)
Period margin	(Gallo and Peccati 1993; Pressacco and Stucchi 1997)
Profit	(Scitovszky 1943, Anthony 1975; Reynolds 1963)
Pure earnings	(Bodenhorn 1964; Boadway and Bruce 1984; Cnossen 1998)
Pure profit	(Machlap 1942; Bodenhorn 1964; Boadway and Bruce 1984; Cnossen 1998)
Quasi-rent	(Peasnell 1981)
Residual earnings	(Ohlson 2003; Pope and Wang 2003; Penman 2010)
Residual (capital) return	(Grant 1998)
Residual income	(Scitovszky 1943, Solomons 1965; Amey 1969, 1975; Tomkins 1975a,b; Emmanuel and Otley 1976; Kay 1976; Mepham 1980; Peasnell 1981, 1982a, b; Ohlson 1989, 1995; Grant 1996; Bromwich and Walker 1998; Ehrbar, 1998; Biddle, Bowen and Wallace 1999; Lee 1999; Arnold 2000; Martin and Petty 2000; Magni 2000a, 2003; Lundholm and O'Keefe 2001; Martin, Petty and Rich 2003; Mowen and Hansen 2008; Brealey, Myers and Allen 2011; Ghiselli Ricci and Magni 2014; O'Hanlon and Peasnell 2014)
Residual operating income	Penman (2010)
Super-profit	(Leake 1921; Edey 1957; Samuels, Wilkes and Brayshaw 1990; Bromwich and Walker 1998; Fernández 2002)
Supernormal profit	(Begg, Fischer, and Dornbusch 1984)
Surplus profit	(Dicksee 1897; Carsberg 1966)
Value added	(Tham and Vélez-Pareja 2004)

[a]Adapted from Magni CA (2009). Splitting up value: A critical review of residual income theories. *European Journal of Operational Research*, 198(1) (October), 1–22 (p. 5)

replacement cost. The authors discounted the stream of excess realizable profits, $\sum_{t=1}^{n}(C_t + F_t - (1+r)C_{t-1})/(1+r)^t$, and, after simple algebraic manipulations, were able to show that it is equal to the NPV of the expected stream of cash flows ('subjective goodwill' in the authors' words): $\text{NPV} = \sum_{t=1}^{n} F_t(1+r)^{-t} - C_0 = \sum_{t=1}^{n}(C_t + F_t - (1+r)C_{t-1})/(1+r)^t$ (see also Lücke 1955).

Some years later, Bodenhorn (1964) defined residual income using the expression *pure earnings* and showed that the discounted sum of pure earnings is equal to the discounted sum of the cash flows for a two-period project (p. 27, footnote 19). It is worth noting that the author acknowledged an important invariance result: The equivalence "is independent of the depreciation pattern" (p. 29); that is, the equivalence holds irrespective of the stream $\boldsymbol{C}$ of capital amounts. This had also been acknowledged by Preinreich in his 1937 paper, where he claimed, in passing, that the equivalence holds "even though both the investment and the excess profits are measured incorrectly" (Preinreich 1937, p. 220). Later, other scholars made it clear that capital depreciation does not affect the equivalence relation: "*any* depreciation method will do" (Peasnell 1982a, p. 365).

In accounting, the notion of residual income came to prominence especially via Solomons (1965) and Amey (1969) who attracted the interest of accounting and managerial scholars on the use of residual income as a performance measure.

In the 80s and 90s, proofs of the equivalence between the discounted sum of residual incomes and NPV flourished in accounting and finance, starting from the notable paper by Peasnell (1982a), which revived the interest on this notion. A special version of residual income was popularized by Stewart (1991) with the label *Economic Value Added* (EVA).

In engineering economics, the notion has entered lately (see Lohmann 1988; Hartman 2000; Shrieves and Wachowicz 2001) and is not frequently employed.
Across the years, a plethora of terms have been attached to the notion of residual income. Table 6.22 collects a multitude of expressions which have been used to refer to the residual income notion. (See Magni 2009c for a historical and critical review of residual income theories. See also O'Hanlon and Peasnell (2014) on residual income and EVA). ▲

Example 6.24 ♦

C^o	C^d
C^l	C^e

(*RI of Mockup project, 80% payout, 3% debt COC*) Suppose we are to calculate all the notions of RI introduced above for the Mockup project with 80% payout ratio. Table 6.23 reports the various incomes, the COCs, the book values of the capitals, and the resulting RIs. For example, the equity RI at time 5 is obtained as $46{,}404 - 11.88\% \cdot 163{,}092 = 27{,}022$.[44] The debt RI is zero in every period because we have assumed $r^d = 0.03 = i^d$, so the project and the replicating portfolio brings the same return in every period. Table 6.24 reports the market values of the RI vectors.

[44] All numbers are rounded for notational convenience, as always.

Table 6.23 Residual income for Mockup project (80% payout, 3% debt COC)

Time			0	1	2	3	4	5
INCOME								
1	Project income	I_t		33,386	37,160	41,036	45,016	49,104
2	NOPAT + after-tax interest income	I_t^u		32,495	36,269	40,145	44,125	48,213
3	Operating income	I_t^o		33,386	36,690	40,072	43,536	47,082
4	NOPAT	I_t^{uo}		32,495	35,954	39,499	43,133	46,858
5	Interest income	I_t^l		0	471	964	1,481	2,022
6	Net income	I_t^e		30,686	34,460	38,336	42,316	46,404
7	Interest expenses	I_t^d		2,700	2,700	2,700	2,700	2,700
8	Net interest	$I_t^{\mathcal{D}}$		2,700	2,229	1,736	1,219	678
COC								
1	asset pre-tax WACC	r_t		9.00%	9.00%	9.00%	9.00%	9.00%
2	asset (after-tax) WACC	r_t^u		8.83%	8.81%	8.79%	8.75%	8.68%
3	operating pre-tax WACC	r_t^o		8.97%	9.28%	9.75%	10.56%	12.39%
4	operating (after-tax) WACC	r_t^{uo}		8.81%	9.12%	9.59%	10.42%	12.27%
5	required return on liquid assets	r_t^l		3.00%	3.00%	3.00%	3.00%	3.00%
6	required return on equity	r_t^e		10.25%	10.41%	10.66%	11.07%	11.88%
7	required return on debt	r_t^d		3.00%	3.00%	3.00%	3.00%	3.00%
8	required return on net debt	$r_t^{\mathcal{D}}$		3.00%	3.00%	3.00%	3.00%	3.00%
CAPITAL								
1	Assets	C_t	500,000	436,537	374,206	313,044	253,092	0
2	Assets	C_t	500,000	436,537	374,206	313,044	253,092	0
3	Operating assets	C_t^o	500,000	413,000	326,000	239,000	152,000	0
4	Operating assets	C_t^o	500,000	413,000	326,000	239,000	152,000	0
5	Liquid assets	C_t^l	0	23,537	48,206	74,044	101,092	0
6	Equity	C_t^e	410,000	346,537	284,206	223,044	163,092	0
7	Debt	C_t^d	90,000	90,000	90,000	90,000	90,000	0
8	Net debt	$C_t^{\mathcal{D}}$	90,000	66,463	41,794	15,956	-11,092	0
RESIDUAL INCOME								
1	project RI (with I_t)	RI_t		−11,614	−2,128	7,358	16,842	26,325
2	project RI (with I_t^u)	RI_t^u		−11,652	−2,196	7,269	16,747	26,248
3	operating RI (with I_t^o)	RI_t^o		−11,489	−1,649	8,290	18,296	28,247
4	operating RI (with I_t^{uo})	RI_t^{uo}		−11,530	−1,712	8,223	18,239	28,215
5	non-operating RI	RI_t^l		0	−235	−482	−740	−1,011
6	equity RI	RI_t^e		−11,336	−1,621	8,042	17,622	27,022
7	debt RI	RI_t^d		0	0	0	0	0
8	net-debt RI	$\text{RI}_t^{\mathcal{D}}$		0	235	482	740	1,011

Table 6.24 Market values of RI for Mockup project (80% payout, 3% debt COC)

Time	0	1	2	3	4	5
MARKET VALUE						
1 $V_t(\text{RI})$	22,277	35,895	41,254	37,609	24,152	0
2 $V_t^u(\text{RI})$	22,277	35,895	41,254	37,609	24,152	0
3 $V_t^o(\text{RI})$	24,469	38,154	43,345	39,281	25,133	0
4 $V_t^{uo}(\text{RI})$	24,469	38,154	43,345	39,281	25,133	0
5 $V_t^l(\text{RI})$	−2,193	−2,259	−2,091	−1,672	−981	0
6 $V_t^e(\text{RI})$	22,277	35,895	41,254	37,609	24,152	0
7 $V_t^d(\text{RI})$	0	0	0	0	0	0
8 $V_t^{\mathcal{D}}(\text{RI})$	2,193	2,259	2,091	1,672	981	0

One can observe that the present values are nothing but the various NPVs. For example, the market value of the operating RI stream at time 0 is \$24,469 and is equal to the operating NPV, while the market value of the non-operating RIs is $-\$2,193$ and is equal to the non-operating NPV; the algebraic sum of the two supplies the market value of the project RI stream, \$22,277, which is in turn equal to the project NPV. The latter can also be found as the sum of market values of the equity RI stream and the debt RI stream (in this case, the latter is zero, as noted above).

Finally, note that, while the operating RI is ambiguous (it depends on whether one complies with the OI notion or the NOPAT notion), the *value* of the operating RI can be unambiguously found either with RI^o or with RI^{uo}, for the differences in the incomes and in the RIs are offset by the discount rate:

$$V_{t-1}^o(\text{RI}) = \frac{V_t^o(\text{RI}) + \text{RI}_t^o}{1 + r_t^o} = \frac{V_t^{uo}(\text{RI}) + \text{RI}_t^{uo}}{1 + r_t^{uo}}.$$

The same reasoning applies to the project RI, so that the value of the project RI is unambiguous and can be found either with RI or RI^u:

$$V_{t-1}(\text{RI}) = \frac{V_t(\text{RI}) + \text{RI}_t}{1 + r_t} = \frac{V_t^u(\text{RI}) + \text{RI}_t^u}{1 + r_t^u}.$$

Example 6.25 ♦

C^o	C^d
C^l	C^e

(*RI of Mockup project, 80% payout, 1% debt COC*) Let us consider Example 6.24 and suppose that the debt COC is 1% instead of 3%: $r^d = 0.01 < 0.03 = i^d$, other things unvaried (in particular, the pre-tax asset WACC still remains at 9%). This means that the debt RI is positive in each period. Table 6.25 reports the RIs and their market values (and, therefore, the NPVs). The debt NPV is positive, as the debt RIs are positive in each period (i.e., debtholders expect to receive, in each period,

Table 6.25 Residual income for Mockup project (80% payout, 1% debt COC)

Time		0	1	2	3	4	5
RESIDUAL INCOME							
1 project RI (with I_t)	RI_t		−11,614	−2,128	7,358	16,842	26,325
2 project RI (with I_t^u)	RI_t^u		−11,652	−2,196	7,269	16,747	26,248
3 operating RI (with I_t^o)	RI_t^o		−11,489	−1,649	8,290	18,296	28,247
4 operating RI (with I_t^{uo})	RI_t^{uo}		−11,530	−1,712	8,223	18,239	28,215
5 non-operating RI	RI_t^l		0	−235	−482	−740	−1,011
6 equity RI	RI_t^e		−13,860	−3,893	5,990	15,750	25,268
7 debt RI	RI_t^d		1,800	1,800	1,800	1,800	1,800
8 net-debt RI	$\mathrm{RI}_t^{\mathcal{D}}$		1,761	1,938	2,102	2,150	3,456
MARKET VALUE							
1 $V_t(\mathrm{RI})$		22,277	35,895	41,254	37,609	24,152	0
2 $V_t^u(\mathrm{RI})$		22,277	35,895	41,254	37,609	24,152	0
3 $V_t^o(\mathrm{RI})$		24,469	38,154	43,345	39,281	25,133	0
4 $V_t^{uo}(\mathrm{RI})$		24,469	38,154	43,345	39,281	25,133	0
5 $V_0^l(\mathrm{RI})$		−2,193	−2,259	−2,091	−1,672	−981	0
6 $V_t^e(\mathrm{RI})$		13,540	28,872	35,960	34,063	22,370	0
7 $V_t^d(\mathrm{RI})$		8,736	7,024	5,294	3,547	1,782	0
8 $V_t^{\mathcal{D}}(\mathrm{RI})$		10,929	9,282	7,385	5,218	2,764	0

higher interest than those they would perceive if they invested the same amounts C_t^d in equivalent-risk alternatives traded in the security market). This has no effect on the market value of the project RI stream, because the pre-tax WACC is assumed to be the same. However, it does have an effect on the equity RI and its market value (the equity NPV): The lower r^d makes the equity COCs be greater than in the previous case, where $r^d = 3\%$ was assumed, so the equity RI decreases in each period. The net effect is that the value of the equity RI stream (i.e., the equity NPV) falls to \$13,540 while the market value of debt RI stream (i.e., the debt NPV) rises to \$8,736. The sum of these two components is equal to the market value of the project RI stream (i.e., the project NPV). Note that the change in the required return on debt does not affect the operating RI (none of the two versions) nor the non-operating RI.

In this situation, the two pairs of components of the project NPV are nonzero. Table 6.26 decomposes the project NPV. The operating assets are expected to create value by \$24,469. However, about 9% of it (\$2,193) is expected to be lost owing to an inefficient policy of financial investments ($r^l > i^l$). Of the remaining \$22,277, only about 61% (\$13,540) is captured by equityholders, who are expected to lose \$8,736,

Table 6.26 Decomposition of Mockup NPV (80% payout, 1% debt COC) investment side and financing side

investment side	$\overbrace{-2,193}^{\text{NPV}^I} + \overbrace{24,469}^{\text{NPV}^o} = \overbrace{22,277}^{\text{NPV}}$
financing side	$\overbrace{8,736}^{\text{NPV}^d} + \overbrace{13,540}^{\text{NPV}^e} = \overbrace{22,277}^{\text{NPV}}$

owing to an inefficient debt policy ($i^d > r^d$). The net result is that the project creates value to shareholders by $13,540, notwithstanding the fact that the debt policy and distribution policy tend to destroy value.

6.7 Valuing Projects with Capitals

In this section, we present a third valuation approach, based on a third benchmark system, whereby the cash flows from $t = 0$ to time $t = n - 1$ are replicated. The transposed Matrix is denoted as $(\mathcal{C}, \boldsymbol{I}^{\mathcal{C}}, \boldsymbol{F}^{\mathcal{C}})$. The comparison of project system and benchmark system boils down to a comparison of terminal alternative capitals, which gives rise to the notion of Value Added. Value added is shown to be equal to the Net Future Value.

6.7.1 The Three Benchmark Systems

Thus far, we have considered two different kinds of benchmark systems, namely, $(\boldsymbol{V}, \boldsymbol{I}^V, \boldsymbol{F}^V)$ and $(\boldsymbol{C}^*, \boldsymbol{I}^*, \boldsymbol{F}^*)$. They trigger the NPV criterion and the RI criterion, respectively. In particular, the two benchmark systems contemplate replication of the project in two different senses:

(i) replication of the cash flows from time 1 to time n (Sect. 6.1)
(ii) replication of the capital amounts from 0 to $n-1$ (Sect. 6.6).

Now, let us consider a third replication strategy:

(iii) replication of the cash flows from time 0 to time $n-1$.

Specifically, the investor invests C_0 in a benchmark portfolio earning the COC, r_t and, at every time t, withdraws/injects the same cash flow F_t which is generated by the project, $t = 1, 2, \ldots, n - 1$.

Portfolio (i) has the following cash-flow and capital streams:

$$\begin{aligned} \boldsymbol{F}^V &= \left(F_0^V, F_1, F_2, \ldots, F_{n-1}, F_n\right) \\ \boldsymbol{V} &= \left(V_0, V_1, V_2, \ldots, V_{n-1}, 0\right) \end{aligned}$$

where we let $F_0^V = -V_0$ denote the first cash flow of the portfolio.

Portfolio (ii) has the following cash-flow and capital streams:

$$\begin{aligned} \boldsymbol{F}^* &= \left(F_0^*, F_1^*, F_2^*, \ldots, F_{n-1}^*, F_n^*\right) \\ \boldsymbol{C}^* &= \left(C_0, C_1, C_2, \ldots, C_{n-1}, 0\right). \end{aligned}$$

with $F_0^* = -C_0$ and $F_t^* = C_{t-1}(1 + r_t) - C_t$ for $t \in \mathbb{N}_n^1$.

Portfolio (iii) has the following cash-flow and capital streams:

$$\begin{aligned} \boldsymbol{F}^{\mathcal{C}} &= \left(F_0, F_1, F_2, \ldots, F_{n-1}, E_n^{\mathcal{C}}\right) \\ \mathcal{C} &= \left(\mathcal{C}_0, \mathcal{C}_1, \mathcal{C}_2, \ldots, \mathcal{C}_{n-1}, 0\right) \end{aligned}$$

where $\mathcal{C}_0 = -F_0$ and $F_t^{\mathcal{C}} = F_t$ for $t \in \mathbb{N}_{n-1}^1$ so that $\mathcal{C}_t = \mathcal{C}_{t-1}(1 + r_t) - F_t$. The final amount $E_n^{\mathcal{C}} = \mathcal{C}_{n-1}(1 + r_n) = F_n^{\mathcal{C}}$ denotes the EOP capital amount (i.e., including the last benchmark profit). We couple the project Matrix and the benchmark Matrix below.

$$\begin{array}{cc} \textbf{Project} & \textbf{Benchmark} \\ \begin{array}{|c|c|} \hline +\Delta C_t^o & +I_t^o \; -F_t^o \\ +\Delta C_t^l & +I_t^l \; -F_t^l \\ \hline +\Delta C_t^d & +I_t^d \; -F_t^d \\ +\Delta C_t^e & +I_t^e \; -F_t^e \\ \hline \end{array} & \begin{array}{|c|c|} \hline +\Delta \mathcal{C}_t^o & +I_t^{o\mathcal{C}} \; -F_t^{o\mathcal{C}} \\ +\Delta \mathcal{C}_t^l & +I_t^{l\mathcal{C}} \; -F_t^{l\mathcal{C}} \\ \hline +\Delta \mathcal{C}_t^d & +I_t^{d\mathcal{C}} \; -F_t^{d\mathcal{C}} \\ +\Delta \mathcal{C}_t^e & +I_t^{e\mathcal{C}} \; -F_t^{e\mathcal{C}} \\ \hline \end{array} \end{array} \tag{6.156}$$

where $I_t^{j\mathcal{C}} = r_t^j \mathcal{C}_{t-1}^j$, $j = o, l, d, e$.

Investors purchasing any one of the three benchmark portfolios earn the market rate:

$$\begin{aligned} V_t^j &= V_{t-1}^j(1 + r_t^j) - F_t^{jV} \\ C_t^j &= C_{t-1}^j(1 + r_t^j) - F_t^{j*} \\ \mathcal{C}_t^j &= \mathcal{C}_{t-1}^j(1 + r_t^j) - F_t^{j\mathcal{C}} \end{aligned} \tag{6.157}$$

where $j = o, l, d, e$ and $C_n^j = V_n^j = \mathcal{C}_n^j = 0$. In other words, any benchmark portfolio is an equilibrium asset:

(i) $\sum_{t=0}^n F_t^{jV} \mathrm{d}_{t,0}^j = 0$
(ii) $\sum_{t=0}^n F_t^{j*} \mathrm{d}_{t,0}^j = 0$
(iii) $\sum_{t=0}^n F_t^{j\mathcal{C}} \mathrm{d}_{t,0}^j = 0$.

Equality (i) is obvious, since $F_0^{jV} = -V_0^j = -\sum_{t=1}^n F_t^{jV} \mathrm{d}_{t,0}^j = -\sum_{t=1}^n F_t^j \mathrm{d}_{t,0}^j$. Equality (ii) derives from the fact that $F_t^j - F_t^{j*} = \mathrm{RI}_t^j$, whence

$$\sum_{t=0}^{n} F_t^{j*} \cdot \mathrm{d}_{t,0}^j = \sum_{t=0}^{n} F_t^j \cdot \mathrm{d}_{t,0}^j - \sum_{t=0}^{n} \mathrm{RI}_t^j \cdot \mathrm{d}_{t,0}^j$$
$$= \mathrm{NPV}^j - \sum_{t=0}^{n} \mathrm{RI}_t^j \cdot \mathrm{d}_{t,0}^j = 0$$

Equality (iii) holds because the benchmark's terminal value is $F_n^{j\mathcal{C}} = E_n^{j\mathcal{C}} = \mathcal{C}_{n-1}^j (1 + r_n^j)$ and $\mathcal{C}_{n-1}^j = -\sum_{t=0}^{n-1} F_t^j \mathrm{u}_{t,n-1}^j$ whence $F_n^{j\mathcal{C}} = E_n^{j\mathcal{C}} = -\sum_{t=0}^{n-1} F_t^j \mathrm{u}_{t,n}^j$ where, as we remind, $\mathrm{u}_{t,n}^j = (1 + r_{t+1}^j) \cdot (1 + r_{t+2}^j) \cdot \ldots \cdot (1 + r_n^j)$. Dividing by $\mathrm{u}_{0,n}^j$ (i.e., multiplying by $\mathrm{d}_{n,0}^j$), (iii) follows.

As a result, any one of the three benchmark systems is entitled to represent the market.

6.7.2 *Value Added or Net Future Value*

Let us compute the difference between the project's cash-flow stream and each of the three replicating portfolio's cash-flow streams:

(i) $(F_0 - F_0^V, 0, 0, \ldots, 0)$
(ii) $(\mathrm{RI}_0, \mathrm{RI}_1, \mathrm{RI}_2, \ldots, \mathrm{RI}_n)$
(iii) $(0, 0, \ldots, 0, E_n - E_n^{\mathcal{C}})$.

Framing (iii) highlights the difference between the project's ending capital, E_n, and the benchmark portfolio's terminal capital, $F_n^{\mathcal{C}}$. This difference suggests that the project is worth undertaking if and only if $E_n - E_n^{\mathcal{C}} > 0$. In other words, if investors can replicate the project's cash flows from time 0 to time $n - 1$ by investing in the market at rate r_t, then the project is worth undertaking if and only if the investors will be left with a terminal value $E_n = C_{n-1}(1 + i_n)$ which is greater than the terminal value of the replicating portfolio, $E_n^{\mathcal{C}} = \mathcal{C}_{n-1}(1 + r_n)$.

Definition 6.4 (*Value Added*) The difference $E_t - E_t^{\mathcal{C}} = C_{t-1}(1 + i_t) - \mathcal{C}_{t-1}(1 + r_t)$ is the project's *residual capital*. The Value Added (VA) is defined as the residual capital at time n:

$$\mathrm{VA} = E_n - E_n^{\mathcal{C}} = C_{n-1}(1 + i_n) - \mathcal{C}_{n-1}(1 + r_n).$$

It measures the value created over and above the benchmark portfolio's terminal value.

The VA concept naturally triggers the following decision criterion.

Definition 6.5 (*VA criterion*) The project is worth undertaking if and only if the VA is positive

$$\text{VA} > 0.$$

As the benchmark's terminal capital is $\text{VA} = E_n^{\mathcal{C}} = \mathcal{C}_{n-1}(1+r_n) = -\sum_{t=0}^{n-1} F_t \mathrm{u}_{t,n}$, the residual terminal capital is

$$E_n - E_n^{\mathcal{C}} = F_n + \sum_{t=0}^{n-1} F_t \mathrm{u}_{t,n} = \text{NPV} \cdot \mathrm{u}_{0,n}, \tag{6.158}$$

whence $E_n - E_n^{\mathcal{C}} > 0$ if and only if $\text{NPV}\mathrm{u}_{0,n} > 0$ which holds if and only if $\text{NPV} > 0$ (since $r > -1$). Since the NPV criterion leads to economically rational decision-making, we have just shown the VA criterion is rational owing to its NPV-consistency.[45]

Proposition 6.6 (Rationality of the VA criterion) *The VA criterion is rational.*

The amount $\text{NPV}\mathrm{u}_{0,n} = F_n + \sum_{t=0}^{n-1} F_t \mathrm{u}_{t,n}$ is a compounded sum of the project's cash flows. It is also known as *Net Future Value* (NFV) or Net Terminal Value, since it is the same as the NPV, but expressed at a future, terminal date.

Equation (6.158) shows that the concept of NFV is not merely an algebraic operation which turns a present value into a future value, as commonly believed; rather, and more significantly, the **NFV measures the difference between the terminal capital of the project system and the terminal capital of a benchmark system** which replicates all the project's cash flows up to (and excluding) time n.

We hereafter use the expressions *value added* or *net future value* (as well as VA criterion or NFV criterion) interchangeably.

Remark 6.11 (*Equity NFV*) The results found above implicitly assume that either $i_t^d = r_t^d$ for all t or the project is not financed with debt ($C_t^d = 0$ for all t). Otherwise, the project VA/NFV must be replaced by the equity VA/NFV, just like what is done with the NPV approach and RI approach. In general, a rational decision is one where the project is undertaken if and only if the equity NFV is positive: $\text{NFV}^e > 0$.

[45] We remind that a given metric/criterion is NPV-consistent if and only if the decision recommended by that metric/criterion is the same as the decision recommended by the NPV (see also Definition 8.6).

The conditions under which the project NFV has the same sign as the equity NFV are the same conditions under which the project NPV has the same sign as the equity NPV. ▲

Remark 6.12 (*Terminology*) In engineering economics, the expression "Net Future Worth" is often used, whereas in finance the term "value" is more common. In investment performance measurement, "value added" seems to be preferred. It is noteworthy that the notion of "*economic* value added" (EVA), often employed in both corporate finance and engineering economics, is not the NFV/VA but, rather, a special case of residual income. One should bear in mind that EVA is not the economic value created: The former is a single-period accounting measure, the latter is a (multi-period) project measure and is quantified by NPV, $V_0(\text{RI})$, and NFV/VA. At the same time, "market value added", another popular metric in applied finance, refers to something different as well: It is equal to the difference between market value and book value at a given point in time. Terminological ambiguities and misunderstandings are always lurking. ▲

Example 6.26 ♦

C^l | C^e

(*Financial investment—NFV*) Consider a financial investment whereby an investor endows a fund manager \$10 million to be invested in the financial market. At the end of the first month, the investor deposits additional \$2 million and at the end of the third month he withdraws \$4 million, while injecting \$1.5 at the end of the fourth month. After six months he liquidates the investment. Assume the manager' policy of selection and allocation results in time-varying growth rates; precisely, the portfolio's value increases by 2% in the first month, 23% in the second month, 15%

Table 6.27 Value added (NFV) of a financial investment

	Investment				**Benchmark portfolio**			
Time t	Cash flow F_t^l	Rate of return i_t^l	EOP value E_t	BOP value C_t	Cash flow F_t^l	Rate of return r_t^l	EOP value $E_t^{\mathcal{C}}$	BOP value $\mathcal{C}_t$
0	−10.00			10.00	−10			10.00
1	−2.00	2%	10.20	12.20	−2	5%	10.50	12.50
2	0.00	23%	15.01	15.01	0	−10%	11.25	11.25
3	4.00	−8%	13.81	9.81	4	24%	13.95	9.95
4	−1.50	0%	9.81	11.31	−1.5	1%	10.05	11.55
5	0.00	15%	13.00	13.00	0	2%	11.78	11.78
6	13.65	5%	**13.65**	0.00	11.43	−3%	**11.43**	0
	Value Added		2.22					

in the fifth month, and 5% in the sixth and final month, while it decreases by -8% in the third month and remains stable in the fourth period.

Let us compute the NFV of this transaction at liquidation date, assuming that the cost of capital is given by the rate of return of a benchmark index. The holding period rates of the benchmark index are $r_1 = 5\%$, $r_2 = -10\%$, $r_3 = 24\%$, $r_4 = 1\%$, $r_5 = 2\%$, $r_6 = -3\%$. The terminal net asset value is equal to $E_6 = \$13.65 = 13 \cdot 1.05$. The benchmark's terminal value is $\$11.43 = 11.78 \cdot 0.97$ (Table 6.27). The difference between these values, \$2.22, is the value added (NFV), which signals that the investment creates value. The NPV may be found back by discounting the NFV at the benchmark returns:

$$\text{NPV} = 2.22 \cdot [1.05 \cdot 0.9 \cdot 1.24 \cdot 1.01 \cdot 1.02 \cdot 0.97]^{-1} = 1.9.$$

Example 6.27 ♦

C^o	C^d
	C^e

(Mockup project—100% payout—NFV) Consider \$500,000, the same amount invested in Mockup project (Example 6.9). Suppose an agent invests said monetary amount in a benchmark portfolio equivalent in risk to Mockup project and withdraws, at the end of each period, an amount which is equal to Mockup's CFA, up to time $n-1$. The expected rates of return of the benchmark portfolio are, by definition, the (pre-tax) WACCs, which we have already calculated: $\boldsymbol{r}^o = \boldsymbol{r} = (8.97\%, 9.28\%, 9.75\%, 10.56\%, 12.39\%)$. We remind that Mockup's (expected) CFAs, assuming 100% payout ratio, are

$$\boldsymbol{F} = (-500{,}000,\ 120{,}386,\ 123{,}845,\ 127{,}390,\ 131{,}024,\ 199{,}749).$$

The values of the benchmark portfolio are reported in Table 6.28 (both ending values and beginning values are displayed).[46] At time 5, the agent liquidates the investment and withdraws the portfolio's closing value, which is equal to $E_5^{\mathcal{C}} = \mathcal{C}_4(1 + r_5) = 158{,}185$ so the prospective value added is $\text{NFV} = E_5 - E_5^{\mathcal{C}} = 199{,}749 - 158{,}185 = 41{,}564$. This means that Mockup's claimholders and the agent investing in the benchmark portfolio expect to receive the same cash flows up to (and including) time 4 but, at time 5, the benchmark portfolio's closing value is expected to be smaller than Mockup project's terminal value. Therefore, Mockup's claimholders expect to be better off than the agent who invests in the benchmark portfolio. Mockup project is then preferable to a replicating investment in the benchmark portfolio. The differential amount \$41,564 is the value added by the project at time 5. The project is then worth undertaking.[47] The project NPV may be found back by discounting the net future worth by 5 periods:

---

[46] Numbers are rounded, as usual.

[47] We have assumed $i^d = r^d = 0.03$ so $\text{NPV}^d = 0$ and the project NPV coincides with shareholder value created (see also Table 6.10).

Table 6.28 Benchmark portfolio for Mockup project (100% payout)

Time	0	1	2	3	4	5
Cash flows	F_0	F_1	F_2	F_3	F_4	$E_5^{\mathcal{C}}$
	−500,000	120,386	123,845	127,390	131,024	158,185
(pre-tax) WACC		r_1	r_2	r_3	r_4	r_5
		8.97%	9.28%	9.75%	10.56%	12.39%
closing value		$\mathcal{C}_0(1+r_1)$	$\mathcal{C}_1(1+r_2)$	$\mathcal{C}_2(1+r_3)$	$\mathcal{C}_3(1+r_4)$	$\mathcal{C}_4(1+r_5)$
		544,875	463,894	373,201	271,769	158,185
opening value	$\mathcal{C}_0$	$\mathcal{C}_1$	$\mathcal{C}_2$	$\mathcal{C}_3$	$\mathcal{C}_4$	$\mathcal{C}_5$
	500,000	424,489	340,049	245,811	140,745	0
Value Added	41,564					

$$\text{NPV} = \frac{41{,}564}{(1.0897)(1.0928)(1.0975)(1.1056)(1.1239)} = 25{,}592.$$

(numbers are rounded).[48] The RI criterion leads to the same result, as can be easily checked.[49]

6.8 The Absolute Approaches and the Basic Notions

The three replicating systems seen above are but three different ways of undoing the factual scenario and constructing a suitable (value-neutral) counterfactual scenario for measuring value creation. Logically, they are three benchmark systems which, confronted with the project system, are capable of measuring the reciprocal equilibrium or disequilibrium. We report below the pairs of Matrices in the three approaches (see (6.159)–(6.161)). Considering that $F_0^V = -V_0$ and $F_0^{\mathcal{C}} = -\mathcal{C}_0$, the differential cash-flow stream (with respect to the project system) may be rewritten, respectively, as

[48]The NPV of the agent investing in the benchmark portfolio is obviously zero, as the benchmark portfolio is an equilibrium asset.

[49]The RI vector is

$$\mathbf{RI} = (0, -11{,}489, \ -1{,}494, \ 8{,}608, \ 18{,}785, \ 28{,}915)$$

whence, discounting with $\boldsymbol{r} = (8.97\%, 9.28\%, 9.75\%, 10.56\%, 12.39\%)$, one gets $V_0(\text{RI}) = 25{,}592$.

(i) (NPV, 0, 0, ..., 0)
(ii) $(\mathrm{RI}_0, \mathrm{RI}_1, \mathrm{RI}_2, \ldots, \mathrm{RI}_n)$
(iii) (0, 0, ..., 0, VA).

This paves the way for associating the three benchmark portfolios with the three decision criteria:

(i) NPV criterion
(ii) RI criterion
(iii) VA/NFV criterion.

The NPV criterion is based on cash flows (NPV is the discounted sum of all the cash flows), the RI criterion is based on incomes (difference between alternative incomes), the NFV criterion is based on residual capitals (difference between two terminal capital values).

$$\begin{array}{c} \textbf{Benchmark (NPV)} \\ \begin{array}{|l|l|} \hline +\Delta V_t^o & +I_t^{oV} \quad -F_t^{oV} \\ +\Delta V_t^l & +I_t^{lV} \quad -F_t^{lV} \\ \hline +\Delta V_t^d & +I_t^{dV} \quad -F_t^{dV} \\ +\Delta V_t^e & +I_t^{eV} \quad -F_t^{eV} \\ \hline \end{array} \end{array} \iff \begin{array}{c} \textbf{Project} \\ \begin{array}{|l|l|} \hline +\Delta C_t^o & +I_t^o \quad -F_t^o \\ +\Delta C_t^l & +I_t^l \quad -F_t^l \\ \hline +\Delta C_t^d & +I_t^d \quad -F_t^d \\ +\Delta C_t^e & +I_t^e \quad -F_t^e \\ \hline \end{array} \end{array} \tag{6.159}$$

$$\begin{array}{c} \textbf{Benchmark (RI)} \\ \begin{array}{|l|l|} \hline +\Delta C_t^o & +I_t^{o*} \quad -F_t^{o*} \\ +\Delta C_t^l & +I_t^{l*} \quad -F_t^{o*} \\ \hline +\Delta C_t^d & +I_t^{d*} \quad -F_t^{d*} \\ +\Delta C_t^e & +I_t^{e*} \quad -F_t^{*} \\ \hline \end{array} \end{array} \iff \begin{array}{c} \textbf{Project} \\ \begin{array}{|l|l|} \hline +\Delta C_t^o & +I_t^o \quad -F_t^o \\ +\Delta C_t^l & +I_t^l \quad -F_t^l \\ \hline +\Delta C_t^d & +I_t^d \quad -F_t^d \\ +\Delta C_t^e & +I_t^e \quad -F_t^e \\ \hline \end{array} \end{array} \tag{6.160}$$

$$\begin{array}{c} \textbf{Benchmark (VA)} \\ \begin{array}{|l|l|} \hline +\Delta \mathcal{C}_t^o & +I_t^{o\mathcal{C}} \quad -F_t^{o\mathcal{C}} \\ +\Delta \mathcal{C}_t^l & +I_t^{l\mathcal{C}} \quad -F_t^{l\mathcal{C}} \\ \hline +\Delta \mathcal{C}_t^d & +I_t^{d\mathcal{C}} \quad -F_t^{d\mathcal{C}} \\ +\Delta \mathcal{C}_t^e & +I_t^{e\mathcal{C}} \quad -F_t^{\mathcal{C}} \\ \hline \end{array} \end{array} \iff \begin{array}{c} \textbf{Project} \\ \begin{array}{|l|l|} \hline +\Delta C_t^o & +I_t^o \quad -F_t^o \\ +\Delta C_t^l & +I_t^l \quad -F_t^l \\ \hline +\Delta C_t^d & +I_t^d \quad -F_t^d \\ +\Delta C_t^e & +I_t^e \quad -F_t^e \\ \hline \end{array} \end{array} \tag{6.161}$$

The (valuation and) decision-making process is a process which may be expressed in terms of the three basic notions:

- a measurement process based on **cash flows** (NPV criterion)
- a measurement process based on (residual) **income** (RI criterion)
- a measurement process based on (residual) **capital** (NFV criterion).

In Part IV we construct three relative approaches to valuation and decision-making which are the relative counterparts of the NPV, RI, and NFV criteria.

Having presented three benchmark systems, three conceptualization of benchmark income are at stake. We then supply a more general definition of normal profit than the one presented in Definition 5.6 in Sect. 5.4.

Definition 6.6 (*Normal or benchmark profit—comprehensive definition*) The three benchmark systems give rise to benchmark, normal profits. The profit is *normal* in the sense that the growth rate of any such system is the equilibrium rate of a competitive, normal market. For example, in a project perspective, $I_t^* = r_t C_{t-1}$, $I_t^V = r_t V_{t-1}$, and $I_t^{\mathcal{C}} = r_t \mathcal{C}_{t-1}$.

Table 6.29 summarizes the three benchmark systems and the related criteria, associated with cash flow, income, and capital, respectively.

Table 6.29 The three benchmark systems and the three decision criteria

Benchmark system	Criterion	Value is created if	Primary notion
(i) Replicating cash flows from time 1 to time n			
$\boldsymbol{V} = (V_0, V_1, V_2, \ldots, V_{n-1}, 0)$ $\boldsymbol{I}^V = (0, r_1 V_0, r_2 V_1, \ldots, r_n V_{n-1})$ $\boldsymbol{F}^V = (F_0^V, F_1, F_2, \ldots, F_{n-1}, F_n)$	NPV	$\sum_{t=0}^{n} F_t \mathrm{d}_{t,0} > 0$	Cash flow
(ii) Replicating capital amounts			
$\boldsymbol{C}^* = (C_0, C_1, C_2, \ldots, C_{n-1}, 0)$ $\boldsymbol{I}^* = (0, r_1 C_0, r_2 C_1, \ldots, r_n C_{n-1})$ $\boldsymbol{F}^* = (F_0^*, F_1^*, F_2^*, \ldots, F_{n-1}^*, F_n^*)$	RI	$\sum_{t=0}^{n} (I_t - I_t^*) \mathrm{d}_{t,0} > 0$	Income
(iii) Replicating cash flows from time 0 to time n−1			
$\boldsymbol{\mathcal{C}} = (\mathcal{C}_0, \mathcal{C}_1, \mathcal{C}_2, \ldots, \mathcal{C}_{n-1}, 0)$ $\boldsymbol{I}^{\mathcal{C}} = (0, r_1 \mathcal{C}_0, r_2 \mathcal{C}_1, \ldots, r_n \mathcal{C}_{n-1})$ $\boldsymbol{F}^{\mathcal{C}} = (F_0, F_1, F_2, \ldots, F_{n-1}, E_n^{\mathcal{C}})$	NFV	$E_n > E_n^{\mathcal{C}}$	Capital

6.9 Key Points

- The calculation of the intrinsic value of a project and the decision about undertaking a project or not may be made on the basis of (i) the cash flows, (ii) the (residual) income, or (iii) the (residual) capital
- Cash flow, income and capital trigger three rational decision criteria, respectively:

  - the net present value (NPV) criterion
  - the residual income (RI) criterion
  - the value added (VA) or net future value (NFV) criterion.

  They stem from the use of three different benchmark systems to measure the degree of disequilibrium. Respectively,

  - a benchmark system replicating the project's cash flows from time 1 to time n (see also Chap. 5)
  - a benchmark system replicating the project's capitals from time 0 to time n
  - a benchmark system replicating the project's cash flows from time 0 to time $n-1$.

- (**Levered perspective**) The value of a project is the sum of the CFAs discounted at the pre-tax weighted average cost of capital (pre-tax WACC), which is the expected rate of return of an equivalent-risk portfolio which replicates the project's CFAs from time 1 to time n. By the law of one price, in a financing perspective, the project value is equal to the sum of the equity value (discounted sum of CFEs at the cost of equity, r_t^e) and the debt value (discounted sum of the CFD at the cost of debt, r_t^d); in an investment perspective, the project value is the sum of the value of operating assets (discounted sum of CFOs at the operating cost of capital, r_t^o) and the value of non-operating assets (discounted sum of CFLs at the non-operating cost of capital, r_t^l)
- (**Potential dividends**) If the present value of the prospective CFLs, discounted at the cost of equity, is zero (i.e., $\sum_{t=1}^{n} F_t^l \mathrm{d}_{t,0}^e = 0$), then the equity value is equal to the present value of prospective potential dividends, discounted at the cost of equity (r_t^e). A sufficient condition for this to occur is that the cost of equity and the expected non-operating ROI (i_t^l) coincide (i.e., investments of retained cash are made at the cost of equity). In this case, the payout policy is irrelevant, which means that the analyst does not have to worry about modeling liquid assets and forecasting the prospective CFEs. It suffices to pretend that PDs are entirely distributed to equityholders, which implies $(\boldsymbol{C}^l, \boldsymbol{I}^l, \boldsymbol{F}^l) = (\mathbf{0}, \mathbf{0}, \mathbf{0})$, which in turn implies that only forecasts of CFO and CFD are required. Discounting then the PDs at r^e, the equity value is obtained (FCFE method). However, if $\sum_{t=1}^{n} F_t^l \mathrm{d}_{t,0}^e \neq 0$ (the most common case), the payout policy will affect the equity value. In this case, the FCFE method should not be applied. One may use FCFEs and get a correct valuation of equity in any situation only if one discounts the FCFEs at the proper required rate of return, namely r_t^{pd} (a generalized weighted mean of r^o and r^d) and then adds the market value of liquid assets

- (**Unlevered perspective**) The value of a project's operating assets is also equal to the discounted sum of the FCFs at the operating after-tax WACC, r_t^{uo} (WACC method) and the value of the project is equal to the discounted sum of the *asset* FCFs ($F_t^{uo} + F_t^{ul}$) at the asset (after-tax) WACC, r_t^u (financially unlevered method)
- (**Textbook WACC**) A particular case of the WACC method obtains by assuming the cost of debt is equal to the ROD and the leverage ratio (in market value) is predetermined. This assumption eliminates the need of explicitly forecasting net interest, $I_t^{\mathcal{D}}$, which in turn removes the need of explicitly estimating the financing (debt + payout) policy; it suffices to discount the FCFs at the WACC (cost of equity and cost of debt are usually assumed constant, as well as the leverage ratio, so that the resultant after-tax WACC is constant for all t). However, a predetermined leverage ratio requires rebalancing of (market value of) net debt in each period, a possibly stringent assumption about a firm's behavior, especially in relation to a capital asset project
- (**APV**) The Adjusted Present Value (APV) method emphasizes the role of the tax shield in the value of a project detaching it from the value of unlevered assets. Specifically, the value of a project is the sum of the value of the unlevered project (the value the project would have if it were unlevered) and the value of the tax shields. The former is obtained by discounting FCFs at the unlevered cost of assets, the latter is obtained by discounting the tax savings at an appropriate risk-adjusted cost of capital
- (**CCF method**) A particular case of APV obtains by assuming that the tax savings have the same risk as the FCFs, so they are discounted at $r^{\text{TS}} = k^{uo}$ (usually assumed constant). This implies that the required return on operating assets is equal to the required return on unlevered assets: $r^o = k^{uo}$. This valuation scheme is also called the *compressed APV* method
- (**Modified CFE discounting**) The modified discounted-cash-flow approach consists of discounting modified CFEs at a suitable cost of capital. More precisely, the CFEs may be discounted at any unlevered cost of assets (k^u, k^{uo}) or any WACC (r, r^o, r^{uo}, r^u) to directly get the equity value, provided that they are suitably adjusted. In this way, if the market input data consist of some prespecified unlevered cost of assets (or some WACC), the cost of equity need not be calculated for discounting the CFEs
- (**Net Present Value**) In general, a project should be undertaken if and only if it creates value for the firm's equityholders. This occurs if and only if the equity NPV (difference between equity value and equity capital injection) is positive. Managers should then undertake only those projects such that $\text{NPV}^e > 0$. If the project is equity financed or the ROD is equal to the cost of debt ($i_t^d = r_t^d$), the debt NPV is zero and project NPV coincides with equity NPV: $\text{NPV} = \text{NPV}^e$. In this case, making decisions with either project NPV or equity NPV leads to the same result.
- (**NPV decomposition**) Shareholder value creation (NPV^e) may be decomposed into operating NPV, debt NPV, and non-operating NPV: $\text{NPV}^e = \text{NPV}^o + (\text{NPV}^l - \text{NPV}^d)$. The first summand measures the contribution of the operations to value creation; it is the value created by the investment policy. The remaining summands

supply information on the value of the debt and distribution policies. The debt policy is value-neutral if $i_t^d = r_t^d$, the distribution policy is value-neutral if $i_t^l = r_t^l$. If both hold, financing policy is value-neutral and value creation only depends on the investment policy (this also occurs, trivially, whenever the project is unlevered, so that both debt and liquid assets are zero)

- **(Residual income)** The market value of the RIs of a firm/project is the market value of the difference between the firm's/project's incomes and the profits that would be earned on an equivalent-risk portfolio which replicates all the capitals. The economic value created is then captured by the market value of the RI stream: $V_0(\text{RI}) = \sum_{t=0}^{n} \text{RI}_t \text{d}_{t,0}$. Given that $V_0(\text{RI}) = \text{NPV}$, if $V_0(\text{RI}) > 0$ value is created, if $V_0(\text{RI}) < 0$ value is destroyed. For any type of NPV there corresponds a well-determined type of RI value. Therefore, one may use any of the aforementioned perspectives to evaluate a project and make a decision.
- **(Residual capital)** Value creation may be assessed in terms of the terminal value $E_n^{\mathcal{C}}$ of a benchmark portfolio which replicates the project's cash flows up to (and excluding) time n: If the terminal value $E_n^{\mathcal{C}}$ of the replicating portfolio is smaller than the terminal value of the project, E_n, then the project creates value; if it is greater, then the project destroys value. The difference between the residual capitals of the project system and the benchmark system is called value added (VA) and is equal to the net future value (NFV), that is, the compounded value of the NPV, which means that value is created if and only if VA = NFV > 0. For any type of NPV, there corresponds a well-determined VA and NFV. Therefore, one may use any of the aforementioned perspectives to evaluate a project and make a decision.

Tables 6.30 and 6.31 represent a summary of costs of capital, acting as the market counterpart of Table 4.37: While the latter shows the income rates of the project, the former show the income rates of a benchmark portfolio which replicates the project's cash flows (from time 1 on), that is, the project's various COCs. The reader will notice the high number of costs of capital and the intricacy of the reciprocal relations among them. However, to assimilate this labyrinth of interconnections, one should bear in mind that any relation among rates must fulfill the law of conservation for income rates, which manifests itself in the shape of weighted means of rates (any other form is a transformation of a weighted mean of rates). For example, in order to catch the relations among cost of equity, unlevered cost of assets, cost of tax shield, and cost of net debt, one should remember that equity value is equal to unlevered value of operating assets plus value of tax shield minus value of net debt. Hence, the cost of equity is a weighted mean of unlevered costs of operating assets, cost of tax shield, and cost of net debt (the latter with negative sign):

$$r_t^e = k_t^{uo} \cdot \frac{V_{t-1}^{uo}}{V_{t-1}^e} + r_t^{\text{TS}} \cdot \frac{V_{t-1}^{\text{TS}}}{V_{t-1}^e} - r_t^{\mathcal{D}} \cdot \frac{V_{t-1}^{\mathcal{D}}}{V_{t-1}^e}.$$

Tables 6.32, 6.33 and 6.34 show the various NPVs associated with the various notions of (i) cash flows, (ii) incomes, and (iii) capitals (with obvious meaning of the symbols).

Table 6.35 shows the various formulas for the modified discounting approach, assuming different market input data.

Figures 6.2 and 6.3 illustrate the mosaics of intrinsic values. They preserve the structures of the mosaics of incomes and cash flows (see Figs. 4.10 and 4.11).

Equation (6.162) below makes use of the thumbnail Matrices and provides a handy guide to helpful implications in terms of capitals, rates of return and COCs/MARRs related to different financial structures.

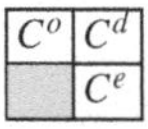

$\Longrightarrow C^o_t = C_t$ and $F^o_t = F_t$ and $i^o_t = i_t$ and $r^o_t = r_t$
and $F^{uo}_t = F^u_t$ and $i^{uo}_t = i^u_t$ and $r^{uo}_t = r^u_t$

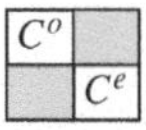

$\Longrightarrow C_t = C^o_t = C^e_t$ and $F_t = F^o_t = F^e_t$ and $i_t = i^o_t = i^e_t$ and $r_t = r^o_t = r^e_t$
and $F_t = F^{uo}_t = F^u_t$ and $i_t = i^{uo}_t = i^u_t$ and $r_t = r^{uo}_t = r^u_t$

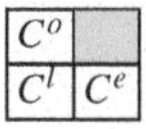

$\Longrightarrow C_t = C^e_t$ and $F_t = F^u_t = F^e_t$ and $i_t = i^u_t = i^e_t$ and $r_t = r^u_t = r^e_t$

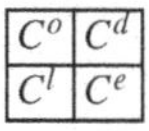

$\Longrightarrow$ no general implications

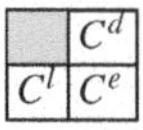

$\Longrightarrow C^l_t = C_t$ and $F^l_t = F_t$ and $i^l_t = i_t$ and $r^l_t = r_t$

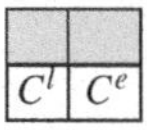

$\Longrightarrow C^l_t = C_t = C^e_t$ and $F^l_t = F_t = F^e_t$ and $i^l_t = i_t = i^e_t$ and $r^l_t = r_t = r^e_t$

(6.162)

Table 6.30 Summary of relations for a project's costs of capital (Part I)

operating pre-tax WACC	$r_t^o = \dfrac{V_t^o + F_t^o}{V_{t-1}^o} - 1$
$r_t^o = \dfrac{r_t V_{t-1} - r_t^l V_{t-1}^l}{V_{t-1} - V_{t-1}^l}$	$r_t^o = \dfrac{r_t^e V_{t-1}^e + r_t^{\mathcal{D}} V_{t-1}^{\mathcal{D}}}{V_{t-1}^e + V_{t-1}^{\mathcal{D}}}$
$r_t^o = k_t^{uo} + (r_t^{\text{TS}} - k_t^{uo}) \cdot \dfrac{V_{t-1}^{\text{TS}}}{V_{t-1}^o}$	$r_t^o = k_t^{uo} \cdot \dfrac{V_{t-1}^{uo}}{V_{t-1}^o} + r_t^{\text{TS}} \cdot \dfrac{V_{t-1}^{\text{TS}}}{V_{t-1}^o}$
asset pre-tax WACC	$r_t = \dfrac{V_t + F_t}{V_{t-1}} - 1$
$r_t = \dfrac{r_t^o V_{t-1}^o + r_t^l V_{t-1}^l}{V_{t-1}^o + V_{t-1}^l}$	$r_t = \dfrac{r_t^e V_{t-1}^e + r_t^d V_{t-1}^d}{V_{t-1}^e + V_{t-1}^d}$
$r_t = k_t^u + (r_t^{\text{GTS}} - k_t^u) \cdot \dfrac{V_{t-1}^{\text{GTS}}}{V_{t-1}}$	$r_t = k_t^u \cdot \dfrac{V_{t-1}^u}{V_{t-1}} + r_t^{\text{GTS}} \cdot \dfrac{V_{t-1}^{\text{GTS}}}{V_{t-1}}$
operating after-tax WACC	$r_t^{uo} = \dfrac{V_t^o + F_t^{uo}}{V_{t-1}^o} - 1$
$r_t^{uo} = r_t^o - \dfrac{\tau I_t^{\mathcal{D}}}{V_{t-1}^o}$	$r_t^{uo} = \dfrac{r_t^e V_{t-1}^e + r_t^{\mathcal{D}}(1-\tau) V_{t-1}^{\mathcal{D}}}{V_{t-1}^e + V_{t-1}^{\mathcal{D}}}$ [assuming $i_t^{\mathcal{D}} = r_t^{\mathcal{D}}$]
$r_t^{uo} = k_t^{uo} + (r_t^{\text{TS}} - k_t^{uo}) \cdot \dfrac{V_{t-1}^{\text{TS}}}{V_{t-1}^o} - \dfrac{\tau I_t^{\mathcal{D}}}{V_{t-1}^o}$	$r_t^{uo} = k_t^{uo} \cdot \dfrac{V_{t-1}^{uo}}{V_{t-1}^o} + \dfrac{r_t^{\text{TS}} V_{t-1}^{\text{TS}} - \tau I_t^{\mathcal{D}}}{V_{t-1}^o}$
asset after-tax WACC	$r_t^u = \dfrac{V_t + F_t^u}{V_{t-1}} - 1$
$r_t^u = r_t - \dfrac{\tau I_t^d}{V_{t-1}}$	$r_t^u = \dfrac{r_t^e V_{t-1}^e + r_t^d(1-\tau) V_{t-1}^d}{V_{t-1}^e + V_{t-1}^d}$ [assuming $i_t^d = r_t^d$]
$r_t^u = k_t^u + (r_t^{\text{GTS}} - k_t^u) \cdot \dfrac{V_{t-1}^{\text{GTS}}}{V_{t-1}} - \dfrac{\tau I_t^d}{V_{t-1}}$	$r_t^u = k_t^u \cdot \dfrac{V_{t-1}^u}{V_{t-1}} + \dfrac{r_t^{\text{GTS}} V_{t-1}^{\text{GTS}} - \tau I_t^d}{V_{t-1}}$

Table 6.31 Summary of relations for a project's costs of capital (Part II)

cost of equity	$r_t^e = \dfrac{V_t^e + F_t^e}{V_{t-1}^e} - 1$
$r_t^e = r_t^{uo} + \dfrac{V_{t-1}^{\mathcal{D}}}{V_{t-1}^e}(r_t^{uo} - r_t^{\mathcal{D}}) + \dfrac{\tau I_t^{\mathcal{D}}}{V_{t-1}^e}$	$r_t^e = r_t^{uo} + \dfrac{V_{t-1}^{\mathcal{D}}}{V_{t-1}^e}(r_t^{uo} - r_t^{\mathcal{D}}(1-\tau))$ [assuming $i_t^{\mathcal{D}} = r_t^{\mathcal{D}}$]
$r_t^e = r_t^u + \dfrac{V_{t-1}^d}{V_{t-1}^e}(r_t^u - r_t^d) + \dfrac{\tau I_t^d}{V_{t-1}^e}$	$r_t^e = r_t^u + \dfrac{V_{t-1}^d}{V_{t-1}^e}(r_t^u - r_t^d(1-\tau))$ [assuming $i_t^d = r_t^d$]
$r_t^e = r_t^o + \dfrac{V_{t-1}^{\mathcal{D}}}{V_{t-1}^e}(r_t^o - r_t^{\mathcal{D}})$	$r_t^e = r_t + \dfrac{V_{t-1}^d}{V_{t-1}^e}(r_t - r_t^d)$
$r_t^e = k_t^{uo} + (k_t^{uo} - r_t^{\mathcal{D}})\dfrac{V_{t-1}^{\mathcal{D}}}{V_{t-1}^e} + (r_t^{\text{TS}} - k_t^{uo})\dfrac{V_{t-1}^{\text{TS}}}{V_{t-1}^e}$	$r_t^e = k_t^u + (k_t^u - r_t^d)\dfrac{V_{t-1}^d}{V_{t-1}^e} + (r_t^{\text{GTS}} - k_t^u)\dfrac{V_{t-1}^{\text{GTS}}}{V_{t-1}^e}$
$r_t^e = k_t^{uo} \cdot \dfrac{V_{t-1}^{uo}}{V_{t-1}^e} + r_t^{\text{TS}} \cdot \dfrac{V_{t-1}^{\text{TS}}}{V_{t-1}^e} - r_t^{\mathcal{D}} \cdot \dfrac{V_{t-1}^{\mathcal{D}}}{V_{t-1}^e}$	$r_t^e = k_t^u \cdot \dfrac{V_{t-1}^u}{V_{t-1}^e} + r_t^{\text{GTS}} \cdot \dfrac{V_{t-1}^{\text{GTS}}}{V_{t-1}^e} - r_t^d \cdot \dfrac{V_{t-1}^d}{V_{t-1}^e}$
cost of net debt	$r_t^{\mathcal{D}} = \dfrac{r_t^d V_{t-1}^d - r_t^l V_{t-1}^l}{V_{t-1}^d - V_{t-1}^l}$
cost of potential capital	$r_t^{pd} = \dfrac{r_t^o V_{t-1}^o - r_t^l V_{t-1}^l}{V_{t-1}^o - V_{t-1}^l}$

Table 6.32 Value creation based on cash flows[a]

TYPES OF CASH FLOW[b]		TYPES OF RATES	TYPES OF NPV
Cash flow from assets (CFA)	F_t^{inv}	asset pre-tax WACC	$\text{NPV} = \sum_{j=0}^{n} \frac{F_j}{\prod_{h=1}^{j}(1+r_h)}$
Cash flow from operations (CFO) Capital cash flow (CCF)	F_t^o	operating pre-tax WACC	$\text{NPV}^o = \sum_{j=0}^{n} \frac{F_j^o}{\prod_{h=1}^{j}(1+r_h^o)}$
Cash flow from liquid assets (CFL)	F_t^l	cost of non-operating assets	$\text{NPV}^l = \sum_{j=0}^{n} \frac{F_j^l}{\prod_{h=1}^{j}(1+r_h^l)}$
Cash flow to equity (CFE)	F_t^e	cost of equity	$\text{NPV}^e = \sum_{j=0}^{n} \frac{F_j^e}{\prod_{h=1}^{j}(1+r_h^e)}$
Cash flow to debt (CFD)	F_t^d	cost of debt	$\text{NPV}^d = \sum_{j=0}^{n} \frac{F_j^d}{\prod_{h=1}^{j}(1+r_h^d)}$
Cash flow to net debt (CF$\mathcal{D}$)	$F_t^{\mathcal{D}}$	cost of net debt	$\text{NPV}^{\mathcal{D}} = \sum_{j=0}^{n} \frac{F_j^{\mathcal{D}}}{\prod_{h=1}^{j}(1+r_h^{\mathcal{D}})}$
Cash flow to capital providers (CFP)	F_t^{fin}	asset pre-tax WACC	$\text{NPV} = \sum_{j=0}^{n} \frac{F_j}{\prod_{h=1}^{j}(1+r_h)}$
Potential dividends (PD) Free cash flow to equity (FCFE)	F_t^{pd}	cost of potential dividends	$\text{NPV}^{pd} = \sum_{j=0}^{n} \frac{F_j^{pd}}{\prod_{h=1}^{j}(1+r_h^{pd})}$
Modified CFE		SEE TABLE 6.35	
UNLEVERED			
Free cash flow (FCF)	F_t^{uo}	operating after-tax WACC	$\text{NPV}^o = \sum_{j=0}^{n} \frac{F_j^{uo}}{\prod_{h=1}^{j}(1+r_h^{uo})}$
Free cash flow (FCF)	F_t^{uo}	unlevered cost of assets	$\text{NPV}^{uo} = \sum_{j=0}^{n} \frac{F_j^{uo}}{\prod_{h=1}^{j}(1+k_h^{uo})}$
Asset free cash flow	F_t^u	asset after-tax WACC	$\text{NPV} = \sum_{j=0}^{n} \frac{F_j^u}{\prod_{h=1}^{j}(1+r_h^u)}$
Asset free cash flow	F_t^u	financially unlevered cost of assets	$\text{NPV}^u = \sum_{j=0}^{n} \frac{F_j^u}{\prod_{h=1}^{j}(1+k_h^u)}$

[a] $\prod_{h=1}^{0}(1+r_h^{(\cdot)}) := 1$
[b] $F_t^{\text{inv}} = F_t^{\text{fin}} = F_t$

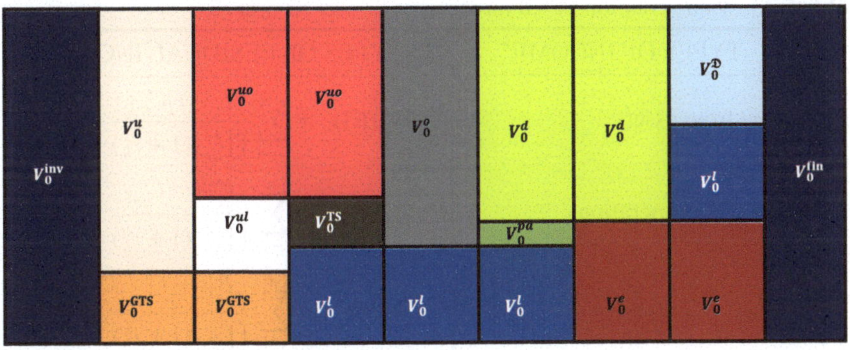

Fig. 6.2 The Mosaic of values of a project (firm) with extra distribution ($\alpha > 1$)

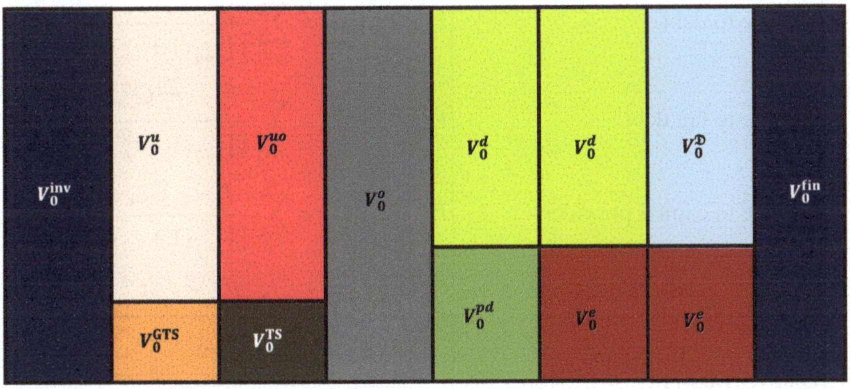

Fig. 6.3 The Mosaic of values of a project (or firm) with full payout ($\alpha = 1$)

Table 6.33 Value creation based on incomes

TYPES OF INCOME[a]		TYPES OF RESIDUAL INCOME
Income from assets	I_t^{inv}	$V_0(\text{RI}) = \sum_{j=0}^{n} \frac{I_j - r_j C_{j-1}}{\prod_{h=1}^{j}(1+r_h)}$
Income from operations	I_t^o	$V_0^o(\text{RI}) = \sum_{j=0}^{n} \frac{I_j^o - r_j^o C_{j-1}^o}{\prod_{h=1}^{j}(1+r_h^o)}$
Income from liquid assets	I_t^l	$V_0^l(\text{RI}) = \sum_{j=0}^{n} \frac{I_j^l - r_j^l C_{j-1}^l}{\prod_{h=1}^{j}(1+r_h^l)}$
Income to equity	I_t^e	$V_0^e(\text{RI}) = \sum_{j=0}^{n} \frac{I_j^e - r_j^e C_{j-1}^e}{\prod_{h=1}^{j}(1+r_h^e)}$
Income to debt	I_t^d	$V_0^d(\text{RI}) = \sum_{j=0}^{n} \frac{I_j^d - r_j^d C_{j-1}^d}{\prod_{h=1}^{j}(1+r_h^d)}$
Income to net debt	$I_t^{\mathcal{D}}$	$V_0^{\mathcal{D}}(\text{RI}) = \sum_{j=0}^{n} \frac{I_j^{\mathcal{D}} - r_j^{\mathcal{D}} C_{j-1}^{\mathcal{D}}}{\prod_{h=1}^{j}(1+r_h^{\mathcal{D}})}$
Income to capital providers	I_t^{fin}	$V_0(\text{RI}) = \sum_{j=0}^{n} \frac{I_j - r_j C_{j-1}}{\prod_{h=1}^{j}(1+r_h)}$
Potential income	I_t^{pd}	$V_0^{pd}(\text{RI}) = \sum_{j=0}^{n} \frac{I_j^{pd} - r_j^{pd} C_{j-1}^{pd}}{\prod_{h=1}^{j}(1+r_h^{pd})}$
UNLEVERED		
NOPAT (Operationally unlevered income)	I_t^{uo}	$V_0^o(\text{RI}) = \sum_{j=0}^{n} \frac{I_j^{uo} - r_j^{uo} C_{j-1}^o}{\prod_{h=1}^{j}(1+r_h^{uo})}$
Financially unlevered income	I_t^u	$V_0(\text{RI}) = \sum_{j=0}^{n} \frac{I_j^u - r_j^u C_{j-1}}{\prod_{h=1}^{j}(1+r_h^u)}$

[a] $I_t^{\text{inv}} = I_t^{\text{fin}} = I_t$

Table 6.34 Value creation based on capitals

TYPES OF CAPITAL[a]		TYPES OF VALUE ADDED
Investments	C_t^{inv}	$\text{NFV} = E_n - E_n^{\mathcal{C}}$
Operating assets	C_t^o	$\text{NFV}^o = E_n^o - E_n^{o\mathcal{C}}$
Liquid assets	C_t^l	$\text{NFV}^l = E_n^l - E_n^{l\mathcal{C}}$
Equity	C_t^e	$\text{NFV}^e = E_n^e - E_n^{e\mathcal{C}}$
Debt	C_t^d	$\text{NFV}^d = E_n^d - E_n^{d\mathcal{C}}$
Net debt	$C_t^{\mathcal{D}}$	$\text{NFV}^{\mathcal{D}} = E_n^{\mathcal{D}} - E_n^{\mathcal{D}\mathcal{C}}$
Financings	C_t^{fin}	$\text{NFV} = E_n - E_n^{\mathcal{C}}$
Potential capital	C_t^{pd}	$\text{NFV} = E_n^{pd} - E_n^{pd\mathcal{C}}$
UNLEVERED		
Operating assets	C_t^o	$\text{NFV}^o = E_n^{uo} - E_n^{uo\mathcal{C}}$
Assets	C_t	$\text{NFV} = E_n^u - E_n^{u\mathcal{C}}$

[a] $C_t^{\text{inv}} = C_t^{\text{fin}} = C_t$

Table 6.35 Modified discounting approach

Known parameters	Equity value
(1) $r^o, r^{\mathcal{D}}$	$V_{t-1}^e = \dfrac{V_t^e + F_t^e - (r_t^o - r_t^{\mathcal{D}}) \cdot V_{t-1}^{\mathcal{D}}}{1 + r_t^o}$
(2) r, r^d	$V_{t-1}^e = \dfrac{V_t^e + F_t^e - (r_t - r_t^d) \cdot V_{t-1}^d}{1 + r_t}$
(3) $r^{uo}, r^{\mathcal{D}}$	$V_{t-1}^e = \dfrac{V_t^e + F_t^e - (r_t^{uo} - r_t^{\mathcal{D}}(1 - \tau)) \cdot V_{t-1}^{\mathcal{D}}}{1 + r_t^{uo}}$
(4) r^u, r^d	$V_{t-1}^e = \dfrac{V_t^e + F_t^e - (r_t^u - r_t^d(1 - \tau)) \cdot V_{t-1}^d}{1 + r_t^u}$
(5) $k^{uo}, r^{\mathrm{TS}}, r^{\mathcal{D}}$	$V_{t-1}^e = \dfrac{V_t^e + F_t^e - (k_t^{uo} - r_t^{\mathcal{D}})V_{t-1}^{\mathcal{D}} - (r_t^{\mathrm{TS}} - k_t^{uo})V_{t-1}^{\mathrm{TS}}}{1 + k_t^{uo}}$
(6) $k^u, r^{\mathrm{GTS}}, r^d$	$V_{t-1}^e = \dfrac{V_t^e + F_t^e - (k_t^u - r_t^d)V_{t-1}^d - (r_t^{\mathrm{GTS}} - k_t^u)V_{t-1}^{\mathrm{GTS}}}{1 + k_t^u}$

Part IV
Valuation and Decision-Making: The Relative Approaches

In this part of the book we complete the accounting-and-finance engineering system, presenting three relative approaches to valuation and decision-making. They are based on the notion of **rate of return** and represent the relative counterparts of the three absolute approaches presented in Part III. The three relative approaches are

Average Internal Rate of Return (AIRR), derived from the RI approach
Internal Average Rate of Return (IARR), derived from the NPV approach
Aggregate Return On Investment (AROI), derived from the NFV approach.

We show how to evaluate a multiperiod asset (project, firm, etc.) and make rational (valuations and) decisions with the relative approaches.

The key concept linking absolute approaches and relative approaches is the notion of **capital**. Specifically, the relative approaches refine the absolute approaches in that they decompose the economic value created in **economic (financial) efficiency** and **project scale** and enable identifying the financial nature of a project and, therefore, capture the source of the shareholders' wealth increase or decrease.

By completing the final step of linking absolute approaches and relative approaches, a unified theory of investment valuation and decision-making is created, which is

- transparent (it uses explicit data present in the pro forma financial statements which reflect the key business assumptions)
- analytical (it is based on a logical system which "engineers" the accounting and the finance of a project)
- intuitive (it rests on intuitive, basic algebraic operations)
- simple (no equation is required to be solved and it is easily comprehensible by non-specialists)
- ready-to-use (it may be implemented on any spreadsheet with no additional data or assumptions).

Chapter 7
The Quest for a Relative Approach

For abstruse reasons that are beyond the scope of this article, college professors prefer NPV analysis … But all surveys indicate that lenders find it more appealing to analyze potential investments in terms of percentage rates of return than by comparing dollars of NPV

Yung and Sherman (1995, p. 18)

Scholar: "What's the point of computing a rate of return when you can use the NPV for making good decisions?".
Practitioner: "What's the point of computing the NPV when I can use a rate of return for making good decisions?"

Private conversation

C. A. Magni, *Investment Decisions and the Logic of Valuation*,
https://doi.org/10.1007/978-3-030-27662-1_7

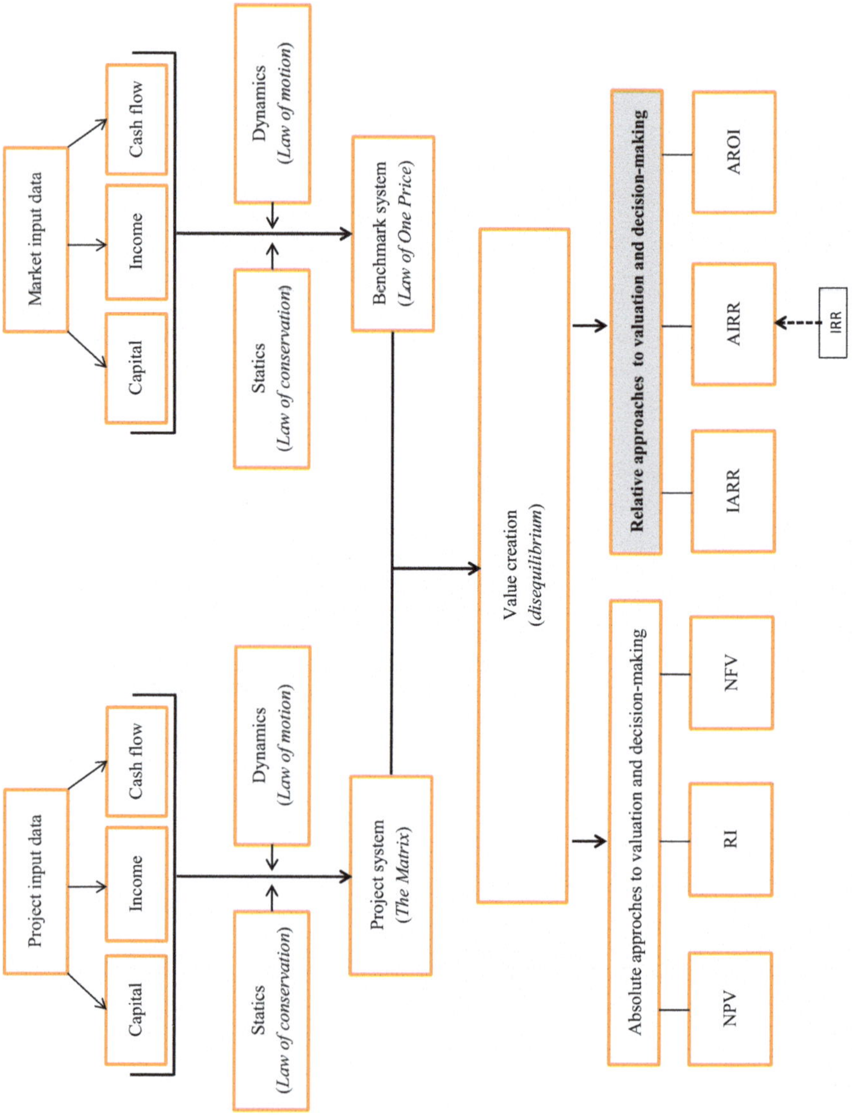

Fig. 7.1 Chapter 7: Relative measures of worth and the capital

The previous chapters have described economic value creation in the form of **absolute measures** of worth: Net present value, residual income, or value added (net future value). Rational decision-making is one where the absolute measure of worth is positive, signaling a positive direction of a disequilibrium. However, to compare alternatives in terms of percentage points is particularly intuitive and is a common habit among practitioners.

Relative measures of worth are widely employed to measure **financial efficiency** in any scientific field and real-life applications, ranging from engineering projects to public policy schemes, health interventions, environmental designs, natural resources conservation plans, educational programs, and many others, where the relevant objective function is expressed in terms of rate of return, benefit-cost ratio, social return on investment, and the like. The **need for linking** relative approaches and absolute approaches is therefore cogent.

Under the traditional perspective, relative measures and absolute measures clash and are in opposition. We argue that the traditional perspective is unsatisfactory and incomplete and that the two classes may be integrated into a sound, comprehensive theory.

The accounting-and-finance engineering system presented in Parts I-III is able to naturally harmonize relative approaches with absolute approaches. Specifically, the NPV decision criterion may be supplemented by an intuitive relative criterion whereby a **rate of return**, expressing the project system's financial efficiency, is compared with a **minimum acceptable rate of return**, based on a benchmark system. How do the absolute perspectives and the relative perspectives relate and reconcile via the accounting-and-finance engineering system? This chapter supplies some relevant questions and anticipates some answers presented in the following chapters, suggesting that the **amount of capital** is the missing, long sought-after link.

After completing this chapter, the reader should be able to:

- feel the need of a more refined economic analysis establishing clear relations among rates of return, capital amounts and value creation
- understand that, in a one-period project, the economic value created is a function of two elements: The project's economic efficiency, expressed in terms of rates of return, and the project scale, expressed in terms of overall amount of capital invested
- appreciate the simple product structure which links efficiency and economic value created in a one-period project and be aware of the role of the capital (scale) for bridging the gap between relative measures and absolute measures
- follow the hints and anticipate the results of the following chapter, namely, that value creation depends on the economic efficiency and the project scale even in a multiperiod project

7.1 Absolute or Relative Measures of Worth?

Generally speaking, there are two camps, one of which recommends the use of absolute measure of worth such as the NPV for detecting value creation and make economic analysis of a project, the other one that favors the use of relative measures of worth and, in particular, of a rate of return susceptible of capturing value creation and being comparable with the COC (MARR in an investment project, MAFR in a financing project).

While many academics belong to the former, many practitioners belong to the latter. Despite some evidence of an increasing use of NPV, many managers, firms, professionals, financial advisors still make substantial use of relative measures of worth (Gitman and Forrester 1977; Stanley and Block 1984; Remer et al. 1993; Burns and Walker 1997; Graham and Harvey 2001; Ryan and Ryan 2002) and, when using the NPV, they often use it in conjunction with some other criteria, including residual income and rates of return (Sandahl and Sjögren 2003; Lindblom and Sjögren 2009; Magni 2009d).

One of the reasons of the appeal of the rate-of-return notion seems to lie with the greater intuitive appeal of a relative (i.e., percentage) measure as opposed to an absolute measure (Evans and Forbes 1993). To compare two rates of return (project rate of return versus cost of capital) is more intuitive and natural to many than to compute an absolute amount (NPV, NFV, RI) which in turn depends on a relative amount (the COC). Consider the two sentences about an investment project:

- "the project's NPV is $500 at 6% MARR"
- "the project's rate of return is 8%, the MARR is 6%".

The second sentence compares homogenous quantities (two rates of return) and immediately identifies a clearly favorable situation. It is a simple sentence, more intuitive than the first one, which involves an absolute amount (NPV) which is a (nontrivial) function of a relative measure of worth (MARR). The second sentence is understandable by anyone, even if unacquainted with basic finance theory. A rate of return

> fills a need that the NPV does not. In analyzing investments, people in general, and financial analysts in particular, seem to prefer talking about rates of return rather than dollar values …In a similar vein, [a rate of return] appears to provide a simple way of communicating information about a proposal. One manager might say to another, 'Remodeling the clerical wing has a 20 percent return.' This may somehow be simpler than saying 'At a 10 percent discount rate, the net present value is 4,000' (Ross et al. 2011, p. 248)

Further, the traditional perspective on NPV seems to be lacking from several points of view. First, it does not identify the project's value drivers, that is, the key factors that create or destroy value. In particular, the NPV, as such, is not capable of distinguishing investment projects from financing projects, nor is it able to identify changes in the firm's financial position (investment versus financing) over time. This implies that it cannot tell us whether value is created out of an investment or out of a financing.

And hence, it cannot tell us whether it creates value because funds are invested at a rate of return which is greater than the (investment) COC or because funds are raised at a financing rate which is smaller than the (financing) COC, a piece of information which is relevant to a firm's manager. (For example, the NPV of $(15, -120, 40, -35, 110, 15, -12)$ at $r = 3\%$ is positive, but it is not clear whether it is an investment project or a financing project.) Also, it does not distinguish the value created or destroyed as a result of raising funds from debtholders from the value created or destroyed as a result of absorbing funds from the assets. Further, it does not provide a measure of (marginal) efficiency of capital for the project at hand; nor does it supply an unambiguous definition of project scale: Initial outlay? Net cash flow? Sum or average of invested amounts? As a result, it is not clear whether a given NPV is the effect of a large amount of capital which is invested at a small rate of return or the effect of a small amount of capital which is invested at a high rate of return. In other words, **the traditional NPV analysis does not inform about the economic efficiency of the capital** and how it relates to investment scale to create value. For example, consider firm A undertaking a one-year project with cash-flow stream $(-94.5, 113.4)$ while firm B undertakes a one-year project whose cash-flow stream is $(-74.2, 91.3)$. Assuming a COC equal to 3% for both projects, the NPVs are $\text{NPV}^A = 15.60$ and $\text{NPV}^B = 14.44$. As such, the numbers do not disclose the reasons why A creates more value than B; specifically, it is not clear whether firm A's managers manage money in a more efficient way than firm B's managers or whether they are less efficient but the less efficiency is more than compensated by the higher scale ($94.5 > 74.2$). The question is easily answered by a rate-of-return analysis. The rate of return of each project is the income rate, i, expressing the growth rate of capital: $i^A = 20\%$ $(= 113.4/94.5 - 1)$ for A and $i^B = 23.05\%$ $(= 91.3/74.2 - 1)$ for B. Therefore, B is more efficient than A. The economic value created by A is greater than the value created by B because, notwithstanding a smaller efficiency, A has a greater scale.[1]

In general, the information about **how investment scale and economic efficiency interact synergically to generate a given NPV** is an important piece of information which a comprehensive value theory should provide. While this is easily derived from a single-period project (the project scale is given by the first cash flow[2] and the economic efficiency is the single-period rate of return), it is not clear how to do this kind of analysis for a multiperiod project without a reliable notion of rate of return. For example, consider firm A undertaking a five-year project with cash-flow stream $(-90, -10, 40, 50, 35)$ while firm B undertakes a five-year project whose cash-flow stream is $(-50, -52, 20, 60, 45)$. Assuming a COC equal to 5% for both projects, the NPVs are $\text{NPV}^A = 8.74$ and $\text{NPV}^B = 7.47$. However, it is not so clear whether

[1]Note that, at a 15% COC, things are reversed: The NPVs are $\text{NPV}^A = 4.11$ and $\text{NPV}^B = 5.19$. Therefore, in this case, the higher economic efficiency of B compensates its smaller scale, to such an extent that the value created by B becomes greater than the value created by A.

[2]As long as $I_0 = 0$.

- the scale of A is greater or smaller than the scale of B
- the economic efficiency of A is greater or smaller than that of B.

We have not yet provided the reader with a well-defined criterion for measuring the project's scale and its marginal efficiency, and NPV alone is of no help. Even more so when a principal-agent relationship arises, whereby inflows and outflow are determined by a principal while management of the capital is upon the agent. Therefore, in this case, value creation is affected by two effects: The agent affects value creation via economic efficiency (how good or bad the investment policy is, given the capital available) whereas the principal affects value creation via deposits or withdrawals and, therefore, via the amounts available to the agent for investment. This is the reason why managers of firms or business units or fund managers are often compensated on the basis of rates of return. For example, in decentralized organizations, managers of business units may not (completely) control the amount of invested capital. Assessing performance of a business unit or a management team necessitates splitting up the economic value created into an efficiency component and a scale component. Only the former measures the managing skills of the manager.[3]

More generally, whenever capital is rationed and managers need to choose small-scale investments or are bound to undertake the same investment at a smaller scale (so that the NPV will be reduced), information on the marginal efficiency of the capital is essential to appraise a business unit's or a manager's investment policy. Comparing different units or different managers on the basis of the value created when capital is scarce would be inappropriate. In these cases, a rate of return measuring the ability of a project of increasing the capital is more informative than the NPV. (Investment decisions in decentralized organizations belong to this class.)

It is also worthy of consideration the fact that **a decision about acceptance or rejection of a project depends on the economic efficiency of the project, not on its scale**. If the project's rate of return is greater than the COC, the project is worth undertaking, regardless of the project scale; since an absolute measure of worth somewhat incorporates both project scale and economic efficiency, it might even be deemed redundant in accept/reject decision-making.

The financial efficiency of a project brings about quick-and-dirty information on the 'safety level' of a project: The greater the efficiency, the greater the resilience of the project to a decrease in the expectations or to adverse realizations. When comparing A (initial investment of 1,000,000 and 1% efficiency, yielding NPV of 10,000) with B (initial investment of 100,000 with 10% efficiency and same NPV), the investor is more protected with B than with A if forecasts are not fulfilled. Otherwise stated, the amount of efficiency is a rule of thumb informing about the resistance of the project to possible unexpected realizations.

[3]See also Magni (2014b, 2015b).

Precisely because, in traditional analysis, every aspect of the economic analysis is condensed in one single number, the traditional NPV (as well as other absolute measures) is a black box (albeit a theoretically correct one) which hides more detailed aspects in the analysis of the sources of value creation, namely, the way money is managed.

In the previous chapter, we have decomposed the equity NPV into operating NPV, non-operating NPV and debt NPV, a relevant decomposition in itself, richer in content than a traditional NPV analysis, for it captures the role of equity and debt in creating value. However, we need to improve on and make an advancement for completing the analysis: It is still unclear how the cost of debt, r_t^d, the cost of equity, r_t^e, the ROD, i_t^d, and the ROE, i_t^e interact to create or destroy value. Nor is there any information about how the return on the operations, i_t^o, and the return on the non-operating assets, i_t^l, interact with their required rates of return in creating value. Likewise, the traditional NPV analysis does not distinguish between the economic efficiency of the project and that of the equity.

A further element is that a negative NPV informs that a suboptimal decision is made if the project is undertaken, but does not say whether the investor loses part of his capital or not. A relative measure of worth such as a rate of return does supply such information: A negative investment rate means that part of the capital invested is lost; a positive investment rate means that the capital grows (see Table 1.4 in Sect. 1.3). For example, if NPV $= -10$, the project destroys value; however, this number does not necessarily mean that investors are expected to lose (part of or all) the money they invest in the project. Consider the following cases: Investment A consists of \$100 outlay at a 3% rate of return with a COC equal to 7.25%. Investment B consists of the same outflow, invested at a -3% with a COC equal to 1%. The NPVs of A and B are equal: NPV $= -3.96$, but investors may find it relevant to know that investment B is expected to lose 3% of the capital, whereas investment A is expected to increase the capital by 3%. If either project must be chosen, investment A may be deemed preferable to investment B, notwithstanding the fact that the NPV is the same. Or, alternatively, consider the investment of \$100 for one period at 4.8% with an associated COC equal to 3.9%. In this case, the project is value-creating: Its NPV is 0.87. However, the same NPV is obtained by a project B with same capital invested and rate of return equal to -2% and a COC equal to -2.84%. The former may be preferable to practitioners, given that they are not expected to give up money if they accept it.[4]

It is rather usual for practitioners to speak about an investment that 'loses money' to refer to projects where the difference between inflows and outflows is negative. Other things equal, if a project loses money, investors may be reluctant to invest in it. Likewise, if a project "earns" money (inflows exceed outflows), then investors may be more willing to invest in the project.

[4] A negative COC is by no means meaningless. It means that a replicating portfolio is expected to earn a negative rate of return, which may occur for several reasons, either relating to macroeconomic factors or to a negative correlation of the project's rate of return to the return of the whole market (see also footnote 16 in Chap. 1).

This means that, implicitly, investors fix an additional, unstated hurdle rate equal to zero. If another project is available which does not lose money and has the same (or similar) NPV, it may make sense that the latter will be preferred. Shareholders of a company which invests in positive-NPV project that lose money may be less satisfied than shareholders of a company which invests in positive-NPV projects that do not lose money. Likewise, equity owners will regard acceptance of a negative-NPV investment with a positive rate of return less disadvantageous than acceptance of a negative-NPV investment with a negative rate of return.

In general, practitioners often compare the project rate of return with several different benchmarks, including zero, the return on Treasury bonds, the shareholder return in companies of the same industry, the return on the stock market index (see Fernández 2002, pp. 19–20), which testifies that relative measures of worth are regarded as extremely important to real-life decision makers. Different measures provide different information: "Comparing different benchmarks enables different questions to be answered" (Fernández 2002, p. 19).

In addition, consider that, for financial investments, the notion of NPV is of limited use because the amount of capital invested can be freely chosen by the investor in financial markets. For example, consider firm A and firm B introduced above with their one-year projects with cash-flow streams $(-94.5, 113.4)$ and $(-74.2, 91.3)$. Now, suppose these cash-flow streams are two financial securities. As such, they may be arbitrarily scaled in a normal market; for example, one might expand the scale of B by $94.5/74.2 = 1.2736$ so that \$94.5 is invested. Given equal amount invested in asset A and asset B, the two rates of return, 20% and 23.05% suffice to conclude that the "expanded B" creates more value than A. In this context, NPV is not so important for investors, precisely because investors may freely choose the scale of the investment; more meaningful is the calculation of the return generated by every single dollar invested.[5] More generally, whenever a project is scalable, the notion of *efficiency* becomes predominant.[6]

On another note, the traditional perspective on NPV is simplistic in terms of informational content: It does not distinguish between the project perspective and the equity perspective, since it does not separate equity investment from the project's invested capital, and does not make it clear that the equity investment has a different story from the project in terms of rate of return and in terms of capital invested, since the project also includes the debtholders' investment.

Suppose, for example, that the firm invests \$100 in a one-year project and let \$140 be the end-of-period cash flow generated by the project. Let $r = 10\%$ be the MARR. The project's economic value created is $\text{NPV} = V_0 - C_0 = 140/1.1 -$

---

[5]This does not diminish the universal validity of NPV: If one computes the NPV of the expanded B, whose cash-flow stream is $B' = (-94.5, 116.28)$, one finds $\text{NPV}^{B'} = -94.5 + 116.28/(1+r) > -94.5 + 113.4/(1+r) = \text{NPV}^A$ for every $r > -1$.

[6]The amount of capital committed in an economic asset does play a major role in determining the amount of economic value created (see Sect. 7.2); but the amount of capital invested in a financial asset (traded securities, funds, portfolios, etc.) is arbitrary.

$100 = 127.27 - 100 = 27.27$. Suppose that the project is financed by a one-year loan for an amount of \$40 and that the ROD is $i^d = 2\%$, equal to the cost of debt, $r^d = 2\%$. As we know, this implies that the debt NPV is zero, so that $\text{NPV}^e = \text{NPV} = 27.27$. However, this analysis does not clarify that this value is the result of two investments which have different economic profitabilities. Debtholders invest \$40 at 2%, while equity holders invest $C_0^e = 60 = 100 - 40$ at $i^e = 65.33\% = (140 - 40 \cdot 1.02)/60 - 1$. Since the equity value is $V_0^e = \text{NPV}^e + C_0^e = 27.27 + 60 = 87.27$, then the required return on equity is $r^e = (0.1 \cdot 127.27 - 0.02 \cdot 40)/87.27 = 13.67\%$. Then, the economic efficiency of the equity investment is $51.67\% = 65.33\% - 13.67\%$, which is much greater than the economic efficiency of the project ($= 30\% = 40\% - 10\%$). The residual income of the project is $\text{RI}_1 = 100 \cdot 0.3 = 30$, whereas the residual income of the equity investment is $\text{RI}_1^e = 60 \cdot 0.5167 = 31$. The reconciliation of the two perspectives occurs when the RIs are discounted at the respective COCs: $\text{NPV} = 30/1.1 = 27.27$ and $\text{NPV}^e = 31/1.1367 = 27.27$. Overall, the equity investment is expected to earn more, in relative terms, than the project, but the capital invested is smaller, to such an extent that the two effects counterbalance.

Should the interest rate on the loan be different from the required rate of return, NPV and NPV^e would not coincide. For example, if $i^d = 8\% > 2\% = r^d$, the equity rate of return would be smaller, equal to $i^e = 61.33\% = (140 - 40 \cdot 1.08)/60 - 1$; the value of debt would be higher, equal to $V_0^d = 42.35 = 40 \cdot 1.08/1.02$; and the required return on equity would be higher, equal to $r^e = 13.99\% = (0.1 \cdot 27.27 - 0.02 \cdot 42.35)/84.92$, thereby decreasing the economic efficiency of the equity investment to $47.34\% = 61.33\% - 13.99\%$. Hence, given the fixed scale of equity ($C_0^e = 60$), the equity NPV would to $\text{NPV}^e = \$24.9 = 60 \cdot 0.4734/1.1399$. The project NPV would not change, and the remaining \$2.35 would be the net gain to debtholders, caused by the increase in the interest rate.

In general, the objectives of human resources initiatives, public policy schemes, health interventions, environmental analyses, natural resources conservation plans and infrastructures, educational programs and even many industrial projects are increasingly set in terms of relative measures such as productivity, efficiency, cost-benefit ratio, ROI, social ROI, conservation ROI, human capital ROI, etc.[7] The use of relative, efficiency measures of worth in operations research is routine for managerial purposes, mainly including the internal rate of return (IRR) and the ROI.[8] Overall, the use of rates of return and relative metrics is pervasive in any scientific field and in real-life applications (even more so whenever the scale of the investment is discretional, at least to some extent). Official institutions sometimes recommend as

[7] For example, see Echols (2005), Fitz-Enz (2009), Cascio and Boudreau (2011), Auerbach et al. (2014), Kumar (2014), Boyd et al. (2015), Donlan et al. (2015), Leck et al. (2016), Gargani (2017), Masters et al. (2017), Michaud et al. (2017), Pattison-Williams et al. (2017), Walker et al. (2017), Yates and Marra (2017), Cordes (2017), Cantarelli et al. (2018), De Risi et al. (2018), McMahon (2018).

[8] For example, see Rapp (1980), Nauss (1988), Danaher and Rust (1996), Myung et al. (1997), Brimberg and ReVelle (2000), Biezma and San Cristóbal (2006), Brimberg et al. (2008), Li et al. (2008), Talavera et al. (2010), Kumbhakar (2011), Menezes et al. (2015), Mørch et al. (2017), Dhavale and Sarkis (2018).

best practice (or even set as a requirement for co-financing) the calculation of rates of return in association with the NPV. For example, infrastructure and productive investments are eligible for EU grants. The EU Regulations for Structural Funds, Cohesion Fund and Instrument For Structural Policies in Pre-Accession recommend that three distinct rates of return be calculated, namely, the financial rate of return of the investment, the financial rate of return of the equity capital, and the economic rate of return. These rates are recommended for the financial evaluation of the project and are used for judging the quality of the appraisal process and for determining the co-financing rate, that is, the percentage of the eligible cost that is covered by EU grants (see Sartori et al. 2014).

As a result, there is a strong need of precisely pinpointing the relations between NPV and rate of return and, more generally, between relatives measures and absolute measures.

7.2 The Capital: A Link Between Absolute and Relative Measures of Worth

While the advantages of computing an additional measure such as a rate of return should now be apparent, it should be clear that a rate of return should complement (not replace) an absolute measure. A rate of return *alone* is unhelpful, misleading, and insufficient for a deep analysis of capital asset investments. First, it does not measure value creation. While NPV/VA/RI do measure value creation, a rate of return alone is only a measure of the growth rate of the capital committed. Which is irrelevant for value creation. What is relevant is the comparison of two rates of return: The project's rate of return and the COC. Therefore, the rate of return should always be accompanied by the corresponding COC. Second, the comparison of rate of return and COC is mute if the analyst has no idea whether the project is an investment project or a borrowing project: If the project is a borrowing project, a project's rate of return is not a rate of return at all: It is a financing rate. And shareholders will be better off if and only if the rate at which they borrow from the project is smaller than the rate at which they might borrow the same funds in the market (MAFR). Therefore, the *use* of a rate requires a precise definition of what is an investment or a borrowing. Without such a definition, a rate may not be used for assessing economic performance. (On the contrary, the NPV does not need any such definition to capture value creation.) Furthermore, the traditional concept of rate of return, so-called *internal rate of return* (widely employed in practice), is flawed from several points of view, to such an extent that it does not enrich the traditional NPV analysis; rather, it obscures its meaningfulness (see Chap. 9 for a detailed analysis).

Moreover, contrary to NPV, any rate of return as such may not be used for ranking capital budgeting projects, precisely because it is a relative measure, neglecting the role of the project scale for decision-making. To understand this, we make use of a simple example. Suppose an investor has the opportunity of investing in one between two mutually single-period investments:

- investment of \$200 at a rate of return of 50%.
- investment of \$10 at a rate of return of 100%.

We assume they have the same risk and that the MARR is $r = 0\%$. This means that both investments create value but the first one creates more value. The NPV of the first one is \$100 ($= -200 + 300/(1+0)$), the NPV of the second one is \$10 ($= -10 + 20/(1+0)$). Note that the two NPVs just express the incomes of the two alternatives, which may be computed as the product of the rate of return and the respective invested amount:

Project	Rate of return	×	Amount of capital	=	Value created
A	50%	×	200	=	\$100
B	100%	×	10	=	\$10.

(7.1)

A rate-of-return comparison incorrectly pinpoints B as more economically profitable than A (100% is preferable than 50%) and, therefore, a decision maker resting on a rate-of-return comparison is induced to choose B. This choice is suboptimal: The wealth increase caused by the first project is ten times that of the second project.

If it were possible to invest \$200 at 100%, then the comparison of rates of return would indeed be significant (see the expanded project B and footnote 5). However, capital asset investments are not easily scalable as is a security traded in a financial market. Which implies that the *amount* of capital invested is relevant for value creation no less than the marginal *efficiency* of that amount. The notion of capital just represents the link between the rate-of-return notion and the NPV notion (and explains the reason why rates of return alone are unreliable for project ranking, unless they have the same scale). This is also confirmed by the notion of residual income, presented in Sect. 6.6. For one-period projects, (6.135) becomes

$$\mathrm{RI}_1^e = \overbrace{C_0^e}^{\text{investment size}} \cdot \overbrace{(i_1^e - r_1^e)}^{\text{economic efficiency}}.$$

whence the economic value created is $\mathrm{NPV}^e = \mathrm{RI}_0^e = C_0^e(i_1^e - r_1^e)/(1 + r_1^e)$. The reader should notice that the value created, in a one-period project, enjoys a product structure where the key drivers are both the economic efficiency (rate of return net of the COC) and the amount of invested capital. This structure provides managerial guidelines in the sense that the objective of wealth maximization consists of two sub objectives: Maximization of the project scale and maximization of the (excess) rate of return.

The product structure in one-period projects shows that the economic value created is a residual income and the capital is the notion that connects the rate of return and the economic value created.[9] In the following chapter, we will see that this RI structure is

[9] Many textbooks devote much space to the estimation of the COC. Yet, some of them devote little, if any, attention to the notion of rate of return and its relevance for a complete economic analysis of projects. However, the COC is itself a rate of return. Therefore, it sounds a bit paradoxical to devote considerable amount of space to the rate of return of an asset which replicates the project

the key to value creation and that the product structure holds for multiperiod projects as well: A project's economic value created will be shown to be equal to the total invested capital (project scale) multiplied by the overall excess return rate (economic efficiency).

The product structure also makes it clear that capital is irrelevant in value creation: It is the financial efficiency (rate of return net of COC) that determines the sign of the NPV and, therefore, determines whether wealth is increased or decreased. The capital only determines the project size and, hence, the size of the economic value created (NPV). Otherwise stated, **whether a project should be accepted or not depends on how efficient the use of capital is**, that is, on how high the project's rate of return is, as opposed to the MARR. Which confirms the paramount importance of the rate-of-return notion. At the same time, this does not diminish the role of the NPV notion and the other absolute measures, for they incorporate the interaction between capital (i.e., the scale) and efficiency and, **in project selection, capital is of paramount importance whereas rate of return may be misleading**. Also, information about the absolute amount of wealth increase is as important as information about the relative amount of wealth increase.

To wrap things up: The traditional perspective on NPV and rates of return is unsatisfactory and creates an artificial opposition, whereby neither NPV nor rate of return are put at full use. A new, more refined methodology is needed which should encompass absolute and relative measures in a consistent and transparent way. The notion of capital precisely represents the link between the two measures of worth. We will investigate and exploit such a link in the following chapters.

NPV is the gold standard for capital budgeting but requires acquaintance with finance to understand it in full. The rate-of-return notion is a tricky one but is intuitive and informs about financial efficiency.

An NPV takes account of both financial efficiency and scale but does so in an implicit way, while a rate of return makes efficiency explicit but does not take into account the scale for decision-making purposes. **Acceptance or rejection** of a project depends on how efficient is the use of capital, not on how amount of capital is tied-up in the project. **Project ranking and selection** depends on the amount of capital employed in the projects to be ranked, not only on how efficient is the use of the capital. Therefore, both the traditional treatment of absolute measures and the traditional treatment of relative measures are unsatisfactory.

A more refined methodology should integrate absolute measures (NPV, RI, VA) and relative measures (rates of return, profitability indices) in a coherent, exhaustive picture, where scale and efficiency are explicitly measured and combined.

---

and dismiss or minimize the discussion about the rate of return of the project itself, which is the subject of the analysis.

Analysts, professionals, financial advisors, managers need a **comprehensive theory which connects the world of the absolute measures and the world of relative measures**. We proceed in bridging the gap between these worlds, armed with the intuition that **capital is the key concept** for linking them.

Remark 7.1 (*Rate of return and glycemic index*) The *glycemic index* (GI) of a nutrient measures how rapidly a carbohydrate is digested and released as glucose into the blood stream. By multiplying the amount of carbohydrate in the food and the food's GI (divided by 100) one finds the *glycemic load* (GL).

$$\frac{\text{GI}}{100} \times \text{ amount of carbohydrates} = \text{GL}$$

For example, suppose a food A has a GI of 50 and the amount of carbohydrates in that food is 200 g. Another food B has a glycemic index of 100 and the amount of carbohydrates in that food is 10. This means that glucose raises much more quickly by eating the second food than by eating the first food. However, the amount of carbohydrates in A is 200 g, much more than the content of carbohydrates in food B. The net effect is that the GL of food B is much smaller: The GLs of the two foods are

Food	**Glycemic index**	×	**Amount of carbohydrates**	=	**Glycemic load**
A	50%	×	200	=	100
B	100%	×	10	=	10.

(7.2)

Equations (7.1)–(7.2) have the same structure (and, in this case, the same numbers). It should be evident that GI is a rate measuring the velocity at which the glucose enters the blood stream, in the same way as a rate of return measures the "velocity" at which a given capital grows. Also, one might say that GI is to GL what a rate of return is to wealth increase. The link between glycemic index and glycemic load is the amount of carbohydrates in the food. The latter is to GL what the amount of capital is to wealth increase.

(The GL is evidently the objective function for people with problems of hyperglycemia, just as wealth increase is the objective function of investors. Yet, physicians do not dismiss the GI as being non-significant: The decomposition of the GL into its constituents enriches the informative structure and provides more guidelines for minimizing the GL.)[10] ▲

[10] A fundamental conceptual difference between glycemic index and rate of return is that glycemic index alone does not determine whether a portion of food is acceptable or not, whereas rate of return (net of MARR) does determine whether a project is acceptable or not. Which makes rate of return much more important to managers and investors than glycemic index is to physicians and hyperglicemic people.

Absolute and relative measures of worth: An alliance – not a clash

Some respectable authors suggest to compute NPV and recommend not using rates of return by resting on a peculiar argument:

There is no point of computing the rate of return
if you can compute the NPV.

However, as long as a reliable rate of return can be identified, the argument might be reversed in favor of the opposite stance:

There is no point of computing NPV if you can compute the rate of return.

Both stances betray a limited view of the logic of project appraisal, for absolute measures of worth and relative measure of worth complement each other. We will then replace both claims with the following one:

There is no point of choosing between NPV and a rate of return
if you can compute both and harmonize them
into a comprehensive, consistent approach.

We will show that the two classes of measures are allies, not competitors. They are both essential elements of a unified view of project appraisal and decision-making.

7.3 Key Points

- Scholars recommend the use of NPV for valuing projects and making decisions. Yet, practitioners often use rates of return as well, which are felt as more intuitive
- The traditional perspective on NPV is insufficient for several reasons, including the fact that it

  1. does not clarify the nature of a project (investment or financing)
  2. does not clarify whether the project creates value owing to a high economic efficiency of the capital committed or a high amount of that capital
  3. is not of much help if capital is rationed (as well as in decentralized organizations). In these cases a rate of return might be more significant
  4. does not clarify the role of the various income rates and required rates of return in value creation
  5. does not inform whether the project "loses" or "earns" money,
  6. does not provide the causes of value creation and, in particular, does not distinguish the value created or destroyed as a result from investing/absorbing resources in/from the assets from the value created as a result of debt financing

- The rate of return (does not replace but) completes the information provided by the NPV and the other absolute measures of worth, precisely because it measures the marginal financial efficiency of the project
- The capital is the link between rate of return and NPV: The latter is intuitively conceptualized as the product of the capital invested and the project's rate of return (net of the MARR)
- A debate about whether rate of return is superior or inferior to an absolute measure of worth for valuation and decision is an idle issue: The two approaches should never be put in reciprocal opposition. Rather, a serious, comprehensive approach to valuation and decision-making uses any possible information to carry out an economic analysis which is as complete as possible. The lack of either the rate of return or the NPV makes the analysis suboptimal for an analyst willing to make informed decisions.

Chapter 8
Average Internal Rate of Return

To the extent that cash-flow capital budgeting measures appear to managers incapable of being arithmetically reconciled with the now familiar accounting measures, then there is a danger that they will not receive the attention they deserve. In these circumstances, there is a lot to be said for presenting the project data in a manner such that the arithmetical connections with accounting profits is made obvious.

Peasnell (1982a, p. 376)

The correct procedure in choosing investments involves in effect comparing the whole schedule of pairs of short-term rates [i_t] with the equilibrium pair of short-term rates [r_t], regardless of whether the equilibrium rates are equal to each other.

Bailey (1959, p. 479)

The project rate is an average rate of return over the project term, not an annual one; ...The book yield is a figure based only on the data related to a given year, and it has no reference to any other part of the project than that year to which it applies. The project rate is the average of the book yields.

Vatter (1966, pp. 695–696)

[T]he rate of return is an *average* concept. It takes a series of n numbers and converts them into an average value.

Haley and Schall (1979, p. 68)

C. A. Magni, *Investment Decisions and the Logic of Valuation*,
https://doi.org/10.1007/978-3-030-27662-1_8

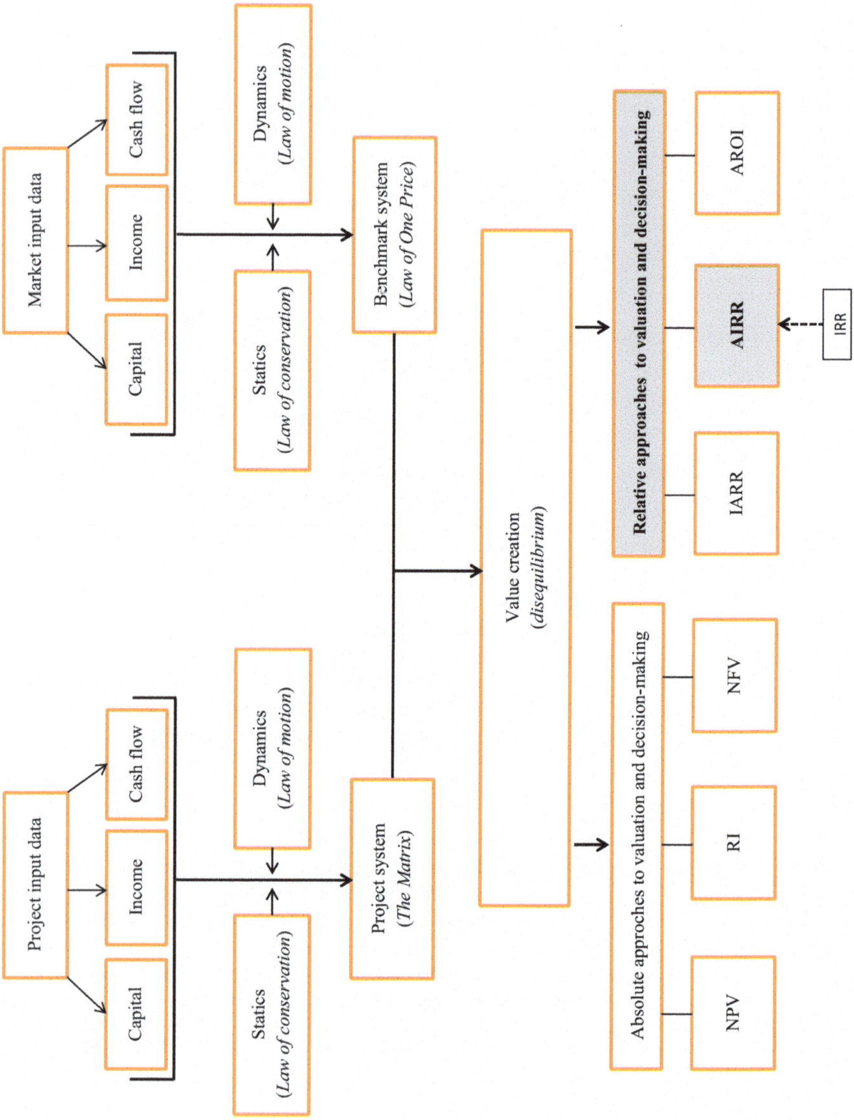

Fig. 8.1 Chapter 8: The average-internal-rate-of-return approach and the book rates of return

In this chapter we investigate the notion of rate of return for multiperiod projects. Literally, the rate of return of an economic entity is an **amount of return per unit of total capital employed**. We first use this intuition to engineer a reliable notion of rate of return. Then, we show that it may also be conceptualized, in a rather natural way, as a **capital-weighted mean of the project's income rates**, called Average Internal Rate of Return (AIRR). Then, we corroborate the intuition with a logical approach and show that the economic value created enjoys a product structure which exposes the project economic efficiency (excess rate of return) and the project scale.

We show that the AIRR notion naturally reconciles with the RI approach, presented in Part III. We show how to compute the AIRR in four different ways, how to use it for rational decision-making and how to practically decompose the economic value created in the project's marginal efficiency (excess AIRR) and the project scale (total capital). In particular, we show how to draw and use the (levered and unlevered) **average ROI** and the **average ROE** from the pro forma financial statements, how to assess the investors' economic profitability (rate of return of the CFA stream) as distinguished from the equityholders' economic profitability (rate of return of the CFE stream), how to break down the project NPV into investment-period NPV and financing-period NPV whenever the project is a mixed transaction.

After completing this chapter, the reader should be able to:

- have a strong intuition of what a rate of return should be for a multiperiod economic asset
- understand that a rate of return is, by definition, a capital-based notion, not a cash-flow based one, and, therefore, it must be expressed as a ratio of return (income) to capital
- learn the notion of Chisini mean and appreciate its usefulness for building the notion of Average Internal Rate of Return (AIRR)
- interpret the rate of return of a project as a generalized mean of the holding period rates, weighted by the capital employed in the respective period
- understand why the AIRR approach leads to rational decisions
- calculate a project's or firm's rate of return in four different-but-logically-equivalent ways
- compute a suitable MARR when the (period) COC is time-varying
- appreciate the major role of accounting rates as key parameters of a project's rate of return and learn how to use them to measure the economic value created
- grasp the relations among various rates of return such as the average ROI, the average ROE, the average ROD
- learn the notion of *Economic AIRR* and interpret it as representative of the project's rate of return from the point of view of the capital market
- understand the notion of excess AIRR and its relation with *Profitability Index* and *Benefit-Cost Ratio*
- break down the project NPV into investment-period NPV and financing-period NPV for a mixed project, one where the firm invests in the project in some periods while borrowing from the project in some other periods.

8.1 Building the Rate-of-Return Notion Intuitively

A rate is, by definition, a "quantity, amount, or degree of something measured as per unit of something else" (Webster's Third New International Dictionary 1961). Consistently, a rate of return is, by definition, an amount of return measured as per unit of invested capital. In other words, it is a ratio of the amount of return generated to the total capital invested. Note that cash flow does not appear in this definition. Rate of return does not depend on cash flow, it depends on *return* (i.e., income) and *capital*. For the time being, we assume $I_0 = 0$ and forget about cash flows. Consider an n-period project with income sequence $\boldsymbol{I}$ and capital sequence $\boldsymbol{C}$ as follows:

$$\boxed{\begin{array}{r} I_1\ I_2\ \ldots\ I_{n-1}\ I_n \\ \hline C_0\ C_1\ C_2\ \ldots\ C_{n-1}\ 0 \end{array}} \tag{8.1}$$

Let $\bar{\imath}$ be the rate of return of the project. How should we calculate $\bar{\imath}$? As a first step, consider $n = 1$. The layout simplifies to

$$\boxed{\begin{array}{r} I_1 \\ \hline C_0\ \ \ \end{array}}$$

and the rate of return is, by definition, the ratio of the amount of return, I_1, to total capital, C_0:

$$\bar{\imath} = \frac{I_1}{C_0} = i_1. \tag{8.2}$$

The project rate of return coincides with the project ROI. Consider now $n = 2$. If a project lasts two periods, it means that some capital C_1 remains invested at time 1, which generates some return I_2 at time 2:

$$\boxed{\begin{array}{r} I_1\ I_2 \\ \hline C_0\ C_1\ \ \ \end{array}}$$

How is the rate of return computed for this project? Starting from (8.2) above, one may simply add the additional return I_2 to the numerator of (8.2) and the additional capital C_1 to the denominator, after due discounting by one period, given that there is a lag of one period between first income and second income as well as between first capital and second capital (so that the time value of money is complied with). Hence,

$$\bar{\imath} = \frac{I_1 + \dfrac{I_2}{1+r}}{C_0 + \dfrac{C_1}{1+r}}. \tag{8.3}$$

For example, consider the following data for a one-year project:

$$\begin{array}{|l|}\hline \quad\quad\$375 \\ \hline \$600 \quad\quad \\ \hline \end{array} \tag{8.4}$$

and let $r = 15\%$ be the MARR. The project rate of return is

$$\bar{\imath} = \frac{375}{600} = 62.5\%$$

which is to be interpreted as the ratio of the overall return to the overall capital, consistently with the primitive notion of rate of return. The RI of this one-year project measures the value created by the project as referred to time 1:

$$\text{RI}_1 = \overbrace{600}^{\text{investment size}} \cdot \overbrace{(62.5\% - 15\%)}^{\text{economic efficiency}} = 285.$$

Note that, consistently with what we have anticipated in Chap. 7, the economic value created is decomposed into investment scale and investment's economic efficiency: The project scale is \$600 and the economic efficiency is 47.5% (= 62.5% − 15%). Reframing it in present terms, the economic efficiency is 47.5%/1.15 = 41.3%, whence

$$V_0(\text{RI}) = \frac{\text{RI}_1}{1.15} = \overbrace{600}^{\text{investment size}} \cdot \overbrace{41.3\%}^{\text{economic efficiency}} = 247.8.$$

As the value of the RI coincides with the NPV, one may write NPV = 247.8. Next, suppose that the project lasts two years and that, in particular, the residual capital invested at the beginning of the second period is \$400, bringing about a second-period profit equal to \$225:

$$\begin{array}{|l|}\hline \quad\quad\$375\ \$225 \\ \hline \$600\ \$400 \quad\quad \\ \hline \end{array}$$

To find the project rate of return we add the additional income and the additional capital (both discounted by one period):

$$\bar{\imath} = \frac{375 + \dfrac{225}{1.15}}{600 + \dfrac{400}{1.15}} = \frac{\overbrace{570.7}^{\text{total return}}}{\underbrace{947.8}_{\text{total capital}}} = 60.2\%.$$

The total return of this project is \$570.7, generated by a total invested capital of \$947.8. Dividing the total return by the total capital one gets the project rate of return. The resulting 60.2% return, greater than the MARR, suggests that the project is worth undertaking. More precisely, the economic efficiency is 60.2% − 15% = 45.2%. As the project scale is \$947.8, the wealth increase, as measured at time 1, is $45.2\% \cdot 947.8 = 428.5$. In present terms, $428.5/1.15 = 372.6$. The intuition is confirmed by the market value of the RIs, which, as we know, coincides with the project NPV:

$$\begin{aligned}\text{NPV} &= V_0(\text{RI})\\ &= \frac{375 - 0.15 \cdot 600}{1.15} + \frac{225 - 0.15 \cdot 400}{1.15^2}\\ &= 247.8 + 124.8\\ &= 372.6\\ &= \underbrace{947.8}_{\text{investment size}} \cdot \underbrace{\frac{60.2\% - 15\%}{1.15}}_{\text{economic efficiency}}\end{aligned}$$

(numbers are rounded).

Next, suppose a 3-period project is at hand, with three incomes and three capitals:

	I_1	I_2	I_3
C_0	C_1	C_2	

(8.5)

We may compute the rate of return as before by dividing the total return by the total capital. With respect to the two-period case, one only has to add the third income to get the total return and the third capital to get the total capital. Taking into account that there is a lag of two periods between I_1 and I_3 and between C_0 and C_2, one may write

$$\bar{\imath} = \frac{I_1 + \dfrac{I_2}{1+r} + \dfrac{I_3}{(1+r)^2}}{C_0 + \dfrac{C_1}{1+r} + \dfrac{C_2}{(1+r)^2}}.$$

Generalization is then straightforward. For an n-period project, $n > 2$, such that

	I_1	I_2	$\dots$	I_{n-1}	I_n
C_0	C_1	C_2	$\dots$	C_{n-1}	

one may divide the total return by the total capital:

$$\bar{\imath} = \frac{\overbrace{I_1 + \dfrac{I_2}{1+r} + \dfrac{I_3}{(1+r)^2} + \dots + \dfrac{I_n}{(1+r)^{n-1}}}^{\text{total return}}}{\underbrace{C_0 + \dfrac{C_1}{1+r} + \dfrac{C_2}{(1+r)^2} + \dots + \dfrac{C_{n-1}}{(1+r)^{n-1}}}_{\text{total capital}}}. \tag{8.6}$$

Essentially, the simple idea is that if, for a single-period project, the rate of return is the ratio of return to capital, the same must be true for a multi-period project.

A project's rate of return is a ratio of
(overall) return to (overall) invested capital.

A different-but-equivalent way of looking at the rate of return is to consider the project as a portfolio of one-period projects generating n rates of return. For example, consider the two-period project analyzed above. It may be interpreted as a portfolio of two one-period investments with the following cash-flow streams:

Time	**0**	**1**	**2**
first project's cash flows	−600	975	0
second project's cash flows	0	−400	625

Precisely, \$600 are invested at $i_1 = (975 - 600)/600 = 62.5\%$ and \$400 are invested at $i_2 = (625 - 400)/400 = 56.3\%$. The rate of return of a portfolio is the mean of the rates of return of the constituent assets, weighted by the respective amounts invested. However, from the point of view of time 0, the invested amount in the second period is (not \$400 but) $\$400/1.15 = \$347.8$, so the overall investment is $\$947.8 = \$600 + \$347.8$. The weights of the portfolio are

$$w_1 = \frac{600}{947.8} = 63.3\%$$
$$w_2 = \frac{347.8}{947.8} = 36.7\%.$$

This means that 63.3% of the capital is invested at 62.5% while 36.7% is invested at 56.3%. Overall, the portfolio's rate of return is $\bar{\imath} = 63.3\% \cdot 62.5\% + 36.7\% \cdot 56.3\% = 60.2\%$. This is exactly the same rate of return we have obtained earlier by dividing the total return by the total capital. In symbols,

$$w_1 = \frac{C_0}{C_0 + C_1(1+r)^{-1}}$$
$$w_2 = \frac{C_1(1+r)^{-1}}{C_0 + C_1(1+r)^{-1}}$$

are the weights of the mean and the rate of return is $\bar{\imath} = w_1 i_1 + w_2 i_2$.[1] More generally, for an n-period project, $n > 2$, the rate of return of a project is the capital-weighted

[1] Apparently, the idea of splitting up a project into a sequence of n one-period projects and using the whole sequence of one-period rates as a measure of economic profitability traces back to Hirshleifer (1958) and Bailey (1959). See Magni (2014a, 2016b) for details.

mean of the period rates:

$$\bar{\imath} = w_1 i_1 + \cdots + w_n i_n \qquad \text{where } w_t = \frac{C_{t-1}(1+r)^{-(t-1)}}{\sum_{k=1}^{n} C_{k-1}(1+r)^{-(k-1)}}, \quad t \in \mathbb{N}_n^1. \quad (8.7)$$

Intuition also suggests that this rate of return should indeed be tied to value creation by an RI structure such as

Total Capital × Economic Efficiency = Economic Value Created.

Also, intuition suggests that, if a project is levered, one should be able to compute, with the same arguments, a rate of return for the four basic classes of incomes and capital, thereby finding

- the rate of return generated by the operations
- the rate of return generated by the non-operating assets
- the rate of return generated by the debt
- the rate of return generated by the equity.

We will devote Sect. 8.4 to corroborating such an intuition with a logically consistent approach.

Example 8.1 ♦

C^o	
	C^e

(Rate of return of a two-period project, assuming constant WACC). Consider the data for a two-period project in the table below.

t	Year	**0**	**1**	**2**
GP_t	Gross profit		\$700.0	\$700.0
$-\mathrm{SGA}_t$	Less: SGA		\$100.0	\$250.0
$-\mathrm{Dep}_t$	Less: Depreciation		\$400.0	\$200.0
$= \mathrm{EBIT}_t$	EBIT		\$200.0	\$250.0
$-\tau\mathrm{EBIT}_t$	Less: Taxes on EBIT (25%)		\$50.0	\$62.5
$= I_t^{uo}$	**NOPAT**		**\$150.0**	**\$187.5**
F_t^{capex}	Capital Expenditure	\$600.0		
C_t^o	**Invested Capital**	**\$600.0**	**\$200.0**	
r_t^{uo}	WACC		15%	15%

Using (8.6), the rate of return of this project is

$$\bar{\imath} = \frac{150 + 187.5(1.15)^{-1}}{600 + 200(1.15)^{-1}} = 40.45\%.$$

Alternatively, one may reason in terms of weighted mean of the ROIs. The ROIs are $i_1 = 150/600 = 25\%$ and $i_2 = 187.5/200 = 93.8\%$. The project may be viewed as a portfolio of two investments: Investment of \$600 at time 0 at 25% return and investment of \$200 at time 1 at 93.8% return. The first investment brings about an income of \$150, so that the EOP capital is \$750 = 600 + 150. The second investment brings about an income of \$187.5, so that the EOP capital is \$387.5 = 200 + 187.5:

Year	0	1	2
First investment's cash flows	−\$600	\$750	
Second investment's cash flows		−\$200	\$387.5
Rate of return		25%	93.8%

As the (present value of the) overall capital is $600 + 200(1.15)^{-1} = 773.9$, one gets the average ROI from (8.7):

$$\begin{aligned}\bar{\imath} &= 25\% \cdot \frac{600}{773.9} + 93.8\% \cdot \frac{200(1.15)^{-1}}{773.9} \\ &= 25\% \cdot 77.5\% + 93.8\% \cdot 22.5\% = 40.45\%.\end{aligned}$$

In tabular format,

Year		1	2
ROI (i_t)		25%	93.8%
weight (w_t)		77.5%	22.5%
Rate of return ($\bar{\imath}$)	40.45%		

We can now interpret the project as a portfolio of two one-year investments. Overall, the investor invests a total \$773.9 capital at a rate of return equal to 40.45%, thereby foregoing the opportunity of investing the same amount of capital in an equivalent-risk asset earning 15%. The overall RI is

$$\text{overall RI} = \overbrace{773.9}^{\text{investment size}} \; \overbrace{(40.45\% - 15\%)}^{\text{economic efficiency}} = 197.$$

In this way, one not only captures value creation, but also decomposes value creation into an investment-scale component (i.e., the capital) and an economic-efficiency component (i.e., the excess rate of return). Discounting the two annual rates, 40.45% and 15%, one gets the present value of the overall RI: $171.3 = 197/1.15$. This is confirmed by explicitly discounting the two RIs:

$$V_0(\text{RI}) = \frac{600 \cdot (25\% - 15\%)}{1 + 0.15} + \frac{200 \cdot (93.8\% - 15\%)}{(1 + 0.15)^2} = 171.3.$$

It should be noted that this approach enables aggregating the pieces of information already provided by the RI model:

$$\frac{600\cdot(25\%-15\%)}{1+0.15}+\frac{200\cdot(93.8\%-15\%)}{(1+0.15)^2}=\frac{773.9\cdot(40.45\%-15\%)}{1+0.15}.$$

The RI notion enables retrieving separate information about scale and profitability of every period, while the approach now introduced enables fleshing out information about *overall* scale and *overall* profitability. It "averages out" the pieces of information already contained in the RI approach, generates a reliable notion of rate of return, and connects rate of return and value creation.

Example 8.2 ♦

C^o	
	C^e

Consider a three-year unlevered project with the following capital stream and income stream:

$$\boldsymbol{C}=(200,100,50,0),\qquad \boldsymbol{I}=(0,80,-10,20).$$

As we know, an unlevered project is such that $r_t^{uo}=r_t^o=r_t=r_t^e$ as well as NPV $=$ NPV^o $=$ NPV^e. In particular, let us assume that the WACC is constant and equal to $r=3\%$; the overall invested capital is $200+100(1.03)^{-1}+50(1.03)^{-2}=344.2$ while the overall return is $80-10(1.03)^{-1}+20(1.03)^{-2}=89.1$. Therefore, the rate of return is

$$\bar{\imath}=\frac{89.1}{344.2}=25.9\%.$$

The excess rate of return is $22.9\%=25.9\%-3\%$. The project may be viewed as an investment of \$344.2 at an excess 22.9% (annual) gain. The instantaneous gain is $344.2\cdot 22.9\%/1.03=76.5=V_0(\text{RI})=\text{NPV}$.

8.2 Coherent Rates, AIRR, and Rational Decisions

A coherent rate of return is one which is true to its name: An amount of return per unit of capital invested. Therefore, it must be suited to be expressed as a ratio.

Definition 8.1 (*Coherence of rate of return*) A rate of return of a multiperiod asset is coherent if it is an amount of return measured per unit of capital employed, that is, if it may be expressed as a ratio:

$$\text{Rate of return}=\frac{\text{Return}}{\text{Capital}}.\tag{8.8}$$

In other words, a coherent rate of return is a rate of growth for capital enjoying the following product structure:

$$\text{Rate of return}\cdot\text{Capital}=\text{Return}.$$

Definition (8.8) simply formalizes Hirshleifer's (1958) idea that a rate of return should be a ratio if it is to be coherent with its purpose: "The idea of a rate of growth involves a *ratio*" (Hirshleifer 1958, p. 347, italics added).

In a multiperiod economic activity, there are $n > 1$ period rates of return (income rates), each of which is obtained as a ratio of the period return I_t to the BOP capital, C_{t-1}. More precisely, given a profit stream equal to $\boldsymbol{I} = (0, I_1, I_2, \dots, I_n)$ and a capital stream equal to $\boldsymbol{C} = (C_0, C_1, \dots, C_{n-1}, 0)$, then the single-period income rates are $i_t = I_t/C_{t-1}$, $t = 1, 2, \dots, n$. However, as $I_t = E_t - C_{t-1}$, one may write $i_t = (E_t - C_{t-1})/C_{t-1}$ (see Eq. (1.9)).

Any multiperiod project may be interpreted as a portfolio of n one-period assets such that asset t is an investment of C_{t-1} at time $t-1$ (or a borrowing, if $C_{t-1} < 0$) with an EOP capital equal to E_t with (investment or financing) rate equal to $i_t = (E_t - C_{t-1})/C_{t-1}$. Graphically, the project may then be broken down into n assets, as reported below, where we have used Eq. (1.3) to find the portfolio's cash flows:

Time	**0**	**1**	**2**	**3**	$\cdots$	$\boldsymbol{n-1}$	$\boldsymbol{n}$	rate of return
Asset 1	$-C_0$	$+E_1$	0	0	$\cdots$	0	0	i_1
Asset 2	0	$-C_1$	$+E_2$	$\cdots$	0	0	0	i_2
Asset 3	0	0	$-C_2$	$+E_3$	$\cdots$	0	0	i_3
$\vdots$	$\vdots$	$\vdots$		$\vdots$	$\vdots$	$\vdots$	$\vdots$	$\vdots$
Asset n	0	0	0	0	$\cdots$	$-C_{n-1}$	$+E_n$	i_n
Net cash flows	$-C_0$	F_1	F_2	F_3	$\cdots$	F_{n-1}	F_n	$\bar{\imath} = ?$

The problem of finding a rate of return, $\bar{\imath}$, for the portfolio boils down to finding a measure which summarizes the pieces of information conveyed by the rates of return i_t of the n constituent assets, with the understanding that the resulting measure is a coherent rate of return according to Definition 8.1.

A solution to this problem can be found by adopting the general notion of mean introduced by Chisini (1929) and apply it to the notion of RI. Oscar Chisini (1889–1967), an Italian mathematician, introduced the general notion of a mean by making use of a simple, pragmatic argument which rests on the operational meaning of the concept underlying it. Specifically, he suggested to simplify the problem of summarizing n observable data by replacing them with a unique number which does not alter the framework of the problem under consideration. This way, it is *as if* the n data all had the same value. Under this approach, it makes no sense to speak of the mean of two or more quantities, but it makes sense to speak of the mean of those quantities *with respect to a given function which remains unvaried*:

> The search for a mean has the purpose of simplifying a given issue, by substituting, to many values, a single value which summarizes them, while leaving the overall picture of the problem under consideration unchanged. [...] It makes no sense to speak of *the mean of two or more values*, but it makes sense to speak of the mean of those values *with reference to the synthetical evaluation of a quantity that depends on them*. (Chisini 1929, pp. 107–108, our translation, adjusted from Muliere and Parmigiani 1993, p. 423; italics in original)

The crucial point here is to select an appropriate function of the quantities that one wants to remain unvaried, while replacing said quantities with a unique value. This unique value is the "Chisini mean".

Definition 8.2 (*Chisini mean*) The Chisini mean of n homogeneous values $x_1, \ldots, x_n$, *with respect to* function g, is the unique number (if it exists) $\bar{x}$ such that

$$g(\bar{x}, \bar{x}, \ldots, \bar{x}) = g(x_1, \ldots, x_n). \quad (8.9)$$

Equation (8.9) is the (Chisini) *invariance requirement.*

Example 8.3 ♦
The arithmetic mean, the geometric mean, the harmonic mean, the power mean are all special cases of Chisini mean. For example, consider the function $g(x_1, x_2, x_3) = x_1^2 \cdot x_2^4 \cdot x_3^6$. The Chisini mean of x_1, x_2, x_3 with respect to g is that unique value $\bar{x}$ such that

$$x_1^2 \cdot x_2^4 \cdot x_3^6 = \bar{x}^2 \cdot \bar{x}^4 \cdot \bar{x}^6$$

which implies $\bar{x} = x_1^{1/6} \cdot x_2^{1/3} \cdot x_3^{1/2}$. This is the weighted geometric mean of x_1, x_2, x_3.

Example 8.4 ♦
Consider the function $g(x_1, x_2, x_3) = x_1 + 2x_2 - 4x_3$. The Chisini mean of x_1, x_2, x_3 with respect to g is that unique value $\bar{x}$ such that

$$x_1 + 2x_2 - 4x_3 = \bar{x} + 2\bar{x} - 4\bar{x},$$

which implies $\bar{x} = -(x_1 + 2x_2 - 4x_3)$.

Example 8.5 ♦
Consider the function $g(x_1, x_2, x_3, x_4) = \ln x_1 + \ln x_2 - e^{x_3/x_4}$. The Chisini mean of x_1, x_2, x_3, x_4 with respect to g is that unique value $\bar{x}$ such that

$$\ln x_1 + \ln x_2 - e^{x_3/x_4} = 2 \ln \bar{x} - e$$

which implies

$$\bar{x} = \sqrt{x_1 \cdot x_2} \cdot e^{\frac{e(x_4 - x_3)}{2x_4}}$$

It should be stressed that, for a given situation, there may be different Chisini means that fit different perspectives. For example, suppose Rental LP rents a car to Mrs. Williams, who covers a distance of 120 km at the speed of x_1 km/h and a distance of 235 km at the speed of x_2 km/h. Suppose Mrs. Williams is interested to know the average speed. Then, let $g(x_1, x_2) = 120/x_1 + 235/x_2$ be the total traveling time. The Chisini mean is that unique value $\bar{x}^*$ such that

$$\frac{120}{x_1} + \frac{235}{x_2} = \frac{120}{\bar{x}^*} + \frac{235}{\bar{x}^*}$$

whence

$$\bar{x}^* = \frac{120 + 235}{\dfrac{120}{x_1} + \dfrac{235}{x_2}}.$$

In contrast, suppose Rental LP is interested to know the average speed with respect to the total consumption. Assume the consumption, for 1 km, depends on the speed x according to the function $c(x) = a + b(x - 60)^2$, where a and b are suitable constant coefficients. Then, the invariant function for Rental LP is

$$\begin{aligned} g(x_1, x_2) &= c(x_1) \cdot 120 + c(x_2) \cdot 235 \\ &= 120 \cdot \big(a + b(x_1 - 60)^2\big) + 235 \cdot \big(a + b(x_2 - 60)^2\big) \end{aligned}$$

which implies that the Chisini mean is that unique value $\bar{x}^{**}$ such that

$$\begin{aligned} &120 \cdot \big(a + b(x_1 - 60)^2\big) + 235 \cdot \big(a + b(x_2 - 60)^2\big) \\ &= 120 \cdot \big(a + b(\bar{x}^{**} - 60)^2\big) + 235 \cdot \big(a + b(\bar{x}^{**} - 60)^2\big) \end{aligned}$$

whence

$$\bar{x}^{**} = 60 + \sqrt{\frac{120(x_1 - 60)^2 + 235(x_2 - 60)^2}{120 + 235}}$$

(see also Chisini 1929, pp. 106–107). As a result, $\bar{x}^*$ is the average speed that leaves the total distance unchanged, while $\bar{x}^{**}$ is the average speed that leaves the total consumption unchanged. In other words, in terms of distance, it is as if the car had covered the same distance at a constant speed $\bar{x}^*$; in terms of consumption, it is as if the car had consumed the same amount of fuel at a constant speed $\bar{x}^{**}$.

Remark 8.1 (*ROI, WACC, and spot rates as Chisini means*) It may be observed that the ROI and the WACC are Chisini means. For example, consider the value of the operating assets, V_t^o. It may be viewed as a function $g = g(r_t^e, r_t^{\mathcal{D}})$ of r_t^e and $r_t^{\mathcal{D}}$:

$$g(r_t^e, r_t^{\mathcal{D}}) = V_{t-1}^e(1 + r_t^e) - F_t^e + V_{t-1}^{\mathcal{D}}(1 + r_t^{\mathcal{D}}) - F_t^{\mathcal{D}}.$$

The Chisini mean of r_t^e and $r_t^{\mathcal{D}}$ with respect to $g(r_t^e, r_t^{\mathcal{D}})$ is the unique number $\bar{x}$ such that $g(r_t^e, r_t^{\mathcal{D}}) = g(\bar{x}, \bar{x})$. Thus,

$$V_{t-1}^e(1 + r_t^e) - F_t^e + V_{t-1}^{\mathcal{D}}(1 + r_t^{\mathcal{D}}) - F_t^{\mathcal{D}} = V_{t-1}^e(1 + \bar{x}) - F_t^e + V_{t-1}^{\mathcal{D}}(1 + \bar{x}) - F_t^{\mathcal{D}}$$

whence

$$\bar{x} = \frac{r_t^e V_{t-1}^e + r_t^{\mathcal{D}} V_{t-1}^{\mathcal{D}}}{V_{t-1}^e + V_{t-1}^{\mathcal{D}}}.$$

Therefore, $\bar{x} = r_t^o$ is the pre-tax (operating) WACC. The same result holds for the other WACCs and holds for any ROI as well: It suffices to replace the market rates with the project income rates in the above invariant function.

The spot rates of the term structure of interest rates, defined in Remark 5.1, are Chisini means as well. Specifically, for any fixed t, consider the function $g(r_1, r_2, \ldots, r_t) = (1+r_1) \cdot (1+r_2) \cdot \ldots \cdot (1+r_t)$. The Chisini mean of r_1, r_2, ..., r_t with respect to g is that unique number $\bar{x}$ such that $g(r_1, r_2, \ldots, r_t) = g(\bar{x}, \bar{x}, \ldots, \bar{x})$ or, equivalently,

$$(1+r_1) \cdot (1+r_2) \cdot \ldots \cdot (1+r_t) = (1+\bar{x})^t.$$

The unique solution of this equation is precisely $\bar{x} = s_t$, the spot rate for a financial transaction (lending or borrowing) starting at time 0 and ending up at time t.

The law of conservation applied to the period rates, introduced in Sect. 2.1.4, may then be interpreted as a Chisini invariance requirement. The same can be said for the law of one price, which guarantees consistency in the term structure of interest rates and in the structure of asset prices. In general, the law of conservation, the law of one price, and Chisini invariance requirement are special cases of a general conservation principle. ▲

The choice of the function g depends on the problem object of the analysis: Different choices of g lead to different kinds of mean (arithmetic, geometric, harmonic, etc.) ...See de Finetti (1931) *for more details and Graziani and Veronese* (2009) *for some applications ...Equation* (8.9) *may have no solutions, so that the mean does not exist, or it may have several solutions and each one may be used as a mean. Indeed the solution exists and is unique if* $g(x, \ldots, x) = q(x)$, *with* $q(\cdot)$ *continuous and strictly monotone. In this case, the mean is given by* $\bar{x} = q^{-1}(g(x_1, \ldots, x_n))$. *Note that the Chisini mean is always consistent, in the sense that when all values* $x_1, \ldots, x_n$ *are equal to a, then* $\bar{x} = a$. *Due to its general definition, it has no other specific properties. In particular it is not necessarily internal, i.e. included between the minimum and the maximum values of the observations, or associative. The latter property which, roughly speaking, states that a mean of n data can be computed from the means of partial non overlapping subgroups of values, is however satisfied when the function q meets suitable conditions. This happens, for example, when the Chisini mean generates an arithmetic or a geometric mean.*

Magni et al. (2018). Chisini means and rational decision-making: Equivalence of investment criteria. *Mathematics and Financial Economics*, 12(2), 193–217 (Sect. 2 and Appendix A). https://doi.org/10.1007/s11579-017-0201-4

In our context, we want to summarize the single-period income rates of the vector $\boldsymbol{i}$. Since the rate of return must be coherent and, at the same time, be consistent with the notion of value creation (and since the rates i_t are linked with economic value created via the RI notion), it seems natural to select the economic value created $V_0(\mathrm{RI})$ as invariance requirement. Specifically, one picks

$$g(i_1, i_2, \ldots, i_n) = \sum_{t=1}^{n} C_{t-1}(i_t - r)(1+r)^{-t} = V_0(\mathrm{RI})$$

and the invariance requirement (8.9) becomes

$$\sum_{t=1}^{n} C_{t-1}(\bar{\imath} - r)(1+r)^{-t} = \sum_{t=1}^{n} C_{t-1}(i_t - r)(1+r)^{-t}$$

which implies that the mean rate of return is given by

$$\bar{\imath} = \frac{\sum_{t=1}^{n} i_t C_{t-1}(1+r)^{-(t-1)}}{\sum_{t=1}^{n} C_{t-1}(1+r)^{-(t-1)}} \tag{8.10}$$

(see also Magni 2010b).

Expression (8.10) is a (generalized) weighted arithmetic mean with weights represented by the present values of the capitals involved, $\mathrm{PV}[C_{t-1}] = C_{t-1}(1+r)^{-(t-1)}$. Being a mean of the income rates i_t, it is called *Average Internal Rate of Return* (AIRR).[2] As $i_t \cdot C_{t-1} = I_t$, Eq. (8.10) implies that the AIRR is indeed an amount of return measured as per unit of capital. Therefore, it expresses the ratio of total return to total capital. This implies that it is coherent according to Definition 8.1. We may then formalize the result in the following definition.

Definition 8.3 (*AIRR*) Given an economic asset with streams of profits and capitals equal to $\boldsymbol{I}$ and $\boldsymbol{C}$, respectively, its rate of return, named *Average Internal Rate of Return* (AIRR), is equal to the ratio of the project's overall return to the project's overall capital:

$$\bar{\imath} = \frac{\overbrace{I_1 + I_2(1+r)^{-1} + \ldots + I_n(1+r)^{-(n-1)}}^{\text{total return}}}{\underbrace{C_0 + C_1(1+r)^{-1} + \ldots + C_{n-1}(1+r)^{-(n-1)}}_{\text{total capital}}}. \tag{8.11}$$

[2]Actually, the adjective "internal" is somewhat redundant. It refers to the fact that the income rates i_t are internal to the project (they do not depend on r) and that AIRR is an average of those rates (note that AIRR is not internal, that is, it does depend on r). See also Sect. 9.2 on this point.

If $C_{t-1} \neq 0$ for every $t \in \mathbb{N}_n^1$, the AIRR may be framed as a (generalized) weighted mean:

$$\bar{\imath} = w_1 i_1 + w_2 i_2 + \cdots + w_n i_n \tag{8.12}$$

where $w_t = C_{t-1}(1+r)^{-(t-1)} / \sum_{k=1}^{n} C_{k-1}(1+r)^{-(k-1)}$. Let $\text{PV}[I] = \sum_{k=1}^{n} I_k$ $d_{k,0}$ be the present value of the period returns and $\text{PV}[C] = \sum_{k=1}^{n} C_{k-1} d_{k-1,0}$ the present value of the capital amounts. The AIRR may also be framed as

$$\bar{\imath} = \overbrace{\frac{\text{PV}[I]}{\text{PV}[C]}}^{\text{instantaneous AIRR}} \cdot \overbrace{(1+r)}^{\text{conversion factor}}. \tag{8.13}$$

The first factor of (8.13) is the ratio $\text{PV}[I]/\text{PV}[C]$: It is an instantaneous return, so to say, for it is a ratio between two present values. The factor $(1 + r)$ is a conversion factor, which turns the instantaneous rate of return into a discrete (e.g., annual, if r is annual) rate of return. Therefore, the ratio $\text{PV}[I]/\text{PV}[C]$ provides the same information as $\bar{\imath}$, albeit in terms of an instantaneous return per unit of invested capital.

Consider the numerical example of a one-year project illustrated in Sect. 8.1: The present value of the return is $375/1.15 = 326.1$. Dividing this present value by 600 one gets the instantaneous rate of return: $\bar{\imath}_0 = 326.1/600 = 54.35\%$. In essence, the investor invests $\text{PV}[C] = 600$ getting an instantaneous 54.35% return. Multiplying by 1.15 one gets the annual rate of return: $\bar{\imath} = 54.35\% \cdot 1.15 = 62.5\%$. We will henceforth use the symbol

$$\bar{\imath}_0 = \frac{\text{PV}[I]}{\text{PV}[C]} \tag{8.14}$$

to denote the instantaneous AIRR, so that $\bar{\imath} = \bar{\imath}_0(1 + r)$. Analogously, one may introduce the instantaneous COC, which is equal to the ratio of the total normal profit to the total capital employed

$$\bar{r}_0 = \frac{r}{1+r} = \frac{\text{PV}[I^*]}{\text{PV}[C]}. \tag{8.15}$$

where $\text{PV}[I^*] = \sum_{t=1}^{n} I_t^*(1+r)^{-t} = \sum_{t=1}^{n} rC_{t-1}(1+r)^{-t}$.[3] Note that (8.13) (as well as (8.11)) does not require the condition $C_{t-1} \neq 0$ for all $t \in \mathbb{N}_n^1$, which is instead required by (8.12). This indicates that Chisini mean serves the purpose of *finding* a coherent rate of return by properly aggregating the period rates *provided that* it exists (i.e., if capitals are all nonzero). But the general definition is precisely given by (8.11) or (8.13).

---

[3] In spreadsheet modelling, (8.14)–(8.15) are particularly easy to use, especially when the COC is time-varying (see Sect. 8.3): One just has to use the already available SUM PRODUCT for calculating the present value of incomes (income stream $\boldsymbol{I}$ times discount-factor stream $\mathbf{d}$) and then copy the same formula for capitals.

If $C_{t-1} < 0$ for some t, then the project is a borrowing in some period, which means that it has a *mixed* financial nature (see Definition 1.1). In order to unambiguously determine the financial nature of $\bar{\imath}$, we need a criterion for identifying, unequivocally, the financial nature of any asset. Building upon the definition of invested capital and borrowed capital given in Sect. 1.3, we supply a definition of investment/financing grounded upon the sign of the overall capital committed, PV$[C]$, which summarizes the overall financial position of the economic agent: If PV$[C] > 0$ (< 0), it means that the investment (financing) positions exceed the financing (investment) positions, so, overall, the asset is a (net) investment (financing).[4]

Definition 8.4 (*Investment or financing*) Let A be a mixed asset, such that $C_j \cdot C_h < 0$ for some $j, h \in \mathbb{N}^0_{n-1}$. Then,

- if PV$[C] > 0$, A is a (net) investment
- if PV$[C] < 0$, A is a (net) financing or borrowing.

Consistently with the financial meaning of any income rate, the AIRR is a financing rate if the asset is a financing (see also Table 1.4 in Sect. 1.3).

The condition of being coherent is not sufficient for a rate of return to be reliable. Whether or not it is rational depends on whether the associated decision criterion is rational according to Definition 5.3. We first supply the AIRR decision-making criterion and then show that it is rational.

Definition 8.5 (*AIRR criterion—accept/reject decisions*) Let A be any economic activity. Then,

- if A is an investment, it is worth undertaking if and only if $\bar{\imath} > r$
- if A is a financing, it is worth undertaking if and only if $\bar{\imath} < r$.

Equivalently,

- if A is an investment, it is worth undertaking if and only if $\bar{\imath}_0 > \bar{r}_0$
- if A is a financing, it is worth undertaking if and only if $\bar{\imath}_0 < \bar{r}_0$

where $\bar{r}_0 = r/(1+r)$.

Further, the investor is indifferent about accepting or rejecting A if and only if $\bar{\imath} = r$ or, equivalently, if and only if $\bar{\imath}_0 = \bar{r}_0$.

[4]See also Magni (2010b, 2013a).

We know that, given a sequence of incomes $\boldsymbol{I}$ and a sequence of capitals $\boldsymbol{C}$, the asset NPV is equal to the market value of RIs: NPV $= V_0(\text{RI}) = \sum_{t=1}^{n}(I_t - I_t^*)(1 + r)^{-t} = \sum_{t=1}^{n}(I_t - rC_{t-1})(1+r)^{-t}$ (see Eq. (6.146)). The latter equality may be reframed as

$$\begin{aligned}\text{NPV}(1+r) &= \sum_{t=1}^{n}(I_t - rC_{t-1})(1+r)^{-(t-1)} \\ &= \text{PV}[I](1+r) - r\text{PV}[C] \\ &= \text{PV}[C]\left(\frac{\text{PV}[I]}{\text{PV}[C]}(1+r) - r\right) \\ &= \text{PV}[C](\bar{\imath} - r)\end{aligned} \tag{8.16}$$

whence the following proposition holds.

Proposition 8.1 (Decomposition of economic value created) *The economic value created is decomposed into economic efficiency and project scale (size):*

$$\text{NPV}(1+r) = \overbrace{\text{PV}[C]}^{\text{project scale}} \cdot \overbrace{(\bar{\imath} - r)}^{\text{economic efficiency}} \tag{8.17}$$

or, using instantaneous economic efficiency,

$$\text{NPV} = \overbrace{\text{PV}[C]}^{\text{project scale}} \cdot \overbrace{(\bar{\imath}_0 - \bar{r}_0)}^{\substack{\text{economic efficiency} \\ \text{(instantaneous)}}} \tag{8.18}$$

(See the various derivations in Magni 2010b, 2013a, 2016b, and Magni et al. 2018.) Equations (8.17)–(8.18) prove, for multiperiod projects, what we have already proved for one-period projects (Chap. 7): The economic value created is an excess profit over and above the profit that would be earned if investors invested in the market at the rate r instead of investing in the asset at the rate $\bar{\imath}$. Equation (8.17) expresses the idea in discrete terms, while Eq. (8.18) expresses the same idea in instantaneous terms: It splits up the instantaneous gain, NPV, into two variables: The instantaneous capital, PV$[C]$, and the instantaneous above-normal rate of return, $\bar{\imath}_0 - \bar{r}_0$.

Using Proposition 5.2, a decision criterion is rational if and only if it is **NPV-consistent**.

Definition 8.6 (*NPV consistency*) A metric and its associated decision criterion is NPV-consistent if and only if the decision recommended by the criterion is the same as the decision recommended by the NPV.

(See Magni 2016b; Magni et al. 2018; Marchioni and Magni 2018 for formal definitions of NPV-consistency and strong NPV-consistency). A project is worth undertaking if and only if $V_0(\text{RI}) = \text{NPV} > 0$; considering (8.17) above, this occurs if and only if $\bar{\imath} > r$ with $\text{PV}[C] > 0$ (or $\bar{\imath} < r$ with $\text{PV}[C] < 0$). Therefore, rationality of AIRR is proved owing to **NPV-consistency**.[5]

Proposition 8.2 (Rationality of AIRR) *The AIRR criterion for accept/reject decision is rational: An investment creates value if and only if* $\bar{\imath} > r$*; a financing creates value if and only if* $\bar{\imath} < r$*. A project is value-neutral if and only if* $\bar{\imath} = r$*.*

Equations (8.17)–(8.18) even provide a shortcut to compute a project's rate of return. Specifically, while the rate-of-return notion depends, by definition, on return and capital (not on cash flows), we know that, in the mechanics of an economic system, incomes and capitals unambiguously determine cash flows via the law of motion: $F_t = I_t - \Delta C_t$. Therefore, one may well calculate a project's AIRR using cash flows. Reverse-engineering Eqs. (8.17)–(8.18), one gets

$$\begin{aligned} \bar{\imath} &= r + \frac{\text{NPV}}{\text{PV}[C]}(1+r) \\ &= r + \frac{F_0 + \frac{F_1}{1+r} + \frac{F_2}{(1+r)^2} + \cdots + \frac{F_n}{(1+r)^n}}{\text{PV}[C]} \cdot (1+r), \end{aligned} \tag{8.19}$$

and, in instantaneous terms,

$$\begin{aligned} \bar{\imath}_0 &= \frac{r}{1+r} + \frac{\text{NPV}}{\text{PV}[C]} \\ &= \frac{r}{1+r} + \frac{F_0 + \frac{F_1}{1+r} + \frac{F_2}{(1+r)^2} + \cdots + \frac{F_n}{(1+r)^n}}{\text{PV}[C]}. \end{aligned} \tag{8.20}$$

[5]NPV consistency is also called *goal congruence* in the accounting and managerial literature: "Goal congruence requires that the managers have an incentive to accept all positive NPV projects" (Gow and Reichelstein 2007, p. 115. See also Peasnell 1995a, b; Hajdasinski 1995, 1997; Hartman 2000; Pfeiffer 2004; Pfeiffer and Velthuis 2005; Mohnen and Bareket 2007; Lindblom and Sjögren 2009; Magni 2009c, 2016b; Chiang et al. 2010; Pasqual et al. 2013 on NPV-consistency). A stronger definition of NPV consistency has been recently introduced in Marchioni and Magni (2018), using Spearman's (1904) correlation coefficient and Iman and Conover's (1987) top-down coefficient.

Four methods for calculating a project's rate of return[6]

A project's rate of return (AIRR) may be conceptualized and computed in four different-but-equivalent ways:

$$① \quad \bar{\imath} = \frac{\sum_{t=1}^{n} I_t(1+r)^{-(t-1)}}{\sum_{t=1}^{n} C_{t-1}(1+r)^{-(t-1)}}$$

$$② \quad \bar{\imath} = w_1 i_1 + w_2 i_2 + \ldots + w_n i_n$$

$$③ \quad \bar{\imath} = \frac{\mathrm{PV}[I]}{\mathrm{PV}[C]}(1+r)$$

$$④ \quad \bar{\imath} = r + \frac{\mathrm{NPV}}{\mathrm{PV}[C]} \cdot (1+r)$$

Example 8.6 ♦

C^o C^e Consider an unlevered project, summarized in the following concise transposed Matrix:

(*C*, *I*, *F*)

Time		0	1	2	3	4	5	6
Capital	C	120	60	10	−80	6	4	0
Income	I		24	−6	5	12	−2	5
Cash Flow	F	−120	84	44	95	−74	0	9

The project is a mixed one, because, in the fourth period, the investor is in a borrowing position. To check whether, overall, the project is a net investment or a net financing, one computes the total capital. Assuming a MARR equal to $r = 10\%$, $\mathrm{PV}[C] = 120 + 60/1.1 + 10/1.1^2 - 80/1.1^3 + 6/1.1^4 + 4/1.1^5 = 129.29 > 0$. Hence, the investment positions prevail over the borrowing position and the project is a net investment. We compute the AIRR using the four methods presented above.

[6]For tutorials see related YouTube Channel: https://www.youtube.com/channel/UC9hfAZ8Sj5DwxUO_gIZUwxg.

① Applying (8.11),

$$\bar{\imath} = \frac{24 - 6/1.1 + 5/1.1^2 + 12/1.1^3 - 2/1.1^4 + 5/1.1^5}{120 + 60/1.1 + 10/1.1^2 - 80/1.1^3 + 6/1.1^4 + 4/1.1^5} = \frac{33.43}{129.3} = 25.86\%.$$

② The weights are

$$\begin{aligned}
w_1 &= \frac{120}{129.3} = 92.8\% \\
w_2 &= \frac{60/1.1}{129.3} = 42.2\% \\
w_3 &= \frac{10/1.1^2}{129.3} = 6.4\% \\
w_4 &= \frac{-80/1.1^3}{129.3} = -46.5\% \\
w_5 &= \frac{6/1.1^4}{129.3} = 3.2\% \\
w_6 &= \frac{4/1.1^5}{129.3} = 1.9\%.
\end{aligned}$$

The income rates are

$$\begin{aligned}
i_1 &= \frac{24}{120} = 20\% \\
i_2 &= \frac{-6}{60} = -10\% \\
i_3 &= \frac{5}{10} = 50\% \\
i_4 &= \frac{12}{-80} = -15\% \\
i_5 &= \frac{-2}{6} = -33.3\% \\
i_6 &= \frac{5}{4} = 125\%.
\end{aligned}$$

(Note that one weight, w_4, is negative, owing to the borrowing position. In this borrowing period, the income rate is a financing rate and is negative, meaning that the investor earns \$12 out of the borrowing. Applying (8.12),

$$\begin{aligned}
\bar{\imath} = 92.8\% \cdot 20\% + 42.2\% \cdot (-10\%) + 6.4\% \cdot 50\% - 46.5\% \\
\cdot (-15\%) + 3.2\%(-33.3\%) + 1.9\% \cdot 125\% = 25.86\%
\end{aligned}$$

③ Applying (8.13),

$$\begin{aligned}\bar{\imath} &= \frac{24/1.1 - 6/1.1^2 + 5/1.1^3 + 12/1.1^4 - 2/1.1^5 + 5/1.1^6}{120 + 60/1.1 + 10/1.1^2 - 80/1.1^3 + 6/1.1^4 + 4/1.1^5} \cdot 1.1 \\ &= 23.51\% \cdot 1.1 \\ &= 25.86\%\end{aligned}$$

④ Applying (8.19),

$$\bar{\imath} = 0.1 + \frac{-120 + 84/1.1 + 44/1.1^2 + 95/1.1^3 - 74/1.1^4 + 9/1.1^6}{129.3} = 25.86\%.$$

Overall, the project consists of investing \$129.3 at 25.86%, while the same amount might be invested at 10%. The project scale is \$129.3, the economic efficiency is 15.86% (= 25.86% − 10%). Hence, the economic value created is NPV $= V_0(\text{RI}) = 129.3 \cdot 15.86\%(1.1)^{-1} = 18.64$. In terms of present value, the instantaneous AIRR is $\bar{\imath}_0 = \bar{\imath}/(1+r) = 0.2586/1.1 = 23.51\%$, the instantaneous COC is $\bar{r}_0 = r/(1+r) = 0.10/1.1 = 9.1\%$. Hence, the (instantaneous) economic efficiency is 23.51% − 9.09% = 14.42%. NPV is decomposed into scale and instantaneous economic efficiency:

$$\text{NPV} = 129.3 \cdot 14.42\%. = 18.64.$$

8.3 Coping with Time-Varying COCs

If the COC is time-varying, Definition 8.3 is generalized by replacing r with r_t. Using the discount factor $\mathrm{d}_{t,j} = \prod_{h=j+1}^{t}(1+r_h)^{-1}$, we give the following general definition of AIRR.

Definition 8.7 (*Average Internal Rate of Return—general*) Given an asset A with streams of profits and capitals equal to $\boldsymbol{I}$ and $\boldsymbol{C}$, the asset's rate of return, *Average Internal Rate of Return* (AIRR), is the ratio of total return to total capital. It may be computed in three different ways:

① as the ratio of total return to total capital:

$$\bar{\imath} = \frac{I_1 + I_2\mathrm{d}_{2,1} + \ldots + I_n\mathrm{d}_{n,1}}{C_0 + C_1\mathrm{d}_{1,0} + \ldots + C_{n-1}\mathrm{d}_{n-1,0}} \tag{8.21}$$

② as linear combination of the income rates i_t (as long as $C_{t-1} \neq 0$ for every $t \in \mathbb{N}_n^1$):

$$\bar{\imath} = w_1 i_1 + w_2 i_2 + \ldots + w_n i_n \tag{8.22}$$

where $w_t = C_{t-1}\mathrm{d}_{t,1}/\mathrm{PV}[C]$
③ as product of the instantaneous rate of return and the converting factor $(1 + r_1)$:

$$\bar{\imath} = \frac{\mathrm{PV}[I]}{\mathrm{PV}[C]}(1 + r_1) \tag{8.23}$$

where $\mathrm{PV}[I] = \sum_{t=1}^{n} I_t \mathrm{d}_{t,0}$.

However, whenever COC is time-varying, an additional puzzling issue arises, which has baffled scholars long since: How should one compute the MARR? The standard answer in finance is that, in this case, it is not possible to derive any meaningful overall COC to be used as a MARR. Contrary to this view, it suffices to apply the same argument to the benchmark system described in Sect. 6.6.1 and define the MARR as the ratio of the (overall) benchmark profit to the (overall) invested capital. Recalling that the benchmark's transposed Matrix is $(\boldsymbol{C}, \boldsymbol{I}^*, \boldsymbol{F}^*)$ such that $I_t^* = r_t C_{t-1}$, one only needs to reframe Definition 8.7 with r_t replacing i_t and I_t^* replacing I_t.

Definition 8.8 (*Cost of capital—general*) Given an economic asset with streams of profits and capitals equal to $\boldsymbol{I}$ and $\boldsymbol{C}$, the COC or MARR is the ratio of replicating portfolio's total return to total capital. It may be computed in three different ways:

① as the ratio of total normal profit to total capital:

$$\bar{r} = \frac{\overbrace{I_1^* + I_2^*\mathrm{d}_{2,1} + \ldots + I_n^*\mathrm{d}_{n,1}}^{\text{benchmark total return}}}{\underbrace{C_0 + C_1\mathrm{d}_{1,0} + \ldots + C_{n-1}\mathrm{d}_{n-1,0}}_{\text{total capital}}} \tag{8.24}$$

② as linear combination of the normal profit rates r_t:

$$\bar{r} = w_1 r_1 + w_2 r_2 + \ldots + w_n r_n, \tag{8.25}$$

where $w_t = C_{t-1}\mathrm{d}_{t,1}/\mathrm{PV}[C]$
③ as product of the instantaneous COC and the converting factor $(1 + r_1)$:

$$\bar{r} = \frac{\mathrm{PV}[I^*]}{\mathrm{PV}[C]}(1 + r_1) \tag{8.26}$$

where $\mathrm{PV}[I^*] = \sum_{t=1}^{n} I_t^* d_{t,0}$.

The equality

$$\mathrm{NPV}(1+r_1) = (\mathrm{PV}[I] - \mathrm{PV}[I^*])(1+r_1) = \mathrm{PV}[I](1+r_1) - \bar{r}\mathrm{PV}[C] \quad (8.27)$$

generalizes (8.16).

We still use the symbol $\bar{r}_0$ to denote the instantaneous COC: $\bar{r}_0 = \mathrm{PV}[I^*]/\mathrm{PV}[C]$ so that $\bar{r} = \bar{r}_0(1+r_1)$, but $\mathrm{PV}[I^*]$ and $\mathrm{PV}[C]$ are now obtained by discounting profits and capitals at time-varying COCs. Equation (8.27) may be written as

$$\mathrm{NPV} = \mathrm{PV}[I] - \mathrm{PV}[I^*] = \mathrm{PV}[I] - \bar{r}\mathrm{PV}[C]/(1+r_1) = \mathrm{PV}[C] \cdot (\bar{\imath}_0 - \bar{r}_0). \quad (8.28)$$

The following proposition generalizes Proposition 8.1 and states that the economic value created is an overall RI which is decomposed into two components: Project scale (how much money one invests) and economic efficiency (how efficiently one invests money).

Proposition 8.3 (Decomposition of economic value created—general) *The economic value created is decomposed into economic efficiency and project scale (size):*

$$\mathrm{NPV}(1+r_1) = \overbrace{\mathrm{PV}[C]}^{\text{project scale}} \cdot \overbrace{(\bar{\imath} - \bar{r})}^{\text{economic efficiency}} \quad (8.29)$$

or, using instantaneous economic efficiency,

$$\mathrm{NPV} = \overbrace{\mathrm{PV}[C]}^{\text{project scale}} \cdot \overbrace{(\bar{\imath}_0 - \bar{r}_0)}^{\substack{\text{economic efficiency} \\ \text{(instantaneous)}}} \quad (8.30)$$

Reverse-engineering (8.29) and (8.30), one finds a fourth generalized method for computing the AIRR and the MARR, based on cash flows:

$$\bar{\imath} = \bar{r} + \frac{\mathrm{NPV}}{\mathrm{PV}[C]} \cdot (1+r_1) \quad (8.31)$$

$$\bar{r} = \bar{\imath} - \frac{\mathrm{NPV}}{\mathrm{PV}[C]} \cdot (1+r_1) \quad (8.32)$$

$$\bar{\imath}_0 = \bar{r}_0 + \frac{\mathrm{NPV}}{\mathrm{PV}[C]} \quad (8.33)$$

$$\bar{r}_0 = \bar{\imath}_0 - \frac{\mathrm{NPV}}{\mathrm{PV}[C]} \quad (8.34)$$

Definitions 8.4 and 8.5 keep on holding, with the caveat that the discount factor $\mathrm{d}_{t,0}$ is read $\prod_{j=1}^{t}(1+r_j)^{-1}$ instead of $(1+r)^{-t}$. Hence, we state the generalized AIRR criterion for accept/reject decisions.

Definition 8.9 (*AIRR criterion—accept/reject decisions—general*) Let A be any economic activity. Then,

- if A is an investment, it is worth undertaking if and only if $\bar{\imath} > \bar{r}$
- if A is a financing, it is worth undertaking if and only if $\bar{\imath} < \bar{r}$.

Equivalently,

- if A is an investment, it is worth undertaking if and only if $\bar{\imath}_0 > \bar{r}_0$
- if A is a financing, it is worth undertaking if and only if $\bar{\imath}_0 < \bar{r}_0$

where incomes and capitals are discounted with $d_{t,0} = \prod_{j=1}^{t}(1+r_j)^{-1}$.

Further, the investor is indifferent about accepting or rejecting A if and only if $\bar{\imath} = \bar{r}$.

Owing to Proposition 8.3, the rationality of the AIRR criterion holds for time-varying equilibrium rates as well as constant equilibrium rates.

Proposition 8.4 (Rationality of AIRR—general) *The AIRR criterion for accept/reject decisions is rational: An investment creates value if and only if* $\bar{\imath} > \bar{r}$ ($\bar{\imath}_0 > \bar{r}_0$)*; a financing creates value if and only if* $\bar{\imath} < \bar{r}$ ($\bar{\imath}_0 < \bar{r}_0$)*. A project is value-neutral if and only if* $\bar{\imath} = \bar{r}$ ($\bar{\imath}_0 = \bar{r}_0$)*.*

Remark 8.2 (*AIRR, RI, and benchmark portfolio*) The relation of AIRR approach and RI notion is strict. The AIRR approach is based on the difference between project system $(\boldsymbol{C}, \boldsymbol{I}, \boldsymbol{F})$ and the benchmark system $(\boldsymbol{C}, \boldsymbol{I}^*, \boldsymbol{F}^*)$, expressed in relative terms. Such a benchmark system is the one which gives rise to the standard RI notion. Therefore, the AIRR is the relative counterpart of the RI approach to value creation and decision-making (see more on the pair (AIRR, RI) in Sect. 10.6). ▲

We finally recollect below the four methods that the analyst may use for computing a project AIRR and the related project COC (MARR).

Four methods for calculating a project's rate of return and the corresponding MARR

A project's rate of return (AIRR) and a project's MARR may be conceptualized and computed in four different-but-equivalent ways:

①	$\bar{\imath} = \dfrac{\sum_{t=1}^{n} I_t\, \mathrm{d}_{t,1}}{\sum_{t=1}^{n} C_{t-1}\, \mathrm{d}_{t-1,0}}$	$\bar{r} = \dfrac{\sum_{t=1}^{n} I_t^*\, \mathrm{d}_{t,1}}{\sum_{t=1}^{n} C_{t-1}\, \mathrm{d}_{t-1,0}}$
②	$\bar{\imath} = w_1 i_1 + w_2 i_2 + \ldots + w_n i_n$	$\bar{r} = w_1 r_1 + w_2 r_2 + \ldots + w_n r_n$
③	$\bar{\imath} = \dfrac{\mathrm{PV}[I]}{\mathrm{PV}[C]}(1+r_1)$	$\bar{r} = \dfrac{\mathrm{PV}[I^*]}{\mathrm{PV}[C]}(1+r_1)$
④	$\bar{\imath} = \bar{r} + \dfrac{\mathrm{NPV}}{\mathrm{PV}[C]} \cdot (1+r_1)$	$\bar{r} = \bar{\imath} - \dfrac{\mathrm{NPV}}{\mathrm{PV}[C]} \cdot (1+r_1)$

Remark 8.3 ($\bar{\imath}_0$ *and* $\bar{r}_0$ *as Chisini means*) Formula ② is not an arithmetic mean (the sum of the weights is not 1). However, the instantaneous rate of return in formula ③ may indeed be framed as a weighted arithmetic mean:

$$\bar{\imath}_0 = \frac{\mathrm{PV}[I]}{\mathrm{PV}[C]} = \frac{i_1' C_0 + i_2' C_1 d_{1,0} + i_3' C_2 d_{2,0} + \ldots + i_n' C_{n-1} d_{n-1,0}}{C_0 + C_1 d_{1,0} + C_2 d_{2,0} + \ldots + C_{n-1} d_{n-1,0}} \tag{8.35}$$

where $i_t' = i_t/(1+r_t)$ is the BOP value of i_t. Analogously for the instantaneous COC:

$$\bar{r}_0 = \frac{\mathrm{PV}[I^*]}{\mathrm{PV}[C]} = \frac{r_1' C_0 + r_2' C_1 d_{1,0} + r_3' C_2 d_{2,0} + \ldots + r_n' C_{n-1} d_{n-1,0}}{C_0 + C_1 d_{1,0} + C_2 d_{2,0} + \ldots + C_{n-1} d_{n-1,0}} \tag{8.36}$$

where $r_t' = r_t/(1+r_t)$ is the BOP value of r_t. Both are Chisini means. Precisely, picking $g = g(i_1', i_2', \ldots, i_n') = \sum_{t=1}^{n} C_{t-1}(i_t' - r_t')d_{t-1,0}$:

$$\overbrace{\sum_{t=1}^{n} C_{t-1}(i_t' - r_t')d_{t-1,0}}^{V_0(\mathrm{RI})} = \sum_{t=1}^{n} C_{t-1}(\bar{\imath}_0 - r_t')d_{t-1,0}$$

$$\Longrightarrow \bar{\imath}_0 = \frac{\sum_{t=1}^{n} C_{t-1} i_t' d_{t-1,0}}{\sum_{t=1}^{n} C_{t-1} d_{t-1,0}} = \frac{\mathrm{PV}[I]}{\mathrm{PV}[C]}.$$

Likewise, picking $g = g(r_1', r_2', \ldots, r_n') = \sum_{t=1}^{n} C_{t-1}(i_t' - r_t')d_{t-1,0}$:

$$\overbrace{\sum_{t=1}^{n} C_{t-1}(i_t' - r_t')d_{t-1,0}}^{V_0(\mathrm{RI})} = \sum_{t=1}^{n} C_{t-1}(i_t' - \bar{r}_0)d_{t-1,0}$$

$$\Longrightarrow \bar{r}_0 = \frac{\sum_{t=1}^{n} C_{t-1} r_t' d_{t-1,0}}{\sum_{t=1}^{n} C_{t-1} d_{t-1,0}} = \frac{\mathrm{PV}[I^*]}{\mathrm{PV}[C]}.$$

▲

Example 8.7 ♦

C^o	
	C^e

Let us pick up Example 8.6, but suppose the COC is time-varying, namely

$$\boldsymbol{r} = (0.2, 0.3, 0.4, 0.25, 0.15, 0.1).$$

This means that the investor might pursue an alternative investment strategy and invests in an equivalent-risk portfolio that replicates the project's capital stream $\boldsymbol{C}$ (see Sects. 6.6.1 and 6.7.1). Therefore, the benchmark portfolio generates a vector of normal capitals $\boldsymbol{C}^*$ $(= \boldsymbol{C})$, a vector of normal profits $\boldsymbol{I}^*$ and a vector of normal cash flows $\boldsymbol{F}^*$, and it may be summarized with a concise transposed Matrix:

$(\boldsymbol{C}^*, \boldsymbol{I}^*, \boldsymbol{F}^*)$

Time		0	1	2	3	4	5	6
Capital	$\boldsymbol{C}^*$	120	60	10	−80	6	4	0
Income	$\boldsymbol{I}^*$		24.0	18.0	4.0	−20.0	0.9	0.4
Cash Flow	$\boldsymbol{F}^*$	−120	84.0	68.0	94.0	−106.0	2.9	4.4

We now compute the AIRR and the MARR using each of the four methods presented and make a decision on whether to accept or reject this project. To this end, we need to compute the relevant discount factors:

$$\mathrm{d}_{0,0} = 1$$

$$\begin{array}{llll}
\mathrm{d}_{1,1} & = 1 & \Longrightarrow & \mathrm{d}_{1,0} = \frac{1}{1.2} = 0.8333 \\
& \swarrow & & \\
\mathrm{d}_{2,1} = \dfrac{1}{1.3} & = 0.7692 & \Longrightarrow & \mathrm{d}_{2,0} = \dfrac{0.7692}{1.2} = 0.6410 \\
& \swarrow & & \\
\mathrm{d}_{3,1} = \dfrac{0.7692}{1.4} & = 0.5495 & \Longrightarrow & \mathrm{d}_{3,0} = \dfrac{0.5495}{1.2} = 0.4579 \\
& \swarrow & & \\
\mathrm{d}_{4,1} = \dfrac{0.5495}{1.25} & = 0.4396 & \Longrightarrow & \mathrm{d}_{4,0} = \dfrac{0.4396}{1.2} = 0.3663 \\
& \swarrow & & \\
\mathrm{d}_{5,1} = \dfrac{0.4396}{1.15} & = 0.3822 & \Longrightarrow & \mathrm{d}_{5,0} = \dfrac{0.3822}{1.2} = 0.3185 \\
& \swarrow & & \\
\mathrm{d}_{6,1} = \dfrac{0.3822}{1.1} & = 0.3475 & \Longrightarrow & \mathrm{d}_{6,0} = \dfrac{0.3475}{1.2} = 0.2896
\end{array}$$

① Applying (8.21),

$$\begin{aligned}
\bar{\imath} &= \frac{24 \cdot 1 - 6 \cdot 0.7692 + 5 \cdot 0.5495 + 12 \cdot 0.4396 - 2 \cdot 0.3822 + 5 \cdot 0.3475}{120 \cdot 1 + 60 \cdot 0.8333 + 10 \cdot 0.6410 - 80 \cdot 0.4579 + 6 \cdot 0.3663 + 4 \cdot 0.3185} \\
&= \frac{28.38}{143.25} = 19.81\%.
\end{aligned}$$

Applying (8.24),

$$\bar{r} = \frac{24 \cdot 1 + 18 \cdot 0.7692 + 4 \cdot 0.5495 - 20 \cdot 0.4396 + 0.9 \cdot 0.3822 + 0.4 \cdot 0.3475}{120 \cdot 1 + 60 \cdot 0.8333 + 10 \cdot 0.6410 - 80 \cdot 0.4579 + 6 \cdot 0.3663 + 4 \cdot 0.3185}$$
$$= \frac{31.73}{143.25} = 22.15\%.$$

② The weights are

$$w_1 = \frac{120}{143.25} = 83.8\%$$
$$w_2 = \frac{60 \cdot 0.7692}{143.25} = 32.2\%$$
$$w_3 = \frac{10 \cdot 0.5495}{143.25} = 3.8\%$$
$$w_4 = \frac{-80 \cdot 0.4396}{143.25} = -24.5\%$$
$$w_5 = \frac{6 \cdot 0.3822}{143.25} = 1.6\%$$
$$w_6 = \frac{4 \cdot 0.3475}{143.25} = 1.0\%.$$

Applying (8.22),

$$\bar{\imath} = 83.8\% \cdot 20\% + 32.2\% \cdot (-10\%) + 3.8\% \cdot 50\% - 24.5\% \cdot (-15\%)$$
$$+ 1.6\% \cdot (-33.3\%) + 1.0\% \cdot 125\% = 19.81\%.$$

Applying (8.25),

$$\bar{r} = 83.8\% \cdot 20\% + 32.2\% \cdot 30\% + 3.8\% \cdot 40\% - 24.5\% \cdot 25\%$$
$$+ 1.6\% \cdot 15\% + 1.0\% \cdot 10\% = 22.15\%$$

③ Applying (8.23),

$$\bar{\imath} = \frac{24 \cdot 0.8333 - 6 \cdot 0.6410 + 5 \cdot 0.4579 + 12 \cdot 0.3663 - 2 \cdot 0.3185 + 5 \cdot 0.2896}{120 \cdot 1 + 60 \cdot 0.8333 + 10 \cdot 0.6410 - 80 \cdot 0.4579 + 6 \cdot 0.3663 + 4 \cdot 0.3185} \cdot 1.2$$
$$= \frac{23.65}{143.25} \cdot 1.2$$
$$= 19.81\%.$$

Applying (8.26),

$$\bar{r} = \frac{24 \cdot 0.8333 + 18 \cdot 0.6410 + 4 \cdot 0.4579 - 20 \cdot 0.3663 + 0.9 \cdot 0.3185 + 0,4 \cdot 0.2896}{120 \cdot 1 + 60 \cdot 0.8333 + 10 \cdot 0.6410 - 80 \cdot 0.4579 + 6 \cdot 0.3663 + 0.4 \cdot 0.3185} \cdot 1.2$$
$$= \frac{26.45}{143.25} \cdot 1.2$$
$$= 22.15\%.$$

④ Applying (8.31)

$$\bar{\imath} = 22.15\% + \frac{-2.8}{143.25} \cdot 1.2 = 19.81\%$$

where $-2.8 = -120 + 84 \cdot 0.8333 + 44 \cdot 0.6410 + 95 \cdot 0.4579 - 74 \cdot 0.3663 + 9 \cdot 0.2896 = \text{NPV}$. Applying (8.32),

$$\bar{r} = 19.81\% - \frac{-2.8}{143.25} \cdot 1.2 = 22.15\%.$$

Therefore, the project scale is 143.25 and the economic efficiency is $-2.34\% = 19.81\% - 22.15\%$. The assumptions are such that the project destroys value: NPV $= 143.25 \cdot (-2.34\%)/1.2 = -2.8$. The investor overall invests \$143.25 at an annual rate which is 2.34% smaller than the normal rate of profit (expected rate of return of an equivalent-risk equilibrium asset).

8.4 The Book AIRRs

Each pair $(\boldsymbol{C}, \boldsymbol{I})$ of capital stream and income stream generates an AIRR. In particular, consider the usual set of transposed Matrices (see Eq. (4.19)), where we remove the cash-flow component (not necessary to build the rate of return):

0	1	2	...	$n-1$	n
C_0^o	C_1^o	C_2^o	...	C_{n-1}^o	
C_0^l	C_1^l	C_2^l	...	C_{n-1}^l	
C_0^d	C_1^d	C_2^d	...	C_{n-1}^d	
C_0^e	C_1^e	C_2^e	...	C_{n-1}^e	
	I_1^o	I_2^o	...	I_{n-1}^o	I_n^o
	I_1^l	I_2^l	...	I_{n-1}^l	I_n^l
	I_1^d	I_2^d	...	I_{n-1}^d	I_n^d
	I_1^e	I_2^e	...	I_{n-1}^e	I_n^e

We may rearrange the rows so as to make the visual association of an income stream with the corresponding capital stream more direct:

0	1	2	...	$n-1$	n
C_0^o	C_1^o	C_2^o	...	C_{n-1}^o	
	I_1^o	I_2^o	...	I_{n-1}^o	I_n^o
C_0^l	C_1^l	C_2^l	...	C_{n-1}^l	
	I_1^l	I_2^l	...	I_{n-1}^l	I_n^l
C_0^d	C_1^d	C_2^d	...	C_{n-1}^d	
	I_1^d	I_2^d	...	I_{n-1}^d	I_n^d
C_0^e	C_1^e	C_2^e	...	C_{n-1}^e	
	I_1^e	I_2^e	...	I_{n-1}^e	I_n^e

Four AIRRs are generated by dividing total income by total capital for each pair of streams, namely the *operating* AIRR, the *non-operating* AIRR, the *debt* AIRR, the *equity* AIRR, which represent, respectively, the *average operating ROI*, the *average non-operating ROI*, the *average ROD*, and the *average ROE*:

$$\bar{\imath}^o = \frac{\sum_{t=1}^n I_t^o \mathrm{d}_{t,1}^o}{\sum_{t=1}^n C_{t-1}^o \mathrm{d}_{t-1,0}^o} = \frac{\mathrm{PV}[I^o]}{\mathrm{PV}[C^o]}(1+r_1^o) \tag{8.37}$$

$$\bar{\imath}^l = \frac{\sum_{t=1}^n I_t^l \mathrm{d}_{t,1}^l}{\sum_{t=1}^n C_{t-1}^l \mathrm{d}_{t-1,0}^l} = \frac{\mathrm{PV}[I^l]}{\mathrm{PV}[C^l]}(1+r_1^l) \tag{8.38}$$

$$\bar{\imath}^d = \frac{\sum_{t=1}^n I_t^d \mathrm{d}_{t,1}^d}{\sum_{t=1}^n C_{t-1}^d \mathrm{d}_{t-1,0}^d} = \frac{\mathrm{PV}[I^d]}{\mathrm{PV}[C^d]}(1+r_1^d) \tag{8.39}$$

$$\bar{\imath}^e = \frac{\sum_{t=1}^n I_t^e \mathrm{d}_{t,1}^e}{\sum_{t=1}^n C_{t-1}^e \mathrm{d}_{t-1,0}^e} = \frac{\mathrm{PV}[I^e]}{\mathrm{PV}[C^e]}(1+r_1^e) \tag{8.40}$$

or, in terms of linear combinations,

$$\bar{\imath}^o = i_1^o \cdot w_1^o + i_2^o \cdot w_2^o + \ldots + i_n^o \cdot w_n^o \tag{8.41}$$

$$\bar{\imath}^l = i_1^l \cdot w_1^l + i_2^l \cdot w_2^l + \ldots + i_n^l \cdot w_n^l \tag{8.42}$$

$$\bar{\imath}^d = i_1^d \cdot w_1^d + i_2^d \cdot w_2^d + \ldots + i_n^d \cdot w_n^d \tag{8.43}$$

$$\bar{\imath}^e = i_1^e \cdot w_1^e + i_2^e \cdot w_2^e + \ldots + i_n^e \cdot w_n^e \tag{8.44}$$

where $w_t^j = C_{t-1}^j \mathrm{d}_{t,1}^j / \mathrm{PV}[C^j]$ for $j = o, l, d, e$. Correspondingly, the COCs are

$$\bar{r}^o = \frac{\sum_{t=1}^n I_t^{o*} \mathrm{d}_{t,1}^o}{\sum_{t=1}^n C_{t-1}^o \mathrm{d}_{t-1,0}^o} = \frac{\mathrm{PV}[I^{o*}]}{\mathrm{PV}[C^o]}(1+r_1^o) \tag{8.45}$$

$$\bar{r}^l = \frac{\sum_{t=1}^n I_t^{l*} \mathrm{d}_{t,1}^l}{\sum_{t=1}^n C_{t-1}^l \mathrm{d}_{t-1,0}^l} = \frac{\mathrm{PV}[I^{l*}]}{\mathrm{PV}[C^l]}(1+r_1^l) \tag{8.46}$$

$$\bar{r}^d = \frac{\sum_{t=1}^n I_t^{d*} \mathrm{d}_{t,1}^d}{\sum_{t=1}^n C_{t-1}^d \mathrm{d}_{t-1,0}^d} = \frac{\mathrm{PV}[I^{d*}]}{\mathrm{PV}[C^d]}(1+r_1^d) \tag{8.47}$$

$$\bar{r}^e = \frac{\sum_{t=1}^n I_t^{e*}\,\mathrm{d}_{t,1}^e}{\sum_{t=1}^n C_{t-1}^e\mathrm{d}_{t-1,0}^e} = \frac{\mathrm{PV}[I^{e*}]}{\mathrm{PV}[C^e]}(1+r_1^e) \tag{8.48}$$

or, in terms of linear combinations,

$$\bar{r}^o = r_1^o \cdot w_1^o + r_2^o \cdot w_2^o + \ldots + r_n^o \cdot w_n^o \tag{8.49}$$

$$\bar{r}^l = r_1^l \cdot w_1^l + r_2^l \cdot w_2^l + \ldots + r_n^l \cdot w_n^l \tag{8.50}$$

$$\bar{r}^d = r_1^d \cdot w_1^d + r_2^d \cdot w_2^d + \ldots + r_n^d \cdot w_n^d \tag{8.51}$$

$$\bar{r}^e = r_1^e \cdot w_1^e + r_2^e \cdot w_2^e + \ldots + r_n^e \cdot w_n^e. \tag{8.52}$$

The average RO$\mathcal{D}$ ($\bar{\imath}^{\mathcal{D}}$) is also defined, consistently with the average ROD, by replacing C^d with $C^{\mathcal{D}}$:

$$\bar{\imath}^{\mathcal{D}} = \frac{\sum_{t=1}^n I_t^{\mathcal{D}}\mathrm{d}_{t,1}^{\mathcal{D}}}{\sum_{t=1}^n C_{t-1}^{\mathcal{D}}\mathrm{d}_{t-1,0}^{\mathcal{D}}} = \frac{\mathrm{PV}[I^{\mathcal{D}}]}{\mathrm{PV}[C^{\mathcal{D}}]}(1+r_1^{\mathcal{D}}) \tag{8.53}$$

with obvious meaning of the symbols. Its affiliated MARR is

$$\bar{r}^{\mathcal{D}} = \frac{\sum_{t=1}^n I_t^{\mathcal{D}*}\mathrm{d}_{t,1}^{\mathcal{D}}}{\sum_{t=1}^n C_{t-1}^{\mathcal{D}}\mathrm{d}_{t-1,0}^{\mathcal{D}}} = \frac{\mathrm{PV}[I^{\mathcal{D}*}]}{\mathrm{PV}[C^{\mathcal{D}}]}(1+r_1^{\mathcal{D}}). \tag{8.54}$$

(Analogously with the instantaneous AIRRs, $\bar{\imath}_0^j$, and MARRs, $\bar{r}_0^j$, $j = o, l, d, e, \mathcal{D}$.) Applying (8.29)–(8.30) for each notion of profit and capital, one gets

$$\begin{aligned}
\mathrm{NPV}^o(1+r_1^o) &= \mathrm{PV}[C^o]\cdot(\bar{\imath}^o - \bar{r}^o) &&\Longleftrightarrow \mathrm{NPV}^o = \mathrm{PV}[C^o]\cdot(\bar{\imath}_0^o - \bar{r}_0^o)\\
\mathrm{NPV}^l(1+r_1^l) &= \mathrm{PV}[C^l]\cdot(\bar{\imath}^l - \bar{r}^l) &&\Longleftrightarrow \mathrm{NPV}^l = \mathrm{PV}[C^l]\cdot(\bar{\imath}_0^l - \bar{r}_0^l)\\
\mathrm{NPV}^d(1+r_1^d) &= \mathrm{PV}[C^d]\cdot(\bar{\imath}^d - \bar{r}^d) &&\Longleftrightarrow \mathrm{NPV}^d = \mathrm{PV}[C^d]\cdot(\bar{\imath}_0^d - \bar{r}_0^d)\\
\mathrm{NPV}^e(1+r_1^e) &= \mathrm{PV}[C^e]\cdot(\bar{\imath}^e - \bar{r}^e) &&\Longleftrightarrow \mathrm{NPV}^e = \mathrm{PV}[C^e]\cdot(\bar{\imath}_0^e - \bar{r}_0^e)\\
\mathrm{NPV}^{\mathcal{D}}(1+r_1^{\mathcal{D}}) &= \mathrm{PV}[C^{\mathcal{D}}]\cdot(\bar{\imath}^{\mathcal{D}} - \bar{r}^{\mathcal{D}}) &&\Longleftrightarrow \mathrm{NPV}^{\mathcal{D}} = \mathrm{PV}[C^{\mathcal{D}}]\cdot(\bar{\imath}_0^{\mathcal{D}} - \bar{r}_0^{\mathcal{D}}).
\end{aligned} \tag{8.55}$$

One can equivalently employ an unlevered perspective. Specifically, let $\mathrm{PV}[I^{uo}] = \sum_{t=1}^n I_t^{uo}\mathrm{d}_{t,0}^{uo}$ be the present value of the NOPATs and $\mathrm{PV}[C^{uo}] = \sum_{t=1}^n C_{t-1}^o\mathrm{d}_{t-1,0}^{uo}$ the present value of the operating assets, where, as we remind, $\mathrm{d}_{t,k}^{uo} = \prod_{j=k+1}^t(1+r_j^{uo})^{-1}$. Hence, one finds

$$\mathrm{NPV}^o(1+r_1^{uo}) = \mathrm{PV}[C^{uo}]\cdot(\bar{\imath}^{uo} - \bar{r}^{uo}) \tag{8.56}$$

where

$$\bar{\imath}^{uo} = \frac{\mathrm{PV}[I^{uo}]}{\mathrm{PV}[C^{uo}]}(1+r_1^{uo}) \qquad \bar{r}^{uo} = \frac{\mathrm{PV}[I^{uo*}]}{\mathrm{PV}[C^{uo}]}(1+r_1^{uo}) \tag{8.57}$$

denote, respectively, the unlevered (operating) average ROI and the associated MARR, with $\mathrm{PV}[I^{uo*}] = \sum_{t=1}^{n} r_t^{uo} \cdot C_{t-1}^{o}\, \mathrm{d}_{t,0}^{uo}$ being the foregone profit.

The unlevered average ROI (and associated MARR) may be obviously computed from $\boldsymbol{i^{uo}} = (i_1^{uo}, i_2^{uo}, \ldots, i_n^{uo})$ and $\boldsymbol{r^{uo}} = (r_1^{uo}, r_2^{uo}, \ldots, r_n^{uo})$ as linear combinations:

$$\begin{aligned} \bar{\imath}^{uo} &= i_1^{uo} \cdot w_1^{uo} + i_2^{uo} \cdot w_2^{uo} + \ldots + i_n^{uo} \cdot w_n^{uo} \\ \bar{r}^{uo} &= r_1^{uo} \cdot w_1^{uo} + r_2^{uo} \cdot w_2^{uo} + \ldots + r_n^{uo} \cdot w_n^{uo} \end{aligned} \tag{8.58}$$

where $w_t^{uo} = C_{t-1}^{o} \mathrm{d}_{t,1}^{uo} / \mathrm{PV}[C^{uo}]$. The unlevered average ROI may also be expressed in an asset perspective (or, in other words, in a financially unlevered perspective) along with its associated MARR:

$$\bar{\imath}^{u} = \frac{\mathrm{PV}[I^{u}]}{\mathrm{PV}[C^{u}]}(1 + r_1^{u}), \qquad \bar{r}^{u} = \frac{\mathrm{PV}[I^{u*}]}{\mathrm{PV}[C^{u}]}(1 + r_1^{u}) \tag{8.59}$$

where $\mathrm{PV}[I^{u}] = \sum_{t=1}^{n} I_t^{u} \mathrm{d}_{t,0}^{u}$, $\mathrm{PV}[I^{u*}] = \sum_{t=1}^{n} r_t^{u} C_{t-1} \mathrm{d}_{t,0}^{u}$, $\mathrm{PV}[C^{u}] = \sum_{t=1}^{n} C_{t-1}$ $\mathrm{d}_{t-1,0}^{u}$ Equivalently, in terms of linear combinations, one may write

$$\begin{aligned} \bar{\imath}^{u} &= i_1^{u} \cdot w_1^{u} + i_2^{u} \cdot w_2^{u} + \ldots + i_n^{u} \cdot w_n^{u} \\ \bar{r}^{u} &= r_1^{u} \cdot w_1^{u} + r_2^{u} \cdot w_2^{u} + \ldots + r_n^{u} \cdot w_n^{u} \end{aligned} \tag{8.60}$$

where $w_t^{u} = C_{t-1}^{o} \mathrm{d}_{t,1}^{u} / \mathrm{PV}[C^{u}]$. If $C_t^{l} = 0$ for every t, then $C_t = C_t^{o}$ and $C_t^{u} = C_t^{uo}$. Likewise, $\bar{\imath}^{uo} = \bar{\imath}^{u}$ and $\bar{r}^{uo} = \bar{r}^{u}$.

Each of the AIRRs presented is, so to say, a *book* AIRR (BAIRR), that is, derived from pro-forma book values.[7]

The project's BAIRRs			**Symbol**	**Associated COC (MARR)**
Asset/project BAIRR	or	average ROI	$\bar{\imath}$	$\bar{r}$
Equity BAIRR	or	average ROE	$\bar{\imath}^{e}$	$\bar{r}^{e}$
Operating BAIRR	or	(operating) average ROI	$\bar{\imath}^{o}$	$\bar{r}^{o}$
Non-operating BAIRR	or	non-operating average ROI	$\bar{\imath}^{l}$	$\bar{r}^{l}$
Unlevered (operating) BAIRR	or	unlevered (operating) average ROI	$\bar{\imath}^{uo}$	$\bar{r}^{uo}$
Unlevered asset BAIRR[8]	or	unlevered average asset ROI	$\bar{\imath}^{u}$	$\bar{r}^{u}$
Debt BAIRR	or	average ROD	$\bar{\imath}^{d}$	$\bar{r}^{d}$
Net debt BAIRR	or	average RO$\mathcal{D}$	$\bar{\imath}^{\mathcal{D}}$	$\bar{r}^{\mathcal{D}}$

[7] See also Magni and Martin (2018).

[8] Unlevered asset BAIRR might also be labeled *financially* unlevered BAIRR (see Definition 2.2).

The goal of a firm's management is to undertake projects that increase equityholders' wealth. Thus, they should undertake projects such that the equity AIRR is greater than the equity COC. Therefore, the equity AIRR is the relevant BAIRR for shareholders.

Proposition 8.5 (Rationality of average ROE and average ROI) *Let P be any project associated with an equity stream of net incomes equal to* $\boldsymbol{I}^e$ *and an equity capital stream equal to* $\boldsymbol{C}^e$. *Shareholder value is created if and only if the average ROE is greater than the average cost of equity:* $\bar{\imath}^e > \bar{r}^e$.[9]

If $i_t^d = r_t^d$, *then* $\bar{\imath}^d = \bar{r}^d$ *and* $\text{NPV}^d = 0$. *Therefore, project NPV and equity NPV coincide:* $\text{NPV} = \text{NPV}^e$. *In this case, shareholder value is created if and only if the average ROI is greater than the average project COC:* $\bar{\imath} > \bar{r}$.

Proposition 8.5 shows that accounting rates of return are economically significant. It gives formal expression to the idea that "for positive NPV investments, the firm is expected to report 'above normal' profitability in the sense that the book rate of return generally exceeds the cost of capital." (Feltham and Ohlson 1996, p. 210).

Using (8.55) in a financing perspective, one may decompose the NPV according to the sources of funds (debt and equity) and according to scale and marginal efficiency. Using instantaneous efficiency,

$$\text{NPV} = \overbrace{\underbrace{\text{PV}[C^e]}_{\substack{\text{scale of}\\ \text{equity}}} \cdot \underbrace{(\bar{\imath}_0^e - \bar{r}_0^e)}_{\substack{\text{economic efficiency of}\\ \text{equity capital}}}}^{\text{NPV}^e} + \overbrace{\underbrace{\text{PV}[C^d]}_{\substack{\text{scale of}\\ \text{debt}}} \cdot \underbrace{(\bar{\imath}_0^d - \bar{r}_0^d)}_{\substack{\text{economic efficiency of}\\ \text{debt capital}}}}^{\text{NPV}^d}. \tag{8.61}$$

Analogously, in an investment perspective,

$$\text{NPV} = \overbrace{\underbrace{\text{PV}[C^o]}_{\substack{\text{scale of}\\ \text{operating assets}}} \cdot \underbrace{(\bar{\imath}_0^o - \bar{r}_0^o)}_{\substack{\text{economic efficiency of}\\ \text{operating assets}}}}^{\text{NPV}^o} + \overbrace{\underbrace{\text{PV}[C^l]}_{\substack{\text{scale of}\\ \text{non-operating assets}}} \cdot \underbrace{(\bar{\imath}_0^l - \bar{r}_0^l)}_{\substack{\text{economic efficiency of}\\ \text{non-operating assets}}}}^{\text{NPV}^l} \tag{8.62}$$

One may frame (8.61) and (8.62) in terms of excess rates:

$$\begin{aligned} \text{NPV} &= \text{PV}[C^e] \cdot \xi^e + \text{PV}[C^d] \cdot \xi^d \\ \text{NPV} &= \text{PV}[C^o] \cdot \xi^o + \text{PV}[C^l] \cdot \xi^l \end{aligned} \tag{8.63}$$

where $\xi^h = \bar{\imath}_0^h - \bar{r}_0^h$, $h = o, l, d, e$. Therefore, one may write

[9] We assume $\text{PV}[C^e] > 0$ (i.e., equity is invested). In case where $\text{PV}[C^e] < 0$, the equity capital is not *raised* from equityholders but, rather, *lent* to equityholders (i.e., equity is financed); then, the sign of the inequality is reversed.

Table 8.1 Shareholder value creation and analysis of economic efficiency—AIRR

Component	**Scale**	**Economic efficiency**
Operating assets	$\text{PV}[C^o]$	$\xi^o = \bar{\imath}^o_0 - \bar{r}^o_0$
Non-operating assets	$\text{PV}[C^l]$	$\xi^l = \bar{\imath}^l_0 - \bar{r}^l_0$
Debt	$\text{PV}[C^d]$	$\xi^d = \bar{\imath}^d_0 - \bar{r}^d_0$
Equity	$\text{PV}[C^e]$	$\xi^e = \bar{\imath}^e_0 - \bar{r}^e_0$

$$\text{NPV}^e = \text{PV}[C^o] \cdot \overbrace{(\bar{\imath}^o_0 - \bar{r}^o_0)}^{\xi^o} + \text{PV}[C^l] \cdot \overbrace{(\bar{\imath}^l_0 - \bar{r}^l_0)}^{\xi^l} - \text{PV}[C^d] \cdot \overbrace{(\bar{\imath}^d_0 - \bar{r}^d_0)}^{\xi^d}. \quad (8.64)$$

(See Table 8.1). For given (positive) scales, the higher the efficiencies of the operating and non-operating assets, the higher the shareholder value created; also, the smaller the efficiency of the debt, the higher the shareholder value created (a less efficient debt means that a greater part of the project NPV is captured by equityholders and lost by debtholders). Symmetrically, for given (positive) efficiencies, the higher the scales of the operating and non-operating assets, the higher the shareholder value created; also, the smaller the debt's scale, the higher the shareholder value created (a smaller scale of debt means that a greater part of the project NPV is captured by equityholders and lost by debtholders). Equation (8.64) may also be framed in terms of net debt: $\text{NPV}^e = \text{PV}[C^o] \cdot (\bar{\imath}^o - \bar{r}^o) - \text{PV}[C^{\mathcal{D}}] \cdot (\bar{\imath}^{\mathcal{D}} - \bar{r}^{\mathcal{D}})$ (Table 8.1).

It is worth noting that, if $i^d_t = r^d_t$, NPV and NPV^e are equal. However, the decomposition of NPV and NPV^e is different:

$$\text{NPV} = \overbrace{\text{PV}[C]}^{\text{project scale}} \cdot \overbrace{(\bar{\imath}_0 - \bar{r}_0)}^{\text{project's efficiency}} = \overbrace{\text{PV}[C^e]}^{\text{equity capital scale}} \cdot \overbrace{(\bar{\imath}^e_0 - \bar{r}^e_0)}^{\text{efficiency of equity}} = \text{NPV}^e. \quad (8.65)$$

In such a way, we have two decompositions of NPV, which separately measure, on one hand, the size of the project and the size of the equity capital and, on the other hand, the marginal efficiency of the capital invested in the project by the firm and the marginal efficiency of the capital invested by equityholders alone.

Remark 8.4 (***Skepticism on book rates of return***) Accounting scholars have always been puzzled by the fact that practitioners often use book rates of return instead of the more appropriate NPV for capital budgeting decisions: "the use of accounting numbers (alongside, in competition with, even to the exclusion of the economic measures) has persisted to an extent which has both surprised and frustrated many financial theorists" (Peasnell 1982a, p. 376). The unreliability of book rates of return for making decisions has been the focus of several decades of academic research, both theoretical and empirical (e.g. Harcourt 1965; Solomon 1966; Livingstone and Salamon 1970; Gordon 1974; Kay 1976; Peasnell 1982a, b; Fisher and McGowan 1983; Luckett 1984; Salamon 1985; Kay and Mayer 1986; Edwards et al. 1987; Gordon and Stark 1989; Whittington 1988; Bar-Yosef and Lustgarten 1994; Peasnell 1996; Feenstra and Wang 2000; Stark 2004). Many accounting and finance scholars still believe that book rates of return are not capable of measuring economic profitability. In a well-known paper, Fisher and McGowan (1983) refused the idea that accounting rates of return may capture economic profitability:

> Accounting rates of return, even if properly and consistently measured, provide almost no information about economic rates of return ... Accounting rates of return are useful only insofar as they yield information as to economic rates of return (Fisher and McGowan 1983, p. 82)

> There is no way in which one can look at accounting rates of return and infer anything about relative economic profitability. (Fisher and McGowan 1983, p. 90)

The idea of comparing accounting rates of return with the cost of capital seemed "clearly like comparing apples with oranges" (Rappaport 1986, p. 31). Solomon (1966) attributed the problem to accounting rates: "It is the book-yield measure which is in error" (Solomon 1966, pp. 233–234, as quoted in Vatter 1966, p. 683). Fisher and McGowan even seemed to extend their criticism to rates of return in general (including but not limited to accounting rates):

> While present value calculations using the cost of capital remain the correct method for analyzing profitability, profitability cannot be summarized by *any rate of return*, including accounting rates of return. (Fisher and McGowan 1983, p. 83, italics added)

Some scholars have acknowledged the need of reconciling book rates of return and value creation:

> For reasons mostly of familiarity and convenience, accounting measures of return continue to be widely used. It is therefore desirable to have an analysis which pinpoints precisely when, how and why the accounting measures differ from the economic measures. (Franks and Hodges 1984, p. 133)

> To the extent that cash-flow capital budgeting measures appear to managers incapable of being arithmetically reconciled with the now familiar accounting measures, then there is a danger that they will not receive the attention they deserve. In these circumstances, there is a lot to be said for presenting the project data in a manner such that *the arithmetical connections with accounting profits is made obvious*. (Peasnell 1982a, p. 376, italics supplied)

As we have seen, the book AIRRs make the arithmetical connections between accounting profits and NPV obvious and natural, while providing a more refined analysis than the traditional NPV analysis (see Magni 2009e, 2016a; Magni and Peasnell 2012, 2015, for a thorough treatment of usefulness of accounting rates of return). ▲

Remark 8.5 (*Zero capital*) Suppose $\text{PV}[C] = 0$. While, from a strictly mathematical viewpoint, the rate of return is not defined (denominator is zero), it is commonplace for economists and practitioners to say that, whenever the denominator is zero, the rate of return is "infinite".[10]

[10] For example, see Krugman (1979, p. 320), Shemin (2004, p. 15), Nosal and Wang (2004, p. 23), Smithers (2013, p. 66).

Although it is just a matter of limited relevance (and only a theoretical one), we may complete the definition of AIRR. First, suppose the COC is constant. Then,

$$\bar{\imath} = \begin{cases} \frac{\mathrm{PV}[I]}{\mathrm{PV}[C]}(1+r_1), & \text{if}\,\mathrm{PV}[C] \neq 0; \\ +\infty, & \text{if}\,\mathrm{PV}[C] = 0\,\text{and}\,\mathrm{PV}[I] > 0; \\ -\infty, & \text{if}\,\mathrm{PV}[C] = 0\,\text{and}\,\mathrm{PV}[I] < 0. \end{cases} \tag{8.66}$$

If $\mathrm{PV}[C] \neq 0$, an investment (financing) project is worth undertaking if and only if $\bar{\imath} > r$ (<). If $\mathrm{PV}[C] = 0$, a project is worth undertaking if and only if $\bar{\imath} = +\infty$.

Now, suppose the COC is time-varying. The MARR is defined as

$$\bar{r} = \begin{cases} \frac{\mathrm{PV}[I^*]}{\mathrm{PV}[C]}(1+r_1), & \text{if}\,\mathrm{PV}[C] \neq 0; \\ +\infty, & \text{if}\,\mathrm{PV}[C] = 0\,\text{and}\,\mathrm{PV}[I^*] > 0; \\ -\infty, & \text{if}\,\mathrm{PV}[C] = 0\,\text{and}\,\mathrm{PV}[I^*] < 0. \end{cases} \tag{8.67}$$

If $\mathrm{PV}[C] \neq 0$, an investment (financing) project is worth undertaking if and only if $\bar{\imath} > \bar{r}$ (<). If $\mathrm{PV}[C] = 0$, a project is worth undertaking if and only if $\mathrm{PV}[I] > \mathrm{PV}[I^*]$. More interesting is the fact that, if $\mathrm{PV}[C] = 0$, the borrowing positions offset the investment positions. It is possible to compute an investment rate for the investment positions and a financing rate for the borrowing positions and find the infinite rate of return as a "weighted mean", so to say, of the investment rate and the financing rate. See Sect. 8.9 for details (see also Magni 2014a, 2015a). ▲

Remark 8.6 (*Copying with nonzero time-0 income, $I_0 \neq 0$*) We have so far assumed that $I_0 = 0$. It is now easy to generalize and include the case where $I_0 \neq 0$. It suffices to add I_0 as an additional instantaneous return. In particular, formula ③ on Sect. 8.3 becomes

$$\bar{\imath} = \frac{\mathrm{PV}[I]}{\mathrm{PV}[C]} \cdot (1+r_1) = \frac{\overset{\downarrow}{\textcircled{I_0}} + I_1 d_{1,0} + I_2 d_{2,0} + \cdots + I_n \mathrm{d}_{n,0}}{C_0 + C_1 d_{1,0} + C_2 d_{2,0} + \cdots + C_{n-1} \mathrm{d}_{n-1,0}} \cdot (1+r_1) \tag{8.68}$$

Formula ④ does not change: $\bar{\imath} = \bar{r} + \mathrm{NPV}(1+r_1)/\mathrm{PV}[C]$ (the difference only lies in the fact that initial capital and initial cash flow do not equate in absolute value: $-C_0 \neq F_0 = I_0 - \Delta C_0$). Formulas ① and ② are modified by adding I_0 to the numerator and by adding $I_0/\mathrm{PV}[C]$ to the linear combination, respectively. The COC/MARR does not change, for $I_0^* = 0$. (Analogously with $\bar{\imath}_0$ and $\bar{r}_0$.) ▲

Example 8.8 ♦

C^o	C^d
	C^e

(*Average ROE and ROD of levered project with time-varying COCs*) Consider the following input data for a two-period project:

	Time	0	1	2
C_t^e	Equity capital	100	60	0
C_t^d	Debt capital	40	20	
I_t^e	Net Income		60	90
I_t^d	Interest expense		2	5
r_t^e	cost of equity		8%	10%
r_t^d	cost of debt		2%	4%

Using (8.39) and (8.40), one computes the present values of the net incomes and the interest expense (i.e., $\text{PV}[I^e] = 131.3$ and $\text{PV}[I^d] = 6.7$) and divide by the respective capital amounts (i.e., $\text{PV}[C^e] = 155.56$ and $\text{PV}[C^d] = 59.61$). Then, after due multiplication for the respective factor (i.e., 1.08 and 1.02), one gets the equity AIRR (average ROE) and the debt AIRR (average ROD). More explicitly,

$$\bar{\imath}^e = \frac{\dfrac{60}{1.08} + \dfrac{90}{1.08 \cdot 1.1}}{100 + \dfrac{60}{1.08}} \cdot (1.08) = \frac{131.3}{155.56} \cdot (1.08) = 91.2\%$$

$$\bar{\imath}^d = \frac{\dfrac{2}{1.02} + \dfrac{5}{1.02 \cdot 1.04}}{40 + \dfrac{20}{1.02}} \cdot (1.02) = \frac{6.7}{59.61} \cdot (1.02) = 11.4\%.$$

As for the equity MARR and debt MARR, they are computed via (8.47) and (8.48). Consider the respective market returns:

$$I_1^{e*} = 0.08 \cdot 100 = 8, \;\; I_2^{e*} = 0.1 \cdot 60 = 6$$
$$I_1^{d*} = 0.02 \cdot 40 = 0.8, \;\; I_2^{d*} = 0.04 \cdot 20 = 0.8$$

whence

$$\bar{r}^e = \frac{\dfrac{8}{1.08} + \dfrac{6}{1.08 \cdot 1.1}}{100 + \dfrac{60}{1.08}} \cdot (1.08) = \frac{12.46}{155.56} \cdot (1.08) = 8.6\%$$

$$\bar{r}^d = \frac{\dfrac{0.8}{1.02} + \dfrac{0.8}{1.02 \cdot 1.04}}{40 + \dfrac{20}{1.02}} \cdot (1.02) = \frac{1.54}{59.61} \cdot (1.02) = 2.6\%.$$

The equity (instantaneous) economic efficiency is $\xi^e = (91.2\% - 8.6\%)/1.08 = 76.48\%$. Therefore the equity NPV is

$$\text{NPV}^e = 155.56 \cdot 76.48\% = 118.9.$$

The efficiency of the debt capital is $\xi^d = (11.4\% - 2.6\%)/1.02 = 8.63\%$ so the debt NPV is

$$\text{NPV}^d = 59.61 \cdot 8.63\% = 5.1.$$

The project NPV is then NPV $= 118.9 + 5.1 = 124$.

The equity NPV is smaller than the project NPV because the firm borrows at an average interest rate, $\bar{\imath}^d = 11.5\%$, which is greater than the average required rate of return on debt, $\bar{r}^d = 2.6\%$. Shareholders lose \$0.089 = \$0.115 − \$0.026 on each of the \$59.61 overall borrowed. Notwithstanding this unfavorable financing policy, the project creates value for shareholders. The reason is that the average ROE, $\bar{\imath}^e = 91.2\%$, is greater than the average cost of equity, $\bar{r}^e = 8.6\%$, to such an extent that only a small part of the value created is given up by equityholders; most part of the project NPV is generated by the equity investment (equity NPV represents about 95.8% of the project NPV).

Example 8.9 ♦

C^o	C^d
	C^e

(Average ROE of levered two-period project, assuming constant WACC and $r^d \neq i^d$*)* Consider the two-year project illustrated in Example 8.1. Assume the project is levered (but $C_t^l = 0$ for all t, which implies $I_t^{uo} = I_t^u$ and $F_t^{uo} = F_t^u$); in particular, suppose the firm borrows from \$250 a lender, repayable with two level payments at a constant ROD equal to $i^d = 2\%$. This implies that the instalments (CFDs) are equal to $F_1^d = F_2^d = 250 \cdot 0.02/(1 - (1.02)^{-2}) = \$128.8$. The outstanding debts are $C_0^d = 250$, $C_1^d = 126.2 = 250 \cdot 1.02 - 128.8$, so that the book value of equity is $C_0^e = 600 - 250 = 350$, $C_1^e = 200 - 126.2 = 73.8$. We assume that the cost of debt is $r^d = 3\% > 2\% = i^d$ while the WACC is constant and equal to $r_t^{uo} = r_t^u = 15\%$. We find the NI from the NOPAT, by subtracting the after-tax interest (i.e., subtracting interest and adding the tax shield): $I_t^e = I_t^u - I_t^d(1 - \tau)$ or $I_t^e = I_t^u - I_t^d + \tau I_t^d$. Therefore, $I_1^e = 150 - 5 + 1.25 = 146.3$ and $I_2^e = 187.5 - 2.5 + 0.6 = 185.6$.

	Year	**0**	**1**	**2**
GP_t	Gross profit		\$700.0	\$700.0
$-\text{SGA}_t$	Less: SGA		−\$100.0	−\$250.0
$-\text{Dep}_t$	Less: Depreciation		−\$400.0	−\$200.0
EBIT_t	EBIT		\$200.0	\$250.0
$-\tau\text{EBIT}_t$	Less: Taxes on EBIT (25%)		−\$50.0	−\$62.5
I_t^u	**NOPAT**		**\$150.0**	**\$187.5**
$-I_t^d$	Less: Interest		−\$5.0	−\$2.5
$+\tau I_t^d$	Plus: Taxes on Interest		\$1.3	\$0.6
I_t^e	**Net Income**		146.3	185.6
C_t	**Invested Capital**	**\$600.0**	**\$200.0**	**\$0.0**
C_t^d	Debt	\$250.0	\$126.2	\$0.0
C_t^e	Equity	\$350.0	\$73.8	\$0.0

As the cost of equity is not exogenously given, one has to draw it from the WACC and the intrinsic values via Eq. (6.88). This implies that cash flows are needed. Specifically, the CFEs may be found as

$$F_t^e = F_t - F_t^d = F_t^u + \tau I_t^d - F_t^d.$$

Substituting the numbers, $F_1^e = 550 + 1.25 - 128.8 = 422.5$, $F_2^e = 387.5 + 0.63 - 128.8 = 259.4$, and, obviously, $F_0^e = -C_0^e = -350$:

	Year	0	1	2
I_t^u	NOPAT		\$150.0	\$187.5
$-\Delta C_t^o$	Less: Change in capital	−\$600.0	\$400.0	\$200
F_t^u	FCF	−\$600.0	\$550.0	\$387.5
$+\tau I_t^d$	Plus: Taxes on Interest		\$1.25	\$0.6
$-F_t^d$	Less: CFD	+\$250.0	−\$128.8	−\$128.8
F_t^e	CFE	−\$350.0	\$422.5	\$259.4

Alternatively, one may use changes in capital, $F_0^e = 0 - 350 = -350$, $F_1^e = 146.3 + 276.2 = 422.5$, $F_2^e = 185.6 + 73.8 = 259.4$.

For greater clarity, we report below the transposed Matrix of this project, where the CFO stream are easily computed from the FCF stream by adding the tax shield, $\tau I_t^{\mathcal{D}} = \tau I_t^d$ (same for the operating income):

$(\boldsymbol{C}, \boldsymbol{I}, \boldsymbol{F})$

Time		0	1	2
Capital				
Operating assets	$\boldsymbol{C^o}$	600	200	0
Non-operating assets	$\boldsymbol{C^l}$	0	0	0
Debt	$\boldsymbol{C^d}$	250	126.24	0
Equity	$\boldsymbol{C^e}$	350	73.76	0
Income				
Operating income	$\boldsymbol{I^o}$	0	151.25	188.13
Non-operating income	$\boldsymbol{I^l}$	0	0	0
Interest expense	$\boldsymbol{I^d}$	0	0	0
Net Income	$\boldsymbol{I^e}$	0	146.25	185.61
Cash flow				
Cash flow from operations	$\boldsymbol{F^o}$	−600	551.25	388.13
Cash flow from non-operating assets	$\boldsymbol{F^l}$	0	0	0
Cash flow to debt	$\boldsymbol{F^d}$	−250	128.76	128.76
Cash flow to equity	$\boldsymbol{F^e}$	−350	422.49	259.37

The debt's economic values are $V_1^d = 128.8/1.03 = 125$ and $V_0^d = (125 + 128.8)/1.03 = 246.4$. Therefore, the debt NPV is $\text{NPV}^d = 246.4 - 250 = -3.6$.

The project's intrinsic values are $V_1 = 387.5/1.15 = 337$ and $V_0 = (337 + 550)/1.15 = 771.3$. Therefore, the project NPV is $\text{NPV} = 771.3 - 600 = 171.3$.

The equity values are $V_1^e = V_1 - V_1^d = 337 - 125 = 212$ and $V_0^e = V_0 - V_0^d = 771.3 - 246.4 = 524.9$. This implies that the equity NPV is $\text{NPV}^e = 524.9 - 350 = 174.9$, higher than the project NPV by \$3.6, the value lost by debtholders.[11]

Considering that $r_t^{uo} = r_t^u$, one may equivalently use the second equation or the fifth equation of (6.88) to find the cost of equity. Specifically, remembering that the WACC is $r_1^u = r_2^u = 15\%$, using the second equation of (6.88), one gets

$$r_1^e = 0.15 + \frac{246.4}{524.9}(0.15 - 0.03) + \frac{1.25}{524.9} = 0.2087$$
$$r_2^e = 0.15 + \frac{125}{212}(0.15 - 0.03) + \frac{0.6}{212} = 0.2238.$$

The market returns are $I_t^{e*} = r_t^e C_{t-1}^e$. Substituting, $I_1^{e*} = 0.2087 \cdot 350 = 73$ and $I_2^{e*} = 0.2238 \cdot 73.8 = 16.5$. Also,

$$\text{PV}[I^e] = \frac{146.3}{1.2087} + \frac{185.6}{1.2087 \cdot 1.2238} = 246.5$$
$$\text{PV}[I^{e*}] = \frac{73}{1.2087} + \frac{16.5}{1.2087 \cdot 1.2238} = 71.6$$

and

$$\text{PV}[C^e] = 350 + \frac{73.8}{1.2087} = 411$$

which implies that the average ROE and the related average cost of equity are

$$\bar{\imath}^e = \frac{246.5}{411} \cdot 1.2087 = 72.48\%, \qquad \bar{r}^e = \frac{71.6}{411} \cdot 1.2087 = 21.05\%.$$

Therefore, the firm's shareholders earn an annual 72.48% on an investment of $\text{PV}[C^e] = 411$ while giving up 21.05% on that same amount. The economic efficiency is $51.43\% = 72.48\% - 21.05\%$, the equity investment scale is \$411, whence we get back to $\text{NPV}^e = 411 \cdot 51.43\%/1.2087 = 174.9$, as opposed to $\text{NPV} = \$171.3$. The difference between equity NPV and project NPV reflects the difference between the levered case and the unlevered case and lies in the economic value subtracted from the debtholders: $\text{NPV}^d = -3.6$. This is the result of a financing policy whereby the firm succeeds in borrowing at a smaller rate than the prevailing market rate for equivalent-risk borrowings.[12] The reader may easily check that the following decomposition holds:

---

[11] We have already computed the project NPV in Example 8.1 as $V_0(\text{RI})$. If the project were unlevered, then project NPV and equity NPV would coincide: $\text{NPV} = \text{NPV}^e = 173.1$. Accordingly, the cost of equity and the WACC would coincide: $r_t^{uo} = r_t^e$. Since the project is levered, the cost of equity is not equal to the WACC.

[12] If, conversely, the cost of debt were equal to the ROD, then the equity NPV would be equal to the project NPV.

$$\text{NPV} = \underbrace{\overbrace{411}^{\text{scale of equity investment}} \cdot \frac{\overbrace{51.43\%}^{\text{efficiency of equity capital}}}{1.2087}}_{\text{NPV}^e=174.88} + \underbrace{\overbrace{372.6}^{\text{scale of debt}} \cdot \frac{\overbrace{-1\%}^{\text{efficiency of debt capital}}}{1.03}}_{\text{NPV}^d=-3.6} = 171.3$$

Example 8.10 ♦

C^o	
	C^e

(*Average ROI of unlevered two-period project, assuming time-varying WACC*) Consider the same figures used in Example 8.1 but let the WACC be time-varying, such that $r_1 = 0.15$, $r_2 = 0.2$. The overall capital is still $\text{PV}[C] = 600 + 200(1.15)^{-1} = 773.9$ but the overall return (total NOPAT) is, in present terms,

$$\text{PV}[I^u] = \text{PV}[I] = 266.3 = \frac{150}{1.15} + \frac{187.5}{1.15 \cdot 1.2}.$$

Hence, the average ROI, from (8.23), is

$$\bar{\imath} = 266.3 \cdot \frac{1.15}{773.9} = 39.57\%,$$

smaller than 40.45%, which is the average ROI computed with a constant WACC equal to 15% (see Example 8.1). This is due to the fact that the value of the overall profit is smaller (due to the higher second-year WACC) while the invested capital remains the same (the second-year WACC does not affect the value of C_1).

The corresponding market returns are $I_1^* = 0.15 \cdot 600 = 90$ and $I_2^* = 0.2 \cdot 200 = 40$. The overall replicating portfolio's return is

$$\text{PV}[I^*] = \frac{90}{1.15} + \frac{40}{1.15 \cdot 1.2} = 107.2.$$

Hence, the MARR is, from (8.26),

$$\bar{r} = 107.2 \cdot \frac{1.15}{773.9} = 15.94\%.$$

The project may be interpreted as an overall investment of \$773.9 at an annual 39.57% average ROI, with a foregone rate of return (average WACC) equal to 15.94%. Hence,

$$\text{NPV} = 773.9 \cdot (39.57\% - 15.94\%)/1.15 = 159.$$

Consistently, we find, via RIs,

$$V_0(\text{RI}) = \frac{150 - 15\% \cdot 600}{1.15} + \frac{187.5 - 20\% \cdot 200}{1.15 \cdot 1.2} = 159.$$

Example 8.11 ♦

C^o	C^d
	C^e

(*Unlevered average ROI of Mockup project, 100% payout*) Let us go back to Example 6.11, where we have computed the time-varying WACCs and the operating value of Mockup project, assuming a 100% distribution policy (this implies $C^l_t = 0$ and $C^o_t = C_t$ for all t). We report the WACCs below (with two decimal digits only), along with the relative discounting factors:

$$\boldsymbol{r}^{uo} = \boldsymbol{r}^{u} = (8.81\%, 9.09\%, 9.51\%, 10.24\%, 11.89\%)$$
$$\mathbf{d}^{uo} = \mathbf{d}^{u} = (1,\ 0.9191,\ 0.8425,\ 0.7694,\ 0.6979,\ 0.6237).$$

We now compute the average ROI (unlevered) and the corresponding average WACC (either via (8.57) or via (8.59), which coincide in this case).

We remind that the EBIT stream is **EBIT** = (0, 48,500, 53,663, 58,954, 64,378, 69,937) (see Table 4.25). Since the tax rate is $\tau = 33\%$ the NOPAT stream is

$$\mathbf{NOPAT} = (1 - 0.33) \cdot \mathbf{EBIT} = (0,\ 32{,}495\ 35{,}954\ 39{,}499\ 43{,}133\ 46{,}858).$$

Table 8.2 reports the computation of average ROI and average WACC. The present values are found by multiplying the NOPATs (and operating capitals) by the respective discount factors (see vector $\mathbf{d}^u$ above). The average ROI is then

$$\bar{\imath}^u = \bar{\imath}^u_0 \cdot (1 + r^u_1) = \frac{149,875}{1,444,196} \cdot 1.0881 = 11.29\%,$$

while the average WACC is

$$\bar{r}^u = \bar{r}^u_0 \cdot (1 + r^u_1) = \frac{124,283}{1,444,196} \cdot 1.0881 = 9.36\%.$$

Therefore, the project creates value for the claimholders (equityholders + debtholders), who earn 11.29% on the invested capital whereas, in equilibrium, an equivalent-risk asset earns an average 9.36% on the same total invested capital. The project NPV is $1,444,196 \cdot (11.29\% - 9.36\%)/1.0881 = 25,592.3$.[13]

The unlevered average ROI may also be computed as a linear combination of ROIs, using Eq. (8.58) (or (8.60), given that, in this case, they coincide). Dividing each NOPAT by the operating capital, one finds the stream of ROIs:

$$\boldsymbol{i}^u = (6.50\%, 8.71\%, 12.12\%, 18.05\%, 30.83\%).$$

[13] Numbers are rounded for illustrative purposes, as usual. Therefore, the manual computations do not produce an exact result. For a better approximation, one should consider more decimal digits. For example, the total capital is, more precisely, \$1,444,195.889, the average ROI is 11.29158%, the average WACC is 9.36347%, the first-period WACC is 8.8053897%, so that multiplying 1,444,195.889 by (11.29158% − 9.36347%) and dividing by (1 + 8.8053897%) one indeed gets 25,592.3. The use of a spreadsheet (e.g., Excel) makes this issue an idle one.

Table 8.2 Mockup project (100% payout)—(unlevered) average ROI and average WACC

	Year	0	1	2	3	4	5
I_t^u	NOPAT		32,495	35,954	39,499	43,133	46,858
C_t	Invested capital	500,000	413,000	326,000	239,000	152,000	
$\text{PV}[I_t^u]$	PV[NOPAT]	149,875					
$\text{PV}[C_t^u]$	PV[Invested capital]	1,444,196					
$\bar{\imath}^u$	**Average ROI (unlevered)**	**11.29%**					
$I_t^{u*} = r_t^u \cdot C_{t-1}$	Market return		44,027	37,525	30,998	24,477	18,072
C_t	Invested capital	500,000	413,000	326,000	239,000	152,000	
$\text{PV}[I^{u*}]$	PV[Market return]	124,283					
$\text{PV}[C_t^u]$	PV[Invested capital]	1,444,196					
$\bar{r}^u$	**Average WACC**	**9.36%**					
$\bar{\imath}^u - \bar{r}^u$	Economic efficiency	1.93%					

Table 8.3 Mockup project (100% payout) (Unlevered) average ROI and average WACC—recalculated

	Year	0	1	2	3	4	5
I_t^u	NOPAT		32,495	35,954	39,499	43,133	46,858
C_t	Investedcapital	500,000	413,000	326,000	239,000	152,000	0
i_t^u	ROI		6.50%	8.71%	12.12%	18.05%	30.83%
w_t^u	weight		34.62%	26.22%	18.90%	12.57%	7.14%
$w_t^u i_t^u$	weight × ROI		2.25%	2.28%	2.29%	2.27%	2.20%
$\bar{\imath}^u = \sum_{t=1}^n w_t^u i_t^u$	**SUM (weight × ROI)**	**11.29%**					
r_t^u	WACC		8.81%	9.09%	9.51%	10.24%	11.89%
w_t^u	weight		34.62%	26.22%	18.90%	12.57%	7.14%
$w_t^u r_t^u$	weight × WACC		3.05%	2.38%	1.80%	1.29%	0.85%
$\bar{r}^u = \sum_{t=1}^n w_t^u r_t^u$	**SUM (weight × WACC)**	**9.36%**					

The weights are obtained as $w_t^u = C_{t-1}\text{d}_{t,1}^u/\text{PV}[C^u]$.[14] Applying the weights to the ROIs and summing, one gets 11.29%, as before. Analogous computations are made to find the MARR (see Table 8.3).

[14] $C_{t-1}\text{d}_{t,1}^u$ and $\text{PV}[C^u]$ are calculated by discounting capital values at WACC (r_t^u), not at r_t (in a levered project, $r_t \neq r_t^u$).

Table 8.4 Mockup project (100% payout) Average ROE and average cost of equity

	Time	0	1	2	3	4	5
I_t^e	Net Income		30,686	34,145	37,690	41,324	45,049
C_t^e	Equity capital	410,000	323,000	236,000	149,000	62,000	0
$\text{PV}[I_t^e]$	PV[Net Income]	133,496.3					
$\text{PV}[C_t^e]$	PV[Equity capital]	1,044,980.4					
$\bar{\imath}^e$	**Average ROE**	**14.08%**					
r_t^e	Cost of equity		10.21%	10.84%	11.91%	14.16%	22.03%
$I_t^{e*}=r_t^e C_{t-1}^e$	Market return		41,858.6	35,024.3	28,118.0	21,092.0	13,655.8
C_t^e	Equity capital	410,000	323,000	236,000	149,000	62,000	0
$\text{PV}[I_t^{e*}]$	PV[Market return]	107,904.0					
$\text{PV}[C_t^e]$	PV[Equity capital]	1,044,980.4					
$\bar{r}^e$	**Average cost of equity**	**11.38%**					
$\bar{\imath}^e - \bar{r}^e$	Economic efficiency	2.7%					

Note that 11.29% is the project's rate of return, not the shareholder rate of return. The latter is equal to the equity AIRR (i.e., average ROE). Table 8.4 reports the computations and the result. Remembering that $r_1^e = 10.2\%$, average ROE is $\bar{\imath}^e = (133,496.3/1,044,980.4) \cdot 1.102 = 14.08\%$, considerably higher than the average ROI. The equity COC is $\bar{r}^e = (107,904/1,044,980.4) = 11.38\%$, higher than the average WACC. The economic efficiency of the equity investment is then $14.08\% - 11.38\% = 2.7\%$. The equity NPV, except for rounding errors, is $\text{NPV}^e = 1,044,980.4 \cdot 2.7\%/1.102 = 25,592.3$. While the decomposition of the equity NPV is different from the decomposition of the project NPV, the NPVs are equal because it is assumed that the efficiency component of the debt is zero (i.e., $i^d = r^d = 3\%$), which implies that $\text{NPV}^d = 0$.

After computing the unlevered average ROI and the average ROE, let us now compute a third BAIRR. Specifically, we compute the (levered) operating average ROI and related (pre-tax) operating average WACC. As we know, the difference between the operating ROI, i_t^o, and the unlevered operating ROI, i_t^{uo}, lies in the fact that the former makes use of the operating profit net of actual taxes, (i.e., $i_t^o = I_t^o/C_{t-1}^o$), whereas the latter employs the operating profit net of taxes on EBIT, thereby excluding taxes on net interest (i.e., $i_t^{uo} = I_t^{uo}/C_{t-1}^o = (I_t^o - \tau I_t^{\mathcal{D}})/C_{t-1}^o$). We use (8.37) and (8.45) to derive the results. Table 8.5 summarizes the computations.

Since non-operating assets do not exist, the operating perspective and the asset perspective coincide, that is, $i_t = i_t^o$ and $r_t = r_t^o$. Hence, the operating average ROI is equal to the asset average ROI and the operating average (pre-tax) WACC is equal to the asset average (pre-tax) WACC: $\bar{\imath} = \bar{\imath}^o = 11.5227\%$ and $\bar{r} = \bar{r}^o = 9.587\%$. Evidently, the resulting rates are (slightly) different from the rates computed under the previous (unlevered) framing, owing to the different treatment of the tax shield, and, therefore, the decomposition of the economic value created is slightly different: $\text{NPV} = 1,440,773 \cdot (11.5227\% - 9.587\%)/1.08975 = 25,592.3$.

Table 8.5 Mockup project (100% payout)—operating average ROI and (pre-tax) average WACC

	Time	**0**	**1**	**2**	**3**	**4**	**5**
I_t^o	Operating Income		33,386	36,845	40,390	44,024	47,749
C_t^o	Invested capital	500,000	413,000	326,000	239,000	152,000	
$\text{PV}[I^o]$	PV[OI]	152,344					
$\text{PV}[C^o]$	PV[Invested capital]	1,440,773					
$\bar{\imath}^o$	**Average ROI**	**11.523%**					
r_t^o	Operating pre-tax WACC		8.97%	9.28%	9.75%	10.56%	12.39%
$I_t^{o*} = r_t^o C_{t-1}^o$	Market return		44,875	38,339	31,782	25,239	18,835
C_t^o	Invested capital	500,000	413,000	326,000	239,000	152,000	
$\text{PV}[I^{o*}]$	PV[Market return]	126,751					
$\text{PV}[C^o]$	PV[Invested capital]	1,440,773					
$\bar{r}^o$	**Average (pre-tax) WACC**	**9.587%**					
$\bar{\imath}^o - \bar{r}^o$	Economic efficiency	1.93%					

Example 8.12 ♦

C^o	C^d
	C^e

(Average ROI and ROE of BHP Billiton, 100% payout) Let us take up Billiton project again, introduced in Example 4.4. In particular, let us consider the case where NOWC is not present, so that only NFA are employed. We adopt an unlevered perspective and compute the unlevered average ROI, $\bar{\imath}^u = \bar{\imath}^{uo}$.

We assume that the WACC is constant and equal to $r^u = 7.5\%$ (thus, $\bar{r}^u = 7.5\%$). Using the financial statements (see Tables 4.13–4.14) we compute the streams of NOPATs, $\boldsymbol{I}^{uo} = \boldsymbol{I}^u$, and invested capitals, $\boldsymbol{C}^o$, as well as the respective total present values (Table 8.6).

From (8.59), the (unlevered) average ROI is

$$\bar{\imath}^u = \overbrace{\left(\frac{\sum_{t=1}^n I_t^u \mathrm{d}_{t,0}^u}{\sum_{t=1}^n C_{t-1} \mathrm{d}_{t-1,0}^u}\right)}^{\bar{\imath}_0^u} \cdot (1 + r_1^u) = \frac{2,249.2}{3,865} \cdot 1.075 = 62.56\%$$

which is greater than the MARR, $\bar{r}^u = 7.5\%$. This immediately identifies the project as a value-creating investment project. Overall, the firm invests \$3,865 at a 62.56% rate of return. The marginal economic efficiency is rather high: The project generates a $55.06\% = 62.56\% - 7.5\%$ excess return. The NPV is $1,979.6 = 3,865 \cdot 55.06\%/1.075$. We assume that the ROD is equal to the cost of debt, so the project NPV is also the equity NPV.

To find the shareholder's rate of return, $\bar{\imath}^e$, one needs to compute the costs of equity, which are endogenous to the input data. The sequence of computations is as follows:

$$\text{project values } (V_t) \Longrightarrow \text{equity values } (V_t^e)$$
$$\Longrightarrow \text{costs of equity } (r_t^e) \Longrightarrow \text{equity AIRR } (\bar{\imath}^e).$$

The project values are computed by discounting FCFs at the WACC: $V_t = (V_{t+1} + F_{t+1}^u)/(1 + r_t^u)$; the equity values are found by deducting the market values of debt (which coincide, by assumption, with the book values): $V_t^e = V_t - V_t^d = V_t - C_t^d$. Then, the cost of equity for each period may be easily computed by the usual formula $r_t^e = (V_t^e + F_t^e)/V_{t-1}^e$ or, alternatively, from the WACC using (6.88) (second equation or fifth equation). Once computed the costs of equity, (8.40) or (8.44) is applied. Using the former,

$$\bar{\imath}^e = \frac{2,147.7}{2,369.9}(1.0766) = 97.57\%.$$

Likewise, one may apply (8.48) or (8.52) for computing the average cost of equity, $\bar{r}^e$. Using the former,

$$\bar{r}^e = \frac{168.1}{2,369.9}(1.0766) = 7.64\%.$$

Table 8.6 reports the relevant data for equity, which result in an overall equity investment of \$2,369.9, an equity AIRR (average ROE) equal to $\bar{\imath}^e = 97.57\%$ and an equity MARR (average cost of equity) equal to $\bar{r}^e = 7.64\%$. The equity NPV is

Table 8.6 BHP Billiton (100% payout, no NOWC)—Rates of return

Time	I_t^u	C_t^o	I_t^e	C_t^e	V_t	V_t^e	F_t^e	r_t^e	I_t^{e*}
0		550.0		350.0	2529.6	2329.6	−350.0		
1	222.3	522.5	211.1	322.5	2469.6	2269.6	238.6	7.66%	26.8
2	222.3	495.0	211.1	295.0	2405.0	2205.0	238.6	7.67%	24.7
3	222.3	467.5	211.1	267.5	2335.7	2135.7	238.6	7.67%	22.6
4	222.3	440.0	211.1	240.0	2261.1	2061.1	238.6	7.68%	20.5
5	222.3	412.5	211.1	212.5	2180.9	1980.9	238.6	7.68%	18.4
6	222.3	385.0	211.1	185.0	2094.7	1894.7	238.6	7.69%	16.3
7	222.3	357.5	211.1	157.5	2002.1	1802.1	238.6	7.70%	14.2
8	222.3	330.0	211.1	130.0	1902.5	1702.5	238.6	7.71%	12.1
9	222.3	302.5	211.1	102.5	1795.4	1595.4	238.6	7.72%	10.0
10	222.3	275.0	211.1	275.0	1680.3	1680.3	38.6	7.74%	7.9
11	222.3	247.5	222.3	247.5	1556.6	1556.6	249.8	7.50%	20.6
12	222.3	220.0	222.3	220.0	1423.6	1423.6	249.8	7.50%	18.6
13	222.3	1925	222.3	192.5	1280.6	1280.6	249.8	7.50%	16.5
14	222.3	165.0	222.3	165.0	1126.9	1126.9	249.8	7.50%	14.4
15	222.3	137.5	222.3	137.5	961.7	961.7	249.8	7.50%	12.4
16	222.3	110.0	222.3	110.0	784.1	784.1	249.8	7.50%	10.3
17	222.3	82.5	222.3	82.5	593.1	593.1	249.8	7.50%	8.2
18	222.3	55.0	222.3	55.0	387.9	387.9	249.8	7.50%	6.2
19	222.3	27.5	222.3	27.5	167.2	167.2	249.8	7.50%	4.1
20	152.3	0.0	152.3	0.0	0.0	0.0	179.8	7.50%	2.1
PV	2,249.2	3,865	2,147.7	2,369.9					168.1

project AIRR (average ROI)	$\bar{\imath}^u = 62.56\%$	project MARR $\bar{r}^u = 7.50\%$	
equity AIRR (average ROE)	$\bar{\imath}^e = 97.57\%$	equity MARR $\bar{r}^e = 7.64\%$	

$$\text{NPV}^e = \overbrace{2,369.9}^{\text{equity scale}} \cdot \frac{\overbrace{97.57\% - 7.64\%}^{\text{efficiency of equity}}}{1.0766} = 1,979.6.$$

Example 8.13 ♦

(*Average ROI of Solar PhV project*) Consider the Solar PhV project introduced in Example 1.13. In that case, $I_0 = 100 \neq 0$, and the stream of incomes is such that $I_t = 100$ for all $t \in \mathbb{N}_{11}^0$, $I_{12} = 900$, $I_t = -100$ for all $t \in \mathbb{N}_{20}^{13}$ (see also Table 1.9). As for the capital base consider that $C_t = 0$ for all $t \in \mathbb{N}_{11}^0$; then, $C_{12} = 8,000$, $C_{13} = 7,000$, $C_{14} = 6,000$, $C_{15} = 5,000$, $C_{16} = 4,000$, $C_{17} = 3,000$, $C_{18} = 2,000$, $C_{19} = 1,000$. Assuming the COC is constant and equal to $r = 3\%$, one may compute the BAIRR using any of the four methods presented. For example, using (8.13),

$$\begin{aligned}\bar{\imath} &= \frac{\text{PV}[I]}{\text{PV}[C]} \cdot (1+r)\\ &= \frac{100 + 100 \cdot \frac{1-(1.03)^{-11}}{0.03} + \frac{900}{1.03^{12}} - 100 \cdot \frac{1-(1.03)^{-8}}{0.03} \cdot \frac{1}{1.03^{12}}}{\frac{8,000}{1.03^{12}} + \frac{7,000}{1.03^{13}} + \frac{6,000}{1.03^{14}} + \frac{5,000}{1.03^{15}} + \frac{4,000}{1.03^{16}} + \frac{3,000}{1.03^{17}} + \frac{2,000}{1.03^{18}} + \frac{1,000}{1.03^{19}}} \cdot 1.03\\ &= \frac{1,164}{23,606} \cdot 1.03\\ &= 5.08\%.\end{aligned}$$

The project is economically profitable ($5.08\% > 3\%$) and the NPV is $23,606 \cdot (5.08\% - 3\%)/1.03 = 477$.

If we were to compute the BAIRR via (8.19), we should employ the cash flows. Using Table 1.9, one may easily compute the NPV directly from cash flows and then use the shortcut (8.19) (which is formula ④):

$$\begin{aligned}\bar{\imath} &= r + \frac{\text{NPV}}{\text{PV}[C]} \cdot (1+r)\\ &= 0.03 + \frac{477}{23,606} \cdot 1.03\\ &= 5.08\%.\end{aligned}$$

Example 8.14 ♦

(*Average ROI of abandonment project*) Consider the abandonment project illustrated in Example 1.11. Suppose the COC is constant and equal to $r = 6\%$. For convenience of the reader, we report below the concise transposed Matrix:

(*C*, *I*, *F*)

Time		0	1	2	3	4	5
Capital	C	−6,000	−4,800	−3,600	−2,400	−1,200	0
Income	I	500	−400	−200	−100	−50	−10
Cash Flow	F	6,500	−1,600	−1,400	−1,300	−1,250	−1,210

Notably, this project is a financing project, according to Definition 8.4, for

$$\text{PV}[C] = -6,000 - 4,800(1.06)^{-1} - 3,600(1.06)^{-2} - 2,400(1.06)^{-3} - 1,200(1.06)^{-4} = -16,697.9 < 0.$$

This means that the resulting AIRR is a financing rate. Specifically, considering that $\text{PV}[I] = 500 - 400(1.06)^{-1} - 200(1.06)^{-2} - 100(1.06)^{-3} - 50(1.06)^{-4} - 10(1.06)^{-5} = -186.4$,

$$\bar{\imath} = \frac{-186.4}{-16,697.9} \cdot 1.06 = 1.18\%.$$

The firm borrows (subtracts) \$16,697.9 from the project at a cost of 1.18% while giving up the opportunity of borrowing the same amount from the market at 6%. Since 1.18% < 6%, the project is economically profitable and worth undertaking. The excess AIRR is then $\bar{\imath} - \bar{r} = 1.18\% - 6\% = -4.82\%$. That is, the firm save financing costs at a rate of 4.82%.

In present terms, the marginal financial efficiency of this borrowing is $-4.82\%/1.06 = -4.55\%$. Applying the latter to the capital base one finds the NPV:

$$\text{NPV} = -16,697.9 \cdot (-4.55\%) = 759.$$

8.5 Equal Cash Flows, Different Rates of Return

As implied by the previous sections, cash flows are not relevant in building the notion of rate of return. Only returns (i.e., incomes) and capitals matter. From this point of view, cash flows are second-order variables.[15] To appreciate this point, we show that, in general, **different economic activities have different rates of return even though their cash-flow stream is the same**. Consider, for example, the following transactions:

- a bank lends \$7,000, repayable with 5 instalments equal to \$4,000, \$3,000, \$1,000, \$1,000, \$196, respectively. The loan rate is time-varying and equal to $\boldsymbol{i}^{\text{loan}} = (11\%, 25\%, 15\%, 20\%, 20\%)$. The COC for this transaction is $r^{\text{loan}} = 10\%$
- a firm is incorporated with \$7,000 capital, which is invested in operating assets for the purpose of launching a new product, whose estimated life is 5 years. This initiative is unlevered and the NOPATs are $\boldsymbol{I}^u = \boldsymbol{I}^{uo} = \boldsymbol{I} = (0, 2{,}200, 1{,}300, -300, 300, -1{,}304)$. Pro forma book values of operating assets are $\boldsymbol{C}^o = (7{,}000, 5{,}200, 3{,}500, 2{,}200, 1{,}500)$. The firm is liquidated after 5 years. The WACC is assumed to constant and equal to $r^{uo} = r^u = r = 10\%$.

[15]This does not diminish the practical usefulness of (8.19)–(8.20) for computing the rate of return and of (8.31)–(8.34) in case of time-varying COCs.

Consider the amortization plan of the loan:

Time	Debt	Interest	Instalment	Interest rate
0	7,000.0		−7,000	
1	3,770.0	770.0	4,000	11%
2	1,712.5	942.5	3,000	25%
3	969.4	256.9	1,000	15%
4	163.3	193.9	1,000	20%
5	0.0	32.7	196	20%
t	C_t^{loan}	I_t^{loan}	F_t^{loan}	i_t^{loan}

Now consider the data for the project:

Time	Operating assets	NOPAT	FCF	ROI
0	7,000		−7,000	0
1	5,200	2,200.0	4,000	31.4%
2	3,500	1,300.0	3,000	25.0%
3	2,200	−300.0	1,000	−8.6%
4	1,500	300.0	1,000	13.6%
5	0	−1,304	196	−86.9%
t	C_t^{firm}	I_t^{firm}	F_t^{firm}	i_t^{firm}

These two economic activities are different in several respects: The former is a financial activity, the latter is a real asset; in the former, the interest rates are exogenously selected by the lender, in the latter the return rates (ROIs) are derived from the estimates of profits (revenues and costs) and capitals. The cash flows in the former are contractually established, the cash flows in the latter depend on the demand of the new product. These differences result in different patterns of profits, different patterns of invested capitals, and different patterns of income rates. It is thus natural to expect that the rates of return of the two activities will be different. Indeed, the rate of return for the bank's loan is

$$
\begin{aligned}
\bar{\imath}^{\text{loan}} &= \frac{\text{PV}[I^{\text{loan}}]}{\text{PV}[C^{\text{loan}}]} \cdot (1+r) \\
&= \frac{770(1.1)^{-1} + 942.5(1.1)^{-2} + 256.9(1.1)^{-3} + 193.9(1.1)^{-4} + 32.7(1.1)^{-5}}{7,000 + 3,770(1.1)^{-1} + 1,712.5(1.1)^{-2} + 969.4(1.1)^{-3} + 163.3(1.1)^{-4}} \cdot 1.1 \\
&= \frac{1,824.61}{12,682.37} \cdot 1.1 \\
&= 15.83\%.
\end{aligned}
$$

The rate of return for the project is

$$
\begin{aligned}
\bar{\imath}^{\text{project}} &= \frac{\text{PV}[I^{\text{project}}]}{\text{PV}[C^{\text{project}}]} \cdot (1+r) \\
&= \frac{2,200(1.1)^{-1} + 1,300(1.1)^{-2} - 300(1.1)^{-3} + 300(1.1)^{-4} - 1,304(1.1)^{-5}}{7,000 + 5,200(1.1)^{-1} + 3,500(1.1)^{-2} + 2,200(1.1)^{-3} + 1,500(1.1)^{-4}} \cdot 1.1 \\
&= \frac{2,244.15}{17,297.25} \cdot 1.1 \\
&= 14.27\%.
\end{aligned}
$$

The only element that the two assets have in common is the cash-flow stream. However, cash flows alone do not define a project and do not suffice to assess its economic rate of return. The latter depends on the relation between return (i.e., profit) and capital. This means that two projects may create the same economic value through investment of different amounts of capital at different degrees of efficiency. For the case at hand, given that cash flows and COCs are equal, the loan and the project generate the same NPV for the bank and the firm:

$$\text{NPV}^{\text{loan}} = 12,682.37 \cdot \frac{5.83\%}{1.1} = 672$$
$$\text{NPV}^{\text{project}} = 17,297.25 \cdot \frac{4.27\%}{1.1} = 672$$

However, the way this NPV is generated is different. Specifically, the bank manages money in a more efficient way than the firm. The stream of the loan's interest rate is $\boldsymbol{i}^{\text{loan}} = (0.11, 0.25, 0.15, 0.2, 0.2)$ as opposed to the equilibrium vector $\boldsymbol{r}^{\text{loan}} = (0.1, 0.1, 0.1, 0.1, 0.1)$. Overall, the bank earns an average 5.83% over and above the market rate on a total invested capital of $12,682.37. In contrast, the firm manages money in a less efficient way: The vector of profit rates is $\boldsymbol{i}^{\text{project}} = (31.4\%, 25\%, -8.6\%, 13.6\%, -86.9\%)$ some of which are greater than the equilibrium rate, while some others are smaller: $\boldsymbol{r}^{\text{project}} = (0.1, 0.1, 0.1, 0.1, 0.1)$. Overall, the firm earns an average 4.27% over and above the market rate, which is less than the 5.83% average excess return earned by the bank. The smaller efficiency is exactly compensated by the project's greater capital base, so that the NPV is the same for the two assets.

Consider now the same pair of assets, but let the NOPAT in the second year be $1,596 instead of $1,300. This implies that the FCF in the second year changes to $\$3,269 = 1,596 - (3,500 - 5,200)$ and the second-year ROI increases to 35.3%, while everything else remains unvaried.

Time	Operating assets	NOPAT	FCF	ROI
0	7,000		−7,000	
1	5,200	2,200.0	4,000	31.4%
2	3,500	**1,596.0**	**3,296**	**30.7%**
3	2,200	−300.0	1,000	−8.6%
4	1,500	300.0	1,000	13.6%
5	0	−1,304	196	−86.9%
t	C_t^{firm}	I_t^{firm}	F_t^{firm}	i_t^{firm}

The rate of return is now 15.83% for both the loan and the project, while the NPVs are different. Specifically, the project's total income is now greater, to such an extent that the firm is just as efficient as the bank, while the invested capital does not vary:

$$
\begin{aligned}
\bar{\imath}^{\text{project}} &= \frac{\text{PV}[I^{\text{project}}]}{\text{PV}[C^{\text{project}}]} \cdot 1.1 \\
&= \frac{2,488.77}{17,297.25} \\
&= 15.83\% \\
&= \frac{1,824.61}{12,682.37} \cdot 1.1 \\
&= \frac{\text{PV}[I^{\text{loan}}]}{\text{PV}[C^{\text{loan}}]} \cdot 1.1 \\
&= \bar{\imath}^{\text{loan}}.
\end{aligned}
$$

As a result, the project creates more value than the loan, given that more capital is invested at the same degree of economic efficiency as the loan. In this case, where a given economic efficiency is shared by the two alternatives, the comparison of the total capital constitutes a rational decision criterion: $17,297.25 > 12,682.37$. This should be rather intuitive, for a project's NPV depends on its size and its economic efficiency. Thus, for a fixed size, the greater the economic efficiency, the greater the economic value created; vice versa, for a fixed economic efficiency, the greater the size, the greater the economic value created.

8.6 The AIRR and Excess AIRR Functions

As seen, the average-based approach naturally triggers a product structure such that the economic value created is the product of a capital base and the excess rate of return, expressing the marginal economic efficiency of the capital. Assuming constant COC,

$$
\text{NPV}(1+r) = \overbrace{\text{PV}[C]}^{\text{project scale}} \cdot \overbrace{(\bar{\imath} - r)}^{\text{economic efficiency}}.
$$

$$
\text{NPV} = \overbrace{\text{PV}[C]}^{\text{project scale}} \cdot \overbrace{(\bar{\imath}_0 - \bar{r}_0)}^{\substack{\text{economic efficiency} \\ \text{(instantaneous)}}}
$$

where, as we remind, $\bar{\imath}_0 = \bar{\imath}/(1+r)$ and $\bar{r}_0 = r/(1+r)$.

We have also seen that, reversing the relation and solving for $\bar{\imath}$, the AIRR may be calculated as

$$
\bar{\imath} = r + \frac{\text{NPV}}{\text{PV}[C]}(1+r) \qquad \bar{\imath}_0 = r_0 + \frac{\text{NPV}}{\text{PV}[C]}. \tag{8.69}
$$

Interestingly, for any fixed cash-flow stream, $\boldsymbol{F} = (F_0, F_1, \ldots, F_n)$, the equality $\text{NPV} = \text{PV}[C](\bar{\imath} - r)/(1 + r)$ holds whatever the sequence $\boldsymbol{C} = (C_0, C_1, \ldots, C_{n-1}, 0)$. This means that, for any pair of capital vectors $\boldsymbol{C}^j \neq \boldsymbol{C}^h$, the equality

$$\text{NPV}(1 + r) = \text{PV}[C^j] \cdot (\bar{\imath}(C^j) - r) = \text{PV}[C^h] \cdot (\bar{\imath}(C^h) - r)$$

holds, even if $\bar{\imath}(C^j) \neq \bar{\imath}(C^h)$. The corresponding AIRRs may be calculated as $\bar{\imath}(C^j) = r + \text{NPV}(1 + r)/\text{PV}[C^j]$ and $\bar{\imath}(C^h) = r + \text{NPV}(1 + r)/\text{PV}[C^h]$.[16] In general, dividing the NPV by some capital base x, one gets a relative measures of excess return (in present terms): NPV/x. Evidently, there may be as many choices of x as the analyst is willing to consider. Considering that $\xi = \bar{\imath}_0 - \bar{r}_0 = (\bar{\imath} - r)/(1 + r)$, one can build two rate-of-return functions,

$$\begin{aligned} \bar{\imath}(x) &= r + \frac{\text{NPV}(1 + r)}{x} \\ \xi(x) &= \frac{\text{NPV}}{x}, \end{aligned} \tag{8.70}$$

which we call, respectively, the (discrete) *AIRR function* and the *excess AIRR function*, with $x \in \mathbb{R} - \{0\}$. The excess AIRR, $\xi(x)$, expresses the instantaneous economic efficiency with respect to the capital base x:[17]

$$\xi(x) = \frac{\text{NPV}}{x} = \bar{\imath}_0 - \bar{r}_0 = \frac{\bar{\imath}(x) - r}{1 + r} \Longleftrightarrow \bar{\imath}(x) = r + \xi(x)(1 + r). \tag{8.71}$$

The AIRR function is a monotonic transformation of the excess AIRR function. Both are homographic functions on the (capital, return)-plane, graphically described by the arcs of a hyperbola. Figure 8.2 represents the AIRR function for a value-creating project. Such arcs delimit an **iso-NPV line**, in the sense that any combination $(x, \bar{\imath}(x))$ or $(x, \xi(x))$ generates the same economic value: $\text{NPV} = x \cdot \xi(x)$ for every x. The AIRR function is then representative of a class of rates of return, $\bar{\imath}(x)$, the ***AIRR class***. The BAIRR belongs to such a class and is the value taken on by the AIRR function whenever x is equal to the overall book value. In the next sections, we investigate other rates belonging to the AIRR class.

---

[16] See details in Magni (2010b, 2013a, 2016b), Magni et al. (2018).

[17] In a thoughtful paper, Keane (1979) realized that the ratio of NPV to the overall capital employed brings about an economically significant measure of efficiency: "If any ratio can be perceived as being the correct 'cost benefit' index it would be the NPV expressed as a ratio of the total number of units of capital employed" (Keane 1979, p. 54, footnote 7).

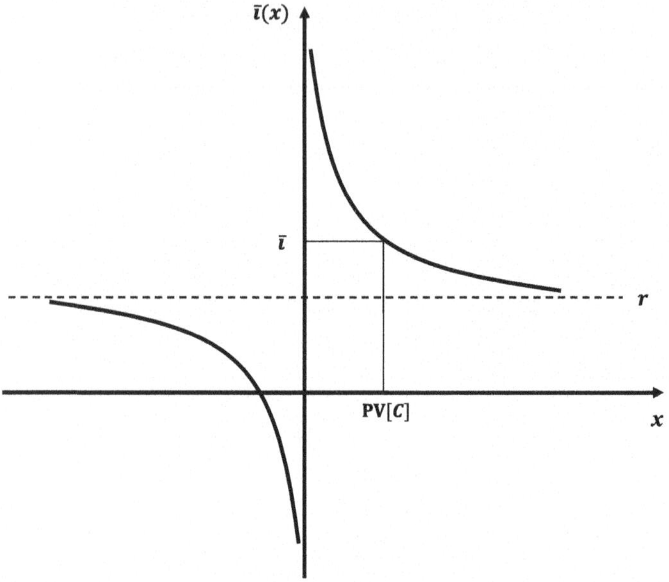

Fig. 8.2 The AIRR function for a value-creating project (NPV > 0). The area of any rectangle with basis $x \neq 0$ and height $\bar{\imath}(x) - r$ is equal to NPV$(1+r)$

8.7 The Profitability Index and the Benefit-Cost Ratio

The Profitability Index is a well-known relative measure of worth, employed by practitioners as a bang-for-the-buck measure. It is equal to the ratio of NPV to initial investment:

$$\mathrm{PI} = \frac{\mathrm{NPV}}{C_0}. \tag{8.72}$$

In the light of what we have seen, the PI may now be reconceptualized as a special case of excess AIRR, where the capital base considered is the initial investment: $x = C_0$, whence $\mathrm{PI} = \xi(C_0)$. The project is worth undertaking if and only if $\mathrm{PI} = \xi(C_0) > 0$, with $C_0 > 0$ (or $\mathrm{PI} < 0$ with $C_0 < 0$). The corresponding rate of return, $\bar{\imath}(C_0) = r + \mathrm{PI} \cdot (1 + r)$, belongs to the AIRR class and expresses the project's return as measured per unit of initial investment.

The so-called *benefit-cost (BC) ratio* is the ratio of the discounted sum of inflows to discounted sum of outflows:

$$\mathrm{BC} = \frac{\sum_{t=0}^{n} F_t^+ \cdot \mathrm{d}_{t,0}}{\sum_{t=0}^{n} F_t^- \cdot \mathrm{d}_{t,0}} \tag{8.73}$$

where $F_t^+ = F_t$ if $F_t > 0$ or $F_t^+ = 0$ otherwise, while $F_t^- = -F_t$ if $F_t < 0$ or $F_t^- = 0$ otherwise. Equivalently,

$$\text{BC} = 1 + \frac{\text{NPV}}{\text{PV}[F^-]} \tag{8.74}$$

where $\text{PV}[F^-] = \sum_{t=0}^{n} F_t^- \text{d}_{t,0}$. As $\text{NPV}/\text{PV}[F^-] = \xi(\text{PV}[F^-])$, the BC ratio is equal to

$$\text{BC} = 1 + \xi(\text{PV}[F^-]). \tag{8.75}$$

The project is worth undertaking if and only if $\xi(F^-) > 0$, that is, if and only if $\text{BC} > 1$. The corresponding rate of return, $\bar{\imath}(\text{PV}[F^-]) = r + \xi(\text{PV}[F^-]) = r + (\text{BC} - 1)$, belongs to the AIRR class and is the project's return as measured per unit of total outflow.

If the only outflow is the initial investment (so that $C_0 = \text{PV}[F^-]$), BC is equal to $1 + \text{PI}$. However, the PI itself may be redefined in terms of total outflow, so that $\text{PI} = \text{NPV}/\text{PV}[F^-]$, thereby employing the same capital base as BC. In general, a profitability index may be defined in terms of any preferred capital base by dividing the NPV by the selected capital base (from this point of view, the excess AIRR function is indeed a PI function).

Remark 8.7 (*Terminology*) The BC ratio and PI are sometimes exchanged for one another or considered synonyms; other times, slightly different definitions are given. For example, Blank and Tarquin (2012) define the PI as $\text{PI} = \sum_{t=1}^{n} F_t d_{t,0}/C_0 = 1 + \text{NPV}/C_0$; Kellison (2009) seems to consider the two measures economically equivalent, given that the term "profitability index" is used to denote both (8.72) and (8.73). In contrast, Ross et al. (2011) use both expressions for (8.72), as well as Hartman (2007), who reserves the term "benefit-cost ratio" to investment made by government entities. Rao (1992) refers to (8.73) as "profitability index." Differently, Park (2011) define profitability index as the ratio of benefits net of operating costs to capital expenditure (see also Brealey et al. 2011; Newnan et al. 2009). In some fields such as human resource management, environmental engineering, and natural sciences, the term "return on investment" is often used to mean either BC ratio or PI (e.g., Gargani 2017; Pattison-Williams et al. 2017; Michaud et al. 2017; De Risi et al. 2018). The different linguistic habits may well cause some misunderstanding. ▲

8.8 The Economic AIRR

Suppose an analyst wants to grasp a piece of information that the BAIRR does not supply. In particular, the analyst wonders what the project's rate of return is *from the point of view of a normal capital market.*

As we have seen in Sect. 5.3, at time 0, there is a temporary state of disequilibrium, reflected in the difference between the value of the project, V_0, and the investment cost, C_0. As a consequence of it, there is "an immediate increase in the value of the company's stock as soon as the company is committed to the investment" (Robichek and Myers 1965, p. 11. See also Ross et al. 2011, p. 327 and Fig. 10.13 therein). This results in

> a "windfall gain", which accrues to the owners of the firm as a result of their being able to invest in a project that is more profitable than the standard market rate. (Robichek and Myers 1965, p. 11).

Formally, after the windfall gain $V_0 - C_0$ has accrued to shareholders instantaneously at time 0, the incremental return generated in the period by the project will be the equilibrium return, $r \cdot V_0$. Adding the two returns, the total first-period income is found:

$$\overbrace{V_0 - C_0}^{\text{instantaneous gain}} + \overbrace{r \cdot V_0}^{\text{equilibrium return}} = \text{NPV} + r \cdot V_0. \tag{8.76}$$

Dividing by the initial investment C_0 one gets the first-period (disequilibrium) rate of return:

$$\begin{aligned} i_1 = i^{\text{dis}} &= \frac{\text{NPV}}{C_0} + r \cdot \frac{V_0}{C_0} \\ &= \text{PI} + r \cdot \frac{V_0}{C_0} \end{aligned} \tag{8.77}$$

or, equivalently, exploiting the equality $\text{NPV} + rV_0 = V_0(1+r) - C_0 = V_1 + F_1 - C_0$,

$$i^{\text{dis}} = \frac{V_1 + F_1 - C_0}{C_0}. \tag{8.78}$$

After earning the disequilibrium rate of return, i_1^{dis}, in the first period, "there is no tendency for subsequent increases and decreases" (Ross et al. 2011, p. 327). In particular,

> after the windfall gain is realized through the increase in value of the owners' stock, income will continue to be realized at a rate of exactly [r] …for the remainder of the project's life (Robichek and Myers 1965, p. 11).

This means that, from time 1 on, the capital value is V_t. Formally, shareholders invest C_0 at time 0 and the economic value V_t at time $t \in \mathbb{N}^1_{n-1}$. Which means that the vector of invested capital is $\boldsymbol{C}^{\text{ec}} = (C_0, V_1, V_2, \dots, V_{n-1}, 0)$.[18] The stream of single-period return rates is then $(i_1^{\text{dis}}, r, r, \dots, r)$. The corresponding rate of return is the so-called *economic AIRR* (EAIRR), such that[19]

$$\begin{aligned} \text{EAIRR} &= \bar{\imath}\Big(\text{PV}[\boldsymbol{C}^{\text{ec}}]\Big) \\ &= \frac{i_1^{\text{dis}} C_0 + rV_1(1+r)^{-1} + rV_2(1+r)^{-2} + \dots + rV_{n-1}(1+r)^{-(n-1)}}{C_0 + V_1(1+r)^{-1} + V_2(1+r)^{-2} + \dots + V_{n-1}(1+r)^{-(n-1)}} \end{aligned} \tag{8.79}$$

or, equivalently,

[18] Do not confuse with $\boldsymbol{V} = (V_0, V_1, \dots, V_{n-1}, 0)$.

[19] Lindblom and Sjögren (2009) call this pattern of capital depreciation "strict market depreciation schedule". See also Magni (2013a, 2014a), Magni et al. (2018).

$$\text{EAIRR} = \frac{\text{PV}[I^{\text{ec}}]}{\text{PV}[C^{\text{ec}}]} \cdot (1+r) \tag{8.80}$$

where $\text{PV}[I^{\text{ec}}] = i_1^{\text{dis}} C_0(1+r)^{-1} + \sum_{t=2}^{n} r_t V_{t-1}(1+r)^{-t}$ and $\text{PV}[C^{\text{ec}}] = C_0 + \sum_{t=1}^{n} V_{t-1} \text{d}_{t,0}$ (see Magni 2013a). Using (8.69), a computational shortcut is

$$\text{EAIRR} = r + \frac{\text{NPV}}{\text{PV}[C^{\text{ec}}]}(1+r). \tag{8.81}$$

If the COC is time-varying, the same formulas hold, with the due adjustments (see Sect. 8.3).[20]

The related decision criterion states that an investment project is worth undertaking if and only if EAIRR $> r$ or, in general, EAIRR $> \bar{r}$ where

$$\bar{r} = \frac{\text{PV}[I^{\text{ec},*}]}{\text{PV}[C^{\text{ec}}]} \cdot (1+r_1)$$

with $I_1^{\text{ec},*} = i^{\text{dis}} C_0$ and $I_t^{\text{ec},*} = r_t V_{t-1}$ for $t > 1$. The decision criterion is rational (the EAIRR is the value taken on by the AIRR function when $x = \text{PV}[C^{\text{ec}}]$, namely, $\bar{\imath}(\text{PV}[C^{\text{ec}}]) = \text{EAIRR}$).

The condition EAIRR $> r$ (in general, EAIRR $> \bar{r}$) is equivalent to $i^{\text{dis}} > r$ (in general $i^{\text{dis}} > r_1$), which is based on the comparison of the first-period rates. This exactly fulfills Bailey's (1959) insight:

> the criterion for multiperiod investments can center on a short-term rate when all other short-term rates are assumed to be equal to the equilibrium rates (Bailey 1959, p. 479. See also Magni 2014a).

While Bailey (1959) did not indicate which rate should be allowed to differ from the equilibrium rate, the choice is unequivocally provided by a normal market; in it, all but the first holding period rates are equal to the equilibrium rates ($i_t = r_t$ for all $t > 1$), so the criterion may center on the disequilibrium rate, i^{dis}.

The EAIRR is also a suitable measure of economic profitability for those people (i.e., external analysts) who, for any reason, have only access to forecasted cash flows and MARR but not to the prospective book values.

Remark 8.8 (***Equivalence class***) Many other metrics may be built, expressing different values taken on by the AIRR function, $\bar{\imath}(x)$. If the COC is constant, then each capital base is associated with an equivalence (or depreciation) class of capital streams and each such equivalence class is associated with a well-defined AIRR. A *K-equivalence class* or *K-depreciation class* is defined as the set of capital streams $\boldsymbol{S} = (S_0, S_1, S_2, \ldots, S_{n-1}, 0)$ such that $\text{PV}[S] = K$. If two capital streams $\boldsymbol{S}^1$ and $\boldsymbol{S}^2$ belong to the same class, then the overall capital coincide: $\text{PV}[S^1] = \text{PV}[S^2] = K$.

[20] See also the use of EAIRR in Magni (2013a, b, 2014a), Cuthbert and Magni (2016), Bosch-Badia et al. (2014), Barry and Robison (2014).

For example, let $r = 0\%$: The capital streams $\boldsymbol{S}^1 = (100, 80, 10, 0)$ and $\boldsymbol{S}^2 = (90, 100, 0, 0)$ belong to the same depreciation class, since $\text{PV}[S^1] = \text{PV}[S^2] = 190$. If any two projects belong to the same equivalence class, then they have the same scale (they entail the same total invested capital). As we will see, given any two projects with the same scale, ranking projects with rates of return is acceptable (see Chap. 11).

The notion of equivalence/depreciation class in the context of rates of return and NPV has been introduced and developed by Magni (2009e, 2010b, 2011a,c). A review of this concept and a discussion of several metrics belonging to AIRR class may be found in Magni (2016b) (see also Mørch et al. 2017; Jiang 2017; Cuthbert and Magni 2018; Lima e Silva et al. 2018 for other metrics belonging to the AIRR class). ▲

Example 8.15 ♦
Consider a project described in the following concise transposed Matrix:

$(\boldsymbol{C}, \boldsymbol{I}, \boldsymbol{F})$							
Time		0	1	2	3	4	5
Capital	C	100	90	60	40	20	0
Income	I	0	−10	20	8	4	5
Cash Flow	F	−100	0	50	28	24	25

Suppose the analyst wants to compute the project's rate of return from the point of view of the market. That is, the analyst aims at computing the EAIRR. It suffices to compute the NPV and the intrinsic values and then use (8.81). Assuming $r = 10\%$, the NPV is $\text{NPV} = -100 + 50(1.1)^{-2} + 28(1.1)^{-3} + 24(1.1)^{-4} + 25(1.1)^{-5} = -5.73$. The intrinsic values are obtained, as usual, via the backward relation $V_t = (V_{t+1} + F_{t+1})/(1 + r)$: $V_5 = 0$, $V_4 = 22.73$, $V_3 = 42.48$, $V_2 = 64.07$, $V_1 = 103.7$. Therefore, $C^{\text{ec}} = (100, 103.7, 64.07, 42.48, 22.73)$ and the capital base is $\text{PV}[C^{\text{ec}}] = 100 + 103.7(1.1)^{-1} + 64.07(1.1)^{-2} + 42.48(1.1)^{-3} + 22.73(1.1)^{-4} = 294.67$. Hence,

$$\text{EAIRR} = \bar{\imath}(294.67) = 0.1 + \frac{-5.73}{294.67} \cdot 1.1 = 7.86\%.$$

The project destroys value, since NPV < 0. In particular, the project's economic efficiency is negative and, in present terms, equal to $\xi(C^{\text{ec}}) = (7.86\% - 10\%)/1.1 =$

-1.94%. Applied to a capital base equal to \$294.67, the NPV is found again: $-5.73 = 294.67 \cdot (-1.94\%)$.[21]

For this project, the PI is

$$\text{PI} = \xi(100) = \frac{-5.73}{100} = -0.0573 < 0.$$

Hence, the rate of return on initial investment is

$$\bar{\imath}(C_0) = \bar{\imath}(100) = 0.1 - 0.0573(1 + 0.1) = 3.7\%.$$

The BC ratio is

$$\text{BC} = \frac{94.27}{100} = 0.9427 < 1$$

and the rate of return on total outflow is

$$\bar{\imath}(F^-) = \bar{\imath}(100) = 0.1 + (0.9427 - 1)(1 + 0.1) = 3.7\%.$$

In this case, $\bar{\imath}(F^-) = \bar{\imath}(C_0) = 3.7\%$ because $C_0 = F^- = 100$.

Example 8.16 ♦

C^o C^d C^e

(*EAIRR of Mockup project, 100% payout*) Consider Mockup project (100% payout) and suppose the analyst wants to draw the EAIRR for the project's CFA stream (which coincides with the CFO stream). We report the concise transposed Matrix below:

$(\boldsymbol{C}, \boldsymbol{I}, \boldsymbol{F})$ Mockup project (100% payout)

Time		0	1	2	3	4	5
Capital	C	500,000	413,000	326,000	239,000	152,000	0
Income	I		33,386	36,845	40,390	44,024	47,749
Cash Flow	F	−500,000	120,386	123,845	127,390	131,024	199,749

The stream of pre-tax WACC is $\boldsymbol{r} = (8.97\%, 9.28\%, 9.75\%, 10.56\%, 12.39\%)$ (see Example 6.9). This enables computing the economic values and, hence, the capital base. The stream of income rates is $\boldsymbol{i}^{\text{ec}} = (i_1^{\text{dis}}, r_2, r_3, r_4, r_5)$ where

[21] It is easy to verify that the BAIRR is $\bar{\imath}\big(\text{PV}[\boldsymbol{C}]\big) = \bar{\imath}(275.12) = 7.71\%$, slightly smaller than the EAIRR. This is due to the fact that, for $t \geq 1$, every book value is slightly smaller than the respective intrinsic value (with a negative numerator, a smaller denominator makes the ratio smaller).

Table 8.7 Mockup project (100% payout)–EAIRR

Time	0	1	2	3	4	5
Income		72,764	41,994	36,123	29,491	22,022
Invested capital	500,000	452,378	370,527	279,260	177,727	0
PV[Economic income]	163,639.4					
PV[Invested capital]	1,562,903.8					
EAIRR	**11.41%**					
Pre-tax WACC		8.97%	9.28%	9.75%	10.56%	12.39%
Market return	44,875	41,994	36,123	29,491	22,022	
PV[Market return]	138,047.1					
PV[Invested capital]	1,562,903.8					
MARR	**9.63%**					
Economic efficiency	1.78%					

$$i_1^{\text{dis}} = \frac{V_1 + F_1 - C_0}{C_0} = \frac{452,378 + 120,386 - 500,000}{500,000} = 14.55\%.$$

The stream of invested capitals is given by $\boldsymbol{C}^{\text{ec}} = (C_0, V_1, V_2, V_3, V_4)$, with $V_t = (V_{t+1} + F_{t+1})/(1 + r_{t+1})$, as usual. The economic incomes are computed by multiplying the income rates by the respective invested capitals. Table 8.7 collects the computations for the EAIRR and its associated COC. The EAIRR is $11.41\% = (163,639.4/1,562,903.8) \cdot 1.0897$ and the associated COC is $9.63\% = (138,047.1/1,562,903.8) \cdot 1.0897$, signaling value creation.

The EAIRR is (slightly) different from the BAIRR computed in Example 8.11 (see the average ROI in Table 8.5). They are close to each other, because the book values are sufficiently close to the economic values.

8.9 Mixed Projects: Investment-Period NPV and Financing-Period NPV

If the project is a mixed transaction, there exist some periods where the firm invests in the project and some other periods where the firm borrows from the project. One may separate the investment region (i.e., the set of those periods where the firm invests in the project) and the financing region (the set of those periods where the firm subtracts resources from the project). Aggregating the RIs of each region, one gets

$$V_0(\text{RI}) = \sum_{t \in T_L} C_{t-1}(i_t - r_t)\text{d}_{t,0} + \sum_{t \in T_B} C_{t-1}(i_t - r_t)\text{d}_{t,0} = \text{NPV} \tag{8.82}$$

where $T_L = \{t \in \mathbb{N}_n^1 : C_{t-1} \geq 0\}$ denotes the set of lending/investment periods and $T_B = \{t \in \mathbb{N}_n^1 : C_{t-1} < 0\}$ denotes the set of borrowing/financing periods. This means that the economic value created is partitioned into an investment-period (or lending-period) NPV, denoted as $\text{NPV}_L = \sum_{t\in T_L} C_{t-1}(i_t - r_t)\text{d}_{t,0}$ and a financing-period (or borrowing-period) NPV, denoted as $\text{NPV}_B = \sum_{t\in T_B} C_{t-1}(i_t - r_t)\text{d}_{t,0}$.

This may easily be reframed as

$$V_0(\text{RI}) = \overbrace{\text{PV}[C_L](\bar{\imath}_0^L - \bar{r}_0^L)}^{\text{NPV}_L} + \overbrace{\text{PV}[C_B](\bar{\imath}_0^B - \bar{r}_0^B)}^{\text{NPV}_B} = \text{NPV} \tag{8.83}$$

where $\text{PV}[C_L] = \sum_{t\in T_L} C_{t-1}d_{t-1,0}$ is the total capital *invested* in the project and $\text{PV}[C_B] = \sum_{t\in T_B} C_{t-1}d_{t-1,0}$ is the total capital *borrowed* from the project, while

$$\begin{aligned}
\bar{\imath}_0^L &= \frac{\text{PV}[I_L]}{\text{PV}[C_L]} = \frac{\sum_{t\in T_L} I_t d_{t,0}}{\sum_{t\in T_L} C_{t-1}d_{t-1,0}} && \text{lending-period ROC/ROI}\\
\bar{\imath}_0^B &= \frac{\text{PV}[I_B]}{\text{PV}[C_B]} = \frac{\sum_{t\in T_B} I_t d_{t,0}}{\sum_{t\in T_B} C_{t-1}d_{t-1,0}} && \text{borrowing-period ROC/ROI}\\
\bar{r}_0^L &= \frac{\text{PV}[I_L^*]}{\text{PV}[C_L]} = \frac{\sum_{t\in T_L} I_t^* d_{t,0}}{\sum_{t\in T_L} C_{t-1}d_{t-1,0}} && \text{lending-period MARR}\\
\bar{r}_0^B &= \frac{\text{PV}[I_B^*]}{\text{PV}[C_B]} = \frac{\sum_{t\in T_B} I_t^* d_{t,0}}{\sum_{t\in T_B} C_{t-1}d_{t-1,0}} && \text{borrowing-period MARR (MAFR)}
\end{aligned} \tag{8.84}$$

are the corresponding relative measures of worth. Alternatively,

$$V_0(\text{RI}) = \text{PV}[C_L] \cdot \frac{\bar{\imath}^L - \bar{r}^L}{1 + r_1} + \text{PV}[C_B] \cdot \frac{\bar{\imath}^B - \bar{r}^B}{1 + r_1} \tag{8.85}$$

with $\bar{\imath}^L = \bar{\imath}_0^L(1 + r_1), \bar{\imath}^B = \bar{\imath}_0^B(1 + r_1), \bar{r}^L = \bar{r}_0^L(1 + r_1), \bar{r}^B = \bar{r}_0^B(1 + r_1)$, as usual.

The rates $\bar{\imath}_0^L$ and $\bar{\imath}_0^B$ are the (instantaneous) investment-period AIRR and financing-period AIRR, respectively. They express an average ROC in the investment periods and an average ROC in the financing periods, respectively (where the latter is an average return on a *negative* investment, that is, a rate of cost). The rates $\bar{r}_0^L$ and $\bar{r}_0^B$ are the corresponding MARR and MAFR in the investment periods and in the financing periods, respectively.

The project's ROC, $\bar{\imath}_0 = \text{PV}[I]/\text{PV}[C]$, may be written, as one should expect, as the weighted mean of the project investment-period ROC, $\bar{\imath}_0^L$ and the project financing-period ROC, $\bar{\imath}_0^B$. Likewise, the project MARR, $\bar{r}_0 = \text{PV}[I^*]/\text{PV}[C]$, may be written, as one should expect, as the weighted mean of the investment-period MARR, $\bar{\imath}_0^L$ and the financing-period MARR (or MAFR), $\bar{r}_0^B$:

$$\bar{\imath}_0 = \frac{\bar{\imath}_0^L \cdot \mathrm{PV}[C_L] + \bar{\imath}_0^B \cdot \mathrm{PV}[C_B]}{\mathrm{PV}[C_L] + \mathrm{PV}[C_B]} \qquad \bar{r}_0 = \frac{\bar{r}_0^L \cdot \mathrm{PV}[C_L] + \bar{r}_0^B \cdot \mathrm{PV}[C_B]}{\mathrm{PV}[C_L] + \mathrm{PV}[C_B]}$$
$$\bar{\imath} = \frac{\bar{\imath}^L \cdot \mathrm{PV}[C_L] + \bar{\imath}^B \cdot \mathrm{PV}[C_B]}{\mathrm{PV}[C_L] + \mathrm{PV}[C_B]} \qquad \bar{r} = \frac{\bar{r}^L \cdot \mathrm{PV}[C_L] + \bar{r}^B \cdot \mathrm{PV}[C_B]}{\mathrm{PV}[C_L] + \mathrm{PV}[C_B]} \tag{8.86}$$

One should not confuse the financing-period ROC ($\bar{\imath}_0^B$) with the project ROD ($\bar{\imath}_0^d$) nor confuse the financing-period NPV (NPV_B) with the debt NPV (NPV^d). The financing-period ROC is the average cost of subtracting funds from the assets, whereas the project ROD is the average interest rate paid by the firm to debtholders. Likewise, the financing-period NPV is the NPV generated by the project in the financing periods as a result of the firm's ability of subtracting resources from the project at a cost which is smaller than the cost of borrowing from the market, whereas the debt NPV is the part of the NPV generated by the project which is captured by debtholders as a result of the debtholders' capacity of lending their money at a rate of return which is greater than the market lending rate. The relations among these concepts may be easily grasped by the following twofold NPV decomposition. For a given NPV,

$$\overbrace{\mathrm{PV}[C^e](\bar{\imath}_0^e - \bar{r}_0^e)}^{\text{equity NPV}} + \overbrace{\mathrm{PV}[C^d](\bar{\imath}_0^d - \bar{r}_0^d)}^{\text{debt NPV}} = \mathrm{NPV} = \overbrace{\mathrm{PV}[C_L](\bar{\imath}_0^L - \bar{r}_0^L)}^{\text{investment-period NPV}} + \overbrace{\mathrm{PV}[C_B](\bar{\imath}_0^B - \bar{r}_0^B)}^{\text{borrowing-period NPV}}. \tag{8.87}$$

The second summand of the left-hand side is the share of project NPV resulting from borrowing from debtholders (if negative, shareholders capture more than the project NPV); the second summand of the right-hand side is the share of project NPV resulting from borrowing from the project (see also discussion in Sect. 2.3). The impact of the two borrowings on the equity NPV is then straightforward:

$$\overbrace{\mathrm{PV}[C^e](\bar{\imath}_0^e - \bar{r}_0^e)}^{\text{equity NPV}} = \overbrace{\mathrm{PV}[C_L](\bar{\imath}_0^L - \bar{r}_0^L)}^{\text{investment-period NPV}} + \overbrace{\mathrm{PV}[C_B](\bar{\imath}_0^B - \bar{r}_0^B)}^{\text{borrowing-period NPV}} - \overbrace{\mathrm{PV}[C^d](\bar{\imath}_0^d - \bar{r}_0^d)}^{\text{debt NPV}}. \tag{8.88}$$

The two NPVs generated by borrowing from debtholders or borrowing from the project may contribute in a positive or negative direction, depending on the signs. If the project is a pure investment project, the financing-period NPV is zero; if the project is a pure financing, the investment-period NPV is zero; if the project is financially unlevered (i.e., equity financed) or the nominal value of debt is equal to its market value, the debt NPV is zero.

It is also worth noting that the breakdown of the equity NPV holds even if $\mathrm{PV}[C] = 0$ because $\mathrm{PV}[C_L] > 0$, $\mathrm{PV}[C_B] > 0$, and $\mathrm{PV}[C^d] > 0$, which implies that all six rates are well-defined.

Example 8.17 ♦

C^o	C^d
	C^e

Consider an expansion project with an estimated life equal to five years, whereby the firm produces a new product. The firm manages the NOWC in such a way that, in the third and fourth period, it becomes negative and greater, in

Table 8.8 Mixed project: Pro forma Balance Sheet and Income Statement

Time	0	1	2	3	4	5
BALANCE SHEET						
Net Fixed Assets	1,000	800	600	400	200	0
Net operating WC	2,000	1000	−1200	−700	100	0
INVESTMENTS	3,000	1,800	−600	−300	300	0
Debt	1,200	800	300	200	100	0
Equity	1,800	1,000	−900	−500	200	0
FINANCINGS	3,000	1,800	−600	−300	300	0
INCOME STATEMENT						
EBITDA		1,000	0	195	140	300
– Depreciation		−200	−200	−200	−200	−200
EBIT		800	−200	−5	−60	100
– Interest		−96	−56	−15	−6	−4
– Taxes		0	0	0	0	0
NET INCOME		704	−256	−20	−66	96

absolute value, than the net fixed assets, which implies that, in those periods, the firm borrows from the project. The project is levered and the ROD is time-varying: $\boldsymbol{i}^d = (8\%, 7\%, 5\%, 3\%, 4\%)$. The other project input data are reported and processed in Table 8.8 (zero tax rate is assumed).

The market input data are exogenously given: The WACC ($= k^{uo}$, since $\tau = 0$) is assumed to be constant and equal to $r = 8\%$ while the required return on debt is assumed to be constant and equal to $r^d = 5\%$.

From the BS and IS the transposed Matrix is derived immediately (Table 8.9). As anticipated, there are investment periods and financing periods, so the project is a mixed transaction. There are also positive and negative incomes; in particular, in the second period, the negative income represents a loss from investing, whereas in the third and fourth periods the negative incomes represent losses from financing.

Discounting the positive and negative capitals separately, one gets a total capital employed in the investment periods equal to $\text{PV}[C_L] = 4,887.2$ and a total capital employed in the borrowing periods equal to $\text{PV}[C_B] = -752.6$. This implies a net invested capital equal to $\text{PV}[C] = 4,134.6$. Discounting the incomes generated in the investment and financing periods, one gets $\text{PV}[I_L] = 637.3$, $\text{PV}[I_B] = -48.1$. This implies that the project's investment-period ROI and the project's financing-period ROI are

$$\bar{\imath}_0^L = \frac{637.3}{4,887.2} = 13.04\% \qquad \bar{\imath}_0^B = \frac{-48.1}{-752.6} = 6.39\%.$$

Applying the WACC to the BOP capitals, the stream of market returns are computed: $\boldsymbol{I}^* = (0, 240, 144, -48, -24, 24)$. Hence, $\text{PV}[I_L^*] = 362$, $\text{PV}[I_B^*] = -55.7$. Hence,

Table 8.9 Mixed project—see Table 8.8 $(\boldsymbol{C}, \boldsymbol{I}, \boldsymbol{F})$

Time		0	1	2	3	4	5
The firm		invests	invests	borrows	borrows	invests	
Capital							
Operating assets	C^o	3,000	1,800	−600	−300	300	0
Non-operating assets	C^l	0	0	0	0	0	0
Debt	C^d	1,200	800	300	200	100	0
Equity	C^e	1,800	1,000	−900	−500	200	0
Income							
Operating income	I^o	0	800	−200	−5	−60	100
Non-operating income	I^l	0	0	0	0	0	0
Interest expense	I^d	0	96	56	15	6	4
Net Income	I^e	0	704	−256	−20	−66	96
Cash flow							
Cash flow from operations	F^o	−3,000	2,000	2,200	−305	−660	400
Cash flow from non-operating assets	F^l	0	0	0	0	0	0
Cash flow to debt	F^d	−1,200	496	556	115	106	104
Cash flow to equity	F^e	−1,800	1,504	1,644	−420	−766	296

the project's investment COC and the project's financing COC are

$$\bar{r}_0^L = \frac{362}{4,887.2} = 7.41\% \qquad \bar{r}_0^B = \frac{-55.7}{-752.6} = 7.41\%$$

(they are necessarily equal, since, $r = 8\%$ is constant). Therefore,

$$\text{NPV} = \overbrace{4,887.2 \cdot (13.04\% - 7.41\%)}^{\text{NPV}_L=275.3} + \overbrace{(-752.6) \cdot (6.39\% - 7.41\%)}^{\text{NPV}_B=7.7} = 283.$$

This means that, overall, the project creates value in both the investment periods and in the financing periods. More precisely, the firm invests \$4,887 in the project at an overall 13% while renouncing to invest them in the market at 7.41%. At the same time, the firm subtracts \$752.6 from the project's assets at a financing cost of 6.41% instead of borrowing the same amount in the market at 7.41%.

The lending-period NPV (\$275.3) is generated in the first, second, and fifth period. Specifically, in the first and fifth period, value is created, for the project income exceeds the market return: $\text{RI}_1 = I_1 - I_1^* = 560$, $\text{RI}_5 = I_5 - I_5^* = 76$. In the second period, value is destroyed: $\text{RI}_2 = I_2 - I_2^* = -344$. The positive RIs outweigh the negative RI and the net market value is precisely \$275.3.

The borrowing-period NPV (\$7.7) is generated in the financing region. Specifically, in the third period value is created, $\text{RI}_3 = I_3 - I_3^* = 43$, whereas in the fourth period value is destroyed, $\text{RI}_4 = I_4 - I_4^* = -36$. Overall, the positive RI outbalances the negative RI and the net market value is precisely \$7.7. To sum up,

$$\begin{array}{ll} \text{RI}_L & = (0, 560, -344, 0, 0, 76) \\ \text{RI}_B & = (0, 0, 0, 43, -36, 0) \\ \text{RI} & = (0, 560, -344, 43, -36, 76) \\ \hline \text{PV}[\text{RI}_L] & = 275.3 \\ \text{PV}[\text{RI}_B] & = 7.7 \\ \hline \text{PV}[\text{RI}] & = 283 = V_0(\text{RI}) \end{array}$$

with obvious meaning of the symbols.

We have used the instantaneous-AIRR framing. We may also use annual rates: $\bar{\imath}_L = 0.01304 \cdot 1.08 = 0.1408$ and $\bar{\imath}_B = 0.0639 \cdot 1.08 = 0.069$. Multiplying and dividing by $(1 + 0.08)$, one gets

$$\text{NPV} = \overbrace{4,887.2 \cdot \frac{14.08\% - 8\%}{1.08}}^{\text{NPV}_L=275.3} + \overbrace{(-752.6) \cdot \frac{6.9\% - 8\%}{1.08}}^{\text{NPV}_B=7.7} = 283$$

as expected. Now, the project ROI may be computed as $\bar{\imath} = \text{PV}[I]/\text{PV}[C]$ or as the weighted mean of the investment-period and borrowing-period ROI:

$$\bar{\imath} = \frac{14.08\% \cdot 4,887.2 + 6.9\% \cdot (-752.6)}{4,887.2 - 752.6} = 15.39\%.$$

Hence, one may aggregate the investment region and the financing region of the project and get

$$4,134.6 \cdot \frac{15.39\% - 8\%}{1.08} = 283.$$

To find the equity NPV (and, therefore, shareholders' wealth increase), one may subtract the debt NPV. From the transposed Matrix one finds $\text{PV}[I^d] = 163.3$ and $\text{PV}[C^d] = 2,489.1$, so that the project ROD is $\bar{\imath}_0^d = 163.3/2,489.1 = 6.56\%$. The benchmark system's income stream is found by multiplying the debt COC with the BOP debt value: $I^{d*} = (0, 60, 40, 15, 10, 5)$ whence $\text{PV}[I^{d*}] = 118.53$ and, therefore, $\bar{r}_0^d = 118.53/2,489.1 = 4.76\%$. Hence,

$$\begin{aligned} \text{NPV}^e = {} & \overbrace{4,887.2 \cdot (13.04\% - 7.41\%)}^{\text{NPV}_L=275.3} + \overbrace{(-752.6) \cdot (6.39\% - 7.41\%)}^{\text{NPV}_B=7.7} \\ & - \overbrace{2,489.1 \cdot (6.56\% - 4.76\%)}^{\text{NPV}^d=44.7} = 238.3. \end{aligned}$$

The equity NPV is broken down into

- investment-period NPV
- financing-period NPV
- debt NPV.

The debt NPV is positive, which means that the firm's financing policy (borrowing from debtholders) is value-destroying. However, this is compensated by the investment policy, part of which consists of borrowing from the project, with a value creation of \$7.7. The project ROC may be computed as $\bar{\imath}_0 = (\text{PV}[I]/\text{PV}[C])(1+r)$ or as the weighted mean of investment-period ROC and borrowing-period ROC:

$$\bar{\imath}_0 = \frac{13.04\% \cdot 4,887.2 + 6.39\% \cdot (-752.6)}{4,887.2 - 752.6} = 14.25\%.$$

The project MARR is computed analogously. As the investment-period MARR and the MAFR are equal, $\bar{r}_0 = \bar{r}_0^L = \bar{r}_0^B = 7.41\%$. Hence, the project's value created may be found by merging investment region and borrowing region:

$$\text{NPV} = 4,134.6 \cdot (14.25\% - 7.41\%) = 283.$$

Example 8.18 ♦

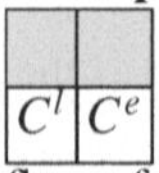

An economic agent has the opportunity of depositing and withdrawing cash flows from an account balance with predetermined (time-varying) borrowing rates and lending rates.[22] The borrowing rates apply when the account balance is negative and the lending rates apply when the account balance is positive.

period	borrowing rates	lending rates
1	23%	16%
2	13%	10%
3	8%	6%
4	20%	19%

We aim at analyzing the financial performance of this transaction in some detail, assuming that the cost of capital is time-varying, namely, $\boldsymbol{r} = (0.21, 0.1, 0.16, 0.12)$.

Suppose the investor deposits \$2 in the account, withdraws \$20 after one period, deposits \$5 and \$75 after two and three periods, respectively, and, finally, withdraws \$70 at the end of the fourth period. The cash-flow stream is then $\boldsymbol{F} = (-2, 20, -5, -75, 70)$ (see Table 8.10 for the relevant data).

---

[22]This example is adapted from Magni (2015a), Investment, financing and the role of ROA and WACC in value creation, *European Journal of Operational Research*, 244(3) (August), 855-866 (pp. 860–861).

Table 8.10 Account balances, lending and borrowing positions

Time	F_t	r_t	borrowing rate	lending rate	C_t^L	C_t^B	C_t
0	-2.00				2.00	0.00	2.00
1	20.00	21.0%	23.00%	16.00%	0.00	−17.68	−17.68
2	-5.00	10.00%	13.00%	10.00%	0.00	−14.98	−14.98
3	-75.00	16.00%	8.00%	6.00%	58.82	0.00	58.82
4	70.00	12.00%	20.00%	19.00%	0.00	0.00	0.00
PV	2.68				40.10	-25.87	14.23

The lending periods are the first one and the fourth one, so the lending rates $i_1 = 16\%$ and $i_4 = 19\%$ are applied to the account balances $C_0 = 2 > 0$ and $C_3 = 58.82 > 0$. The financing periods are the second one and the third one, so the borrowing rates $i_2 = 13\%$ and $i_3 = 8\%$ are applied to the account balances $C_1 = -17.68 < 0$ and $C_2 = -14.98 < 0$. The vector of income rates is then $\boldsymbol{i} = (0.16, 0.13, 0.08, 0.19)$.[23] The investment region consists of two periods: The first period and the fourth period. The latter is a value-creating period, where the investor renounces to investing funds at 12% in exchange of a 19% return from the project; therefore, in this period, the investor earns an above-normal 7% return. Conversely, the first period is a value-destroying period, because the investor gains a 16% return but foregoes a 21% return; therefore, in this period, the investor secures a return which falls short of the normal return by 5%.

To measure the net impact of these opposing results, we amalgamate the 19% and 16% investment rates into a single rate which condenses the overall performance of the project in the investment periods: From (8.84) (first and third formula), $\bar{\imath}_0^L = 16.78\% = 6.73/40.1$ (or $\bar{\imath}^L = 20.3\% = 16.78\% \cdot 1.21$); analogously, the 12% and 21% costs of capital are combined into a single benchmark rate which summarizes the overall performance of the benchmark portfolio in the investment periods: $\bar{r}_0^L = 11.05\% = 4.43/40.1$ (or $\bar{r}^L = 13.37\% = 11.05\% \cdot 1.21$). On average, the investor invests at 16.78% return in two periods while foregoing the opportunity of investing at 11.05% return. The net effect is positive and the investment region creates value. As for the financing region, the second-period borrowing position destroys value, for funds are borrowed at $i_2 = 13\%$ while the market only requires $r_2 = 10\%$. In the third period, value is created, for funds are borrowed at $i_3 = 8\%$ whereas the market interest rate is $r_3 = 16\%$. We amalgamate the project borrowing rates and the costs of capital by applying (8.84) (second and fourth formula). This brings about $\bar{\imath}_0^B = 9.68\% = (-2.5)/(-25.87)$ (or $\bar{\imath}^B = 11.71\% = 9.68\% \cdot 1.21$) and $\bar{r}_0^B = 11.14\% = (-2.88)/(-25.87)$ (or $\bar{r}^B = 13.48\% = 11.14\% \cdot 1.21$). Overall, the financing periods create value, since, on average, funds are borrowed at 9.68% while the market requires 11.14%. By (8.87) (right-hand side), the NPV of the project is

[23] As we know, the vector of incomes rates is such that $-2 + 20 \cdot 1.16^{-1} - 5 \cdot (1.16 \cdot 1.13)^{-1} - 75 \cdot (1.16 \cdot 1.13 \cdot 1.08)^{-1} + 5 \cdot (1.16 \cdot 1.13 \cdot 1.08 \cdot 1.19)^{-1} = 0$.

$$\text{NPV} = \overbrace{40.1}^{\text{PV}[C_L]} \cdot \overbrace{(16.78\% - 11.05\%)}^{(\bar{\imath}_0^L - \bar{r}_0^L)} + \overbrace{(-25.87)}^{\text{PV}[C_B]} \cdot \overbrace{(9.68\% - 11.14\%)}^{(\bar{\imath}_0^B - \bar{r}_0^B)}$$
$$= \underbrace{2.3}_{\text{NPV}_L} + \underbrace{0.38}_{\text{NPV}_B} = 2.68.$$

In turn, coupling the lending-period rate and the borrowing-period rate, as well as the lending-period and borrowing-period costs of capital, the project rate of return and the project cost of capital are calculated. Specifically, from (8.86), $\bar{\imath}_0 = 29.68\%$ and $\bar{r}_0 = 10.88\%$ (in discrete terms, $\bar{\imath} = 35.92\%$ and $\bar{r} = 13.16\%$). As $\text{PV}[C_L] + \text{PV}[C_B] = \text{PV}[C] = 14.23$, the investor invests, overall, a net \$14.23 at an average return rate of 29.68% with a MARR of 10.88%. The NPV may then be rederived as $14.23 \cdot (29.68\% - 10.88\%) = 2.68$).

rate of return	project	benchmark	efficiency	capital	NPV
lending-period	16.78%	11.05%	5.73%	40.10	2.30
borrowing-period	9.68%	11.14%	−1.46%	−25.87	0.38
average	29.68%	10.88%	18.80%	14.23	2.68

In this example, the project is a financial transaction (no production of goods is involved) consisting in a contractual relationship between two parties. The economic agent, who deposits and withdraws cash amounts, changes its position from lender to borrower and then from borrower to lender. Assuming the transaction is equity financed (the economic agent does not raise funds from any debtholder), the debt NPV is zero and the borrowing-period NPV generated in this transaction arises as a result of borrowing from the transaction itself at a borrowing rate which is smaller than the market rate by 1.46% points.

8.10 Key Points

- The rate of return of a multiperiod asset (project, firm, etc.) is a relative measure of worth. It is defined as an amount of return per unit of invested capital
- The AIRR approach defines a project's rate of return as the total (discounted) return to total (discounted) capital:

$$\text{AIRR} = \frac{\text{Total discounted return}}{\text{Total discounted capital}}$$

- From the point of view of the income rates, an AIRR may be viewed as the mean of the income rates, i_t, weighted by the present value of the capitals invested (or borrowed) in each period:

$$\text{AIRR} = \text{Capital-weighted mean of the period return rates}$$

- Computationally, a third, fruitful way for calculating the AIRR is to divide the present value of the incomes by the present value of the capitals. In such a way, an instantaneous rate of return is obtained. Multiplying by the compounding factor (1 + cost of capital), the discrete (e.g., annual) rate of return is obtained:

$$\text{AIRR} = \text{Instantaneous AIRR} \times (1 + \text{COC})$$

- A fourth equivalent way to compute the AIRR is to start from the NPV: Dividing the NPV by the total capital employed, a relative instantaneous measure of value creation is obtained. Multiplying the NPV by the compounding factor (1 + cost of capital), it is turned to a discrete (e.g., annual) excess rate of return (excess AIRR). Adding the COC, the rate of return is calculated:

$$\text{AIRR} = (1 + \text{COC}) \times \text{NPV/Total capital} + \text{COC}$$

- Given pro-forma financial statements collecting estimated income streams and capital streams, a rate of return exists for each pair (income, capital). We have analyzed 8 kinds of capitals and incomes. Therefore, 8 types of AIRRs are properly associated with a project, called *book AIRRs*:
  - asset average ROI
  - average ROE
  - operating average ROI
  - non-operating average ROI
  - unlevered operating average ROI
  - unlevered asset average ROI
  - average ROD
  - average RO$\mathcal{D}$
- The economic value created reconciles with the AIRR via a product structure, according to which the economic value created is equal to the product of the total capital (project size or scale) and the project's excess AIRR (marginal economic efficiency of the capital):

$$\text{Economic Value Created} = \text{Project size} \times \text{Excess AIRR}$$

- In capital asset investments (and, in general, in any situation where pro forma financial statements are drawn) book rates of return play a major role: Their weighted means signal whether the project/firm is efficient or not from several points of view (levered or unlevered perspective, shareholders' or debtholders' perspective, operating or non-operating perspective) and whether the project should be undertaken or not. The connection with the NPV is explicit and simple to understand (see point above). This result represents a refutation of the long-standing belief that book rates of return are not capable of providing information on a firm's/project's economic profitability

- Profitability Index and Benefit-Cost ratio are relative measures of worth linked to the AIRR approach. The rates of return associated with them measure a project's return per unit of initial investment and total outflow, respectively.
- Economic AIRR is the rate of return from the point of view of the market. It may be also interpreted as the expected rate of return of the project if it were traded in the market
- A project is associated with a rate-of-return function (AIRR function) and an excess rate-of-return function (excess AIRR function), whose independent variable represents a capital base on which the project's return is measured. The BAIRRs above mentioned, the EAIRR, the AIRRs associated with PI and BC are just values taken on by the function for different capital bases
- In a mixed project, the firm invests in the project in some periods and borrows from the project in some other periods. The economic value created may be decomposed into investment-period (or lending-period) NPV and financing-period (or borrowing-period) NPV. The investment-period ROI is the rate of return generated in the investment periods while the borrowing-period ROI is the cost of financing borne by the firm in the financing periods. Accordingly, the investment COC/MARR is the benchmark investment rate in the investment periods, the borrowing COC/MARR (or MAFR) is the benchmark financing rate in the financing periods.

8.11 Appendix: Underdetermination of Rate of Return by Cash Flows

Consider a window. What is the area of the window? For example, suppose the height of the window is $\xi = 1.5$ m and its width is $x = 2.8$ m. Then, its area is $A = \xi \times x = 1.5 \times 2.8 = 4.2\,\text{m}^2$. Now, let us reverse the problem. Suppose we know that the area of a window is 4.2. What is the window's height? To answer this question, one needs to know the width, for there are infinitely many combinations of width and height that generate the same areas: $\xi = A/x = 4.2/x$. The area alone is not sufficient to derive the height of the window. In other words, the area *underdetermines* the height of the window. In general, the area of a rectangle underdetermines the height of the rectangle.

The same phenomenon of underdetermination is observable in the notion of rate of return: The (excess) rate of return is to NPV what the height of the window is to its area. Hence, the NPV underdetermines the rate of return and, given that NPV depends on cash flows, a project's cash flows underdetermine the rate of return. Knowledge of cash flows does enable one to get the NPV (as long as the COC is available) but is not sufficient to get the marginal economic efficiency of the capital employed. For example, for a given NPV of \$4.2, the marginal economic efficiency, ξ, depends on the capital base, x, which is invested in (or borrowed from) the project: $\xi = 4.2/x$. If the capital invested is \$10, then the economic efficiency is $4.2/10 = 0.42$; if the capital invested is \$30, then the economic efficiency is $4.2/30 = 0.14$; if the capital is \$50, then the economic efficiency is $4.2/50 = 0.084$, and so on. The NPV is the area of a rectangle whose height is the (excess) rate of return and the width is the total capital invested: NPV $= x \cdot \xi$.

In general, the shape of the NPV rectangle, $\xi \cdot x$, is underdetermined by the project's cash flows. The excess AIRR function, $\xi(x)$ determines infinitely many NPV rectangles with the same area (see Fig. 8.3). Only after selecting the appropriate capital base the rectangle (and its height) will be determined. This insight was anticipated by Vatter (1966) half a century ago: "*unless capital recovery process is specified*, there is no single way to measure the annual productivity of the investment" (Vatter 1966, p. 687, italics added). As pro forma book values are the capital values that determine the cash flows, the overall book value is a privileged choice for capital asset investments.[24] As a result, while it is usual to claim in finance that "cash is king" in reference to the economic value created by a project, it is also appropriate to say that, in reference to a project's economic efficiency, "capital is king" (as well as income). A cash-flow stream does not determine any economic efficiency.[25]

The underdetermination of the rate of return by cash flows is an example of *underdetermination of theory by data,* a widely known phenomenon in science. The Stanford Encyclopedia of Philosophy defines this phenomenon in intuitive terms:

> At the heart of the underdetermination of scientific theory by evidence is the simple idea that the evidence available to us at a given time may be insufficient to determine what beliefs we should hold in response to it. In a textbook example, if all I know is that you spent \$10 on apples and oranges and that apples cost \$1 while oranges cost \$2, then I do not know whether you bought one orange and eight apples, two oranges and six apples, and so on.
>
> (Stanford Encyclopedia of Philosophy: https://plato.stanford.edu/entries/scientific-underdetermination)

This problem has first been raised by Duhem (1914) and Schlick (1948), and later became associated with Quine's (1951) analysis. Formally, consider a set of empirical data plotted on an xy-plane: These data may be connected with infinitely many functions, so there is no way of detecting the "true" analytical relation explaining those data. In terms of our issue, consider the time-capital plane, (t, C_t); the only empirical data available are $(0, C_0)$ and $(n, 0)$. Any function connecting those two

---

[24]Determination of *ex post* rate of return in financial investments and portfolios is better served by the investment's observed market value.

[25]Capital and income are "kings" for a further reason: The estimates of cash flows in a capital asset investment or in a firm depend on the determination of the capital's depreciation, according to the law of motion $F_t = I_t - \Delta C_t$: "the more general point is all changes in the balance sheet are relevant in the determination of cash flows because they come, in an important sense, logically prior to the estimate of cash flows." (Magni and Peasnell 2012, p. 14). One may object that, if liquid assets are present, CFLs are not zero for all t. which means that the latter depend on decisions about payout policy as well as on accounting magnitudes. Furthermore, when dealing with financial investments, the principal (client) has full control on cash drawdowns and injections (as well as the length of the investment); in all these cases, accounting magnitudes are not at stake.

We sustain this objection. In actual facts, the project system consists of a network of projected accounting magnitudes and a set of decisions regarding the financing and distribution policy, which jointly determine the distributed cash flow (and, therefore, the value created). We then finally claim that the whole basic trinity $(\boldsymbol{C}, \boldsymbol{I}, \boldsymbol{F})$ is the only true "king".

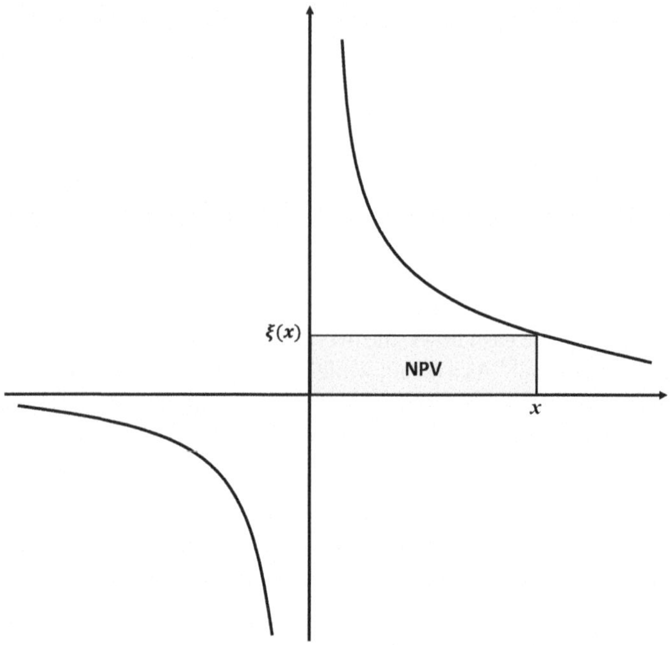

Fig. 8.3 The NPV rectangle is underdetermined by the cash flows. The excess AIRR function, $\xi(x)$, determines infinitely many NPV rectangles. Economic efficiency is determined by the selection of a given capital base, x

points expresses a depreciation function for the project's capital. And to each such depreciation there corresponds a given rate of return: $\bar{\imath} : \boldsymbol{C} \rightarrow \bar{\imath}(\boldsymbol{C})$. Which implies that knowledge of cash flows is not sufficient to identify the correct rate of return: A cash-flow stream underdetermines the capital stream and, hence, the rate of return.

Chapter 9
Internal Rate of Return

> These instances of failure of the multi-period internal-rate-of-return rule ... are, of course, merely the symptom of an underlying erroneous conception
>
> Hirshleifer (1958, p. 349).

> *This is a crucial criticism of the IRR – even though it may be unique, and real in the mathematical sense, this in itself is not a sufficient condition for it to be a correct measure of return on investment*
>
> Herbst (1978, p. 367, italics in original)

> Recognition of the correct general solution of the investment problem has been hindered by the habit of thinking in terms of a single, long-term rate of interest ... This is an example of the 'paradox' that has attracted so much attention in connection with investment decision criteria. It should be evident, however, that this paradox is merely an accident of the simplifying device of dealing with a single long-term rate of interest, and that it has no special importance.
>
> Bailey (1959, pp. 477–479)

> It is better to use accountant's estimates of what we really want to know than to use numbers which bear no relationship to the problems we need to solve.
>
> Vatter (1966, p. 697)

C. A. Magni, *Investment Decisions and the Logic of Valuation*,
https://doi.org/10.1007/978-3-030-27662-1_9

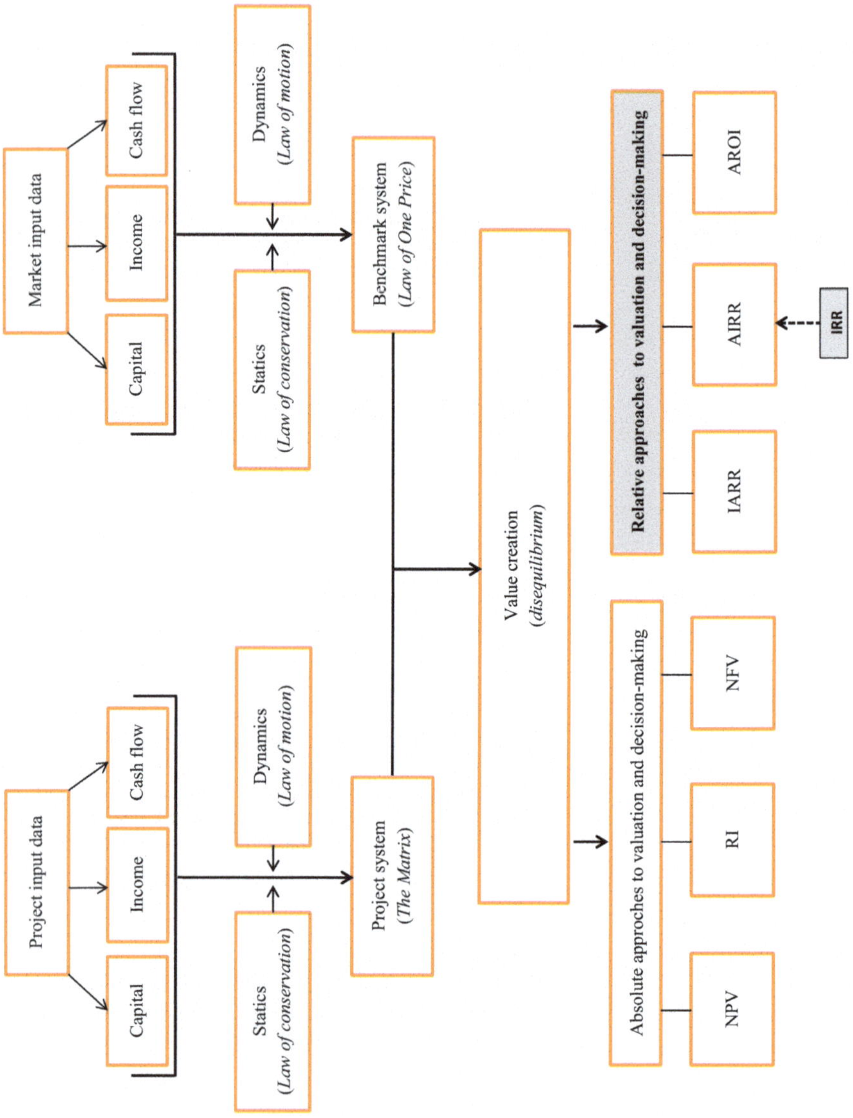

Fig. 9.1 Chapter 9: The internal-rate-of-return approach and its pitfalls

This chapter is devoted to the multiperiod Internal Rate of Return (IRR). Notwithstanding its venerable ancestry and its predominance among practitioners, it is marred by several **pitfalls** that make it unsuitable in many circumstances. We show in some detail its pitfalls, including the fact that it is inconsistent with the mechanics of the project as presented in Chaps. 1–2 and is not economically meaningful, in the sense that it bears **no direct relationship to the business transactions** underlying the project's activities and, therefore, no relationship to the capital actually employed. It is advisable for practitioners to dispense with it (or, at the very least, use it with extreme caution).

We illustrate two alternative approaches that were advanced in the past decades for overcoming some of the classical difficulties of the IRR: The **Modified-Internal-Rate-of-Return** (MIRR) approach and the **Teichroew-Robichek-Montalbano** (TRM) approach. These approaches remedy some of the IRR's pitfalls but do not provide the *project*'s rate of return and are based on assumptions which are even more stringent than the IRR's assumptions. As a matter of fact, the AIRR approach, introduced in Magni (2010b) and presented in Chap. 8, provides a **complete resolution** to the long-lasting IRR conundrum.

After completing this chapter, the reader should be able to:

- understand the notion of (multiperiod) *Internal Rate of Return* (IRR)
- encompass the IRR in the AIRR approach
- recognize the IRR's anomalies and its theoretical difficulties for a reliable project analysis and assessment of economic profitability
- understand how to use IRR for accept-reject decisions
- be aware of the harmful effects of IRR in terms of valuation logic and in terms of economic significance
- understand the notion of *Modified Internal Rate of Return* (MIRR), its underlying assumptions, its pros and cons
- understand the so-called *Teichroew-Robichek-Montalbano* (TRM) approach, its underlying assumptions, its pros and cons
- grasp the reasons why the AIRR approach is not marred by the IRR approach's limitations and, rather, provides a complete resolution to the IRR conundrums.

9.1 IRR and Chisini Mean

In Chap. 8 we have defined the project rate of return as the ratio of total income to total capital. We have shown that, if $C_t \neq 0$ for all $t \in \mathbb{N}^0_{n-1}$ (and $I_0 = 0$), it may be interpreted as a Chisini mean where the proposed invariant function is $g(i_1, i_2, \ldots, i_n) = \sum_{t=1}^{n} i_t C_{t-1}(1+r)^{-(t-1)}$.

However, suppose one selects a different invariant function, namely

$$g(i_1, i_2, \ldots, i_n) = \sum_{t=1}^{n} \frac{F_t}{(1+i_1)(1+i_2)\cdot \ldots \cdot (1+i_t)}. \tag{9.1}$$

The associated Chisini mean is that rate σ such that

$$\sum_{t=1}^{n} \frac{F_t}{(1+i_1)(1+i_2)\cdot \ldots \cdot (1+i_t)} = \sum_{t=1}^{n} \frac{F_t}{(1+\sigma)^t}.$$

However, the left-hand side of the above invariance requirement represents the project cost, C_0 (see Eq. (1.15). Therefore, since $C_0 = -F_0$,[1]

$$F_0 + \frac{F_1}{1+\sigma} + \frac{F_2}{(1+\sigma)^2} + \ldots + \frac{F_n}{(1+\sigma)^n} = 0$$

Definition 9.1 (*Internal rate of return*) Consider the discounted-cash-flow (DCF) function $\phi(x) = \sum_{t=0}^{n} F_t \cdot (1+x)^{-t}$ and the equation

$$\phi(x) = \sum_{t=0}^{n} \frac{F_t}{(1+x)^t} = 0. \tag{9.2}$$

An internal rate of return (IRR) is a solution $\sigma \in (-1, +\infty)$ of (9.2). That is, $\phi(\sigma) = 0$.

The IRR notion is strictly connected with the notions of rate of return over cost (Fisher 1930) and marginal efficiency of capital (Keynes 1936/1967) (see also Alchian 1955). However, it was apparently Boulding (1935) who first qualified an economic asset's rate of return as "internal". Dealing with multiperiod projects, he explicitly assumed that

[1] With no loss of generality, we assume $I_0 = 0$.

> there is some rate of return [. . .] which is characteristic of the investment as a whole. This is of course a rate of interest, or a rate of discount. But it must be emphasized that it is a rate of interest which the enterprise itself produces [. . .]. That is to say, it is an internal rate." (p. 478)

Essentially, Boulding (1935) assumed that firm's capital is remunerated at a constant (and exogenously fixed) remuneration rate, such that

$$C_t - C_{t-1} = i \cdot C_{t-1} - F_t \tag{9.3}$$

(see Boulding 1935, Eqs. (3)–(5)). It should be clear that (9.3) is nothing but the fundamental law of motion with constant income rate (see Eq. (1.16)). Explicitly noting that C_n must be zero, the author solved the dynamical system for $t = n$ finding $\sum_{t=0}^{n} F_t(1+i)^{n-t} = 0$ and divided by $1/(1+i)^n$ to get the present-value relation $\sum_{t=0}^{n} F_t(1+i)^{-t} = 0$. Boulding then noted that, while he assumed i as exogenously fixed, the same relation $\sum_{t=0}^{n} F_t(1+i)^{-t} = 0$ can be reverse engineered, that is, it can be used as "an equation from which we can calculate the internal rate of return of the enterprise itself, treating the net revenue series as given and i as an unknown". (Boulding 1935, p. 481).

Across the years, a plethora of labels have been used to denote the IRR, depending on the economic milieu where the investment is undertaken or the academic domain where it is discussed, including *money-weighted rate of return, effective rate, yield to maturity, multiperiod IRR, implied rate of return, economic rate of return, true yield. marginal rate of return on investment.*

The IRR brings about the following decision criterion.

Definition 9.2 (*IRR acceptability criterion*) Consider a project P and let $\boldsymbol{F}$ be the associated cash-flow stream. If P is an investment, then it is economically profitable (acceptable) if and only if $\sigma > r$ (IRR is greater than the COC); if P is a financing, then it is acceptable if and only if $\sigma < r$.

The essence of the IRR criterion is that an investment is worth undertaking if its IRR exceeds the cost of capital and that a financing is worth undertaking if its IRR is smaller than the cost of capital.

There can be as many IRRs as are the types of cash flows. For example, the equity IRR, σ^e, is the IRR of the CFE stream, such that $\sum_{t=0}^{n} F_t^e(1+\sigma^e)^{-t} = 0$. The debt IRR, σ^d, is the IRR of the CFD stream, such that $\sum_{t=0}^{n} F_t^d(1+\sigma^d)^{-t} = 0$. An IRR may be defined for any relevant cash-flow stream: CFO, CF$\mathcal{D}$, CFL, FCF, CFA, etc.

Since the goal of a firm is to maximize shareholders' wealth, the accept/reject decision should be based on the equity IRR alone and $\sigma^e > r^e$ (<) is the relevant acceptance condition for investments (financings). However, in the special case where $i_t^d = r_t^d$, the project NPV coincides with the equity NPV, which in turn implies that $\sigma > r$ if and only if $\sigma^e > r^e$, so that either condition may be used for making a decision.[2] Also, if the project is equity-financed, then the notion of project IRR and equity IRR coincide: $\sigma = \sigma^e$, as well as $r = r^e$, and the two conditions become the same.

Example 9.1 ♦
Consider the Opening project's cash-flow stream in the base-case (unlevered project with no taxes):

$$\boldsymbol{F} = \boldsymbol{F}^e = (-400, 175, 179.5, 184.1, 188.8, 193.5) \qquad \text{(in thousands)}$$

(see Table 4.6). The project and equity IRR coincide, obtained as the solution of

$$-400 + \frac{175}{1+x} + \frac{179.5}{(1+x)^2} + \frac{184.1}{(1+x)^3} + \frac{188.8}{(1+x)^4} + \frac{193.5}{(1+x)^5} = 0.$$

The root is $\sigma = 35.4\%$. Assume the COC is $r = 25\%$. The IRR criterion recommends acceptance, for $35.4\% > 25\%$. This is rational, because it is consistent with the NPV criterion:

$$-400 + \frac{175}{1+0.25} + \frac{179.5}{(1+0.25)^2} + \frac{184.1}{(1+0.25)^3} + \frac{188.8}{(1+0.25)^4} + \frac{193.5}{(1+0.25)^5} = 89.88 > 0.$$

It should not escape the reader's notice that, once computed the IRR from a cash-flow series, Eq. (9.3) automatically establishes a stream of capital amounts C_t as functions of σ:

$$C_t = C_t(\sigma) = C_{t-1}(\sigma)(1+\sigma) - F_t. \tag{9.4}$$

In other words, the application of the IRR notion to the dynamics of an asset *automatically* implies the determination of capital values $C_t(\sigma)$, which, in general, are not equal to book values nor to market values.

Definition 9.3 (*Internal capital*) Let $C_t(\sigma) = C_{t-1}(\sigma)(1+\sigma) - F_t$ be the capital implied by the IRR, with $C_0(\sigma) = -F_0$. It is called the *internal* value or capital; $\boldsymbol{C}(\sigma) = (C_0(\sigma), C_1(\sigma), C_2(\sigma), \ldots, C_{n-1}(\sigma), 0)$ is the vector of the internal capitals $C_t(\sigma)$.

As a result, the multiperiod IRR may be interpreted as the rate of return of an economic asset whose capital grows at a constant rate of return and whose cash-flow stream is the same as the project's. Or, equivalently, the IRR may be said to measure what the rate of return of the project would be if the project's capital were enforced to grow at a constant pace. This interpretation derives from the fact that the notion of

[2]This is not strictly true, because the equity capital might in principle have a sign which differs from the sign of the invested capital. We will not dwell on this.

IRR is akin to the notion of interest rate for a loan. If the loan has a constant interest rate, then its IRR coincides with the interest rate. Indeed, IRR may also be defined as that interest rate that would be contractually stipulated if the project were a financial transaction with constant interest rate.[3]

9.2 IRR and AIRR

The term "internal" in this context means "not affected by parameters which are exogenous to the project". However, this expression is quite a loose and nonrigorous expression, for it is not clear what is internal to a project and what is external to it. Usually, scholars refer to the fact that the IRR is not affected by the COC, r. From this point of view, the moniker "Average Internal Rate of Return" introduced in Chap. 8, might be considered a misnomer, for AIRR does depend on r. However, each project's income rate i_t is internal, since it does not depend on r. The vector $\boldsymbol{i} = (i_1, i_2, \ldots, i_n)$ is then an *internal* vector (Weingartner 1966). The AIRR is a mean of such single-period rates, so it is indeed an average of rates which are internal. Yet, the weights of the mean depend on the COC, so the mean itself is not internal.[4]

At the same time, the multiperiod IRR is a rate of return belonging to the AIRR class, where the capital stream belongs to the same equivalence class as $\boldsymbol{C}(\sigma)$ (so-called Hotelling or internal class).

The internal class is the equivalence class of those capital streams $\boldsymbol{S} = (S_0, S_1, S_2, \ldots, S_{n-1}, 0)$ such that $S_0 = -F_0$ and $\text{PV}[S] = \text{PV}[C(\sigma)]$. The internal stream $\boldsymbol{C}(\sigma)$ is the only stream of that class which has constant income rates: $\boldsymbol{\sigma} = (\sigma, \sigma, \ldots, \sigma)$ while, in general, a member of the internal class has time-varying rates. In other words, IRR is a generalized arithmetic mean of (generally time-varying) period rates, denoted as $\sigma_t = (S_t + F_t - S_{t-1})/S_{t-1}$, weighted by present values of capitals, $\text{PV}[S_{t-1}]$, whose discounted sum amounts to $\text{PV}[S] = \text{PV}[C(\sigma)]$. Formally, for any $\boldsymbol{S}$,

$$\sigma = \frac{\sigma_1 S_0 + \sigma_2 S_1 (1+r)^{-1} + \ldots + \sigma_n S_{n-1}(1+r)^{-(n-1)}}{S_0 + S_1(1+r)^{-1} + \ldots + S_{n-1}(1+r)^{-(n-1)}}.$$

As a result, if one interprets the multiperiod IRR as a constant growth rate, as is usual, one is implicitly selecting the stream $\boldsymbol{C}(\sigma)$ as the representative stream of the internal class (for further details and illustrative examples on this issue, see Magni 2010b, 2010e, 2016b; Magni and Peasnell 2012; Magni et al. 2018. See also Magni 2009e).

[3]Financial transactions include (but are not limited to) loans. In a generic financial transaction, lending positions and borrowing positions may change over time.

[4]In Chap. 10 we will introduce another relative metric which is genuinely internal. To pinpoint the difference with AIRR, we will switch "average" and "internal" and label it "Internal Average Rate of Return".

As a result, the IRR is also the value taken on by the AIRR function, $\bar{\imath}(x) = r + (1+r)\cdot \mathrm{NPV}/x$, when one picks $x = \mathrm{PV}[C(\sigma)]$:

$$\sigma = \bar{\imath}\Big(\mathrm{PV}[C(\sigma)]\Big) = r + \frac{\mathrm{NPV}}{\mathrm{PV}[C(\sigma)]}\cdot(1+r) \tag{9.5}$$

whence the product structure

$$\mathrm{NPV}(1+r) = \mathrm{PV}[C(\sigma)](\sigma - r) \tag{9.6}$$

(see also Hazen 2003, 2009).[5]

The IRR is interpretable as

- the rate of return of an asset whose capital growth is time-invariant and whose cash flows mimic the project's cash flows
- the rate of return the project would have if its capital grew at a constant pace
- the interest rate that would be contractually established if the project were a savings-and-credit account
- a weighted mean of time-varying holding period rates associated with any capital stream belonging to the internal class and, therefore,
- the AIRR associated with the internal capital base $\mathrm{PV}[C(\sigma)]$

Another aspect is that the IRR is a third-order variable with respect to the book rates of return. The reason is that IRR depends on cash flows (second-order variable) and cash flows depend on incomes and capitals (first-order variables). Specifically, for every $t > 0$, the expected cash flow may be viewed as a function of (capital and) book rate of return: $F_t = F_t(i_1, i_2, \ldots, i_n) = i_t C_{t-1} - \Delta C_t$, $t \in \mathbb{N}_n^1$. Therefore, the IRR itself is a function of the book rates of return:

[5]Equation (9.5) may be reframed as

$$\sigma - r - \frac{\overbrace{\sum_{t=0}^{n} F_t(1+r)^{-t+1}}^{\mathrm{NPV}\cdot(1+r)}}{\underbrace{\sum_{t=1}^{n}\underbrace{\sum_{k=t}^{n} F_k(1+\sigma)^{t-k-1}}_{C_{t-1}(\sigma)}\cdot(1+r)^{-t+1}}_{\mathrm{PV}[C(\sigma)]}} = 0.$$

The latter may well be considered an alternative definition of IRR (note that the solutions of this equation are necessarily invariant under changes in r).

$$\phi(x) = F_0 + \sum_{t=1}^{n} \frac{i_t C_{t-1} - \Delta C_t}{(1+x)^t} = 0. \tag{9.7}$$

As a result, the IRR and its built-in internal values are affected by book rates of return and book values. As noted in Remark 8.4, skepticism is prevalent among accounting scholars about the role of book rates of return and some believe that the IRR notion is a better, albeit imperfect, measure of economic profitability. In fact, the IRR depends on cash flows and cash flows depend on book rates of return and book values. As a result, the use of invariance requirement (9.7) for aggregating the book rates of return is unfortunate: A deluge of theoretical and practical shortcomings spring up.

9.3 The Classical Pitfalls of IRR

Although the IRR is widely used in practice, its flaws are numerous. Three of them were detected rather soon after Boulding's (1935) paper and have puzzled scholars for many years. They may be classified as follows:

- multiplicity of IRR
- non-existence of IRR
- ambiguous financial nature of IRR.[6]

The following examples illustrate the first two cases.

Example 9.2 ♦
Consider a two-period project such that $\boldsymbol{I} = (0, 11, -12)$ and $\boldsymbol{C} = (10, 5, 0)$. Let 15% be the COC and suppose an analyst wants to find the IRR and the resulting economic efficiency. As usual, one applies the fundamental law of motion (1.19) and gets an expected cash-flow stream equal to $\boldsymbol{F} = (-10, 16, -7)$. The IRR equation has no solution in $(-1, +\infty)$ (i.e., $\phi(x) < 0$ for every $x \in (-1, +\infty)$). No IRR exists, so it is not clear what its economic profitability is.

No such problem arises with AIRR:

$$\bar{i} = \frac{11 \cdot (1.15)^{-1} - 12 \cdot (1.15)^{-2}}{10 + 5 \cdot (1.15)^{-1}} \cdot 1.15 = \frac{0.491}{14.348} \cdot 1.15 = 3.94\%.$$

The firm invests \$14.348 at an average rate of 3.94%, so giving up a 15% return. Therefore, value is destroyed at an average rate of $11.06\% = |3.94\% - 15\%|$, which represents the project's economic (in)efficiency.

---

[6]See also Magni (2013a) for a compendium of 18 flaws of the IRR (many of which are treated in the next section).

Example 9.3 ♦
Consider the following cash flow stream:

$$\boldsymbol{F} = (-4, 3, 2.25, 1.5, 0.75, 0, -0.75, -1.5, -2.25).$$

The project has two IRRs in the relevant interval $(-1, +\infty)$, namely $\sigma^1 = 10.43\%$ and $\sigma^2 = 26.31\%$.[7] It is not clear which one is the 'correct' or 'relevant' IRR. The IRR criterion is not applicable. For example, assuming a COC equal to $r = 18\%$, the IRR criterion suggests to undertake the project if the larger IRR is the relevant IRR or reject the project if the smaller IRR is the relevant IRR.

No problem arises with the AIRR approach. One simply makes use of the book value estimates which have generated the project's cash-flow stream. However, for the sake of illustration, let us calculate the EAIRR, either assuming that book values are not available or assuming that the analyst wants to draw information about the project's rate of return as established by a normal market whose equilibrium rate for equivalent-risk assets is 18%. First, one computes the economic values, discounting prospective cash flows at 18%. This results in a capital stream equal to

$$\boldsymbol{C}^{\text{ec}} = (4, 1.85, -0.07, -1.58, -2.61, -3.08, -2.89, -1.91, 0).$$

It is clear that, in the first two periods, the project is an investment (the firm invests in the project) whereas in the remaining periods it is a financing (the firm borrows from the project). Overall, it may be checked that the investment positions exceed the borrowing positions: $\text{PV}[\boldsymbol{C}^{ec}] = 0.2$. This means that, overall, the project is a net investment (see Definition 8.4). Using the shortcut, one gets the EAIRR, which is an investment rate:

$$\begin{aligned}\text{EAIRR} &= 0.18 + \frac{-4 + \frac{3}{1.18} + \frac{2.25}{1.18^2} + \frac{1.5}{1.18^3} + \frac{0.75}{1.18^4} - \frac{0.75}{1.18^6} - \frac{1.5}{1.18^7} - \frac{2.25}{1.18^8}}{0.2} \cdot 1.18 \\ &= 84.09\%.\end{aligned}$$

Overall, the project undertaking boils down to borrowing a net \$0.2 at an average investment rate equal to 84.09%, which is greater than the expected investment rate that the market pays on an equivalent-risk portfolio. The efficiency of the project is then 66.09%. Assuming the project is unlevered (so that $\text{NPV} = \text{NPV}^e$), shareholders wealth is increased. The financing scale is \$0.2 and the efficiency, expressed in present terms, is $66.09\%/1.18 = 56.01\%$. The wealth increase is then equal to $\text{NPV} = 0.2 \cdot 56.01\% = 0.11$.

It is evident, from the above counterexamples, that the IRR manifests problems of existence and uniqueness. When IRRs are multiple or no IRR exists, the IRR criterion may not be applied. In this respect, there is a fundamental difference between the

[7] This cash-flow stream was presented in Eschenbach (1995, Sect. 7.6), associated with a mineral extraction project, and later analyzed by Hazen (2003) and Magni (2010b).

way AIRR and IRR are defined. AIRR may be defined abstracting from the Chisini mean notion. In particular, (8.21), (8.23), and (8.31) are more general than (8.22) (they do not require existence of the income rates). This implies that there is no need of solving any equation to calculate an AIRR. In contrast, the IRR is necessarily tied to a Chisini mean and the analyst must necessarily solve an equation for calculating it. This is precisely the source of the problems of existence and uniqueness.

These two problems were soon recognized since 1930s (Boulding 1936; Samuelson 1937) but only two decades later the problems started to attract scholars' attention (e.g., Lorie and Savage 1955; Solomon 1956). Since then, a growing amount of contributions started to appear in the literature in order to cope with these two difficulties. Many scholars focused on conditions for an IRR to exist and be unique and strived to provide more and more general conditions under which an IRR exists and is unique. For example, Pitchford and Hagger (1958) (PH) identified a class of projects with a unique IRR in the interval $(-1, +\infty)$. They showed that, if the stream of cash flows $\boldsymbol{F}$ possesses one and only one change of sign, then the IRR, σ, exists and is unique in the interval $(-1, +\infty)$. Cash-flow streams that satisfy this condition are often called ***conventional***.

Soper (1959) extended the class of projects with unique IRRs, and his results were later generalized by Teichroew et al. (TRM) (1965b) and Gronchi (1986). Their results can be combined and stated as follows. If σ is an IRR of $\boldsymbol{F}$ and the amounts $\sum_{k=0}^{t} F_k(1+\sigma)^{t-k}$ are all non-negative or all non-positive for all $t \in \mathbb{N}_{n-1}^0$, then an IRR, σ, exists and is unique in the interval $(-1, +\infty)$.

Owing to (1.17), the Soper-TRM-Gronchi (STRMG) condition can be reframed in terms of internal values, since (9.4) implies $C_t(\sigma) = -\sum_{k=0}^{t} F_k(1+\sigma)^{t-k}$. Specifically, if an IRR exists and $C_t(\sigma) \geq 0$ for every t or $C_t(\sigma) \leq 0$ for every t, then the IRR is unique. In these cases, the project is ***internally pure***, namely, an investment project in the former case, a financing project in the latter case.

Furthermore, it may be proved that, if the cash-flow stream is conventional, then the project is internally pure. Put it in other words, if PH condition holds, the project is associated with internal values that do not change sign (i.e., PH conditions imply STRMG condition). The reverse is not true: A cash-flow stream may be internally pure even if it is not conventional. For example, $\boldsymbol{F} = (-100, 50, -10, 30, 40, 40)$ is nonconventional (its cash-flow stream changes sign three times) but it is internally pure. Specifically, the IRR is 14.3% and the internal values are non-negative: $\boldsymbol{C}(\sigma) = (100, 64.3, 83.6, 65.6, 35, 0)$.

STRMG's findings unearthed the third classical problem of the IRR, related to nonconventional cash-flow streams: If the cash-flow stream is not conventional, then it may occur that $C_t(\sigma) < 0$ for some $t \in \mathbb{N}_{n-1}^0$ and $C_t(\sigma) > 0$ for some other $t \in \mathbb{N}_{n-1}^0$. In this case, the project and the IRR are financially ambiguous: In those periods where $C_t(\sigma) > 0$, the project might be considered an investment project and the IRR might be considered an investment rate; in those periods where $C_t(\sigma) < 0$, the project might be considered a financing project and the IRR might be considered a financing rate (see also Sect. 1.3). In this case, the traditional IRR criterion is not applicable, for it does not distinguish investment projects from financing projects. Such problems of ambiguity may arise even if the project possesses a unique IRR (see Example 9.4

below). This means that, while existence and uniqueness of σ in (9.2) is a necessary condition for the IRR to have an unambiguous nature (and for the IRR criterion to be employed), the condition is by no means sufficient.

Remark 9.1 (*Other conditions for existence or uniqueness of IRR*) A further extension of the class of projects with unique IRR in $(-1, +\infty)$ was accomplished by Kaplan (1965, 1967) by making use of the Sturm's sequence. It may be proved that, if STRMG condition is fulfilled, then Kaplan's conditions are fulfilled as well. In other words, the set of projects with unique IRR detected by Kaplan is greater than the set of projects with unique IRRs identified by STRMG.

Other scholars focused on the interval $(0, +\infty)$. Jean (1968) showed that, assuming $F_0 < 0$, if $\sum_{t=0}^{n} F_t > 0$ and the cash-flow stream $\boldsymbol{F}$ possesses one or two sign changes, then a unique IRR exists in the interval $(0, +\infty)$ (more precisely, this is a generalization of Jean's original result, provided by Gronchi 1984/1987). Norstrøm (1972) studied the cumulative cash flow, defined as $X_t = \sum_{j=0}^{t} F_j$. He showed that, if the sequence $\boldsymbol{X} = (X_0, X_1, \ldots, X_n)$ possesses one and only one change of sign and $X_n \neq 0$, then, a unique IRR exists in the interval $(0, +\infty)$.[8] Bernhard (1979, 1980), later generalized by Gronchi (1984/1987), considered the sequence $\boldsymbol{m} = (m_0, m_1, \ldots, m_n)$ such that $m_t = \sum_{j=0}^{t} \binom{n-j}{t-j}$. If said sequence possesses one and only one change in sign, then a unique IRR exists in the interval $(0, +\infty)$. Other conditions were set by Aucamp and Eckardt (1976), De Faro (1978), Pratt and Hammond (1979) (see also Magni 2016b for a review. See Gronchi 1984/1987 for a detailed formal treatment of these findings).

Another strand of literature dealt with the concept of *truncated project*. Notable were the contributions by Karmel (1959) and Arrow and Levhari (1969). Karmel (1959) assumed that a project can be truncated at any time and considered the truncated cash-flow streams $\boldsymbol{F}^{(t)} = (F_0, F_1, \ldots, F_t, 0, 0, \ldots, 0), t \in \mathbb{N}_n^0$. The author recommended to select the "maximum" IRR according to the following procedure:

1. exclude all truncated cash-flow streams $\boldsymbol{F}^{(t)}$ which possess no IRR
2. if there exists some t such that $\boldsymbol{F}^{(t)}$ possesses multiple IRRs, then pick the greatest one and neglect the other ones
3. take the IRRs selected in the previous steps and choose the greatest one. Let $\sigma^{\max}$ be such IRR and $t^{\max}$ be the associated truncation date.

The author proved that, if the investor truncates the project at time $t^{\max}$, then the IRR is unique. Arrow and Levhari (1969) utilized the truncated cash-flow stream alongside their DCF functions $G_t = \sum_{j=0}^{t} F_j(1+x)^{-j}$ and defined the maximum present value as the function $G(x) = \max[G_t(x) : t = 0, 1, 2, n]$. Assuming $F_0 < 0$, the authors showed that function $G(x)$ is continuous, such that $\lim_{x \to -1^+} G(x) = +\infty$ and $\lim_{x \to +\infty} G(x) = F_0 < 0$. Further, the function is monotonically decreasing, which implies that the IRR associated with $G(x)$ is unique. Norstrøm (1970) proved

[8]This does not exclude that other IRRs exist in the interval $(-1, 0)$.

that the IRR derived via Karmel's procedure is equal to the IRR obtained by Arrow and Levhari.[9]

Example 9.4 ♦
Consider the cash-flow stream $\boldsymbol{F} = (-10, 27, -27, 17)$. The IRR exists and is unique: $\sigma = 0.7$. However, the internal (i.e., IRR-implied) capital amounts are

$$\begin{aligned} C_0(0.7) &= 10 > 0 \\ C_1(0.7) &= 10(1+0.7) - 27 = -10 < 0 \\ C_2(0.7) &= -10(1+0.7) + 27 = 10 > 0 \\ C_3(0.7) &= 10(1+0.7) - 17 = 0. \end{aligned}$$

In the first and third periods, the project seems to be an investment (the firm invests funds in the project at the IRR), whereas in the second period the project is a financing (the firm subtracts funds from the project at the IRR). The internal values identify this project as a mixed one. Therefore, IRR is an investment rate in the first and third period, whereas it is a financing rate in the second period. Hence, the IRR as such does not supply any useful information about the project's economic profitability. Also, it is not clear how the traditional IRR criterion should be applied. For example, assuming the COC is $r = 65\%$, it is not clear whether $70\% > 65\%$ signals value creation or, rather, value destruction. As we know, the NPV criterion, while not distinguishing between the two types of projects, immediately reveals that the project creates value: $\text{NPV} = 0.23 > 0$. As we also know, this may be decomposed into project scale, $\text{PV}[C]$, and economic efficiency, $\xi_0 = \bar{\imath}_0 - r_0$. If $\text{PV}[C] > 0$, the AIRR is an investment rate, if $\text{PV}[C] < 0$, the AIRR is a financing rate.

Hence, uniqueness of IRR does not guarantee that the IRR has a straight financial meaning.

Example 9.5 ♦
Consider Example 9.3, whose cash-flow stream is

$$\boldsymbol{F} = (-4, 3, 2.25, 1.5, 0.75, 0, -0.75, -1.5, -2.25).$$

There are two sequences of internal values, one associated with the smaller IRR, 10.43%, the other one associated with the larger IRR, 26.31%:

$$\begin{aligned} \boldsymbol{C}(0.1043) &= (4, 1.42, -0.68, -2.26, -3.24, -3.58, -3.20, -2.04, 0) \\ \boldsymbol{C}(0.2631) &= (4, 2.05, 0.34, -1.07, -2.10, -2.65, -2.60, -1.78, 0) \end{aligned}$$

Both IRRs signal that the project starts as an investment and ends as a financing. It is not clear whether the two IRRs are, overall, investment rates or financing rates and which one is the relevant IRR.

---

[9] See also Soper (1959), Hicks (1973), Nuti (1974), Sen (1975), Ross et al. (1980), Cuthbert (2018) on truncation.

Example 9.6 ♦
Consider a cash-flow stream $\boldsymbol{F} = (-1, 6, -11, 6)$. The IRR equation

$$-1 + 6(1+x)^{-1} - 11(1+x)^{-2} + 6(1+x)^{-3} = 0$$

brings about three solutions: $\sigma^1 = 0\%, \sigma^2 = 100\%, \sigma^3 = 200\%$. Each of these IRRs implies a description of the project as a mixed (not pure) one:

$$\boldsymbol{C}(0) = (1, -5, 6, 0) \quad \boldsymbol{C}(1) = (1, -4, 3, 0) \quad \boldsymbol{C}(2) = (1, -3, 2, 0).$$

The problem of ambiguity of IRR afflicted scholars for decades, until Hazen (2003) adjusted the traditional IRR criterion taking into explicit account the sign of the overall internal capital $\text{PV}[C(\sigma)] = \sum_{t=1}^{n} C_{t-1}(\sigma)(1+r)^{-(t-1)}$. Specifically, if $\text{PV}[C(\sigma)] > 0$, then the project is described as a net investment and the IRR is an investment rate, if $\text{PV}[C(\sigma)] < 0$, the IRR is described as a net financing and the IRR is a financing rate.[10] Hence, the IRR criterion can be applied as a rational criterion. With his notable contribution, Hazen (2003) showed that

i. if a project possesses multiple IRRs, the project may be arbitrarily described as a net investment or a net financing, depending on which IRR is chosen to be representative of the project
ii. any one IRR may be equivalently used for rational decision-making.

These two results guarantee that any IRR supplies rational decision-making.[11] However, this still leaves the question open as to which one IRR correctly measures the project's economic efficiency among the multiple ones (and, therefore, whether the project is an investment or a financing).

Among others, Hartman and Schafrick (2004) is worth mentioning. The authors made use of the graph of the DCF function, $\psi(x) = \sum_{t=0}^{n} F_t/(1+x)^t$, and the sign of the NPV derivative to single out the *relevant* IRR (see also Cannaday et al. 1986 and Hajdasinski 1987 on relevant rates of return).

In sharp contrast with this view, Osborne (2010) recommended the use of *all* the (real-valued and complex-valued) solutions to the IRR equation, $\sum_{t=0}^{n} F_t/(1+x)^t = 0$, thereby denying that one IRR is more relevant than any other one.[12]

Most recently, Cuthbert (2018) turned the issue on its head and showed an important result: Any cash-flow stream may be partitioned into consecutive, non-overlapping sequences of conventional cash-flow streams possessing (unique) strictly decreasing IRRs, associated with internally pure capital streams. This means that, to any project, there corresponds a *unique* sequence of implicit IRRs which have the same financial nature. The rule for finding Cuthbert partition is as follows. Consider a cash-flow

---

[10]This is in line with Definition 8.4 in this book. Indeed, Hazen's (2003) definition may be interpreted as a special case of it (see Magni 2010b).

[11]Hazen's adjusted IRR criterion also implies a third result: The IRR's financial nature is not internal, for it depends on the sign of $\text{PV}[C(\sigma)]$ which in turn depends on the COC, r (see also Chap. 10).

[12]Complex-valued IRRs are also dealt with in Hazen (2003) and Pierru (2010).

stream $F = (F_0, F_1, \ldots, F_n)$ and let a *simple transaction* be a cash-flow stream where the ouflows precede the inflows (i.e., a conventional transaction). Then,

1. split up F into a sequence of simple transactions
2. find the IRRs of the simple transactions
3. amalgamate any runs of simple transactions with increasing IRRs
4. find the IRRs of the resulting transactions
5. amalgamate any runs of transactions with increasing IRRs so as to obtain a partition with fewer transactions
6. repeat steps 4. and 5. until this process terminates in a partition of F into pure transactions with strictly decreasing IRRs. This is Cuthbert partition (see Cuthbert 2018 for details).

The resultant IRRs of the partition may be aggregated to give rise to an IRR-implied rational rate of return called *pure-investment Average Internal Rate of Return* (PIAIRR) (see Cuthbert and Magni 2018).[13]

9.4 The Practical Relevance of the Classical Pitfalls

The above mentioned problems reveal the inadequacy of the IRR to represent a meaningful rate of return in real-life applications. Indeed, the practical relevance of these problems should not be overlooked. Following are some types of practical applications that manifest the problems (the classes of projects below are not mutually exclusive: A project may well belong to more than one class).

---

[13]The literature on multiple IRRs is enormous. Authors from various fields advanced methods for choosing the relevant IRR among the multiple ones or methods for avoiding the multiple-IRR issue. To cite just (very) few contributions,

- Ramsey (1970) distinguished IRR from marginal efficiency of capital and provided a redefinition of internal rate of return
- Oakford et al. (1977) proposed a pseudo-IRR for mixed transactions
- Cantor and Lippman (1983, 1995) recommended the use of the smallest nonnegative IRR
- Cannaday et al. (1986) and Colwell (1995) selected the relevant IRR on the basis of the effect of a marginal increase in a cash flow in the IRRs (their solution is limited to two roots)
- Howe (1991) advanced the use of the ratio of an equivalent perpetuity return to the initial outlay
- Bidard (1999) recommended the use of maximal root
- Zhang (2005) described an IRR parity technique (based on the number even or odd IRRs which are greater than the cost of capital) to choose the relevant IRR
- Bosch et al. (2007) used a payback coefficients to derive an NPV-compatible normalized index
- Chiu and Garza Escalante (2012) provided a Generalized Relative Rate of Return as a mark-up to the time-varying costs of capital
- Shestopaloff and Shestopaloff (2013) endorsed the rule of the largest root
- and many others ...

(The results found in Hartman and Schafrick 2004; Magni 2010b; and Cuthbert 2018 have been awarded the "Eugene L. Grant" award by ASEE in 2005, 2011, and 2019, respectively.) .

Projects with Disposal/Cleanup Costs

From PH, we know that a necessary condition for a cash-flow stream to have multiple IRRs or no IRR is that the cash-flow sequence is nonconventional, that is, it has zero, two or more than two changes in sign.

If disposal and/or remedial, clean-up costs are sustained at the end of a project, a typical sign pattern for cash flows is $(-, +, +, \ldots, +, -)$ or, if the construction phase lasts more than one period, $(-, -, -, \ldots, +, +, \ldots, +, -)$. Several engineering economy textbooks and some finance textbooks describe examples where the problem of multiple IRRs (or even no IRR) arises because of a terminal outflow (e.g., Brealey et al. 2011, p. 110; Berk and DeMarzo, 2014, pp. 210–214; Blank and Tarquin 2012, Example 7.3 and pp. 180–184). These cases are more frequent than one might believe. Manufacturing sites, utility plants, extraction of natural resources and any case where a facility is to be dismantled or equipment is to be disposed of may lead to multiple IRRs.

Same sign pattern occurs, *ex post*, whenever an unexpected event terminates the project abruptly and the investor experiences one or more losses as a consequence of hurricanes, wildfires, floods and other natural disasters. As unexpected events may not be foreseen, these situations arise in an ex post performance measurement, rather than in an ex ante setting.

Phased Expansion

The capacity of a project may be increased over a number of periods. In this case, multiple changes of sign may arise in the cash-flow stream, as a result of the additional capital expenditures made by the firm over time. The resulting sign pattern is $(-, +, +, \ldots, +, -, +, +, \ldots, -, +, +, \ldots, +)$ which may lead to multiple IRRs (see also Hartman 2007, pp. 393–394). If disposal costs are associated with phased expansion, then the pattern becomes $(-, +, +, \ldots, +, -, +, +, \ldots, -, +, +, \ldots, +, -)$ which makes it even possible that no IRR exists. (In general, a necessary condition for no IRR to occur is that the initial cash flow and the last nonzero cash flow have the same sign.)

Replacement Projects

While the expansion projects add a new course of action to the status quo (i.e., the firm's current operations), replacement projects consist of replacing part of the status quo with a different (or even equivalent) course of action: The decision maker must decide whether the status quo should be retained or replaced by such an alternative (see Sect. 1.4.3). As seen, the current operation which is challenged by the new one is often called the ***defender*** while the new course of action is called the ***challenger***. The replacement project is then described, in terms of cash flows, as the difference between the challenger's cash-flow stream and the defender's cash-flow stream. Examples of this class are decisions regarding replacement of a piece of equipment, a process, a system, a technology, a supplier, etc. This kind of projects may be subject to multiple IRRs or even no IRR.

No Payout

The IRR concept, based as it is on cash flows, collapses when the cash-flow stream consists of a single cash flow. This occurs whenever the payout ratio is zero and negative cash flows are covered by internal financing. For example, consider a project with initial operating investment equal to $C_0^o = -F_0^o > 0$. If the operating investment is supported by internal financing, the firm raises funds from the existing assets instead of raising funds from claimholders (debtholders and equityholders) (see Sect. 2.2). The firm might finance the capital expenditure with excess cash or by selling marketable securities, especially if the project is not a large scale one (even more so if the firm has a substantial amount of excess cash). This implies $F_0^l = -C_0^l = C_0^o = -F_0^o > 0$, so that a decant occurs from liquid assets to operating assets and $F_0 = F_0^o + F_0^l = 0$ (see Remark 3.1 on decant). Since the payout ratio is $\alpha = 0\%$, the prospective positive operating cash flows are absorbed internally (e.g., deposited back in the liquid assets to replenish the balance) and the prospective negative operating cash flows are covered by withdrawals from the liquid assets, so that $F_t^o = -F_t^l$ and $F_t = F_t^o + F_t^l = 0$ for every $t \in \mathbb{N}_{n-1}^0$.

These situations logically entail that only one cash flow is generated by the project's assets, namely, the terminal one, which may be positive or negative: $\boldsymbol{F} = (0, 0, \ldots, 0, F_n)$. In all such situations, the IRR does not exist.

Cash-Flow Streams with No Change of Sign

The above case is a special case of a more general one where the cash-flow pattern has no change of sign. Far from being a peculiar situation, this case may be frequently encountered. For example, consider a firm which borrows from the project and succeeds in managing the funds so that the project's cash flows are always nonnegative. This may be done in several ways. For instance, the negative operating cash flows are absorbed internally while the positive operating cash flows are wholly or partially distributed. In this case, the IRR of the project (and of the equity, since $F_t = F_t^e$) does not exist.

Alternatively, one may think of a firm that suitably manages the working capital: If the cash cycle is negative (customers pay for the goods earlier than the firm pays for supplies), negative assets may arise, which means that the firm borrows from the project. If the income is positive and greater than (or equal to) the change in assets in every period ($I_t \geq \Delta C_t$ for every t), then the cash-flow stream does not contain any negative cash flow ($F_t \geq 0$ for every t), and the IRR does not exist.

A third possibility is the case were the negative cash flows are covered entirely by debt. In this case, all equityholders' cash flows will be nonnegative (as long as the last CFO exceeds the last CFD), and the equity IRR will not exist.

Projects with Upfront Payments

A firm may be involved in contract-based work, whereby it receives a purchase order from a customer, accompanied by one or more upfront payments for the delivery of some good (machinery, equipment, fixed asset, services, etc.) at a certain date. After the down payment(s), the firm incurs costs associated with the production or construction of the good or service. The balance is settled at or after the delivery date. The pattern is then either $(+, -, -, \ldots, -, -)$ or $(+, -, -, \ldots, -, +)$. While the former case is a conventional project (with unique IRR), the latter pattern is symmetric to the pattern of a project with disposal/cleanup costs, with opposite signs. For this reason, this kind of cash-flow streams are open to multiple IRRs or even no IRR. Examples of this kind also include construction, modernization, overhaul, innovation, as well as maintenance costs for real assets such as building, roads, highways, infrastructures, etc. whereby the concessionaire increases tolls for financing the project, then pays the costs, and keeps on receiving the increased tolls after the last payment has been made.

Equity Perspective

The three classical problems lurk even in projects having conventional cash-flow streams. The reason is that the relevant IRR for equityholders is not the IRR of the CFA stream (nor the FCF stream) but the equity IRR, that is, the IRR of the CFE stream. If the project is levered, the CFE stream may well be a non-conventional one, even if the CFA (or FCF) stream is a conventional one. Let $\boldsymbol{F} = (F_0, F_1, \dots, F_n)$ be a conventional stream of CFAs (e.g., $F_0 < 0$ and $F_t \geq 0$ for $t \in \mathbb{N}^1_n$ with some positive cash flow) so that the sign pattern is $(-, +, +, \dots, +)$. This implies that the project IRR exists and is unique and is an investment rate ($C_t(\sigma) \geq 0$ for all t). Let $\boldsymbol{F}^d = (F^d_0, F^d_1, \dots, F^d_n)$ be a conventional streams of CFDs (e.g., $F^d_0 < 0$ and $F^d_t \geq 0$ for $t \in \mathbb{N}^1_n$ with some positive cash flow) so that the sign pattern is $(+, -, -, \dots, -)$: The IRR of the debt's stream exists and is unique as well. The equity cash-flow stream is equal to the difference between the CFA stream and the CFD stream:

$$\boldsymbol{F}^e = \boldsymbol{F} - \boldsymbol{F}^d = (F_0 - F^d_0, F_1 - F^d_1, \dots, F_n - F^d_n).$$

It should be evident that, even though both CFA and CFD streams are conventional streams, $\boldsymbol{F}^e$ might turn out to be an unconventional stream, one which generates multiple IRRs or no one at all. A unique project IRR is of limited use to the analyst if the corresponding equity IRR does not exist or multiple equity IRRs arise.

Old Shareholders

Let us consider Table 6.1, whereby the firm raises C^e_0 by issuing new shares for undertaking the project. Consider an existing shareholder, say Mrs. Old, and suppose she does not purchase any new share. After the issuance of the new shares, Mrs. Old owns θ^{Old} percent of the firm's equity, where $\theta^{\text{Old}} = n^{\text{Old}}/(n_0 + \Delta n_0)$ and n^{Old} is the number of shares owned by Mrs. Old, n_0 is the number of existing shares before issuance of new shares, Δn_0 is the number of shares issued.

Remembering that f^e_t is the equity cash flow of the firm-with-no-project and F^e_t is the project's equity cash flow, she expects to receive a cash flow of $F^{\text{Old}}_t = \theta^{\text{Old}} F^e_t - (1 - \theta^{\text{Old}}) f^e_t$ at every date from $t = 1$ to $t = n$, incurring no outlay at $t = 0$: $F^{\text{Old}}_0 = 0$. If $\theta^{\text{Old}} F^e_t > (1 - \theta^{\text{Old}}) f^e_t$ for every t, Mrs. Old's cash-flow stream is non-negative, her cash-flow pattern being $(0, +, +, \dots, +)$. The IRR of this cash-flow stream does not exist.

The AIRR approach can handle this case effectively. Mrs. Old's outflow is zero ($F^{\text{Old}}_0 = 0$) but, being a shareholder, she owns a quota of the incremental equity, C^e_t, for every $t \in \mathbb{N}^0_{n-1}$. Let us denote such a quota as $C^{\text{Old}}_t = \theta^{\text{Old}} C^e_t$. Her capital stream is

$$\boldsymbol{C}^{\text{Old}} = (C_0^{\text{Old}}, C_1^{\text{Old}}, C_2^{\text{Old}}, \ldots, C_{n-1}^{\text{Old}}, 0).$$

Thus, she is entitled to a quota of the project's incomes and cash flows. Her income stream will be

$$\boldsymbol{I}^{\text{Old}} = (I_0^{\text{Old}}, I_1^{\text{Old}}, I_2^{\text{Old}}, \ldots, I_{n-1}^{\text{Old}}, I_n^{\text{Old}})$$

where $I_t^{\text{Old}} = F_t^{\text{Old}} + \Delta C_t^{\text{Old}}$ and, in particular,

$$I_0^{\text{Old}} = F_0^{\text{Old}} + \Delta C_0^{\text{Old}} = F_0^{\text{Old}} + C_0^{\text{Old}} = C_0^{\text{Old}}.$$

Her AIRR may then be calculated as usual. For example, using the third method (Eq. (8.23)),

$$\bar{\imath}^{\text{Old}} = \frac{\text{PV}[\boldsymbol{I}^{\text{Old}}]}{\text{PV}[\boldsymbol{C}^{\text{Old}}]}(1 + r_1).$$

Rotating Savings and Credit Associations (ROSCAs)

Another compelling example is offered by the so-called Rotating Savings and Credit Association (ROSCA), a vast class of financial tools for saving and borrowing which are widely employed throughout the world. The social function of ROSCAs has been studied by anthropologists since 1960s (Geertz 1962; Ardener 1964; Kurtz 1973) while their economic role and performance has attracted the interest of economists only since 1990s, starting with the well-known contribution by Besley et al. (1993) (see also Bouman 1995; Hevener 2006; Dupas and Robinson 2013; Acquah and Dahal 2018; Bauchet and Larsen 2018). There are numerous variants of ROSCA. In its basic form, participants of a ROSCA meet periodically and deposit a fixed amount of money that creates a "pot" which is distributed to one of the participants according to some selection criteria (e.g., a lottery). Once a member of the pool has received the pot, he keeps on depositing the fixed amount in the next periods until the ROSCA ends (the number of periods being equal to the number of the members). For example, if a ROSCA has m participants depositing x each month, then the pot is mx and the cash-flow stream of, say, the fifth member receiving the pot is equal to

$$(-x, -x, -x, -x, mx - x, -x, -x, \ldots, -x).$$

It is easy to see that, usually, two IRRs arise for all members (except the first and last member, who have a unique IRR).[14] This makes it impossible to establish the rate of return of the deal for each member.

Further, it is not even clear whether the m deals are investments or financings, given that, as noted, Definition 9.1 does not come along with an unambiguous criterion for distinguishing an investment from a financing. In general, one may say that the case

[14]More precisely, it may be proved that a unique IRR is possible only for the intermediate member and only if n is odd. All other members have two IRRs.

of a unique IRR for the intermediate members (2 through n) is extremely infrequent and, whenever it occurs, the unique IRR has no clear financial meaning.

ROSCAs may be organized by a financial institution, in which case things get even worse. For example, suppose a ROSCA with m members is organized by a bank, according to which it periodically distributes an amount of xm' (thousands) to one member, with $m' < m$, while all the participants contribute an amount of x. In this case, all members (except the first one and the last one) will have multiple IRRs or even no IRR at all. For example, assuming the number of members is $m = 10$, the sixth member's cash-flow stream is

$$(-x, -x, -x, -x, -x, xm' - x, -x, -x, -x, -x).$$

It may well occur that multiple IRRs arise for this cash-flow stream, or even no IRR at all and these cases occur with very high frequency.[15] (Lima e Silva et al. 2018 illustrate some real-life ROSCAs in Brazil where these problems have appeared. See also Ahn et al. 2017.)

The difficulties of IRR are further amplified whenever an individual purchases multiple shares of a ROSCA and searches for a relative measure of economic profitability of its committed capital. In this case, there may be more than two changes in sign in the sequence of cash flows, with unpredictable results in terms of multiple IRRs or no IRR.

Finally, there is an additional aspect: From the point of view of any financial institution organizing ROSCAs, every ROSCA is a project with no IRR. The reason is that, from the point of view of the organizer, the cash flow is positive for every $t \in \mathbb{N}^0_m$, since $m > m'$ implies

$$\overbrace{xm}^{\text{cash flow received from the members}} - \overbrace{xm'}^{\text{cash flow distributed to one member}} > 0.$$

This in turn implies that, from the point of view of the financial institution, the IRR equation has no solution, since its NPV is positive:

[15] In the previous ROSCA, existence of IRR is guaranteed by the fact that

$$\phi(0) = \sum_{t=0}^{n} F_t = -(\overbrace{x + x + \cdots + x}^{m \text{ summands}}) + mx = 0,$$

so the solution $\sigma = 0$ always exists. Conversely, in this case, where $m' < m$,

$$\phi(0) = \sum_{t=0}^{n} F_t = -(\overbrace{x + x + \cdots + x}^{m \text{ summands}}) + m'x < 0$$

inexistence of IRR may indeed occur.

$$\sum_{j=1}^{m} \frac{x(m-m')}{(1+r)^j} > 0$$

for all $r \in (-1, +\infty)$.

In any ROSCA, all three problems are simultaneously at play: Multiple IRRs, no IRR, ambiguous IRR (Lima e Silva et al. 2018 use the AIRR approach to solve the problems).

Principal-Agent Relationship

Consider the situation where a principal entrusts a manager or a management team an initial endowment of $C_0 > 0$ to be invested in an economic activity. The principal periodically makes a decision on the amount of cash flow that is additionally contributed in the asset or withdrawn from the asset. This setting is typical of portfolio management, where a client entrusts a given capital amount to a fund manager, who actively makes decisions on asset allocation and selection. In this case, the client determines the amount of contributions and distributions (as well as the liquidation date). Therefore, large contributions may alternate with large distributions, which often result in a cash-flow stream that has more than one sign change, which increases the likelihood of multiple IRRs.

Another domain where the principal-agent relation arises regards investment decisions in decentralized organizations. In this case, the principal may be the board of directors of a company, which delegates the execution of an economic activity to an agent, represented by a new business unit or a subsidiary firm which is managed by a manager (or a management team). In this case, the board of directors retains the right of periodically withdrawing funds from or investing additional funds into the business unit. In such a way, the overall business unit's performance depends on both the investment policy followed by the business unit's management, and the principal's decisions about injections and withdrawals of funds.

In these situations, there may be considerable sways in the amount and sign of cash flows: Contributions may alternate to distributions, resulting in irregular, nonconventional cash-flow streams. The sways of the cash flows may be such that multiple IRRs arise (non-existence of IRR is excluded, as long as the terminal value is nonnegative).

Portfolios of Projects

Whenever a portfolio of projects is considered, multiple IRRs may easily arise even if the individual projects have conventional cash-flow streams, as long as they are sufficiently dissimilar in the timing and magnitude of outflows and inflows (even more so if an equity perspective is taken. See above).[16] Even in those cases where

[16]The distinction between a project and a portfolio of projects is not clear-cut: A project might be itself split into lower-order components (e.g., as we have seen, a project is a portfolio of equity and debt as well as a portfolio of operating assets and non-operating assets). Whether an asset should be considered an individual project or a portfolio is a matter of perspective and, therefore, of convention: Whether several courses of action should be aggregated into a single one or a single

each individual component is described by a conventional cash-flow stream, the resulting portfolio may be a nonconventional cash-flow stream with large sways, which make it more likely to find multiple solutions or no solution at all of the IRR equation.

Examples of portfolios include (but are not limited to) a bundle of projects undertaken by a firm, a multiperiod trading strategy with long and short positions taken on different financial assets, the purchase of multiple shares of a ROSCA, the equity part of a project,[17] simultaneous transactions made by a bank raising funds from some agents (e.g., customers) and using those funds for lending money to some other agents (e.g., a firm).

Spillover Effects

As we recall, the project is a differential (i.e., incremental) set of activities as opposed to the firm's status quo. As a byproduct of the project undertaking, several kinds of additional costs or revenues may arise. For example, *cannibalization* implies that the introduction of a new product by a firm reduces the sales of other goods manufactured by the firm and, at the same time, reduces the associated production costs.

The use of an existing resource which is currently utilized for other activities in the firm is an additional opportunity cost (by undertaking the project, the firm foregoes the income related to the use of that resources). The more substantial these effects, the higher the likelihood that no IRR or multiple IRRs will arise. Other externalities may augment costs or revenues stemming from other areas of the firm.

Example 9.7 ♦
(*Replacement project*) The costs associated with a presently owned machine (defender) with an economic life of four periods are $\boldsymbol{f}^{\text{def}} = (-35, -70, -125, -165)$ (in thousands), whereas a possible replacement with the same lifespan will generate costs equal to $\boldsymbol{f}^{\text{ch}} = (-36, -64, -136, -159)$. Therefore, the project's cash-flow stream is $\boldsymbol{F} = \boldsymbol{f}^{\text{ch}} - \boldsymbol{f}^{\text{def}} = (-1, 6, -11, 6)$ which gives rise to three IRRs: 0%, 100%, and 200%.

Example 9.8 ♦
(*Phased expansion and disposal costs*) A facility costs \$200 and will generate three inflows equal to \$75. In two years the firm expects to have the opportunity of expanding the project with an additional capital expenditure of \$350 which will give rise to inflows equal to \$120 for three years. After other two years, a further capital increment of \$1,000 may be injected into the project, followed by inflows equal to \$350 for three years. Remedial costs equal to \$75 are due at the end of the eighth year.

---

course of action should be partitioned into multiple ones depends on the conceptualization/framing of the situation, which may differ depending on several variables such as the domain and the purpose of the analysis, the pieces of information available to the analyst, the easiness and intuitiveness of the conceptual framework, etc.

[17]The equity may be viewed as a portfolio of an investment position (project) and a financing position (debt). See also above on the equity IRR.

Table 9.1 Phased expansion—multiple IRRs

Time	1st phase	2nd phase	3rd phase	remedial costs	project
	Cash flows				
0	−200				−200
1	75				75
2	75	−350			−275
3	75	120			195
4		120	−1000		−880
5		120	350		470
6			350		350
7			350		350
8				−75	−75
IRR	6.1%	1.4%	2.5%		0.35% −82.24%

This project's cash-flow vector is

$$\boldsymbol{F} = (-200, 75, -275, 195, -880, 470, 350, 350, -75)$$

and there are two IRRs: $\sigma^1 = 0.35\%$ and $\sigma^2 = -82.24\%$ (see Table 9.1).

Example 9.9 ♦

(*Replacement project*) Consider the solar PhV system illustrated in Example 1.13, the defender being represented by the current system (utility) and the challenger being represented by the solar PhV system. The IRR equation for this project may be framed as

$$100(1+x)\cdot\frac{1-(1+x)^{-12}}{x} - 7,100(1+x)^{-12} + 900\frac{1-(1+x)^{-8}}{x}\cdot(1+x)^{-12} = 0.$$

No IRR exists.

In contrast, the AIRR is easily calculated. As shown in Example 8.13, it is equal to the average ROI, $\bar{\imath} = 5.03\%$. It is an investment rate ($\mathrm{PV}[C] > 0$) and is greater than the MARR.[18]

Example 9.10 ♦

C^o	
C^l	C^e

(*No payout*) A firm invests \$20,000 for purchasing some machinery which will be used for the production of a new good. The project lasts 7 years, after which production will cease. The firm uses some internal cash to finance the project and establishes that the cash flow realized by the sales will be deposited back in the liquid asset account. The excess cash will be invested in risk-free assets, and the risk-free rate is $r_f = 1\% = r^l$. Therefore, $i^l = r^l = 1\%$ and the NPV of liquid assets will be zero. The operating pre-tax WACC is $r^o = 12\%$. The tax rate is 33%.

Below we describe the BS and IS, based on the analysts' estimations. They imply a cash-flow stream equal to $\boldsymbol{F} = \boldsymbol{F}^e = (0, 0, 0, 0, 0, 0, 0,\ 28{,}882.61)$. This is a one-cash-flow project. The IRR equation is

$$\frac{28{,}882.61}{(1+x)^7} = 0$$

and it has no solution. Hence, this project has no IRR.

On the contrary, with the AIRR approach one may easily compute the average ROE, after computing the equity COC. It may be checked that $\boldsymbol{r}^e = (29.9, 20.4, 14.8, 11.0, 8.1, 5.6, 3.4\%)$. Hence, $\mathrm{PV}[I^e]$=14,841.51, $\mathrm{PV}[C^e] = 36{,}774.09$, $\mathrm{PV}[I^{e*}] = 2{,}566.7$, whence the average ROE is $\bar{\imath}^e_0 = 40.36\%$ and the equity MARR is $\bar{r}^e_0 = 6.98\%$. The project creates value at a rate of $\xi^e = 33.38\%$, which, applied to the capital base of \$36, 774.09, makes a total value creation of $\mathrm{NPV} = \mathrm{NPV}^e = \$12{,}275$. The latter is also obtained by discounting the last CFE: $F^e_7 \cdot \mathrm{d}^e_{7,0} = 28{,}882.61 \cdot 0.424989791 = 12{,}275$.

---

[18]In Magni and Marchioni (2019), a more refined example of PhV plant with no IRR is illustrated.

No payout—Balance Sheets

Year	**0**	**1**	**2**	**3**	**4**	**5**	**6**	**7**
NFA	20,000.00	14,285.71	10,204.08	7,288.63	5,206.16	3,718.69	2,656.210	0.00
NOWC	0.00	0.00	0.00	0.00	0.00	0.00	0.00	0.00
Liquid assets	−20,000.00	−12,218.29	−4,923.21	2,035.90	8,766.76	15,346.36	21,829.80	0.00
Investments	0.0	2,067.43	5,280.87	9,324.53	13,972.92	19,065.05	24,486.01	0.00
Debt	0.00	0.00	0.00	0.00	0.00	0.00	0.00	0.00
Equity	0.00	2,067.43	5,280.87	9,324.53	13,972.92	19,065.05	24,486.01	0.00
Financings	0.00	2,067.43	5,280.87	9,324.53	13,972.92	19,065.05	24,486.01	0.00

No payout—Income Statements and CFEs

Year	**1**	**2**	**3**	**4**	**5**	**6**	**7**
Sales	14,000.0	14,000.0	14,000.0	14,000.0	14,000.0	14,000.0	14,000.0
−Operating costs	−5,000.0	−5,000.0	−5,000.0	−5,000.0	−5,000.0	−5,000.0	−5,000.0
−Depreciation	−5,714.29	−4,081.63	−2,915.45	−2,082.47	−1,487.48	−1,062.48	−2,656.21
EBIT	3,285.71	4,918.37	6,084.55	6,917.53	7,512.52	7,937.52	6,343.79
−Interest expense	0.00	0.00	0.00	0.00	0.00	0.00	0.00
+Interest income	−200.00	−122.18	−49.23	20.36	87.67	153.46	218.30
EBT	3,085.71	4,796.18	6,035.32	6,937.89	7,600.19	8,090.98	6,562.09
−Taxes	−1,018.29	−1,582.74	−1,991.65	−2,289.50	−2,508.06	−2,670.02	−2,165.49
NI	2,067.43	3,213.44	4,043.66	4,648.39	5,092.13	5,420.96	4,396.60
$-\Delta$ Equity	−2,067.43	−3,213.44	−4,043.66	−4,648.39	−5,092.13	−5,420.96	24,486.01
CFE	0.00	0.00	0.00	0.00	0.00	0.00	28,882.61

(See also Bumps example in Chap. 12.)

Example 9.11 ♦
(*Upfront payment*)A firm receives a purchase order from a customer, accompanied by an upfront payment of \$20,000 for the delivery of some goods after 3 months. The production begins immediately; related payments for additional costs of \$22,000 and \$80,000 will be due, respectively, after one month and two months. At the end of the third month, the finished product is delivered. The customer settles the balance with a payment of \$88,000. This implies $\boldsymbol{F} = (20, -22, -80, 88)$ (in thousands). As already seen in Example 1.10, the upfront payment is a non-interest bearing liability (customers loan to the firm at a 0% interest rate), so, in the first period, the project is a financing ($C_0 = -20 < 0$). The IRR equation is

$$20 - \frac{22}{1+x} - \frac{80}{(1+x)^2} + \frac{88}{(1+x)^3} = 0.$$

Two IRRs arise: $\sigma^1 = 10\%$ and $\sigma^2 = 100\%$. It is easy to check that the internal capital vectors are

$$\boldsymbol{C}(0.1) = (-20, 0, 80, 0), \qquad \boldsymbol{C}(1) = (-20, -18, 44, 0).$$

According to the second IRR, 100%, the project is a financing in the first two periods while turning to an investments in the third period. According to the first IRR, 10%, a peculiar situation arises: The project is a financing in the first period and an investment in the third period. However, in the second period, it is neither an investment nor a financing. No capital is either invested in the project nor absorbed from the project. This creates a situation of further ambiguity: If nothing is invested or borrowed in that period, then, how can 10% be a growth rate for the capital? The growth rate is, simply, not defined.

Example 9.12 ♦
(*Old shareholder*) Consider a firm whose expected cash flow to equity is $f_t^e = \$150$ for every $t > 0$. The firm's managers decide to undertake a project whose initial cost is $C_0^e = \$14$ and whose prospective CFEs are equal to $F_t^e = \$6$ for every $t \in \mathbb{N}_8^1$. The firm issues new shares to raise funds. The required return on equity is assumed to be $r^e = 15\%$, which means that funds will be raised at 15%.

After the issuance, Mrs. Old, a long-standing equityholder, owns 98% of the firm's shares, whereas Mrs. New owns the remaining 2%. At time 0, Mrs. New incurs an outlay equal to $-\$14$, whereas Mrs. Old's cash position remains unvaried (no inflow nor outflow). The incremental cash flow expected by Mrs' Old is $98\% \cdot \$6 - 2\% \cdot \$150 = \$2.88$, whereas the incremental cash flow expected by Mrs' New is $2\% \cdot (\$150 + \$6) = \$3.12$.

Table 9.2 reports the cash flows of Mrs. Old and Mrs. New. The IRR of Mrs. New coincides with the required return on equity (15%), whereas Mrs. Old's IRR does not exist (see also Remark 6.3 and, in particular, Table 6.1).

Table 9.2 Old shares, new shares, and IRR[a]

Time	CFE (old shares)	CFE (new shares)	Total
0	\$0	−\$14	−\$14
1	\$2.88	\$3.12	\$6
2	\$2.88	\$3.12	\$6
3	\$2.88	\$3.12	\$6
4	\$2.88	\$3.12	\$6
5	\$2.88	\$3.12	\$6
6	\$2.88	\$3.12	\$6
7	\$2.88	\$3.12	\$6
8	\$2.88	\$3.12	\$6
NPV^e	\$12.92	\$0.00	\$12.92
IRR	NONE	15%	39.94%

[a] See also Table 6.1

Example 9.13 ♦

(*ROSCA*) Consider a ROSCA with 10 participants, each one paying \$1 each month. The pot is $\$10 = \$1 \cdot 10$. Table 9.3 reports the cash flows for each participant. The first's and last member's cash-flow streams are conventional, which implies that the IRRs exist and are unique, both equal to 0%, one representing an investment rate (last member) while the other is a financing rate (first member). As for members 2 through 9, their cash-flow streams are nonconventional and two IRRs arise for each member's deal. Specifically, all deals share an IRR equal to 0%; the second IRR is decreasing from deal 2 to deal 9: 787.3%, 142.4%, 50.6%, 13.1%, −11.6%, −33.6%, −58.7%, −88.7%. For these members, it is not clear whether they invest or borrow. For example, let us suppose the cost of capital is $r = 5\%$ and let us focus on deal 5. The IRRs are $\sigma^1 = 0\%$ and $\sigma^2 = 13.1\%$. It is easy to check that both sequences of internal capitals alternate in sign. If one considers 0% as the relevant IRR, the deal is a net financing and the IRR is a financing rate, because $\text{PV}[C(0\%)] = -2.38\% < 0$. If one considers 13.1% as the relevant IRR, then the deal is a net investment and the IRR is an investment rate, because $\text{PV}[C(13.1\%)] = 1.48\% > 0$.[19] (Analogously with the other deals.)

[19] We assume the first meeting will take place in one period. In this respect, it is irrelevant whether the first cash flow occurs at time 1 or at time 0 (no matter what unit of time has been selected): The solution of the IRR equation and the sign of $\text{PV}[C(\sigma)]$ are not affected.

Table 9.3 Rotating savings and credit association

TIME	PARTICIPANTS									
	1	**2**	**3**	**4**	**5**	**6**	**7**	**8**	**9**	**10**
0	0	0	0	0	0	0	0	0	0	0
1	9	−1	−1	−1	−1	−1	−1	−1	−1	−1
2	−1	9	−1	−1	−1	−1	−1	−1	−1	−1
3	−1	−1	9	−1	−1	−1	−1	−1	−1	−1
4	−1	−1	−1	9	−1	−1	−1	−1	−1	−1
5	−1	−1	−1	−1	9	−1	−1	−1	−1	−1
6	−1	−1	−1	−1	−1	9	−1	−1	−1	−1
7	−1	−1	−1	−1	−1	−1	9	−1	−1	−1
8	−1	−1	−1	−1	−1	−1	−1	9	−1	−1
9	−1	−1	−1	−1	−1	−1	−1	−1	9	−1
10	−1	−1	−1	−1	−1	−1	−1	−1	−1	9
IRR	0%	0%	0%	0%	0%	0%	0%	0%	0%	0%
IRR		787.3%	142.4%	50.6%	13.1%	−11.6%	−33.6%	−58.7%	−88.7%	

Example 9.14 ♦

(*ROSCA*) Suppose a 10-member ROSCA is organized by a bank according to which it periodically distributes an amount of \$178 (thousands) to one member while all the other participants contribute an amount of \$22 (thousands) (see Table 9.4).

The first and last deal are described by conventional cash-flow streams, so IRR is unique, 2.2% and −2.1% respectively. Of the other 8 participants, the fifth and the sixth ones have no IRR. The remaining 6 participants have two IRRs each: (2.9%, 694.7%), (4.2%, 124.6%), (8.5%, 36.6%), (−7.9%, −26.78%), (−4%, −55.5%), (−2.78%, −87.42%) for member 2, 3, 4, 7, 8, 9 respectively.

As seen, the member receiving the pot at time 1 has a unique IRR, as well as the member receiving the pot at time 10. However, assuming that one participant

Table 9.4 Rotating savings and credit association organized by a bank

TIME	PARTICIPANTS									
	1	**2**	**3**	**4**	**5**	**6**	**7**	**8**	**9**	**10**
0	0	0	0	0	0	0	0	0	0	0
1	178	−22	−22	−22	−22	−22	−22	−22	−22	−22
2	−22	178	−22	−22	−22	−22	−22	−22	−22	−22
3	−22	−22	178	−22	−22	−22	−22	−22	−22	−22
4	−22	−22	−22	178	−22	−22	−22	−22	−22	−22
5	−22	−22	−22	−22	178	−22	−22	−22	−22	−22
6	−22	−22	−22	−22	−22	178	−22	−22	−22	−22
7	−22	−22	−22	−22	−22	−22	178	−22	−22	−22
8	−22	−22	−22	−22	−22	−22	−22	178	−22	−22
9	−22	−22	−22	−22	−22	−22	−22	−22	178	−22
10	−22	−22	−22	−22	−22	−22	−22	−22	−22	178
IRR	2.2%	2.9%	4.2%	8.5%	NONE	NONE	−7.9%	−4.0%	−2.8%	−2.1%
IRR		694.7%	124.6%	36.6%			−26.8%	−55.5%	−87.4%	

purchases both shares, the relevant cash-flow stream of the combined deal will be

$$(-156, 44, 44, 44, 44, 44, 44, 44, 44, 44, -156)$$

and two IRRs will be generated. Specifically, the IRR equation is

$$-\frac{156}{1+x}+\sum_{t=2}^{9}\frac{44}{(1+x)^t}-\frac{156}{(1+x)^{10}}=0$$

which has two solutions: 14.05 and −12.32%.

Analyzing the ROSCA from the point of view of the organizing bank, its expected cash-flow stream boils down to the sequence (20, 20, 20, 20, 20, 20, 20, 20, 20, 20). Therefore, the IRR does not exist.

Example 9.15 ♦

(*Upfront payment*) The concessionaire of a highway announces the plan for rehabilitation and repair of a stretch of motorway. Work will start in one year and will cost \$3,000 (in thousands) a year for three years, to be paid at the end of each year. Toll charges will be immediately increased by \$0.1. We assume the concession will end after 10 years and that the current traffic amounts to 40 thousands cars a day (expected to remain stable). Thus, the firm will collect an additional amount of \$4 thousands (0.1 × 40) a day, which means \$1,460 thousands a year for 10 years, starting from the end of this year. Therefore, the vector of cash flows is

$$\boldsymbol{F}=(0, 1{,}460, -1{,}540, -1{,}540, -1{,}540, 1{,}460, 1{,}460, 1{,}460, 1, 460, 1{,}460, 1{,}460).$$

Barring the initial zero (which is irrelevant), the sign pattern is (+, − − −, + + + + ++) which makes it likely that the project possesses zero or more than one solution. Indeed, the IRR equation,

$$\frac{1}{1+x}\left(1,460-1,540\cdot\frac{1-(1+x)^{-3}}{x}+1,460\cdot\frac{1-(1+x)^{-6}}{x}\cdot\frac{1}{(1+x)^3}\right)=0,$$

has no solution in the relevant interval $(-1, +\infty)$.

Example 9.16 ♦

(*Equity perspective*) Consider the case where a project's CFA stream is $\boldsymbol{F} = (-100, 40, 40, 40)$. The project IRR is $\sigma = 9.7\%$. Suppose that 70% of the project is financed with a loan, repayable with a lump sum after three periods equal to $F_3^d = 91$. Hence, the debt's cash-flow stream is $\boldsymbol{F}^d = (-70, 0, 0, 91)$. Obviously, the debt's IRR exists and is unique, equal to $\sigma^d = 9.14\%$.[20] However, the equity stream is non-

[20] The debt's IRR is not necessarily the ROD. The ROD is a single-period concept (the ratio of the interest expense to the BOP debt), while the IRR is a multiperiod concept. ROD and debt's IRR coincide if and only if the ROD is time-invariant (a rather unrealistic occurrence unless the debt consists of one single liability bearing interest at a constant rate of interest).

conventional: $\boldsymbol{F}^e = (-30, 40, 40, -51)$. The IRR equation gives rise to two equity IRRs: 3.5% and 43.9%. The internal values associated with the IRRs are, respectively, $\boldsymbol{C}(3.5\%) = (30, -9, -49, 0)$ and $\boldsymbol{C}(43.9\%) = (30, 3, -35, 0)$. The first IRR signals a financing position in the second period ($C_1(3.5\%) < 0$) whereas the second IRR signals an investment position for the same period ($C_1(43.9\%) > 0$), thereby conveying contradictory information.

Example 9.17 ♦

(*Portfolio of projects*) A company undertakes three projects, say α, β, and γ, whose cash flow streams are

$$\begin{aligned} \boldsymbol{F}^{\alpha} &= (-1800, 1000, 500, 2300) \\ \boldsymbol{F}^{\beta} &= (-2050, 1200, 0, 1100) \\ \boldsymbol{F}^{\gamma} &= (2850, 1600, -5235, -1465). \end{aligned}$$

The (unique) IRRs are $\sigma^{\alpha} = 40.3\%$, $\sigma^{\beta} = 6.2\%$, $\sigma^{\gamma} = 24.5\%$. The IRR-implied capital is positive in each period for α and β: $\boldsymbol{C}^{\alpha} = (1800, 1525.1, 1639.5)$, $\boldsymbol{C}^{\beta} = (2050, 976.2, 1036.2)$. This signals that the IRRs of α and β describe them as (internally) pure investments. As for γ, the internal values are negative, $\boldsymbol{C}^{\gamma} = (-2850, -5148.9, -1176.5)$, which signals that it is interpreted, according to its IRR, as a pure financing.

Despite this (seemingly) clear situation, mixing the three transactions, a portfolio with multiple IRRs is generated. Precisely, the portfolio's cash-flow stream is $\boldsymbol{F} = \boldsymbol{F}^{\alpha} + \boldsymbol{F}^{\beta} + \boldsymbol{F}^{\gamma} = (-1{,}000,\ 3{,}800,\ -4{,}735,\ 1{,}935)$. It is easy to check that this portfolio gives rise to three IRRs: $\sigma^1 = 0\%$, $\sigma^2 = 24.2\%$, and $\sigma^3 = 55.8\%$. The portfolio's internal values, as implied by each IRR, alternate in sign:

$$\begin{aligned} \boldsymbol{C}^{\alpha+\beta+\gamma}(0) &= (1{,}000, -2{,}800, 1{,}935.0) \\ \boldsymbol{C}^{\alpha+\beta+\gamma}(0.242) &= (1{,}000,\ -2{,}558.1,\ 1{,}558.1) \\ \boldsymbol{C}^{\alpha+\beta+\gamma}(0.558) &= (1{,}000,\ -2{,}241.9,\ 1{,}241.9). \end{aligned}$$

This means that, according to an IRR-based approach, this is a mixed project with no clear capital amounts and no clear rate of return/cost.

Example 9.18 ♦

C^l | C^e

(*Principal-agent relationship*) In portfolio management, interim cash flows are the effect of decisions made by the investor. Specifically, the investor decides how much to contribute into the fund or distribute from the fund in each and every period. Large sways of contributions and distributions increase the likelihood of multiple IRRs. For example, consider an insurance company (principal) which has endowed a money management firm (agent) with \$400 (in thousands) at time 0. The company withdraws \$200 at the end of the first period and \$1,000 at the end of the third period while it contributes \$800 at the end of the fourth period, respectively. The investment is liquidated at the end of the fifth period. Assume that the growth rates of the fund's value are $\boldsymbol{i} = (10\%, 100\%, 150\%, -50.0\%, -91.33\%)$, thereby signaling a sharp increase in the first three periods and a sharp decrease in the last two periods. Using the retrospective relation Eq. (1.11), the capital at the beginning of the fifth period is

$$C_4^l = 400 \cdot (1.1)(2)(1.5)(0.5) - 200 \cdot (2)(1.5)(0.5) + 0 \cdot (1.5)(0.5) \\ - 1,000 \cdot (0.5) + 800 = 900.$$

Therefore, at the end of the fifth period, the fund's value is $E_5 = 900 \cdot (1 - 0.913) = 78$ whence the last cash flow is $F_5 = E_5 = 78$. The concise transposed Matrix is as follows:

$(\boldsymbol{C}, \boldsymbol{I}, \boldsymbol{F})$

Time		0	1	2	3	4	5
Capital	$\boldsymbol{C}^l$	400	240	480	200	900	0
Income	$\boldsymbol{I}^l$	0	40	240	720	−100	−822
Cash Flow	$\boldsymbol{F}^l$	−400	200	0	1,000	−800	78

The cash-flow stream produces three IRRs, $\sigma^1 = 17.9\%$, $\sigma^2 = -27.2\%$, $\sigma^3 = -88.6\%$, and it is therefore unclear whether the rate of return of this transaction is positive or negative and whether it is indeed an investment or, rather, a financing. Yet, a quick glance at the capital stream makes it clear that the investor has invested money. In the AIRR approach, PV[C] will unambiguously be positive, so AIRR will correctly signal the investment nature of this (and any other) transaction.

Example 9.19 ♦

(*Spillover*) A company must decide whether a new product should be merchandized. The product life is expected to be 5 years and the new product requires investment in working capital by \$30. The amount will be recovered at the end of the fourth year. Increased sales are expected, as well as increased costs for this product. However, the sale of the new good will cannibalize a product already sold by the company. This will cause a reduction in the old product's sales and, at the same time, a reduction in its associated costs. Below are reported the amount of expected incremental sales and costs, taking account of the lost sales and the avoided costs. The corporate marginal tax rate is 35% and the (after-tax) WACC is $r^{uo} = 10\%$.

The FCF stream is $\boldsymbol{F}^{uo} = (-30, -20, 10, 40, -20, 0)$ which brings about a negative NPV. This means that the effect of cannibalization is too heavy and the addition of the new product to the existing range of products is not profitable. However, the analyst will not find how inefficient this project is if the IRR concept is used. The IRR equation

$$-30 - \frac{20}{1+x} + \frac{10}{(1+x)^2} + \frac{40}{(1+x)^3} - \frac{20}{(1+x)^4} = 0$$

has no solution. In contrast, as we know, the AIRR approach measures the project's inefficiency by comparing average ROI and WACC. One discounts the NOPATs at 10% and divide it by the present values of the NOWCs and easily get the project's average ROI: $\bar{\iota}_0^{uo} = -9.71\%$, which signals a loss of \$0.0971 out of any dollar invested in the project. The economic inefficiency is high, considering that the same amount might be invested in the market producing at an instantaneous $\bar{r}^{uo} = 9.09\% (= 0.1/1.1)$. Precisely, the project would lead to an underperformance of $\xi_0^{uo} = \bar{\iota}_0^{uo} - \bar{r}_0^{uo} = -18.8\%$.[21]

Year	0	1	2	3	4	5
Sales of new product		200.00	200.00	323.08	230.77	184.62
−Lost sales		−246.15	−200.00	−230.77	−184.62	−76.92
Incremental sales		−46.15	0.00	92.31	46.15	107.69
−Costs of new product		−46.15	−30.77	−61.54	−153.856	−200.00
+Cost savings		61.54	46.15	30.77	76.92	46.15
−Incremental costs		15.38	15.38	−30.77	−76.92	−153.85
EBIT		−30.77	15.38	61.54	−30.77	−46.15
−Taxes (35%)		10.77	−5.38	−21.54	10.77	16.15
NOPAT		−20.00	10.00	40.00	−20.00	−30.00
NOWC	30.00	30.00	30.00	30.00	30.00	0.00
−Change in NOWC	−30.00	0.00	0.00	0.00	0.00	30.00
FCF	−30.00	−20.00	10.00	40.00	−20.00	0.00

WACC	10.00%
NPV	−\$23.53
IRR	does not exist
PV[NOPAT]	−\$12.15
PV[NOWC]	\$125.1
average ROI	−9.71% (instantaneous)
	−10.69% (discrete)
MARR	9.09% (instantaneous)
	10.00% (discrete)

These figures show that the negative NPV is caused by a high inefficiency, one which is not captured by IRR (the value destruction would be even higher if, other things unvaried, the scale of the project were greater).

In general, the more complex a project, the higher the likelihood that spillovers have substantial effects, thereby augmenting in turn the likelihood of no IRR or multiple IRRs.

[21] In discrete terms, $\bar{\iota}^{uo} = -10.69\% = -9.71\% \cdot 1.1$ and the economic inefficiency is $\xi^{uo} = -20.69\% (= -10.69\% - 10\%)$.

Example 9.20 ♦

(*Cash-flow stream with no change in sign*) The Rainbow Educational Program, which will be illustrated in Chap. 12, consists of an educational project made by a brick-and-mortar bookstore. While we crossrefer to Sect. 12.3 for more details on the project, the estimates for the educational program are such that the owners of the bookstore are able to distribute positive payouts (full distribution policy) from year 4 through 7 with no previous disbursement, because the bookstore raises funds via internal financing. The project's concise transposed Matrix is

$(\boldsymbol{C}, \boldsymbol{I}, \boldsymbol{F})$ – Rainbow Educational Program

Time	0	1	2	3	4	5	6	7
C	0.0	−27,978.0	−47,477.8	−58,698.1	−47,182.6	−35,299.0	−23,036.1	0.0
I	0.0	−27,978.0	−19,499.8	−11,220.3	21,920.5	27,520.1	33,287.9	39,228.9
F	0.0	0.0	0.0	0.0	10,405.0	15,636.6	21,024.9	16,192.8

Clearly, the IRR does not exist. In such situations, the IRR concept lays bare its inadequacy.

Example 9.21 ♦

(*Equity cash-flow stream with no change in sign*) The equity cash-flow stream may have nonnegative cash flows if all negative cash flows of the project are covered

Table 9.5 Equity cash-flow stream with no change in sign

Time		0	1	2	3	4	5
Capital							
Operating assets	C^o	1,200.00	1,822.00	1,000.00	250.00	100.00	0.00
Non-operating assets	C^l	0.00	0.00	0.00	0.00	0.00	0.00
Debt	C^d	1,200.00	1,786.00	908.26	200.00	100.00	0.00
Equity	C^e	0.00	36.00	91.74	50.00	0.00	0.00
Income							
Operating income	I^o		72.00	109.32	60.00	15.00	6.00
Non-operating income	I^l		0	0	0	0	0
Interest expense	I^d		36.00	53.58	27.25	6.00	3.00
Net Income	I^e		36.00	55.74	32.75	9.00	3.00
Cash flow							
Cash flow from operations	F^o	−1,200.00	−550.00	931.32	810.00	165.00	106.00
Cash flow from non-operating assets	F^l	0.00	0.00	0.00	0.00	0.00	0.00
Cash flow to debt	F^d	−1,200.00	−550.00	931.32	735.51	106.00	103.00
Cash flow to equity	F^e	0.00	0.00	0.00	74.49	59.00	3.00

by debt. For example, consider a project with two initial capital expenditures equal to \$1,200 and \$550 in $t = 0$ and $t = 1$, respectively. Suppose the two outlays are covered by debtholders and let it be repaid with four cash flows, starting at $t = 2$, amounting to $F_2^d = 931.32$, $F_3^d = 735.51$, $F_4^d = 106$, $F_5^d = 103$. This means that the (assumed constant) interest rate is $i^d = 3\%$. Equityholders will be distributed the PDs (full payout policy) as soon as the project will start generating positive PDs. The transposed Matrix is reported in Table 9.5.

In the first three periods, $F_t^o = F_t^d$ so the CFE is zero. Starting from time $t = 3$, the CFO exceeds the CFD, and the PDs are fully distributed to equityholders. As a result, the CFE vector is nonnegative. The equity IRR does not exist.

9.5 Other Mishaps of IRR

Economists, mathematicians, finance scholars have been puzzled by the problems of existence and uniqueness mentioned in the previous section. They are a direct consequence of an unfortunate choice of Chisini's invariance requirement. This choice brings about other difficulties which have largely gone unnoticed by scholars and practitioners. Following is a non-exhaustive compendium (see also Magni 2013a).

Non-additivity of Internal Values

As we have seen, the IRR may be interpreted as a time-invariant income rate which self-produces the project's interim values. Once solved the IRR equation, the interim values are obtained with the following law of motion: $C_t(\sigma) = C_{t-1}(\sigma)(1 + \sigma) - F_t$. Unfortunately, this automatism in generating interim values breaks the additivity principle (i.e., the law of conservation). Specifically, consider a portfolio of assets/projects with cash-flow streams $\boldsymbol{F}^j$, $j \in \mathbb{N}_m^1$ and suppose they are conventional cash-flow streams so that they possess a unique IRR. Suppose that the portfolio of the m projects, denoted as $P = \sum_{j=1}^m P_j$, is itself described by a conventional cash-flow stream. Then, the portfolio's IRR exists and is unique. However, the portfolio's internal values are not equal to the sum of the individual assets' internal values. Therefore, there exists some t such that

$$\sum_{j=1}^{m} C_t^{P_j}(\sigma) \neq C_t^{\sum_{j=1}^m P_j}(\sigma). \tag{9.8}$$

In general, the inequality occurs for every $t \in \mathbb{N}^1_{n-1}$, equality being a mere happenstance (in the AIRR approach, the capital value of a portfolio, at a given date t, is equal to the sum of its constituent assets, by law of conservation: $C^P_t = C^{P_1}_t + C^{P_2}_t + \ldots + C^{P_n}_t$).

An implication of this is that the project's internal value, at time t, is not equal to the sum of the equity's internal value and the debt's internal value at the same date: $C^e_t(\sigma^e) + C^d_t(\sigma^e) \neq C_t(\sigma)$, whereas, in contrast, the AIRR approach gives no problem: $C^e_t + C^d_t = C_t$.

Example 9.22 ♦

Consider Table 9.6, where five projects' cash-flow streams are described. They are all projects with unique IRRs. Consider the portfolio of these projects; its cash-flow stream, denoted as P, has a unique IRR as well, equal to $\sigma^P = 5.5\%$. However, the portfolio's internal value is ambiguous for every $t \in \mathbb{N}^1_{n-1}$. Table 9.7 collects the interim values generated by the IRR of each cash-flow stream. The reader will note that the sum of the internal values of the individual assets is sharply different from the portfolio's internal values for every $t \in \mathbb{N}^1_{n-1}$. For example, at time 2, the internal value of the portfolio is either 748.1 or 657.9, respectively, depending on whether one derives it from the portfolio's IRR or from the sum of the IRR-implied values of each constituent asset.

Table 9.6 Portfolio of projects: Cash flows and IRRs

Time	F^{P_1}	F^{P_2}	F^{P_3}	F^{P_4}	F^{P_5}	F^P
0	−100	−50	−20	−250	−50	−470
1	50	10	0	−150	0	−90
2	−150	10	100	−50	−40	−130
3	−50	40	−120	300	0	170
4	40	20	60	0	50	170
5	0	0	10	0	500	510
IRR	−62.0%	18.3%	59.4%	−15.5%	53.9%	5.5%

Table 9.7 Portfolio of projects: Value non-additivity

Time	$C^{P_1}_t(\sigma)$	$C^{P_2}_t(\sigma)$	$C^{P_3}_t(\sigma)$	$C^{P_4}_t(\sigma)$	$C^{P_5}_t(\sigma)$	$C^P_t(\sigma)$	$\sum_{j=1}^5 C^{P_j}_t(\sigma)$
0	100	50	20	250	50	470	470.0
1	−12.0	49.1	31.9	361.2	76.9	585.9	507.1
2	145.4	48.1	−49.2	355.1	158.4	748.1	657.9
3	105.3	16.9	41.6	0.0	243.7	619.3	407.4
4	0.0	0.0	6.3	0.0	325.0	483.4	331.2
5	0.0	0.0	0.0	0.0	0.0	0.0	0.0

Table 9.8 BHP Billiton (90% payout) non-additivity of internal values

	INTERNAL VALUES			
Time	Equity	Debt	Equity + Debt	Assets
0	400.0	200.0	600.0	600.0
1	393.8	200.0	593.8	594.4
2	391.6	200.0	591.6	593.9
3	388.5	200.0	588.5	593.5
4	383.9	200.0	583.9	593.2
5	377.1	200.0	577.1	592.9
6	366.8	200.0	566.8	592.8
7	351.1	200.0	551.1	592.8
8	327.2	200.0	527.2	593.0
9	290.6	200.0	490.6	593.6
10	414.6	0.0	414.6	574.6
11	415.5	0.0	415.5	574.3
12	417.0	0.0	417.0	574.2
13	419.7	0.0	419.7	574.2
14	424.1	0.0	424.1	574.6
15	431.2	0.0	431.2	575.4
16	442.3	0.0	442.3	576.8
17	459.8	0.0	459.8	579.0
18	487.1	0.0	487.1	582.4
19	529.3	0.0	529.3	587.4
20	0.0	0.0	0.0	0.0
PV[$C^j(\sigma)$]	2,798.6	1,449.4	4,248.0	4,422.5

Example 9.23 ♦

C^o	C^d
	C^e

(BHP Billiton, 90% payout, IRR) Consider BHP Billiton project, as described in Example 4.13. The equity IRR is $\sigma^e = 53.9\%$, and the debt IRR is $\sigma^d = i^d = 8.0\%$. The project's IRR associated with the CFA stream is $\sigma = 38.7\%$. Table 9.8 collects the debt's internal value and the equity's internal value. For every $t > 0$, the project's internal value calculated from the project IRR is different from the value obtained by summing the debt's internal value and the equity's internal value. (The last line of the table shows the problem of global non-additivity. See Example 9.37 for details.)

Time-varying COCs

The IRR criterion does not contemplate the possibility of time-varying COCs, so an IRR, σ, may not be compared with a vector $\boldsymbol{r} = (r_1, r_2, \ldots, r_n)$ of COCs. This criticism was first addressed by such scholars as Hirshleifer (1958) and Bailey (1959). While it is usual in corporate finance to use a constant COC, this is not always appropriate, for two reasons: (i) the term structure of interest rate may be non-flat (e.g., the risk-free rate is time-varying), and/or (ii) the risk of a project may change during its life and, therefore, the risk premium, π_t, may be time-varying. Furthermore, in investment performance measurement, where it is usual to match ex post performance of a given portfolio with a benchmark portfolio, the case of a constant benchmark rate is highly unrealistic. Notwithstanding its importance, few textbooks mention this pitfall, with some notable exceptions (e.g., Brealey et al. 2011, pp. 113-14, pp. 230–233; Pike et al. 2012, p. 98).

This issue is even more compelling if one considers that, even if the project WACC, r, is constant, the cost of equity, r_t^e, is time-varying. Indeed, if r is constant, then the equity COC is equal to

$$r_t^e = r + \frac{V_{t-1}^d}{V_{t-1}^e}(r - r_t^d)$$

(see Eq. (6.22)). If the debt/equity ratio V_{t-1}^d/V_{t-1}^e is constant for all t and the cost of debt, r_t^d, is constant for all t, then the cost of equity is constant. If one of these conditions are not satisfied, as is usual in capital asset investments,[22] the cost of equity is time-varying, which makes the equity IRR not comparable with the stream $\boldsymbol{r}^e = (r_1^e, r_2^e, \ldots, r_n^e)$.

A possible remedy is to use the IRR concept and carve out a constant COC from the equation

$$\overbrace{\sum_{t=1}^{n} \frac{F_t}{\prod_{j=1}^{t}(1+r_j)}}^{V_0} = \sum_{t=1}^{n} \frac{F_t}{(1+x)^t}. \tag{9.9}$$

However, this equation does not necessarily have a solution or multiple solutions may arise. In Chap. 8 we have shown that the correct approach to find the MARR is to calculate the weighted mean of the COCs.[23]

Example 9.24 ♦

C^o	C^d
	C^e

Consider Mockup project with 100% payout. We have already computed the CFEs (see Example 6.9), namely, $\boldsymbol{F}^e = (-410{,}000,\ 117{,}686,\ 121{,}145,\ 124{,}690,\ 128{,}324,\ 107{,}049)$. The IRR equation is then

$$-410{,}000 + \frac{117{,}686}{1+x} + \frac{121{,}145}{(1+x)^2} + \frac{124{,}690}{(1+x)^3} + \frac{128{,}324}{(1+x)^4} + \frac{107{,}049}{(1+x)^5} = 0.$$

[22] Even assuming that $r_t^d = r^d$ is constant, the debt-to-equity ratio (V_{t-1}^d/V_{t-1}^e) will most frequently be time-varying.

[23] The AIRR approach may help technically solve this problem by computing the arithmetic mean of the COCs, weighted by the discounted internal capitals. A net investment is worth undertaking if and only if this mean is smaller than the IRR (see Magni 2013a, pp. 105–108).

The (unique) IRR is $\sigma^e = 14.22\%$. We have also computed the cost of equity, period by period: $\boldsymbol{r}^e = (10.2\%, 10.8\%, 11.9\%, 14.2\%, 22.0\%)$. The first three COCs are smaller than the IRR, the fourth COC is almost the same as the IRR and the fifth COC is greater than the IRR. The IRR criterion is not applicable. (To make it applicable, one needs to use the AIRR approach. See also footnote 23).

Signaling Erroneous Financial Position

The IRR may signal that a project is a mixed transaction when, in fact, it is a pure investment or a pure financing. In particular, it may signal a financing position when, in actual facts, it is not.

Indeed, it may well occur that the firm is investing resources in a project and, yet, the IRR generates internal capitals that are negative, thereby erroneously signaling that the firm absorbs resources from the project. To illustrate, we make use of the following simple example.

Example 9.25 ♦

	C^d
C^l	C^e

Consider a three-period investment in a portfolio of financial assets made by an investor. The firm has invested $C_0^l = 10,000$ at time 0. Assume that the investment's value increases by 80% in the first year. At the end of the year, the firm withdraws \$15,000 from the portfolio. In the second year, the investment loses 10% of its value; at the end of that year, the firm additionally deposits \$18,000. In the third year, the investment's value grows by 5% and the firm liquidates the investment. This investment implies the following returns: $I_1^l = \$8,000$, $I_2^l = -\$300$, and $I_3^l = \$1,035$. The concise transposed Matrix is

$(\boldsymbol{C}^l, \boldsymbol{I}^l, \boldsymbol{F}^l)$

Time		0	1	2	3
Capital	$\boldsymbol{C}^l$	10,000	3,000	20,700	0
Income	$\boldsymbol{I}^l$	0	8,000	−300	1,035
Cash Flow	$\boldsymbol{F}^l$	−10,000	15,000	−18,000	21,735

The (unique) IRR for this cash-flow stream is $\sigma = 35.52\%$. If the IRR is a meaningful rate of return, then the rate of change of the capital is indeed a constant 35.52%. However, if this were true, the BOP capital at time 1 would be negative and equal to $C_1(\sigma) = 10{,}000(1 + 35.52\%) - 15{,}000 = -1{,}448$ (this value is the only compatible value with a growth rate of 35.52%). But this would imply that, in the second period, the firm borrows \$1,448. Yet, in the second period the firm does not borrow at all; it invests \$3,000 in the market: $C_1 = 3{,}000$. Therefore, the IRR forces the economic asset to have capital values consistent with the IRR itself, thereby departing from the true economic transactions involved in the investment and the actual values. Therefore, 35.52% is not an appropriate rate of return for this project.

As we know, with the AIRR approach, the actual capital values are explicitly used, with no fictitious assumption on their values. Assuming, for mere simplicity, that the COC is constant and equal to $r = 10\%$ and, using the shortcut,

$$\bar{\imath}^l = \overbrace{0.1}^{r} + \frac{\overbrace{-10,000 + \frac{15,000}{1.1} - \frac{18,000}{1.1^2} + \frac{21,735}{1.1^3}}^{\text{NPV}}}{\underbrace{10,000 + \frac{3,000}{1.1} + \frac{20,700}{1.1^2}}_{\text{PV}[C]}} \cdot 1.1 = 28.8\%.$$

Improper Capital Amounts

Even in those cases where the IRR correctly signals the investor's financial position (lending vs. borrowing), the magnitude of the capital amount is, in general, inappropriate, for an economic asset does not grow at a constant force of interest, unless the growth rate is explicitly enforced to be time-invariant. Certainly, this is not the case for capital asset investments or for going concerns such as firms. As a result, the internal values express an unrealistic pattern of capital depreciation which has no explicit link with the estimated data of a project (let alone with the transactions involved in the project). Nor does it have any relation with intrinsic/economic depreciation, that is, with the change in value as determined by investors in a normal market. The IRR concocts, so to say, its own built-in values, and, therefore, it is improper to interpret it as the return on the capital which is actually invested in the project. Rather, it may be interpreted, as already seen, as the rate of return of a project which has the same cash flows as the project at hand and whose capital grows at a constant income rate, or as the hypothetical return the project would have if its capital grew at a constant rate of return (see also Magni and Peasnell 2012, 2015; Altshuler and Magni 2012, 2015 on this point).

Example 9.26 ♦
Consider the investment of \$100 in a single-period project whose income rate is 40%, followed after one period by another investment of \$250 in a single-period project whose income rate is 5%, which is in turn followed after another period by a final investment of \$90 in a single-period project with income rate 50%. The cash-flow streams of the three investments are $\boldsymbol{F}^1 = (-100, 140, 0, 0)$, $\boldsymbol{F}^2 = (0, -250, 262.5, 0)$, and $\boldsymbol{F}^3 = (0, 0, -90, 135)$. Considering the entire span of three periods, the cash-flow stream is

$$\boldsymbol{F} = \boldsymbol{F}^1 + \boldsymbol{F}^2 + \boldsymbol{F}^3 = (-100, -110, 172.5, 135)$$

(this cash-flow stream may well be associated with a single project with phased expansion) and the interim values are equal to the cash flows invested in the three one-period projects:

$$\begin{aligned} C_0 &= 100 \\ C_1 &= 100 \cdot 1.4 + 110 = 250 \\ C_2 &= 250 \cdot 1.05 - 172.5 = 90. \end{aligned}$$

The (unique) IRR is $\sigma = 22\%$ and, assuming the MARR is $r = 10\%$ the decision triggered by the IRR criterion is correct. In this case, the internal values correctly signal that the firm finds itself in a lending position in each period. However, the IRR devises its own internal values, which do not seem to have to do with the actual invested capital amounts: $\boldsymbol{C}(22\%) = (100, 232, 110.6, 0)$. The interim values at time 1 and 2 seem to distort the actual representation of the projects, because, at those dates, the investor injects \$250 $\neq$ \$232 and \$90 $\neq$ \$110.6 in the second and third period, respectively. Rather, the internal values are the hypothetical capital amounts that the investor would earn if the project's value grew at a constant rate of return.

Example 9.27 ♦

C^d C^l C^e Consider Example 9.25. As seen, the project IRR is 35.52%, so the sequence of internal values concocted by the IRR is $\boldsymbol{C}(0.3552)$=(10, 000, −1, 448, 16, 037.95). Apart from the issue of a financing position incorrectly signalled by the second internal capital, $C_1(0.3552)$, the other interim capital amount is not tied to the project's actual economic referents and, therefore, it is not related with the mechanics of the project. Only if the capital amounts had grown at a constant rate of return, then the internal values would be the appropriate capital values and IRR would be the appropriate rate of return.

Example 9.28 ♦

C^d C^l C^e Consider a fund whose value at time 0 is \$1,000. Suppose it grows by 20% in the first period so that the end-of period market value is \$1,200. The fund manager invests this amount in the second period in a portfolio of assets. Unfortunately, the capital invested decreases by −40%, so only 60% of the \$1,200 (= \$720) remains invested in the fund. In the third period, the capital grows by 70%, which implies that the market value of the fund at time 3 is \$1,224. We assume that, at time 3, the investor liquidates the fund. This trivial investment may be described in terms of the following concise transposed Matrix:

$(\boldsymbol{C}^l, \boldsymbol{I}^l, \boldsymbol{F}^l)$

Time		0	1	2	3
Capital	$\boldsymbol{C}^l$	1, 000	1, 200	720	0
Income	$\boldsymbol{I}^l$	0	200	−480	504
Cash Flow	$\boldsymbol{F}^l$	−1, 000	0	0	1, 224

However, under an IRR approach, the same deal is described as an investment of \$1,000 growing at a constant 6.97% (solution of $1{,}000 + 1{,}224(1+x)^{-3} = 0$) so that the investment's value is described as being equal to $C_1(0.0697) = 1{,}000(1.0697) = 1{,}070$ and $C_2(0.0697) = 1{,}070(1.0697) = 1{,}144$. Therefore, the sequence of internal values contradicts the true asset's values.

Example 9.29 ♦
The deviation of IRR from AIRR may be considerable, owing to a substantial deviation of the internal capital from the actual capital employed by the investor. For example, consider the mixed transaction in Example 8.18. Its cash-flow stream is $\boldsymbol{F} = (-2, 20, -5, -75, 70)$. We remind that the vector of interest rates applied to the BOP balances is $\boldsymbol{i} = (16\%, 13\%, 8\%, 19\%)$. The AIRR has been shown to be equal to $\bar{\imath} = 35.92\%$. However, the (unique) IRR is $\sigma = 834.6\%$, which may be interpreted as the rate of growth of the project's capital if it were enforced to increase at a constant rate of return (which, in fact, is not). This value seems nothing more than a mathematical artifact and brings about fictitious capitals which are substantially lower, in absolute value, than the actual BOP capitals: $\boldsymbol{C}(8.346) = (2, -1.31, -7.22, 7.49, 0)$, as opposed to $\boldsymbol{C} = (2, -17.68, -14.98, 58.82, 0)$. With an assumed MARR vector equal to $\boldsymbol{r} = (21\%, 10\%, 16\%, 12\%)$, the two metrics are associated with a capital base equal to $\text{PV}[C(\sigma)] = 0.343$ and $\text{PV}[C] = 14.23$, respectively. The respective excess rates are then

$$\xi(0.343) = \frac{\text{NPV}}{\text{PV}[C(\sigma)]} = \frac{2.68}{0.343} = 781.3\%$$
$$\xi(14.23) = \frac{\text{NPV}}{\text{PV}[C]} = \frac{2.68}{14.23} = 18.83\%.$$

Use of IRR in Association with the NPV Profile

Some believe that the IRR may reliably provide insights on value creation if it is used in association with the profile of the DCF function, $\phi(x) = \sum_{t=0}^{n} F_t(1+x)^{-t}$. However, this is not a general case. For example, consider the following project:

$(\boldsymbol{C}, \boldsymbol{I}, \boldsymbol{F})$

Time		0	1	2
Capital	$\boldsymbol{C}$	350	80	0
Income	$\boldsymbol{I}$	0	230	−260
Cash Flow	$\boldsymbol{F}$	−350	500	−180

The DCF function is negative for every discount rate: $\phi(x) = -350 + 500/(1+x) - 180/(1+x)^2 < 0$ for every $x > -1$. The IRR does not exist because the NPV graph lies below the horizontal axis. As the NPV is negative no matter what the discount rate is, it seems that the project is a value-destroying one. However, suppose that the COC is time-varying and equal to $\boldsymbol{r} = (1\%, 30\%)$. This implies $\text{NPV} = -350 + 500(1.01)^{-1} - 180[(1.01)(1.3)]^{-1} = 7.96$, which means that value is created. Therefore, not even the IRR in association with NPV profile is always reliable, for the mere fact that the DCF function $\phi(x)$ is not itself reliable in general.

As a matter of fact, this is a criticism of the use of the traditional DCF function as well as of the use of IRR. Instead of $\phi(x)$, one should more properly study the function

$$\phi(x_1, x_2, \ldots, x_n) = \sum_{t=0}^{n} \frac{F_t}{\prod_{j=1}^{t}(1 + x_j)}$$

which may not be plotted on a graph if $n > 2$.

With the AIRR approach, the rate of return and the MARR is easily computed. Using, for example, the instantaneous AIRR:

$$\bar{\imath}_0 = \frac{\text{PV}[I]}{\text{PV}[C]} = \frac{29.7}{429.21} = 6.92\% \qquad \bar{r}_0 = \frac{\text{PV}[I^*]}{\text{PV}[C]} = \frac{21.74}{429.21} = 5.06\%.$$

The situation clearly defines a financial efficiency equal to $\xi = 1.85\%$. Since the scale is 429.21, the economic value created is $429.21 \cdot 0.0186 = 7.98$, as expected.

Ex Post Performance and Sensitivity Analysis

Assessment of ex post performance of a project consists of measuring the deviations of a given metric (NPV, NFV, RI, AIRR, etc.) with respect to the expected value calculated ex ante on the basis of the input data. Let $\boldsymbol{F}^{\text{ex ante}}$ be the ex ante project's cash-flow stream; even if it enjoys a unique IRR, the ex post cash-flow stream, $\boldsymbol{F}^{\text{ex post}}$, may result in no IRR or multiple IRRs, making it impossible to compare the expected value of the IRR with the actual, realized IRR.

The same problems may take place in risk management whenever, ex ante, the analyst considers one or more different scenarios, in order to assess the impact of a deviation of the base-case parameters on the project IRR. This analysis is called *sensitivity analysis*: It measures the impact of uncertainty of one or more inputs of a model upon the model output and investigates how the output variance can be apportioned to the various inputs (see Saltelli et al. 2004. See also Pianosi et al. 2016 for a review). Sensitivity analysis has been conducted in recent years on various economic measures including IRR, NPV, and AIRR (see Borgonovo and Peccati 2006; Borgonovo et al. 2010; Percoco and Borgonovo 2012; Marchioni and Magni 2018).[24] Even if $\boldsymbol{F}^{\text{ex ante}}$ possesses a unique IRR in the base case, a perturbation of one or more inputs may well cause the DCF function of a different scenario change in such a way that no IRR exists or multiple IRRs arise (see Marchioni and Magni 2016). This makes it impossible to derive significant information on the sensitivity of the project's economic profitability to changes in the input data. (These kinds of problems do not arise with AIRR.)

[24] See Joy and Bradley (1973) and Whisler (1976) for less recent contributions on this issue.

Example 9.30 ♦
Consider a natural-resource project and let $\boldsymbol{F}^{\text{ex ante}} = (-750, 450, 350, 360, 50)$ be the estimated cash-flow stream. Suppose that expectations are fulfilled but, at time 4, the firm is requested to pay remedial costs equal to \$500. The realized cash-flow stream is then $\boldsymbol{F}^{\text{ex post}} = (-750, 450, 350, 360, -450)$. The analyst would like to know what impact this event has had on IRR. However, the IRR of the expected cash-flow stream is 28.52%, while the IRR of the ex post cash-flow stream does not exist. Ex post analysis cannot be conducted.

Now, consider a different case. Let $\boldsymbol{F}^{\text{ex ante}} = (-800, 1{,}000, -100, -200, 700)$ be a project's ex ante cash-flow stream. Suppose that the ex post case is described as $\boldsymbol{F}^{\text{ex post}} = (-800, 2{,}360, -940, -2{,}350, 1{,}750)$: Both inflows and outflows have turned out to be much greater than expected. What is the impact on the IRR? The (unique) ex ante IRR is $\sigma = 36.72\%$, while three ex post IRRs arise: $\sigma^1 = 8.07\%$, $\sigma^2 = 25\%$, and $\sigma^3 = 61.93\%$. It is then unclear whether ex post IRR is greater or smaller than ex ante IRR.

Project's Operating Life[25]

The existence of a project is evidenced empirically by the existence of an ongoing operating structure (plants, equipment, working capital, labor), which has relation with several categories of economic agents (suppliers, customers, government, etc.), not by the appearance of cash flows. The IRR approach, since it is based on the (nonzero) cash flows, overlooks the operating and economic life of a project.

A simple example may help understand. Suppose two firms undertake two different projects at time 0. The operations of firm A's project stop at time n, while the operations of firm B's project continue for other m periods, so the operating activity cease at time $n+m$. Suppose the two projects have the same cash flows up to time n; from time $n+1$ to time $n+m$, firm B's project generates an income that is equal to the change in capital value: that is, $I_t = \Delta C_t$. This means that cash flows are zero in the last m periods: $F_t = I_t - \Delta C_t = 0, t \in \mathbb{N}^n_{n+m}$. The respective cash-flow streams are then

$$\boldsymbol{F}^A = (F_0^A, F_1^A, \ldots, F_n^A) \quad \text{and} \quad \boldsymbol{F}^B = (F_0^B, F_1^B, \ldots, F_n^B, \underbrace{0, 0, \ldots, 0}_{m \text{ zeros}}).$$

The IRR approach approach disregards the operating activity of firm B's project in the last m periods because the cash flows are zero; the analyst who uses the IRR approach to measure the economic efficiency implicitly assumes that the operating activity of project B stops at time n and nothing economically significant occurs between n and $n+m$. The internal capital, $C_t(\sigma)$, is zero between time n and time $n+m$ so, allegedly, no capital is invested and no return is generated: $\sigma \cdot C_t(\sigma) = \sigma \cdot 0 = 0$ for $t > n$. In other words, the IRR "forgets", so to speak, that the activities of firm B's project continue: The firm uses labor, materials, plants, equipment, to sell goods,

---

[25] This paragraph and Example 9.31 are adapted from Magni (2013a), The Internal Rate of Return Approach and the AIRR paradigm: A refutation and a corroboration, *The Engineering Economist*, 58(2), 73–111 (p. 82 and p. 100).

to pay suppliers; in other words, the firm keeps on with the project for other m periods during which it invests, in each period, some capital which generates a return. Therefore, there is a clash between the IRR approach to rate of return and the operations and economic transactions that take place in actual facts. The IRR takes account of cash flows but neglects the actual returns that derive from the operations and the economic transactions associated with them. The AIRR does not suffer from this problem, since it takes account of the incomes and the capitals until the end of the project, regardless of whether cash flows are zero or nonzero.

Example 9.31 ♦

Consider two firms, A and B, incorporated for undertaking a project. Firm A ceases operations at the end of the second year, firm B ceases operations at the end of the fifth year. Suppose A and B have interim capitals equal to $\boldsymbol{C}^A = (100, 70, 0)$ and $\boldsymbol{C}^B = (100, 70, 40, 50, 70, 0)$, respectively, while the profits are $\boldsymbol{I}^A = (0, 35, -10)$ and $\boldsymbol{I}^B = (0, 35, 30, 10, 20, -70)$. The single-period return rates of firms A and B are $\boldsymbol{i}^A = (35\%, 14.3\%)$ and $\boldsymbol{i}^B = (35\%, 42.9\%, 25\%, 40\%, -100\%)$.

The two projects/firms are clearly different in many respects, namely, in the structure of the operations, in the earning power, in the scale of the operations, in the lengths. Yet, the IRR treats them as if they were the same project, for the only reason that they (seem to) have the same cash flows: $\boldsymbol{F}^A = (-100, 65, 60)$, $\boldsymbol{F}^B = (-100, 65, 60, 0, 0, 0)$. Under an IRR perspective, it is as if firm B did not operate in the last three periods, because investors do not receive or invest any more cash flow. Therefore, the rate of return would be $\sigma = 16.5\%$ for both firms. However, in the interval [2, 5] firm B does operate by deploying specific economic resources that generate specific returns (ones which are easily processed by the AIRR approach).

Framing Effect: Present Value Versus Future Value[26]

A rational decision maker is one whose decisions comply with the principle of description invariance, according to which a decision does not change if a decision problem is presented in different (but logically equivalent) formats. Violations of such a principle creates *cognitive illusions* known as "framing effects" (Tversky and Kahneman 1981; Kahneman and Tversky 1984). In this respect, the IRR equation is a present-value equation, that is, all elements of the equation are referred to time 0. However, the future-value format of the same equation was taken into account by Boulding (1935) and may well be considered by any analyst. In principle, one expects that the solution of the equation should not change under changes in the framing, namely, the IRR should not change if cash flows are referred to any date $T > 0$. Multiplying both sides of $\sum_{t=0}^{n} F_t(1+x)^{-t} = 0$ by $(1+x)^T$, one gets the future-value equation $\sum_{t=0}^{n} F_t(1+x)^{T-t} = 0$. If $T > n$, the solution $\sigma = -1$ holds whatever the cash flows, which signals the loss of 100% of the invested capital, *whatever the project* the firm will undertake. Further, if $F_0 < 0$ and $F_t = 0$ for $t \in \mathbb{N}_n^1$ (i.e., the investor loses the entire capital initially invested) the present-format IRR

---

[26]This paragraph and Example 9.32 are adapted from Magni (2013a), The Internal Rate of Return Approach and the AIRR paradigm: A refutation and a corroboration, *The Engineering Economist*, 58(2), 73–111 (pp. 79–80).

equation becomes $F_0 + \sum_{t=1}^{n} 0/(1+x)^t = 0$, which has no solution. As a result, IRR either never or always signals a loss of the entire capital, depending on how the equation is framed.

From this point of view, the definition of AIRR is unambiguous:

$$\bar{i} = \frac{\sum_{t=1}^{n} I_t(1+r)^{-(t-1)}}{\sum_{t=1}^{n} C_{t-1}(1+r)^{-(t-1)}} = \frac{\sum_{t=1}^{n} I_t(1+r)^{T-(t-1)}}{\sum_{t=1}^{n} C_{t-1}(1+r)^{T-(t-1)}}$$

for every $T \in \mathbb{R}$, and the rate of return coherently depends on the incomes generated and the invested capitals.

Example 9.32 ♦
Consider the cash-flow stream $\boldsymbol{F} = (-200, 0, 0, 266.2, 0)$. From $-200 + 266.2/(1+x)^3 = 0$, one obtains $\sigma = 10\%$ as the unique solution; from $-200(1+x)^4 + 266.2(1+x) = 0$ one gets two solutions, $\sigma^1 = 10\%$ and $\sigma^2 = -100\%$. If, instead, $\boldsymbol{F} = (-200, 0, 0, 0, 0)$, the IRR will or will not signal the loss of the invested capital, depending on whether the future-value or the present-value framing is used. For example, for $T = 1$, the future-value equation is $-200(1+x)^4 + \sum_{t=1}^{4} 0 \cdot (1+x)^{4-t} = 0$ whose solution is $\sigma = -1$; in contrast, the present-value equation, $-200 + \sum_{t=1}^{4} 0/(1+x)^4 = 0$, has no solution.

Example 9.33 ♦
Consider a project whose initial cost is \$100. At the end of the first period, the cash flow is \$120. The firm continues operations until the end of the second year, and the terminal cash flow is zero. Then, the IRR future-value equation, $-100(1+x)^2 + 120(1+x) + 0 = 0$ provides two solutions: $\sigma^1 = -100\%$ and $\sigma^2 = 20\%$. Which one of the two is correct?[27]

Framing Effect: Expected Value of Stochastic IRR Versus IRR of Expected Value of Stochastic Cash-Flow Stream

Another ambiguity regards the way the IRR may be computed starting from the (uncertain) cash flows. Whenever the cash-flow stream is stochastic, as is the case of capital asset investments, the IRR is an *expected* IRR. However, such an expected IRR depends on how IRR is computed. To understand this statement, let the vector $\tilde{\boldsymbol{F}} = (F_0, \tilde{F}_1, \tilde{F}_2, \ldots, \tilde{F}_n)$ be the stochastic cash flow stream of a project (F_0 may be assumed to be nonrandom). Let $\mathbb{E}(\tilde{F}_t) = F_t$ be the expected value of $\tilde{F}_t$ so that

$$\boldsymbol{F} = (F_0, F_1, F_2, \ldots, F_n) = \mathbb{E}(\tilde{\boldsymbol{F}}) = (F_0, \mathbb{E}(\tilde{F}_1), \mathbb{E}(\tilde{F}_2), \ldots, \mathbb{E}(\tilde{F}_n))$$

---

[27] The answer is: None of them. The project's rate of return is the weighted mean of the income rates that are generated in the two periods. For example, if the vector of capitals is $\boldsymbol{C} = (100, 50, 0)$, then the vector of incomes is $\boldsymbol{I} = (0, 70, -50)$ and the holding period rates are $i_1 = 70\%$ and $i_2 = -100\%$. This means that, in the first period, \$100 is invested and the capital grows by $i_1 = 70\%$. In the second period, the firm loses the whole capital that remains invested in the project, so the period rate of return is $i_2 = -100\%$. An appropriate rate of return is a mean of these income rates, properly weighted by the respective capitals (i.e., the average ROI in the AIRR approach).

denotes the expected value of $\tilde{F}$ and let $\tilde{\sigma}(\tilde{F})$ be the (stochastic) IRR of $\tilde{F}$. Therefore, the rate $\sigma(\mathbb{E}(\tilde{F}))$ is the IRR of the vector of expected values of $\tilde{F}$ and the rate $\mathbb{E}(\tilde{\sigma}(\tilde{F}))$ is the expected value of the IRR of $\tilde{F}$. However, since it is generally the case that $\sigma(\mathbb{E}(\tilde{F})) \neq \mathbb{E}(\tilde{\sigma}(\tilde{F}))$ it is not clear which the correct order of computation is. Does it make sense first to form expectations of future cash flows and then compute IRR? Alternatively, is it better first to compute the IRR of each possible stream of cash flows and then take the expectation of these possible IRRs? The IRR can be different depending on the choice (see also "Jensen's Inequality" in Brown 2006, pp. 197–199).

Example 9.34 ♦
Consider a 4-year project whose stochastic income, $\tilde{I}_t$, is based on three different scenarios, namely the base-case state, the pessimistic state, and the optimistic state. Table 9.9 illustrates the various incomes at the four dates for each state of nature. Suppose the investment cost is \$100 and the assets depreciate uniformly: $C_0 = 100$, $C_1 = 75$, $C_2 = 50$, $C_3 = 25$. The cash flows are then computed from the income as $\tilde{F}_t = \tilde{I}_t - \Delta C_t$. For example, at time 1, the cash flow in the pessimistic state is $21 = -4 + 25$. Table 9.10 shows the cash flows for each scenario and the two framings for the IRR. If the IRR is computed for each scenario and then expectation is taken, the resulting IRR is

$$\mathbb{E}(\tilde{\sigma}(\tilde{F})) = 0.3 \cdot 51.3\% + 0.4 \cdot 15.3\% + 0.3 \cdot (-7.6\%) = 19.2\%;$$

Table 9.9 Framing effect: Stochastic income

Probability	Scenario	0	1	2	3	4
0.3	optimistic		29	47	65	11
0.4	base		5	15	25	−5
0.3	pessimistic		−4	3	10	−25

Table 9.10 Framing effect: Stochastic cash flows and expected IRR

	Probability	Scenario	0	1	2	3	4	IRR	Expected IRR
$\tilde{F} = \{$	0.3	optimistic	−100	54	72	90	36	51.3%	
	0.4	base	−100	30	40	50	20	15.3%	$\mathcal{E}(\tilde{\sigma}(\tilde{F})) = 19.2\%$
	0.3	pessimistic	−100	21	28	35	0	−7.6%	
		$\mathcal{E}(\tilde{F})$	−100	34.5	46	57.5	18.8		$\sigma(\mathcal{E}(\tilde{F})) = 21.6\%$

Table 9.11 Framing effect: Stochastic ROI

Probability	Scenario	0	1	2	3	4
0.3	optimistic		29%	62.7%	130%	44%
0.4	base		5%	20%	50%	−20%
0.3	pessimistic		−4%	4%	20%	−100%

if, instead, the expected value of the cash-flow stream is computed first and then its IRR is calculated, then the resulting IRR is $\sigma(\mathbb{E}(\tilde{\boldsymbol{F}})) = 21.6\%$. Which is the "true" IRR?

In the AIRR approach, no such framing effect arises, for the order of computations is irrelevant. Specifically, in this case, one starts from the random ROIs of each period, displayed in Table 9.11. Assuming $r = 10\%$, the vector of the four weights, $w_t = C_{t-1}(1+r)^{-(t-1)}/\text{PV}[C]$, is easily calculated: $\boldsymbol{w} = (43.8\%, 29.9\%, 18.1\%, 8.2\%)$. Hence, one computes the average ROI for each scenario, getting the random average ROI, described by the vector

$$(\overbrace{58.6\%}^{\text{optimistic}}, \overbrace{15.6\%}^{\text{base}}, \overbrace{-5.2\%}^{\text{pessimistic}}).$$

Weighting each random ROI by the respective probability, one gets the expected average ROI:

$$\bar{\imath} = 0.3 \cdot 58.6\% + 0.4 \cdot 15.6\% + 0.3 \cdot (-5.2\%) = 22.3\%.$$

Now, let us revert the order of computations. First, one multiplies, for each period, the probability of each scenario by the respective random ROI (obtained, as usual, by dividing the income by the BOP capital). This results in four expected ROIs:

$$\begin{aligned}
i_1 &= 9.5\% &&= 0.3 \cdot (29\%) + 0.4 \cdot (5\%) + 0.3 \cdot (-4\%)\\
i_2 &= 28\% &&= 0.3 \cdot (62.7\%) + 0.4 \cdot (20\%) + 0.3 \cdot (4\%)\\
i_3 &= 65\% &&= 0.3 \cdot (130\%) + 0.4 \cdot (50\%) + 0.3 \cdot (20\%)\\
i_4 &= -24.8\% &&= 0.3 \cdot (44\%) + 0.4 \cdot (-20\%) + 0.3 \cdot (-100\%).
\end{aligned}$$

Then, one uses the weights w_t of the four periods for finding the average of these expected ROIs:

$$\bar{\imath} = 43.8\% \cdot 9.5\% + 29.9\% \cdot 28\% + 18.1\% \cdot 65\% + 8.2\% \cdot (-24.8\%) = 22.3\%,$$

as expected.

Nonzero Initial Income

The IRR approach does not distinguish between income and capital. This is a further source of distortion. For example, in the IRR approach, the first internal capital is equal to the first cash flow changed in sign even if $I_0 \neq 0$. This is inconsistent with the law of motion of any economic asset, which requires that the first incremental cash flow is equal to the incremental income less the incremental capital committed by the firm in the project: $F_0 = I_0 - \Delta C_0 = I_0 - C_0$. This implies that the IRR approach is not able to properly detect the project's initial investment in situations where an immediate income arises (e.g. replacement projects or abandonment projects). In other words, part of the first cash flow is not an invested capital, but an (incremental) profit. In these cases, the IRR may not even be interpreted as a constant growth rate for the invested capital, since it fails to remove the income component from the initial cash flow.

Example 9.35 ♦
Consider the replacement project illustrated in Example 1.12. We report its concise transposed Matrix below:

$(\boldsymbol{C}, \boldsymbol{I}, \boldsymbol{F})$

Time		0	1	2	3	4	5
Capital	C	28	44	60	40	20	0
Income	I	8	2	1.5	1	0.5	0.2
Cash Flow	F	−20	−14	−14.5	21	20.5	20.2

The (unique) IRR is 8.1%. The first internal capital is $C_0(0.081) = -F_0 = 20$ and, on that capital, the firm allegedly earns 8.1% in the first period. This means that, according to the IRR approach, the firm invests an incremental \$20 for replacing the old equipment. In actual facts, the firm increments its own assets by $C_0 = \$28$, obtained as the difference between the price of the new equipment (\$100) and the salvage value of the old equipment (\$72). However, the firm succeeds in selling the old equipment at \$80, which is \$8 more than the salvage value. Therefore, an incremental income occurs at time 0 equal to $I_0 = \$8$.

Put it differently, without the replacement, the firm's income is \$25; by replacing the equipment, the firm's income is \$33. The difference is \$8, which the firm properly records in the income statement as a gain on asset disposal (see also Table 1.8). In other words, by purchasing the new asset the firm increments the income by $I_0 = \$8$. This is not the result of the investment in the new asset but, rather, the result of the old asset's sale. The project may then be viewed as a portfolio of two courses of actions: Sale of old asset + purchase of new assets. The latter entails a capital commitment of \$100, the former entails a reduction in capital of \$72.

This implies that the actual cash outlay of the firm is $|F_0| = \$20$, which is \$8 smaller than the capital invested. As a result, to represent that \$20 as the capital invested in the project means to distort the actual economic meaning of the transaction. And to assume that, in the first period, the firm earns 8.1% on \$20 means to delineate a biased description of the dynamics of the project.

Differences in Scale: A Non-pitfall

When comparing different projects, the IRR criterion is not generally consistent with the NPV. However, this one case should not be considered a pitfall of the IRR notion. The fact is that IRR is a rate of return, and any rate of return (including AIRR) is a relative measure of worth. A relative measure of worth provides information on the project acceptability and on its economic efficiency, not on the project size. The latter is an essential ingredient of the amount of value created. In other words, the different ranking provided by an absolute measure of worth as opposed to a rate of return is caused by differences in the project scales (see Sect. 11.2. See also Keane 1979).

Therefore, two or more competing projects should never be ranked by comparison of respective rates of return, unless they share the same capital base. NPV or V_0(RI) are the right tools to accomplish project ranking (but see Chap. 11 on rational project ranking with relative measures of worth).

Makeham's Formula

William Makeham was an English actuary and mathematician, living in the 19th century. He found a formula for calculating the price of a bond in terms of the coupon rate (or modified coupon rate) and the valuation rate. The formula was then generalized for any type of loans.

Owing to the unified framework we have introduced (namely, the statics and the dynamics of any economic system), Makeham's formula may be applied to any class of economic assets, be it a loan, a bond, an operating asset, a portfolio of financial assets, etc. Consider an economic asset such that $i = I_t/C_{t-1}$ is constant. The present value of the period returns, PV[I], is, by definition, equal to $\text{PV}[I] = \sum_{t=1}^{n} I_t(1+r)^{-t}$.[28] For a loan or bond, this is the value of interest. Makeham's formula reframes the value of interest as a function of the interest/income rate i and the benchmark rate r:

$$\text{PV}[I] = \frac{i}{r}\Big(C_0 - \text{PV}[-\Delta C]\Big) \tag{9.10}$$

where

$$\text{PV}[-\Delta C] = -\sum_{t=1}^{n} \Delta C_t(1+r)^{-t}$$

is the present value of the principal repayments (or, in case of an operating asset, the present value of the capital depreciation charges). Therefore, the market value of the asset is

$$V_0 = \text{PV}[I] + \text{PV}[-\Delta C] = \frac{i}{r}\Big(C_0 + \text{PV}[\Delta C]\Big) - \text{PV}[\Delta C] \tag{9.11}$$

[28] For simplicity, we assume $I_0 = 0$.

(Makeham 1874; Glen 1893; Broverman 2008; Magni 2013b). In a loan/bond with constant interest rate, the IRR coincides with the interest rate: $\sigma = i$. Therefore, Makeham's formula expresses the value of an asset in terms of its internal rate of return. One may then write

$$\text{PV}[I] = \frac{\sigma}{r}\Big(C_0 + \text{PV}[\Delta C]\Big) \quad \text{and} \quad V_0 = \frac{\sigma}{r}\Big(C_0 + \text{PV}[\Delta C]\Big) - \text{PV}[\Delta C]. \tag{9.12}$$

However, (9.12) holds only if the income rate $i = \sigma$ is constant, which means that, for loans with time-varying interest rates (as well as for any operating investment), it is not valid.

This pitfall is healed if one replaces IRR with the AIRR, $\bar{\imath}$. It may be shown that

$$\text{PV}[I] = \frac{\bar{\imath}}{r}\Big(C_0 + \text{PV}[\Delta C]\Big) \quad \text{and} \quad V_0 = \frac{\bar{\imath}}{r}\Big(C_0 + \text{PV}[\Delta C]\Big) - \text{PV}[\Delta C] \tag{9.13}$$

for any vector of income rates $\boldsymbol{i} = (i_1, i_2, \ldots, i_n)$. If the benchmark rates are time-varying, Makeham formula in (9.12) may not even be applied, whereas (9.13) is easily generalized by replacing r with the average benchmark rate, $\bar{r}$:

$$\text{PV}[I] = \frac{\bar{\imath}}{\bar{r}}\Big(C_0 + \text{PV}[\Delta C]\Big) \quad \text{and} \quad V_0 = \frac{\bar{\imath}}{\bar{r}}\Big(C_0 + \text{PV}[\Delta C]\Big) + \text{PV}[\Delta C] \tag{9.14}$$

(see Magni 2013b, Proposition 3.1).

Example 9.36 ♦

Consider a 6-period loan with time-varying interest rates, i_t and let $r = 1.5\%$ be the benchmark rate (MARR). Table 9.12 reports the relevant data: Principal outstanding (book value), interest earned, instalments, principal repayments, loan rates and market rate, as well as the respective present values (PV). The unique IRR is $\sigma = 10.7\%$. Equation (9.12) is not fulfilled:

$$\text{PV}[I] = 809.6 \neq 738.83 = \frac{10.7\%}{1.5\%}(1,487.16 - 1,383.63).$$

In our case, the average rate is $\bar{\imath} = 809.6(1 + 0.015)/7,005.76 = 11.73\%$, so that (9.13) is indeed satisfied:

$$\text{PV}[I] = 809.6 = \frac{11.73\%}{1.5\%}(1,487.16 - 1,383.63).$$

Global (non)Additivity: A Shared Difficulty

We say that a metric is globally additive if, in overall terms, the capital committed in a portfolio of assets during its lifetime is equal to the sum of the capitals committed in

Table 9.12 Amortization schedule of a 6-period loan

Time	Book value C	Interest earned I	Cash flows F	Principal repayments $-\Delta C$	Interest rates i	Market rate r	Discount factor $\mathbf{d}$
0	1,487.16		−1,487.16				1.0000
1	1,316.91	29.74	200.00	170.26	2.00%	1.50%	0.9852
2	1,222.26	105.35	200.00	94.65	8.00%	1.50%	0.9707
3	1,144.48	122.23	200.00	77.77	10.00%	1.50%	0.9563
4	1,066.16	171.67	250.00	78.33	15.00%	1.50%	0.9422
5	1,008.06	191.91	250.00	58.09	18.00%	1.50%	0.9283
6	0.00	241.94	1250.00	1,008.06	24.00%	1.50%	0.9145
PV	7,005.76	809.60	706.07	1,383.63			

its constituent assets during their lifetimes. The IRR is globally nonadditive: Given any bundle of m projects, consider the portfolio $P = \sum_{j=1}^{m} P_j$. Then,

$$\text{PV}[C^{P_1}(\sigma)] + \text{PV}[C^{P_2}(\sigma)] + \ldots + \text{PV}[C^{P_m}(\sigma)] \neq \text{PV}[C^P(\sigma)]$$

(this is straightforward from the nonadditivity at the period level illustrated on Sect. 9.5). As for AIRR, if the COCs are equal across the constituent assets, then the AIRR is globally additive because the capital of the portfolio is, by definition, the sum of the capitals of the constituent assets: $C_t^P = \sum_{j=1}^{m} C_t^{P_j}$ for every t and the equality of the COCs implies the equality of all discount factors: $\text{d}_{t,0}^P = \text{d}_{t,0}^{P_1} = \text{d}_{t,0}^{P_2} = \ldots = \text{d}_{t,0}^{P_m}$ for every t. Hence, $C_t^P \text{d}_{t,0}^P = \sum_{j=1}^{m} C_t^{P_j} \text{d}_{t,0}^{P_j}$, which in turn implies $\sum_{j=1}^{m} \text{PV}[C^{P_j}] = \text{PV}[C^P]$. (See also Magni 2018.) It is preciesly this global additivity that guarantees that, for a mixed project, the equality $\text{PV}[C] = \text{PV}[C_L] + \text{PV}[C_B]$ holds. However, if the COC is different across assets, then the AIRR is not globally additive. It is easy to verify that, under this assumption, in general, $C_t^P \text{d}_{t,0}^P \neq \sum_{j=1}^{m} C_t^{P_j} \text{d}_{t,0}^{P_j}$, whence $\sum_{j=1}^{m} \text{PV}[C^{P_j}] \neq \text{PV}[C^P]$. The problem lies in the discounting process and in the subtle relations linking the COCs of the individual projects and the COCs of the portfolio.[29]

Example 9.37 ♦

Consider Example 9.23 again and, in particular, Table 9.8. The last line illustrates the global non-additivity implied by the non-additivity of internal values. Specifically, the project is a portfolio of equity and debt, and $\text{PV}[C^e(\sigma^e)] = 2{,}798.6$, $\text{PV}[C^d(\sigma^d)] = 1{,}449.4$ such that $\text{PV}[C^e(\sigma^e)] + \text{PV}[C^d(\sigma^d)] = 4{,}248$. However, the total internal value of the portfolio is different: $\text{PV}[C(\sigma)] = 4{,}422.5$.

[29] We do not dwell on it, but, in Chap. 10, we will introduce a class of rates of return which do not incur this difficulty.

Now, let us consider Example 6.5 where we have assumed that BHP Billiton project's cost of equity is equal to $r^e = 15\%$ and the cost of debt is equal to $r^d = 8\%$. Discounting all the equity book values at 15% and all the debt book values at 8%, one gets $\text{PV}[C^e] = 3{,}084.2$ and $\text{PV}[C^d] = 1{,}449.4$ so that $\text{PV}[C^e] + \text{PV}[C^d] = 4{,}533.6$. To find $\text{PV}[C]$ one needs to calculate the project's pre-tax WACC. As we know, this is realized by calculating the arithmetic mean of the equity COC and the debt COC: $r_t = (r^e C^e_{t-1} + r^d C^d_{t-1})/(C^e_{t-1} + C^d_{t-1})$. In the first 10 periods, the project is partially financed with debt and the pre-tax WACC is time-varing:

$$r_1 = 14.10\%,\ r_2 = 14.08\%,\ r_3 = 14.07\%,\ r_4 = 14.06\%,\ r_5 = 14.05\%,$$
$$r_6 = 14.04\%,\ r_7 = 14.02\%,\ r_8 = 14.00\%,\ r_9 = 13.97\%,\ r_{10} = 13.94\%.$$

In the last 10 periods, the project is unlevered; therefore, $r_{11} = r_{12} = \ldots = r_{20} = r^e = 15\%$ for $t \in \mathbb{N}^{11}_{20}$. Discounting the project's capital, C_t, with the pre-tax WACCs, one finds $\text{PV}[C] = 4{,}408.2$. Therefore,

$$\text{PV}[C^e] + \text{PV}[C^d] = 4,533.6 \neq 4{,}408.2 = \text{PV}[C].$$

9.6 The Modified IRR (MIRR) and the Sinking Fund Methods (SFMs)

Attempts to provide solutions to the three classical problems have been presented since 1950s. However, instead of modifying the invariant function (9.1) to get a more reliable notion of project rate of return, researches have addressed their efforts to modifying the cash-flow stream while keeping the same invariance requirement.

In this respect, a well-known approach is the Modified Internal Rate of Return (MIRR). This approach was anticipated by the French actuary Duvillard (1755–1832) (see Biondi 2006), then re-discovered and studied since 1950s by several authors.[30] The idea is to modify the original cash-flow stream in some way, so as to obtain a new, conventional cash-flow stream. As such, the IRR of the modified cash-flow stream exists, is unique and its internal values do not change sign, so that financial meaning of IRR as either investment rate or financing rate is unambiguous. The modification of the original cash-flow stream is usually obtained by discounting and/or compounding some cash flows at some auxiliary rate y^{ext} (also known as *external* rate) from some date to some other date.

The notion of MIRR is strictly connected with the notion of Sinking Fund Method (SFM). These are methods for adjusting the cash flows of the original project so as to guarantee that the adjusted cash-flow stream possesses a unique IRR. The modification is obtained by assuming that funds are set aside in a (hypothetical) sinking fund which accrues interests.

---

[30] Among others, see Solomon (1956), Baldwin (1959), Kirshenbaum (1965), Lin (1976), Athanasopoulos (1978), Chang and Owens (1999), Kierulff (2008).

Following are some methods for modifying a nonconventional cash-flow stream resulting in a conventional one with unique IRR.

1. **Compounding approach**. All positive cash flows are compounded to time n.
2. **Discounting approach**. All negative cash flows are discounted to time 0.
3. **Combination approach**. All positive cash flows are compounded to time n and all negative cash flows are discounted to time 0.
4. **Accumulating approach**. All cash flows except F_0 are compounded to time n[31]
5. **SFM approach (first version)**. Later negative cash flows are discounted and subtracted from preceding cash flows. The procedure is iterated to the extent required to zero out the negative flows following some positive cash flow.
6. **SFM approach (second version)**. Later negative cash flows are discounted and subtracted from preceding cash flows. The procedure is iterated to the extent required to zero out the negative flows following some positive cash flow *and* the negative flows following the first negative flow.[32]

Example 9.38 ♦
Consider $\boldsymbol{F} = (-1, 0, -11, 6, -4, 1.8)$. The cash-flow stream has three IRRs: $\sigma^1 = -39.6\%$, $\sigma^2 = 49.8\%$, and $\sigma^3 = 401.7\%$. Let us apply the six methods described above, by assuming that the auxiliary rate is $y^{\text{ext}} = 20\%$. Table 9.13 collects the resulting modified cash-flow streams data for the six adjustments made to the original cash-flow stream and the corresponding IRRs (i.e., the MIRRs).

1. **Compounding approach**. In this approach, one compounds at 20% all positive cash flows to time n, which results in a terminal amount of \$25. The negative cash flows are unvaried.
2. **Discounting approach**. In this approach, one discounts at 20% the negative cash flows to time 0, resulting in an initial revised outflow equal to \$10.6; the positive cash flows are unvaried.
3. **Combination approach**. In this approach, one combines the former two approaches and therefore considers only two nonzero cash flows: The initial revised amount of the discounting approach and the compounded amount of the compounding approach.
4. **Accumulating approach**. In this approach, every positive and negative cash flow (except F_0) is compounded at 20% to time 5, resulting in a terminal amount equal to \$1.15. The initial investment is left unvaried.

---

[31] This method is applicable only if the compounded sum of the cash flows has a different sign from F_0.

[32] More specifically, let t be the last date where a negative cash flow occurs. Then, one discounts back the outflow by one period at rate y^{ext} and subtracts it from the preceding cash flow. If the resulting amount is nonnegative, the procedure ends, otherwise it is iterated, until some date τ, where the resulting amount is nonnegative. At this point, if there is another negative cash flow occurring prior to τ, the same iterative procedure is reinstantiated, until no negative cash flow occurs after some positive cash flow (first version) or until no negative cash flow occurs after some positive cash flow and an no negative cash flow occurs after the first negative flow (second version). Other variants of the SFM are possible (e.g., later positive cash flows may be discounted to pay off earlier negative cash flows (see Herbst 2002 on SFMs).

Table 9.13 Six MIRR/SFM methods for a nonconventional cash-flow stream

Time	F	Modified cash-flow stream					
		1.	**2.**	**3.**	**4.**	**5.**	**6.**
0	−1.0	−1.0	−10.6	−10.6	−1.0	−1.0	−2.8
1	7.0	0.0	7.0	0.0	0.0	−2.2	0.0
2	−11.0	−11.0	0.0	0.0	0.0	0.0	0.0
3	6.0	0.0	6.0	0.0	0.0	2.7	2.7
4	−4.0	−4.0	0.0	0.0	0.0	0.0	0.0
5	1.8	25.0	1.8	25.0	1.15	1.8	1.8
IRR	−39.6% 49.8% 401.7%	17.8%	16.9%	18.8%	2.8%	11.8%	13.2%

5. **SFM approach (first version)**. In the first version of SFM, the iterated procedure is applied until no negative cash flow occurs after some positive cash flow. Specifically, one starts from the last outflow, −\$4, which is discounted by one period and added to \$6 so as to generate a modified inflow of \$2.7 at time 3. Then, the procedure is repeated for the outflow at time 2, equal to −\$11: It is discounted back by one period and added to \$7, resulting in a modified cash flow of −\$2.2 at time 1. The procedure ends because all the negative cash flows are now separated from the positive cash flows.
6. **SFM approach (second version)**. In this case, the procedure is iterated until no negative cash flow occurs after the initial one. This boils down to first applying approach 5. Then, after noticing that the cash flow at time 1 is still negative (= −\$2.2), one discounts it back by one period and subtracts \$1, the initial original outflow. The revised initial cash flow is then $-\$2.2/1.2 - \$1 = -\$2.8$. In such a way, all negative cash flows following the first one are zeroed out.

The MIRRs are quite different, spanning from 2.8% to 18.8%. They all represent investment rates, since every method generates a sequence of positive internal capital values, so signaling permanent lending positions. For example, considering approach 5. (SFM, first version), the internal values are

$$0 < C_0(0.118) = 1.0$$
$$0 < C_1(0.118) = 3.3 = 1(1.2) + 2.2$$
$$0 < C_2(0.118) = 3.7 = 3.3(1.2) - 0$$
$$0 < C_3(0.118) = 1.4 = 3.7(1.2) - 2.7$$
$$0 < C_4(0.118) = 1.6 = 1.4(1.2) - 0.$$

The MIRR/SFM approach has pros and cons. As for the pros, it solves the problem of uniqueness, and supplies an IRR with a well-defined financial nature (the internal values do not change sign). As for the cons, these are of two types. The first type has to do with the fact that the MIRR does not always exist (e.g., projects in Examples 9.10 and 9.20 have no MIRR) and does not solve the remaining pitfalls of the IRR illustrated in Sect. 9.5. The second type of cons consists of additional problems that the original project's IRR does not incur. Following is a list of these problems.

(i) The approach presupposes integrative (hypothetical) operations that do not have anything to do with the actual transactions implied by the project. For example:

- the discounting of negative cash flows may be interpreted as the accumulation of a positive amount in a sinking fund at the interest rate y^{ext} which offsets a future negative cash flow[33]
- the compounding of a positive cash flow may be conceived of as the result of the reinvestment in some other economic activity at the income rate y^{ext} (the reinvestment zeroes out the positive cash flow and the accumulated amount is added to an original cash flow of a future date)[34]
- the compounding of a negative cash flow may be interpreted as the result of raising funds from capital providers at the financing rate y^{ext}: The raising of funds zeroes out the negative cash flow and the accumulated amount is subtracted from an original cash flow of a future date.[35]

However, these operations are merely hypothetical and have no real economic content in the sense that they are not associated with real underlying business transactions. They only represent simplifying mathematical devices aimed at generating a unique IRR, whose relation with the actual project is then obfuscated.

(ii) Even if the integrative operations underlined real transactions, they would nonetheless have nothing to do with the project: Any investment or financing associated with the project is explicitly reported in the pro forma financial statements. For example, if a negative cash flow appears in the CFE stream, then it means that the firm is expected to issue shares or to distribute to equityholders an amount of CFE which is smaller than it would be if the project were not undertaken. If the firm decided to borrow the amount from external sources

---

[33] For example, consider the SFM in the example above: At time 4 an outflow of \$4 occurs. Assuming that \$3.33 is invested for one year in a sinking fund earning 20%, the accumulated amount is \$4, which zeroes out the outflow. At time 3, the investment of \$3.33 is subtracted from \$6 to get a net inflow of \$2.7.

[34] For example, in the compounding approach, the \$7 is zeroed out by assuming a reinvestment of \$7 at 20% for 4 periods, and the \$6 is zeroed out by assuming a reinvestment of \$6 at 20% for two periods. At time 5, the accumulated amount of the two reinvestments will be added to the terminal cash flow: $7(1.2)^4 + 6(1.2)^2 + 1.8 = 25$. In this way, at time 1 and 3, the net cash flow is zero.

[35] For example, in the accumulating approach, the outflow of \$11 at time 2 is zeroed out by assuming that \$11 is borrowed at 20%. The accumulated amount is then subtracted from the accumulated amount of the other integrative operations realized at the various dates.

at some auxiliary rate y^{ext}, then the loan would be recorded in the pro forma financial statements, along with its instalments at the various dates. Likewise, if a positive CFE appears at some date in the CFE stream, it means it is distributed to equityholders. If it were reinvested at some rate y^{ext}, the retained cash would appear in the liquid assets along with the subsequent interest incomes. If integrative operations are tied to the project and they really need to be taken into account, it means that they have not already been included in the pro forma financial statements, as they should have been, which implies that the project's law of conservation has not been complied with.[36] As a result, the integrative operations, if really dependent on the project, should be already incorporated in the project's pro forma financial statements (and, in such a way, encompassed in the triplet $(\boldsymbol{C}, \boldsymbol{I}, \boldsymbol{F})$); if, alternatively, they are independent of the project (i.e., they would be undertaken regardless of project acceptance/rejection), then they should not be taken into account in the modelling system (only differential amounts should be considered for decision-making):

> The prospective return on another independent investment should never be allowed to influence the investment decision. (Brealey et al. 2011, p. 141)

> The reinvestment opportunities for the cash flows of a project ...are not relevant for the purposes of assessing the project's worth. (Keane 1979, p. 50)

> The reinvestment opportunities are irrelevant to the choice of a criterion provided the cash flows are correctly specified. (Keane 1979, p. 55)

(iii) An implication of the point above is that a MIRR does not measure the project's rate of return; it merely measures the IRR of a course of action which includes the project being analyzed and other transactions (sinking funds, reinvestments of inflows, financing of outflows, etc.) introduced for the sole purpose of adjusting the project's cash-flow stream and render it a conventional cash-flow stream. As such, the MIRR is an internal rate of return "on a modified set of cash flows, not the project's actual cash flows" (Ross et al. 2011, p. 250). As a result, one might say that, at the very best, MIRR answers a question which has not been asked, one which is not relevant for decisions regarding the project undertaking.

---

[36] Keane (1979) expressed the same concept with the following words:

> If the satellite projects are physically or economically dependent, then the original cash flow specifications are incorrect because they should have included the effects of the entire programme of interdependent projects. (Keane 1979, p. 50)

> If a manager measures the NPV of a project's cash flows without including the reinvestment flows, the only implicit assumption in the decision to do so is that the reinvestment opportunities are not relevant. If they are in fact relevant there is no benefit in omitting the additional flows and speculating about the arithmetic relationship between the NPV of the incomplete set and the NPV that would have obtained had the additional flows been included. (Keane 1979, p. 53).

(iv) It is not clear how the original cash-flow stream should be modified. We have shown six methods and, actually, there are infinitely many ways of adjusting a cash-flow stream into a conventional one which supplies a unique IRR, and there is no clear reason to prefer one adjustment over the other ones (e.g., which of the six MIRRs computed in the example above is the "correct" or "relevant" MIRR?).

(v) In general, the approach does not capture value creation. For example, if $r = 1\%$ is the COC in Example 9.38, the project's NPV is negative (NPV $= -1.16$), but all methods incorrectly suggest project acceptance (all the MIRRs exceed 1%). More generally, if $-39.6\% < r < 2.8\%$, the NPV is negative but all methods incorrectly suggest project acceptance. If $2.8\% < r < 18.8\%$, the NPV is still negative and some of the MIRRs recommend acceptance, while some other ones recommend rejection. If $18.8\% < r < 49.8\%$ or $r > 401.7\%$, all the MIRRs correctly signal value destruction. If $49.8\% < r < 401.7\%$, the NPV is positive and all the MIRRs incorrectly suggest to reject the project (see also the graph of the DCF function of the project in Fig. 9.2).

(vi) The MIRR/SFM methods requires an additional piece of information with respect to the IRR, which is the auxiliary rate. Sometimes, two (or even more) auxiliary rates are employed (e.g., one for discounting, one for compounding) The presence of two auxiliary rates makes the methodology inconsistent with rational decision-making, as can be easily checked.

(vii) If a unique IRR exists for the original cash-flow stream and internal values are nonnegative, the MIRR/SFM approach is useless. Indeed, in this case, the IRR is preferable than the MIRR/SFM, because it does not require any modification of the cash-flow stream nor the determination of any additional auxiliary rate. So, this methodology is appropriate whenever no IRR exists or multiple IRRs

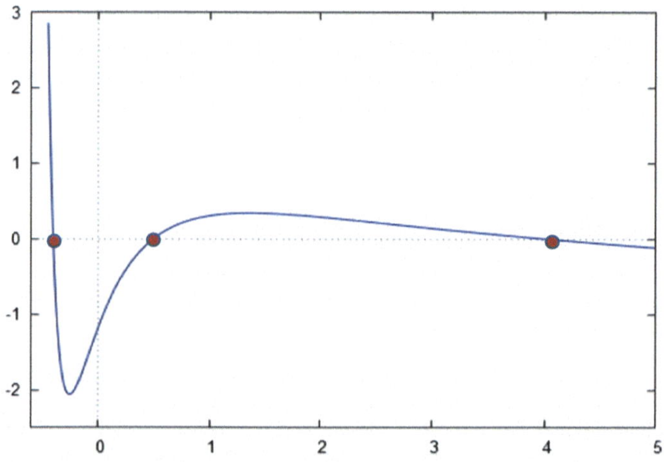

Fig. 9.2 Graph of the DCF function, $\phi(x)$, of the cash-flow stream $(-1, 7, -11, 6, -4, 1.8)$. The IRRs are $\sigma^1 = -39.6\%$, $\sigma^2 = 49.8\%$, $\sigma^3 = 401.7\%$

arise but unhelpful whenever the cash-flow stream has a unique, financially unambiguous IRR. A paradox then arises, according to which the MIRR is economically meaningful depending on whether the original cash-flow has a unique, unambiguous IRR or not.

9.7 The TRM Model[37]

A well-known proposal for overcoming the classical problems of IRR was advanced by Teichroew et al. (TRM) (1965b, 1965a). The authors proposed a generalization of the IRR notion and derived (not one, but) two rational rate-of-return-based decision rules.

The situation is described by TRM (1965b, 1965a) in terms of *project balance*, which they define in the following way:

$$S_t(i_B, i_L) = \begin{cases} S_{t-1}(i_B, i_L)(1+i_B) + F_t & \text{if } S_{t-1} > 0 \\ S_{t-1}(i_B, i_L)(1+i_L) + F_t & \text{otherwise} \end{cases} \tag{9.15}$$

which may be written as

$$C_t(i_B, i_L) = \begin{cases} C_{t-1}(i_B, i_L)(1+i_B) - F_t & \text{if } C_{t-1} < 0 \\ C_{t-1}(i_B, i_L)(1+i_L) - F_t & \text{otherwise} \end{cases} \tag{9.16}$$

where $C_t(i_B, i_L) = -S_t(i_B, i_L)$. Therefore, TRM's project balance is equal to the project's capital *under the assumption* that the capital grows at a constant rate, which differs depending on whether the BOP capital is positive or negative. Specifically, whenever $C_{t-1}(i_B, i_L) > 0$, the firm invests resources in the project at the investment/lending rate i_L; whenever $C_{t-1}(i_B, i_L) < 0$, the firm absorbs resources from the project at the financing/borrowing rate i_B. The terminal boundary condition for a project may then be written as $C_n(i_B, i_L) = 0$ (see TRM 1965b, p. 401; TRM, 1965a, p. 169). TRM called i_B the *project financing (or borrowing) rate* and i_L the *project investment (or lending) rate*. The rate i_B is only active in the borrowing periods, that is, when $C_{t-1}(i_B, i_L) < 0$, while the rate i_L is only active in the investment periods, that is, when $C_{t-1}(i_B, i_L) > 0$.[38] Mathematically, the internal pair (i_B, i_L) may also be framed as an internal vector of holding period rates $\boldsymbol{i} = (i_1, \ldots, i_n)$ such that

[37] The first part of this section is adapted from Magni (2015a), Investment, financing and the role of ROA and WACC in value creation, *European Journal of Operational Research,* 244(3) (August), 855–866 (Sect. 2). See See also Magni (2014a) for an illustration of this approach in greater detail.

[38] If $C_t(i_B, i_L) \geq 0$ for every t, then the project is internally pure and i_B does not apply: The rate i_L is the traditional project's IRR (which exists and is unique by STRMG): $i_L = \sigma$. If $C_t(i_B, i_L) \leq 0$ for every t, then the project is internally pure and i_L does not apply: The rate i_B is the traditional project's IRR (which exists and is unique by STRMG): $i_B = \sigma$.

$$i_t = \begin{cases} i_B & \text{if } C_{t-1} < 0 \\ i_L & \text{otherwise} \end{cases}$$

TRM showed the following result connecting i_B and i_L.

Proposition 9.1 *The boundary condition $C_n(i_B, i_L) = 0$ generates an implicit function $i_B = i_B(i_L)$ and an implicit function $i_L = i_L(i_B)$, which is the inverse function of the former.*

(See TRM 1965b, Theorem IV, Corollary IVB; TRM 1965a, p. 169).

Reminding that $\phi(x) = \sum_{j=0}^{n} F_j(1+x)^{-j}$ is the DCF function, such that NPV $= \phi(r)$ and using Proposition 9.1, TRM proved the following result.

Proposition 9.2 *For any acceptable interest rate x (i.e., belonging to the domain of the implicit functions),*

$$\phi(x) > 0 \quad \text{if and only if} \quad i_L(x) > x \tag{9.17a}$$

$$\phi(x) > 0 \quad \text{if and only if} \quad i_B(x) < x. \tag{9.17b}$$

(See TRM 1965b, Theorem V, TRM 1965a, p. 176).

Therefore, considering that the project creates value if and only if $\phi(r) = \text{NPV} > 0$, the following accept/reject decision rules can be stated.

Proposition 9.3 *Given the project COC, r,*

$$\textit{accept the project if and only if} \quad i_L(r) > r \tag{9.18a}$$

$$\textit{accept the project if and only if} \quad i_B(r) < r. \tag{9.18b}$$

(TRM 1965b, p. 403; TRM 1965a, section VI and p. 177).

Proposition 9.3 may be expressed in graphical terms as follows: Move along the locus of points (i_B, i_L) which fulfill $C_n(i_B, i_L) = 0$ and focus on the points $(r, i_L(r))$ and $(i_B(r), r)$. The comparison of abscissa and ordinate in either pair establishes project acceptability (see Fig. 9.3).

Textbooks dealing with the TRM approach report Rule (9.18a) while forgetting Rule (9.18b). Actually, there is no logical rationale why either rule should be preferred; the two rules are logically equivalent, and it is only a matter of preference

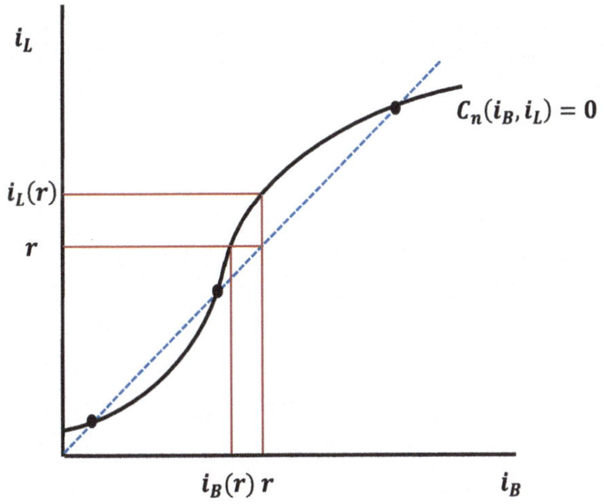

Fig. 9.3 The TRM rules. The black line shows $i_L(i_B)$. Its inverse function is $i_B(i_L)$. The dashed line describes the set of points such that $i_L = i_B$, so the intersection of black and dashed line identifies the project IRRs (in the figure, three IRRs are illustrated). The figure describes a value-creating project; this is signaled by both $i_L(r) > r$ (investment setting) and $i_B(r) < r$ (borrowing setting)

whether the analyst should choose the investment setting (Rule (9.18a)) or the borrowing setting (Rule (9.18b)).[39]

TRM did not provide any explicit relation between a project NPV and the internal pair (i_B, i_L). Such a link is straightforward with the AIRR approach, which includes TRM approach as a (very) special case. More precisely, in Sect. 8.9 we showed that the NPV may be split up into investment-period NPV and borrowing-period NPV:

$$V_0(\text{RI}) = \overbrace{\sum_{t \in T_L} C_{t-1}(i_t - r_t)\text{d}_{t,0}}^{\text{NPV}_L} + \overbrace{\sum_{t \in T_B} C_{t-1}(i_t - r_t)\text{d}_{t,0}}^{\text{NPV}_B} = \text{NPV} \tag{9.19}$$

where T_L denotes the set of lending periods and T_B denotes the set of borrowing periods, or, equivalently,

$$\begin{aligned} V_0(\text{RI}) &= \overbrace{\text{PV}[C_L] \cdot (\bar{\imath}_0^L - \bar{r}_0^L)}^{\text{NPV}_L} + \overbrace{\text{PV}[C_B] \cdot (\bar{\imath}_0^B - \bar{r}_0^B)}^{\text{NPV}_B} \\ &= \underbrace{\text{PV}[C_L] \cdot \frac{\bar{\imath}^L - \bar{r}^L}{1 + r_1}}_{\text{NPV}_L} + \underbrace{\text{PV}[C_B] \cdot \frac{\bar{\imath}^B - \bar{r}^B}{1 + r_1}}_{\text{NPV}_B} = \text{NPV}. \end{aligned}$$

[39] There may however be a cognitive rationale in preferring the investment setting. Practitioners and academics seem to prefer to think in terms of rates of return rather than rates of cost.

TRM's smart idea of setting either $i_B = r$ or $i_L = r$ implies that one of the two summands above is forced to be equal to zero. This is an artificial construct, for the period return rate of an engineering project (and, in general, any corporate project) has no direct relation with the rate of return of financial assets traded in a capital market; rather, it is tied to the response of customers and, therefore, to the incremental sales, as well as to the cost savings obtained by the undertaking of the project. Borrowing from the project (or investing in the project) is not like borrowing from (or lending to) agents operating in the capital market (see related discussion in Sect. 8.9).

We now show how TRM's decision rules are easily derived from (9.19) as follows. First, note that, if $i_t = i_B$ for $t \in T_B$ and $i_t = i_L$ for $t \in T_L$, then (9.19) becomes

$$\text{NPV} = \sum_{t \in T_L(i_B, i_L)} C_{t-1}(i_B, i_L) \cdot (i_L - r_t)\text{d}_{t,0} + \sum_{t \in T_B(i_B, i_L)} C_{t-1}(i_B, i_L) \cdot (i_B - r_t)\text{d}_{t,0} \tag{9.20}$$

where $T_L(i_B, i_L)$ denotes the periods where $C_{t-1}(i_B, i_L) \geq 0$ and $T_B(i_B, i_L)$ denotes the periods where $C_{t-1}(i_B, i_L) < 0$.

Investment Setting

Assume that

1. the COC is constant, $r_t = r$ for every t
2. in the borrowing periods, the period rate i_t is constant, $i_t = i_B$ for every $t \in T_B(i_B, i_L)$
3. the period rate i_B is equal to the COC, $i_B = r$
4. in the investment periods, the period rate i_t is constant, $i_t = i_L$ for every $t \in T_L(i_B, i_L)$.

Under these assumptions, the capital amount is $C_t(r, i_L(r))$ and (9.20) becomes

$$\begin{aligned} \text{NPV} &= \sum_{t \in T_L(i_B, i_L)} C_{t-1}(r, i_L(r)) \cdot (i_L(r) - r)\text{d}_{t,0} + \overbrace{\sum_{t \in T_B(i_B, i_L)} C_{t-1}(r, i_L(r)) \cdot (r - r)\text{d}_{t,0}}^{=0} \\ &= \sum_{t \in T_L(i_B, i_L)} C_{t-1}(r, i_L(r)) \cdot (i_L(r) - r)\text{d}_{t,0} \end{aligned}$$

which implies NPV > 0 if and only if $i_L(r) > r$ whence Rule (9.18a) follows at once.

Borrowing Setting

Assume that

1. the COC is constant, $r_t = r$ for every t
2. in the investment periods, the period rate i_t is constant, $i_t = i_L$ for every $t \in T_L(i_B, i_L)$
3. the period rate i_L is equal to the COC, $i_L = r$
4. in the borowing periods, the period rate i_t is constant, $i_t = i_B$ for every $t \in T_B(i_B, i_L)$.

Under these assumptions, the capital amount is $C_t(i_B(r), r)$ and (9.20) becomes

$$\begin{aligned}\text{NPV} &= \overbrace{\sum_{t\in T_L(i_B,i_L)} C_{t-1}(i_B(r), r)\cdot(r-r)\mathrm{d}_{t,0}}^{=0} + \sum_{t\in T_B(i_B,i_L)} C_{t-1}(i_B(r), r)\cdot(i_B(r)-r)\mathrm{d}_{t,0}\\ &= \sum_{t\in T_B(i_B,i_L)} C_{t-1}(i_B(r), r)\cdot(i_B(r)-r)\mathrm{d}_{t,0}\end{aligned}$$

which implies NPV > 0 if and only if $i_B(r) < r$ whence Rule (9.18b) follows.

With the former set of assumptions (investment setting), the RIs in the financing periods are zeroed out; with the latter set of assumptions (borrowing setting), the RIs in the investment periods are zeroed out. In both cases, value creation is simply signaled by the comparison of the two remaining rates.

However, as said, such assumptions are largely inadequate in real-life applications, to such an extent that TRM's lending rate and borrowing rate are, in either setting, beautiful-but-economically-unacceptable artifacts. On one hand, they do ameliorate the crude assumption of the IRR approach of a constant rate of return but, on the other hand, they provide a result which is economically significant only insofar as one accepts the idea that the ROI is equal to the WACC, r, either in the investment periods or in the financing periods. We have already seen that, in real-life projects, the project's income rate, i_t, is time-varying and is not chained to the interest rate prevailing in the financial markets.

As a result, TRM's lending rate and borrowing rate may be said to be just special cases of investment-period AIRR and financing-period AIRR, respectively (see Sect. 8.9). Specifically, under the assumptions made by TRM, the investment-period AIRR and the financing-period AIRR are

$$\bar{\imath}^L = i_L \qquad \bar{\imath}^B = r \qquad \text{(investment setting)}$$

$$\bar{\imath}^L = r \qquad \bar{\imath}^B = i_B \qquad \text{(borrowing setting)}$$

and the investment-period MARR coincides with the financing-period MARR (MAFR), which is constant:

$$\bar{r}^L = \bar{r}^B = \bar{r} = r.$$

Moreover, for a given setting, neither i_B nor i_L represents the project's rate of return, not even under the stated assumptions. Indeed, i_L and i_B are income rates which apply only in the investment region and the financing region, respectively, not to the whole project. Therefore, at the very best, each of them describes only *one side* of the project's economic profitability under unrealistic assumptions.

TRM did not advance any project's rate of return. Using the AIRR approach with TRM's assumptions, the project's rate of return might indeed be obtained by weighting TRM's lending rate and borrowing rate with the respective implied capital. The result would be a rate of return belonging to the AIRR class, associated with a capital base equal to $x = \text{PV}[C_L(i_B, i_L)] + \text{PV}[C_B(i_B, i_L)]$ where $\text{PV}[C_L(i_B, i_L)] = \sum_{t \in T_L(i_B, i_L)} C_{t-1}(i_B, i_L) d_{t-1,0}$ is the present value of the TRM's capitals in the lending periods and $\text{PV}[C_B(i_B, i_L)] = \sum_{t \in T_B(i_B, i_L)} C_{t-1}(i_B, i_L) d_{t-1,0}$ is the present value of the TRM's capitals in the borrowing periods. As there are two settings, two project rates of return arise. Specifically, using Eq. (8.86), if the investment setting is selected, the rate of return would be

$$\frac{r \cdot \text{PV}[C_B(r, i_L(r))] + i_L(r) \cdot \text{PV}[C_L(r, i_L(r))]}{\text{PV}[C_L(r, i_L(r))] + \text{PV}[C_B(r, i_L(r))]}.$$

If the borrowing setting is selected, the rate of return would be

$$\frac{i_B(r) \cdot \text{PV}[C_B(i_B(r), r)] + r \cdot \text{PV}[C_L(i_B(r), r)]}{\text{PV}[C_L(i_B(r), r)] + \text{PV}[C_B(i_B(r), r)]}.$$

However, why should one compute either rate of return, when another (unambiguous and simpler) one is readily available that is more properly associated with the project's actual economic referents, namely, the average ROI, $\bar{\imath} = \bar{\imath}(\text{PV}[C])$?

In addition, consider that equation $C_n(i_B, i_L) = 0$ may be rewritten as $\sum_{j=0}^{n} F_j(1 + i_B)^{-\alpha_j}(1 + i_L)^{-\beta_j} = 0$, where α_j represents the number of financing periods and β_j represents the number of investment periods between time 0 and time j, so that $\alpha_j + \beta_j = j$, $j = 1, 2, \ldots, n$, and $\alpha_0 = \beta_0 = 0$. In the investment setting, such that $i_B = r$, this implies

$$F_0 + \frac{F_1^{\alpha}}{(1 + i_L)^{\beta_1}} + \frac{F_2^{\alpha}}{(1 + i_L)^{\beta_2}} + \ldots + \frac{F_n^{\alpha}}{(1 + i_L)^{\beta_n}} = 0$$

where $F_j^{\alpha} = F_j(1 + r)^{-\alpha_j}$. Therefore, i_L is the IRR of a cash-flow stream which releases F_j^{α} at time β_j. Likewise, in the borrowing setting, such that $i_L = r$,

$$F_0 + \frac{F_1^{\beta}}{(1 + i_B)^{\alpha_1}} + \frac{F_2^{\beta}}{(1 + i_B)^{\alpha_2}} + \ldots + \frac{F_n^{\beta}}{(1 + i_B)^{\alpha_n}} = 0$$

where $F_j^{\beta} = F_j(1 + r)^{-\beta_j}$. Therefore, i_B is the IRR of a cash-flow stream which releases F_j^{β} at time α_j. Therefore, TRM's lending rate and financing rate are themselves IRRs. Which implies that the pitfalls of the IRR approach reverberate on the TRM approach.

Finally, the TRM model implicitly adopts an unlevered perspective. This implies that it does not address the role of debt and equity in the investment and financing periods, thereby obscuring the notable difference between borrowing from the project and borrowing from debtholders. This has (mis)led some scholars to believe that i_B is indeed the rate at which the firm borrows from debtholders (instead of the rate at which the firm subtracts resources from the assets in the borrowing period) and, therefore, to neglect that, in general, the equilibrium market rate, r, has not to do with the operations of a firm/project (see Sect. 2.3).

To sum up, the TRM approach does not suffer from any problem of existence and the two TRM criteria do lead to rational decisions. However, the model

- is based on two sets of artificial (and mutually exclusive) assumptions
- does not supply any information about the project's rate of return
- suffers from many of the IRR pitfalls (since both i_B and i_L are IRRs)
- does not cope with the firm's debt policy, thereby blurring the distinction between borrowing from the project and borrowing from debtholders.

Example 9.39 ♦
Consider $\boldsymbol{F} = (55, -50, -48, -50, 100)$ and assume the COC is $r = 7\%$.[40] If one sets $i_B = r = 7\%$ (investment setting), then $C_n(i_B, i_L) = 0$ becomes $C_4(7\%, i_L(7\%)) = 0$ whose solution is $i_L(7\%) = 8.8\%$. Therefore, in the borrowing periods, the firm borrows from the project's assets at 7%, which is exactly the same financing rate available in the market (the borrowing periods are then value-neutral). In the investment periods, the firm invests in the project's assets at 8.8%, which is greater than the market lending rate: $i_L(r) = 8.8\% > 7\% = r$. The investment periods are then value-creating and the project is then accepted (Rule (9.18a)).

If, alternatively, one sets $i_L = r = 7\%$ (borrowing setting), then $C_n(i_B, i_L) = 0$ becomes $C_4(i_B(7\%), 7\%) = 0$ whose solution is $i_B(7\%) = 3.8\%$. Under this assumption, in the investment periods, the firm invests in the project's assets at 7%, which is the same return rate expected for investing in the capital market (the investment periods are value-neutral). In the borrowing periods, the firm borrows from the project's assets at 3.8% which is smaller than the market financing rate: $i_B(r) = 3.8\% < 7\% = r$. The borrowing periods are then value-creating and project is then accepted (Rule (9.18b)).

Note that the recommendation is the same regardless of the setting chosen, although the reason is symmetric: In the investment setting, it is implicitly assumed that borrowing from the project has the same cost as borrowing in the capital market, whereas in the borrowing setting, it is implicitly assumed that investment in the project has the same rate of return as an investment in the capital market.

9.8 Popularity of IRR with Practitioners

The IRR is still very popular in practice, notwithstanding the considerable amount of evidence about its theoretical and practical inadequacy to measure an asset's rate of return and the repeated warnings of many academics. There may be tons of reasons why the IRR is still widely employed. We tentatively provide a nonexhausting list of possible reasons.

---

[40] This example is adapted from Magni (2015a), Investment, financing and the role of ROA and WACC in value creation, *European Journal of Operational Research*, 244(3) (August), 855–866 (p. 857).

1. **First move advantage**. The IRR has been devised in a period where no theory in economics linked the notions of capital, cash flow, and rate of return. The field was open for any proposal, there was no competitor to the IRR notion and the proposal of IRR was a "first move" (the first move has always the highest chance of being the winning move).
2. **Venerable ancestry**. The IRR was introduced by three foremost 20th-century economists: Fisher (1930), Boulding (1935), Keynes (1936/1967). This means that the IRR has a venerable academic ancestry and is therefore regarded to hold a stamp of scientific approval.
3. **Effective rate**. The IRR notion extends to firms and projects the concept of "effective rate" or "yield", well-rooted in actuarial mathematics since 19th century (e.g., see Glen 1893).
4. **Loan and savings accounts**. A (constant) interest rate in a loan or savings account is experienced by people everyday. It is easier to use this as a metaphor for a project's rate of return than thinking in terms of a sequence of incomes and a sequence of capitals, which requires an understanding of the mechanics of an economic system.
5. **Objectivity**. The IRR seems to be "objective": It is the solution of an equation, therefore something that magically pops up from a cash-flow stream.
6. **Automatism**. The IRR is automatic in the sense that it does not require the analyst to make any effort of reasoning about capital (many people are not even aware that IRR necessarily devises its own capital values).
7. **Acceptance in accounting**. Accounting scholars do know that IRR essentially means a constant growth rate for the capital, but have somewhat acritically accepted this metric. They have given a firm's IRR the privileged status of "gold standard" and have given the internal capital stream the honorific label of "neutral depreciation", while any other depreciation method is considered "biased". A vast literature has been devoted in accounting to criticizing accounting rates because they are not close enough to IRR, which is regarded as the only meaningful "economic rate of return" (see Feenstra and Wang 2000. See also Remark 8.4).
8. **Imprinting in schools**. Scholars and practitioners have been exposed to IRR at school, so they have a mental habit (an "imprinting") that makes it difficult to conceptualize a rate of return differently.
9. **Unfamiliarity with IRR's pitfalls**. Many people are aware of the multiplicity issue and no-IRR issue, but they are not aware of the other pitfalls.
10. **Do-what-the-other-do strategy**. As testified by the bounded-rationality literature, individuals often use the *do-what-others-do* (or *do-what-successful-people-do*) heuristic: "If everyone uses IRR, then it means it is OK". See Laland (2001).
11. **Use of IRR for healing IRR**. All the attempts of healing the IRR's pitfalls have been based on the notion of IRR itself (e.g., MIRR, TRM), thereby enhancing the idea that the IRR equation cannot be abandoned. This means that, unawarely, scholars and practitioners have sticked with the invariance requirement (9.1), thereby refusing to change it in a more robust and economically significant one.

12. **Excel**. IRR is a canned function in Excel. Anyone can use it with a simple click. This favors the habit of using it without addressing the problem of its reliability.
13. **Internality**. Practitioners seem to prefer relative measures of worth which are internal. Indeed, the IRR seems to be genuinely internal.[41]

All these (and possibly many other) reasons makes IRR to be culturally transmitted in an effective and powerful way. Nowadays, IRR is not a mere rate of return, but a formidable mechanism for storage and transmission by communication, imitation, teaching and learning.

9.9 Key Points

- An internal rate of return (IRR), σ, is defined as a rate that makes the NPV equal to zero. It is then a solution of the equation $\phi(x) = \sum_{t=0}^{n} F_t(1+x)^{-t} = 0$
- The IRR may be interpreted into four ways:
  - the rate that would be contractually stipulated if the project were a loan with constant interest rate
  - the hypothetical rate of return that the project would have if its capital grew at a constant rate of return
  - the actual rate of return of an asset having the same cash flows as the project but a different capital, growing at a constant rate of return
  - a weighted mean of generally time-varying income rates such that the discounted sum of the capital values is equal to the discounted sum of the internal (i.e., IRR-self-produced) values
  - a rate of return belonging to the AIRR class.
- The IRR may not exist or multiple IRRs may arise. These cases may occur in projects with disposal costs, replacement projects, phased expansion projects. It may also crop up when computing the equityholders' IRR of a levered project. And it may also occur in projects with one or more upfront payments as well as in those financial investments where the investor has full control on interim deposits and withdrawals, and in case of internal financing. This problem may also spring up in portfolios with projects having different timing and magnitude. It *always* occurs in Rotating Savings and Credit Associations (ROSCAs).
- Other anomalies of the IRR have to do with (non)additivity of internal values and with the use of the traditional IRR criterion with time-varying COCs. The IRR is not economically meaningful in those cases where the internal (i.e., IRR-implied) capital signals a financing position when, in actual facts, the firm is investing resources in the project. Even when the financial position is correctly signaled, the internal capital amounts are incorrect, for they are self-generated by the IRR

[41]The financial nature of an IRR is not genuinely internal: It depends on the cost of capital. See Chap. 10.

and, therefore, have no actual referents in the business transactions underlying the project activities. Furthermore, the IRR is subject to framing effects, so that a different IRR is calculated depending on how the computations are made; sensitivity analysis with the IRR may prove problematic; the IRR disregards the actual project's operating life. Finally, its use along with the NPV profile, sometimes recommended, is not always reliable.

- Historically, a modified version of IRR has been advanced by some scholars since 1950s, whereby the original cash-flow stream is distorted by assuming integrative investments/financings, in such a way that the modified cash-flow stream possesses one and only one IRR. This approach is called *modified IRR* (MIRR) and generates a class of rates of return. While it solves the problem of uniqueness, it does not solve the problem of existence and it does not properly measure the rate of return of the project, for the integrative operations it rests on are fictitious and, anyway, have not to do with the project. Also, the MIRR concept is ambiguous, since there are infinitely many ways of modifying a cash-flow stream and, therefore, infinitely many MIRRs; scholars do not agree about what the "relevant" MIRR should be. In addition, being an IRR, the MIRR is also subject to all the other mishaps of IRR. The AIRR is not subject the IRR's pitfalls.[42]
- The TRM model, developed in the 1960s, addressed the classical difficulties of the IRR by assuming that the capital growth is based on two rates, a lending-period rate, i_L, and a borrowing-period rate, i_B, one of which must be set equal to the COC in order to get a rational decision rule. The assumptions of this approach imply that, in any project, economic value is created or destroyed either in the investment periods or in the financing periods, but not in both. TRM's investment rate and financing rate are two rates which induce two decision rules; both are rational decision criteria but it is not clear which is the "relevant" rate. Both may be viewed as members of the AIRR class, associated with the total capital invested in the investment periods and with the total capital borrowed in the financing periods, respectively. TRM's implicit assumptions are unrealistic, the actual economic referents are distorted and forced to comply with artificial magnitudes. Finally, neither rate describes the project's performance; rather, at the very best, they describe only part of the project's performance, the one which is generated in the investment region (project investment rate) or, in the alternative setting, the one which is generated in the financing region (project financing rate). The TRM model did not provide the rate of return for the project as a whole.

[42] An exception is the global (non)additivity of AIRR when the COCs of a portolio's constituent assets are not equal.

Chapter 10
Internal Average Rate of Return and Aggregate Return on Investment

> Accounting measurements are, at this writing, the only basic sources of data which establish (however imperfectly) the income for a period, the amount of investment, and the bases of classification and matching which establish the rate of return currently being realized by operations or projects in which we are interested
>
> Vatter (1966, p. 682)

C. A. Magni, *Investment Decisions and the Logic of Valuation*,
https://doi.org/10.1007/978-3-030-27662-1_10

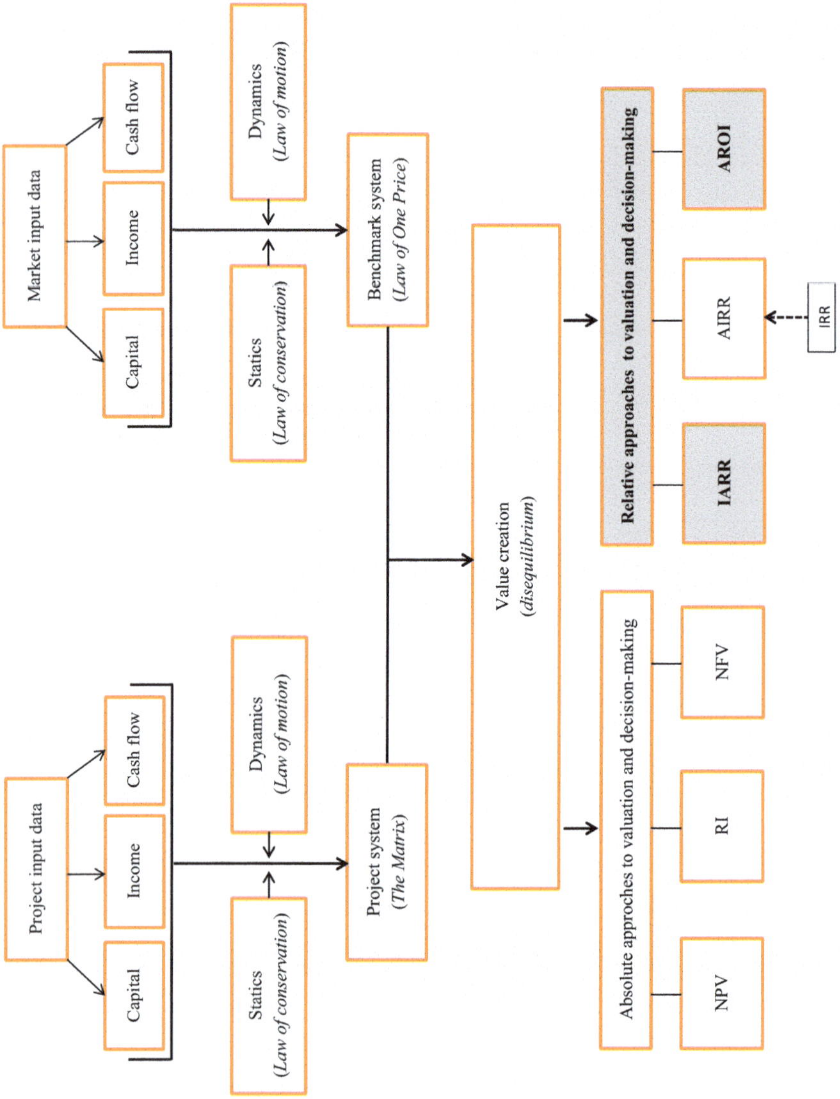

Fig. 10.1 Chapter 10: The internal-average-rate-of-return approach and the aggregate-return-on-investment approach

In Chap. 8 we have presented the AIRR approach, associated with a well-defined benchmark system representing the equilibrium in the capital market. In this chapter we present two other relative approaches, which use different benchmark systems to represent the same market equilibrium.

The chapter begins with the presentation of the **Internal Average Rate of Return** (IARR) approach. The IARR does not depend on the COC; nor does its financial nature depend on the COC; as such, it is a **purely internal** rate of return (as opposed to the IRR, whose financial nature is not internal). Also, the IARR does not incur the IRR's fallacies, so it may be used in place of IRR by those practioners who are willing to calculate a reliable, internal relative measure of worth.

Then, we illustrate the **Aggregate Return On Investment** (AROI) approach. While it shares with IARR the same project's rate of return, its associated MARR is different, being based on a different benchmark system.

We show that IARR and the AROI are conveniently associated with cash flows and capitals, respectively. This implies that the three absolute measures of worth and the three relative measures of worth may be coupled, giving rise to three twofold valuation systems. Specifically, we show that

- the pair **(AIRR, RI)** may be conveniently conceptualized as an income-based methodology
- the pair **(IARR, NPV)** may be conveniently conceptualized as a cash-flow-based methodology
- the pair **(AROI, NFV)** may be conveniently conceptualized as a capital-based methodology.

The three methodologies are logically equivalent; they produce the same value and the same decision.

As a result, the analyst has a complete toolkit of approaches for appraising projects, conducting refined economic analyses, and making rational decisions. This unified theory, based on an explicit link between project scale and financial efficiency, provides a **complete reconciliation between absolute valuation methods and relative valuation methods**.

After completing this chapter, the reader will be able to:

- compute an *internal* project rate of return which does not suffer from the pitfalls of the multi-period IRR
- understand the difference between the Internal Average Rate of Return (IARR) approach and the AIRR approach
- learn the notion of Aggregate Return on Investment (AROI) and its relation with IARR
- use internal average ROI and internal average ROE to appraise a project and make rational decisions
- calculate the project MARR and the equity MARR
- reconcile absolute measures of worth and relative measures of worth
- understand that the differences among the three appraisal methods rest on the different benchmark systems employed as counterfactual alternatives
- associate the three basic notions to the three valuation methodologies: The (AIRR, RI) method is associated with the notion of income, the (IARR, NPV) is associated with the notion of cash flow, and the (AROI, NFV) method is associated with capital
- understand that all the methods are different ways of conceptualizing value creation and rational decision-making, thereby acting as three pairs of glasses which shed different lights on the same project.

10.1 IARR

Practitioners seem to appreciate relative measures of worth that are *internal*, which is one of the reasons why the multiperiod IRR still has its mentors in academia and is extensively used in practice (see Sect. 9.8). However, we have shown that IRR suffers from (theoretical and practical) limitations that make it inappropriate as a general notion of rate of return, while the AIRR approach and, in particular, the average ROI and the average ROE, are more reliable relative measures of worth.

However, AIRR is not internal and one may ask whether a reliable, genuinely internal rate of return possessing the same favorable features as the AIRR exists which, at the same time, is not blemished by the shortcomings of the IRR.

The answer is positive. However, before presenting the internal measure, it is worth clarifying why the IRR is not *genuinely* internal. More precisely, IRR is internal but its financial nature is not. Indeed, while it is true that the IRR equation $\sum_{t=0}^{n} F_t(1+x)^t = 0$ does not contain the COC, the nature of σ depends on the internal values, and the latter depend on the COC. Consider a stream of CFAs, $\boldsymbol{F} = (F_0, F_1, \ldots, F_n)$ and its associated internal stream, $\boldsymbol{C}(\sigma) = (C_0(\sigma), C_1(\sigma), C_2(\sigma), \ldots, C_{n-1}(\sigma), 0)$. We remind that

$$\text{NPV}(1+r) = \text{PV}[C(\sigma)](\sigma - r) \tag{10.1}$$

(see Eq. (9.6)). The IRR is an investment rate if and only if $\text{PV}[C_{t-1}(\sigma)] > 0$, while it is a financing rate if and only if $\text{PV}[C(\sigma)] < 0$. Given that $\text{PV}[C(\sigma)] = \sum_{t=1}^{n} C_{t-1}(\sigma)(1+r)^{-(t-1)}$ is a function of r as well as σ, the financial nature of the IRR changes under changes in the COC. For example, 10% is an IRR of $\boldsymbol{F} = (-100, 150, 100, -158.4)$. However, the financial nature of IRR depends on the COC. Specifically, for a given r, $\text{PV}[C(\sigma)] = 100 - 40/(1+r) - 144/(1+r)^2$, whence

$$\begin{aligned} &\text{PV}[C(\sigma)] > 0 \quad \text{for all } r > 0.416553 \\ \text{and} \quad &\text{PV}[C(\sigma)] < 0 \quad \text{for all } r < 0.416553. \end{aligned}$$

Therefore, the IRR is an investment rate if $r > 0.416553$, it is a financing rate if $r < 0.416553$. (If $r = 0.416553$, the IRR's financial nature is not clear.)

This implies that IRR may not be considered a genuinely internal valuation metric. By no means we intend that this is a drawback. Quite the contrary, this fact is rather natural from Definition 8.4. The reader should not be surprised that monetary amounts available at different dates have different present values (and, therefore, different signs) depending on *external* rates (i.e., market rates). After all, the choice of a relative measure of worth should not be selected on the basis of whether it is or it is not internal. Internality itself may even lose any significance in some domains. For example, the COC is itself an IRR, and its internal values are, at the same, time market values: There is no separation between internal and external factors, so it is indeed an internal measure but, at the same time, it depends on market forces. Also, consider a financial project consisting in investing in a portfolio of n securities traded in the capital market which are held in some proportion. Given a benchmark index based on (some or all of) those assets held in different proportions, the holding period rates $(r_1, r_2, \ldots, r_n)$ of the constituent assets are the same for the project and the benchmark index. Hence, the project's period rates (which depend on the constituent asset's period rates) are, at the same time, internal (internal to the project) and external (i.e. they are market rates). As a result, the portfolio's IRR will be, at the same time, internal and external (since it depends on the market rates). Therefore, in some cases, the notion of *internal* rate as opposed to an *external* rate does not make much sense.[1]

That said, some practitioners seem to be particularly concerned about finding a rate of return which may be genuinely internal. We show that it is possible to work out a rate of return from income and capital which

- is coherent (i.e., it is a ratio of return to capital)
- has the same nice features as the AIRR
- does not suffer from the IRR's pitfalls
- is purely internal, in the sense that it does not depend on the COC and its financial nature does not depend on the COC.

To begin with, let us assume $I_0 = 0$ and let us compare the project system and the benchmark system associated with the NPV approach. Specifically, let us pick up Matrix (6.18), which matches the project's change in capital against a replicating portfolio's change in capital:

$$\begin{array}{c}\textbf{Project}\\ \begin{array}{|c|ccc|}\hline +\Delta C_t & +i_t C_{t-1} & -F_t \\ \hline +\Delta C_t^d & +i_t^d C_{t-1}^d & -F_t^d \\ \hline +\Delta C_t^e & +i_t^e C_{t-1}^e & -F_t^e \\ \hline \end{array}\end{array} \quad \begin{array}{c}\textbf{Benchmark}\\ \begin{array}{|c|ccc|}\hline +\Delta V_t & +r_t V_{t-1} & -F_t \\ \hline +\Delta V_t^d & +r_t^d V_{t-1} & -F_t^d \\ \hline +\Delta V_t^e & +r_t^e V_{t-1} & -F_t^e \\ \hline \end{array}\end{array} \tag{10.2}$$

Subtracting the elements of the benchmark Matrix from the corresponding elements of the project Matrix, one gets an economically significant incremental Matrix which informs, in each period, about the project's return over and above the profit of the replicating portfolio:

$$\begin{array}{c}\textbf{Incremental Matrix}\\ \begin{array}{|c|ccc|}\hline +\Delta C_t - \Delta V_t & +I_t - r_t \cdot V_{t-1} & 0 \\ \hline +\Delta C_t^d - \Delta V_t^d & +I_t^d - r_t^d \cdot V_{t-1}^d & 0 \\ \hline +\Delta C_t^e - \Delta V_t^e & +I_t^e - r_t^e \cdot V_{t-1}^e & 0 \\ \hline \end{array}\end{array} \tag{10.3}$$

The economic profitability of the whole project is embedded in the set of n such incremental Matrices, $t = 1, 2, \ldots, n$. More precisely, focusing on the first row and summing through 1 to n,

[1]This is also true whenever fair-value accounting is applied. Fair-value accounting means that assets are evaluated at their fair (market) value, $C_t = V_t$, and, therefore, the capital amounts (and the relative income rates) are, at the same time, internal and external.

$$\sum_{t=1}^{n}(\Delta C_t - \Delta V_t) = \sum_{t=1}^{n} I_t - \sum_{t=1}^{n} I_t^V \tag{10.4}$$

where, as we remind, $I_t^V = r_t V_{t-1}$ is the profit of the benchmark portfolio. Unfolding the elements of the left-hand side,

$$\begin{aligned}\sum_{t=1}^{n}(\Delta C_t - \Delta V_t) &= \big(C_1 - C_0 + C_2 - C_1 + \cdots + C_n - C_{n-1}\big)\\ &\quad - \big(V_1 - V_0 + V_2 - V_1 + \cdots + V_n - V_{n-1}\big)\\ &= -C_0 + C_n - (-V_0 + V_n)\\ &= V_0 - C_0 = [\text{since} I_0 = 0] = \text{NPV}.\end{aligned} \tag{10.5}$$

This implies

$$\text{NPV} = C \cdot (\bar{\jmath} - \bar{\rho}) \tag{10.6}$$

where C is defined as the undiscounted sum of the invested capital, $C = \sum_{t=1}^{n} C_{t-1}$, and

$$\bar{\jmath} = \frac{I}{C} = \frac{I_1 + I_2 + \ldots + I_n}{C_0 + C_1 + \ldots + C_{n-1}} = \frac{\text{Total income}}{\text{Total capital}} \tag{10.7}$$

is called *Internal Average Rate of Return* (IARR) while

$$\bar{\rho} = \frac{I^V}{C} = \frac{I_1^V + I_2^V + \ldots + I_n^V}{C_0 + C_1 + \ldots + C_{n-1}} = \frac{\text{Total benchmark return}}{\text{Total capital}} \tag{10.8}$$

is the related project COC/MARR. If $C_t \neq 0$ for all $t \in \mathbb{N}_{n-1}^1$, the income may be framed as $I_t = i_t C_{t-1}$, so the IARR may be framed as a weighted mean of income rates:

$$\bar{\jmath} = \frac{i_1 C_0 + i_2 C_1 + \ldots + i_n C_{n-1}}{C_0 + C_1 + \ldots + C_{n-1}} \tag{10.9}$$

while the associated MARR is

$$\bar{\rho} = \frac{r_1 V_0 + r_2 V_1 + \ldots + r_n V_{n-1}}{C_0 + C_1 + \ldots + C_{n-1}}. \tag{10.10}$$

The interpretation of $\bar{\jmath}$ and $\bar{\rho}$ should be clear: The investors invest C at a rate of return equal to $\bar{\jmath}$; if, alternatively, one were to invest the same amount in a market (value-neutral) portfolio replicating the project's cash flows (from time 1 to time n), one would earn an overall rate of return of $\bar{\rho}$. Note that the IARR and the MARR are coherent so the product structure required by the notion of rate of return is fulfilled, with the investment scale being expressed by C and the economic efficiency being expressed by the excess IARR, here denoted as $\epsilon = \bar{\jmath} - \bar{\rho}$.

The following definition mirrors Definition 8.4.

Definition 10.1 (*Investment or financing*) Let A be a mixed asset, such that $C_j \cdot C_h < 0$ for some $j, h \in \mathbb{N}^0_{n-1}$. Then, A is an investment if $C > 0$, A is a financing (borrowing) if $C < 0$.

We can now supply the following decision criterion, which mirrors Definition 8.5.

Definition 10.2 (*IARR criterion for accept/reject decisions*) Let A be any multiperiod economic activity associated with a profit stream equal to $\boldsymbol{I}$ and a capital stream equal to $\boldsymbol{C}$. Then,

- if A is an investment, it is worth undertaking if and only if $\bar{\jmath} > \bar{\rho}$
- if A is a financing, it is worth undertaking if and only if $\bar{\jmath} < \bar{\rho}$.

Further, the investor is indifferent about accepting or rejecting A if and only if $\bar{\jmath} = \bar{\rho}$.

From (10.6), the rationality of IARR is straightforward.

Proposition 10.1 (Rationality of IARR) *The IARR criterion for accept/reject decision is rational: An investment creates value if and only if $\bar{\jmath} > \bar{\rho}$; a financing creates value if and only if $\bar{\jmath} < \bar{\rho}$. A project is value-neutral if and only if $\bar{\jmath} = \bar{\rho}$.*

If $C_{t-1} \neq 0$ for all t, the IARR may be viewed as a Chisini mean where the invariant function is $g(i_1, i_2, \ldots, i_n) = \sum_{t=1}^{n} i_t C_{t-1}$. As for $\bar{\rho}$, it may be noted that it is an adjusted COC which takes into account the fact that, if the investor invested in the replicating portfolio, the invested capital would be V_t, not C_t. More precisely, one can rewrite $\bar{\rho}$ as

$$\bar{\rho} = \bar{r}_V \cdot \frac{V}{C} \tag{10.11}$$

where $V = \sum_{t=1}^{n} V_{t-1}$ and $\bar{r}_V$ is the undiscounted weighted mean of the (pre-tax) average WACCs:[2]

$$\bar{r}_V = \frac{r_1 V_0 + r_2 V_1 + \ldots + r_n V_{n-1}}{V_0 + V_1 + \ldots + V_{n-1}}. \tag{10.12}$$

[2]Interestingly, (10.12) holds even if there exists some t for which $V_{t-1} = 0$. The reason is that r_t is determined by demand and supply in the market, not by V_{t-1}. In mathematical terms, r_t is not a function of V_{t-1} (in contrast, i_t is a function of C_{t-1} so it is not defined if $C_{t-1} = 0$).

The average WACC, $\bar{r}_V$, is itself a Chisini mean with invariant function equal to $g(r_1, r_2, \ldots, r_n) = \sum_{t=1}^{n} r_t V_{t-1}$. Therefore, $\bar{\rho}$ is a weighted arithmetic mean of COCs adjusted for the value-to-book ratio (the latter expresses how more or less capital should be invested in order to replicate the project's prospective cash flows in a market portfolio). Note that, if $r_t = r$ for all t, the MARR becomes, simply, $\bar{\rho} = r \cdot (V/C)$.

The MARR may itself be viewed as a Chisini mean. To see it, consider the adjusted (period) COC, denoted as $\rho_t = r_t(V_{t-1}/C_{t-1})$ and pick $g(\rho_1, \rho_2, \ldots, \rho_n) = \sum_{t=1}^{n} \rho_t C_{t-1}$ as the invariant function. Then, the equation $g(\rho_1, \rho_2, \ldots, \rho_n) = g(\bar{\rho}, \bar{\rho}, \ldots, \bar{\rho})$ brings

$$\bar{\rho} = \frac{\rho_1 C_0 + \rho_2 C_1 + \ldots + \rho_n C_{n-1}}{C_0 + C_1 + \ldots + C_{n-1}}. \tag{10.13}$$

Given the definition of ρ_t and the definition of V, (10.11) and (10.13) are equal.

Hence, the excess IARR may be found as the weighted mean of the excess holding period rates, $\epsilon_t = i_t - \rho_t$:

$$\epsilon = \bar{\jmath} - \bar{\rho} = \frac{\epsilon_1 C_0 + \epsilon_2 C_1 + \ldots + \epsilon_n C_{n-1}}{C_0 + C_1 + \ldots + C_{n-1}}.$$

The reader should not confuse $\bar{r}$, $\bar{\rho}$, and $\bar{r}_V$. They have related-but-different economic meaning.

- The rate $\bar{r}$ is a **linear combination of the period COCs weighted by the project's invested amounts** $C_{t-1}\mathrm{d}_{t-1,1}$. It represents the overall rate of return that investors would earn if they invested, in each period, the amount C_{t-1} at the rate r_t. This scenario involves the purchase, on the market, of a portfolio replicating the capital amounts (not the cash flows) of the project from time 0 to time $n-1$. This scenario is expressed by the triplet $(\boldsymbol{C}, \boldsymbol{I}^*, \boldsymbol{F}^*)$ and gives rise to the RI and AIRR benchmark system. For decision-making, $\bar{r}$ must be compared with $\bar{\imath}$.
- The rate $\bar{r}_V$ (average COC) is a **mean of the COCs weighted by the project's market values**, V_{t-1}. It represents the overall rate of return that investors would earn if they invested, in each period, the amount V_{t-1} at the rate r_t. This scenario involves the purchase, on the market, of a portfolio replicating the cash flows (not the capital amounts) from time 1 to time n and is affiliated to the NPV and IARR benchmark system, characterized by the triplet $(\boldsymbol{V}, \boldsymbol{I}^V, \boldsymbol{F}^V)$.
- The rate $\bar{\rho}$ is a **mean of adjusted (time-varying) COCs weighted by the project's invested amounts** C_{t-1} and expresses the same return as the average COC, $\bar{r}_V$, but is measured in units of the project's invested capital C rather than in units of the replicating portfolio's capital V. In other words, a portfolio whereby the investor invests V_{t-1} at the rate r_t is financially equivalent to a portfolio whereby the investor invests C_{t-1} at the rate ρ_t. For decision-making, $\bar{\rho}$ must be compared with $\bar{\jmath}$.

$$\bar{r} = \frac{r_1 C_0 + r_2 C_1 d_{2,1} + \ldots + r_n C_{n-1} d_{n-1,1}}{C_0 + C_1 d_{1,0} + \ldots + C_1 d_{1,0}}$$

$$\bar{r}_V = \frac{r_1 V_0 + r_2 V_1 + \ldots + r_n V_{n-1}}{V_0 + V_1 + \ldots + V_{n-1}}$$

$$\bar{\rho} = \frac{\rho_1 C_0 + \rho_2 C_1 + \ldots + \rho_n C_{n-1}}{C_0 + C_1 + \ldots + C_{n-1}}.$$

Finally, note that the IARR is indeed coherent, for it is an amount of return (numerator) measured per unit of capital employed (denominator).

Remark 10.1 (*More on the opportunity cost of capital,* $\bar{\rho}$). The return rate ρ_t takes into consideration the fact that the capital invested in the replicating portfolio is V_{t-1}, not C_{t-1}. Therefore, the return on V_{t-1} may be viewed as the result of the combined effect of two different components, namely, (i) a rate component (the return foregone on the capital actually employed in the project); (ii) a capital component (the return foregone on the excess capital invested in the replicating portfolio):

$$I_t^V = r_t V_{t-1} = \overbrace{r_t C_{t-1}}^{\text{rate component}} + \overbrace{r_t (V_{t-1} - C_{t-1})}^{\text{capital component}}.$$

For this reason, the cost of capital is a multiple of the equilibrium rate r_t:

$$\rho_t = r_t + r_t \cdot \frac{V_{t-1} - C_{t-1}}{C_{t-1}} = r_t \cdot \frac{V_{t-1}}{C_{t-1}}.$$

The ratio V_{t-1}/C_{t-1} is a correction factor which adjusts for the different capital, so ρ_t may be expressed as per unit of project's invested capital. If the market rate is time-invariant, by aggregating the various adjusted COCs one gets the overall COC: $\bar{\rho} = r(V/C)$. If the market rates are time-varying, then their average $\bar{r}_V$ is recruited: $\bar{\rho} = \bar{r}_V(V/C)$. For example, if $\bar{r}_V = 10\%$ and $V/C = 1.2$, the firm is not merely renouncing to earn, on average, 10% on the capital C invested in the project; it is also renouncing to invest an extra 20% of capital at a 10% return. Therefore, the cost of capital is 10%(1.2) = 12%. The excess 2% is due to the additional return foregone on the excess capital $V - C = 0.2C$. ▲

A useful shortcut may be used for computing the project COC, $\bar{\rho}$, which avoids the computation of the economic value, V. It directly stems from (10.6) by solving the equation for $\bar{\rho}$:

$$\bar{\rho} = \bar{\jmath} - \frac{\text{NPV}}{C}. \tag{10.14}$$

Vice versa, the IARR may be computed as

$$\bar{\jmath} = \bar{\rho} + \frac{\text{NPV}}{C}. \tag{10.15}$$

We have so far presented the IARR as a rate of return derived from profits and capital amounts or via a shortcut which makes use of the NPV. However, the link with cash flows may be made more interesting recalling that the algebraic sum of the profits is always equal to the algebraic sum of the cash flows. More precisely, let $F = \sum_{t=0}^{n} F_t$ be the project's net cash flow. Then, exploiting the usual law of motion, $F_t = I_t - \Delta C_t$, one gets

$$F = \sum_{t=0}^{n}(I_t - \Delta C_t) = \sum_{t=0}^{n} I_t - \overbrace{\sum_{t=0}^{n} \Delta C_t}^{=0} = I$$

(see also Eq. (1.7)). Then, one may write

$$\bar{\jmath} = \frac{I}{C} = \frac{F}{C} = \frac{F_0 + F_1 + \ldots + F_n}{C_0 + C_1 + \ldots + C_{n-1}}. \tag{10.16}$$

As for the MARR, we make use of the law of motion again: $F_t = I_t^V - \Delta V_t$ for $t \geq 1$. Therefore,

$$\sum_{t=1}^{n} I_t^V = \sum_{t=1}^{n}(F_t + \Delta V_t) = F_1 + F_2 + \ldots + F_n - V_0.$$

Hence, by (10.13), $\sum_{t=1}^{n} \rho_t C_{t-1} = \bar{\rho} C = \sum_{t=1}^{n} r_t V_{t-1} = \sum_{t=1}^{n} I_t^V$, which implies

$$\bar{\rho} = \frac{F_1 + F_2 + \ldots + F_n - V_0}{C_0 + C_1 + \ldots + C_{n-1}}. \tag{10.17}$$

It is then worth underlining the difference between the IARR and its MARR in terms of the latest framings. Since $F_0 = -C_0$ (we are assuming $I_0 = 0$), the NPV springs up from the difference between the two numerators:

numerator of $\bar{\jmath}$	$+F_1$	$+F_2$	$+\ldots$	$+F_n$	$-C_0$
numerator of $\bar{\rho}$	$+F_1$	$+F_2$	$+\ldots$	$+F_n$	$-V_0$
difference	$+0$	$+0$	$+\ldots$	$+0$	$+$NPV

Since $F - \text{NPV} = F_1 + \ldots + F_n - V_0$, one may also rewrite (10.17) as

$$\bar{\rho} = \frac{F - \text{NPV}}{C} = \frac{\sum_{t=0}^{n} F_t(1 - d_{t,0})}{C} \tag{10.18}$$

(which is also straightforward from (10.14) and (10.16)).

Four methods for calculating a project's Internal Average Rate of Return and the corresponding COC/MARR

A project's rate of return (IARR) and a project COC may be conceptualized and computed in four different-but-equivalent ways:

① $\bar{\jmath} = \frac{I}{C} = \frac{\sum_{t=1}^{n} I_t}{\sum_{t=1}^{n} C_{t-1}}$ $\qquad \bar{\rho} = \frac{I^V}{C} = \frac{\sum_{t=1}^{n} I_t^V}{\sum_{t=1}^{n} C_{t-1}}$

② $\bar{\jmath} = \alpha_1 i_1 + \alpha_2 i_2 + \ldots + \alpha_n i_n$ $\quad \bar{\rho} = \alpha_1 \rho_1 + \alpha_2 \rho_2 + \ldots + \alpha_n \rho_n$ $(\alpha_t = \frac{C_{t-1}}{C})$

③ $\bar{\jmath} = \frac{F}{C} = \frac{\sum_{t=1}^{n} F_t - C_0}{\sum_{t=1}^{n} C_{t-1}}$ $\qquad \bar{\rho} = \frac{F - \text{NPV}}{C} = \frac{\sum_{t=1}^{n} F_t - V_0}{\sum_{t=1}^{n} C_{t-1}}$

④ $\bar{\jmath} = \bar{\rho} + \frac{\text{NPV}}{C}$ $\qquad \bar{\rho} = \bar{\jmath} - \frac{\text{NPV}}{C}$

(Compare with the four methods for AIRR computations in Chap. 8).

Remark 10.2 (*Copying with nonzero time-0 income,* $I_0 \neq 0$) The IARR is easily generalized for coping with the case $I_0 \neq 0$. Considering that

- $\sum_{t=1}^{n}(\Delta C_t - \Delta V_t) = V_0 - C_0$ (see (10.5))
- $\text{NPV} = V_0 + I_0 - C_0$ (see Eq. (5.14))

then, $\sum_{t=1}^{n}(\Delta C_t - \Delta V_t) = \text{NPV} - I_0$. This implies, owing to (10.4), $\text{NPV} - I_0 = \sum_{t=1}^{n} I_t - \sum_{t=1}^{n} I_t^V$ whence $\text{NPV} = \sum_{t=0}^{n} I_t - \sum_{t=0}^{n} I_t^V$ where $I_0^V = 0$. Dividing by C the right-hand side of the equality, $\text{NPV} = C(\bar{\jmath} - \bar{\rho})$ holds where $\bar{\jmath}$ is generalized as $\bar{\jmath} = \sum_{t=0}^{n} I_t / \sum_{t=0}^{n} C_t$. This is formula ① where I is redefined as $I = I_0 + I_1 + \ldots + I_n$.

Formula ② needs only to be integrated with the summand I_0/C.

As for formula ③, the first ratio, namely $\bar{\jmath} = F/C$, remains unvaried since $\sum_{t=0}^{n} I_t = \sum_{t=0}^{n}(F_t + \Delta C_t) = \sum_{t=0}^{n} F_t$ (remember that $\Delta C_0 = C_0$). However, considering that $F_0 = I_0 - \Delta C_0 = I_0 - C_0$, the numerator of the second ratio must be integrated by I_0: $\bar{\jmath} = (F_1 + F_2 + \ldots + F_n + I_0 - C_0)$.

Formula ④ remains unvaried. (The formulas for MARR do not change in any case.) Hence,

$$
\begin{aligned}
&① \quad \bar{\jmath} = \frac{I}{C} = \frac{\sum_{t=0}^{n} I_t}{C} \\
&② \quad \bar{\jmath} = (I_0/C) + \alpha_1 i_1 + \alpha_2 i_2 + \ldots + \alpha_n i_n \\
&③ \quad \bar{\jmath} = \frac{F}{C} = \frac{\sum_{t=1}^{n} F_t + \overbrace{(I_0 - C_0)}^{F_0}}{C} \\
&④ \quad \bar{\jmath} = \bar{\rho} + \frac{\text{NPV}}{C}.
\end{aligned}
\tag{10.19}
$$

▲

Example 10.1 ♦

C^o C^e (*Applying the four methods, assuming time-varying COCs*) Consider an unlevered project whose concise transposed Matrix is as follows:

$(\boldsymbol{C}, \boldsymbol{I}, \boldsymbol{F})$

Time		0	1	2	3	4	5	6
Capital	$\boldsymbol{C}$	120	60	10	−80	6	4	0
Income	$\boldsymbol{I}$		24	−6	5	12	−2	5
Cash Flow	$\boldsymbol{F}$	−120	84	44	95	−74	0	9

The vector of income rates is obtained by dividing each income by the respective BOP capital:

$$\boldsymbol{i} = (0.2, -0.1, 0.5, -0.15, -0.333, 1.25).$$

We assume the COC is time-varying and equal to

$$\boldsymbol{r} = (0.2, 0.3, 0.4, 0.25, 0.15, 0.1)$$

(this is the same project described in Examples 8.6–8.7).

The investor might pursue an alternative investment strategy and invests V_0 in an equivalent-risk portfolio whose cash-flow stream $\boldsymbol{F}^V$ replicates the project's cash-flow stream $\boldsymbol{F}$ from time 1 to time n. The concise transposed Matrix of the benchmark system is

$(\boldsymbol{V}, \boldsymbol{I}^V, \boldsymbol{F}^V)$

Time		0	1	2	3	4	5	6
Capital	$\boldsymbol{V}$	117.2	56.64	29.64	−53.51	7.11	8.18	0
Income	$\boldsymbol{I}^V$		23.44	16.99	11.85	−13.38	1.07	0.82
Cash Flow	$\boldsymbol{F}^V$	−117.2	84	44	95	−74	0	9

The values of the benchmark portfolio are the project's economic values and are obtained by the usual recursive relation $V_t = V_{t-1}(1 + r_t) - F_t$, for $t \in \mathbb{N}^1_n$ and the profits are obtained as $I^V_t = r_t V_{t-1}$.[3] We now compute the IARR and the related MARR using each of the four methods presented.

[3] More precisely (and with a pinch of pedantry): $V_t = V_{t-1}(1 + r_t) - F^V_t$ for $t \in \mathbb{N}^0_n$ where $F^V_t = F_t$ for $t \in \mathbb{N}^1_n$ and $F^V_0 = -V_0$.

① Applying (10.7),

$$\bar{\jmath} = \frac{24 - 6 + 5 + 12 - 2 + 5}{120 + 60 + 10 - 80 + 6 + 4} = \frac{38}{120} = 31.67\%.$$

Applying (10.8)

$$\bar{\rho} = \frac{23.44 + 16.99 + 11.85 - 13.38 + 1.07 + 0.82}{120 + 60 + 10 - 80 + 6 + 4} = \frac{40.79}{120} = 34\%.$$

② The weights are

$$\begin{aligned}
\alpha_1 &= \frac{120}{120} &&= 100\% \\
\alpha_2 &= \frac{60}{120} &&= 50\% \\
\alpha_3 &= \frac{10}{120} &&= 8.33\% \\
\alpha_4 &= \frac{-80}{120} &&= -66.67\% \\
\alpha_5 &= \frac{6}{120} &&= 5\% \\
\alpha_6 &= \frac{4}{120} &&= 3.33\%.
\end{aligned}$$

and the adjusted COCs are

$$\begin{aligned}
\rho_1 &= 0.2 \cdot \frac{117.2}{120} &&= 19.5\% \\
\rho_2 &= 0.3 \cdot \frac{56.64}{60} &&= 28.3\% \\
\rho_3 &= 0.4 \cdot \frac{29.64}{10} &&= 118.5\% \\
\rho_4 &= 0.25 \cdot \frac{-53.51}{-80} &&= 16.7\% \\
\rho_5 &= 0.15 \cdot \frac{7.11}{6} &&= 17.8\% \\
\rho_6 &= 0.1 \cdot \frac{8.18}{4} &&= 20.5\%.
\end{aligned}$$

Applying (10.9)–(10.10),

$$\begin{aligned}
\bar{\jmath} &= 20\% \cdot 100\% - 10\% \cdot 50\% + 50\% \cdot 8.33\% \\
&\quad - 15\% \cdot (-66.67\%) - 33.3\% \cdot 5\% + 125\% \cdot 3.33\% = 31.67\%
\end{aligned}$$

and

$$\bar{\rho} = 19.5\% \cdot 100\% + 28.3\% \cdot 50\% + 118.5\% \cdot 8.33\% + 16.7\% \cdot (-66.67\%) + 17.8\% \cdot 5\% + 20.5\% \cdot 3.33\% = 34\%.$$

③ Applying (10.16),

$$\bar{j} = \frac{84 + 44 + 95 - 74 + 9 - 120}{120 + 60 + 10 - 80 + 6 + 4} = \frac{38}{120} = 31.67\%.$$

Applying (10.17),

$$\bar{\rho} = \frac{84 + 44 + 95 - 74 + 9 - 117.2}{120 + 60 + 10 - 80 + 6 + 4} = \frac{40.8}{120} = 34\%.$$

④ Applying (10.15),

$$\bar{j} = 0.34 + \frac{117.2 - 120}{120} = 31.67\%.$$

Applying (10.14),

$$\bar{r} = 0.3167 - \frac{117.2 - 120}{120} = 34\%.$$

(Compare with the results of AIRR in Example 8.7.)

Therefore, the project scale is \$120 and the economic efficiency is $\epsilon = \bar{j} - \bar{\rho} = 31.67\% - 34\% = -2.33\%$. The assumptions are such that the project destroys value: $\text{NPV} = 120 \cdot (-2.33\%) = -2.8$. The investor overall invests \$120 at a rate which is smaller than the MARR by 2.33%. As the project is unlevered, $\text{NPV} = \text{NPV}^e$, the project decreases shareholders' wealth and, therefore, it should not be undertaken.

Example 10.2 ♦
(*Constant COC*) Consider the following project:

$(\boldsymbol{C}, \boldsymbol{I}, \boldsymbol{F})$

Time		0	1	2
Capital	C	300	150	0
Income	I	0	90	80
Cash Flow	F	−300	240	230

and let $r = 0.1$ be the (assumed constant) COC. Then, the IARR is easily computed as

$$\bar{j} = \frac{90 + 80}{300 + 150} = 37.78\%.$$

If one uses the cash flows to compute the IARR, the calculation is

$$\bar{j} = \frac{-300 + 240 + 230}{300 + 150} = 37.78\%.$$

For illustrative purposes, we now compute the MARR using (10.11). The market values are $V_1 = 230/1.1 = 209.09$ and $V_0 = (209.09 + 240)/1.1 = 408.26$. Therefore, the MARR is

$$\bar{\rho} = \bar{r}_V \cdot \frac{V}{C} = 0.1 \cdot \frac{408.26 + 209.09}{300 + 150} = 13.72\%.$$

If investors invested in the replicating portfolio, they would invest a total $\$617.35 = 408.26 + 209.09$ at 10%, which is financially equivalent to an investment of $\$450 = 300 + 150$ at 13.72%. More precisely, investors would invest in the benchmark portfolio \$167.35 ($= 617.35 - 450$) more than what they invest in the project. That additional capital would be invested at 10%. Therefore, they would get 10% on \$450 and 10% on an additional \$167.35 or, equivalently, they would earn a 10% return on a total capital which is 37.2% ($= 167.35/450$) greater than the project overall capital. This results in a COC of $10\% \cdot (1 + 37.2\%) = 13.72\%$. The financial efficiency is $\epsilon = 37.78\% - 13.72\% = 24.06\%$. The NPV may be decomposed into project scale and financial efficiency:

$$\text{NPV} = \overbrace{450}^{\text{scale}} \cdot \overbrace{24.06\%}^{\text{efficiency}} = 108.3.$$

The project COC might also be computed very easily with no recourse to economic values using the shortcut (10.14):

$$\bar{\rho} = 37.78\% - \frac{-300 + 240/1.1 + +230/1.1^2}{450} = 13.72\%.$$

This shortcut is particularly time-saving from a computational point of view.[4]

[4]For this project, if one uses the AIRR approach, one gets

$$\bar{\imath} = \frac{90/1.1 + 80/1.1^2}{300 + 150/1.1} \cdot 1.1 = 37.3\%.$$

The economic efficiency is $\xi = (37.3\% - 10\%)/1.1 = 24.82\%$ and the economic value created is $\text{NPV} = 436.36 \cdot 24.82\% = 108.3$.

10.2 The Book IARRs

Consider the four basic pairs of capital and income, in the investment perspective (operating, non-operating) and in the financing perspective (debt, equity). Each such pair generates an IARR. More precisely, we define I^h as total (undiscounted) income and C^h as total (undiscounted) capital for $h = o, l, d, e$. One then finds the *operating* IARR, the *non-operating* IARR, the *debt* IARR, the *equity* IARR as ratio of the respective total incomes to total capitals:

$$\bar{j}^o = \frac{I^o}{C^o} = \frac{\sum_{t=1}^n I_t^o}{\sum_{t=1}^n C_{t-1}^o} \tag{10.20}$$

$$\bar{j}^l = \frac{I^l}{C^l} = \frac{\sum_{t=1}^n I_t^l}{\sum_{t=1}^n C_{t-1}^l} \tag{10.21}$$

$$\bar{j}^d = \frac{I^d}{C^d} = \frac{\sum_{t=1}^n I_t^d}{\sum_{t=1}^n C_{t-1}^d} \tag{10.22}$$

$$\bar{j}^e = \frac{I^e}{C^e} = \frac{\sum_{t=1}^n I_t^e}{\sum_{t=1}^n C_{t-1}^e} \tag{10.23}$$

(with the understanding that, if $I_0^j \neq 0$, $j = o, l, d, e$, it must be added to the respective numerators). Equivalently, if the capitals are all nonzero,

$$\bar{j}^o = \frac{\sum_{t=1}^n i_t^o C_{t-1}^o}{\sum_{t=1}^n C_{t-1}^o} \tag{10.24}$$

$$\bar{j}^l = \frac{\sum_{t=1}^n i_t^l C_{t-1}^l}{\sum_{t=1}^n C_{t-1}^l} \tag{10.25}$$

$$\bar{j}^d = \frac{\sum_{t=1}^n i_t^d C_{t-1}^d}{\sum_{t=1}^n C_{t-1}^d} \tag{10.26}$$

$$\bar{j}^e = \frac{\sum_{t=1}^n i_t^e C_{t-1}^e}{\sum_{t=1}^n C_{t-1}^e}. \tag{10.27}$$

Using cash flows,

$$\bar{j}^o = \frac{F^o}{C^o} = \frac{\sum_{t=0}^n F_t^o}{\sum_{t=1}^n C_{t-1}^o} \tag{10.28}$$

$$\bar{j}^l = \frac{F^l}{C^l} = \frac{\sum_{t=0}^n F_t^l}{\sum_{t=1}^n C_{t-1}^l} \tag{10.29}$$

$$\bar{j}^d = \frac{F^d}{C^d} = \frac{\sum_{t=0}^n I_t^d}{\sum_{t=1}^n C_{t-1}^d} \tag{10.30}$$

$$\bar{j}^e = \frac{F^e}{C^e} = \frac{\sum_{t=0}^n F_t^e}{\sum_{t=1}^n C_{t-1}^e}. \tag{10.31}$$

Correspondingly, the MARRs are equal to the benchmark portfolio's total profit measured per unit of total capital employed in the project:

$$\bar{\rho}^o = \frac{I^{Vo}}{C^o} = \frac{\sum_{t=1}^n I_t^{Vo}}{\sum_{t=1}^n C_{t-1}^o} \tag{10.32}$$

$$\bar{\rho}^l = \frac{I^{Vl}}{C^l} = \frac{\sum_{t=1}^n I_t^{Vl}}{\sum_{t=1}^n C_{t-1}^l} \tag{10.33}$$

$$\bar{\rho}^d = \frac{I^{Vd}}{C^d} = \frac{\sum_{t=1}^n I_t^{Vd}}{\sum_{t=1}^n C_{t-1}^d} \tag{10.34}$$

$$\bar{\rho}^e = \frac{I^{Ve}}{C^e} = \frac{\sum_{t=1}^n I_t^{Ve}}{\sum_{t=1}^n C_{t-1}^e} \tag{10.35}$$

or

$$\bar{\rho}^o = \frac{\sum_{t=1}^n r_t^o V_{t-1}^o}{\sum_{t=1}^n C_{t-1}^o} = \bar{r}_V^o \cdot \frac{V^o}{C^o} \tag{10.36}$$

$$\bar{\rho}^l = \frac{\sum_{t=1}^n r_t^l V_{t-1}^l}{\sum_{t=1}^n C_{t-1}^l} = \bar{r}_V^l \cdot \frac{V^l}{C^l} \tag{10.37}$$

$$\bar{\rho}^d = \frac{\sum_{t=1}^n r_t^d V_{t-1}^d}{\sum_{t=1}^n C_{t-1}^d} = \bar{r}_V^d \cdot \frac{V^d}{C^d} \tag{10.38}$$

$$\bar{\rho}^e = \frac{\sum_{t=1}^n r_t^e V_{t-1}^e}{\sum_{t=1}^n C_{t-1}^e} = \bar{r}_V^e \cdot \frac{V^e}{C^e} \tag{10.39}$$

where $\bar{r}_V^h = (\sum_{t=1}^n r_t^h V_{t-1}^h) / \sum_{t=1}^n V_{t-1}^h$, $h = o, l, d, e$ represents the average cost of capital of item h. Therefore, the respective MARR is obtained by multiplying the average COC by the market-to-book ratio (V^h/C^h). Equivalently, one may reframe the above equality with adjusted (period) COCs as follows:

$$\bar{\rho}^o = \frac{\sum_{t=1}^n \rho_t^o C_{t-1}^o}{\sum_{t=1}^n C_{t-1}^o} \tag{10.40}$$

$$\bar{\rho}^l = \frac{\sum_{t=1}^n \rho_t^l C_{t-1}^l}{\sum_{t=1}^n C_{t-1}^l} \tag{10.41}$$

$$\bar{\rho}^d = \frac{\sum_{t=1}^n \rho_t^d C_{t-1}^d}{\sum_{t=1}^n C_{t-1}^d} \tag{10.42}$$

$$\bar{\rho}^e = \frac{\sum_{t=1}^n \rho_t^e C_{t-1}^e}{\sum_{t=1}^n C_{t-1}^e} \tag{10.43}$$

with $\rho_t^h = r_t^h (V_{t-1}^h / C_{t-1}^h)$, $h = o, l, d, e$.

From a financing perspective, one may write

$$\text{NPV} = I - \bar{\rho}C = \overbrace{I^e - \bar{\rho}^e C^e}^{\text{NPV}^e} + \overbrace{I^d - \bar{\rho}^d C^d}^{\text{NPV}^d} \tag{10.44}$$

with $C^e = \sum_{t=1}^n C_{t-1}^e$ and $C^o = \sum_{t=1}^n C_{t-1}^o$. This enables accomplishing a twofold decomposition, in terms of the sources of funds (debt and equity) and in terms of scale and marginal efficiency:

$$\text{NPV} = \overbrace{\underbrace{C^e}_{\substack{\text{scale of}\\\text{equity}}} \cdot \underbrace{(\bar{\jmath}^e - \bar{\rho}^e)}_{\substack{\text{economic efficiency of}\\\text{equity capital}}}}^{\text{NPV}^e} + \overbrace{\underbrace{C^d}_{\substack{\text{scale of}\\\text{debt}}} \cdot \underbrace{(\bar{\jmath}^d - \bar{\rho}^d)}_{\substack{\text{economic efficiency of}\\\text{debt capital}}}}^{\text{NPV}^d}. \tag{10.45}$$

Analogously, in an investment perspective,

$$\text{NPV} = I - \bar{\jmath}C = \overbrace{I^o - \bar{\jmath}^o C^o}^{\text{NPV}^o} + \overbrace{I^l - \bar{\jmath}^l C^l}^{\text{NPV}^l} \tag{10.46}$$

whence

$$\text{NPV} = \overbrace{\underbrace{C^o}_{\substack{\text{scale of}\\\text{operating assets}}} \cdot \underbrace{(\bar{\jmath}^o - \bar{\rho}^o)}_{\substack{\text{economic efficiency of}\\\text{operating assets}}}}^{\text{NPV}^o} + \overbrace{\underbrace{C^l}_{\substack{\text{scale of}\\\text{non-operating assets}}} \cdot \underbrace{(\bar{\jmath}^l - \bar{\rho}^l)}_{\substack{\text{economic efficiency of}\\\text{non-operating assets}}}}^{\text{NPV}^l} \tag{10.47}$$

Mingling the investment and financing perspectives, and exploiting the fact that the law of conservation keeps on holding for total capitals (i.e., global additivity holds: $C^o + C^l = C = C^e + C^d$), one may formalize the equilibrium of investments and financings in terms of IARRs and MARRs:

$$\frac{\bar{\jmath}^o C^o + \bar{\jmath}^l C^l}{C^o + C^l} = \bar{\jmath} = \frac{\bar{\jmath}^e C^e + \bar{\jmath}^d C^d}{C^d + C^e} \tag{10.48}$$

$$\frac{\bar{\rho}^o C^o + \bar{\rho}^l C^l}{C^o + C^l} = \bar{\rho} = \frac{\bar{\rho}^e C^e + \bar{\rho}^d C^d}{C^d + C^e}. \tag{10.49}$$

One may frame (10.45) and (10.47) in terms of excess rates, denoted as ϵ:

$$\begin{aligned} \text{NPV} &= C^e \cdot \epsilon^e + C^d \cdot \epsilon^d \\ \text{NPV} &= C^o \cdot \epsilon^o + C^l \cdot \epsilon^l \end{aligned} \tag{10.50}$$

Table 10.1 Shareholder value creation and analysis of economic efficiency

Component	Scale	Economic efficiency
Operating assets	C^o	ϵ^o
Non-operating assets	C^l	ϵ^l
Debt	C^d	ϵ^d
Equity	C^e	ϵ^e

where $\epsilon^h = \bar{\jmath}^h - \bar{\rho}^h$, $h = o, l, d, e$. Hence, NPV $= C \cdot \epsilon$, where

$$\frac{\epsilon^o C^o + \epsilon^l C^l}{C^o + C^l} = \epsilon = \frac{\epsilon^e C^e + \epsilon^d C^d}{C^e + C^d}. \tag{10.51}$$

Therefore, one may write

$$\text{NPV}^e = C^o \cdot \overbrace{(\bar{\jmath}^o - \bar{\rho}^o)}^{\epsilon^o} + C^l \cdot \overbrace{(\bar{\jmath}^l - \bar{\rho}^l)}^{\epsilon^l} - C^d \cdot \overbrace{(\bar{\jmath}^d - \bar{\rho}^d)}^{\epsilon^d} \tag{10.52}$$

(see also Table 10.1). This decomposition refines the traditional economic analysis of a project and represents a substantial improvement upon the traditional NPV analysis.

The notion of net-debt IARR may be derived as a proper (generalized) mean of debt IARR and non-operating IARR:

$$\bar{\jmath}^{\mathcal{D}} = \frac{\bar{\jmath}^d C^d - \bar{\jmath}^l C^l}{C^d - C^l}. \tag{10.53}$$

Analogously for the respective MARR:

$$\bar{\rho}^{\mathcal{D}} = \frac{\bar{\rho}^d C^d - \bar{\rho}^l C^l}{C^d - C^l} \tag{10.54}$$

so that NPV$^o = C^{\mathcal{D}}\epsilon^{\mathcal{D}} + C^e\epsilon^e$ where $\epsilon^{\mathcal{D}} = (\epsilon^d C^d - \epsilon^l C^l)/(C^d - C^l)$. Hence,

$$\text{NPV}^e = C^o \cdot \overbrace{(\bar{\jmath}^o - \bar{\rho}^o)}^{\epsilon^o} - C^{\mathcal{D}} \cdot \overbrace{(\bar{\jmath}^{\mathcal{D}} - \bar{\rho}^{\mathcal{D}})}^{\epsilon^{\mathcal{D}}}. \tag{10.55}$$

We have hitherto followed a levered perspective. As for the unlevered perspective, let us denote the total NOPAT as $I^{uo} = \sum_{t=1}^n I_t^{uo}$ and the net unlevered cash flow as $F^{uo} = \sum_{t=0}^n F_t^{uo}$. One finds

$$\text{NPV}^o = C^o(\bar{\jmath}^{uo} - \bar{\rho}^{uo}) \tag{10.56}$$

where

$$\bar{j}^{uo} = \frac{F^{uo}}{C^o} = \frac{I^{uo}}{C^o} = \frac{\sum_{t=1}^{n} i_t^{uo} C_{t-1}^o}{C^o} \tag{10.57}$$

and

$$\bar{\rho}^{uo} = \frac{\sum_{t=1}^{n} \rho_t^{uo} C_{t-1}^o}{C^o} = \bar{r}_V^{uo} \cdot \frac{V^o}{C^o} \tag{10.58}$$

with $\bar{r}_V^{uo} = \sum_{t=1}^{n} r_t^{uo} \cdot V_{t-1}^o / V^o$. Alternatively, one may use the shortcut (10.18):

$$\bar{\rho}^{uo} = \bar{j}^{uo} - \frac{\text{NPV}^o}{C^o}. \tag{10.59}$$

Therefore, one may write

$$\begin{aligned} \text{NPV} = I - \bar{\rho}C &= \overbrace{I^{uo} - \bar{\rho}^{uo} C^o}^{\text{NPV}^o} + \overbrace{I^l - \bar{\rho}^l C^l}^{\text{NPV}^l} \\ &= \overbrace{C^o(\bar{j}^{uo} - \bar{\rho}^{uo})}^{\text{NPV}^o} + \overbrace{C^l(\bar{j}^l - \bar{\rho}^l)}^{\text{NPV}^l}. \end{aligned} \tag{10.60}$$

The decomposition of NPV in terms of scale and efficiency provided by the unlevered perspective is equal to the decomposition provided by the investment (levered) perspective. More precisely, considering (10.47) and (10.60), their only seeming difference lies in the excess returns of the operating assets. However,

$$\begin{aligned} \epsilon^o &= \bar{j}^o - \bar{\rho}^o \\ &= \frac{\sum_{t=1}^{n} I_t^o}{\sum_{t=1}^{n} C_{t-1}^o} - \frac{\sum_{t=1}^{n} \rho_t^o C_{t-1}^o}{\sum_{t=1}^{n} C_{t-1}^o} \\ &= \frac{\sum_{t=1}^{n} I_t^o - \sum_{t=1}^{n} r_t^o V_{t-1}^o}{C^o} \\ &= \frac{\sum_{t=1}^{n} I_t^{uo} - \sum_{t=1}^{n} (r_t^o V_{t-1}^o - \tau I_t^{\mathcal{D}})}{C^o} \\ &= \frac{I^{uo}}{C^o} - \frac{\sum_{t=1}^{n} \left(r_t^o - \frac{\tau I_t^{\mathcal{D}}}{V_{t-1}^o}\right) V_{t-1}^o}{C^o} \\ &= \bar{j}^{uo} - \frac{\sum_{t=1}^{n} r_t^{uo} \cdot V_{t-1}^o}{C^o} \\ &= \bar{j}^{uo} - \frac{\sum_{t=1}^{n} \rho_t^{uo} C_{t-1}^o}{C^o} \\ &= \bar{j}^{uo} - \bar{\rho}^{uo}. \end{aligned} \tag{10.61}$$

This means that, while the IARRs and the MARRs are different (i.e., $\bar{j}^o \neq \bar{j}^{uo}$ and $\bar{\rho}^o \neq \bar{\rho}^{uo}$), the excess rates of return coincide in both perspectives. Conceptually, the difference only lies in the treatment of the tax shield: In the former perspective, the total tax shield, $\sum_{t=1}^{n} \tau I_t^{\mathcal{D}}$, is incorporated in $\bar{\jmath}^o$ while in the latter perspective the tax shield is incorporated in $\bar{\rho}^{uo}$.

In an asset (financially unlevered) perspective, one may also make use of the financially unlevered incomes $I_t^u = (\text{EBIT}_t + I_t^l)(1-\tau) = \text{NOPAT}_t + I_t^l(1-\tau)$. The sum of the unlevered incomes is $I^u = \sum_{t=1}^{n} I_t^u$, which is equal to the algebraic sum of the asset FCFs, $F^u = \sum_{t=0}^{n} F_t^u$. Hence, one may decompose the project NPV as

$$\text{NPV} = C(\bar{\jmath}^u - \bar{\rho}^u)$$

where

$$\bar{\jmath}^u = \frac{F^u}{C} = \frac{I^u}{C} = \frac{\sum_{t=1}^{n} i_t^u C_{t-1}}{C} \tag{10.62}$$

and

$$\bar{\rho}^u = \frac{\sum_{t=1}^{n} \rho_t^u C_{t-1}}{C} = \bar{r}_V^u \cdot \frac{V}{C} \tag{10.63}$$

with $\bar{r}_V^u = \sum_{t=1}^{n} r_t^u V_{t-1}/V$. Alternatively, one may use the shortcut (10.18):

$$\bar{\rho}^u = \bar{\jmath}^u - \frac{\text{NPV}}{C}. \tag{10.64}$$

As in the previous case, the NPV decomposition in the unlevered perspective is equal to the decomposition provided by the levered perspective, that is, $\text{NPV} = C(\bar{\jmath} - \bar{\rho}) = C(\bar{\jmath}^u - \bar{\rho}^u)$. (The proof may also be obtained from (10.61); it suffices to delete the superscript "o", replace the superscript "$\mathcal{D}$" with "d", and replace the superscript "uo" with "u".) Conceptually, the difference lies in the fact that, in the former perspective, the total gross tax shield, $\sum_{t=1}^{n} \tau I_t^d$, is incorporated in $\bar{\jmath}$ while in the latter perspective it is incorporated in $\bar{\rho}^u$.

We have found eight book IARRs (BIARRs). Below is the terminology we will use to refer to them.

The project's BIARRs		**Symbol**	**Associated COC (MARR)**
Asset/project IARR	or internal average ROI	$\bar{\jmath}$	$\bar{\rho}$
Equity IARR	or internal average ROE	$\bar{\jmath}^e$	$\bar{\rho}^e$
Operating IARR	or internal (operating) average ROI	$\bar{\jmath}^o$	$\bar{\rho}^o$
Non-operating IARR	or internal non-operating average ROI	$\bar{\jmath}^l$	$\bar{\rho}^l$
Unlevered IARR	or internal (unlevered) average ROI	$\bar{\jmath}^{uo}$	$\bar{\rho}^{uo}$
Unlevered asset IARR	or internal (unlevered) average asset ROI	$\bar{\jmath}^u$	$\bar{\rho}^u$
Debt IARR	or internal average ROD	$\bar{\jmath}^d$	$\bar{\rho}^d$
Net debt IARR	or internal average RO$\mathcal{D}$	$\bar{\jmath}^{\mathcal{D}}$	$\bar{\rho}^{\mathcal{D}}$

The goal of a firm's management is to undertake projects that increase equityholders' wealth. Thus, firms should undertake projects such that the equity IARR is greater than equity COC.[5] Therefore, the equity IARR (not the project IARR) is the relevant IARR for shareholders.

Proposition 10.2 (Rationality of average ROE) *Let P be any project associated with a stream of net incomes equal to* $\boldsymbol{I}^e$ *and an equity capital stream equal to* $\boldsymbol{C}^e$*. Then, shareholder value is created (and the project is worth undertaking) if and only if the internal average ROE is greater than the overall cost of equity,* $\bar{\jmath}^e > \bar{\rho}^e$ *or, equivalently, if the excess equity IARR is positive,* $\epsilon^e > 0$.[6]

If the project is levered but $i_t^d = r_t^d$ for all t, then $C_t^d = V_t^d$ for all t, which implies $\text{NPV}^d = 0$ and $\text{NPV} = \text{NPV}^e$. In this case, decision can be made with either the internal average ROI or the internal average ROE, because $\bar{\jmath} > \bar{\rho}$ if and only if $\bar{\jmath}^e > \bar{\rho}^e$ (as long as C and C^e have the same sign). Note that the decomposition of the NPV is different in an asset perspective and in an equity perspective:

$$\text{NPV} = C(\bar{\jmath} - \bar{\rho}) = C^e(\bar{\jmath}^e - \bar{\rho}^e) = \text{NPV}^e$$

(compare with (8.65).

If the project is financially unlevered (i.e., equity-financed), the internal average ROI is equal to the internal average ROE: $\bar{\jmath} = \bar{\jmath}^e$ and the respective MARR coincide: $\bar{\rho} = \bar{\rho}^e$ (see an example in Magni amd Marchioni 2019).

Remark 10.3 (*Zero capital*) While scarcely relevant practically, one may complete the definition of IARR as follows, analogously to what we have made with AIRR (see Remark 8.5):

$$\bar{\jmath} = \begin{cases} \frac{I}{C}, & \text{if } C \neq 0; \\ +\infty, & \text{if } C = 0 \text{ and } I > 0; \\ -\infty, & \text{if } C = 0 \text{ and } I < 0. \end{cases} \tag{10.65}$$

If $C \neq 0$, an investment (financing) project is worth undertaking if and only if $\bar{\jmath} > rV/C$ (<); if $C = 0$, the project is worth undertaking if and only if $I > r\sum_{t=1}^{n} V_{t-1}$.

In case of time-varying COCs, the MARR is defined as

$$\bar{\rho} = \begin{cases} \bar{r}_V \cdot \frac{V}{C}, & \text{if } C \neq 0; \\ +\infty, & \text{if } C = 0 \text{ and } \sum_{t=1}^{n} r_t V_{t-1} > 0; \\ -\infty, & \text{if } C = 0 \text{ and } \sum_{t=1}^{n} r_t V_{t-1} < 0. \end{cases} \tag{10.66}$$

[5]Unlevered asset BIARR might also be said to called *financially* unlevered BIARR (see Definition 2.2).

[6]In the case where $C^e < 0$, that is, the equity capital is not raised from equityholders but, rather, lent to equityholders, the sign of the inequality is reversed.

If $C \neq 0$, an investment (financing) project is worth undertaking if and only if $\bar{\jmath} > \bar{\rho}$ (<); if $C = 0$, the project is worth undertaking if and only if $I > \sum_{t=1}^{n} r_t V_{t-1}$. ▲

Remark 10.4 (*Pitfall-free rate of return*) In Chap. 9 we have shown that the IRR suffers from several notable shortcomings. We have also shown that the AIRR approach does not incur the IRR's drawbacks, with the exception of global non-additivity of AIRR in case the COCs of a portfolio's constituent assets are not equal. Specifically, given any two assets j and k, global additivity holds for AIRR only if the COC is the same for the two assets. Therefore, if the COC is not the same, then $\text{PV}[C^j] + \text{PV}[C^k] \neq \text{PV}[C^{j+k}]$ (see discussion on global additivity in Chap. 9). Given that, barring exceptional cases, equity and debt have different costs of capital, the following inequality holds:

$$\text{PV}[C] \neq \text{PV}[C^o] + \text{PV}[C^l] \neq \text{PV}[C^e] + \text{PV}[C^d] \neq \text{PV}[C]$$

(As we know, this has no impact on valuation, since $\text{PV}[C](\bar{\imath}_0 - \bar{r}_0) = \text{PV}[C^e](\bar{\imath}_0^e - \bar{r}_0^e) + \text{PV}[C^d](\bar{\imath}_0^d - \bar{r}_0^d)$.) This implies that the project's AIRR is not equal to the mean of the average ROE and average ROD nor the mean of the average ROI and the average non-operating ROI:

$$\bar{\imath} \neq \frac{\bar{\imath}^e \text{PV}[C^e] + \bar{\imath}^d \text{PV}[C^d]}{\text{PV}[C^e] + \text{PV}[C^d]} \neq \frac{\bar{\imath}^o \text{PV}[C^o] + \bar{\imath}^l \text{PV}[C^l]}{\text{PV}[C^o] + \text{PV}[C^l]} \neq \bar{\imath}. \quad (10.67)$$

As seen, the IRR suffers from this problem as well, but in a more severe way, for it has the additional problem that the same inequality arises in reference to a single period: $C_t(\sigma) \neq C_t^e(\sigma^e) + C_t^d(\sigma^d)$ (whereas $C_t = C_t^e + C_t^d$).

It is to be noted that the **IARR does not suffer from *any* of the difficulties associated with IRR, including global non-additivity**. In particular, $C = C^j + C^k$ for any pair of assets j and k, and $\bar{\jmath} = (\bar{\jmath}^j C^j + \bar{\jmath}^k C^k)/(C^j + C^k)$. Therefore,

$$\frac{\bar{\jmath}^o C^o + \bar{\jmath}^l C^l}{C^o + C^l} = \bar{\jmath} = \frac{\bar{\jmath}^e C^e + \bar{\jmath}^d C^d}{C^e + C^d}.$$

Analogously for the MARRs and the excess rates of return (see also (10.48)–(10.49) and (10.51)). ▲

Remark 10.5 (*Intrinsic Rate of Return*) Consider the IARR function, defined as $\bar{\jmath}(x) = I/x = F/x$. The NPV may be framed as $\text{NPV} = x \cdot (\bar{\jmath}(x) - \bar{\rho}(x))$ for every $x \neq 0$, with $\bar{\rho}(x) = \bar{r}_V \cdot (V/x)$, which means that IARR gives rise to a class of rates of return, the *IARR class*, differentiated by the capital base. If, in the IARR approach, one measures the return per unit of total economic value (i.e., $x = V$), one gets the *Intrinsic Rate of Return* (IROR). This measure is not internal, precisely because the capital base is based on intrinsic values, which depend on the COCs:

$$\text{IROR} = \frac{I}{V} = \frac{F}{V} = \frac{\sum_{t=0}^{n} F_t}{\sum_{t=0}^{n} V_t}. \quad (10.68)$$

In this case, the capital base coincides with the replicating portfolio's total value, so no adjustment factor is needed and the MARR coincides with r. Hence, the direct comparison of IROR and r captures value creation:

The project creates value if and only if IROR $> r$.

In case of time-varying COCs, the MARR is calculated as the value-weighted mean of the COCs, and the inequality IROR $> \bar{r}_V$ replaces IROR $> r$ for detecting value creation (see Magni and Marchioni 2018 for details on intrinsic average ROI and intrinsic average ROE for decision-making). The IROR in the IARR class plays the same role as EAIRR in the AIRR class. ▲

Remark 10.6 (*Mixed projects*) If there are lending periods and borrowing periods, as described in Sect. 8.9, one may easily accomplish a twofold breakdown of the NPV. In the IARR approach, Eq. (8.87) becomes

$$\overbrace{C^e(\bar{\jmath}^e - \bar{\rho}^e)}^{\text{equity NPV}} + \overbrace{C^d(\bar{\jmath}^d - \bar{\rho}^d)}^{\text{debt NPV}} = \text{NPV} = \overbrace{C_L(\bar{\jmath}^L - \bar{\rho}^L)}^{\text{investment-period NPV}} + \overbrace{C_B(\bar{\jmath}^B - \bar{\rho}^B)}^{\text{borrowing-period NPV}} \tag{10.69}$$

with $C_B = \sum_{t \in T_B} C_{t-1}$, $C_L = \sum_{t \in T_L} C_{t-1}$, and obvious meaning of the rate-of-return symbols. As opposed to the AIRR, the IARR fulfills global additivity. Therefore, $C^e + C^d = C = C_L + C_B$.

From (10.69) one gets the equity NPV:

$$\overbrace{C^e(\bar{\jmath}^e - \bar{\rho}^e)}^{\text{equity NPV}} = \overbrace{C_L(\bar{\jmath}^L - \bar{\rho}^L)}^{\text{investment-period NPV}} + \overbrace{C_B(\bar{\jmath}^B - \bar{\rho}^B)}^{\text{borrowing-period NPV}} - \overbrace{C^d(\bar{\jmath}^d - \bar{\rho}^d)}^{\text{debt NPV}}. \tag{10.70}$$

Example 10.3 ♦

(*Project analysis*) We analyze shareholder value creation for a levered project in 7 different scenarios, corresponding to 7 different projects (Table 10.2). The table is divided into three panels: Panel (a) describes shareholder value creation by disaggregation into capital and efficiency components of (i) operations, (ii) non-operating assets, and (iii) debt. In Panel (b) shareholder's value creation is viewed in terms of scale of equity capital and efficiency of equity capital. Panel (c) summarizes the role of net debt and, therefore, the role of capital structure.

Scenario 1

The input data for the analysis are $C^o = 120$, $\epsilon^o = 2\%$, $C^l = 70$ $\epsilon^l = 0\%$, $C^d = 40$, $\epsilon^d_0 = 0\%$. Hence, $\text{NPV}^e = 120 \cdot 2\% + 70 \cdot 0\% - 40 \cdot 0\% = 2.4 > 0$. The operating NPV and the equity NPV coincide, because the borrowing and payout policies are both value-neutral. Shareholder value is created out of an overall investment of $C^e = 150 = 120 + 70 - 40$ which earns an excess average ROE equal to $\epsilon^e = 1.6\%$ ($\text{NPV}^e = 150 \cdot 1.6\% = 2.4$).

Table 10.2 Project analysis

SCENARIO		**1**	**2**	**3**	**4**	**5**	**6**	**7**
Panel (a)								
	C^o	120	120	150	105	105	140	140
	ϵ^o	2.0%	2.0%	2.0%	2.0%	2.0%	1.5%	1.5%
[1]	NPV^o	2.4	2.4	3	2.1	2.1	2.1	2.1
	C^l	70	70	70	70	70	70	0
	ϵ^l	0.0%	−2.0%	−2.0%	−2.0%	−2.0%	−2.0%	−2.0%
[2]	NPV^l	0	−1.4	−1.4	−1.4	−1.4	−1.4	0
	C^d	40	40	40	40	175	175	120
	ϵ^d	0.0%	3.0%	3.0%	3.0%	3.0%	3.0%	0.0%
[3]	NPV^d	0	1.2	1.2	1.2	5.25	5.25	0
[1] + [2] − [3]	NPV^e	2.4	−0.2	0.4	−0.5	−4.55	−4.55	2.1
Panel (b)								
	C^e	150	150	180	135	0	35	20
	ϵ^e	1.6%	−0.13%	0.22%	−0.37%	$-\infty$	−13%	10.5%
	NPV^e	2.4	−0.2	0.4	−0.5	−4.55	−4.55	2.1
Panel (c)								
	$C^{\mathcal{D}}$	−30	−30	−30	−30	105	105	120
	$\epsilon^{\mathcal{D}}$	0.0%	−8.67%	−8.67%	−8.67%	6.33%	6.33%	0%
[3] − [2]	$\text{NPV}^{\mathcal{D}}$	0	2.6	2.6	2.6	6.65	6.65	0

Scenario 2

Consider the same input data, except for the non-operating excess IARR, which is now assumed to be equal to $\epsilon^l = -2\%$, and the excess debt IARR, which is now assumed to be equal to $\epsilon^d = 3\%$:

$$\text{NPV}^e = 120 \cdot 2\% + 70 \cdot (-2\%) - 40 \cdot 3\% = 2.4 - 1.4 - 1.2 = -0.2 < 0.$$

The project is not worth undertaking, despite the fact that the NPV of the operations is positive. That is, the investment policy is value-creating (\$2.4), but the financing policy is value-destroying, both in terms of borrowing (−\$1.2) and in terms of investment in liquid assets (−\$1.4).

This raises the issue of possibly changing the financing policy for this project so as to nullify or, at least, minimize, its negative effect. The analyst should make some sensitivity analysis to measure the benefit of a reduction in liquid assets and/or debt. Note that liquid assets destroy value more than the debt although, for each dollar, the use of debt implies a higher loss of value than the liquid asset's (3% > 2%). The reason is that the scale of the borrowing is smaller than the scale of the liquid assets. Managers may then investigate the effect of shrinking the capital invested in liquid assets (and, therefore, the effect of an increase in the payout policy) or, rather, the effect of a reduction in the capital borrowed from debtholders.

The total net debt is negative: $C^{\mathcal{D}} = 40 - 70 = -30$, and its excess IARR is $\epsilon^{\mathcal{D}} = -8.67\%$. Overall, the course of action may be viewed, from the point of view of equityholders, as a portfolio of an operating investment of \$120 at an excess 2% return and a financial investment of \$30 at an excess −8.67%. The operating investment has an NPV of \$2.4, the financial investment has an NPV of −\$2.6. Merging these figures, shareholders invest a net \$150 at a −0.13% excess return: $\text{NPV}^e = 150 \cdot (-0.13\%) = -0.2$.

Scenario 3

Same inputs as scenario 2, except for a higher scale of operations: $C^o = 150$ instead of $C^o = 120$.

In this case, the project is worth undertaking, because the adverse financing policy is more than compensated by the greater scale of the operations. Earning an excess 2% on \$150 makes the operating NPV increase by $2\% \cdot 30 = 0.6$ with respect to scenarios 1 and 2, while the NPV of the financing policy is unvaried with respect to scenario 2 and equal to $\text{NPV}^{\mathcal{D}} = 2.6$. Hence, $\text{NPV}^e = \text{NPV}^o - \text{NPV}^{\mathcal{D}} = 3 - 2.6 = 0.4$, obtained with a scale of \$180 equity capital and economic efficiency equal to \$0.22%.

Scenario 4

Suppose $C^o = 105$, other figures unvaried with respect to scenario 3. In this case, the scale of the operations are too small to compensate the value lost by the financing policy. Overall, the equity capital is \$135, which is invested at $\epsilon^e = -0.37\%$. Multiplying the equity scale and the efficiency of equity, the equity NPV is obtained: $\text{NPV}^e = 135 \cdot (-0.37\%) = -0.5$.

Scenario 5

It is the same as scenario 4, but $C^d = 175$ so that $C^{\mathcal{D}} = 105 = C^o$ and $C^e = 0$. In this case, total equity invested is zero, because the total net debt is equal to the total operating invested capital. The operating NPV is \$2.1, and the net-debt NPV is \$6.65. This means that the equity NPV is $\text{NPV}^e = 2.1 - 6.65 = -4.55$. The equity rate of return is *minus infinity*. The reason is that, while equityholders's overall investment is zero, their wealth decreases by \$4.55. This is equivalent to an *arbitrage loss*, so the excess rate of return is $-\infty$. More precisely, equity may be viewed as a portfolio of an operating investment of \$105 at an excess 2% return, totally financed by net debt at an excess 6.33% financing rate: Nothing is invested by equityholders but much is lost!

Scenario 6

Same as scenario 5, except that the scale of operations is higher ($C^o = 140$) and the economic efficiency is smaller ($\epsilon^o = 1.5\%$). Everything remains equal in terms of NPVs. However, n terms of equity, the same NPV is obtained with a positive scale ($C^e = 35$) and a negative marginal efficiency ($\epsilon^e = -13\%$).

Scenario 7

In scenario 7, we assume that a full payout policy is employed, so liquid assets are zero in each period. Also, the project is financed with debt by a total of \$120, that is, the overall leverage ratio (in book value) is $C^d/C = 120/140 = 85.7\%$. The excess debt IARR is assumed to be equal to $\epsilon^d = 0\%$. This means that the project NPV is equal to equity NPV: $\text{NPV} = \text{NPV}^o = \text{NPV}^e = 2.1$. However, the decomposition of NPV and NPV^e is different. The economic efficiency of the equity investment is $\epsilon^e = 10.5\%$, significantly greater than the economic efficiency of the operations, which is equal to $\epsilon = \epsilon^o = 1.5\%$. The reason is that shareholders finance a very small fraction of the project (\$20 out of \$140). Given that the NPVs of equity and project are equal, then the higher economic efficiency of the equity must be offset by a much smaller scale of the equity.

Example 10.4 ♦

C^o	C^d
	C^e

(Average ROE, ROD and ROI of levered project with time-varying COCs)

Consider the input data used in Example 8.8, reported below:

Time	**0**	**1**	**2**	Total
Equity capital (C_t^e)	100	60	0	160
Debt capital (C_t^d)	40	20	0	60
Net Income (I_t^e)		60	90	150
Interest expense (I_t^d)		2	5	7
Cost of equity (r_t^e)		8%	10%	
Cost of debt (r_t^d)		2%	4%	

Hence, the internal average ROE and the internal average ROD are

$$\bar{j}^e = \frac{150}{160} = 93.8\%, \quad \bar{j}^d = \frac{7}{60} = 11.7\%$$

The CFEs and CFDs are, respectively, $\boldsymbol{F}^e = (-100, 100, 150)$ and $\boldsymbol{F}^d = (-40, 22, 25)$. Hence, discounting the cash flows at the respective discount rates, the economic values of equity, V_t^e, and debt, V_t^d, are

Time	**0**	**1**	**2**	Total
Equity value (V_t^e)	218.9	136.4	0	355.2
Debt value (V_t^d)	45.1	24.0	0	69.2

Therefore, the average cost of equity and cost of debt are

$$\bar{r}_V^e = \frac{8\% \cdot 218.9 + 10\% \cdot 136.4}{355.2} = 8.8\%, \quad \bar{r}_V^d = \frac{2\% \cdot 45.1 + 4\% \cdot 24}{69.2} = 2.7\%.$$

The equity MARR and the debt MARR are then found by adjusting the two average COCs:

$$\bar{\rho}^e = 8.8\% \cdot \frac{355.2}{160} = 19.5\% \qquad \bar{\rho}^d = 2.7\% \cdot \frac{69.2}{60} = 3.1\%.$$

The equity MARR and the debt MARR may also be computed in one shot:

$$\bar{\rho}^e = \frac{8\% \cdot 218.9 + 10\% \cdot 136.4}{160} = 19.5\% \qquad \bar{\rho}^d = \frac{2\% \cdot 45.1 + 4\% \cdot 24}{60} = 3.1\%.$$

Shareholder value creation is $\text{NPV}^e = 160 \cdot (93.8\% - 19.5\%) = 118.9$.

Owing to (10.48) and (10.49), the average ROI may be found as the mean of the average ROE and the average ROD and the overall (pre-tax) WACC may be found as the mean of the average cost of equity and the average cost of debt:

$$\bar{\jmath} = \frac{93.8\% \cdot 160 + 11.7\% \cdot 60}{160 + 60} = 71.4\%$$
$$\bar{\rho} = \frac{19.5\% \cdot 160 + 3.1\% \cdot 60}{160 + 60} = 15\%.$$

The economic efficiency of the project is smaller than the economic efficiency of the equity investment: $\epsilon = 71.4\% - 15\% = 56.4\% < 74.3\% = 93.8\% - 19.5\% = \epsilon^e$. This is because ϵ is a weighted mean of $\epsilon^e = 74.3\%$ and $\epsilon^d = 8.6\%$. However, this is applied to the project scale, which is greater than the scale of the equity investment: $C = 220 > 160$. The smaller scale of the equity investment outweighs the higher economic efficiency of the equity investment and thus make shareholder value creation be smaller than the value created by the project:

$$\text{NPV}^e = \underbrace{\overbrace{220}^{\text{project scale}} \cdot \overbrace{(71.4\% - 15\%)}^{\text{project's economic efficiency}}}_{123.99} - \underbrace{\overbrace{60}^{\text{debt scale}} \cdot \overbrace{(11.7\% - 3.1\%)}^{\text{debt's economic efficiency}}}_{5.14} = 118.9.$$

Out of \$123.99 value created, shareholders grasp \$118.9. The remaining \$5.14 is captured by debtholders. (Compare with the results obtained in Example 8.8.) With a different framing,

$$\text{NPV}^e = \underbrace{\overbrace{220 \cdot 71.4\% - 60 \cdot 11.7\%}^{\text{Total return from equity } (I^e)}}_{150.06} - \underbrace{\overbrace{(220 \cdot 15\% - 60 \cdot 3.1\%)}^{\text{Total equity benchmark return } (I^{Ve})}}_{31.14} = 118.9.$$

The project yields absolute returns for a total of \$150.06, while an equivalent-risk benchmark portfolio would yield \$31.14; hence, the excess return of \$118.9.

Example 10.5 ♦

C^o	C^d
	C^e

Consider Example 10.4. For the sake of illustration, we recompute the average ROE using the cash flows and recompute the respective equity COC using the shortcut. The average ROE is

$$\bar{j}^e = \frac{F^e}{C^e} = \frac{-100 + 100 + 150}{100 + 60} = 93.8\%.$$

In other words, the investors invest, overall, \$160 and receive, in exchange for it, a net \$150 cash flow, which makes 93.8% of the total investment. The equity COC is obtained by subtracting from it the ratio of the equity NPV to the total equity employed:

$$\bar{\rho}^e = 93.8\% - \frac{118.9}{160} = 19.5\%.$$

Alternatively, one might have used one of the other methods. For example, using (10.17):

$$\bar{\rho}^e = \frac{F_1^e + F_2^e - V_0^e}{C^e} = \frac{100 + 150 - 218.9}{160} = 19.5\%.$$

In this case, only the first market value of equity is required. Analogously with average ROD, average ROC and their respective COCs.

Example 10.6 ♦

C^o	C^d
C^l	C^e

(*Average ROI and ROE of Mockup*, 80% *payout*, $i^d = 3\%$, $r^d = 1\%$) Let us take up Mockup project again, assuming 80% payout. We also assume that the cost of debt is different from the ROD, namely, $i^d = 3\% > 1\% = r^d$ and the cost of asset (pre-tax WACC) is $r = 9\%$. These are the assumptions made in Example 6.25, where we have computed the project NPV (= 22,277), the equity NPV (= 13,540), and the debt NPV (= 8,736). We now compute the internal IARRs and the related COCs.

For the benefit of the reader, we report below the concise transposed Matrix (see also Table 4.24):

(C, I, F)

Time		0	1	2	3	4	5
Capital	C	500,000	436,537	374,206	313,044	253,092	0
Income	I	0	33,386	37,160	41,036	45,016	49,104
Cash Flow	F	−500,000	96,849	99,492	102,198	104,968	302,196

The shortest way to compute the rate of return is to divide the net cash flow (or total income) by the total capital:

$$\begin{aligned}\bar{j} = \frac{F}{C} &= \frac{-500{,}000 + 96{,}849 + 99{,}492 + 102{,}198 + 104{,}968 + 302{,}196}{500{,}000 + 436{,}537 + 374{,}206 + 313{,}044 + 253{,}092} \\ &= \frac{205{,}703}{1{,}876{,}879} \\ &= 10.96\%.\end{aligned}$$

To compute the MARR, several ways are available. For example, one may use (10.18):

$$\bar{\rho} = 0.1096 - \frac{22{,}277}{1{,}876{,}879} = 9.773\%.$$

The investors invest $C = 1{,}876{,}879$ at an excess rate of return equal to $\epsilon = 10.96\% - 9.773\% = 1.187\%$, so gaining $1{,}876{,}879 \cdot 1.187\% = 22{,}277$ (the error is due to the rounding of the numbers).

Taking an equity perspective, consider Mockup's basic streams:

$(\boldsymbol{C}^e, \boldsymbol{I}^e, \boldsymbol{F}^e)$

Time		0	1	2	3	4	5
Capital	C^e	410,000	346,537	284,206	223,044	163,092	0
Income	I^e		0	30,686	34,460	38,336	42,316	46,404
Cash Flow	F^e	−410,000	94,149	96,792	99,498	102,268	209,496

The average ROE is

$$\begin{aligned}\bar{\jmath}^e &= \frac{F^e}{C^e} = \frac{-410{,}000 + 94{,}149 + 96{,}792 + 99{,}498 + 102{,}268 + 209{,}496}{410{,}000 + 346{,}537 + 284{,}206 + 223{,}044 + 163{,}092} \\ &= \frac{192{,}203}{1{,}426{,}879} \\ &= 13.47\%.\end{aligned}$$

Using (10.18), the equity MARR is

$$\bar{\rho}^e = \bar{\jmath}^e - \frac{\text{NPV}^e}{C^e} = 0.1347 - \frac{13{,}540}{1{,}426{,}879} = 12.521\%.$$

While the total invested capital is \$1,876,879, equityholders only finance \$1,426,879 with an economic efficiency equal to $\epsilon^e = 13.47\% - 12.521\% = 0.949\%$. The net gain, 13,540, is then decomposed into the scale of the equity capital, $C^e = 1{,}426{,}879$, and its marginal efficiency, $\epsilon^e = 0.949\%$.

As for the debt, considering that the ROD is constant, the average ROD is $\bar{\jmath}^d = i^d = 3\%$ and the total debt capital is $C^d = C - C^e = 1{,}876{,}879 - 1{,}426{,}879 = 450{,}000$. The related MARR is then

$$\bar{\rho}^d = \bar{\jmath}^d - \frac{\text{NPV}^d}{C^d} = 0.03 - \frac{8{,}736}{450{,}000} = 1.059\%$$

so the debt's economic efficiency is $\epsilon^d = 0.03 - 0.01059 = 0.01941$. Hence, we can split up the equity NPV into project component and debt component (numbers are rounded, as usual):

$$\text{NPV}^e = \overbrace{1{,}876{,}879}^{\text{project scale}} \cdot \overbrace{1.187\%}^{\text{project's economic efficiency}} - \overbrace{450{,}000}^{\text{scale of debt}} \cdot \overbrace{1.941\%}^{\text{debt's efficiency}} = 13{,}540.$$

(In turn, the first summand may be split up into operating component and non-operating component. We leave it to the reader.)

The project's rates are weighted average of the equity rates and the debt rates. Specifically,

- the average ROI, $\bar{\jmath}$ is the weighted average of the average ROE, $\bar{\jmath}^e$, and the average ROD, $\bar{\jmath}^d$
- the project COC, $\bar{\rho}$, is the weighted average of the equity COC, $\bar{\rho}^e$ and the debt COC, $\bar{\rho}^d$
- the efficiency of the project, ϵ, is the weighted average of the efficiency of equity, ϵ^e, and the efficiency of debt, ϵ^d.

For example, we verify the latter:

$$\epsilon = \frac{\epsilon^e \cdot C^e + \epsilon^d \cdot C^d}{C^e + C^d} = \frac{0.949\% \cdot 1{,}426{,}879 + 1.941\% \cdot 450{,}000}{1{,}426{,}879 + 450{,}000} = 1.187\%$$

Example 10.7 ♦

C^o C^e (*Average ROI and MARR of unlevered project with constant WACC*) Consider the following project, which we assume to be unlevered (so that $I_t = I_t^{uo} = I_t^u$ and $F_t = F_t^{uo} = F_t^u$ for all t, as well as $\bar{\jmath} = \bar{\jmath}^{uo} = \bar{\jmath}^u$, and $\bar{\rho} = \bar{\rho}^{uo} = \bar{\rho}^u$):

(C, I, F)

Time		0	1	2
Capital	C	600	200	0
Income	I	0	150	187.5
Cash Flow	F	−600	550	387.5

(same data as Example 8.1). The average ROI is

$$\bar{\jmath} = \frac{\overbrace{337.5}^{I=F}}{\underbrace{800}_{C}} = 42.19\%.$$

Assuming the WACC is constant and equal to $r = 15\%$ (as in Example 8.1), the project WACC, $\bar{\rho}$, is easily computed with the shortcut:

$$\bar{\rho} = \overbrace{42.19\%}^{\bar{j}} - \frac{\overbrace{-600 + 550/1.15 + 387.5/1.15^2}^{\text{NPV}}}{\underbrace{800}_{C}} = 20.78\%.$$

Alternatively, if one used the economic values, one would have

$$V = V_1 + V_0 = \frac{387.5}{1.15} + \frac{337 + 550}{1.15} = 1{,}108.2$$

whence, multiplying the WACC by the market-to-book ratio,

$$\bar{\rho} = \overbrace{0.15}^{r} \cdot \frac{\overbrace{1{,}108.2}^{V}}{\underbrace{800}_{C}} = 20.78\%.$$

This means that, if one invested in the replicating portfolio, one would earn 15% on a total \$1,108.2, which is financially equivalent to 20.78% on a total of \$800. The comparison of 42.19 and 20.78% captures value creation, indicating that the project is worth undertaking. The difference between the two rates, equal to $\epsilon = 21.41\%$, expresses the economic efficiency of the project. Applied to a scale of $C = 800$, the economic value created is obtained: $\text{NPV} = 800 \cdot 21.41\% = 171.3$. (Compare with the values obtained in Example 8.1.)

Example 10.8 ♦

C^o C^e *(Average ROI and MARR of unlevered project with time-varying WACCs)* Let us consider the same project above (Example 10.7) but let us assume that the WACC is time-varying, such that $r_1 = 15\%, r_2 = 20\%$ (as in Example 8.10). The average ROI does not change (it is an internal measure, so it is not affected by the WACC). However, the project MARR does change, since it takes account that the replicating portfolio now produces profit at a rate of 20% in the second period, as opposed to 15%. The replicating portfolio's values (i.e., the market values of the project) are $V_1 = 387.5/1.2 = 322.9$, $V_0 = (322.9 + 550)/1.15 = 759.1$. Hence, the project's market-to-book ratio is $1{,}082/800 = 1.352$, which means that the replicating portfolio's total investment is 35.2% greater than the project's total investment. Using (10.11), the project WACC/MARR is equal to the mean of the WACC multiplied by the market-to-book ratio V/C:

$$\bar{\rho} = \frac{\overbrace{0.15 \cdot 759.1 + 0.2 \cdot 322.9}^{r_1 V_0 + r_2 V_1}}{\underbrace{759.1 + 322.9}_{V_0 + V_1}} \cdot \overbrace{1.352}^{V/C} = 22.31\%.$$

The higher WACC in the second period results in a higher cutoff rate. The excess return rate is then $\epsilon = \bar{\jmath} - \bar{\rho} = 42.19\% - 22.3\% = 19.89\%$. For each dollar invested, the project returns \$0.1989 over and above the cutoff rate, which means that value is created. How much value is created? Multiplying the excess rate by the investment scale, one gets $\text{NPV} = 800 \cdot 19.89\% = 159.1$.

Scenario	WACC	Average ROI	Project WACC	NPV
Time-invariant	$r_1 = 15\%$ $r_2 = 15\%$	$\bar{\jmath} = 42.19\%$	$\bar{\rho} = 20.78\%$	$\text{NPV} = 171.3$
Time-varying	$r_1 = 15\%$ $r_2 = 20\%$	$\bar{\jmath} = 42.19\%$	$\bar{\rho} = 22.31\%$	$\text{NPV} = 159.1$

(Compare the results with those of Example 8.10.) As opposed to the traditional NPV analysis, which disregards invested capital amounts and rate of return, it is rather easy to understand why the difference occurs in the two scenarios, for it explicitly highlights the sources of value creation: Financial efficiency ($\bar{\jmath} - \bar{\rho}$) and project scale ($C$). Specifically, the reduction in the NPV is due to a decrease in the efficiency, while the project scale remains unvaried. More precisely, the project has a scale of \$800 in both scenarios (constant WACC and time-varying WACC). However, in the scenario with time-varying WACCs, the second-year WACC is higher than the first-year WACC, which triggers two conflicting effects:

- the average WACC ($\bar{r}$) increases from 15% to 16.49%
- the relative project scale (V/C) is reduced from $1{,}108.3/800 = 1.39$ to $1{,}082/800 = 1.35$.

The first effect prevails, so the net effect is an increase of the project COC. Therefore, the economic efficiency decreases from 21.41% to 19.89%, which makes the NPV smaller. The difference of the NPV with constant WACC and the NPV with time-varying WACCs is equal to the product of the project scale and the difference in the economic efficiencies:

$$800 \cdot (21.41\% - 19.89\%) = 12.2 = 171.3 - 159.1.$$

Example 10.9 ♦

C^o	C^d
	C^e

(Unlevered/levered average ROI and ROE of Mockup, 100% *payout,* $i^d = 3\% = r^d$) Consider Mockup project again, but assume that the payout ratio is 100% (so that $C^o_t = C_t$, $F^{uo}_t = F^u_t$, $i^{uo}_t = i^u_t$, and $r^{uo}_t = r^u_t$ for all t) and that $i^d = r^d = 3\%$ so that $\text{NPV}^d = 0$ and $\text{NPV} = \text{NPV}^o = \text{NPV}^e$.

Suppose the analyst wants to calculate the internal (unlevered) average ROI, $\bar{\jmath}^u = \bar{\jmath}^{uo}$. In this example, we exploit (10.63), which requires the computations of the economic values for calculating the MARR.

Table 10.3 reports the necessary input data (NOPATs, capitals, and WACCs) and the computations for the internal average ROI and the associated COC.

Table 10.3 Mockup project (100% payout) (unlevered) average ROI and associated MARR

Time	0	1	2	3	4	5
NOPAT (I_t^u)		32,495	35,954	39,499	43,133	46,858
Invested capital (C_t)	500,000	413,000	326,000	239,000	152,000	
Total NOPAT (I^u)	197,939					
Total Invested capital (C)	1,630,000					
average ROI ($\bar{j}^u$)	12.14%					
WACC (r^u)		8.81%	9.09%	9.51%	10.24%	11.89%
Intrinsic value (V_t)	525,592	452,378	370,527	279,260	177,727	
Market return ($I_t^{Vu} = r_t^u V_{t-1}$)		46,280	41,103	35,232	28,600	21,131
Total market return ($\sum_{t=1}^n I_t^{Vu}$)	172,347					
Total intrinsic value (V)	1,805,484					
project WACC ($\bar{\rho}^u$)	10.57%					
NPV	25,592					

The (unlevered) average ROI is found by dividing the overall NOPAT by the overall invested capital: $\bar{j}^u = 12.14\%$, while the MARR is found by dividing the total market return (computed as $\sum_{t=1}^n I_t^{Vu} = \sum_{t=1}^n r_t^u \cdot V_{t-1}$) by the total invested capital: $\bar{\rho}^u = 172{,}347/1{,}630{,}000 = 10.57\%$.

As an alternative, one may also compute the MARR as a multiple of the average WACC. The latter is $\bar{r}_V^u = 172{,}347/1{,}805{,}484 = 9.545\%$, and the scale-adjustment factor (market-to-book ratio) is $V/C = 1{,}805{,}484/1{,}630{,}000 = 1.1077$. By investing in the replicating portfolio, the investors would invest 10.77% more than they actually invest in the project at a rate of 9.545%, which is equivalent to $\bar{\rho}^u = 9.545\% \cdot 1.1077 = 10.57\%$ on a \$1,630,000 investment. The project creates value for the claimholders as a whole (equityholders + debtholders), since $12.14\% > 10.57\%$. The project NPV is equal to

$$\text{NPV} = \overbrace{1,630,000}^{\text{investment scale}} \cdot \overbrace{(12.14\% - 10.57\%)}^{\text{project's economic efficiency}} = 25,592.$$

The average ROI might also be computed from the stream of ROIs,

$$\boldsymbol{i}^u = (6.50\%, 8.71\%, 12.12\%, 18.05\%, 30.83\%)$$

where $i_t^u = i_t^{uo} = \text{NOPAT}_t/C_{t-1}^o$ (see Eq. (10.57)). The weights are obtained by dividing C_{t-1} by C. Applying the weights to the ROIs and summing, one gets 12.14%, as before (see Table 10.4).

Table 10.4 Mockup project (100% payout) (unlevered) average ROI—recalculated

	0	1	2	3	4	5
NOPAT (I_t^u)		32,495	35,954	39,499	43,133	46,858
(Operating) assets (C_t)	500,000	413,000	326,000	239,000	152,000	0
ROI (j_t^u)		6.50%	8.71%	12.12%	18.05%	30.83%
weight (C_{t-1}/C)		30.67%	25.34%	20.00%	14.66%	9.33%
weight × ROI		1.99%	2.21%	2.42%	2.65%	2.87%
SUM (weight × ROI)	12.14%					

It is important to remind that 12.14% is the unlevered average ROI, $\bar{j}^u$, not the (levered) average ROI, $\bar{j}$. The former does not incorporate the tax shield, the latter does incorporate the tax shield. It is easy to compute $\bar{j}$: It suffices to divide the net CFA, F, by the total capital invested C. We remind that the cash-flow stream is $\boldsymbol{F} = (-500{,}000, 120{,}386, 123{,}845, 127{,}390, 131{,}024, 199{,}749)$ while the capital stream is $\boldsymbol{C} = (500{,}000, 413{,}000, 326{,}000, 239{,}000, 152{,}000, 0)$ (see Table 4.27). Hence, the levered average ROI is

$$\bar{j} = \frac{F}{C} = \frac{-500{,}000 + 120{,}386 + 123{,}845 + 127{,}390 + 131{,}024 + 199{,}749}{500{,}000 + 413{,}000 + 326{,}000 + 239{,}000 + 152{,}000}$$
$$= 12.42\%.$$

This figure is higher than $\bar{j}^u = 12.14\%$ by 0.28% for it shows the effects of the tax shield on the investors' rate of return. In other words, if the project were unlevered, the rate of return would be smaller by 0.28% points. Therefore, the rate of return actually earned by the claimholders is $\bar{j} = 12.42\%$, while $\bar{j}^u = 12.14\%$ is the rate of return net of the tax shield contribution:[7]

$$\bar{j} = \bar{j}^u + \frac{\tau I^d}{C} = 12.14\% + \overbrace{\left(\frac{0.33 \cdot (5 \cdot 2{,}700)}{1{,}630{,}000}\right)}^{0.28\%} = 12.42\%.$$

The levered project COC associated with 12.42% is $\bar{\rho} = 10.85\%$, which does not include the tax shield. The difference between $\bar{\rho} = 10.85\%$ and $\bar{\rho}^u = 10.57\%$ is just 0.28%, confirming, on one hand, the impact of the tax shield on the rate of return and, on the other hand, the fact that the decomposition of the project NPV is the same no matter whether one uses a levered perspective or an unlevered perspective ($\epsilon = \epsilon^u = 1.57\%$).

[7] A more precise value of the tax shield contribution is 0.273%. The error is due to rounding the average ROI and its MARR to the second decimal. More precisely, rounding to the third decimal, one gets $\bar{j} = 12.417\%$, $\bar{\rho} = 12.144\%$ and their difference is precisely 0.273%.

Table 10.5 Mockup project (100% payout) average ROE and equity MARR

Time	0	1	2	3	4	5
Net Income (I_t^e)		30,686	34,145	37,690	41,324	45,049
Equity capital (C_t^e)	410,000	323,000	236,000	149,000	62,000	0
Total Net Income (I^e)	188,894.3					
Total equity capital (C^e)	1,180,000					
Average ROE ($\bar{\jmath}^e$)	16%					
Cost of equity (r_t^e)		10.21%	10.84%	11.91%	14.16%	22.03%
Equity value (V_t^e)	435,592	362,378	280,527	189,260	87,727	0
Market return ($r_t^e V_{t-1}^e$)		44,471	39,294	33,423	26,791	19,322
Total market return ($\sum_{t=1}^n r_t^e V_{t-1}^e$)	163,302					
Equity MARR ($\bar{\rho}^e$)	13.8%					
Equity NPV (NPV^e)	25,592					

The shareholder rate of return is the average ROE, equal to $\bar{\jmath}^e = 16\% = 188{,}894/1{,}180{,}000$, while the equity COC is $\bar{\rho}^e = 13.8\% = 163{,}302/1{,}180{,}000$ (see Table 10.5). The economic efficiency of the equity investment is then $16\% - 13.8\% = 2.2\%$. The equity NPV is then $\text{NPV}^e = 1{,}180{,}000 \cdot 2.2\% = 25{,}592.3$. This is exactly equal to the project NPV, because it is assumed $i^d = r^d = 3\%$, so that $\text{NPV}^d = 0$. It is easy to check that the 1.57% efficiency we have computed for the project is the weighted mean of the 2.2% efficiency of the equity capital and the 0% efficiency of the debt capital. (Needless to say, the use of the shortcuts would have led to the same results.)

Example 10.10 ♦

C^o	C^d
	C^e

(Unlevered average ROI and average ROE of BHP Billiton, 100% payout) Let us take up BHP Billiton project again, assuming that NOWC is not present, so that $C_t = \text{NFA}_t$ for every t. As in Example 8.12, we assume that the WACC is constant: $r^u = 7.5\%$. We compute the internal (unlevered) average ROI by dividing the sum of NOPATs by the sum of operating capitals (see columns 1 and 2 in Table 10.6): $\bar{\jmath}^u = 4{,}375/5{,}775 = 75.76\%$. The market-to-book ratio is $V/C = 31{,}938.7/5{,}775 = 5.53$. Multiplying by the (unadjusted) WACC one gets the project WACC: $\bar{\rho}^u = 0.075 \cdot 5.53 = 41.48\%$. The project's economic efficiency is $\epsilon^u = 34.28\%$. Multiplying the scale and the excess return one gets the economic value created by the project: $\text{NPV} = 5{,}775 \cdot 34.28\% = 1{,}979.6$.

As we know, the rate of return for equityholders is different. To compute it, we report the necessary data in Table 10.7. The average ROE is $\bar{\jmath}^e = 4{,}263/3{,}775 = 112.93\%$ while the equity COC is $\bar{\rho}^e = 2{,}283.4/3{,}775 = 60.49\%$. The equity NPV is

$$\text{NPV}^e = 3{,}775 \cdot \overbrace{(112.93\% - 60.49\%)}^{\epsilon^e = 52.44\%} = 1{,}979.6.$$

Table 10.6 BHP Billiton (100% payout, no NOWC, no liquid assets) internal average ROI and MARR

Time	NOPAT $(I_t^u = I_t^{uo})$	Invested capital $(C_t = C_t^o)$	Project value $(V_t = V_t^o)$
0		550.0	2529.6
1	222.3	522.5	2469.6
2	222.3	495.0	2405.0
3	222.3	467.5	2335.7
4	222.3	440.0	2261.1
5	222.3	412.5	2180.9
6	222.3	385.0	2094.7
7	222.3	357.5	2002.1
8	222.3	330.0	1902.5
9	222.3	302.5	1795.4
10	222.3	275.0	1680.3
11	222.3	247.5	1556.6
12	222.3	220.0	1423.6
13	222.3	192.5	1280.6
14	222.3	165.0	1126.9
15	222.3	137.5	961.7
16	222.3	110.0	784.1
17	222.3	82.5	593.1
18	222.3	55.0	387.9
19	222.3	27.5	167.2
20	152.3	0.0	0.0
TOTAL	4,375	5,775	31,938.7
Average ROI $(\bar{\jmath}^u)$	75.76%		
MARR $(\bar{\rho}^u)$	41.48%		

This is the same as the project NPV, which should come as no surprise, given that $r^d = i^d$ (which implies $\text{NPV}^d = V_0^d - C_0^d = 0$). Precisely for this reason, the same economic value is generated for equityholders with a higher economic efficiency (52.44% > 34.28%) and a smaller investment scale (3,775 < 5,775) than the project.

Example 10.11 ♦
(*Average ROI of abandonment project*) Consider the abandonment project illustrated in Example 1.11. Suppose the COC is constant and equal to $r = 6\%$. For convenience of the reader, we report below the concise transposed Matrix:

Table 10.7 BHP Billiton (100% payout, no NOWC) (internal) average ROE

Time	Net Income (I_t^e)	Equity capital (C_t^e)	Equity value (V_t^e)	Cost of equity (r_t^e)	Market return ($r_r^e V_{t-1}^e$)
0		350.0	2329.6		
1	211.1	322.5	2269.6	7.66%	178.5
2	211.1	295.0	2205.0	7.67%	174.0
3	211.1	267.5	2135.7	7.67%	169.2
4	211.1	240.0	2061.1	7.68%	164.0
5	211.1	212.5	1980.9	7.68%	158.4
6	211.1	185.0	1894.7	7.69%	152.4
7	211.1	157.5	1802.1	7.70%	145.9
8	211.1	130.0	1702.5	7.71%	139.0
9	211.1	102.5	1595.4	7.72%	131.5
10	211.1	275.0	1680.3	7.74%	123.5
11	222.3	247.5	1556.6	7.50%	126.0
12	222.3	220.0	1423.6	7.50%	116.7
13	222.3	192.5	1280.6	7.50%	106.8
14	222.3	165.0	1126.9	7.50%	96.0
15	222.3	137.5	961.7	7.50%	84.5
16	222.3	110.0	784.1	7.50%	72.1
17	222.3	82.5	593.1	7.50%	58.8
18	222.3	55.0	387.9	7.50%	44.5
19	222.3	27.5	167.2	7.50%	29.1
20	152.3	0.0	0.0	7.50%	12.5
Total	4,263	3,775			2,283.4
Average ROE	112.93%				
Equity MARR	60.49%				

$(\boldsymbol{C}, \boldsymbol{I}, \boldsymbol{F})$

Time		0	1	2	3	4	5
Capital	C	−6,000	−4,800	−3,600	−2,400	−1,200	0
Income	I	500	−400	−200	−100	−50	−10
Cash Flow	F	6,500	−1,600	−1,400	−1,300	−1,250	−1,210

This project is a financing project, according to Definition 10.1, for

$$C = -6{,}000 - 4{,}800 - 3{,}600 - 2{,}400 - 1{,}200 = -18{,}000 < 0.$$

This means that the resulting IARR is a financing rate. Specifically, considering that

$$I = 500 - 400 - 200 - 100 - 50 - 10 = -260,$$

one gets

$$\bar{\jmath} = \frac{-260}{-18{,}000} = 1.44\%.$$

The investors borrow (subtract) $|-\$18{,}000| = \$18{,}000$ from the project at a cost of 1.44% while giving up the opportunity of borrowing the same amount from the market. The corresponding COC (MAFR, since this is a financing project) is

$$\bar{\rho} = 0.0144 - \frac{6{,}500 - \frac{1{,}600}{1.06} - \frac{1{,}400}{1.06^2} - \frac{1{,}300}{1.06^3} - \frac{1{,}250}{1.06^4} - \frac{1{,}210}{1.06^5}}{-18{,}000} = 5.66\%.$$

Since $\bar{\jmath} = 1.44\% < \bar{\rho} = 5.66\%$, the project is worth undertaking and, in particular, the project's efficiency, measured by the excess IARR, is $\bar{\jmath} - \bar{r} = 1.44\% - 5.66\% = -4.22\%$. That is, the firm saves financing costs at a rate of 4.22% (compare these results with those obtained for AIRR in Example 8.14).

10.3 The Average ROE and the Equity COC

We know that the ROE bears a significant relation to the various ROIs which passes through the debt/equity ratios. We report below the four expressions for the ROE, presented in Eqs. (4.48) and (4.61):

$$i_t^e = i_t^o + \frac{C_{t-1}^{\mathcal{D}}}{C_{t-1}^e} \cdot (i_t^o - i_t^{\mathcal{D}}) \tag{10.71}$$

$$i_t^e = i_t^{uo} + \frac{C_{t-1}^{\mathcal{D}}}{C_{t-1}^e} \cdot \left(i_t^{uo} - i_t^{\mathcal{D}}(1-\tau)\right) \tag{10.72}$$

$$i_t^e = i_t + \frac{C_{t-1}^d}{C_{t-1}^e} \cdot (i_t - i_t^d) \tag{10.73}$$

$$i_t^e = i_t^u + \frac{C_{t-1}^d}{C_{t-1}^e} \cdot \left(i_t^u - i_t^d(1-\tau)\right). \tag{10.74}$$

Owing to the linear structure of the IARR approach, analogous relations holds between the average ROE and the average ROI. In particular, the average ROE may be computed as

$$\bar{\jmath}^e = \bar{\jmath}^o + \frac{C^{\mathcal{D}}}{C^e} \cdot (\bar{\jmath}^o - \bar{\jmath}^{\mathcal{D}}) \tag{10.75}$$

$$\bar{\jmath}^e = \bar{\jmath}^{uo} + \frac{C^{\mathcal{D}}}{C^e} \cdot \left(\bar{\jmath}^{uo} - \bar{\jmath}^{\mathcal{D}}(1-\tau)\right) \tag{10.76}$$

$$\bar{\jmath}^e = \bar{\jmath} + \frac{C^d}{C^e} \cdot (\bar{\jmath} - \bar{\jmath}^d) \tag{10.77}$$

$$\bar{\jmath}^e = \bar{\jmath}^u + \frac{C^d}{C^e} \cdot \left(\bar{\jmath}^u - \bar{\jmath}^d(1-\tau)\right). \tag{10.78}$$

(The proof is straightforward: It suffices to multiply each sides of the three equations in (10.71) by C^e_{t-1}, sum through 1 to n, divide by C^e and make simple algebraic manipulations.)

As for the equity MARR, we know that

$$\bar{\rho}^e = \frac{r^e_1 V^e_0 + r^e_2 V^e_1 + \ldots + r^e_n V^e_{n-1}}{C^e_0 + C^e_1 + \ldots + C^e_{n-1}} \tag{10.79}$$

$$\bar{\rho}^e = \frac{\rho^e_1 C^e_0 + \rho^e_2 C^e_1 + \ldots + \rho^e_n C^e_{n-1}}{C^e_0 + C^e_1 + \ldots + C^e_{n-1}} \tag{10.80}$$

$$\bar{\rho}^e = \bar{r}^e_V \cdot \frac{V^e}{C^e} \tag{10.81}$$

$$\bar{\rho}^e = \bar{\jmath}^e - \frac{\text{NPV}^e}{C^e} \tag{10.82}$$

with $\rho^e_t = r^e_t(V^e_{t-1}/C^e_{t-1})$. However, we also know that

$$\begin{aligned}
r^e_t &= r^o_t + \frac{V^{\mathcal{D}}_{t-1}}{V^e_{t-1}} \cdot (r^o_t - r^{\mathcal{D}}_t) \\
r^e_t &= r^{uo}_t + \frac{V^{\mathcal{D}}_{t-1}}{V^e_{t-1}} \cdot \left(r^{uo}_t - r^{\mathcal{D}}_t\right) + \frac{\tau I^{\mathcal{D}}_t}{V^e_{t-1}} \\
r^e_t &= r^{uo}_t + \frac{V^{\mathcal{D}}_{t-1}}{V^e_{t-1}} \cdot \left(r^{uo}_t - r^{\mathcal{D}}_t(1-\tau)\right) \quad [\text{if } I^{\mathcal{D}}_t = r^{\mathcal{D}}_t \cdot V^{\mathcal{D}}_{t-1}] \\
r^e_t &= r_t + \frac{V^d_{t-1}}{V^e_{t-1}} \cdot (r_t - r^d_t) \\
r^e_t &= r^u_t + \frac{V^d_{t-1}}{V^e_{t-1}} \cdot \left(r^u_t - r^d_t\right) + \frac{\tau I^d_t}{V^e_{t-1}} \\
r^e_t &= r^u_t + \frac{V^d_{t-1}}{V^e_{t-1}} \cdot \left(r^u_t - r^d_t(1-\tau)\right) \quad [\text{if } I^d_t = r^d_t \cdot V^d_{t-1}]
\end{aligned} \tag{10.83}$$

(see Eq. (6.88)). Exploiting again the linear structure of the IARR approach, one finds the same formal structure for the relations between average cost of equity and average cost of asset:

$$
\begin{aligned}
\bar{r}^e_V &= \bar{r}^o_V + \frac{V^{\mathcal{D}}}{V^e} \cdot (\bar{r}^o_V - \bar{r}^{\mathcal{D}}_V) \\
\bar{r}^e_V &= \bar{r}^{uo}_V + \frac{V^{\mathcal{D}}}{V^e} \cdot \left(\bar{r}^{uo}_V - \bar{r}^{\mathcal{D}}_V\right) + \frac{\tau I^{\mathcal{D}}}{V^e} \\
\bar{r}^e_V &= \bar{r}^{uo}_V + \frac{V^{\mathcal{D}}}{V^e} \cdot \left(\bar{r}^{uo}_V - \bar{r}^{\mathcal{D}}_V(1-\tau)\right) \ \ [\text{if } I^{\mathcal{D}}_t = r^{\mathcal{D}}_t \cdot V^{\mathcal{D}}_{t-1}] \\
\bar{r}^e_V &= \bar{r}_V + \frac{V^d}{V^e} \cdot (\bar{r}_V - \bar{r}^d_V) \\
\bar{r}^e_V &= \bar{r}^u_V + \frac{V^d}{V^e} \cdot \left(\bar{r}^u_V - \bar{r}^d_V\right) + \frac{\tau I^d}{V^e} \\
\bar{r}^e_V &= \bar{r}^u_V + \frac{V^d}{V^e} \cdot \left(\bar{r}^u_V - \bar{r}^d_V(1-\tau)\right) \ \ [\text{if } I^d_t = r^d_t \cdot V^d_{t-1}]
\end{aligned}
\tag{10.84}
$$

where $I^{\mathcal{D}} = \sum_{t=1}^{n} I^{\mathcal{D}}_t$ (the proof is the same as for the average ROE, with the costs of capital replacing the income rates and the economic values replacing the book values).

Hence, considering that the equity COC is a multiple of the average cost of equity, $\bar{\rho}^e = \bar{r}^e_V(V^e/C^e)$, one may also write

$$
\begin{aligned}
\bar{\rho}^e &= \bar{r}^o_V \cdot \frac{V^e}{C^e} + \frac{V^{\mathcal{D}}}{C^e} \cdot (\bar{r}^o_V - \bar{r}^{\mathcal{D}}_V) \\
\bar{\rho}^e &= \bar{r}^{uo}_V \cdot \frac{V^e}{C^e} + \frac{V^{\mathcal{D}}}{C^e} \cdot \left(\bar{r}^{uo}_V - \bar{r}^{\mathcal{D}}_V\right) + \frac{\tau I^{\mathcal{D}}}{C^e} \\
\bar{\rho}^e &= \bar{r}^{uo}_V \cdot \frac{V^e}{C^e} + \frac{V^{\mathcal{D}}}{C^e} \cdot \left(\bar{r}^{uo}_V - \bar{r}^{\mathcal{D}}_V(1-\tau)\right) \ \ [\text{if } I^{\mathcal{D}}_t = r^{\mathcal{D}}_t \cdot V^{\mathcal{D}}_{t-1}] \\
\bar{\rho}^e &= \bar{r}_V \cdot \frac{V^e}{C^e} + \frac{V^d}{C^e} \cdot (\bar{r}_V - \bar{r}^d_V) \\
\bar{\rho}^e &= \bar{r}^u_V \cdot \frac{V^e}{C^e} + \frac{V^d}{C^e} \cdot \left(\bar{r}^u_V - \bar{r}^d_V\right) + \frac{\tau I^d}{C^e} \\
\bar{\rho}^e &= \bar{r}^u_V \cdot \frac{V^e}{C^e} + \frac{V^d}{C^e} \cdot \left(\bar{r}^u_V - \bar{r}^d_V(1-\tau)\right) \ \ [\text{if } I^d_t = r^d_t \cdot V^d_{t-1}]
\end{aligned}
\tag{10.85}
$$

Example 10.12 ♦

C^o	C^d
	C^e

(Unlevered average ROI and average ROE of 2-period levered project, assuming constant WACC and $r^d \neq i^d$) Consider a levered project whose initial cost is $C^o_0 = C_0 = \$600$, which depreciates by \$400 in the first period, so that $C^o_1 = 200$. The firm borrows \$250 from a lender, repayable with two level payments at a ROD of $i^d = 2\%$. Suppose also that the WACC is exogenously given and is equal to $r^u = 15\%$ while the required return on debt is constant and greater than the ROD: $r^d = 3\% > 2\% = i^d$. This implies that the instalments are equal to $\$128.8 = 0.02 \cdot 250/(1 - 1.02^{-2})$ and the outstanding debts are $C^d_0 = 250$, $C^d_1 = 126.2\ (= 250 \cdot 1.02 - 128.8)$

so that the book value of equity is $C_0^e = 600 - 250 = 350$, $C_1^e = 200 - 126.2 = 73.8$ and $C^e = 350 + 73.8 = 423.8$. The transposed Matrix is as follows:

(C, I, F)

Time		0	1	2
Capital				
Operating assets	C^o	600	200	0
Non-operating assets	C^l	0	0	0
Debt	C^d	250	126.24	0
Equity	C^e	350	73.76	0
Income				
Operating income	I^o	0	151.25	188.13
Non-operating income	I^l	0	0	0
Interest expense	I^d	0	5.00	2.52
Net Income	I^e	0	146.25	185.61
Cash flow				
Cash flow from operations	F^o	−600	551.25	388.13
Cash flow from non-operating assets	F^l	0	0	0
Cash flow to debt	F^d	−250	128.76	128.76
Cash flow to equity	F^e	−350	422.49	259.37

With an assumed tax rate of $\tau = 25\%$, the FCFs are obtained from CFOs as $F_t^u = F_t^o - \tau I_t^d$:

$$\begin{aligned} F_0^u &= -600 \\ F_1^u &= 551.25 - 0.25 \cdot 5 = 550 \\ F_2^u &= 388.13 - 0.25 \cdot 2.52 = 387.5. \end{aligned}$$

Discounting the FCFs at $r^{uo} = r^u = 15\%$, one finds the project values: $V_0 = 771.27$ and $V_1 = 336.94$, so that $V = 1{,}108.2$. We now compute the average ROE and its associated MARR. First, we compute the average ROI:

$$\bar{\jmath}^u = \frac{-600 + 550 + 387.5}{600 + 200} = 42.19\%.$$

The average ROD is, obviously, $\bar{\jmath}^d = i^d = 2\%$. Then, using (10.78), one finds the internal average ROE:

$$\bar{\jmath}^e = \overbrace{42.19\%}^{\bar{\jmath}^u} + \frac{\overbrace{376.2}^{C^d}}{\underbrace{423.8}_{C^e}} \cdot \big(\overbrace{42.19\%}^{\bar{\jmath}^u} - \overbrace{2\%}^{\bar{\jmath}^d} \cdot \overbrace{(1 - 0.25)}^{(1-\tau)}\big) = 78.31\%.$$

As for the equity COC, one needs both the equity market values and the debt market value. Discounting the CFDs at $r^d = 3\%$, one finds the market values of debt: $V_0^d = 246.4$ and $V_1^d = 125$ so that $V^d = 246.4 + 125 = 371.4$. The economic values of equity are $V_0^e = 771.27 - 246.4 = 524.88$, $V_1^e = 336.94 - 125 = 211.94$. Hence, the cost of equity is

$$r_1^e = \frac{211.94 + 422.49}{524.88} - 1 = 20.87\%$$
$$r_2^e = \frac{0 + 259.37}{211.94} - 1 = 22.38\%.$$

Using (10.82),

$$\bar{\rho}^e = 78.31\% - \frac{-350 + \dfrac{422.49}{1.2087} + \dfrac{259.37}{(1.2087)(1.2238)}}{423.76} = 37.04\%.$$

The scale of the equity investment is $C^e = 423.76$, and its marginal efficiency is $\epsilon^e = 78.31\% - 37.04\% = 41.27\%$. The equity NPV may be viewed as being derived from assembling scale and efficiency of the equity capital: $\text{NPV}^e = 423.76 \cdot 41.27\% = 174.88$. The debt NPV is nonzero, for the efficiency of the debt is negative. To compute it, consider that the average ROD is $\bar{\jmath}^d = i^d = 2\%$.

The debt's MARR is a multiple of $r^d = 3\%$. Precisely,

$$\bar{\rho}^d = r^d \cdot \frac{V^d}{C^d} = 0.03 \cdot \frac{371.4}{376.24} = 2.96\%.$$

The economic efficiency of the debt is $\epsilon^d = 2\% - 2.96\% = -0.96\%$. The project NPV may be framed as

$$\text{NPV} = \overbrace{423.76 \cdot 41.27\%}^{\text{NPV}^e = 174.88} + \overbrace{376.24 \cdot (-0.96\%)}^{\text{NPV}^d = -3.6} = 171.28.$$

The equityholders gain \$3.6 more than the project NPV, at the expense of the debtholders.[8]

The equity COC might have been computed in many other ways. For the sake of illustration, we use (10.85) (fifth equality) to calculate it. First, consider that $\bar{r}_V^u = r^u = 15\%$ and $\bar{r}_V^d = r^d = 3\%$. Hence,

$$\bar{\rho}^e = \overbrace{15\%}^{r^u} \cdot \frac{\overbrace{736.83}^{V^e}}{\underbrace{423.76}_{C^e}} + \frac{\overbrace{371.4}^{V^d}}{\underbrace{423.76}_{C^e}} \overbrace{(15\% - 3\%)}^{r^u - r^d} + \frac{\overbrace{0.25 \cdot 7.52}^{\tau I^d}}{\underbrace{423.76}_{C^e}} = 37.04\%.$$

---

[8] If one assumes, *ceteris paribus*, $r^d = i^d = 2\%$, then $C_t^d = V_t^d$ for $t = 0, 1, 2$ so that $\text{NPV}^d = 0$ and $\text{NPV} = \text{NPV}^e = 171.28$: Equityholders do not capture any additional value from debtholders.

10.4 Cash Multiple and Modified Dietz

In Chap. 8 we have seen that the PI and the BC are incorporated in the AIRR approach. However, some firms do not discount cash flows when calculating PI and BC. This results in a *cash multiple* (CM), which is often seen as the sign of an incomplete understanding of the time value of money:

> A few companies do not discount the benefits or costs before calculating the profitability index. The less said about these companies the better (Brealey et al. 2011, p. 143).

> A naïve approach that is often used is to divide the inflows by the outflows ...This formula does not work in a multiperiod setting. ...The naïve approach—because it does not properly take into account the timing of the cash flows—is not a correct measure of return (Rao 1992, pp. 74–75).

> Real-world practitioners often use IRR and the *cash multiple* (or multiple of money) as alternative valuation metrics ...The cash multiple (also called the multiple of money or absolute return) is the ratio of the total cash received to the total cash invested ...The cash multiple is a common metric used by investors in transactions such as this one. It has an obvious weakness: The cash multiple does not depend on the amount of time it takes to receive the cash (Berk and DeMarzo 2014, pp. 697–698).

Formally, remembering that $F_t^- \geq 0$ denotes an outflow and $F_t^+ \geq 0$ denotes an inflow, the CM may be defined in various ways. As described by the second and third quotations above, it may be defined as total inflow divided by total outflow:

$$\text{CM} = \frac{F^+}{F^-}. \tag{10.86}$$

This means that CM is an undiscounted Benefit-Cost ratio (see Eq. (8.73)). It may be also defined as net cash flow, $F = F^+ - F^-$, divided by total outflow (or initial outflow, as intended by Brealey et al. 2011. See above). In this case, it is given the moniker *Modified Dietz* (MD):

$$\text{MD} = \frac{F}{F^-} = \text{CM} - 1. \tag{10.87}$$

In this form, it is an undiscounted Profitability Index with capital base equal to F^- (instead of C_0). As seen in Remark 10.5, the IARR function $\bar{\jmath}(x) = I/x = F/x$ gives rise to the IARR class of rates of return, the *IARR class*, differentiated by the capital base. The MD belongs to the IARR class, with $x = F^-$, so that $\text{MD} = \bar{\jmath}(F^-)$. The corresponding MARR is $\bar{\rho}(F^-) = \bar{r}_V \cdot (V/F^-)$. Hence, $\text{NPV} = F^- \cdot (\text{MD} - \bar{\rho}(F^-))$ and the project is worth undertaking if and only if $\text{MD} > \bar{\rho}(F^-)$ or, equivalently, if and only if $\text{CM} > 1 + \bar{\rho}(F^-)$.

This turns a seemingly naïve metric into a rational measure of economic profitability. Even more compellingly, the computation of CM/MD and its MARR are extremely simple. In particular, a shortcut is available for a fast calculation of the MARR. One merely reverse-engineers the product structure described above and gets

$$\bar{\rho}(F^-) = \text{MD} - \frac{\text{NPV}}{F^-} = \frac{F - \text{NPV}}{F^-}.$$

Remark 10.7 Strictly speaking, the definition of MD above is a variant of the Modified Dietz return, which is itself a variant of the Simple Dietz return. These and other variants are widely employed in the practice of portfolio management. In particular, the Modified Dietz is often used as an approximation of the IRR (see Spaulding 2011; Fischer and Wermers 2013):

> the IRR is solved iteratively (i.e., by trial and error). As noted above, a rule-of-thumb is to start with the result you'd obtain by using Modified Dietz formula (which serves as the "first order" approximation to the IRR) (Spaulding 2011, pp. 98–99).

Modified Dietz is also used as an approximation of the so-called *Time-Weighted Rate of Return* by chain-linking a series of Modified Dietz returns (Spaulding 2011, p. 92).[9]

If the CM is defined in terms of net cash flow and initial capital C_0 (instead of F^-), then it becomes an undiscounted version of the PI, as implied by Brealey et al. (2011, p. 143):

$$\text{CM} = \frac{F}{C_0} \tag{10.88}$$

(see Magni et al. 2018. See also Magni 2014c.) ▲

Example 10.13 ♦
Consider the cash-flow stream $\boldsymbol{F} = (-150, 80, -40, 50, 120)$. Then, $F^- = 190$, $F^+ = 250$, $F = 250 - 190 = 60$. Let $r = 10\%$ be the MARR. The CM may be expressed as $\text{CM} = F^+/F^- = 250/190 = 1.3158$ or as $\text{CM} = F/F_0 = 60/150 = 0.4$ (undiscounted PI), depending on the preferences. Alternatively, using the MD version, $\text{MD} = F/F^- = 60/190 = 0.3158$. As for the MARR, considering the MD and taking into account the fact that the total economic value is $V = 508.03$ (as can be checked), then $\bar{\rho}(F^-) = 0.1 \cdot 508.03/190 = 0.2674 < 0.3158$. The investors earn 31.58% on the total outlay, thereby renouncing to earn 26.74% on that amount. The difference is positive and the NPV is $190 \cdot (31.58\% - 26.74\%) = 9.2 > 0$.

10.5 Aggregate Return on Investment

The Aggregate Return On Investment (AROI) is a relative approach which makes use of the same replicating portfolio (and, therefore, the same benchmark system) which is used for deriving the notion of net future value.[10] Therefore, the benchmark system

[9] See Gray and Dewar (1971) for an axiomatization of the Time-Weighted Rate of Return.

[10] It has been introduced in Magni (2009e) and developed in Magni (2011a, 2015b, 2016a) (see an application to real estate assets in Althsuler and Magni 2015).

is described as $(\mathcal{C}, \boldsymbol{I}^{\mathcal{C}}, \boldsymbol{F}^{\mathcal{C}})$. Suppose, for the time being, that the COC is constant. Investors might invest funds in a portfolio replicating the project's cash flows from time 0 to time $n-1$. The law of motion for such a portfolio is $\mathcal{C}_t = \mathcal{C}_{t-1}(1+r) - F_t$ for $t \geq 0$ with $\mathcal{C}_{-1} = 0$. We have already ascertained that $E_n^{\mathcal{C}} = \mathcal{C}_{n-1}(1+r)$ is the EOP value of the benchmark portfolio at time n, so that the value added is NFV $= E_n - E_n^{\mathcal{C}}$ (see Sect. 6.7.2). Consider that

$$\begin{aligned}
\sum_{t=0}^{n}(I_t - I_t^{\mathcal{C}}) &= \sum_{t=0}^{n}(I_t - r_t\mathcal{C}_{t-1}) \\
&= \sum_{t=0}^{n}(C_t - C_{t-1} + F_t) - \sum_{t=0}^{n-1}(\mathcal{C}_t - \mathcal{C}_{t-1} + F_t) - (\mathcal{C}_n - \mathcal{C}_{n-1} + E_n^{\mathcal{C}}) \\
&= F_n - E_n^{\mathcal{C}} \\
&= E_n - E_n^{\mathcal{C}} \\
&= \text{VA} = [\text{by (6.158)}] = \text{NFV}.
\end{aligned} \tag{10.89}$$

(see also next section and Magni 2011a, Eq. (11)). Therefore,

$$\text{NFV} = C(\bar{\jmath} - \bar{\varrho}) \tag{10.90}$$

where $\bar{\varrho} = r\mathcal{C}/C$ and $\mathcal{C} = \sum_{t=1}^{n} \mathcal{C}_{t-1}$. (It is far too evident that AROI is rational, in the sense that $\bar{\jmath} > \bar{\varrho}$ signals value creation.) In case of time-varying COCs, the MARR is generalized as $\bar{\varrho} = \bar{r}_{\mathcal{C}}\mathcal{C}/C$ where $\bar{r}_{\mathcal{C}} = \sum_{t=1}^{n} r_t\mathcal{C}_{t-1}/\mathcal{C}$. The economic efficiency, denoted as $\varepsilon = \bar{\jmath} - \bar{\varrho}$, is equal to the ratio of NFV to total capital employed.

As a result, the rates of return of a project in the IARR and AROI approach coincide. However, with respect to the IARR, the AROI makes use of a different counterfactual alternative and, therefore, a different benchmark system, which is reflected in the different scale-adjustment factor: $\mathcal{C}/C$ replaces V/C. The MARRs are then defined in the same way as in the IARR approach, except that the scale-adjustment is different. The book AROIs (BAROIs) are equal to the BIARRs.

The three approaches (AIRR, IARR, AROI) are rational (i.e., they signal value creation and provide correct guidance for decision-making). The difference lies in the different product structures implied by the three benchmark systems. Specifically,

$$\text{NPV} = C \cdot (\bar{\jmath} - \bar{\rho}) \qquad \text{NFV} = C \cdot (\bar{\jmath} - \bar{\varrho}) \qquad V_0(\text{RI}) = \text{PV}[C](\bar{\iota}_0 - \bar{r}_0).$$

AIRR and RI subsume a benchmark portfolio which replicates the capital amounts; IARR and NPV subsume a benchmark portfolio which replicates the cash flows from time 1 to time n; AROI and NFV subsume a benchmark portfolio which replicates the cash flows from time 0 to time $n-1$. We delve into this issue in the next section, in order to provide a complete framework of absolute and relative approaches to valuation and decision-making.

It is important, in the context of AROI, to note that the relation with NPV may be restated as follows. Specifically, discounting both sides of (10.90) from n to zero, one expresses the same relation in present terms:

$$\text{NPV} = C_{\text{d}} \cdot (\bar{\jmath}_{\text{d}} - \bar{\varrho}_{\text{d}}) \tag{10.91}$$

where $C_{\text{d}} = C \cdot \text{d}_{n,0}$ and

$$\bar{\jmath}_{\text{d}} = \frac{I_{\text{d}}}{C_{\text{d}}} = \frac{I \cdot \text{d}_{n,0}}{C \cdot \text{d}_{n,0}} \qquad \bar{\varrho}_{\text{d}} = r \cdot \frac{\mathcal{C}_{\text{d}}}{C_{\text{d}}} = r \cdot \frac{\mathcal{C} \cdot \text{d}_{n,0}}{C \cdot \text{d}_{n,0}}. \tag{10.92}$$

This may seem overly redundant and, therefore, useless, given that the discount factors neutralize each other and $\bar{\jmath} = \bar{\jmath}_{\text{d}}$ and $\bar{\varrho} = \bar{\varrho}_{\text{d}}$. Yet, when dealing with incremental analysis for project ranking, the latter framing is necessary for providing rational decision-making (see Chap. 11. See also Magni 2015b, Sect. 4, for this definition).

In such a way, the economic efficiency may be described as either $\varepsilon = \bar{\jmath} - \bar{\varrho}$ or $\varepsilon_{\text{d}} = \bar{\jmath}_{\text{d}} - \bar{\varrho}_{\text{d}}$. For single assets, the two are equal:

$$\varepsilon = \varepsilon_{\text{d}} = \begin{cases} \bar{\jmath} - \bar{\varrho} &= \dfrac{\text{NFV}}{C} \\[2ex] \bar{\jmath}_{\text{d}} - \bar{\varrho}_{\text{d}} &= \dfrac{\text{NPV}}{C_{\text{d}}} \end{cases}$$

However, they are not equal when an incremental analysis of two projects is made (see Sect. 11.2).

The AROI procedure is somewhat opposite to the AIRR procedure; namely, the AROI procedure requires summing before discounting, whereas the AIRR procedure requires discounting before summing: $\mathcal{C}_{\text{d}}$ indicates that capitals must be summed and then discounted, PV$[C]$ indicates that capitals must be discounted and then summed. As a result, the AROI is a Sum&Discount approach while the AIRR is a Discount&Sum approach (see Magni 2010a, 2011d).

10.6 The Six Approaches and the Three Pairs of Glasses

The three benchmark systems are three pairs of glasses for looking into a project. In this section we summarize their relations.

(RI, AIRR) These approaches subsume the same benchmark system, whereby the project's capital amounts, C_t, are replicated. The concise transposed Matrix of the replicating portfolio is

$(\boldsymbol{C}^*, \boldsymbol{I}^*, \boldsymbol{F}^*)$

Time		0	1	2	...	$n-1$	n
Capital	$\boldsymbol{C}^*$	C_0	C_1	C_2	...	C_{n-1}	0
Income	$\boldsymbol{I}^*$	0	r_1C_0	r_2C_1	...	$r_{n-1}C_{n-2}$	r_nC_{n-1}
Cash Flow	$\boldsymbol{F}^*$	F_0^*	F_1^*	F_2^*	...	F_{n-1}^*	F_n^*

AIRR and RI reconcile via the coherent structure (capital × efficiency):

$$\begin{aligned} V_0(\text{RI}) &= \sum_{t=0}^{n}(I_t - I_t^*)d_{t,0} \\ &= \text{PV}[I] - \text{PV}[I^*] \\ &= \text{PV}[C](\bar{\imath}_0 - \bar{r}_0) \qquad \big(\text{or PV}[C](\bar{\imath} - \bar{r})/(1+r_1)\big) \end{aligned} \tag{10.93}$$

(with $I_0^* = 0$). Hence, the absolute criterion $V_0(\text{RI}) > 0$ is replaced by the relative counterpart $\bar{\imath}_0 > \bar{r}_0$ (or $\bar{\imath} > \bar{r}$).

The IRR approach is a special case of this approach where the capital amounts are forced to grow at a constant force of interest. That is, using $C_{t-1}(\sigma)$ in place of C_{t-1}, $V_0(\text{RI}) = \text{PV}[C(\sigma)](\sigma - \bar{r}(\sigma))/(1+r_1)$ where $\bar{r}(\sigma) = \sum_{t=1}^{n} r_t C_{t-1}(\sigma) d_{t,1} / \sum_{t=1}^{n} C_{t-1}(\sigma) d_{t-1,0}$.[11]

(**NPV, IARR**) These approaches subsume the same benchmark system, whereby the project's cash flows are replicated from time 1 to time n. The concise transposed Matrix of the replicating portfolio is

$(\boldsymbol{V}, \boldsymbol{I}^V, \boldsymbol{F}^V)$

Time		0	1	2	...	$n-1$	n
Capital	$\boldsymbol{V}$	V_0	V_1	V_2	...	V_{n-1}	0
Income	$\boldsymbol{I}^V$	0	r_1V_0	r_2V_1	...	$r_{n-1}V_{n-2}$	r_nV_{n-1}
Cash Flow	$\boldsymbol{F}^V$	F_0^V	F_1	F_2	...	F_{n-1}	F_n

[11] It is worth reminding that, if $I_0 \neq 0$, the first internal capital does not comply with the mechanics of an economic system, since $-F_0 = C_0(\sigma) \neq C_0 = \Delta C_0 = I_0 - F_0$.

IARR and NPV reconcile via the coherent structure (capital × efficiency):

$$
\begin{aligned}
\text{NPV} &= V_0 + I_0 - C_0 \\
&= V_0 + \overbrace{I_0 - C_0}^{=F_0} + \overbrace{\sum_{t=0}^{n} \Delta C_t}^{=0} - \overbrace{\sum_{t=0}^{n} \Delta V_t}^{=0} \\
&= \overbrace{V_0 - \Delta V_0}^{=0} + \sum_{t=0}^{n}(C_t - C_{t-1} + F_t) - \sum_{t=1}^{n}(V_t - V_{t-1} + F_t) \\
&= \sum_{t=0}^{n} I_t - \sum_{t=1}^{n} I_t^V \\
&= [\text{since } I_0^V = 0] = \sum_{t=0}^{n} I_t - \sum_{t=0}^{n} I_t^V \\
&= C(\bar{\jmath} - \bar{\rho}).
\end{aligned}
\tag{10.94}
$$

Hence, the absolute criterion NPV > 0 is replaced by the relative counterpart $\bar{\jmath} > \bar{\rho}$.

(**NFV, AROI**) These approaches subsume the same benchmark system, whereby the project's cash flows from time 0 to time $n-1$ are replicated. The concise transposed Matrix of the replicating portfolio is

$(\mathcal{C}, \boldsymbol{I}^{\mathcal{C}}, \boldsymbol{F}^{\mathcal{C}})$

Time		0	1	2	...	$n-1$	n
Capital	$\mathcal{C}$	$\mathcal{C}_0$	$\mathcal{C}_1$	$\mathcal{C}_2$	...	$\mathcal{C}_{n-1}$	0
Income	$\boldsymbol{I}^{\mathcal{C}}$	0	$r_1\mathcal{C}_0$	$r_2\mathcal{C}_1$	...	$r_{n-1}\mathcal{C}_{n-2}$	$r_n\mathcal{C}_{n-1}$
Cash Flow	$\boldsymbol{F}^{\mathcal{C}}$	F_0	F_1	F_2	...	F_{n-1}	$E_n^{\mathcal{C}}$

AROI and NFV reconcile via the coherent product structure (capital × efficiency):

$$\begin{aligned}
\text{NFV} = \text{VA} &= E_n - E_n^{\mathcal{C}} \\
&= \overbrace{\sum_{t=0}^{n} \Delta C_t + F_n}^{=0} - \overbrace{\sum_{t=0}^{n} \Delta \mathcal{C}_t - E_n^{\mathcal{C}}}^{=0} \\
&= \sum_{t=0}^{n}(C_t - C_{t-1} + F_t) - \sum_{t=0}^{n-1}(\mathcal{C}_t - \mathcal{C}_{t-1} + F_t) - (\mathcal{C}_n - \mathcal{C}_{n-1} + E_n^{\mathcal{C}}) \\
&= \sum_{t=0}^{n} I_t - \sum_{t=0}^{n} r_t \mathcal{C}_{t-1} \\
&= C(\bar{\jmath} - \bar{\varrho}).
\end{aligned} \tag{10.95}$$

The correspondence between absolute measures and relative measures creates three appraisal pairs, each of which is primarily linked with one of the three basic elements. Specifically, NPV, RI, and NFV have been associated with the respective notions of cash flow, income, and capital (see Sect. 6.8). In this section, we have made explicit the association between absolute measures of worth and relative measures of worth. Therefore, the same association applies between relative approaches and basic notions. One might then speak of the cash-flow based approach (IARR, NPV), the capital-based approach (AROI, NFV), and the income-based approach (AIRR, RI) with the understanding that the three basic notions are reciprocally intertwined and chained one another into a dynamical system where they flow into one another (see Part I), which means that the three methodologies are just three different pairs of glasses for looking into a project/firm, depending on which counterfactual alternative is selected as the normal, benchmark alternative (see Table 10.8).[12]

The (valuation and) decision-making process is a process which may be expressed in terms of the three basic notions:

- a measurement process based on **cash flows** (NPV-IARR criterion)
- a measurement process based on **incomes** (RI-AIRR criterion)
- a measurement process based on **capitals** (NFV-AROI criterion).

---

[12]The three methodologies, as well as the three basic notions, are three in one: They are the same concept that may take on three personas (see also Remark 1.4 on Sect. 1.3). Thus, a debate about which of the three methodologies best serves the needs of practitioners boils down to debating whether capital or income or cash flow is preferable. From a logical point of view, they are equivalent; whether cash flows or incomes or capitals should be used in real life to make decisions is a practical issue. All six methods are now embodied in a unified logically consistent framework. Learning about similarities and differences between the approaches and their links with the three basic notions may bolster the understanding of the mechanics of the project/firm and the appreciation of six perspectives which provide different pieces of economic information but are naturally reconciled one another.

Table 10.8 Absolute and relative approaches to valuation

	Absolute approach	Relative approach	Linking structure	Primary notion
(NPV, IARR)	NPV	IARR	$C \cdot (\bar{\jmath} - \bar{\rho})$	Cash flow
(RI, AIRR)	RI	AIRR	$\text{PV}[C] \cdot (\bar{\imath}_0 - \bar{r}_0)$	Income
(NFV, AROI)	NFV	AROI	$C \cdot (\bar{\jmath} - \bar{\varrho})$	Capital

---

Example 10.14 ♦

(*The three pairs of glasses*) Consider the same investment analyzed in Example 6.24 in Sect. 6.7.2. We apply the three methodologies breaking down the project and the benchmark portfolio. We will look through the lens of each approach.

Table 10.9 reports the relevant data for both the project and the replicating portfolio, for each methodology. The left-hand side presents the data for the project, while the right-hand side presents the data for the benchmark portfolios. The three left-hand sides are equal, the right-hand sides change.

Under the lens of the (AIRR, RI) approach, an investor undertaking the project foregoes, in each period, the opportunity of investing, at rate r_t, the market value of a portfolio which replicates the project's capital amounts.

Under the lens of the (IARR, NPV) approach, an investor undertaking the project foregoes, in each period, the opportunity of investing, at rate r_t, the market value of a replicating portfolio which mimics the project's cash flows from time 1 to time n.

Under the lens of the (AROI, NFV) approach, an investor undertaking the project foregoes, in each period, the opportunity of investing, at rate r_t, the market value of a portfolio which replicates the cash flows from time 0 to time $n-1$.

Computationally, one gets

- the (instantaneous) AIRR and its associated MARR by summing, respectively, the project's incomes and the benchmark returns, discounted at the benchmark rates r_t, and then dividing by the discounted sum of capitals:

$$\bar{\imath}_0 = \frac{4.10}{66.19} = 6.19\% \qquad \bar{r}_0 = \frac{2.20}{66.19} = 3.32\%$$

  whence $V_0(\text{RI}) = 66.19 \cdot (6.19\% - 3.32\%) = 1.9$.
- the IARR and its associated MARR by summing, respectively, the project's incomes and the benchmark returns and then dividing by total capital:

$$\bar{\jmath} = \frac{4.15}{71.32} = 5.82\% \qquad \bar{\rho} = \frac{2.25}{71.32} = 3.16\%$$

  whence $\text{NPV} = 71.32 \cdot (5.82\% - 3.16\%) = 1.9$

Table 10.9 The three coupled approaches

AIRR and RI

	Project system					Benchmark system				
Time t	Cash flow F_t	Rate of return i_t	EOP value $C_{t-1}(1+i_t)$	BOP value C_t	Interest I_t	Cash flow F_t^*	Rate of return r_t	EOP value $C_{t-1}(1+r_t)$	BOP value C_t	Interest r_tC_{t-1}
0	−10.00			10.00		−10.00			10.00	
1	−2.00	2.00%	10.20	12.20	0.20	−1.70	5.00%	10.50	12.20	0.50
2	0.00	23.00%	15.01	15.01	2.81	−4.03	−10.00%	10.98	15.01	−1.22
3	4.00	−8.00%	13.81	9.81	−1.20	8.80	24.00%	18.61	9.81	3.60
4	−1.50	0.00%	9.81	11.31	0.00	−1.40	1.00%	9.90	11.31	0.10
5	0.00	15.00%	13.00	13.00	1.70	−1.47	2.00%	11.53	13.00	0.23
6	13.65	5.00%	13.65	0.00	0.65	12.61	−3.00%	12.61	0	−0.39
			PV	66.19	4.10				PV	2.20

IARR and NPV

	Project system					Benchmark system				
Time t	Cash flow F_t	Rate of return i_t	EOP value $C_{t-1}(1+i_t)$	BOP value C_t	Interest I_t	Cash flow F_t^V	Rate of return r_t	EOP value $V_{t-1}(1+r_t)$	BOP value V_t	Interest r_tV_{t-1}
0	−10.00			10.00		−11.90			11.90	
1	−2.00	2.00%	10.20	12.20	0.20	−2.00	5.00%	12.49	14.49	0.59
2	0.00	23.00%	15.01	15.01	2.81	0.00	−10.00%	13.05	13.05	−1.45
3	4.00	−8.00%	13.81	9.81	−1.20	4.00	24.00%	16.18	12.18	3.13
4	−1.50	0.00%	9.81	11.31	0.00	−1.50	1.00%	12.30	13.80	0.12
5	0.00	15.00%	13.00	13.00	1.70	0.00	2.00%	14.07	14.07	0.28
6	13.65	5.00%	13.65	0.00	0.65	13.65	−3.00%	13.65	0	−0.42
			Total	71.32	4.15				Total	2.25

AROI and NFV

	Project system					Benchmark system				
Time t	Cash flow F_t	Rate of return i_t	EOP value $C_{t-1}(1+i_t)$	BOP value C_t	Interest I_t	Cash flow $F_t^{\mathcal{C}}$	Rate of return r_t	EOP value $\mathcal{C}_{t-1}(1+r_t)$	BOP value $\mathcal{C}_t$	Interest $r_t\mathcal{C}_{t-1}$
0	−10.00			10.00		−10.00			10.00	
1	−2.00	2.00%	10.20	12.20	0.20	−2.00	5.00%	10.50	12.50	0.50
2	0.00	23.00%	15.01	15.01	2.81	0.00	−10.00%	11.25	11.25	−1.25
3	4.00	−8.00%	13.81	9.81	−1.20	4.00	24.00%	13.95	9.95	2.70
4	−1.50	0.00%	9.81	11.31	0.00	−1.50	1.00%	10.05	11.55	0.10
5	0.00	15.00%	13.00	13.00	1.70	0.00	2.00%	11.78	11.78	0.23
6	13.65	5.00%	13.65	0.00	0.65	11.43	−3.00%	11.43	0.00	−0.35
			Total	71.32	4.15				Total	1.93

- the AROI and its associated MARR by summing, respectively, the project's incomes and the benchmark returns and then dividing by total capital:

$$\bar{\jmath} = \frac{4.15}{71.32} = 5.82\% \qquad \bar{\varrho} = \frac{1.93}{71.32} = 2.70\%$$

whence NFV $= 71.32 \cdot (5.82\% - 2.70\%) = 2.22$.

10.7 Key Points

- The IARR approach provides a genuinely internal rate of return. Contrary to the IRR (and, more generally, the AIRR), its financial nature is not affected by the COC
- The IARR approach defines a project's rate of return as the total (undiscounted) return to total (undiscounted) capital:

$$\text{IARR} = \frac{\text{Total return}}{\text{Total capital}}$$

- From the point of view of the income rates, an IARR may be viewed as the mean of the income rates, i_t, weighted by the capitals invested (or borrowed) in each period:

$$\text{IARR} = \text{Capital-weighted mean of the period return rates}$$

- As the total income coincides with the algebraic sum of the cash flows, the IARR may be defined as the ratio of (undiscounted) total cash flow to (undiscounted) total capital

$$\text{IARR} = \frac{\text{Total cash flow}}{\text{Total capital}}$$

- A fourth equivalent way to compute the IARR is to start from the NPV: Dividing the NPV by the total capital employed and adding the MARR, the IARR is obtained:

$$\text{IARR} = \text{MARR} + \text{NPV}/\text{Total capital}$$

- The MARR of the IARR approach is equal to the total profit of a value-neutral replicating portfolio which mimics the project's cash flows from time 1 to time n, divided by the total capital. It is therefore a multiple of the COC, namely, the product of the (average) COC and the overall market-to-book ratio:

$$\text{MARR} = \text{(average) COC} \times \text{market-to-book ratio}$$

- A rate of return exists for each income and each cash-flow stream. We have analyzed 8 kinds of capitals, incomes, and cash flows. Therefore, 8 types of IARRs are properly associated with a project, called *book* IARRs:

- internal asset average ROI
- internal average ROE
- internal operating average ROI
- internal non-operating average ROI
- internal unlevered average ROI
- internal unlevered average asset ROI
- internal average ROD
- internal average RO$\mathcal{D}$

- The economic value created reconciles with the IARR via a product structure, according to which the economic value created is equal to the product of the total undiscounted capital (project size or scale) and the project's excess IARR (marginal economic efficiency of the capital):

$$\text{Economic Value Created} = \text{Project size} \times \text{Excess IARR}$$

- Cash Multiple (Modified Dietz return) are incorporated in the IARR class, whereby the return is measured per unit of total outflow instead of total capital
- A third relative approach to rate of return is the AROI approach. The project's rate of return is equal to the IARR but the MARR is different: It is equal to the total profit of a value-neutral replicating portfolio which mimics the project's cash flows from time 0 to time $n-1$, divided by the total capital
- The three relative approaches, AIRR, IARR, and AROI, represent the relative counterparts of the three absolute approaches, RI, NPV, and NFV, respectively. The AIRR is affiliated with RI, the IARR is affiliated with to NPV, the AROI is affiliated with NFV. Each pair unequivocally identifies a benchmark system, well-defined by a Matrix containing the benchmark capitals, the benchmark incomes, and the benchmark cash flows. Each pair may be connected with one of the three basic notions. Therefore, each pair acts as a pair of glasses which looks into the project with a capital-oriented perspective, income-oriented perspective, and cash-flow-oriented perspective, respectively:

Basic notion	**Absolute criterion**	**Relative criterion**
Capital	Net future value	AROI
Income	Residual income	AIRR
Cash flow	Net present value	IARR

Table 10.10 explicitly associates each (absolute and relative) criterion with its own benchmark portfolio. Table 10.11 summarizes the main features of the three approaches.

Table 10.10 The three benchmark systems and the six decision criteria

Benchmark system	**Criterion**	**Value is created if**	**Primary notion**
(i-a) Replicating cash flows from time 1 to time n $\boldsymbol{V} = (V_0, V_1, V_2, \ldots, V_{n-1}, 0)$ $\boldsymbol{I}^V = (0, r_1 V_0, r_2 V_1, \ldots, r_n V_{n-1})$ $\boldsymbol{F}^V = (F_0^V, F_1, F_2, \ldots, F_{n-1}, F_n)$	NPV	$\sum_{t=0}^{n} F_t \mathrm{d}_{t,0} > 0$	Cash flow
(i-b) Replicating cash flows from time 1 to time n $\boldsymbol{V} = (V_0, V_1, V_2, \ldots, V_{n-1}, 0)$ $\boldsymbol{I}^V = (0, r_1 V_0, r_2 V_1, \ldots, r_n V_{n-1})$ $\boldsymbol{F}^V = (F_0^V, F_1, F_2, \ldots, F_{n-1}, F_n)$	IARR	$\bar{j} > \bar{\rho}$	Cash flow
(ii-a) Replicating capital amounts $\boldsymbol{C}^* = (C_0, C_1, C_2, \ldots, C_{n-1}, 0)$ $\boldsymbol{I}^* = (0, r_1 C_0, r_2 C_1, \ldots, r_n C_{n-1})$ $\boldsymbol{F}^* = (F_0^*, F_1^*, F_2^*, \ldots, F_{n-1}^*, F_n^*)$	RI	$\sum_{t=0}^{n} (I_t - I_t^*) \mathrm{d}_{t,0} > 0$	Income
(ii-b) Replicating capital amounts $\boldsymbol{C}^* = (C_0, C_1, C_2, \ldots, C_{n-1}, 0)$ $\boldsymbol{I}^* = (0, r_1 C_0, r_2 C_1, \ldots, r_n C_{n-1})$ $\boldsymbol{F}^* = (F_0^*, F_1^*, F_2^*, \ldots, F_{n-1}^*, F_n^*)$	AIRR	$\bar{\imath} > \bar{r}$	Income
(iii-a) Replicating cash flows from time 0 to time n−1 $\mathcal{C} = (\mathcal{C}_0, \mathcal{C}_1, \mathcal{C}_2, \ldots, \mathcal{C}_{n-1}, 0)$ $\boldsymbol{I}^{\mathcal{C}} = (0, r_1 \mathcal{C}_0, r_2 \mathcal{C}_1, \ldots, r_n \mathcal{C}_{n-1})$ $\boldsymbol{F}^{\mathcal{C}} = (F_0, F_1, F_2, \ldots, F_{n-1}, E_n^{\mathcal{C}})$	NFV	$E_n > E_n^{\mathcal{C}}$	Capital
(iii-b) Replicating cash flows from time 0 to time n−1 $\mathcal{C} = (\mathcal{C}_0, \mathcal{C}_1, \mathcal{C}_2, \ldots, \mathcal{C}_{n-1}, 0)$ $\boldsymbol{I}^{\mathcal{C}} = (0, r_1 \mathcal{C}_0, r_2 \mathcal{C}_1, \ldots, r_n \mathcal{C}_{n-1})$ $\boldsymbol{F}^{\mathcal{C}} = (F_0, F_1, F_2, \ldots, F_{n-1}, E_n^{\mathcal{C}})$	AROI	$\bar{j} > \bar{\varrho}$	Capital

Table 10.11 The three appraisal pairs: breakdown of absolute and relative measures

Relative approach	**AIRR**	**IARR**	**AROI**
Rate of return	$\bar{\imath}_0 = \dfrac{\mathrm{PV}[I]}{\mathrm{PV}[C]}$	$\bar{\jmath} = \dfrac{F}{C}$	$\bar{\jmath} = \dfrac{F}{C}$
MARR	$\bar{r}_0 = \dfrac{\sum_{t=1}^{n} r_t C_{t-1} d_{t,0}}{\mathrm{PV}[C]}$	$\bar{\rho} = \dfrac{\sum_{t=1}^{n} r_t V_{t-1}}{C}$	$\bar{\varrho} = \dfrac{\sum_{t=1}^{n} r_t \mathcal{C}_{t-1}}{C}$
Economic Efficiency	$\xi = \bar{\imath}_0 - \bar{r}_0$	$\epsilon = \bar{\jmath} - \bar{\rho}$	$\varepsilon = \bar{\jmath} - \bar{\varrho}$
Project scale	$\mathrm{PV}[C] = \sum_{t=1}^{n} \mathrm{PV}[C_{t-1}]$	$C = \sum_{t=1}^{n} C_{t-1}$	$C = \sum_{t=1}^{n} C_{t-1}$
Product structure	$\mathrm{PV}[C] \cdot \xi$	$C \cdot \epsilon$	$C \cdot \varepsilon$
Absolute approach	V_0(**RI**)	**NPV**	**NFV**

Part V
Project Selection and Some Final Examples

This final part includes a chapter which supplies some tips for ranking and selecting projects with the six methods and a chapter illustrating three projects analyzed with the three appraisal pairs: (NPV, IARR), (RI, AIRR), and (NFV, AROI). The first two projects are engineering projects, the third one is a human resource initiative.

Chapter 11
Ranking Projects

> It can be shown that the scale effect is present in all project comparisons even between projects with identical outlay.
>
> Keane (1979, p. 55)

C. A. Magni, *Investment Decisions and the Logic of Valuation*,
https://doi.org/10.1007/978-3-030-27662-1_11

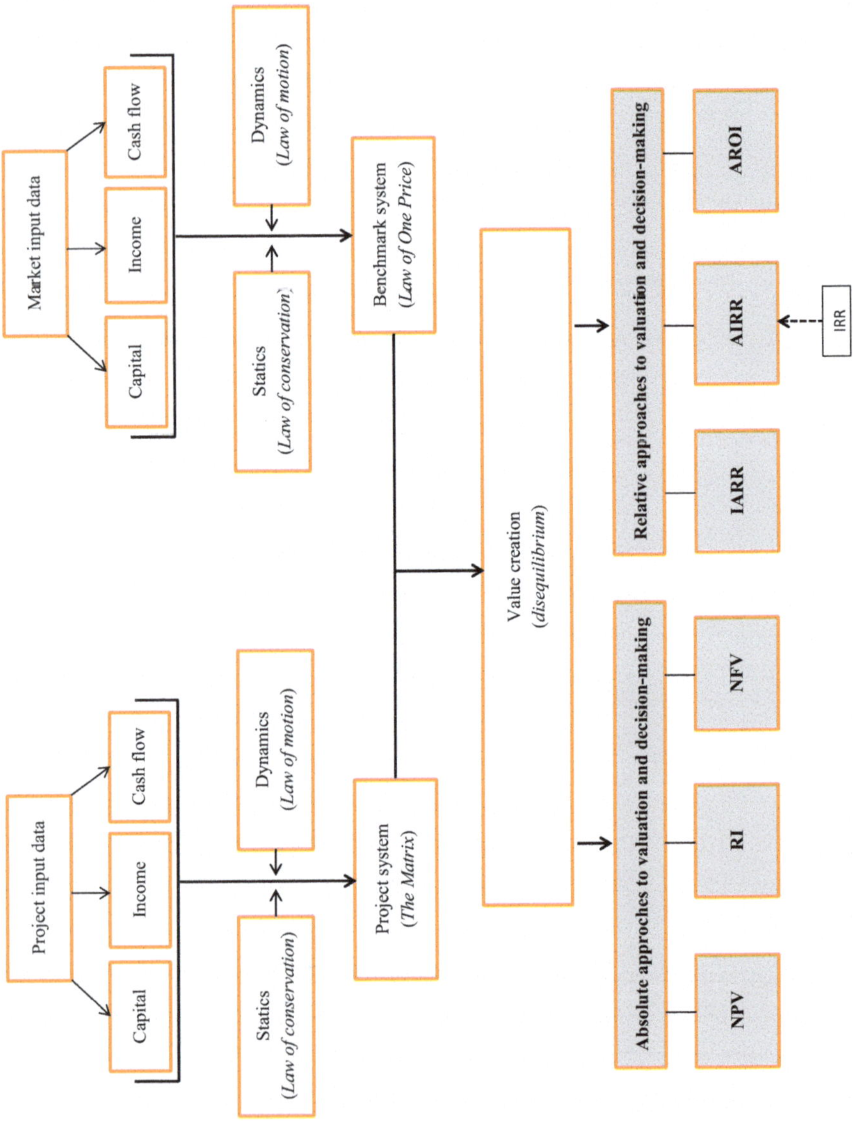

Fig. 11.1 Chapter 11: Choice between mutually exclusive projects and project ranking

This chapter addresses the issue of choosing between mutually exclusive projects and ranking a bundle of projects. We show that ranking projects with RI and NPV is rational; in contrast, ranking projects with NFV is rational if and only if the competing projects have the same cost of capital.

Then, we show that ranking projects with rates of return is not rational. This should now be obvious, for we know that rates of return aimed at measuring economic efficiency, not project scale, which is an essential element for determining the magnitude of the economic value created.

Nevertheless, it is possible to obviate these limitations with two techniques: An **incremental** or **differential** analysis, whereby an iterated pairwise comparison of projects is accomplished, taking into account the differences between project's incomes and projects' capitals; the second method is a **direct** method, which **standardizes** the project scale so that the relative measures of worth become comparable.

The incremental method informs about the **sources of incremental value** of one project over another, something that the absolute measures of worth cannot provide.

After completing this chapter, the reader will be able to

- understand the reason why NPV ranking and RI ranking are rational, whereas NFV ranking is rational if and only if the projects have (the same risk and, therefore) the same costs of capital
- understand the scale effect and the reason why ranking projects with relative measures of worth is problematic
- rationally ranking alternatives with relative measures of worth via incremental analysis, which consists of an iterated pairwise comparison of incremental/differential rates of return (or excess rates of return)
- understand that the incremental method provides additional information that the absolute measures do not provide
- rationally ranking alternatives with a direct method which standardizes the (excess) rates of return by use of a standardized capital.

11.1 Project Ranking with Absolute Measures of Worth

An economically rational project ranking is one where the shareholders' wealth is maximized.

Definition 11.1 (*Rational project ranking*) Given a bundle of m projects, project h is ranked higher than project k if shareholders' wealth increase generated by h is greater than the shareholders' wealth increase generated by k. Two projects have the same rank if they generate the same wealth increase.

Let j denote a generic project. As we know, the equity NPV of project j is

$$\mathrm{NPV}^{e,j} = V_0^{e,j} - C_0^{e,j} = \sum_{t=0}^{n_j} F_t^{e,j} \cdot \mathrm{d}_{t,0}^{e,j}$$

with $\mathrm{d}_{0,0}^{e,j} = 1$ as usual and n_j denoting project j's lifespan.[1] We use the notation $h \succ k$ to designate preference of h over k.

Definition 11.2 (*Ranking via absolute measures*) Consider the independent project h and k.

NPV criterion. Project h is preferable to project k, $h \succ k$, if and only if $\mathrm{NPV}^{e,h} > \mathrm{NPV}^{e,k}$. In general, considering a bundle of $m > 2$ competing projects, they are ranked, by their NPVs.

RI criterion. Project h is preferable to project k, $h \succ k$, if and only if $V_0^{e,h}(\mathrm{RI}) > V_0^{e,k}(\mathrm{RI})$. In general, considering a bundle of $m > 2$ competing projects, they are ranked by the value of their overall RIs.

NFV criterion. Project h is preferable to project k, $h \succ k$, if and only if $\mathrm{NFV}^{e,h} > \mathrm{NFV}^{e,k}$. In general, considering a bundle of $m > 2$ competing projects, they are ranked by their NFVs.

Let us assume, for the moment, that the projects share the same COC structure: $r_t^j = r_t$ for every $j \in \mathbb{N}_m^1$. Project h is preferred to project k if shareholders' wealth increase generated by h is higher than shareholders' wealth increase generated by k. Shareholders' wealth increases is measured by $n_0 \Delta p_0$. Since the latter is equal to NPV (see Sect. 5.3 and Eq. (5.11)), the NPV ranking is rational. The other two absolute measures of worth are equivalent to NPV, for

$$\mathrm{NPV}^{e,h} > \mathrm{NPV}^{e,k} \iff V_0^{e,h}(\mathrm{RI}) > V_0^{e,k}(\mathrm{RI}) \iff \mathrm{NFV}^{e,h} > \mathrm{NFV}^{e,k}. \quad (11.1)$$

Therefore, we may state the following proposition.

[1] If $I_0^{e,j} \neq 0$, $\mathrm{NPV}^{e,j} = V_0^{e,j} + I_0^{e,j} - C_0^{e,j} = \sum_{t=0}^{n} F_t^{e,j} \cdot \mathrm{d}_{t,0}^{e,j}$.

Proposition 11.1 (Rationality of ranking via absolute measures—same COCs) *Project ranking with the three absolute measures (*NPV, NFV, RI*) is rational, according to Definition 11.1, provided that the projects share the same COC structure.*

As seen, one may substitute the equity perspective with the asset perspective (i.e., replace NPV^j for $\text{NPV}^{e,j}$, NFV^j for $\text{NFV}^{e,j}$, and $V_0^j(\text{RI})$ for $V_0^{e,j}(\text{RI})$) whenever $C_t^d = V_t^d$ for every t or whenever the project is unlevered. In these cases, project NPV and equity NPV coincide (this is the implicit or explicit assumption made in many texts) and $h \succ k$ if one of the following holds:

- $\text{NPV}^h > \text{NPV}^k$
- $\text{NFV}^h > \text{NFV}^k$
- $V_0^h(\text{RI}) > V_0^k(\text{RI})$.

Example 11.1 ♦
Consider two unlevered 4-year projects, α and β. The profit streams and capital streams are, respectively, $\boldsymbol{I}^\alpha = (0, 50, 0, 0, 150)$, $\boldsymbol{I}^\beta = (0, -50, 70, -60, 270)$ and $\boldsymbol{C}^\alpha = (250, 200, 150, 100, 0)$, $\boldsymbol{C}^\beta = (250, 250, 200, 0, 0)$. Hence, the cash-flow streams are $\boldsymbol{F}^\alpha = (-250, 100, 50, 50, 250)$, $\boldsymbol{F}^\beta = (-250, -50, 120, 140, 270)$. Assume the projects are equally risky, so that they share the same COC, equal to $r = 0\%$. It is easy to verify that $\text{NPV}^\alpha = \text{NFV}^\alpha = V_0(\text{RI}) = 200$ and $\text{NPV}^\beta = \text{NFV}^\beta = V_0^\beta(\text{RI}) = 230$. The project ranking is then $\beta \succ \alpha$.

If projects have a different COC structure, Proposition 11.1 is not valid. More precisely, the NFV ranking is not rational. The reason lies in the relation between NPV and VA/NFV. The VA/NFV may be written as a compounded NPV, namely,

$$\text{NFV}^j = \text{NPV}^j \overbrace{(1+r_1^j)\cdot(1+r_2^j)\cdot\ldots\cdot(1+r_{n_j}^j)}^{\text{u}_{0,n_j}^j}$$

or, with time-invariant COC, $\text{NFV}^j = \text{NPV}^j(1+r^j)^{n_j}$. Now, when confronting assets h and k, $\text{NPV}^h > \text{NPV}^k$ is not equivalent to $\text{NPV}^h \cdot \text{u}_{0,n_h}^h > \text{NPV}^k \cdot \text{u}_{0,n_k}^k$, due to the different effect of the compounding factors.[2] Then, the order of NFVs may be reversed as opposed to the order of NPVs. For example, if $\text{NPV}^h = 10 > 8 = \text{NPV}^k$, $n_h = n_k = 7$ and $r^h = 2\% \neq r^k = 6\%$, then $\text{NFV}^h = 10 \cdot (1.02)^7 = 11.48 < 12.03 = \text{NFV}^k = 8 \cdot (1.06)^7$. Therefore, only NPV and RI ranking are rational.

[2]The use of $n = \max[n_h, n_k]$ as a common date for the computation of net future value does not solve the problem (possibly, it worsens it).

Proposition 11.2 (Rationality of ranking via absolute measures—different COCs) *Assume the COC structure of the m projects is different. Then, project ranking with* NPV *and* RI *is rational, according to Definition 11.1, whereas, in general, project ranking with* VA/NFV *is not rational.*

Example 11.2 ♦
Consider three projects, A, B, and C, and the following accounting estimates:

Time	I^{A}	I^{B}	I^{C}	C^{A}	C^{B}	C^{C}
0				100	0	30
1	−10	0	20	90	10	150
2	65	−10	90	80	80	60
3	55	−30	−40	70	50	20
4	−10	45	−20	60	30	0
5	−60	65	0	0	0	0

Table 11.1 collects the corresponding streams of CFAs, along with the respective (assumed constant) COCs and resulting discount factors. The projects are assumed to have unequal risks (C is riskier than B and B is riskier than A) and this is reflected in different COCs. Assume, for simplicity, that the projects are levered but the debt NPV is zero for each project. This means that NPV and V_0(RI) correctly capture shareholder value created.[3] The table reports the NPVs of the projects, which trigger the following ranking: B $\succ$ A $\succ$ C. All projects are value-creating transactions. Therefore, all of them are worth undertaking, if they are not mutually exclusive; otherwise, B should be undertaken, because it increases the firm's share price more than the other two projects.[4]

---

[3]In this particular case, NFV ranking is equivalent to NPV and RI ranking only because the magnitudes of the COCs and the number of periods are not sufficiently dissimilar to cause a ranking's reversal.

[4]More precisely, remembering the arguments in Sect. 5.3:

- if the firm announces the undertaking of project A, a normal market will acknowledge that its value is \$130.2 so the share price will increase by $(\$130.2 - \$100)/n_0 = \$30.2/n_0$. The firm's new share price, after the equilibrium is restored, will be $p_0 + \$30.2/n_0$
- if the firm announces the undertaking of project B, the market will acknowledge that its value is \$45.8 so the share price increases by $(\$45.8 - \$0)/n_0 = \$45.8/n_0$. The firm's new share price will be $p_0 + \$45.8/n_0$
- if the firm announces the undertaking of project C, the market will acknowledge that its market value is \$57.9 so the share price will increase by $(\$57.9 - \$30)/n_0 = \$27.9/n_0$. The firms' new share price will be $p_0 + \$27.9/n_0$.

In order to maximize shareholders' wealth, managers must maximize the share price, which is equivalent to maximizing NPV.

Table 11.1 Ranking three projects with NPV—constant COCs

Time	F^{A}	F^{B}	F^{C}	r^{A}	r^{B}	r^{C}	$\mathbf{d}^{\mathrm{A}}$	$\mathbf{d}^{\mathrm{B}}$	$\mathbf{d}^{\mathrm{C}}$
0	−100	0	−30				1	1	1
1	0	−10	−100	3%	5%	10%	0.9709	0.9524	0.9091
2	75	−80	180	3%	5%	10%	0.9426	0.9070	0.8264
3	65	0	0	3%	5%	10%	0.9151	0.8638	0.7513
4	0	65	0	3%	5%	10%	0.8885	0.8227	0.6830
5	0	95	0	3%	5%	10%	0.8626	0.7835	0.6209
NPV	30.2	45.8	27.9						

Suppose now that the COCs of the projects are time-varying, such that

$$
\begin{aligned}
\boldsymbol{r}^{\mathrm{A}} &= (25\%, 15\%, 20\%) \\
\boldsymbol{r}^{\mathrm{B}} &= (5\%, 10\%, 8\%, 15\%, 7\%) \\
\boldsymbol{r}^{\mathrm{C}} &= (1\%, 2\%)
\end{aligned}
$$

so that

$$
\begin{aligned}
\mathbf{d}^{\mathrm{A}} &= (1, 0.8000, 0.6957, 0.5797) \\
\mathbf{d}^{\mathrm{B}} &= (1, 0.9524, 0.8658, 0.8017, 0.6971, 0.6515) \\
\mathbf{d}^{\mathrm{C}} &= (1, 0.9901, 0.9707).
\end{aligned}
$$

Discounting the CFAs of each project with the respective discount factors, one gets

$$
\mathrm{NPV}^{\mathrm{C}} = 45.71 > \mathrm{NPV}^{\mathrm{B}} = 28.42 > \mathrm{NPV}^{\mathrm{A}} = -10.14.
$$

In this scenario, project A destroys value, for the firm might invest in an equivalent-risk asset with expected period returns at 25%, 15%, 20% (so getting zero NPV). Only projects B and C are worth undertaking. However, the ranking is reversed: C is now preferred to B, since its NPV is higher. The ranking is then $C \succ B \succ 0 \succ A$.[5] In case of mutually exclusive projects, C will be undertaken.

There may be situations where a given course of action has a negative NPV, yet the course of action is accepted. There may be strategic, legal, social, ethical considerations that make it necessary to undertake it. In this case, the preferred option is

[5] The zero symbol here discriminates the region of value creation from the region of value destruction. Economically, it means that investing in C and B is better than investing in the respective null alternatives (assets equivalent in risk to C and B, respectively) whereas investing in A is worse than investing in its null alternative.

the one with the highest NPV. If, for example, two negative alternatives are at stake, with NPVs equal to $\text{NPV}^A = -100$ and $\text{NPV}^B = -120$, and if the two options are equivalent from a non-economic point of view, then option A will be preferred, since it destroys less value than B.

In some circumstances, a negative NPV may arise as a consequence of additional cost components associated with a connected positive-NPV project, which could not be foreseen at the time of the project analysis. And it may also occur that such additional cost components, while destroying value, are such that, combining the courses of actions, the resultant effect is still positive. Following is a simple instance of this situation.

Example 11.3 ♦

(*GreenGas Heating System*) The following example takes inspiration from Example 8.1 in Rogers and Duffy (2012, p. 172), suitably adjusted for our purposes. Suppose a company has just purchased a building which will be devoted to the production of a new product. Suppose the cost of the building is \$150,000 and the project value is \$160,000, which implies a project NPV equal to $\text{NPV}^{\text{purchase}} = 160{,}000 - 150{,}000 = 10{,}000 > 0$. Just after the purchase of the building, which has an old heating system tracing back to 1980s, local authorities impose all companies to change heating systems older than 20 years and replace it with a new, less polluting, heating system. The company then needs to consider the effect of a new heating system. The NPV of the new heating system is equal to the discounted sum of the (negative) incremental FCFs involved by the new heating system.

GreenGas heating system- Input data		
Project data		
Cost of plant and equipment	(\$)	1,500
Depreciation	Double-declining balance	
Heat energy demand	(kWh/annum)	10,000
Efficiency of the heating system	(%)	85
Fuel used	(kWh/annum)	$11{,}764.71 = 10{,}000/0.85$
CO_2-equivalent emission intensity	(kg/kWh)	0.275
CO_2-equivalent emissions	(kg/annum)	$3{,}235.29 = 0.275 \times 11{,}764.71$
Unit cost		
Fuel	(\$/kWh)	0.08
CO_2	(\$/kg)	0.03
Life-cycle costs		
Energy	(\$/annum)	$941.18 = 0.08 \times 11{,}764.71$
CO_2	(\$/annum)	$97.06 = 0.03 \times 3{,}235.29$
Maintenance (fixed cost)	(\$/annum)	250
Tax rate	40%	
Market data		
(After-tax) WACC	$r^{uo} = 20\%$	constant
cost of debt	equal to nominal interest rate	

Suppose the company considers the installation of GreenGas system, a new heating system whereby the greenhouse gases produced are fully abated, thanks to a newly developed technology. In particular, the system will produce 0.275 kg CO_2-equivalent emissions per kWh. To eliminate it, the cost for the firm is \$0.03 per kg. Energy costs depend on the demand, the efficiency of the system, and the prospective fuel costs. Maintenance costs are assumed to be constant. The heating system's duration is tuned to the life-cycle of the new product, which is equal to 15 years. Sales of the new product are evidently unaffected by the heating system (production will not change as a consequence of a new heating system). Double-declining depreciation of the heating system is used, which means that the pro forma book value declines by a depreciation rate of $2/15 = 0.1333$, applied to the BOP book value (see Eq. (3.3)).

We assume that the cost of debt and the interest rate on debt coincide: $r_t^d = i_t^d$. Therefore, the project NPV is equal to the equity NPV and it suffices to discount the project FCFs at the WACC to get the relevant NPV.

The cost-volume-profit IS reports the variable costs, the fixed costs, the contribution margin, and the EBIT after deducting the depreciation charge (see Table 11.2). The additional costs make the additional EBIT generated by the heating system negative. However, a partial shelter is given by the tax rate ($\tau = 0.4$), which makes the taxes be negative and, therefore, reduces the loss.[6] For example, the GreenGas system will entail additional projected costs for producing the new product equal to \$1,488.2 in year 1. Therefore, the firm's EBIT will be less than initially forecasted by \$1,488.2. This decrease is partially sheltered by the tax rate, so that the loss implied by the heating system in year 1 is only \$892.9. Adding the depreciation charge, this results in a FCF which is smaller than initially forecasted by \$692.9. The NPV of the GreenGas heating system is negative and equal to $\text{NPV}^{\text{GG}} = -\$4{,}871.13$. Therefore, acceptance of the GG system makes the equityholders worse off. However, this does not imply that the GG system should be rejected. It depends on what alternative course of action the company may pursue. For example, if the new production will not be initiated, the firm's managers decide that the building will be resold. Therefore, by rejecting the project, the firm will gain the selling price from the building's disposal. At the same time it will forego the value of the new production, equal to \$160,000. Let P be the resale price. If the building is disposed, then the NPV is $\text{NPV}^{\text{resale}} = -160{,}000 + P$.

The NPV of this alternative is positive if and only $P > 160{,}000$. For example, if $P = 162{,}000$, then $\text{NPV}^{\text{resale}} = 2{,}000$. This means that resale is a value-creating alternative and should be undertaken. If $P = 160{,}000$, the resale option has zero NPV and if $P < 160{,}000$, the GG option is preferable to the resale.

If $P < 155{,}128.87$, both alternatives have negative NPVs, but the GG option is preferable. If $P > 155{,}128.87$, the resale option is preferable. At $P = 155{,}128.87$, the two alternatives have the same NPV, equal to $-4{,}871.13$; therefore, they are financially equivalent.

[6]Negative taxes for this project means that the firm will pay less taxes than it would pay if it rejected the project.

Table 11.2 GreenGas heating system—Income statement

Year	0	1	2	3	4	5	6	7	8	9	10	11	12	13	14	15
Sales		0.0	0.0	0.0	0.0	0.0	0.0	0.0	0.0	0.0	0.0	0.0	0.0	0.0	0.0	0.0
–Energy costs		–941.2	–941.2	–941.2	–941.2	–941.2	–941.2	–941.2	–941.2	–941.2	–941.2	–941.2	–941.2	–941.2	–941.2	–941.2
–CO_2-eq costs		–97.1	–97.1	–97.1	–97.1	–97.1	–97.1	–97.1	–97.1	–97.1	–97.1	–97.1	–97.1	–97.1	–97.1	–97.1
– Variable costs		–1,038.2	–1,038.2	–1,038.2	–1,038.2	–1,038.2	–1,038.2	–1,038.2	–1,038.2	–1,038.2	–1,038.2	–1,038.2	–1,038.2	–1,038.2	–1,038.2	–1,038.2
Contribution Margin		–1,038.2	–1,038.2	–1,038.2	–1,038.2	–1,038.2	–1,038.2	–1,038.2	–1,038.2	–1,038.2	–1,038.2	–1,038.2	–1,038.2	–1,038.2	–1,038.2	–1,038.2
–Fixed costs		–250.0	–250.0	–250.0	–250.0	–250.0	–250.0	–250.0	–250.0	–250.0	–250.0	–250.0	–250.0	–250.0	–250.0	–250.0
EBITDA		–1,288.2	–1,288.2	–1,288.2	–1,288.2	–1,288.2	–1,288.2	–1,288.2	–1,288.2	–1,288.2	–1,288.2	–1,288.2	–1,288.2	–1,288.2	–1,288.2	–1,288.2
–Depreciation		–200.0	–173.3	–150.2	–130.2	–112.8	–97.8	–84.8	–73.5	–63.7	–55.2	–47.8	–41.4	–35.9	–31.1	–202.3
EBIT		–1,488.2	–1,461.6	–1,438.5	–1,418.4	–1,401.1	–1,386.0	–1,373.0	–1,361.7	–1,351.9	–1,343.4	–1,336.0	–1,329.7	–1,324.1	–1,319.4	–1,490.5
–Taxes		595.3	584.6	575.4	567.4	560.4	554.4	549.2	544.7	540.8	537.4	534.4	531.9	529.7	527.7	596.2
NOPAT		–892.9	–876.9	–863.1	–851.1	–840.6	–831.6	–823.8	–817.0	–811.1	–806.0	–801.6	–797.8	–794.5	–791.6	–894.3
$-\Delta C_t$	–1,500.0	200.0	173.3	150.2	130.2	112.8	97.8	84.8	73.5	63.7	55.2	47.8	41.4	35.9	31.1	202.3
FCF	–1,500.0	–692.9	–703.6	–712.9	–720.9	–727.8	–733.8	–739.0	–743.6	–747.5	–750.9	–753.8	–756.4	–758.6	–760.5	–692.0

NPV of GreenGas Heating System	–4,871.13

Year	0	1	2	3	4	5	6	7	8	9	10	11	12	13	14	15
Invested capital	1,500.00	1,300.00	1,126.67	976.44	846.25	733.42	635.63	550.88	477.43	413.77	358.60	310.79	269.35	233.44	202.31	0

However, the two options might also be described by combining them with the puchase-and-launch option. If the new product is launched, equityholders first gain \$10,000 and then lose \$4,871.13, owing to the new legislation. Hence, the combined NPV of the purchase of the building and the installation of the GG system equal to \$5,128.87. This means that, combining the two decisions, equityholders still have a net gain. If, instead, the building is disposed, the equityholders first gain \$10,000 and then lose (if negative) or gain (if positive) the amount $\text{NPV}^{\text{resale}} = -\$160{,}000 + P$, which makes the combined NPV of purchase and resale be equal to $-160{,}000 + P + 10{,}000 = P - 150{,}000$. At a price of $P = \$160{,}000$, the combined NPV is \$10,000, which is exactly equal to $\text{NPV}^{\text{purchase}}$, the NPV that were generated by the launch of the new product if the new legislation had not been enacted. This is greater than \$5,128.87, so reselling the building is economically preferable to launching the new product.

Now, suppose that, other things unvaried, the WACC of the GG heating system is 2%. This implies $\text{NPV}^{\text{GG}} = -10{,}903.19$. In this case, the installation of the GG system (and the production of the new product) will be undertaken only if $-10{,}903.19 > -160{,}000 + P$, that is, if and only if $P < 149{,}096.81$. At $P = 149{,}096.81$, the two options are equivalent.

If $P = 150{,}000$, then

$$\text{NPV}^{\text{GG}} = -10{,}903.19 < -10{,}000 = \text{NPV}^{\text{resale}}.$$

In terms of the combined framing, the NPVs will be $-\$903.19$ and \$0, respectively.

If $P = 149{,}500$ the resale of the building is still preferable, notwithstanding the fact that the selling price of the building is smaller than the cost sustained for purchasing it:

$$\text{NPV}^{\text{GG}} = -10{,}903.19 < -10{,}500 = \text{NPV}^{\text{sell}}.$$

In terms of combined NPV, the two combined NPVs are both negative and equal to $-\$903.19$ and $-\$500$, respectively.

The following table considers various scenarios for the selling price, associated with two different WACCs, 20% and 2% respectively. Note that the NPV of the resale option is not affected by the WACC, which is the discount rate for the GG heating system. It is evident that, if the WACC is smaller, then the value of the (negative) FCFs generated by the GG option decreases, which makes GG less attractive, other things unvaried. While a 20% WACC determines a break-even price of \$155,128.87 as the indifference threshold, a 2% WACC determines a break-even price much smaller, equal to \$149,096.81. Therefore, even in the case where the building is sold at a loss (i.e., building is resold at a smaller price than the cost), it may be more rational to sell the building, if the price is not too small (i.e., greater than \$149,096.81).

FRAMING 1

	NPV (20% WACC)			NPV (2% WACC)		
Selling price	Resale	GreenGas	choice	Resale	GreenGas	choice
162,000.00	2,000.00	−4,871.13	resale	2,000.00	−10,903.19	resale
160,000.00	0.00	−4,871.13	resale	0.00	−10,903.19	resale
155,128.87	−4,871.13	−4,871.13	indifferent	−4,871.13	−10,903.19	resale
153,000.00	−7,000.00	−4,871.13	GG	−7,000.00	−10,903.19	resale
150,000.00	−10,000.00	−4,871.13	GG	−10,000.00	−10,903.19	resale
149,500.00	−10,500.00	−4,871.13	GG	−10,500.00	−10,903.19	resale
149,096.81	−10,903.19	−4,871.13	GG	−10,903.19	−10,903.19	indifferent
145,000.00	−15,000.00	−4,871.13	GG	−15,000.00	−10,903.19	GG

FRAMING 2 (combined)

	NPV (20% WACC)			NPV (2% WACC)		
Selling price	Purchase and resale	Purchase and GreenGas	choice	Purchase and Resale	Purchase and GreenGas	choice
162,000.00	12,000.00	5,128.87	resale	12,000.00	−903.19	resale
160,000.00	10,000.00	5,128.87	resale	10,000.00	−903.19	resale
155,128.87	5,128.87	5,128.87	indifferent	5,128.87	−903.19	resale
153,000.00	3,000.00	5,128.87	GG	3,000.00	−903.19	resale
150,000.00	0.00	5,128.87	GG	0.00	−903.19	resale
149,500.00	−500.00	5,128.87	GG	−500.00	−903.19	resale
149,096.81	−903.19	5,128.87	GG	−903.19	−903.19	indifferent
145,000.00	−5,000.00	5,128.87	GG	−5,000.00	−903.19	GG

11.2 Project Ranking with Relative Measures of Worth

Projects should not be ranked with rates of return, for reasons that will become apparent later. However, in real-life applications, many practitioners use rates of return, especially the IRR. The ranking provided by the IRR is not consistent with NPV ranking, except under exceptional circumstances.

Scholars have studied the conflict between IRR and NPV, sometimes relating it to specific assumptions about reinvestment/financing of interim cash flows. As we have shown, neither NPV nor IRR is based on any assumption about reinvestment or financing (see Keane 1979; Keef and Roush 2001; Magni and Martin 2017).

The key point of the conflict between the rate-of-return notion and the NPV notion is the concept of *capital*. However, this is also the key point of the reconciliation between the two notions. We have already treated this issue in Sect. 7.2 and have clarified that the economic value created and rate of return are, at the same time, interlocked and differentiated by the ***project scale***. Indeed, rates of return are relative measures, designed for measuring the value of each dollar invested, while NPV, RI, and NFV are absolute measures of worth, designed for measuring the total value created by the project. Economic value created is a medal with two sides: One side is the rate of return and the other side is the project scale:

$$\text{NPV} = \text{Project scale} \times \text{(excess) Rate of return.}$$

In general, given any two projects, α and β, the amount of capital necessary to generate project α's NPV is different from the amount that is required to generate project β's NPV. It is this *scale effect* that prevents rate of return to be used for project ranking (see also Remark 7.1).

A forerunner of this explanation is Keane (1979). The author realized that reinvestment assumptions play no role in the conflicting ranking of an absolute measure of worth and a relative measure of worth. He explicitly acknowledged the scale effect as the source of the conflict:

> a difference in scale necessarily introduces the possibility of conflict between an absolute measure of value and a ratio of profitability ...Conflict between IRR and NPV can be attributed to this fact alone (Keane 1979, pp. 53–54).

> Conflict between IRR and NPV can be attributed entirely to the effects of scale, and it can be shown that the scale effect is present in all project comparisons even between projects with identical outlays. It is concluded that the IRR fails to qualify as a valid investment selection criterion because it is a ratio, and an inadequate ratio at that, in a context where ratios are substantially inappropriate (Keane 1979, p. 55).

> since it is not 'performance' but 'contribution to value' that matters, the scale difference invalidates the direct use of the IRR (Keane 1979, p. 53).

It was then clear to Keane that the missing link is the capital (a link that AIRR, IARR, and AROI explicitly provide for each one of the three absolute measures of worth).[7] For instance, consider Example 11.1 where a zero COC is assumed. The IRRs are $\sigma^\alpha = 23.1\%$ and $\sigma^\beta = 20.4\%$. The IRR ranking is then $\alpha \succ \beta$, opposite to the NPV ranking. The reason lies in the scale effect:

$$\text{NPV}^\alpha = 200 = \overbrace{866.47}^{\text{scale}} \cdot \overbrace{23.1\%}^{\text{(excess) IRR}}$$

$$\text{NPV}^\beta = 230 = \overbrace{1,127.81}^{\text{scale}} \cdot \overbrace{20.4\%}^{\text{(excess) IRR}}.$$

where 866.47 and 1,127.81 are easily calculated as the sum of the internal values of α and β, respectively (discounted at $r = 0\%$), and the excess IRR is equal to the IRR, given that $r = 0\%$. While project α is more efficient, the capital invested in α is smaller than the capital invested in β. The overall capital invested in β is \$261.34 higher than the capital invested in α; the IRR generated by α is 2.7% points greater than the IRR generated by β. The two effects are conflicting. In this case, the scale effect prevails and β creates more value than project α (notwithstanding the smaller rate of return), so that the ranking via IRR does not reflect the correct ranking.

[7] Keane (1979) also realized that the project scale is expressed by the total capital: "Profitability Index fails because it incorrectly equates scale with initial outlay. If any ratio can be perceived as being the correct 'cost-benefit' index it would be the NPV expressed as a ratio of the total number of units of capital employed" (Keane 1979, p. 54, footnote 7). This is exactly what we have called, in this book, economic (or financial) efficiency. Interestingly, the proposal of the author for measuring the overall capital was to consider the discounted sum of $\mathcal{C}_t = \mathcal{C}_{t-1}(1 + r_t) - F_t$. The latter is exactly the capital base of the benchmark system $(\mathcal{C}, \boldsymbol{I}^{\mathcal{C}}, \boldsymbol{F}^{\mathcal{C}})$, associated with AROI (see also the RP-AIRR introduced in Magni 2016a).

Let us then consider the three relative approaches associated with the three absolute measures of worth: AIRR, IARR, and AROI. From the profit and capital estimates of α and β illustrated in Example 11.1 one derives $\bar{\imath}^{\alpha} = \bar{\jmath}^{\alpha} = 200/700 = 28.6\%$ and $\bar{\imath}^{\beta} = \bar{\jmath}^{\beta} = 230/700 = 32.9\%$.[8] The ranking is the same as the NPV: $\beta \succ \alpha$. The reason lies in the fact that, in the AIRR and IARR approach, α and β have the same scale:

$$\text{NPV}^{\alpha} = 200 = \overbrace{700}^{\text{scale}} \cdot \overbrace{28.6\%}^{\text{(excess) average ROI}}$$

$$\text{NPV}^{\beta} = 230 = \overbrace{700}^{\text{scale}} \cdot \overbrace{32.9\%.}^{\text{(excess) average ROI}}$$

The difference in the rates fully reflects the difference in value creation only because the investment scale coincides. However, in general, ranking should not be accomplished with rates of return, because the project scales differ.

Example 11.4 ♦
One should be advised that the IRR, IARR, and AROI rankings are not sensitive to changes in the COC(s), whereas NPV ranking, RI ranking and AIRR ranking are. To illustrate, consider Example 11.2. We report below the IRRs, the average ROIs (AIRRs) and the internal average ROIs (IARRs/AROIs) of projects A, B, and C, in the case of constant COC:

Project	A	B	C
COC (r)	3%	5%	10%
NPV/RI	30.2	45.8	27.9
IRR (σ)	14.7%	23.8%	29.6%
AIRR ($\bar{\imath}$)	11.2%	37.1%	23.3%
IARR/AROI ($\bar{\jmath}$)	10.0%	41.2%	19.2%

The ranking in the five approaches is different:

Ranking	
NPV	$B \succ A \succ C$
IRR(σ)	$C \succ B \succ A$
AIRR($\bar{\imath}$)	$B \succ C \succ A$
IARR/AROI($\bar{\jmath}$)	$B \succ C \succ A$

Assume now the COC of B is 30% instead of 5%, other things unvaried. The IRR and IARR/AROI rankings do not change (they do not depend on the COC), while the AIRR and the NPV/RI ranking change:

[8] In this case, the project's average ROIs, the investment scales, and the MARRs coincide in the three approaches (AIRR, IARR, and AROI) because we have assumed $r = 0\%$.

Project	A	B	C
COC (r)	3%	30%	10%
NPV	30.2	−6.7	27.9
IRR (σ)	14.7%	23.8%	29.6%
AIRR ($\bar{\imath}$)	11.2%	20.2%	23.3%
IARR/AROI ($\bar{\jmath}$)	10.0%	41.2%	19.2%.

NPV ranking and AIRR ranking change:

Ranking	
NPV	$A \succ C \succ B$
IRR(σ)	$C \succ B \succ A$
AIRR($\bar{\imath}$)	$C \succ B \succ A$
IARR/AROI($\bar{\jmath}$)	$B \succ C \succ A$

In general, the ranking of the three relative approaches will not be reciprocally consistent nor NPV/RI-consistent (and, therefore, it will not be an economically rational ranking).

Practitioners (more than scholars) are particularly fond of using rates of return for comparing competing projects. As seen, the scale issue makes the comparison of rates of return as such unreliable for project ranking. However, the problem can be overcome with AIRR, IARR, and AROI. Specifically, we briefly show two procedures that empowers AIRR, IARR, and AROI to produce rational ranking: The *incremental* (or *differential*) method and the *direct* method.

11.3 Incremental Analysis

Consider a pair of projects h and k and let n_h and n_k be the duration of the operations. Let $n = \max[n_h, n_k]$. For the time being, we assume that the COC is invariant across periods and across projects. We know that the following three product structures holds, for AIRR, IARR, and AROI, respectively:

AIRR

$$V_0^h(\text{RI}) = \text{PV}[C^h](\bar{\imath}_0^h - \bar{r}_0^h)$$
$$V_0^k(\text{RI}) = \text{PV}[C^k](\bar{\imath}_0^k - \bar{r}_0^k)$$

IARR

$$\text{NPV}^h = C^h(\bar{\jmath}^h - \bar{\rho}^h)$$
$$\text{NPV}^k = C^k(\bar{\jmath}^k - \bar{\rho}^k)$$

AROI

$$\text{NFV}^h = C^h(\bar{\jmath}^h - \bar{\varrho}^h)$$
$$\text{NFV}^k = C^h(\bar{\jmath}^h - \bar{\varrho}^k)$$

or, in present terms,

$$\text{NPV}^h = C_\text{d}^h(\bar{\jmath}_\text{d}^h - \bar{\varrho}_\text{d}^h)$$
$$\text{NPV}^k = C_\text{d}^h(\bar{\jmath}_\text{d}^h - \bar{\varrho}_\text{d}^k)$$

This implies an incremental product structure for the three relative approaches. Specifically, for AIRR we give the following definitions:

$$
\begin{aligned}
\mathrm{PV}[C^{h-k}] &= \mathrm{PV}[C^h] - \mathrm{PV}[C^k] \\
\bar{\imath}_0^{h-k} &= \frac{\bar{\imath}_0^h \mathrm{PV}[C^h] - \bar{\imath}_0^k \mathrm{PV}[C^k]}{\mathrm{PV}[C^h] - \mathrm{PV}[C^k]} \\
\bar{r}_0^{h-k} &= \frac{\bar{r}_0^h \mathrm{PV}[C^h] - \bar{r}_0^k \mathrm{PV}[C^k]}{\mathrm{PV}[C^h] - \mathrm{PV}[C^k]}.
\end{aligned}
$$

One may then write

$$
V_0^h(\mathrm{RI}) - V_0^k(\mathrm{RI}) = \mathrm{PV}[C^{h-k}](\bar{\imath}_0^{h-k} - \bar{r}_0^{h-k}). \tag{11.2}
$$

For IARR, we give the following definitions:

$$
\begin{aligned}
C^{h-k} &= C^h - C^k \\
\bar{\jmath}^{h-k} &= \frac{\bar{\jmath}^h C^h - \bar{\jmath}^k C^k}{C^h - C^k} \\
\bar{\rho}^{h-k} &= \frac{\bar{\rho}^h C^h - \bar{\rho}^k C^k}{C^h - C^k}.
\end{aligned}
$$

One may then write

$$
\mathrm{NPV}^h - \mathrm{NPV}^k = C^{h-k}(\bar{\jmath}^{h-k} - \bar{\rho}^{h-k}) \tag{11.3}
$$

For AROI, we give the following definitions:

$$
\begin{aligned}
C_{\mathrm{d}}^{h-k} &= C_{\mathrm{d}}^h - C_{\mathrm{d}}^k \\
\bar{\jmath}_{\mathrm{d}}^{h-k} &= \frac{\bar{\jmath}_{\mathrm{d}}^h C_{\mathrm{d}}^h - \bar{\jmath}_{\mathrm{d}}^k C_{\mathrm{d}}^k}{C_{\mathrm{d}}^h - C_{\mathrm{d}}^k} \\
\bar{\varrho}_{\mathrm{d}}^{h-k} &= \frac{\bar{\varrho}^h C_{\mathrm{d}}^h - \bar{\varrho}^k C_{\mathrm{d}}^k}{C_{\mathrm{d}}^h - C_{\mathrm{d}}^k}.
\end{aligned}
$$

One may then write[9]

$$
\begin{aligned}
\mathrm{NPV}^h - \mathrm{NPV}^k &= \mathrm{NFV}^h \cdot d_{n,0}^h - \mathrm{NFV}^k \cdot d_{n,0}^k \\
&= C_{\mathrm{d}}^{h-k} \cdot (\bar{\jmath}_{\mathrm{d}}^{h-k} - \bar{\varrho}_{\mathrm{d}}^{h-k})
\end{aligned} \tag{11.4}
$$

[9]It is not possible to use (10.90), for the NFV ranking is not rational in general. One must necessarily rest on (10.91). (See also Magni 2015b on project ranking with AROI.)

where the rates $\bar{\imath}_0^{h-k}$, $\bar{\jmath}^{h-k}$, and $\bar{\jmath}_{\rm d}^{h-k}$ measure, in their respective approaches, the incremental rate of return of h over k, and they are (generalized) weighted mean of the rates of return of h and k.[10] Likewise, the rates $\bar{r}_0^{h-k}$, $\bar{\rho}^{h-k}$, and $\bar{\varrho}^{h-k}$ measure, in the respective domains, the incremental rate of return required by an investor to accept h over k.

The incremental rate of return may also be computed in different ways. For example, one may rest on the very definition of rate of return (ratio of return to capital) and have

$$\bar{\imath}_0^{h-k} = \frac{\text{PV}[I^h] - \text{PV}[I^k]}{\text{PV}[C^h] - \text{PV}[C^k]} \tag{11.5}$$

for AIRR and

$$\bar{\jmath}^{h-k} = \frac{I^h - I^k}{C^h - C^k} = \frac{F^h - F^k}{C^h - C^k} \tag{11.6}$$

for IARR, and

$$\bar{\jmath}_{\rm d}^{h-k} = \frac{I_{\rm d}^h - I_{\rm d}^k}{C_{\rm d}^h - C_{\rm d}^k} = \frac{F_{\rm d}^h - F_{\rm d}^k}{C_{\rm d}^h - C_{\rm d}^k} \tag{11.7}$$

for AROI, where the subscript "d" of a symbol indicates that the amount is discounted from n to 0.[11] Or, starting from NPV,

$$\bar{\imath}_0^{h-k} = \bar{r}_0^{h-k} + \frac{\text{NPV}}{\text{PV}[C^{h-k}]}$$

$$\bar{\jmath}^{h-k} = \bar{\rho}^{h-k} + \frac{\text{NPV}^{h-k}}{C^{h-k}}$$

$$\bar{\jmath}_{\rm d}^{h-k} = \bar{\varrho}_{\rm d}^{h-k} + \frac{\text{NPV}^{h-k}}{C_{\rm d}^{h-k}}$$

where $\text{NPV}^{h-k} = \text{NPV}^h - \text{NPV}^k$ and

$$\bar{r}_0^{h-k} = \frac{\text{PV}[I^{*h}] - \text{PV}[I^{*k}]}{\text{PV}[C^h] - \text{PV}[C^k]} = \frac{\sum_{t=1}^n r_t^h C_{t-1}^h \text{d}_{t,0}^h - \sum_{t=1}^n r_t^k C_{t-1}^k \text{d}_{t,0}^k}{\text{PV}[C^h] - \text{PV}[C^k]}$$

$$\bar{\rho}^{h-k} = \frac{I^{Vh} - I^{Vk}}{C^h - C^k} = \frac{\sum_{t=1}^n r_t^h V_{t-1}^h - \sum_{t=1}^n r_t^k V_t^k}{C^h - C^k}$$

$$\bar{\varrho}_{\rm d}^{h-k} = \frac{I^{\mathcal{C}h} - I^{\mathcal{C}k}}{C_{\rm d}^h - C_{\rm d}^k} = \frac{\sum_{t=1}^n r_t^h \mathcal{C}_{t-1}^h \text{d}_{n,0}^h - \sum_{t=1}^n r_t^k \mathcal{C}_{t-1}^k \text{d}_{n,0}^k}{C_{\rm d}^h - C_{\rm d}^k}.$$

[10]If the COC is constant across periods and equal across assets (i.e., $r_t^h = r_t^k = r$) then IARR and AROI coincide: $\bar{\jmath}_{\rm d}^{h-k} = I_{\rm d}/C_{\rm d} = I/C = \bar{\jmath}^{h-k}$.

[11]That is, one first sums and then discounts the amounts from n to 0, as opposed to the AIRR approach where amounts are first discounted from t to 0 and then summed. See also Sect. 10.5.

If the COC is constant across periods (i.e., $r_t^h = r^h$ and $r_t^k = r^k$), then the above relations are further simplified:

$$\bar{r}_0^{h-k} = \frac{r^h \sum_{t=1}^n C_{t-1}^h \mathrm{d}_{t,0} - r^k \sum_{t=1}^n C_{t-1}^k \mathrm{d}_{t,0}}{\mathrm{PV}[C^h] - \mathrm{PV}[C^k]}$$

$$\bar{\rho}^{h-k} = \frac{r^h V^h - r^k V^k}{C^h - C^k}$$

$$\bar{\varrho}_{\mathrm{d}}^{h-k} = \frac{r^h \mathcal{C}_{\mathrm{d}}^h - r^k \mathcal{C}_{\mathrm{d}}^k}{C_{\mathrm{d}}^h - C_{\mathrm{d}}^k}.$$

If, in addition, the two assets have (the same risk and, therefore) the same COC, $r^h = r^k = r$, one may write, respectively,

$$\bar{r}_0^{h-k} = \frac{r}{1+r}$$

$$\bar{\rho}^{h-k} = r \cdot \frac{V^{h-k}}{C^{h-k}}$$

$$\bar{\varrho}_{\mathrm{d}}^{h-k} = r \cdot \frac{\mathcal{C}_{\mathrm{d}}^{h-k}}{C_{\mathrm{d}}^{h-k}}.$$

The choice between two mutually exclusive alternatives boils down to accepting or rejecting the incremental project $h - k$. This accept-reject decision is iterated across the various projects under analysis to get a rational ranking.

Proposition 11.3 (Project ranking with the relative approaches—Incremental method) *Consider the independent projects h and k.*

AIRR. *Suppose* $\mathrm{PV}[C^h] > PV[C^k]$; *then, h is ranked higher than k if and only if* $\bar{\imath}_0^{h-k} > \bar{r}_0^{h-k}$. *In general, facing a bundle of m independent projects, they are ranked according to the pairwise comparison of* $\bar{\imath}_0^{h-k}$ *and* $\bar{r}_0^{h-k}$, $h, k \in \{1, 2, \ldots, m\}$.

IARR. *Suppose* $C^h > C^k$. *Then, h is ranked higher than k if and only if* $\bar{\jmath}^{h-k} > \bar{\rho}^{h-k}$. *In general, facing a bundle of m independent projects, they are ranked according to the pairwise comparison of* $\bar{\jmath}^{h-k}$ *and* $\bar{\rho}^{h-k}$. $h, k \in \{1, 2, \ldots, m\}$.

AROI. *Suppose* $C^h > C^k$. *Then, h is ranked higher than k if and only if* $\bar{\jmath}_{\mathrm{d}}^{h-k} > \bar{\varrho}_{\mathrm{d}}^{h-k}$. *In general, facing a bundle of m independent projects, they are ranked according to the pairwise comparison of* $\bar{\jmath}_{\mathrm{d}}^{h-k}$ *and* $\bar{\varrho}_{\mathrm{d}}^{h-k}$. $h, k \in \{1, 2, \ldots, m\}$.

For all methods, the resulting ranking is rational according to Definition 11.2.

Note that the comparison of incremental rate of return and incremental COC boils down to checking the sign of the incremental financial efficiency, which in turn may be framed as the ratio of incremental NPV to the incremental capital base in each respective approach. The inequalities

$$\begin{aligned} \xi^{h-k} &= \bar{\imath}_0^{h-k} - \bar{r}_0^{h-k} = \frac{\text{NPV}^{h-k}}{\text{PV}[C^{h-k}]} > 0 \\ \epsilon^{h-k} &= \bar{\jmath}^{h-k} - \bar{\rho}^{h-k} = \frac{\text{NPV}^{h-k}}{C^{h-k}} > 0 \\ \varepsilon_{\text{d}}^{h-k} &= \bar{\jmath}_{\text{d}}^{h-k} - \bar{\varrho}_{\text{d}}^{h-k} = \frac{\text{NPV}^{h-k}}{C_{\text{d}}^{h-k}} > 0 \end{aligned} \tag{11.8}$$

are the respective conditions for ranking h over k.

It is also worthy of attention that the incremental economic efficiencies may themselves be framed as generalized weighted means of the individual projects' economic efficiencies:

$$\begin{aligned} \xi^{h-k} &= \frac{\xi^h \text{PV}[C^h] - \xi^k \text{PV}[C^k]}{\text{PV}[C^h] - \text{PV}[C^k]} \\ \epsilon^{h-k} &= \frac{\epsilon^h C^h - \epsilon^k C^k}{C^h - C^k} \\ \varepsilon_{\text{d}}^{h-k} &= \frac{\varepsilon_{\text{d}}^h C_{\text{d}}^h - \varepsilon_{\text{d}}^k C_{\text{d}}^k}{C_{\text{d}}^h - C_{\text{d}}^k}. \end{aligned}$$

Notice that

- if $\text{PV}[C^h] - \text{PV}[C^k] = 0$, then $\text{NPV}^h > \text{NPV}^k$ if and only if $\xi^h > \xi^k$
- if $C^h - C^k = 0$, then $\text{NPV}^h > \text{NPV}^k$ if and only if $\epsilon^h > \epsilon^k$
- if $C_{\text{d}}^h - C_{\text{d}}^k = 0$, then $\text{NPV}^h > \text{NPV}^k$ if and only if $\varepsilon_{\text{d}}^h > \varepsilon_{\text{d}}^h$.

Also, if $\text{PV}[C^h] - \text{PV}[C^k] = 0$ and $r_t^h = r_t^k = r$ for every t, then $\text{NPV}^h > \text{NPV}^k$ if and only if $\bar{\imath}_0^h > \bar{\imath}_0^k$ (or, equivalently, $\bar{\imath}^h > \bar{\imath}^k$). In other words, the comparison of the rates of return of the individual assets suffices to provide rational guidance.

True as it is, a pairwise comparison of m projects may be burdensome for the analyst if m is large (although, admittedly, this does not occur often). However, it is important that, theoretically, the incremental procedure provides a rational ranking and that the incremental value created or destroyed is suitably decomposed into its economic efficiency component and its incremental size.

An incremental approach for IRR is also sometimes invoked by scholars. It consists in computing the IRR of the incremental cash-flow stream, σ^{h-k}, such that $\sum_{t=0}^{n} F_t^{h-k}/(1+\sigma^{h-k})^t = 0$ where $F_t^{h-k} = F_t^h - F_t^k$. However, σ^{h-k} is nothing but the IRR of an economic asset, labeled $h-k$, and, as any economic asset, meets all the anomalies described in Chap. 9. With an additional problem: If the competing projects have different COCs/MARRs, then the incremental IRR σ^{h-k} is useless, since it is not clear what the MARR should be as a cutoff rate for σ^{h-k}.[12]

Example 11.5 ♦
Consider the following estimation for two projects, a and b,[13] whose assets are assumed to depreciate uniformly:

$(\boldsymbol{C}^a, \boldsymbol{I}^a, \boldsymbol{F}^a)$

Time		0	1	2	3
Capital	C^a	100,000	66,666.67	33,333.33	0
Income	I^a	0	266.67	38,986.67	6,586.67
Cash Flow	F^a	−100,000	33,600	72,320	39,920

$(\boldsymbol{C}^b, \boldsymbol{I}^b, \boldsymbol{F}^b)$

Time		0	1	2	3
Capital	C^b	104,000	69,333.33	34,666.67	0
Income	I^b	0	15,333.33	15,333.33	15,333.33
Cash Flow	F^b	−104,000	50,000	50,000	50,000

The concise transposed matrix of the incremental project $b - a$ is

$(\boldsymbol{C}^{b-a}, \boldsymbol{I}^{b-a}, \boldsymbol{F}^{b-a})$

Time		0	1	2	3
Capital	C^{b-a}	4,000	2,666.67	1,333.33	0
Income	I^{b-a}	0	15,066.67	−23,653.33	8,746.67
Cash Flow	F^{b-a}	−4,000	16,400	−22,320	10,080

[12]Worshippers of IRR might be relieved in knowing that each of the three relative paradigms (AIRR, IARR, AROI) supply a lifeline for IRR to overcome this problem. Simply as it is, the incremental MARRs associated with incremental IRRs may be computed as weighted means of the market period returns, r_t. We crossrefer the reader to Magni et al. (2018, Sect. 4.1) for details. Needless to say, all the other shortcomings of IRR may not be healed. (See also Ben-Horin and Kroll 2017 on NPV-consistent methods of project ranking. See Foster and Mitra 2003 on dominance conditions independent of the discount rate.)

[13]The two cash-flow streams of these projects have appeared in White et al. (2014, p. 223, Example 6.5).

We note, in passing, that this incremental project has three IRRs: 20%, 40%, and 50% and that both projects are value-creating ($\text{NPV}^a = 37{,}322.35$, $\text{NPV}^b = 37{,}430.57$); if possible, both should be undertaken. Then, we proceed to calculate the incremental efficiency and, specifically, the incremental rate of return and incremental MARR in the three relative approaches. For simplicity, we assume that the COC is equal across periods and across projects: $r = 3\%$. Note also that the incremental project is a pure investment project for all capital amounts are positive. The AIRR is

$$\bar{\imath}_0^{b-a} = \frac{\overbrace{336.74}^{\text{PV}[I^{b-a}]}}{\underbrace{7{,}845.79}_{\text{PV}[C^{b-a}]}} = 4.29\% > \frac{\overbrace{0.03}^{r}}{\underbrace{1.03}_{1+r}} = 2.91\% = \bar{r}_0^{b-a}$$

or, in annual terms,

$$\bar{\imath}^{b-a} = 0.0429 \cdot 1.03 = 4.42\% > 3\% = \bar{r}^{b-a}.$$

The AIRR is greater than the MARR, thereby signaling value creation for the incremental project: Project b is preferred to project a.

The IARR is

$$\bar{j}^{b-a} = \frac{\overbrace{160}^{I^{b-a}}}{\underbrace{8{,}000}_{C^{b-a}}} = 2\%,$$

the corresponding MARR is

$$\bar{\rho}^{b-a} = \overbrace{3\%}^{r} \cdot \frac{\overbrace{4{,}108.22 - 12{,}168.54 + 9{,}786.41}^{V^{h-k}}}{\underbrace{4{,}000 + 2{,}666.67 + 1{,}333.33}_{C^{h-k}}} = 3\% \cdot \frac{1{,}726.1}{8{,}000} = 0.65\%.$$

The investor invests an undiscounted \$8,000 at 2% while foregoing the opportunity of replicating the same prospective cash flows by investing \$1,726.1 at 3%, which is equivalent to investing \$8,000 at 0.65%. The latter is smaller than 2%, which signals that the incremental project is favorable, so b is preferred to a.

As for AROI, it is equal to IARR (see footnote 10 in this chapter):

$$\bar{j}_{\text{d}}^{b-a} = \frac{\overbrace{160 \cdot (1.03)^{-3}}^{I_{\text{d}}^{b-a}}}{\underbrace{8{,}000 \cdot (1.031)^{-3}}_{C_{\text{d}}^{b-a}}} = \frac{146.42}{7{,}321.1} = 2\%.$$

As for the corresponding MARR,

$$\bar{\varrho}_{\mathrm{d}}^{b-a} = \overbrace{3\%}^{r} \cdot \frac{\overbrace{(4{,}000 - 12{,}280 + 9{,}671.6) \cdot 1.03^{-3}}^{\mathcal{C}_{\mathrm{d}}^{b-a}}}{\underbrace{8{,}000 \cdot 1.03^{-3}}_{C_{\mathrm{d}}^{b-a}}} = 3\% \cdot \frac{1{,}273.51}{7{,}321.1} = 0.52\%.$$

The investor invests \$7,321 at 2% while foregoing the opportunity of replicating the same cash flows up to (and excluding) time n by investing \$1,273.5 at 3%, which is equivalent to investing \$7,321.1 at 0.52%. The latter is smaller than 2%, which signals that the incremental project is favorable, so b is preferred to a. The product structure is conserved for the three approaches and the incremental NPV of b over a is

$$\mathrm{NPV}^{b-a} = \overbrace{7{,}845.8}^{\mathrm{PV}[C^{b-a}]} \cdot \overbrace{(4.29\% - 2.91\%)}^{\xi^{b-a}} = 108.22$$

$$\mathrm{NPV}^{b-a} = \overbrace{8{,}000.0}^{C^{b-a}} \cdot \overbrace{(2.00\% - 0.65\%)}^{\epsilon^{b-a}} = 108.22$$

$$\mathrm{NPV}^{b-a} = \overbrace{7{,}321.1}^{C_{\mathrm{d}}^{b-a}} \cdot \overbrace{(2.00\% - 0.52\%)}^{\varepsilon_{\mathrm{d}}^{b-a}} = 108.22.$$

Project b creates \$108.22 more value than project a.

Example 11.6 ♦
Consider Example 11.2, where constant COCs are assumed (see Table 11.1). A pairwise comparison of incremental excess rates of return makes it possible to correctly rank the projects. Using (11.8), one finds the efficiencies of the incremental assets in all the approaches.

Project	A	B	C
NPV	30.18	45.82	27.85
$\mathrm{PV}[C^j]$	380.16	149.96	230.98
C^j	400.00	170.00	260.00
C_{d}^j	345.04	133.20	161.44

pair	(A, B)	(A, C)	(C, B)
AIRR approach			
ξ^{h-k}	−6.80%	1.56%	−22.18%
IARR approach			
ϵ^{h-k}	−6.80%	1.66%	−19.97%
AROI approach			
$\varepsilon_{\mathrm{d}}^{h-k}$	−7.39%	1.27%	−63.64%
preference	$B \succ A$	$A \succ C$	$B \succ C$
ranking	$B \succ A \succ C$		

In particular, for any approach, it is evident that B is preferred to A, A is preferred to C and B is preferred to C. (In this case, the latter comparison is unneeded as it follows from the previous two by transitivity.)[14]

The signs of the excess return rates signal incremental value creation/destruction of h over k. For example, $\epsilon^{A-C} = (\text{NPV}^A - \text{NPV}^C)/(C^A - C^C) = 2.33/140 = 1.66\%$ signals that A is preferred to C. To better understand this figure, consider that the excess (internal) average ROI of A is $\epsilon^A = \bar{\jmath}^A - \bar{\rho}^A = \text{NPV}^A/C^A = 30.18/400 = 7.54\%$, whereas $\epsilon^C = \bar{\jmath}^C - \bar{\rho}^C = \text{NPV}^C/C^C = 27.85/260 = 10.71\%$. This means that A is less efficient than C but has a larger scale. Project A may be viewed as a portfolio of C and the incremental asset $A - C$, and the economic efficiency of A (7.54%) is precisely the weighted average of the two constituent assets' economic efficiencies.

Asset	Scale	Efficiency	Value created
C	260	10.71%	27.85
$A - C$	140	1.66%	2.33
A	400	7.54%	30.18

11.4 Direct Method

A direct method for ranking projects via rates of return is possible as well. Consider a bundle of m projects and assume that the total capital amounts employed is the same in all projects. This means that they have the same scale or, in formal words, that they belong to the same equivalence class (see Remark 8.8).

In the AIRR approach, this means $\text{PV}[C^1] = \text{PV}[C^2] = \ldots = \text{PV}[C^m] = K$ and, hence (assuming $K > 0$), $\text{NPV}^h > \text{NPV}^k$ if and only if $\xi^h > \xi^k$.

In the IARR approach, same project scale means $C^1 = C^2 = \ldots = C^m = K$ and, hence (assuming $K > 0$), $\text{NPV}^h > \text{NPV}^k$ if and only if $\epsilon^h > \epsilon^k$.

In the AROI approach, this means $C^1_{\text{d}} = C^2_{\text{d}} = \ldots = C^m_{\text{d}} = K$ and, hence (assuming $K > 0$), $\text{NPV}^h > \text{NPV}^k$ if and only if $\varepsilon^h_{\text{d}} > \varepsilon^k_{\text{d}}$.

Notice that, in the AIRR approach, if the COC is constant across periods and equal across assets (i.e., $r^1_t = r^2_t = \ldots = r^m_t = r$ for every $t \in \mathbb{N}^1_n$), then $\text{NPV}^h > \text{NPV}^k$ if and only if $\bar{\imath}^h > \bar{\imath}^k$ (i.e., comparison of rates of return suffices to provide the correct ranking. See also Magni 2010b, Theorem 5). This is straightforward from the fact that AIRR may be viewed as an affine transformation of the NPV:

[14] The rates ξ^{A-B} and ϵ^{A-B} appear to be equal, for the numbers are rounded. More precisely, $\xi^{A-B} = -6.797\%$ and $\epsilon^{A-B} = -6.802\%$.

$$\bar{\imath}_0^j = r_0 + \frac{\mathrm{NPV}^j}{K} \quad \text{or}$$
$$\bar{\imath}^j = r + \frac{\mathrm{NPV}^j(1+r)}{K} \qquad j = 1, 2, \ldots, m. \tag{11.9}$$

However, the economic efficiency is more informative than the rates of return as such: $\bar{\imath}^h > \bar{\imath}^k$ informs that h is preferred to k but does not tell whether h and k are value-creating projects. If they are both value-destroying, the comparison is pointless, for both should be rejected (unless, for some non-economic reason, one of them must necessarily be undertaken). In contrast, the comparison of ξ^h and ξ^k also informs about value creation. In particular, both are value-destroying if $0 > \xi^h > \xi^k$; both are value-creating if $\xi^h > \xi^k > 0$; only h creates value if $\xi^h > 0 > \xi^k$.

More problematic is the case where the projects are not in the same equivalence class, that is, they have different size. In this case, one can make an NPV-invariant adjustment and express all excess rates of return in terms of a common capital base K, suitably selected, which we call *standardized* capital. Specifically, for the AIRR approach, one selects the standardized capital K, then calculates the standardized excess rates of return, respectively, $\xi^j(K) = \mathrm{NPV}^j/K$ in the AIRR approach or $\epsilon^j(K) = \mathrm{NPV}^j/K$ in the IARR approach or $\varepsilon_{\mathrm{d}}^j(K) = \mathrm{NPV}^j/K$ in the AROI approach, for $j \in \mathbb{N}_m^1$.

If the COC is constant across periods and assets, and the AIRR approach is employed, then (11.9) enables easily deriving a **standardized AIRR**, which we denote as

$$\bar{\imath}^j(K) = r + \frac{\mathrm{NPV}^j \cdot (1+r)}{K}$$

for $j \in \mathbb{N}_m^1$. The set of m standardized AIRRs are directly comparable, since they refer to the same scale and are associated with the same COC.

Proposition 11.4 (Project ranking with the relative approaches—Direct method) *Consider a bundle of m independent projects and let K be any standardized capital. The projects may be ranked via the standardized excess AIRRs,* $\xi^j(K)$, *or standardized excess IARRs,* $\epsilon^j(K)$, *or standardized excess AROIs,* $\varepsilon_{\mathrm{d}}^j(K)$, $j \in \mathbb{N}_m^1$. *The resulting ranking is rational.*

The selection of the standardized capital is not a mathematical issue. Mathematically, any capital goes. It is, rather, an economic issue, and a nontrivial one, as it raises the question of how a representative capital should be chosen. In some case, the choice is naturally triggered by the specific scenario. For example, when dealing with ROSCAs, the amount distributed is the same for all members, so the standardized capital K may be set equal to such an amount. Lima e Silva et al. (2018) use this standardized capital to rank the deals in pre-purchase financing pool, a type of (Brazilian) ROSCA. (The authors use the AIRR approach to correctly rank the deals).

In other cases, it is less simple. One possibility is that of computing a Chisini mean of the scales of the projects, using some significant function as invariant function. The total NPV generated by the projects might be such a function. Let P be the portfolio of the m projects to be ranked and let $n = \max\{n_1, n_2, \ldots, n_m\}$. Then, assuming that the approach selected is the AIRR approach and the COC is constant and equal across assets, the total available NPV is framed as

$$\text{NPV}^P = \sum_{j=1}^{m} \text{NPV}^j = \sum_{j=1}^{m} \text{PV}[C^j](\bar{\imath}_0^j - r_0). \tag{11.10}$$

Considering NPV^P as a function of the project scales and imposing Chisini invariance requirement upon (11.10),

$$\text{NPV}^P(\text{PV}[C^1], \text{PV}[C^2], \ldots, \text{PV}[C^m]) = \text{NPV}^P(K, K, \ldots, K)$$

one gets

$$K = \frac{\text{PV}[C^1] \cdot \xi^1 + \text{PV}[C^2] \cdot \xi^2 + \ldots + \text{PV}[C^m] \cdot \xi^m}{\xi^1 + \xi^2 + \ldots + \xi^m}. \tag{11.11}$$

Equation (11.11) expresses the standardized capital as an average scale; more precisely, it is a (generalized) weighted mean of the projects scales, where the weights are the excess AIRRs.

As for IARR, the same procedure leads to

$$K = \frac{C^1 \cdot \epsilon^1 + C^2 \cdot \epsilon^2 + \ldots + C^m \cdot \epsilon^m}{C^1 + C^2 + \ldots + C^m}. \tag{11.12}$$

As for AROI, the same procedure leads to

$$K = \frac{C_{\text{d}}^1 \cdot \varepsilon_{\text{d}}^1 + C_{\text{d}}^2 \cdot \varepsilon_{\text{d}}^2 + \ldots + C_{\text{d}}^m \cdot \varepsilon_{\text{d}}^m}{C_{\text{d}}^1 + C_{\text{d}}^2 + \ldots + C_{\text{d}}^m}. \tag{11.13}$$

A more compelling choice of standardized capital is available if the projects have the same initial capital: $C_0^1 = C_0^2 = \ldots = C_0^m = C_0$. In this case, there are two possibilities. The first one is to use C_0 as the standardized capital; the second one is to compute an average rate of change of capital in the following way. Let λ_t^j be the depreciation rate of project j's BOP capital, expressed in terms of C_0^j, that is, $\lambda_t^j = (C_{t-1}^j - C_t^j)/C_0^j$. Then, $C_n^j = C_0^j(1 - \lambda_1^j - \lambda_2^j - \ldots - \lambda_n^j)$. Let us consider the following set of m Chisini invariance requirements:

$$C_0^1(1-\lambda_1^1-\lambda_2^1-\ldots-\lambda_n^1)=C_0^1(1-\lambda^1-\lambda^1-\ldots-\lambda^1)$$
$$C_0^2(1-\lambda_1^2-\lambda_2^2-\ldots-\lambda_n^2)=C_0^2(1-\lambda^2-\lambda^2-\ldots-\lambda^2)$$
$$\ldots\ldots$$
$$C_0^m(1-\lambda_1^m-\lambda_2^m-\ldots-\lambda_n^m)=C_0^m(1-\lambda^m-\lambda^m-\ldots-\lambda^m)$$

Since $C_0^j=C_0$ and $C_n^j=0$ for all $j\in\mathbb{N}_m^1$, this set of equations has a unique solution:

$$\lambda^1=\lambda^2=\ldots=\lambda^m=\lambda=1/n$$

(this derives from the fact that $\sum_{t=1}^n\lambda_t^j=1$ for all $j\in\mathbb{N}_m^1$). Therefore, the average rate of change is the same for all projects. As a result, the average capital is $K_t=K_{t-1}-\lambda C_0$ for all projects which means $K_t=C_0(1-t/n)$ for every $t\in\mathbb{N}_n^1$.[15] Therefore, the project's capital streams belong to the same equivalence (straight-line depreciation) class. The standardized overall capital is then

$$K=\sum_{t=1}^n K_{t-1}(1+r)^{-(t-1)}=\sum_{t=1}^n C_0\cdot\left(1-\frac{(t-1)}{n}\right)(1+r)^{-(t-1)}$$

in the AIRR approach. The resultant rate of return, $\bar{\imath}^j(K)$ may be called ***straight-line rate of return***. Ranking of the projects may then be accomplished via ranking of $\bar{\imath}^j(K)$. If the COC is not equal across projects, then rational ranking is possible in general only via ranking of the *excess* straight-line rates of return, $\xi^j(K)$.[16] As for the IARR and AROI approach, the standardized capital becomes

$$K=\sum_{t=1}^n K_{t-1}=\sum_{t=1}^n C_0\cdot\left(1-\frac{(t-1)}{n}\right)=\frac{C_0\cdot(n+1)}{2}.$$

Hence, the standardized IARR (AROI) is equal to $\bar{\jmath}(K)=2I/(C_0(n+1))$.[17] Project ranking is made with the excess rates of return (respectively, $\epsilon(K)$ and $\varepsilon(K)$).

With time-varying COCs, generalization is straightforward by replacing r with $\bar{r}$, $\bar{\rho}$, and $\bar{\varrho}$, respectively.

Example 11.7 ♦
Consider three mutually exclusive projects, with lifespan equal to 6 years, and assume their initial outlays are equal. We report the basic triplets for the various projects in Table 11.3. We select the average capital as the standardized capital. The average capital depreciation rate is then $\lambda=1/6=0.167$. Therefore, the standardized capital is $\boldsymbol{K}=(100, 83.33, 66.67, 50, 33.3, 16.67)$.

---

[15] This is equivalent to straight-line depreciation.

[16] See Marchioni and Magni (2016) on strong NPV-consistency of this rate of return in the AIRR approach.

[17] Note that $\bar{\jmath}(K)=(2/(n+1))\cdot\bar{\jmath}(C_0)$.

Table 11.3 Three projects with equal initial outlay and their standardized capital

Year	C^A	C^B	C^C	I^A	I^B	I^C	F^A	F^B	F^C	K
0	100	100	100				−100	−100	−100	100.00
1	100	60	0	50	0	−50	50	40	50	83.33
2	100	20	0	20	0	50	20	40	50	66.67
3	100	0	0	30	20	−20	30	40	−20	50.00
4	100	0	0	20	30	−20	20	30	−20	33.33
5	100	0	0	10	30	30	10	30	30	16.67
6	0	0	0	−95	30	60	5	30	60	0.00

Table 11.4 Ranking of the three projects

Project	A	B	C
AIRR approach			
Overall standardized capital	301.54	301.54	301.54
Standardized economic efficiency (ξ^j)	2.39%	18.41%	3.51%
Ranking		$B > C > A$	
IARR approach			
Overall standardized capital	350	350	350
Standardized economic efficiency (ϵ^j)	2.06%	15.86%	3.02%
Ranking		$B > C > A$	
AROI approach			
Overall standardized capital	197.57	197.57	197.57
Standardized economic efficiency (ε^j)	3.65%	28.11%	5.36%
Ranking		$B > C > A$	
NPV			
Net present value	7.21	55.53	10.59
Ranking		$B > C > A$	

Table 11.4 reports the (consistent) ranking in the various approaches, assuming $r^A = r^B = r^C = 10\%$. The NPV is found from each relative approach by multiplying the standardized excess return rates and the overall standardized capital base or, alternatively, by discounting the cash flows at 10%.

Given the assumption of a constant COC across periods and across projects, if the AIRR approach is used, ranking of projects may be accomplished via standardized rates of return (in place of excess rates of return). The (instantaneous) standardized average ROI is found as $\bar{\imath}_0^j = \xi^j + r/(1+r)$. Hence, $\bar{\imath}_0^A = 11.48\%$, $\bar{\imath}_0^B = 27.51\%$, $\bar{\imath}_0^C = 12.6\%$, which preserves the ranking: $B \succ C \succ A$.

11.5 Key Points

- Ranking competing projects with NPV criterion and with RI criterion is rational: The higher the NPV (the overall value of RI), the higher the rank in terms of preference
- Ranking projects with NFV is rational if the projects have the same COC (or same pattern of COCs); otherwise, ranking projects with NFV is not rational, owing to the distortive effect of the compounding factors, which may reverse the NPV/RI ranking
- A project's value creation is the result of two effects: The *scale effect* and the *efficiency effect*. Relative measures of worth quantify the efficiency, not the scale, which is determined by the capital invested. Therefore, in general, to compare two rates of return is just like comparing oranges and apples. Owing to the scale effect, ranking projects with relative measures of worth is non-rational
- It is possible to obviate the ranking problem with relative measures of worth via incremental analysis, which works irrespective of the COC patterns of the competing projects. AIRR, IARR, and AROI may be reliably employed to get rational rankings. Intriguingly, while ranking with NFV is not rational, ranking with AROI (its relative companion) is rational
- It is possible to obviate the ranking problem with relative measures of worth via a direct method whereby the rates of return are standardized with a benchmark capital. In this way, the scale effect is sterilized
- The problem of selecting a significant benchmark capital is not a logical one (any capital goes), but an economic one: It is not trivial to select the most appropriate benchmark capital. This should be done case by case. Chisini approach may be of some help when economic intuition does not supply any tip.

Chapter 12
Three Decisions

> Some of the assets are tangible assets such as plants and machinery; other are intangible assets such as brand names and patents. [Firms] finance these assets by borrowing, by retaining and reinvesting cash flow, and by selling additional shares of stocks to the [firm's] shareholders
>
> Brealey, Myers and Allen (2011, p. 29)

> In manufacturing, engineering is involved in every detail of a product's production, from conceptual design to shipping. In fact, engineering decisions account for majority (some say 85%) of product costs
>
> Park (2011, p. 31)

> The demand for accountability among all business functions has never been greater. Recent events show how vital decisions are about human resources … A key responsibility … is to articulate the logical connections between progressive HR practices and firm performance.
>
> Cascio and Boudreau (2011, Preface, p. xvi)

C. A. Magni, *Investment Decisions and the Logic of Valuation*,
https://doi.org/10.1007/978-3-030-27662-1_12

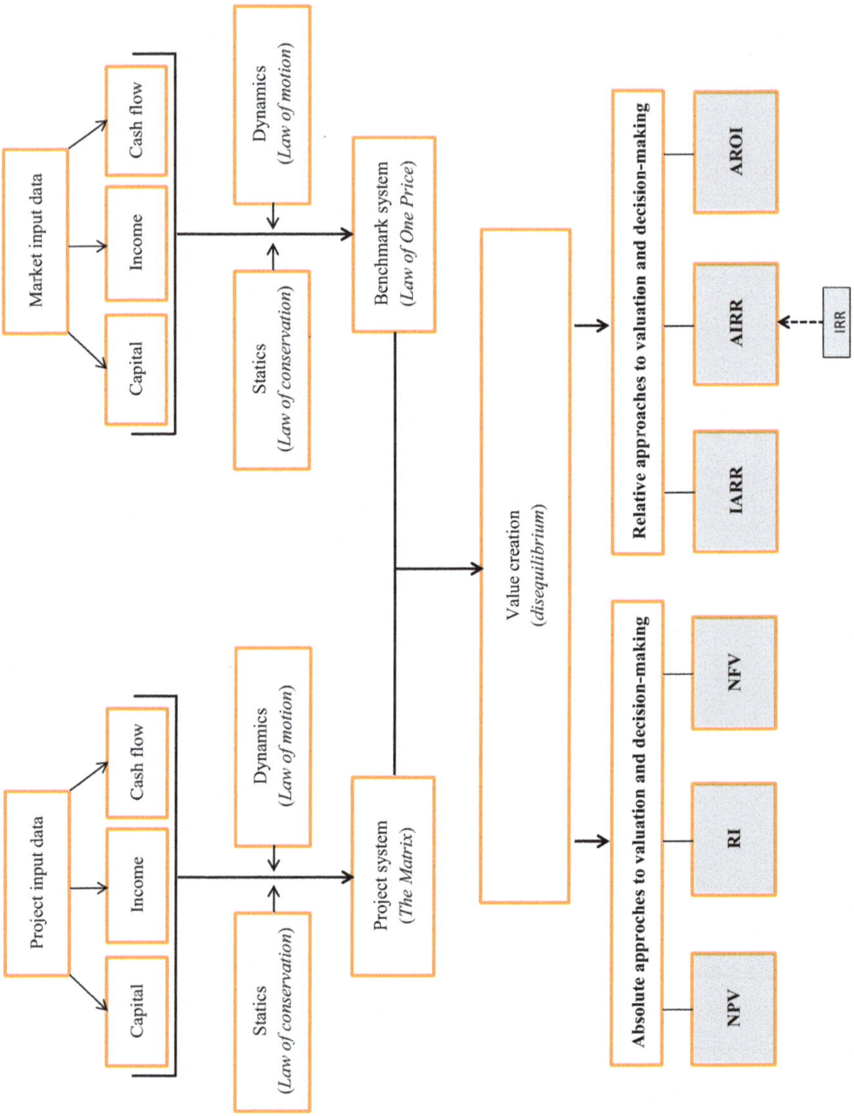

Fig. 12.1 Chapter 12: Three final examples and related decision-making processes analyzed with absolute and relative measures of worth

This final chapter scrutinizes three examples: Two engineering decisions (named Bumps and HomeNet, respectively) and a human resource initiative (named Rainbow Educational Program). Bumps consists in a choice between two mutually exclusive alternatives. HomeNet and Rainbow projects are accept-reject decisions. For illustrative purposes,

- Bumps's economic profitability is analyzed with the (NPV, IARR) method
- HomeNet's economic profitability is analyzed with the (RI, AIRR) method
- Rainbow Educational Program's economic profitability is analyzed with the (NFV, AROI) method.

12.1 Bumps Inc. (IARR and NPV)

This example is inspired by Problem 28 in White et al. (2014 p. 280), suitably adjusted and enriched for our purposes.

Bumps Inc., a highway contractor, investigates the possibility of overhauling a tractor and scraper and, at the same time, explores the opportunity of replacing it with a new one. A decision is required about whether (i) doing nothing (status quo), (ii) overhauling, or (iii) replacing. The difference between (ii) and (i) may be classified as an expansion project (assets are expanded by the cost of overhauling), the difference between (iii) and (i) is a replacement project.

Project input data. The old equipment was purchased 5 years ago for $130,000; it had a 12-year projected life. The current and projected operating and maintenance (O&M) costs are $40,000/year. Overhauling the equipment will cost $10,000, which will be depreciated with double-declining balance for 7 years. If overhauled, operating and maintenance (O&M) costs will decrease to $25,000/year. The old equipment depreciates evenly, so salvage value will be zero in 7 years. If traded for a new tractor and scraper, the old equipment can be now sold for $60,000. If replaced, a new tractor and scraper can be purchased for $150,000. It will depreciate evenly for 7 years and will be sold at an expected price of $35,000. O&M costs will fall down to $7,000/year.

If overhauled, the overhaul cost will be financed internally, that is, by withdrawing cash from the liquid asset-account; the interest rate on the account is 0.5%. If replaced, the cost of new equipment will be partly financed by a loan (40% of the invested capital at time 0) and the remaining part by equity. The loan is repaid with 7 level principal repayments at 2%.

If either overhaul or replacement is selected, the payout ratio of the incremental cash flows from time 1 to time 6 will be set to 30%, which implies that 70% of the cash available for distribution will be retained in the firm's cash account from year 1 to 7. Tax rate is 33%.

Market input data. The risk of the two projects is considered equivalent, and the pre-tax operating WACC is set to $r^o = 15\%$ for both overhaul and replacement. The required return on liquid assets is $r^l = 0.5\%$, equal to the interest rate on liquid assets. The required return on debt is $r^d = 2\%$, equal to the loan rate.

We analyze the two projects separately and then combine the results.

1. Overhaul

C^o	
C^l	C^e

We report the input data in the table below.

Overhaul – Input data

Project data	
Old equipment	
O&M costs	\$40,000/year
Overhaul	
Initial investment	\$10,000
Useful life	7 years
O&M costs	\$25,000/year
Depreciation	double-declining balance
Financing (year 0)	
Liquid assets	100% ($= -C_0^l/C_0^0$) (withdrawal of cash)
return on liquid assets	$i^l = 0.5\%$ (constant and equal to r^l)
Equity financing	0%
Debt financing	0%
Payout ratio	$\alpha = 30\%$ (from time 1 to time 6) if $F_t^{pd} > 0$ and $I_t^e > 0$ ($\alpha = 0\%$ otherwise)
Tax rate	33%
Market data	
Pre-tax (operating) WACC	$r^o = 15\%$ (constant)
Required return on liquid assets	$r^l = 0.5\%$ (constant and equal to i^l)

Consider that:

- depreciation of overhaul is calculated as $\text{Dep}_t^{\text{ovh}} = \text{Dep}_t^{DDB} = (2/N) \cdot Q_{t-1}^{DDB}$ (see Sect. 3.2)
- sales are not affected by the project, so $S_t = 0$ for all t
- operating costs are negative, owing to O&M savings, which amount to $\text{OpC}_t = -\$15{,}000 = \$25{,}000 - \$40{,}000$ per year. Therefore, the (negative) operating costs are positive income components
- as soon as inflows are generated by the project, the balance of liquid assets increases (70% of CFO = PD is retained in the firm). From time 2, the balance is above zero. Therefore, the firm is subtracting funds from liquid assets in the first two periods while investing funds in liquid assets in the remaining periods
- the project IRR does not exist, for all the project's CFAs are positive. Since CFAs and CFEs coincide ($F_t = F_t^e$), the equity IRR does not exist as well
- two IRRs for the non-operating assets arise: $\sigma_1^l = 0.5\%$ (the interest rate on liquid assets) and $\sigma_2^l = 64.51\%$, which has no reference with the transactions involved in the liquid assets
- the operating IRR is unique and equal to $\sigma^o = 107.41\%$.

Bumps Inc. – Overhaul – Balance Sheets

Year	0	1	2	3	4	5	6	7
NFA	10,000.0	7,142.9	5,102.0	3,644.3	2,603.1	1,859.3	1,328.1	0.0
Overhaul	10,000.0	7,142.9	5,102.0	3,644.3	2,603.1	1,859.3	1,328.1	0.0
NOWC	0.0	0.0	0.0	0.0	0.0	0.0	0.0	0.0
Liquid assets	–10,000.0	–2,343.5	5,154.0	12,545.5	19,869.3	27,152.5	34,414.6	0.0
Investments	0.0	4,799.4	10,256.0	16,189.8	22,472.4	29,011.8	35,742.7	0.0
Debt	0.0	0.0	0.0	0.0	0.0	0.0	0.0	0.0
Equity	0.0	4,799.4	10,256.0	16,189.8	22,472.4	29,011.8	35,742.7	0.0
Financings	0.0	4,799.4	10,256.0	16,189.8	22,472.4	29,011.8	35,742.7	0.0

Bumps Inc. – Overhaul – Income Statements

Year	1	2	3	4	5	6	7
Sales	0.0	0.0	0.0	0.0	0.0	0.0	0.0
–OpCosts	+15,000.0	+15,000.0	+15,000.0	+15,000.0	+15,000.0	+15,000.0	+15,000.0
–Depreciation overhaul	–2,857.1	–2,040.8	–1,457.7	–1,041.2	–743.7	–531.2	–1,328.1
EBIT	12,142.9	12,959.2	13,542.3	13,958.8	14,256.3	14,468.8	13,671.9
–Interest expense	0.0	0.0	0.0	0.0	0.0	0.0	0.0
+Interest income	–50.0	–11.7	+25.8	+62.7	+99.3	+135.8	+172.1
EBT	12,092.9	12,947.5	13,568.0	14,021.5	14,355.6	14,604.5	13,844.0
–Taxes	–3,990.6	–4,272.7	–4,477.5	–4,627.1	–4,737.4	–4,819.5	–4,568.5
NI	8,102.2	8,674.8	9,090.6	9,394.4	9,618.3	9,785.0	9,275.5

In contrast, all the book IARRs exist, are unique, and are economically significant. For the sake of illustration, we focus on internal average ROI and ROE, which coincide, given that $C_t = C^e_t$ for all t. Discounting the CFOs and the CFLs, one gets the economic values of operations and liquid assets. The internal average ROE/ROI is simply obtained by dividing the total income to the total invested capital: $\bar{\jmath} = \bar{\jmath}^e = I^e/C^e = 63,940.8/118,472.1 = 53.97\%$. The internal average operating ROI, $\bar{\jmath}^o$ is very high, owing to the fact that the operating incomes are much higher than the operating invested capitals in every period barring the first one: $\bar{\jmath}^o = I^o/C^o = 63{,}506.8/31{,}679.7 = 200.46\%$. This reflects the performance of the operations better than the operating IRR, which underestimates the operating performance by almost 100 percentage points.

The internal average ROI, $\bar{\jmath}$, is much lower than the average operating ROI, $\bar{\jmath}^o$, because of the role of the liquid assets. Specifically, the average ROI is a (generalized) weighted mean of the average operating ROI and the average return on liquid assets (in this case, the latter is constant). Overall, the total capital invested in the liquid assets is substantial, more than the capital invested in the operating assets: $C^l = 86{,}792.4$ as opposed to $C^o = 31{,}679.7$. This boosts the total capital invested. However, the total income from such a corpulent investment is negligible ($I^l = 434$), which explains why the resultant project ROI is much lower than the project operating ROI:

$$\bar{\jmath} = \bar{\jmath}^e = \frac{63{,}506.8 + 434}{31{,}679.7 + 86{,}792.4} = 53.97\% < 200.46\% = \bar{\jmath}^o = \frac{63{,}506.8}{31{,}679.7}.$$

As for the MARR, one first computes the economic value of operations. Specifically, discounting the cash flows at the respective COCs ($r^o = 15\%$) one finds

$$V^o = (43{,}959.5,\ 39{,}544.1,\ 34{,}748.3,\ 29{,}438,\ 23{,}480.8,\ 16{,}740.3,\ 9{,}070.9,\ 0)$$

whence the total intrinsic value of the operating capital is $V^o = 196{,}982$. Also, $V^l = C^l = 86{,}792.4$ because $i^l = r^l$. Hence, one can compute the project MARR aggregating the pre-tax operating WACC and the required return on liquid assets:

$$\bar{\rho} = \frac{\overbrace{r^o \cdot V^o + r^l \cdot V^l}^{I^V}}{\underbrace{C^o + C^l}_{C}} = \frac{15\% \cdot 196{,}982 + 0.5\% \cdot 86{,}792.4}{31{,}679.7 + 86{,}792.4} = 25.31\% \ (= \bar{\rho}^e).$$

Therefore, the project consists of a total investment of $C^o + C^l = 118{,}472.1$ at an excess rate of return of $\epsilon = 53.97\% - 25.31\% = 28.66\%$. The NPV is NPV $=$ NPV$^e = 118{,}472.1 \cdot 28.66\% \simeq 33{,}959.5$.[1] All of the NPV is generated by the operations, since the liquid assets are value-neutral ($C^l_t = V^l_t$ for every t).

Overhaul is then preferable to the null alternative (status quo, doing nothing). Figure 12.2 shows the modeling relationships among project input data, incomes, cash flows, and book values.

[1]Numbers are rounded, as usual. With more digits, one indeed gets $118{,}472.1457 \cdot 53.971129\% - 25.306587\% = 33{,}959.5$.

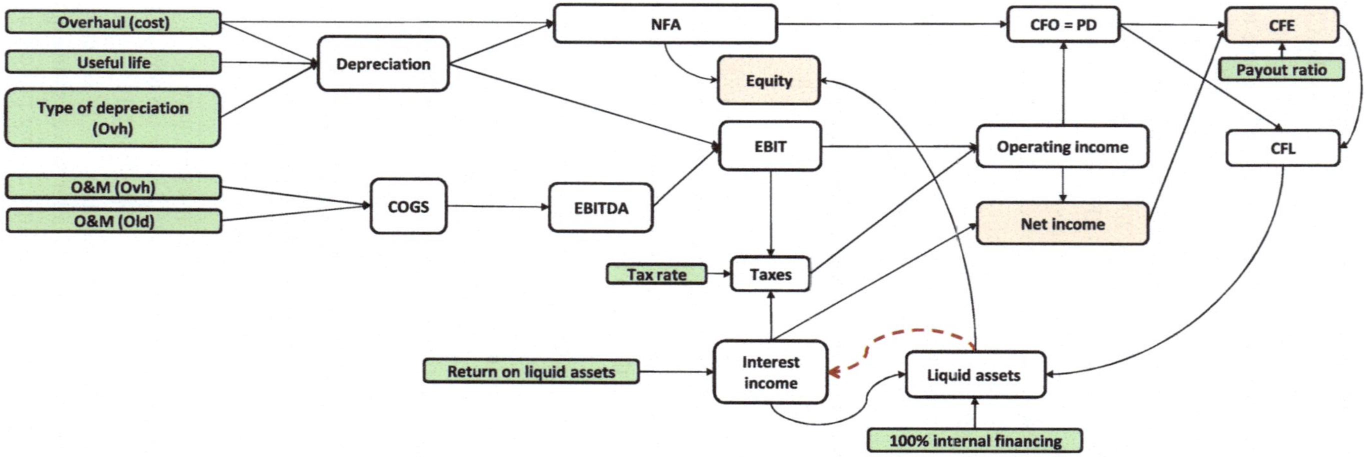

Fig. 12.2 Influence diagram indicating the functional relationships of Bumps Inc. (Overhaul). The dashed line refers to time $t-1$: The balance of liquid assets at time $t-1$ determines the interest income in t via application of the interest rate on liquid assets: $I_t^l = i_t^l C_{t-1}^l$

$(\boldsymbol{C}, \boldsymbol{I}, \boldsymbol{F})$ Bumps Inc.—Overhaul

Year	0	1	2	3	4	5	6	7
C	0.0	4,799.4	10,256.0	16,189.8	22,472.4	29,011.8	35,742.7	0.0
C^o	10,000.0	7,142.9	5,102.0	3,644.3	2,603.1	1,859.3	1,328.1	0.0
C^l	−10,000.0	−2,343.5	5,154.0	12,545.5	19,869.3	27,152.5	34,414.6	0.0
C^d	0.0	0.0	0.0	0.0	0.0	0.0	0.0	0.0
C^e	0.0	4,799.4	10,256.0	16,189.8	22,472.4	29,011.8	35,742.7	0.0
I	0.0	8,102.2	8.674.8	9,090.6	9,394.4	9,618.3	9,785.0	9,275.5
I^o	0.0	8,152.2	8.686.5	9,064.8	9,331.7	9,518.9	9,649.3	9,103.4
I^l	0.0	−50.0	−11.7	25.8	62.7	99.3	135.8	172.1
I^d	0.0	0.0	0.0	0.0	0.0	0.0	0.0	0.0
I^e	0.0	8,102.2	8.674.8	9,090.6	9,394.4	9,618.3	9,785.0	9,275.5
F	0.0	3,302.8	3,218.2	3,156.8	3,111.9	3,078.8	3,054.2	45,018.2
F^o	−10,000.0	11,009.4	10,727.3	10,522.5	10,372.9	10,262.6	10,180.5	10,431.5
F^l	10,000.0	−7,706.6	−7,509.1	−7,365.8	−7,261.0	−7,183.9	−7,126.4	34,586.7
F^d	0.0	0.0	0.0	0.0	0.0	0.0	0.0	0.0
F^e	0.0	3,302.8	3,218.2	3,156.8	3,111.9	3,078.8	3,054.2	45,018.2

2. Replacement

C^o	C^d
C^l	C^e

We report the input data in the table below.

Replacement – Input data

Project data	
Old equipment	
Original cost	\$130,000
Date of acquisition	5 years ago
Residual useful life	7 years
Depreciation	straight-line
O&M costs	\$40,000/year
Selling price	\$60,000
Replacement	
Initial investment	\$150,000
Useful life	7 years
O&M costs	\$7,000/year
Depreciation	straight-line
Expected selling price	\$35,000
Financing policy (year 0)	
Liquid assets	0%
Equity financing	\$60% $(= C_0^e/C_0^o)$
Debt financing	\$40% $(= C_0^d/C_0^o)$
length	7 years
amortization	level principal repayments $(= C_0^d/7)$
loan rate	$i^d = 2\%$ (constant and equal to r^d)
Return on liquid assets	$i^l = 0.5\%$ (constant and equal to r^l)
Payout ratio	$\alpha = 30\%$ (from time 1 to time 6) if $F_t^{pd} > 0$ and $I_t^e > 0$ ($\alpha = 0\%$ otherwise)
Tax rate	33%
Market data	
Pre-tax (operating) WACC	$r^o = 15\%$ (constant)
Required return on liquid assets	$r^l = 0.5\%$ (constant and equal to i^l)
Required return on debt	$r^d = 2\%$ (constant and equal to i^d)

Consider that:

- depreciation of old equipment is $\text{Dep}^{\text{old}} = \$130{,}000/5 = \$10{,}833.3$; therefore, the accumulated depreciation thus far is $\$10{,}833.3 \cdot 5 = \$54{,}166.7$. Hence, the year-0 book value is $\text{NFA}_0^{\text{old}} = \$130{,}000 - \$54{,}166.7 = \$75{,}833.3$
- at time 0, the project's NFA is equal to the difference between the purchasing price of the new tractor and scraper and the carrying amount of the old one: $\text{NFA}_0 = 74{,}166.7 = 150{,}000 - 75{,}833.3$. The firm's NFA decreases by the depreciation expense of the new equipment ($\text{Dep}^{\text{new}} = \$21{,}428.6 = \$150{,}000/7$) but increases by the depreciation expense of the old equipment ($\text{Dep}^{\text{old}} = \$10{,}833.3$). Therefore, $\text{Dep}_t = \$10{,}595.2 = \$21{,}428.6 - \$10{,}833.3$. Hence, the project's NFA decreases by \$10,833.3 in every period from $t = 1$ to $t = 7$
- the firm raises funds from debt for an amount of \$29,666.7, which is equal to 40% of the initial operating capital; the residual \$44,500 (= \$74,166.7 − \$29,666.7), which expresses 60% of the initial operating capital, is invested by equityholders
- the sale of old equipment brings about a loss on disposed asset equal to the difference between the selling price and the residual book value: $z_0 = -\$15{,}833.3 = \$65{,}000 - \$75{,}833.3$; the sale of new equipment produces a gain on disposed asset equal to the difference between the selling price and the salvage value, which is zero: $z_7 = \$35{,}000 = \$35{,}000 - \$0$
- the loss on disposed assets is financed by equity. More precisely, consider that the loss causes the initial (operating and net) income to be nonzero: $I_0^o = I_0^e = -\$10{,}608.3$. This implies that the first (operating and equity) outflow is greater than the initial invested capital:

$$\overbrace{\$10{,}608.3}^{-I_0^o} + \overbrace{\$74{,}166.7}^{+\Delta C_0^o} = \overbrace{\$84{,}775}^{|F_0^o|} > \overbrace{\$74{,}166.7}^{C_0^o}$$

$$\overbrace{\$10{,}608.3}^{-I_0^e} + \overbrace{\$44{,}500.0}^{+\Delta C_0^e} = \overbrace{\$55{,}108.3}^{|F_0^e|} > \overbrace{\$44{,}500.0}^{C_0^e}$$

(in contrast, $C_0^d = -F_0^d$). As a result, equityholders provide \$10,608.3 to cover the loss on sale (in addition to \$44,500 equity capital). Overall, equityholders cover $\$55{,}108.3/\$84{,}775 = 65\%$ of the total initial disbursement, whereas debtholders only cover the remaining 35% (= 29,666.7/84,775)
- sales are not affected by the project, so $S_t = 0$ for all t
- operating costs are negative, owing to O&M savings, which amount to \$33,000 = \$40,000 − \$7,000 per year. Therefore, the (negative) operating costs are positive income components
- as cash flows are generated by the project, part of them is used for servicing the debt; since both PD and NI are positive, 30% of the PD is distributed to equityholders, 70% of the PD is deposited in the liquid assets
- the project IRR is $\sigma = 17.23\%$, the equity IRR is $\sigma^e = 21.32\%$. Note that the IRR concept does not recognize the initial income, thereby providing a further source of bias (see Sect. 9.5)

Bumps Inc. – Replacement – Balance Sheets

Year	0	1	2	3	4	5	6	7
NFA	74,166.7	63,571.4	52,976.2	42,381.0	31,785.7	21,190.5	10,595.2	0.0
New equipment	150,000.0	128,571.4	107,142.9	85,714.3	64,285.7	42,857.1	21,428.6	0.0
–Old equipment	−75,833.3	−65,000.0	−54,166.7	−43,333.3	−32,500.0	−21,666.7	−10,833.3	0.0
NOWC	0.0	0.0	0.0	0.0	0.0	0.0	0.0	0.0
Liquid assets	0.0	14,679.6	29,455.3	44,327.6	59,296.9	74,363.5	89,527.7	0.0
Investments	74,166.7	78,251.0	82,431.5	86,708.6	91,082.6	95,553.9	100,123.0	0.0
Debt	29,666.7	25,428.6	21,190.5	16,952.4	12,714.3	8,476.2	4,238.1	0.0
Equity	44,500.0	52,822.4	61,241.0	69,756.2	78,368.3	87,077.8	95,884.9	0.0
Financings	74,166.7	78,251.0	82,431.5	86,708.6	91,082.6	95,553.9	100,123.0	0.0

Bumps Inc. – Replacement – Income Statements

Year	0	1	2	3	4	5	6	7
Sales		0.0	0.0	0.0	0.0	0.0	0.0	0.0
+ Gains on sales	−15,833.3	0.0	0.0	0.0	0.0	0.0	0.0	+35,000.0
−OpCosts		+33,000.0	+33,000.0	+33,000.0	+33,000.0	+33,000.0	+33,000.0	+33,000.0
$-\text{Dep}^{new}$		−21,428.6	− 21,428.6	−21,428.6	−21,428.6	−21,428.6	−21,428.6	−21,428.6
$+\text{Dep}^{old}$		+10,833.3	+10,833.3	+10,833.3	+10,833.3	+10,833.3	+10,833.3	+10,833.3
EBIT	−15,833.3	22,404.8	22,404.8	22,404.8	22,404.8	22,404.8	22,404.8	57,404.8
−Int. expense		−593.3	−508.6	−423.8	−339.0	−254.3	−169.5	−84.8
+Int. income		0.0	+73.4	+147.3	+221.6	+296.5	+371.8	+447.6
EBT	−15,833.3	21,811.4	21,969.6	22,128.2	22,287.4	22,447.0	22,607.1	57,767.6
−Taxes	+5,225.0	−7,197.8	−7,250.0	−7,302.3	−7,354.8	−7,407.5	−7,460.3	−19,063.3
NI	−10,608.3	14,613.7	14,719.6	14,825.9	14,932.5	15,039.5	15,146.7	38,704.3

$(\boldsymbol{C}, \boldsymbol{I}, \boldsymbol{F})$ Bumps Inc. – Replacement

Year	0	1	2	3	4	5	6	7
$\boldsymbol{C}$	74,166.7	78,251.0	82,431.5	86,708.6	91,082.6	95,553.9	100,123.0	0.0
$\boldsymbol{C^o}$	74,166.7	63,571.4	52,976.2	42,381.0	31,785.7	21,190.5	10,595.2	0.0
$\boldsymbol{C^l}$	0.0	14,679.6	29,455.3	44,327.6	59,296.9	74,363.5	89,527.7	0.0
$\boldsymbol{C^d}$	29,666.7	25,428.6	21,190.5	16,952.4	12,714.3	8,476.2	4,238.1	0.0
$\boldsymbol{C^e}$	44,500.0	52,822.4	61,241.0	69,756.2	78,368.3	87,077.8	95,884.9	0.0
$\boldsymbol{I}$	−10,608.3	15,207.0	15,228.2	15,249.7	15,271.6	15,293.7	15,316.3	38,789.1
$\boldsymbol{I^o}$	−10,608.3	15,207.0	15,154.8	15,102.4	15,049.9	14,997.3	14,944.4	38,341.4
$\boldsymbol{I^l}$	0.0	0.0	73.4	147.3	221.6	296.5	371.8	447.6
$\boldsymbol{I^d}$	0.0	593.3	508.6	423.8	339.0	254.3	169.5	84.8
$\boldsymbol{I^e}$	−10,608.3	14,613.7	14,719.6	14,825.9	14,932.5	15,039.5	15,146.7	38,704.3
$\boldsymbol{F}$	−84,775.0	11,122.7	11,047.7	10,972.6	10,897.6	10,822.4	10,747.2	138,912.0
$\boldsymbol{F^o}$	−84,775.0	25,802.2	25,750.0	25,697.7	25,645.2	25,592.5	25,539.7	48,936.7
$\boldsymbol{F^l}$	0.0	−14,679.6	−14,702.4	−14,725.0	−14,747.6	−14,770.1	−14,792.4	89,975.4
$\boldsymbol{F^d}$	−29,666.7	4,831.4	4,746.7	4,661.9	4,577.1	4,492.4	4,407.6	4,322.9
$\boldsymbol{F^e}$	−55,108.3	6,291.2	6,301.0	6,310.7	6,320.4	6,330.0	6,339.6	134,589.2

The BIARRs are easily computed as

$$\bar{j} = \frac{I}{C} = \frac{119{,}747.2}{608{,}317.3} = 19.68\%$$
$$\bar{j}^o = \frac{I^o}{C^o} = \frac{118{,}189.0}{296{,}666.7} = 39.84\%$$
$$\bar{j}^l = \frac{I^l}{C^l} = \frac{1{,}558.3}{311{,}650.6} = 0.50\% = i^l$$
$$\bar{j}^d = \frac{I^d}{C^d} = \frac{2{,}373.3}{118{,}666.7} = 2\% = i^d$$
$$\bar{j}^e = \frac{I^e}{C^e} = \frac{117{,}373.9}{489{,}650.6} = 23.97\%.$$

As for the MARRs, discounting CFOs at 15%, one finds

$$\boldsymbol{V}^o = (115{,}629.4,\ 107{,}171.6,\ 97{,}497.3,\ 86{,}424.2,\ 73{,}742.7,\ 59{,}211.6,\ 42{,}553.6,\ 0)$$

whence $V^o = \sum_{t=1}^{n} V^o_{t-1} = 582{,}230.4$, while $V^l = C^l = 311{,}650.6$. Hence,

$$\bar{\rho} = \frac{\overbrace{15\% \cdot 582{,}230.4 + 0.5\% \cdot 311{,}650.6}^{r^o V^o + r^l V^l}}{\underbrace{296{,}666.7 + 311{,}650.6}_{C^o + C^l}} = 14.61\%.$$

Therefore, the project consists of a total investment of $C = 608{,}317.3$ at an excess rate of return of $\epsilon = 19.68\% - 14.61\% = 5.07\%$. The value created by the project is NPV $= 608{,}317.3 \cdot 5.07\% = 30{,}854.4$.[2] All of the NPV is generated by the operations, since the liquid assets are value-neutral ($C^l_t = V^l_t$ for every t). All the NPV is grasped by equityholders, since $i^d = r^d$. In particular, the equity MARR is

$$\bar{\rho}^e = \frac{\overbrace{15\% \cdot 582{,}230.4 + 0.5\% \cdot 311{,}650.6 - 2\% \cdot 118{,}666.7}^{r^o V^o + r^l V^l - r^d V^d = r^e V^e}}{\underbrace{296{,}666.7 + 311{,}650.6 - 118{,}666.7}_{C^o + C^l - C^d = C^e}} = 17.67\%.$$

[2]Numbers are rounded. Using more digits, one indeed gets NPV $= 608{,}317.27063 \cdot 5.072093\% = 30{,}854.4$.

(One may also use (10.85) to calculate $\bar{\rho}^e$.) The economic efficiency of the equity is $\epsilon^e = 23.97\% - 17.67\% = 6.3\%$. Since $C^e = 489{,}650.6$, the shareholder created value is $\text{NPV}^e = 489{,}650.6 \cdot 6.3\% = 30{,}854.4$.[3] The difference between the project perspective and equity perspective resides in the debt perspective (debtholders invest \$118,666.7 at a 0% excess return):

$$\begin{array}{rlll} & \text{NPV}^e & = 489{,}650.6 \times 6.3\% & = 30{,}854.4 \\ + & \text{NPV}^d & = 118{,}666.7 \times 0.0\% & = 0 \\ \hline & \text{NPV} & = 608{,}317.3 \times 5.07\% & = 30{,}854.4 \end{array}$$

The 5.07% is the weighted mean of 6.3 and 0%.

Replacement is then preferable to the null alternative (status quo, doing nothing). Figure 12.3 reports an influence diagram for the replacement case, which makes it clear that incomes, capitals and cash flows are intertwined in a nontrivial way.

Overhaul Versus Replacement

Both alternatives are worth undertaking. However, the comparison of equity NPV establishes the preferred alternative: $\text{NPV}^{e,\text{rep}} = 30{,}854.4 < 33{,}959.5 = \text{NPV}^{e,\text{ovh}}$. Therefore, overhaul should be undertaken. Observe that equity NPV and project NPV coincide in both cases but for different reasons (overhaul alternative does not require debt, replacing requires a debt whose interest rate is equal to the required return on debt: $i^d = r^d$).

The incremental equity NPV is negative: $\text{NPV}^{e,\text{rep}} - \text{NPV}^{e,\text{ovh}} = -3{,}105.1$. The reason is that overhauling entails the use of less equity capital at a higher rate of return (sufficiently higher to outweigh the smaller size). To shed some light on this differential NPV, one needs to accomplish an incremental IARR analysis. Below is the summary of project's net incomes, benchmark net incomes, and equity capitals, along with the respective totals.

[3] More precisely, $489{,}650.60397 \cdot 6.301313\% = 30{,}854.4$.

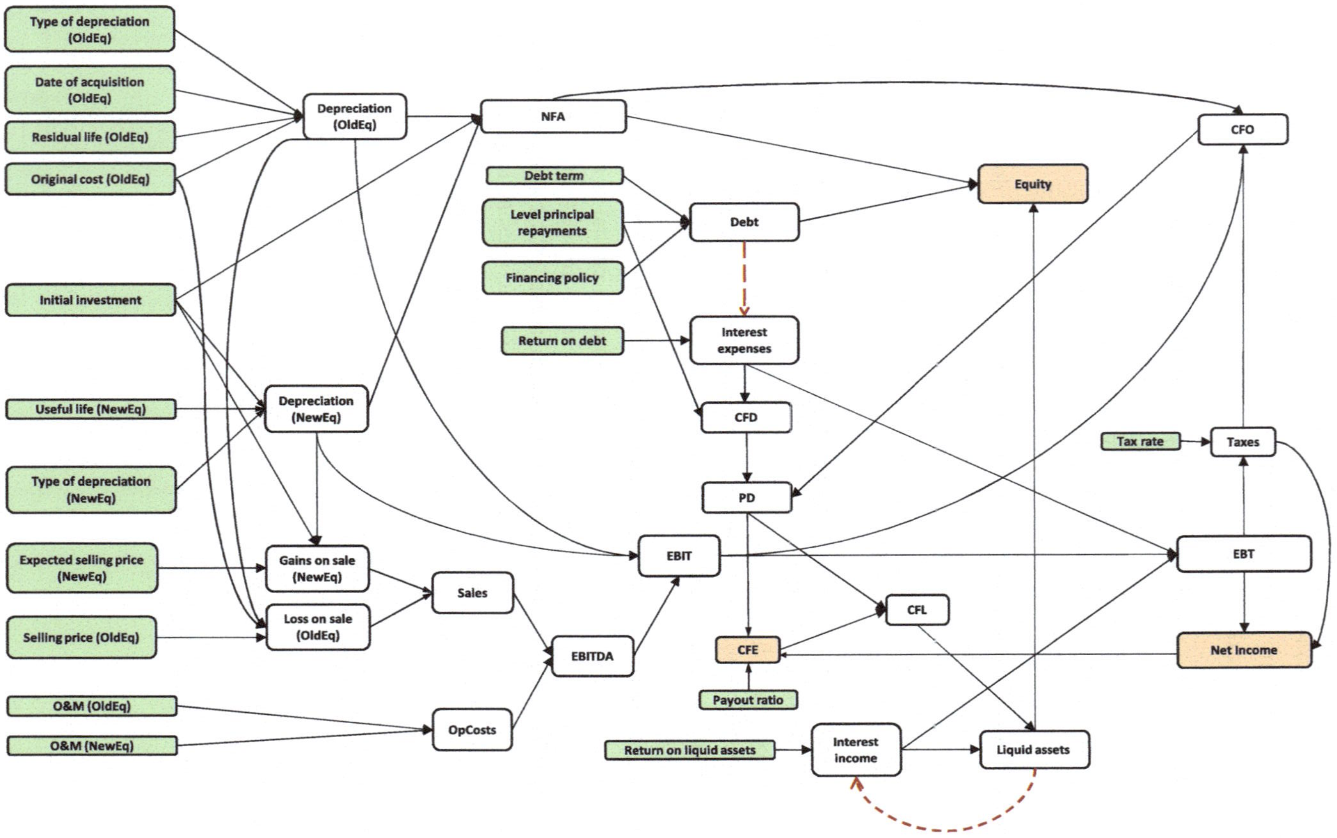

Fig. 12.3 Influence diagram indicating the functional relationships of Bumps Inc. (Replacement). The dashed lines refer to time $t-1$: The balance of liquid asset, C_{t-1}^{l}, along with i_t^l, affects the interest income as of at time t; the outstanding debt, C_{t-1}^{d}, along with i_t^d, affects the interest expenses as of time t

Time	0	1	2	3	4	5	6	7	TOTAL
Replacement									
net income ($I^{e,\mathrm{rep}}$)	−10,608.3	14,613.7	14,719.6	14,825.9	14,932.5	15,039.5	15,146.7	38,704.3	117,373.9
benchmark net income		16,751.1	15,640.6	14,348.1	12,846.2	11,103.6	9,084.0	6,745.9	86,519.5
equity capital	44,500.0	52,822.4	61,241.0	69,756.2	78,368.3	87,077.8	95,884.9	0.0	489,650.6
Overhaul									
net income ($I^{e,\mathrm{ovh}}$)	0.0	8,102.2	8,674.8	9,090.6	9,394.4	9,618.3	9,785.0	9,275.5	63,940.8
benchmark net income		6,543.9	5,919.9	5,238.0	4,478.4	3,621.5	2,646.8	1,532.7	29,981.3
equity capital	0.0	4,799.4	10,256.0	16,189.8	22,472.4	29,011.8	35,742.7	0.0	118,472.1

With the undertaking of the replacement project, it is as if shareholders invested \$118,472.1 in the overhaul and \$371,178.5 in an asset earning an income of \$53,433.1:

$$371{,}178.5 = 489{,}650.6 - 118{,}472.1 = C^{e,\,\text{rep}} - C^{e,\,\text{ovh}}$$
$$53{,}433.1 = 117{,}373.9 - 63{,}940.8 \;\; = I^{e,\,\text{rep}} - I^{e,\,\text{ovh}}.$$

Therefore, the incremental equity rate of return is $\bar{\jmath}^{e,\text{rep}-\text{ovh}} = 53{,}433.1/371{,}178.5 = 14.40\%$ (see Eq. (11.6)). However, in such a way, shareholders renounce to earn a total \$56,538.2 = \$86,519.5 − \$29,981.3 or, in relative terms, give up a rate of return of $\bar{\rho}^{e,\text{rep}-\text{ovh}} = 56{,}538.2/371{,}178.5 = 15.23\%$.

Incremental amounts and resultant rates (Equity perspective)	
incremental scale	\$371,178.5
incremental return	\$53,433.1
incremental benchmark return	\$56,538.2
incremental average ROE	14.40%
incremental MARR	15.23%
incremental efficiency	−0.84%
differential NPV	−\$3,105.1

The economic efficiency of this differential investment is negative and equal to $\epsilon^{e,\,\text{rep}-\text{ovh}} = -0.84\%$; multiplying the latter by the incremental scale, the NPV of the incremental asset is found.[4]

Equity investment	Scale ×	Economic efficiency =	NPV^e	(%)
Overhaul	118,472.1 ×	28.66% =	33,959.5	100.00%
+ Incremental investment	371,178.5 ×	(−0.84%) =	−3,105.1	−9.14%
= Replacement	489,650.6 ×	6.30% =	30,854.4	90.86%

For shareholders, to finance the replacement is financially equivalent to investing in the overhaul and investing additional capital in an economically inefficient asset with a size about three times greater than the overhaul's size. The relatively small loss in value (−\$3,105.1) is due to the fact that, despite the large scale, the economic inefficiency of the incremental investment is low. The net effect is that replacement only subtracts about 9% of the value created by overhauling.

[4]Numbers are rounded, as usual. With more precision, the incremental size of equity is \$371,178.458 and the incremental efficiency is −0.83655%, so that $-0.83655\% \cdot 371{,}178.458 = -3{,}105.1$.

12.2 HomeNet Project (AIRR and RI)

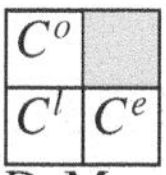

This example is inspired by the HomeNet example presented in Berk and DeMarzo (2014, Chap. 8), suitably adjusted for our purposes.

A company is considering the development of a wireless home networking appliance, called HomeNet, consisting of a 'smart' (hardware and software) system which will enable running the entire home (PCs, heating system, air-conditioning, stereos, telephones, office equipments, etc.) from any Internet connection.

Project input data. For the hardware, total engineering costs and design costs will amount to \$5 million; for the software, 50 software engineers will be working on it in year 0 for one year at a cost of \$220,000 per capita.

HomeNet will be supplied by a manufacturer, and the cost of any unit to the company will be \$115, while the wholesale price will be \$260, which is expected to grow at 5% from year 1 to year 2, 10% from year 2 to year 3, 15% from year 3 to year 4. No product will be sold after the end of year 4. At the end of year 0, the design will be completed and the firm will be ready to start deliveries to customers. Selling and shipping costs in year 1 to 4 will amount to \$2.8 million, 70% of which are salaries (paid in the year where they occur) and 30% are cash expenses for delivery of HomeNet.

Expected sales, on a per year basis, are 100,000 units.

\$8.5 million will be spent in lab equipment, in order to verify compatibility of new Internet-ready appliances with HomeNet system. In order to provide services for existing customers, the equipment will be in use for five years and it will be depreciated on a straight-line basis.

In case of project undertaking HomeNet's new lab will be housed in a rental property owned by the company which is currently rented for \$200,000 a year. If the project is undertaken, the property will be unavailable for rent until the end of year 4. At that date, the property will be rented again for the same amount. The property will keep its value unvaried.

HomeNet will partially cannibalize an existing router sold by the company, for approximately 25% of HomeNet's first-year sales (the router's average price is \$100 and the related operating cost is \$55).

No inventory will be needed (HomeNet will be directly shipped from manufacturer to retail stores). Receivables are expected to account for 40% of HomeNet annual sales while payables are estimated at 25% of HomeNet annual COGS. (We assume the router is sold on a cash-basis only, so its suspension has no effect on receivables and payables.)

The payout ratio is 80% (20% of potential dividends will be invested in cash and cash equivalents). The interest rate on liquid assets is 1%. The project will be equity financed.

The tax rate is 40%.

Market input data. The HomeNet project has the same risk as the firm and the firm's operating (pre-tax) WACC is 21%. The COC of the non-operating assets (including property as well as cash and marketable securities) is 5%.

HomeNet Project – Input Data

Project data	
HomeNet	
Units sold	100,000 (year 1 to 4)
Wholesale price in year 1	\$260
Price increase	5% (year 2), 10% (year 3), 15% (year 4)
Unit cost	\$115
Product life	4 years
Selling and shipping costs	\$2,800,000 (70% salaries, 30% delivery expenses)
R&D	
Hardware engineering and design	\$5,000,000
Software engineers	50
Cost of a software engineer	\$220,000 (annual)
Working time of the software engineering team	1 year
Old product (router)	
Cannibalization rate	25% of fist-year HomeNet sales
Average price	\$100
Unit cost	\$55
Non-operating income	
Lost rent	\$200,000 (year 0 through 4)
Interest rate on liquid assets	1% (constant)
Fixed asset	
Lab equipment	\$8,500,000
Depreciation	straight-line
Useful life	5 years
Inventory	None
Accounts receivable (% HomeNet sales)	40% (year 1 to 4)
Accounts payable (% HomeNet COGS)	25% (year 1 to 4)
Financing	
Equity financing	100% (= C_0^e/C_0)
Debt financing	0% (= C_0^d/C_0)
Payout ratio	$\alpha = 80\%$ (from time 1 to time 4)
Tax rate	40%
Market data	
Operating pre-tax WACC	$r^o = 21\%$ (constant)
Required return on cash and cash equivalents	$r^l = 5\%$

Consider that:

- R&D costs amount to $\$5{,}000{,}000 + 50 \cdot \$220{,}000 = \$16{,}000{,}000$, expensed immediately
- HomeNet sales in the first year are $100{,}000 \cdot \$260 = \$26{,}000{,}000$ and increase at 5%, 10%, 15% in the subsequent years
- lost sales for cannibalization are $25\% \cdot 100{,}000 \cdot \$100 = \$2{,}500{,}000$; related cost savings for lost sales are equal to $25\% \cdot 100{,}000 \cdot \$55 = \$1{,}375{,}000$
- non-operating assets consist of rental property as well as liquid assets. The firm's rental property with or without the project is unvaried, so the incremental rental property is zero. This implies that lost rental income is equal to lost rental payment. Liquid assets are increased by the interest income (at 1%) and by additional deposits (20% of potential dividends) until the end of year 5, when the liquidating terminal CFE takes place, including residual working capital and residual cash
- interest income does not coincide with the return on non-operating assets (i_t^l). The latter is the ratio of the non-operating income (interest income + lost rent) to the book value of non-operating assets (cash + rental property). For example, in the fourth year, the return on non-operating assets is $i_4^l = -175{,}183.4/2{,}481{,}662 = -7.06\%$, which means that non-operating assets generate a loss of \$7.06 for every \$100 invested in the fourth year[5]
- consistently with what stated above, the rate $r^l = 5\%$ is not the required return on liquid assets, but the required return on non-operating assets, which means that it takes account of the risk of the rental income as well as the risk of financial investments
- HomeNet production cost is equal to $\$115 \cdot 100{,}000 = \$11{,}500{,}000$, which is paid by the firm to the manufacturer
- in year 5 the only nonzero income components are the interest income and the depreciation charge, equal to $\$1{,}700{,}000 = \$8{,}500{,}000/5$
- in year 0 and year 5 incremental taxes are negative, thereby representing a positive income component: If the project is undertaken, the company will pay less taxes than it would if the project will not be undertaken
- the project is equity financed ($C_t^d = 0$ for all t). This implies that the overall project RI (and NPV) is equal to the overall equity RI (and NPV): $V_0(\text{RI}) = V_0^e(\text{RI}) = \text{NPV} = \text{NPV}^e$
- the project is positively levered at time 1 ($C_1^{\mathcal{D}} = -C_1^l > 0$) and negatively levered from time 2 to time 4 ($C_t^{\mathcal{D}} = -C_t^l < 0$ for $t \in \mathbb{N}_4^2$)
- the incremental invested capital in year 0 is $C_0 = \$8{,}500{,}000$. However, the incremental cash outflow is $|F_0| = \$18{,}220{,}000$. This is caused by (i) the research and development costs sustained in year 0, (ii) the lost rental income, and (iii) the avoided taxes: $I_0 = -\$16{,}000{,}000 - \$200{,}000 + \$6{,}480{,}000 = -\$9{,}720{,}000$, entirely covered by equity

[5]The superscript "l" refers to non-operating assets. In this case, the non-operating assets include liquid assets and other non-operating assets, namely, the rental property. The latter has a higher degree of liquidity than the HomeNet's operations but a lower degree of liquidity than cash and marketable securities. Hence, C_t^l expresses the sum of liquid assets and rental property, I_t^l is the sum of interest income and rental income, F_t^l is the sum of withdrawal/deposit from/into liquid assets and (lost) rental payments.

– the cash which is retained in liquid assets is calculated as a residue of the whole non-operating cash flow. Precisely, the CFL (non-operating cash flow) is equal to $F_t^l = -(1-0.8)\cdot F_t^{pd}$. The lost cash flow from rental property is $F_t^{\text{other non-op}} = -\$200{,}000$. Therefore, using Eq. (4.84), the cash retained in liquid assets at time t is

$$-F_t^{\text{liquid}} = (1-0.8)\cdot F_t^{pd} - 200{,}000.$$

The balance of liquid account is then computed as

$$C_t^{\text{liquid}} = C_{t-1}^{\text{liquid}} + I_t^{\text{liquid}} - F_t^{\text{liquid}}$$

up to year 4 (see also Remark 4.8 on this issue)

– the project's IRR is $\sigma = 19.92\%$. Note that the initial internal capital is $C_0(\sigma) = \$18{,}220{,}000 = -F_0$, which is not the invested capital, as seen above. Therefore, the subsequent internal capital amounts are concocted and 19.92% does not measure the correct return rate.

As for the AIRR, the average ROE and the average ROI are equal, given that the project is equity financed, as well as the respective MARRs. The intrinsic values of operations and non-operating assets are easily found by discounting the respective cash flows at $r^o = 21\%$ and $r^l = 5\%$. The mean of the rates, weighted by the respective intrinsic values, supplies the asset pre-tax WACC: $r_t = (r^o V_{t-1}^o + r^l V_{t-1}^l)/(V_{t-1}^o + V_{t-1}^l)$. In particular, $\boldsymbol{r} = (21.85\%, 21.79\%, 20.74\%, 19.24\%, 16.11\%)$. This enables calculating the average ROI, $\bar{\imath}_0 = 7{,}279{,}729.2/45{,}895{,}847.3 = 15.86\%$. Then, after calculating the benchmark returns, one finds the project MARR: $\bar{r}_0 = 7{,}680{,}037.9/45{,}895{,}847.3 = 16.73\%$ (see related table on Sect. 12.3). The economic efficiency is -0.87%. This means that the capital has been invested in a (slightly) inefficient way. The size of the investment is notable: \$45,895,847.3. Applied to the economic efficiency, it brings about a shareholders wealth decrease equal to $-\$400{,}308.7 = \$45{,}895{,}847.3\cdot(-0.87\%)$.[6] The negative result is mainly due to a ponderous scale effect.

The same negative net value would result by discounting RIs at the pre-tax WACC, r_t. Positive RIs arise from year 1 to year 4, whereas RIs are negative in year 0 and year 5. In particular, the R&D costs in year 0 are substantial and the later positive RIs are not sufficient to turn the project into an economically profitable system.

The traditional NPV analysis hiddens the relevant information that inefficiency is very small and scale plays a major role in making the project unfavorable. This implies that the project might turn out to be worth undertaking if some efficiency parameters could be adjusted. For example, it may be seen that cannibalization plays a major role in depressing the project value. If the two products were more differentiated (so that no cannibalization would occur), the financial efficiency would increase to 2.71%. With a scale of \$46,745,293, the NPV would rise to \$1,267,020. Even a reduction

[6]Error is due to rounding, as usual. With more digits, $\bar{\imath}_0 = 15.8614115\%$, $\bar{r}_0 = 16.7336226\%$, and $PV[C] = \$45{,}895{,}847.3461908$ so that $\xi = -0.8722111\%$ and shareholder's value destroyed is $-0.8722111\%\cdot 45{,}895{,}847.3461908 = 400{,}308.7$.

of only 7 percentage points in the cannibalization rate would turn the project into a favorable one.

The economic value created is also sensitive to the wholesale price. If the firm could increase the expected price by only \$5 (other things unvaried), the project would turn into a favorable one, with a 0.71% efficiency and an NPV equal to \$330,384 (assuming units sold would not be affected).

A different cash cycle also impacts on value. If the firm were able to reduce the cash cycle and halve the AR in every period (other things unvaried), the project would turn out to be substantially profitable with a 5.71% efficiency and an NPV equal to \$1,970,344 (similar but less substantial result in case of a doubling of AP).

On the contrary, a change in the payout ratio (other things unvaried) would not contribute to turn the value-destroying project into a value-creating one; however, a 100% payout policy would minimize the loss, bringing a financial inefficiency of only −0.18% and an NPV of −\$74,068, as opposed to a greater inefficiency and larger absolute loss assuming a 0% payout ($\xi = -2.56\%$ and NPV $= -\$1,702,278$).

Further, information about RIs is relevant because it shows that the major problems occur in the initial and in the last period:

- In the initial period, the RI is highly negative, owing to the high R&D expenditures. The difference is not sufficiently compensated in the following periods by the increased revenues. However, the firm's managers might investigate the consequence of a smaller commitment in the project, either reducing the R&D expenditures or reducing the number of engineers involved. A deeper analysis might be required to verify whether the number of engineers has been overestimated. If, for example, the same project could be undertaken with 45 engineers instead of 50, the project would be profitable, other things unvaried ($\xi = 0.57\%$, NPV $= 259,691$)
- In the last period, the last depreciation charge has a high impact and managers might wonder whether a different depreciation schedule (e.g., accelerated depreciation) is possible which may have a more favorable effect or, alternatively, whether an equipment with the same characteristics may be purchased from a different supplier at a smaller cost, thereby reducing depreciation charges.

However, the change in a parameter might imply a change in some other parameters (e.g., a reduction in the cannibalization rate may be associated with increasing costs for differentiating the product or with higher marketing expenses), so a deep sensitivity analysis is recommendable to extract more information by changing more interdependent variables simultaneously (see Saltelli et al. 2004 and Borgonovo 2017 for detailed guides to sensitivity analysis).

A detailed split-screen Matrix, splitting up the equity components in its basic constituents, may be drawn up from the input data and may be used for deriving more precise information about accounting and financial relationships. It also acts as a further diagnostic tool, in addition to the concise transposed Matrix, for spotting modeling errors and inconsistencies. We focus on the Matrix as of time $t = 1$ (see next page).

To build the Matrix, we have used the information that 70% of the selling costs is labor cost (salaries), so that labor cost at time $t = 1$ is $0.7 \cdot 2{,}800{,}000 = 1{,}960{,}000$,

$t = 1$	Cash flow	Income	$-\Delta$Capital
Accounts receivable	13,100,000	23,500,000	−10,400,000
(Accounts payable)	−8,090,000	−10,965,000	2,875,000
Inventory	+0	+0	−0
(Salaries & Wages payable)	−1,960,000	−1,960,000	−0
Net Fixed Assets	+0	−1,700,000	+1,700,000
Rental property	−200,000	−200,000	0
(Taxes payable)	−3,470,000	−3,470,000	0
(Debt)	−0	−0	+0
Cash	+284,000	+0	+284,000
Equity	−336,000	5,205,000	−5,541,000

which equates the cash flow to employees. The cost of supplies is obtained by summing the COGS and 30% of the selling expenses. The cash from customers is $F_1^c = 13{,}100{,}00$. Cash to suppliers for purchases is equal to $F_1^{cop} = 8{,}090{,}000$. The CFO may be calculated in four different ways. Using cash flows,

$$F_1^o = F_1^c - F_1^{cop} - F_1^{swp} - F_1^T = 13{,}100 - 8{,}090 - 1{,}960 - 3{,}470 = -420.$$

(in thousands). The CFO is negative and must be covered. In addition, some cash is lost from rental property, by an amount of \$200.[7] This makes a deficit of $420 + 200 = 620$. The firm covers this deficit via cash account (\$284) and equity (\$336). The law of conservation for cash flows is fulfilled:

$$420 + 200 = 620 = 284 + 336.$$

Reminding that the project is intrinsically incremental in nature (see Sect. 1.4), the negative CFL ($F_1^l = -\$284$) does not necessarily mean that the firm withdraws cash from the non-operating assets: It means that cash flow from non-operating assets of the firm-with-the-project will be smaller than cash flow from non-operating assets of the firm-without-the-project. In particular, it will be smaller by \$200 (rent) plus \$84 (cash). For example, if the CFL of the firm-without-the-project will be \$X, then the firm-with-the-project's CFL will be $\$X - \$284$. Likewise, the negative CFE ($F_t^e = -\$336$) does not necessarily mean that the firm issues new shares: It means that the cash flow to equityholders of the firm-with-the-project will be smaller than the cash flow to equityholders of the firm-without-the-project. For example, if the CFE of the firm-without-the-project is \$Y, then the firm-with-the-project's CFE will be $\$Y - \$336$.

Note that, summing by row and by column, the static and dynamic equilibrium is fulfilled, so there is no modeling error. (The whole *split-screen film* of HomeNet is described in Fig. 12.4. The "*OK*" rows and columns check fulfilment of laws of motion and laws of conservation, thereby acting as diagnostic tools for spotting errors and inconsistencies.)

[7]Book value of rental property is the same, regardless of whether the project is undertaken or not, so the lost rental income is, at the same time, a lost cash flow.

HomeNet Project – Balance Sheets

Year	0	1	2	3	4	5
Net Fixed Assets	8,500,000.0	6,800,000.0	5,100,000.0	3,400,000.0	1,700,000.0	0.0
Lab equipment	8,500,000.0	6,800,000.0	5,100,000.0	3,400,000.0	1,700,000.0	0.0
Net Operating WC	0.0	7,525,000.0	8,045,000.0	9,137,000.0	10,938,800.0	0.0
Inventory	0.0	0.0	0.0	0.0	0.0	0.0
+Accounts receivable	0.0	+10,400,000.0	+10,920,000.0	+12,012,000.0	+13,813,800.0	0.0
−Accounts payable	0.0	−2,875,000.0	−2,875,000.0	−2,875,000.0	−2,875,000.0	0.0
Non-operating assets	0.0	−284,000.0	986,387.2	2,481,662.0	4,389,273.3	0.0
Rental property	0.0	0.0	0.0	0.0	0.0	0.0
+Liquid assets	0.0	−284,000.0	986,387.2	2,481,662.0	4,389,273.3	0.0
Investments	8,500,000.0	14,041,000.0	14,131,387.2	15,018,662.0	17,028,073.3	0.0
Debt	0.0	0.0	0.0	0.0	0.0	0.0
Equity	8,500,000.0	14,041,000.0	14,131,387.2	15,018,662.0	17,028,073.3	0.0
Financings	8,500,000.0	14,041,000.0	14,131,387.2	15,018,662.0	17,028,073.3	0.0

HomeNet Project – Income Statements						
Year	0	1	2	3	4	5
Sales		23,500,000.0	24,800,000.0	27,530,000.0	32,034,500.0	0.0
HomeNet sales		26,000,000.0	27,300,000.0	30,030,000.0	34,534,500.0	0.0
–lost sales		−2,500,000.0	−2,500,000.0	−2,500,000.0	−2,500,000.0	0.0
−COGS		−10,125,000.0	−10,125,000.0	−10,125,000.0	−10,125,000.0	0.0
–HomeNet production cost		−11,500,000.0	−11,500,000.0	−11,500,000.0	−11,500,000.0	0.0
+cost savings		+1,375,000.0	+1,375,000.0	+1,375,000.0	+1,375,000.0	0.0
Gross Profit	0.0	13,375,000.0	14,675,000.0	17,405,000.0	21,909,500.0	0.0
−SGA	−16,000,000.0	−2,800,000.0	−2,800,000.0	−2,800,000.0	−2,800,000.0	0.0
–selling and shipping expenses		−2,800,000.0	−2,800,000.0	−2,800,000.0	−2,800,000.0	0.0
–R&D costs	−16,000,000.0	0.0	0.0	0.0	0.0	0.0
EBITDA	−16,000,000.0	10,575,000.0	11,875,000.0	14,605,000.0	19,109,500.0	0.0
−Depreciation		−1,700,000.0	−1,700,000.0	−1,700,000.0	−1,700,000.0	−1,700,000.0
EBIT	−16,000,000.0	8,875,000.0	10,175,000.0	12,905,000.0	17,409,500.0	−1,700,000.0
−Interest expense		0.0	0.0	0.0	0.0	0.0
+Non-operating income	−200,000.0	−200,000.0	−202,840.0	−190,136.1	−175,183.4	+43,892.7
–lost rental income	−200,000.0	−200,000.0	−200,000.0	−200,000.0	−200,000.0	0.0
+interest income		0.0	−2,840.0	+9,863.9	+24,816.6	+43,892.7
EBT	−16,200,000.0	8,675,000.0	9,972,160.0	12,714,863.9	17,234,316.6	−1,656,107.3
−Taxes	+6,480,000.0	−3,470,000.0	−3,988,864.0	−5,085,945.5	−6,893,726.6	+662,442.9
Net Income	−9,720,000.0	5,205,000.0	5,983,296.0	7,628,918.3	10,340,590.0	−993,664.4

$(\boldsymbol{C}, \boldsymbol{I}, \boldsymbol{F})$ HomeNet

Year	0	1	2	3	4	5
$\boldsymbol{C}$	8,500,000.0	14,041,000.0	14,131,387.2	15,018,662.0	17,028,073.3	0.0
$\boldsymbol{C}^o$	8,500,000.0	14,325,000.0	13,145,000.0	12,537,000.0	12,638,800.0	0.0
$\boldsymbol{C}^l$	0.0	-284,000.0	986,387.2	2,481,662.0	4,389,273.3	0.0
$\boldsymbol{C}^d$	0.0	0.0	0.0	0.0	0.0	0.0
$\boldsymbol{C}^e$	8,500,000.0	14,041,000.0	14,131,387.2	15,018,662.0	17,028,073.3	0.0
$\boldsymbol{I}$	−9,720,000.0	5,205,000.0	5,983,296.0	7,628,918.3	10,340,590.0	−993,664.4
$\boldsymbol{I}^o$	−9,520,000.0	5,405,000.0	6,186,136.0	7,819,054.5	10,515,773.4	−1,037,557.1
$\boldsymbol{I}^l$	−200,000.0	−200,000.0	−202,840.0	−190,136.1	−175,183.4	43,892.7
$\boldsymbol{I}^d$	0.0	0.0	0.0	0.0	0.0	0.0
$\boldsymbol{I}^e$	−9,720,000.0	5,205,000.0	5,983,296.0	7,628,918.3	10,340,590.0	−993,664.4
$\boldsymbol{F}$	−18,220,000.0	−336,000.0	5,892,908.8	6,741,643.6	8,331,178.7	16,034,408.9
$\boldsymbol{F}^o$	−18,020,000.0	−420,000.0	7,366,136.0	8,427,054.5	10,413,973.4	11,601,242.9
$\boldsymbol{F}^l$	−200,000.0	84,000.0	−1,473,227.2	−1,685,410.9	−2,082,794.7	4,433,166.0
$\boldsymbol{F}^d$	0.0	0.0	0.0	0.0	0.0	0.0
$\boldsymbol{F}^e$	−18,220,000.0	−336,000.0	5,892,908.8	6,741,643.6	8,331,178.7	16,034,408.9

HomeNet – AIRR and RI calculation

Year	0	1	2	3	4	5
Cost of capital (r_t)		21.85%	21.79%	20.74%	19.24%	16.11%
Discount factor (d)	1	0.82065	0.67384	0.55807	0.46801	0.40308
Income (I_t)	−9,720,000.0	5,205,000.0	5,983,296.0	7,628,918.3	10,340,590.0	−993,664.4
Overall income (PV[I])	7,279,729.2					
Overall capital invested (PV[C])	45,895,847.3					
Average ROI ($\bar{\iota}_0$)	15.86%					
Benchmark return (I_t^*)		1,857,672.0	3,059,032.9	2,931,431.5	2,890,245.3	2,742,942.0
Total benchmark return (PV[I^*])	7,680,037.9					
Total capital invested (PV[C])	45,895,847.3					
project MARR ($\bar{r}_0$)	16.73%					
Economic efficiency (ξ)	−0.87%					
Value created PV[C] $\cdot\, \xi$	−400,308.7					
Residual Income (RI)	−9,720,000.0	3,347,328.0	2,924,263.1	4,697,486.8	7,450,344.7	−3,736,606.4
Value of RIs (V_0(RI))	−400,308.7					

	capital	income	-cash flow	capital	
Time	-1	0	0	0	
+Accounts receivable	0	0	C	0	OK
-Accounts payable	0	-5,000,000	5,000,00C	0	OK
+Inventory	0	0	C	0	OK
-S&W payable	0	-11,000,000	11,000,00C	0	OK
+Equipment	0	0	8,500,000	8,500,000	OK
+Cash	0	0	0	0	OK
+Other non-op. assets	0	-200,000	200,000	0	OK
-Taxes	0	6,480,000	-6,480,000	0	OK
-Debt	0	0	0	0	OK
=Equity	0	-9,720,000	18,220,000	8,500,000	OK
	OK	OK	OK	OK	

	capital	income	-cash flow	capital	
Time	0	1	1	1	
+Accounts receivable	0	23,500,000	-13,100,000	10,400,000	OK
-Accounts payable	0	-10,965,000	8,090,000	-2,875,000	OK
+Inventory	0	0	0	0	OK
-S&W payable	0	-1,960,000	1,960,000	0	OK
+Equipment	8,500,000	-1,700,000	0	6,800,000	OK
+Cash	0	0	-284,000	-284,000	OK
+Other non-op. assets	0	-200,000	200,000	0	OK
-Taxes	0	-3,470,000	3,470,000	0	OK
-Debt	0	0	0	0	OK
=Equity	8,500,000	5,205,000	336,000	14,041,000	OK
	OK	OK	OK	OK	

Fig. 12.4 HomeNet *split-screen film*: two framings. In the first six Matrices, the split screen is such that $C_t = C_{t-1} + I_t - F_t$, whereas in the last six Matrices, the split screen is such that $F_t = I_t + C_{t-1} - C_t$

	capital	income	-cash flow	capital	
Time	1	2	2	2	
+Accounts receivable	10,400,000	24,800,000	-24,280,000	10,920,000	OK
-Accounts payable	-2,875,000	-10,965,000	10,965,000	-2,875,000	OK
+Inventory	0	0	0	0	OK
-S&W payable	0	-1,960,000	1,960,000	0	OK
+Equipment	6,800,000	-1,700,000	0	5,100,000	OK
+Cash	-284,000	-2,840	1,273,227	986,387	OK
+Other non-op. assets	0	-200,000	200,000	0	OK
-Taxes	0	-3,988,864	3,988,864	0	OK
-Debt	0	0	0	0	OK
=Equity	14,041,000	5,983,296	-5,892,909	14,131,387	OK
	OK	OK	OK	OK	

	capital	income	-cash flow	capital	
Time	2	3	3	3	
+Accounts receivable	10,920,000	27,530,000	-26,438,000	12,012,000	OK
-Accounts payable	-2,875,000	-10,965,000	10,965,000	-2,875,000	OK
+Inventory	0	0	0	0	OK
-S&W payable	0	-1,960,000	1,960,000	0	OK
+Equipment	5,100,000	-1,700,000	0	3,400,000	OK
+Cash	986,387	9,864	1,485,411	2,481,662	OK
+Other non-op. assets	0	-200,000	200,000	0	OK
-Taxes	0	-5,085,946	5,085,946	0	OK
-Debt	0	0	0	0	OK
=Equity	14,131,387	7,628,918	-6,741,644	15,018,662	OK
	OK	OK	OK	OK	

Fig. 12.4 Continued

	capital	income	-cash flow	capital	
Time	3	4	4	4	
+Accounts receivable	12,012,000	32,034,500	-30,232,700	13,813,800	OK
-Accounts payable	-2,875,000	-10,965,000	10,965,000	-2,875,000	OK
+Inventory	0	0	0	0	OK
-S&W payable	0	-1,960,000	1,960,000	0	OK
+Equipment	3,400,000	-1,700,000	0	1,700,000	OK
+Cash	2,481,662	24,817	1,882,795	4,389,273	OK
+Other non-op. assets	0	-200,000	200,000	0	OK
-Taxes	0	-6,893,727	6,893,727	0	OK
-Debt	0	0	0	0	OK
=Equity	15,018,662	10,340,590	-8,331,179	17,028,073	OK
	OK	OK	OK	OK	

	capital	income	-cash flow	capital	
Time	4	5	5	5	
+Accounts receivable	13,813,800	0	-13,813,800	0	OK
-Accounts payable	-2,875,000	0	2,875,000	0	OK
+Inventory	0	0	0	0	OK
-S&W payable	0	0	0	0	OK
+Equipment	1,700,000	-1,700,000	0	0	OK
+Cash	4,389,273	43,893	-4,433,166	0	OK
+Other non-op. assets	0	0	0	0	OK
-Taxes	0	662,443	-662,443	0	OK
-Debt	0	0	0	0	OK
=Equity	17,028,073	-993,664	-16,034,409	0	OK
	OK	OK	OK	OK	OK

Fig. 12.4 Continued

	cash flow	income	+capital	-capital	
	0	0	-1	0	Time
OK	0	0	0	0	+Accounts receivable
OK	-5,000,000	-5,000,000	0	0	-Accounts payable
OK	0	0	0	0	+Inventory
OK	-11,000,000	-11,000,000	0	0	-S&W payable
OK	-8,500,000	0	0	-8,500,000	+Equipment
OK	0	0	0	0	+Cash
OK	-200,000	-200,000	0	0	+Other non-op. assets
OK	6,480,000	6,480,000	0	0	-Taxes
OK	0	0	0	0	-Debt
OK	-18,220,000	-9,720,000	0	-8,500,000	=Equity
	OK	OK	OK	OK	

→

	cash flow	income	+capital	-capital	
	1	1	0	1	Time
OK	13,100,000	23,500,000	0	-10,400,000	+Accounts receivable
OK	-8,090,000	-10,965,000	0	2,875,000	-Accounts payable
OK	0	0	0	0	+Inventory
OK	-1,960,000	-1,960,000	0	0	-S&W payable
OK	0	-1,700,000	8,500,000	-6,800,000	+Equipment
OK	284,000	0	0	284,000	+Cash
OK	-200,000	-200,000	0	0	+Other non-op. assets
OK	-3,470,000	-3,470,000	0	0	-Taxes
OK	0	0	0	0	-Debt
OK	-336,000	5,205,000	8,500,000	-14,041,000	=Equity
	OK	OK	OK	OK	

Fig. 12.4 Continued

	cash flow	income	+capital	-capital	
	2	2	1	2	Time
OK	24,280,000	24,800,000	10,400,000	-10,920,000	+Accounts receivable
OK	-10,965,000	-10,965,000	-2,875,000	2,875,000	-Accounts payable
OK	0	0	0	0	+Inventory
OK	-1,960,000	-1,960,000	0	0	-S&W payable
OK	0	-1,700,000	6,800,000	-5,100,000	+Equipment
OK	-1,273,227	-2,840	-284,000	-986,387	+Cash
OK	-200,000	-200,000	0	0	+Other non-op. assets
OK	-3,988,864	-3,988,864	0	0	-Taxes
OK	0	0	0	0	-Debt
OK	5,892,909	5,983,296	14,041,000	-14,131,387	=Equity
	OK	OK	OK	OK	

→

	cash flow	income	+capital	-capital	
	3	3	2	3	Time
OK	26,438,000	27,530,000	10,920,000	-12,012,000	+Accounts receivable
OK	-10,965,000	-10,965,000	-2,875,000	2,875,000	-Accounts payable
OK	0	0	0	0	+Inventory
OK	-1,960,000	-1,960,000	0	0	-S&W payable
OK	0	-1,700,000	5,100,000	-3,400,000	+Equipment
OK	-1,485,411	9,864	986,387	-2,481,662	+Cash
OK	-200,000	-200,000	0	0	+Other non-op. assets
OK	-5,085,946	-5,085,946	0	0	-Taxes
OK	0	0	0	0	-Debt
OK	6,741,644	7,628,918	14,131,387	-15,018,662	=Equity
	OK	OK	OK	OK	

Fig. 12.4 Continued

	cash flow	income	+capital	-capital	
	4	4	3	4	Time
OK	30,232,700	32,034,500	12,012,000	-13,813,800	+Accounts receivable
OK	-10,965,000	-10,965,000	-2,875,000	2,875,000	-Accounts payable
OK	0	0	0	0	+Inventory
OK	-1,960,000	-1,960,000	0	0	-S&W payable
OK	0	-1,700,000	3,400,000	-1,700,000	+Equipment
OK	-1,882,795	24,817	2,481,662	-4,389,273	+Cash
OK	-200,000	-200,000	0	0	+Other non-op. assets
OK	-6,893,727	-6,893,727	0	0	-Taxes
OK	0	0	0	0	-Debt
OK	8,331,179	10,340,590	15,018,662	-17,028,073	=Equity
	OK	OK	OK	OK	

	cash flow	income	+capital	-capital	
	5	5	4	5	Time
OK	13,813,800	0	13,813,800	0	+Accounts receivable
OK	-2,875,000	0	-2,875,000	0	-Accounts payable
OK	0	0	0	0	+Inventory
OK	0	0	0	0	-S&W payable
OK	0	-1,700,000	1,700,000	0	+Equipment
OK	4,433,166	43,893	4,389,273	0	+Cash
OK	0	0	0	0	+Other non-op. assets
OK	662,443	662,443	0	0	-Taxes
OK	0	0	0	0	-Debt
OK	16,034,409	-993,664	17,028,073	0	=Equity
	OK	OK	OK	OK	

Fig. 12.4 Continued

12.3 Rainbow Educational Program (AROI and NFV)

C^o C^l C^e

Human Science Inheritage (HSI) is an independent bookstore chain with brick-and-mortar locations. One of them, Rainbow Bookshop, is located in the centre of the city and hosted in a building owned by HSI. Rainbow is a large bookstore: It extends over the 4 floors of the building and is especially dedicated to human sciences, spreading from literature to philosophy and social sciences. Its customers are highly-educated people with demanding bibliographic needs and interests. HSI has a strong social-and-culture-orientation approach, and HSI's management team is currently evaluating a program of cultural qualification for 10 employees. The initiative consists of a partnership with the local university to develop a HSI-approved curriculum of literature, history, social and philosophical studies, spreading over three years. The *Rainbow Educational Program* has the aim of enhancing the cultural background of the participants in the relevant fields. The educational program will last 3 years and will entail tailor-made lectures by scholars and university professors. Lectures and other related activities will be held at the university.

HSI's managers believe that cultural exposure, serious study, and selective interactions with experts will cause participants to acquire above-average knowledge in the relevant fields, as well as specific skills and relevant abilities, which will largely favor the activities of book promotion and handselling, raise the social effectiveness in public relations, increase customer satisfaction and loyalty, and boost the bookstore reputation (as well as augmenting employees' commitment and satisfaction). HSI's management expects that the program will reverberate positively on the sales for the next 7 years.[8] The participants will lose 15% of their working time for lectures, study, and related activities. Ownership and employees agree upon the following absent time recovery plan: 70% of the total lost worktime will be recovered by hiring substitute employees; the remaining 30% will be recuperated by participants with additional openings of the bookstore in the evening.[9] Participants' annual pay will be kept unvaried. Substitute employees will be paid at a rate of \$18 per hour.

The total working hours of an employee is 1,700 a year. Educational costs amount to \$1,300 a year per participant.

HSI's managers consider that the last year's sales (year 0) have been \$1,000,000 and variables costs have been \$700,000. They estimate that the ratio of variables costs to sales will not change in the following 7 years.

If the educational program is not undertaken, sales are expected to fall by an annual 3% in the next three years, then keep stable in the following 4 years. If the project is undertaken, benefits from the educational advances are expected to materialize in year 2, when a 2% annual sales growth with respect to the preceding year is expected.

[8]Educational programs have enduring effects. We assume that the effects of the program will last only few years for illustrative purposes.

[9]This means that participants will lose a net $10.5\% = 15\% \cdot 70\%$ of their worktime. However, considering the whole time spent for both work and educational activities, they will spend $4.5\% = 30\% \cdot 15\%$ more time if the program is undertaken than they would if it is not undertaken.

The same growth is expected in year 3. From year 4 to 7, the annual growth rate is estimated at 3%.

Additional fixed costs will be tied to additional evening openings, amounting to \$1,500 a year for the next three years. Sales are made on cash and purchases are made on credit. Accounts payable will amount to 10% of the variable costs. Inventory will be 50% of sales. Tax rate is 40%. Outflows generated by the program will be financed internally via existing liquid assets; inflows generated by the project will be entirely distributed to equityholders.

The interest rate on liquid assets is 1.5%, equal to its required rate of return. HSI's owners believe that, considering risk, 15% is appropriate as their minimum attractive rate of return.

On the basis of the sales and variable costs estimates,

- if the program is not undertaken, Rainbow's expected sales in year 1 through 7 are

$$(970{,}000,\ 940{,}900,\ 912{,}673,\ 912{,}673,\ 912{,}673,\ 912{,}673,\ 912{,}673)$$

- if the program is undertaken, Rainbow's expected sales in year 1 through 7 are

$$(970{,}000,\ 989{,}400,\ 1{,}009{,}188,\ 1{,}039{,}463.64,\ 1{,}070{,}647.55,\ 1{,}102{,}766.98,\ 1{,}135{,}849.98)$$

- the expected incremental sales are obtained as the difference of the above streams:

$$\boldsymbol{S} = (0,\ 48{,}500,\ 96{,}515,\ 126{,}790.64,\ 157{,}974.55,\ 190{,}093.98,\ 223{,}176.98)$$

- incremental variables costs are 70% of incremental sales:

$$(0, 33{,}950, 67{,}560.5, 88{,}753.45, 110{,}582.18, 133{,}065.78, 156{,}223.89).$$

Consider that

- since the project does not use any debt, the CFAs coincide with the CFEs:

$$\boldsymbol{F} = \boldsymbol{F}^e = (0, 0, 0, 0, 10{,}405, 15{,}636.6, 21{,}024.9, 16{,}192.8).$$

  The project's IRR and the equity IRR do not exist, so it is not possible to say whether, overall, the project is efficient or not (let alone measure the degree of efficiency). It is not even possible to measure the efficiency of the equity capital
- the project is (negatively) levered, so the FCF stream and the CFO stream does not coincide. Respectively,

$$\boldsymbol{F}^{uo} = (0,\ -27{,}978.0,\ -40{,}103.0,\ -31{,}251.8,\ 9{,}803.8,\ 15{,}026.3,\ 20{,}405.6,\ 121{,}912.3)$$
$$\boldsymbol{F}^{o} = (0,\ -27{,}978.0,\ -39{,}935.1,\ -30{,}841.8,\ 10{,}405.0,\ 15{,}636.6,\ 21{,}024.9,\ 122{,}540.9).$$

Rainbow Educational Program – Input data

Project data	
Participants	10
Duration of the program	3 years
Educational expenses	\$1,300 (per year, per person)
Annual worktime	1,700 hours
Lost worktime	15% (of annual worktime)
Worktime of substitute employees	70% (of lost worktime)
Hourly pay	\$18
Sales	
Last year's sales	\$1,000,000
Annual sales growth with no program	−3% (year 1 to 3), 0% (year 4 to 7)
Annual sales growth with program	−3% (year 1), 2% (year 2 to 3), 3% (year 4 to 7)
Incremental variables costs (% of sales)	70%
Incremental costs for evening openings (light, heating, insurance, etc.)	\$1,500 (year 1 to 3)
Working capital	
Inventory	50% (of incremental sales)
Accounts receivable	0% (of incremental sales)
Accounts payable	10% (of variable costs)
Financing policy	internal (liquid assets)
Distribution policy	full payout
Interest rate on liquid assets	1.5% (constant)
Payout ratio	$\alpha_t = 0$ if $\mathrm{PD}_t < 0$; $\alpha_t = 1$ otherwise
Tax rate	40%
Market data	
Cost of equity	$r^e = 15\%$ (constant)
Required return on liquid assets	$r^l = 1.5\%$ (constant)

Rainbow Educational Program – Balance Sheets (financial structure)

Year	0	1	2	3	4	5	6	7
Net Operating WC	0.0	0.0	20,855.0	41,501.5	54,520.0	67,929.1	81,740.4	0.0
Inventory	0.0	0.0	24,250.0	48,257.5	63,395.3	78,987.3	95,047.0	0.0
+Accounts receivable	0.0	0.0	0.0	0.0	0.0	0.0	0.0	0.0
–Accounts payable	0.0	0.0	−3,395.0	−6,756.1	−8,875.3	−11,058.2	−13,306.6	0.0
Liquid assets	0.0	−27,978.0	−68,332.8	−100,199.5	−101,702.5	−103,228.1	−104,776.5	0.0
Investments	0.0	−27,978.0	−47,477.8	−58,698.1	−47,182.6	−35,299.0	−23,036.1	0.0
Debt	0.0	0.0	0.0	0.0	0.0	0.0	0.0	0.0
Equity	0.0	−27,978.0	−47,477.8	−58,698.1	−47,182.6	−35,299.0	−23,036.1	0.0
Financings	0.0	−27,978.0	−47,477.8	−58,698.1	−47,182.6	−35,299.0	−23,036.1	0.0

Rainbow Educational Program – Balance Sheets (capital structure)

Year	0	1	2	3	4	5	6	7
Net Operating WC	0.0	0.0	20,855.0	41,501.5	54,520.0	67,929.1	81,740.4	0.0
Inventory	0.0	0.0	24,250.0	48,257.5	63,395.3	78,987.3	95,047.0	0.0
+Accounts receivable	0.0	0.0	0.0	0.0	0.0	0.0	0.0	0.0
–Accounts payable	0.0	0.0	−3,395.0	−6,756.1	−8,875.3	−11,058.2	−13,306.6	0.0
Operating investments	0.0	0.0	20,855.0	41,501.5	54,520.0	67,929.1	81,740.4	0.0
Net debt	0.0	27,978.0	68,332.8	100,199.5	101,702.5	103,228.1	104,776.5	0.0
Equity	0.0	−27,978.0	−47,477.8	−58,698.1	−47,182.6	−35,299.0	−23,036.1	0.0
Net financings	0.0	0.0	20,855.0	41,501.5	54,520.0	67,929.1	81,740.4	0.0

Rainbow – Income Statements

Year	0	1	2	3	4	5	6	7
Sales		0.0	48,500.0	96,515.0	126,790.6	157,974.5	190,094.0	223,177.0
−Variable costs		0.0	−33,950.0	−67,560.5	−88,753.4	−110,582.2	−133,065.8	−156,223.9
Contribution margin		0.0	14,550.0	28,954.5	38,037.2	47,392.4	57,028.2	66,953.1
−Fixed costs		−46,630.0	−46,630.0	−46,630.0	0.0	0.0	0.0	0.0
–Educational costs		−13,000.0	−13,000.0	−13,000.0	0.0	0.0	0.0	0.0
–Substitute employees		−32,130.0	−32,130.0	−32,130.0	0.0	0.0	0.0	0.0
–Evening openings		−1,500	−1,500	−1,500	0.0	0.0	0.0	0.0
EBITDA=EBIT		−46,630.0	−32,080.0	−17,675.5	38,037.2	47,392.4	57,028.2	66,953.1
−Interest expense		0.0	0.0	0.0	0.0	0.0	0.0	0.0
+Interest income		0.0	−419.7	−1,025.0	−1,503.0	−1,525.5	−1,548.4	−1,571.6
Earnings before taxes		−46,630.0	−32,499.7	−18,700.5	36,534.2	45,866.8	55,479.8	65,381.4
−Taxes		+18,652.0	+12,999.9	+7,480.2	−14,613.7	−18,346.7	−22,191.9	−26,152.6
Net Income		−27,978.0	−19,499.8	−11,220.3	21,920.5	27,520.1	33,287.9	39,228.9

Potential dividends, cash flow to equity and financing/distribution policy								
Year	0	1	2	3	4	5	6	7
Payout ratio[(a)]		0%	0%	0%	100%	100%	100%	
CFO	0.0	–27,978.0	–39,935.1	–30,841.8	10,405.0	15,636.6	21,024.9	122,540.9
−CFD	0.0	0.0	0.0	0.0	0.0	0.0	0.0	0.0
PD	0.0	–27,978.0	–39,935.1	–30,841.8	10,405.0	15,636.6	21,024.9	122,540.9
−CFE	0.0	0.0	0.0	0.0	–10,405.0	–15,636.6	–21,024.9	–16,192.8
CFL	0.0	27,978.0	39,935.1	30,841.8	0.0	0.0	0.0	-106,348.1

[a]If PDs are negative, $\alpha_t = 0\%$; otherwise, $\alpha_t = 100\%$

They differ by the tax shield, which only depends on interest income (interest expense is zero). More precisely, $F_t^{uo} = F_t^o - \tau I_t^{\mathcal{D}}$ where $\tau I_t^{\mathcal{D}} = -\tau I_t^l$. The CFO is equal to the PD in every period; hence, it represents the cash available for distribution to shareholders. The FCF is not cash available for distribution; it would be the cash available for distribution if the project were unlevered ($C_t^d = C_t^l = 0$ for every t)

- since CFO is negative in the first three periods, it represents the cash that needs to be funded, rather than available cash. The funds are raised internally, from liquid assets
- in the first three periods, the payout ratio is a pay-in ratio. Since PD is negative, α_t represents the degree at which the firm borrows from shareholders. Since HSI's managers set $\alpha_t = 0$, no negative CFA/CFE occurs (funds are raised from liquid assets)
- HSI's managers will distribute cash flows only if the project generates positive cash flows. In the first three years, the CFO is expected to be negative and equal to the PDs, so owners of HSI will not be required to make any contribution: Any financial need will be fulfilled with internal financing. This means that CFLs will be equal to PDs in absolute value and changed in sign, and the CFEs will be zero. From year 4 to 6, $\text{PD}_t > 0$ and, therefore, $\alpha_t = 100\%$. In year 7 (end of project) everything left is distributed, namely, the EOP operating capital,

$$E_7^o = C_6^o + I_7^o = 81{,}740.4 + 40{,}800.5 = 122{,}540.9$$

plus the EOP balance of liquid assets,

$$E_7^l = C_6^l + I_7^l = -104{,}776.5 - 1{,}571.6 = -106{,}348.1,$$

whence $F_7^e = E_7^e = 122{,}540.9 - 106{,}348.1 = 16{,}192.8$

- the IRR of the CFO stream and FCF stream are 12.88% and 12.33%, respectively. They are investment IRRs (the respective internal capitals are positive). However, they do not say anything about the marginal efficiency of the equity capital
- the project is a financing project ($C_t^e = C_t < 0$ for every period). HSI's managers subtract funds from liquid assets (and, partially, from suppliers) to invest in education (and working capital). Education may be considered an intangible asset (see also the related BS under the capital-structure perspective)
- Rainbow Educational Program's IS follows a cost-volume-profit perspective: Costs are divided into variable and fixed. Fixed assets depreciation is unaffected, so EBITDA = EBIT (see Sect. 3.9)
- the project is equity-invested, that is, the equity is negative in every period. It means that equityholders are in a borrowing position with respect to the firm. In turn, the firm is in a borrowing position with respect to the project. The equity rate of return is then a borrowing rate for HSI's shareholders
- The AROI (= internal average ROI) coincides with the internal average ROE, since $C_t^e = C_t$ for all t. To compute it, one divides the total income by the total capital: $\bar{\jmath} = \bar{\jmath}^e = 63{,}259.25/(-239{,}671.58) = -26.39\%$. It is a financing rate, since the

total capital is negative, and it is negative, meaning that shareholders make money out of a borrowing (see Sect. 1.3); out of \$100 borrowed, shareholders earn \$26.39. The MARR is obtained from the benchmark capital, $\mathcal{C}_t = \mathcal{C}_{t-1}(1+0.15) - F_t^{\mathcal{C}}$, $\mathcal{C}_0 = -F_0$, and the benchmark income $I_t^{\mathcal{C}} = 0.15\mathcal{C}_{t-1}$, whence

$$\mathcal{C} = (0,\ 0,\ 0,\ 0,\ -10{,}405,\ -27{,}602.3,\ -52{,}767.6,\ -76{,}875.5, 0)$$

and

$$\boldsymbol{I}^{\mathcal{C}} = (0,\ 0,\ 0,\ 0,\ 0,\ -1{,}560.7,\ -4{,}140.3,\ -7{,}915.1).$$

This results in $\bar{\varrho} = -13{,}616.23/(-239{,}671.58) = 5.68\%$. This means that, if HSI's shareholders borrowed the same amount in the market, they would have to pay a 5.68% return to lenders. The efficiency of the borrowing project is equal to $\varepsilon = -32.08\%$: Shareholders pay borrowing costs at a rate which is 32.08% percentage points smaller than the market borrowing rate. The value added by the educational program after 7 years is then $-239{,}671.58 \cdot (-32.08\%) = 76{,}875.48$ (numbers rounded, as usual). The project is economically profitable and is worth undertaking. The AROI may also be calculated as a weighted average of operating rate of return, $\bar{\jmath}^o$, and non-operating rate of return, $\bar{\jmath}^l$:

$$\bar{\jmath}^o = \frac{\overbrace{70{,}852.5}^{\text{total operating income}}}{\underbrace{266{,}545.9}_{\text{total operating assets}}} = 26.58\%, \qquad \bar{\jmath}^l = i^l = 1.5\%$$

$$\Longrightarrow \bar{\jmath} = \frac{0.266 \cdot 266{,}545.9 + 0.015 \cdot (-506{,}217.5)}{266{,}545.9 - 506{,}217.5} = -0.2639$$

Analogously, the MARR may be calculated as a weighted average of $\bar{\varrho}^o$ and $\bar{\varrho}^l$ (we leave the calculations to the reader).

(C, I, F) Rainbow Educational Program

Year	0	1	2	3	4	5	6	7
C	0.0	−27,978.0	−47,477.8	−58,698.1	−47,182.6	−35,299.0	−23,036.1	0.0
C^o	0.0	0.0	20,855.0	41,501.5	54,520.0	67,929.1	81,740.4	0.0
C^l	0.0	−27,978.0	−68,332.8	−100,199.5	−101,702.5	−103,228.1	−104,776.5	0.0
C^d	0.0	0.0	0.0	0.0	0.0	0.0	0.0	0.0
C^e	0.0	−27,978.0	−47,477.8	−58,698.1	−47,182.6	−35,299.0	−23,036.1	0.0
I	0.0	−27,978.0	−19,499.8	−11,220.3	21,920.5	27,520.1	33,287.9	39,228.9
I^o	0.0	−27, 978.0	−19, 080.1	−10, 195.3	23, 423.5	29, 045.6	34, 836.3	40, 800.5
I^l	0.0	0.0	−419.7	−1,025.0	−1,503.0	−1,525.5	−1,548.4	−1,571.6
I^d	0.0	0.0	0.0	0.0	0.0	0.0	0.0	0.0
I^e	0.0	−27,978.0	−19,499.8	−11,220.3	21,920.5	27,520.1	33,287.9	39,228.9
F	0.0	0.0	0.0	0.0	10, 405.0	15, 636.6	21, 024.9	16,192.8
F^o	0.0	−27, 978.0	−39, 935.1	−30, 841.8	10, 405.0	15, 636.6	21, 024.9	122, 540.9
F^l	0.0	27,978.0	39,935.1	30,841.8	0.00	0.00	0.00	−106,348.15
F^d	0.0	0.0	0.0	0.0	0.0	0.0	0.0	0.0
F^e	0.0	0.0	0.0	0.0	10,405.0	15,636.6	21,024.9	16,192.8

Rainbow Educational Program – AROI and VA/NFV calculation								
Year	0	1	2	3	4	5	6	7
Total income (I)	63,259.25							
Total capital (C)	−239,671.58							
Internal average ROI ($\bar{j}$)	−26.39%							
Benchmark capital ($\mathcal{C}_t$)	0	0	0	−10,405.0	−27,602.3	−52,767.6	−76,875.5	0
Benchmark income ($I_t^{\mathcal{C}}$)	0	0	0	0.0	−1,560.7	−4,140.3	−7,915.1	0
Total benchmark return ($I^{\mathcal{C}}$)	−13,616.2							
Total capital (C)	−239,671.58							
project MARR ($\bar{\varrho}$)	5.68%							
Economic efficiency (ε)	−32.08%							
Value added ($C \cdot \varepsilon$)	76,875.48							

Following is a third representation of Rainbow Educational Program's financial position, which makes use of positive numbers. Shifting the equity row in the investment side and accounts payable and liquid assets in the financing side, one gets a clear picture of what constitutes the source of funds and what constitute the use of funds. As can be gleaned from the statement below, HSI's managers subtract funds from liquid assets and from suppliers (debtholders are not involved). The funds raised are used for investments in inventory and equity. For example, in year 4, suppliers provide \$8,875.3 and liquid assets provide \$101,702.5 for a total of \$110,577.89, part of which is invested in inventory (\$63,395.3) and the remaining part is invested in equity (\$47,182.6). Overall, the investment in equity by the firm has a negative rate of return of −26.39%, which is also the rate at which equityholders borrow from the firm. As seen, this is highly efficient for shareholders, since a financing in the market would cost much more. (If the firm's shares were traded in a normal market, then the equityholders' wealth would increase by $76{,}875.5/1.15^7 = 28{,}900 = \text{NPV}^e$.)

Rainbow Educational Program – Statement of financial position with positive numbers

Year	0	1	2	3	4	5	6	7
Inventory	0.0	0.0	24,250.0	48,257.5	63,395.3	78,987.3	95,047.0	0.0
Accounts receivable	0.0	0.0	0.0	0.0	0.0	0.0	0.0	0.0
Equity	0.0	27,978.0	47,477.8	58,698.1	47,182.6	35,299.0	23,036.1	0.0
Investments	0.0	27,978.0	71,727.8	106,955.6	110,577.9	114,286.3	118,083.1	0.0
Accounts payable	0.0	0.0	3,395.0	6,756.1	8,875.3	11,058.2	13,306.6	0.0
Debt	0.0	0.0	0.0	0.0	0.0	0.0	0.0	0.0
Liquid assets	0.0	27,978.0	68,332.8	100,199.5	101,702.5	103,228.1	104,776.5	0.0
Financings	0.0	27,978.0	71,727.8	106,955.6	110,577.9	114,286.3	118,083.1	0.0

References

Abel, A. B. (1990). Consumption and investment. In B. M. Friedman & F. H. Hahn (Eds.), *Handbook of monetary economics*. Amsterdam: North-Holland.

Acquah, J. F., & Dahal, R. (2018). ROSCAs as lenders of last resort after financial crises: Lessons from Indonesia. *Journal of International Development*, *30*(7), 1223–1239.

Ahn, D., Kang, W., Kim, K.-K., & Shin, H. (2017). Analysis and design of microfinance services: A case of ROSCA. *The Engineering Economist*, *62*(3), 197–230.

Alchian, A. A. (1955). The rate of interest, Fisher's rate of return over cost and Keynes' internal rate of return. *American Economic Review*, *45*, 938–943.

Altshuler, D., & Magni, C. A. (2012). Why IRR is not the rate of return on your investment: Introducing the AIRR to the Real Estate community. *Journal of Real Estate Portfolio Management*, *18*(2), 219–230.

Althsuler, D., & Magni, C. A. (2015). Introducing aggregate return on investment as a solution to the contradiction between some PME metrics and IRR. *Journal of Performance Measurement*, *20*(1), 48–56.

Amey, L. (1969). *The efficiency of business enterprises*. London: George Allen and Unwin.

Amey, L. (1975). Tomkins on 'Residual income'. *Journal of Business Finance and Accounting*, *2*(1), 55–68.

Anthony, R. N. (1975). *Accounting for the cost of interest*. Lexington: D. C. Heath and Company.

Archer, S. H., & D'Ambrosio, A. A. (1972). *Business finance. Theory and management* (2nd ed.). New York: MacMillan Company.

Ardener, S. (1964). The comparative study of rotating credit associations. *The Journal of the Royal Anthropological Institute of Great Britain and Ireland*, *94*(2), 201–229.

Arditti, F. D., & Levy, H. (1977). The weighted average cost of capital as a cutoff rate: A critical analysis of the classical textbook weighted average. *Financial Management*, *6*(3), 24–34.

Arnold, G. (2000). Tracing the development of value-based management. In G. Arnold & M. Davies (Eds.), *Value-based management: Context and application*. Chichester: Wiley.

Arnold, G. (2005). *Corporate financial management*. Harlow: Pearson Education.

Arnold, G. (2007). *Essentials of corporate financial management*. Harlow: Prentice Hall.

Arrow, K. J., & Levhari, D. (1969). Uniqueness of the internal rate of return with variable life of investment. *The Economic Journal*, *79*(315), 560–566.

Arzac, E. R., & Glosten, L. R. (2005). A reconsideration of tax shield valuation. *European Financial Management*, *11*(4), 453–461.

C. A. Magni, *Investment Decisions and the Logic of Valuation*,
https://doi.org/10.1007/978-3-030-27662-1

Athanasopoulos, P. J. (1978). A note on the modified internal rate of return and investment criterion. *The Engineering Economist*, *23*(2), 131–133.

Aucamp, D. L., & Eckardt, W. L. (1976). A sufficient condition for unique nonnegative internal rate of return: Comment. *Journal of Financial and Quantitative Analysis*, *11*, 329–332.

Auerbach, N. A., Tulloch, A. I. T., & Possingham, H. P. (2014). Informed actions: Where to cost effectively manage multiple threats to species to maximize return on investment. *Ecological Applications*, *24*(6), 1357–1373.

Bailey, M. J. (1959). Formal criteria for investment decisions. *Journal of Political Economy*, *67*(5), 476–488.

Baldwin, R. H. (1959). How to assess investment proposals. *Harvard Business Review*, *37*(3), 98–99.

Bao, B.-H., & Bao, D.-H. (1998). Usefulness of value added and abnormal economic earnings: An empirical examination. *Journal of Business Finance & Accounting*, *25*(1/2), 251–264.

Barney, J. B. (1986). Strategic factor markets: Expectations, luck, and business. *Management Science*, *32*, 1231–1241.

Barney, J. B. (1991). Firm resources and sustained competitive advantage. *Journal of Management*, *17*, 99–120.

Barney, J. B. (2001). Is the resource-based 'view' a useful perspective for strategic management research? Yes. *Academy of Management Review*, *26*(1), 41–57.

Bar-Yosef, S., & Lustgarten, S. (1994). Economic depreciation, accounting depreciation and their relation to current cost accounting. *Journal of Accounting, Auditing & Finance*, *9*, 41–60.

Barry, P. J., & Robison, L. J. (2014). Technical note: Economic rates of return and investment analysis. *The Engineering Economist*, *59*(3), 231–236.

Bathala, C. T., Bowlin, O. D., & Dukes, W. P. (2006). Use of debt covenants in small firms. *The Journal of Entrepreneurial Finance*, *11*(2), 49–72.

Bauchet, J., & Larsen, V. (2018). ROSCA composition and repayment: Evidence from Taiwanese bidding ROSCAs. *The Journal of Development Studies*, *54*(9), 1483–1495.

Begg, D., Fischer, S., & Dornbusch, R. (1984). *Economics* (British ed.). Maidenhead: Mc-Graw Hill.

Ben-Horin, M. (1979). The weighted average cost of capital as a cutoff rate. *Financial Management*, *8*(2), 18–21.

Ben-Horin, M., & Kroll, Y. (2017). A simple intuitive NPV-IRR consistent ranking. *The Quarterly Review of Economics and Finance*, *66*, 108–114.

Berg, N., & Lien, D. (2003). Tracking error decision rules and accumulated wealth. *Applied Mathematical Finance*, *10*, 91–119.

Berk, J., & DeMarzo, P. (2014). *Corporate Finance*, Global edition (3rd ed.). Harlow: Pearson.

Bernhard, R. H. (1979). A more general sufficient condition for a unique internal rate of return. *Journal of Financial and Quantitative Analysis*, *14*, 337–341.

Bernhard, R. H. (1980). A simplification and an extension of the Bernhard-De Faro sufficient condition for a unique non-negative internal rate of return. *Journal of Financial and Quantitative Analysis*, *15*(1), 201–209.

Besley, T., Coate, S., & Loury, G. (1993). The economics of rotating savings and credit associations. *American Economic Review*, *83*(4), 792–810.

Bidard, C. (1999). Fixed capital and internal rate of return. *Journal of Mathematical Economics*, *31*, 523–541.

Biddle, G. C., Bowen, R. M., & Wallace, J. S. (1999). Evidence on EVA. *Journal of Applied Corporate Finance*, *12*(2), 69–79.

Biezma, M. V., & San Cristóbal, J. R. (2006). Investment criteria for the selection of cogeneration plants – a state of the art review. *Applied Thermal Engineering*, *26*, 583–588.

Bierman, H., & Hass, J. E. (1973). Capital budgeting under uncertainty: A reformulation. *The Journal of Finance*, *28*, 119–129.

Biondi, Y. (2006). The double emergence of the modified internal rate of return. The neglected contribution of Duvillard (1755–1832) in a comparative perspective. *European Journal of History of Economic Thought*, *13*(3), 311–335.

Blank, L., & Tarquin, A. (2012). *Engineering economy*, International Edition (7th ed.). New York: McGraw-Hill.

Blank, L., & Tarquin, A. (2014). *Basics of engineering economy*, International edition (2nd ed.). New York: McGraw-Hill.

Block, S. (2011). Does the weighted average cost of capital describe the real-world approach to the discount rate?. *The Engineering Economist*, *56*(2), 170–180.

Boadway, R., & Bruce, N. (1984). A general proposition on the design of a neutral business tax. *Journal of Public Economics*, *24*(2), 231–239.

Bodenhorn, D. (1959). On the problem of capital budgeting. *The Journal of Finance*, *14*(4), 473–492.

Bodenhorn, D. (1964). A cash-flow concept of profit. *The Journal of Finance*, *19*(1), 16–31.

Bogue, M. C., & Roll, R. (1974). Capital budgeting of risky projects with "imperfect markets" for physical capital. *The Journal of Finance*, *29*(2), 601–613.

Booth, L. (2007). Capital cash flows, APV and valuation. *European Financial Management*, *13*(1), 29–48.

Borgonovo, E. (2017). *Sensitivity analysis. An introduction for the management scientist*. Berlin: Springer International Publishing.

Borgonovo, E., Gatti, S., & Peccati, L. (2010). What drives value creation in investment projects? An application of sensitivity analysis to project finance transactions. *European Journal of Operational Research*, *205*, 227–236.

Borgonovo, E., & Peccati, L. (2006). Uncertainty and global sensitivity analysis in the evaluation of investment projects. *International Journal of Production Economics*, *104*, 62–73.

Bosch-Badia, M. T., Montllor-Serrats, J., & Tarrazon-Rodon, M. A. (2014). Capital budgeting and shareholders' value: Investment projects versus courses of action. *The Engineering Economist*, *59*(3), 207–230.

Bosch, M. T., Montllor-Serrats, J., & Tarrazon, M. A. (2007). NPV as a function of the IRR: The value drivers of investment projection. *Journal of Applied Finance*, *17*, 41–45.

Boulding, K. E. (1935). The theory of the single investment. *Quarterly Journal of Economics*, *49*, 475–494.

Bouman, F. J. A. (1995). Rotating and accumulating savings and credit associations: A development perspective. *World Development*, *23*(3), 371–384.

Boyd, J., Epanchin-Niell, R., & Siikamäki, J. (2015). Conservation planning: A review of return on investment analysis. *Review of Environmental Economics and Policy*, *9*(1), 23–42.

Brackenborough, S., McLean, T., & Oldroyd, D. (2001). The emergence of discounted cash flow analysis in the Tyneside coal industry c.1700–1820. *British Accounting Review*, *33*, 137–155.

Brealey, R. A., & Myers, S. (2000). *Principles of corporate finance* (6th ed.). Boston: McGraw-Hill/Irwin.

Brealey, R. A., Myers, S., & Allen, F. (2011). *Principles of corporate finance* (Global ed.). New York: McGraw-Hill Irwin.

Brief, R. P. (1996). Using accounting data in present value models. *Journal of Financial Statement Analysis*, *1*, 21–29.

Brief, R. P., & Peasnell, K. V. (Eds.). (1996). *Clean surplus: A link between accounting and finance*. New York: Garland Publishing.

Brigham, E. F. (1975). Hurdle rates for screening capital expenditure proposals. *Financial Management*, *4*(3), 17–26.

Brimberg, J., Hansen, P., Laporte, G., Mladenovic, N., & Urosevic, D. (2008). The maximum return-on-investment plant location problem with market share. *Journal of the Operational Research Society*, *59*(3), 399–406.

Brimberg, J., & ReVelle, C. (2000). The maximum return-on-investment plant location problem. *Journal of the Operational Research Society*, *51*(6), 729–735.

Bromiley, P. (2005). *The behavioral foundations of strategic management*. Oxford: Blackwell Publishing.

Bromwich, M., & Walker, M. (1998). Residual income past and future. *Management Accounting Research, 9*(4), 391–419.

Broverman, S. A. (2008). *Mathematics of investment and credit* (4th ed.). Winsted: ACTEX Publications.

Brown, R. J. (2006). Sins of the IRR. *Journal of Real Estate Portfolio Management, 12*(2), 195–199.

Buchanan, J. (1969). *Cost and choice. An inquiry in economic theory*. Chicago: Markham. Republished as Midway reprint (1977). Chicago: University of Chicago Press.

Burns, R. M., & Walker, J. (1997). Investment techniques among the Fortune 500: A rationale approach. *Managerial Finance, 23*(9), 3–15.

Byrne, R. M. J. (2002). Mental models and counterfactual thoughts about what might have been. *Trends in Cognitive Science, 6*(10), 426–431.

Byrne, R. M. J. (2016a). Counterfactual thought. *Annual Review of Psychology, 67*, 135–157.

Byrne, R. M. J. (2016b). Counterfactual thinking. In A. Kind (Ed.), *The Routledge handbook of philosophy of imagination*. London: Routledge.

Byrne, R. M. J., & Tasso, A. (1999). Deductive reasoning with factual, possible, and counterfactual conditionals. *Memory and Cognition, 27*(4), 726–740.

Cannaday, R. E., Colwell, P. F., & Paley, H. (1986). Relevant and irrelevant internal rates of return. *The Engineering Economist, 32*(1), 17–38.

Cannan, E. (1921). Early history of the term capital. *The Quarterly Journal of Economics, 35*(3), 469–481.

Canning, J. B. (1929). *The economics of accountancy: A critical analysis of accounting theory*. New York: The Ronald Press.

Cantarelli, R., Bausch, G., Murdock, J., & Chyatte, M. R. (2018). Return-on-Investment (ROI) analyses of an inpatient lay health worker model on 30-day readmission rates in a rural community hospital. *Journal of Rural Health, 34*, 411–422.

Cantor, D. G., & Lippman, S. A. (1983). Investment selection with imperfect capital markets. *Econometrica, 51*(4), 1121–1144.

Cantor, D. G., & Lippman, S. A. (1995). Optimal investment selection with a multitude of projects. *Econometrica, 63*(5), 1231–1240.

Carsberg, B. V. (1966). The contribution of P.D. Leake to the theory of Goodwill valuation. *Journal of Accounting, 4*(1), 1–15.

Carter, E. E. (1971). The behavioral theory of the firm and top-level corporate decisions. *Administrative Science Quarterly, 16*(4), 413–428.

Cascio, W., & Boudreau, J. (2011). *Investing in people. Financial impact of human resource initiatives* (2nd ed.). Upper Saddle River: Pearson Education Inc.

Chambers, D. R., Harris, R. S., & Pringle, J. J. (1982). Treatment of financing mix in analyzing investment opportunities. *Financial Management, 11*(2), 24–41.

Chang, C. E., & Owens, R. W. (1999). Modifying the internal rate of return method for real estate investment selection. *Real Estate Review, 29*(3), 36–41.

Chiang, Y., Cheng, E., & Lam, P. (2010). Employing the net present value-consistent IRR methods for PFI contracts. *Journal of Construction Engineering & Management, 136*(7), 811–814.

Chisholm, R. M. (1946). The contrary-to-fact conditional. *Mind, 55*(222), 289–307.

Chisini, O. (1929). Sul concetto di media [On the concept of mean]. *Periodico di Matematiche, 4*, 106–116.

Chiu, S., & Garza Escalante, E. F. (2012). A companion for NPV: The generalized relative rate of return. *The Engineering Economist, 57*, 192–205.

Cnossen, S. (1998). Reform and coordination of company taxes in the European union. In P. B. Sørensen (Ed.), *Public finance in a changing world*. London: MacMillan.

Coase, R. H., (1938). Business organization and the accountant. *The Accountant.* Reprinted as "The nature of costs". In: Solomons D, (Eds.). (1968). *Studies in cost analysis* (2nd ed.). London: Sweet & Maxwell.

Collis, D., & Montgomery, C. (1995). Competing on resources: Strategy in the 1990's. *Harvard Business Review*, *73*(4), 118–128.

Colwell, P. (1995). Solving the dual IRR puzzle. *Journal of Property Management*, 60–61.

Cooper, I. A., & Nyborg, K. G. (2006). The value of tax shields IS equal to the present value of tax shields. *Journal of Financial Economics*, *81*(1), 215–225.

Cooper, I. A., & Nyborg, K. G. (2007). Valuing the debt tax shield. *Journal of Applied Corporate Finance*, *19*(2), 50–59.

Cooper, I. A., & Nyborg, K. G. (2008). Tax-adjusted discount rates with investor taxes and risky debt. *Financial Management*, *37*(2), 365–379.

Cooper, I. A., & Nyborg, K. G. (2018). Consistent valuation of project finance and LBOs using the flows-to-equity method. *European Financial Management*, *24*(1), 34–52.

Copeland, T., Koller, T., & Murrin, J. (2000). *Valuation. Measuring and managing the value of companies*. New York: Wiley.

Copeland, T. E., & Weston, J. F. (1988). Financial theory and corporate finance. Reading: Addison-Wesley Publishing Company Inc.

Cordes, J. J. (2017). Using cost-benefit analysis and social return on investment to evaluate the impact of social enterprise: Promises, implementation, and limitations. *Evaluation and Program Planning*, *64*, 98–104.

Crum, R. L., & Derkinderen, F. G. J. (Eds.). (1981). *Capital budgeting under conditions of uncertainty*. Boston: Martinus Nijhoff Publishing.

Cuthbert, J. R. (2018). Partitioning transaction vectors into pure investments. *The Engineering Economist*, *63*(2), 143–152.

Cuthbert, J. R., & Magni, C. A. (2016). Measuring the inadequacy of IRR in PFI schemes using profitability index and AIRR. *International Journal of Production Economics*, *179*, 130–140.

Cuthbert, J. R., & Magni, C. A. (2018). Some problems of the IRR in measuring PEI performance and how to solve it with the pure-investment AIRR. *Journal of Performance Measurement*, *22*(2), 39–50.

Damodaran, A. (1999). *Applied corporate finance. A user's manual*. New York: Wiley.

Damodaran, A. (2006a). *Damodaran on valuation. Security analysis for investment and corporate finance*. Hoboken: Wiley.

Damodaran, A. (2006b). Valuation approaches and metrics: A survey of the theory and evidence. Available at SSRN. http://www.stern.nyu.edu/adamodar/pdfiles/papers/valuesurvey.pdf. Originally published in *Foundations and Trends© in Finance*, *1*(8), 693 (2005).

Danaher, P. J., & Rust, R. T. (1996). Determining the optimal return on investment for an advertising campaign. *European Journal of Operational Research*, *95*(3), 511–521.

De Angelo, H., & De Angelo, L. (2006). The irrelevance of the MM dividend irrelevance theorem. *Journal of Financial Economics*, *79*(2), 293–315.

De Angelo, H., & De Angelo, L. (2007). Payout policy pedagogy: What matters and why. *European Financial Management*, *13*(1), 11–27.

De Angelo, H., De Angelo, L., & Wruck, K. H. (2002). Asset liquidity, debt covenants, and managerial discretion in financial distress: The collapse of LA Gear. *Journal of Financial Economics*, *64*(1), 3–34.

de Faro, C. (1978). A sufficient condition for a unique non-negative internal rate of return: Further comment. *Journal of Financial and Quantitative Analysis*, *13*, 577–584.

de Finetti, B. (1931). Sul concetto di media. [On the concept of mean]. *Giornale dell'Istituto Italiano degli Attuari* *2*(3), 369–396.

de Finetti, B. (1937). La prévision: ses lois logiques, ses sources subjectives. *Annales de l'Institut Henri Poincaré*, *17*(1), 1–68.

Dempsey, M. (2013). Consistent cash flow valuation with tax-deductible debt: A clarification. *European Financial Management*, *19*(4), 830–836.

De Risi, R., De Paola, F., Turpie, J., & Kroeger, T. (2018). Life cycle cost and return on investment as complementary decision variables for urban flood risk management in developing countries. *International Journal of Disaster Risk Reduction*, *28*, 88–106.

de Scitovszky, T. (1943). A note on profit maximisation and its implications. *The Review of Economic Studies, 11*(1), 57–60.
Dhavale, D. G., & Sarkis, J. (2018). Stochastic internal rate of return on investments in sustainable assets generating carbon credits. *Computers & Operations Research, 89*, 324–336.
Dicksee, L. R. (1897). *Goodwill and its treatment in accounts*. London.
Dixit, A. (1992). Investment and hysteresis. *Journal of Economic Perspectives, 6*(1), 107–132.
Dixit, A., & Pindyck, R. (1994). *Investment under uncertainty*. Princeton: Princeton University Press.
Donlan, C. J., Luque, G. M., & Wilcox, C. (2015). Maximizing return on investment for island restoration and species conservation. *Conservation Letters, 8*(3), 171–179.
Duhem, P. (1914). *La théorie physique: son object et sa structure [The aim and structure of physical theory]*. Paris: Marcel Rivière.
Dulman, S. P. (1989). The development of discounted cash flow techniques in U.S. industry. *The Business History Review 63*(3), 555–587.
Dunbar, K. (2013). Economic value added (EVA^{TM}): A thematic-bibliography. *The Journal of New Business Ideas & Trends, 11*(1), 54–66.
Dupas, P., & Robinson, J. (2013). Why don't the poor save more? Evidence from health savings experiments. *American Economic Review, 103*(4), 1138–1171.
Dybvig, P. H., & Ingersoll, J. E. (1982). Mean-variance theory in complete markets. *The Journal of Business, 55*(2), 233–251.
Echols, M. E. (2005). *ROI on human capital investment*. New York: iUniverse, Inc.
Edey, H. C. (1957). Business valuation, goodwill and the super-profit method. Accountancy. Reprinted. In: Baxter WT, Davidson S, (Eds.). (1962). *Studies in accounting theory*. London: Sweet & Maxwell.
Edwards, E. (1980). The fundamental character of excess current income. *Accounting and Business Research, 10*(40), 375–384.
Edwards, E., & Bell, P. (1961). *The theory and measurement of business income*. Berkeley: University of California Press.
Edwards, E., Kay, J., & Mayer, C. (1987). *The economic analysis of accounting profitability*. New York: Oxford University Press.
Edwards, J. R., & Warman, A. (1981). Discounted cash flow and business valuation in a nineteenth century merger: A note. *The Accounting Historians Journal, 8*(2), 37–50.
Ehrbar, A. (1998). *EVA: The real key to creating value*. New York: Wiley.
Ehrhardt, M. C., & Brigham, E. F. (2016). *Corporate finance. A focused approach*. Boston: Cengage Learning.
Emmanuel, C. R., & Otley, D. T. (1976). The usefulness of residual income. *Journal of Business Finance & Accounting, 3*(4), 43–51.
Eschenbach, T. G. (1995). *Engineering economy: Applying theory to practice*. Chicago: Irwin.
Evans, D. A., & Forbes, S. M. (1993). Decision making in display methods: The case of prescription and practice in capital budgeting. *The Engineering Economist, 39*(1), 87–92.
Esty, B. (2004). Why study large projects? An introduction to research on project finance. *European Financial Management, 10*(2), 213–224.
Fabozzi, F. J., & Grant, J. L. (Eds.). (2000). *Value-based metrics: Foundations and practice*. New Hope: Frank J. Fabozzi Associates.
Fama, E. F. (1965). Random walks in stock market prices. *Financial Analyst Journal, 21*(5), 55–59.
Fama, E. F. (1970). Efficient capital markets: A review of theory and empirical work. *The Journal of Finance, 25*(2), 383–417.
Feenstra, D. W., & Wang, H. (2000). Economic and accounting rates of return. Research report 00E42, University of Groningen, Research Institute SOM (Systems, Organisations and Management). http://irs.ub.rug.nl/ppn/240961587.
Feltham, G., & Ohlson, J. (1995). Valuation and clean surplus accounting for operating and financial activities. *Contemporary Accounting Research, 11*(2), 689–731.

Feltham, G., & Ohlson, J. (1996). Uncertainty resolution and the theory of depreciation measurement. *Journal of Accounting Research, 34*(2), 209–234.

Fernández, P. (2002). *Valuation methods and shareholder value creation*. San Diego: Elsevier Science.

Fernández, P. (2004). The value of tax shields is NOT equal to the present value of tax shields. *The Accounting Review, 73*(1), 145–165.

Fetter, F. A. (1937). Reformulation of the concepts of capital and income in economics and accounting. *The Accounting Review, 12*(1), 3–12.

Finnerty, J. D. (1986). *Corporate financial analysis*. New York: McGraw-Hill.

Finnerty, J. D. (2013). *Project financing: Asset-based financial engineering* (3rd ed.). Hoboken: Wiley.

Fischer, B. R., & Wermers, R. (2013). *Performance evaluation and attribution of security portfolios*. Waltham: Academic Press.

Fisher, F. M., & McGowan, J. J. (1983). On the misuse of accounting rates of return to infer monopoly profits. *American Economic Review, 73*(1), 82–97.

Fisher, I. (1930). *The theory of interest*. New York: The Macmillan Company. Reprint by Augustum, M. (1974). Clifton: Kelley Publishers.

Fitz-Enz, J. (2009). *The ROI of human capital: Measuring the economic value of employee performance* (2nd ed.). New York: Amacom.

Foster, J. E., & Mitra, T. (2003). Ranking investment projects. *Economic Theory, 22*, 469–494.

Francis, J., Olsson, P., & Oswald, D. R. (2000). Comparing the accuracy and explainability of dividend, free cash flow, and abnormal earnings equity value estimates. *Journal of Accounting Research, 38*(1), 45–70.

Franks, J. R., & Hodges, S. D. (1984). The meaning of accounting numbers in target setting and performance measurement: Implications for managers and regulators. In: *Annual meeting of the American finance association*, San Francisco (28–30 December 1983). Reprinted. In R. Brief & K. V. Peasnell (Eds.) (1996). *Clean surplus: a link between accounting and finance*. New York: Garland Publishing.

Gallo, P., & Peccati, L. (1993). The appraisal of industrial investments: A new method and a case study. *International Journal of Production Economics, 30/31*(1), 465–476.

Gargani, J. (2017). The leap from ROI to SROI: Farther than expected? *Evaluation and Program Planning, 64*, 116–126.

Gatti, S. (2013). *Project finance in theory and practice. Designing, structuring, and financing private and public projects* (2nd ed.). Burlington: Elsevier.

Geertz, C. (1962). The Rotating Credit Association: A "middle rung" in development. *Economic Development and Cultural Change, 10*(3), 241–263.

Ghiselli Ricci, R., & Magni, C. A. (2006). Economic value added and systemic value added: Symmetry, additive coherence and differences in performance. *Applied Financial Economics Letters, 2*(3), 151–154.

Ghiselli Ricci, R., & Magni, C. A. (2014). Axiomatization of residual income and generation of financial securities. *Quantitative Finance, 14*(7), 1257–1271.

Gigerenzer, G. (2001). The adaptive toolbox. In G. Gigerenzer & R. Selten (Eds.), *Bounded rationality*. Cambridge: The MIT Press.

Gitman, L. J., & Forrester, J. R. (1977). A survey of capital budgeting techniques used by major U.S. firms. *Financial Management, 6*(3), 66–71.

Gitman, L. J., & Mercurio, V. A. (1982). Cost of capital techniques used by major U.S. firms: Survey and analysis of fortune's 1000. *Financial Management, 11*(4), 21–29.

Gleeson-White, J. (2012). *Double-Entry. How the merchants of Venice created modern finance* (First American ed.). New York: WW Norton & Company.

Glen, N. (1893). *Actuarial science: An elementary manual*. Glasgow: John Smith & Son.

Goetzmann, W. N. (2004). *Fibonacci and the financial revolution* (NBER Working Paper 10352) (March). http://www.nber.org/papers/w10352.

Goetzmann, W. N., & Garstka S. J. (1999). *The development of corporate performance measures: Benchmarks before EVA* (Yale ICF Working Paper No. 99–06). Available at SSRN. http://ssrn.com/abstract=170674.

Goetzmann, W. N., & Rouwenhorst, K. G. (Eds.). (2005). *The origins of value: The financial innovations that created modern capital markets*. New York: Oxford University Press.

Goodman, L. E., & Warner, W. H. (2001). *Dynamics*. Mineola: Dover; Correct republication of third printing (1965). Belmont: Wadsworth Publishing Company Inc.

Goodman, N. (1947). The problem of counterfactual conditionals. *The Journal of Philosophy*, *44*(5), 113–128.

Goodman, N. (1983). *Fact, fiction, and forecast* (4th ed.). Cambridge: Harvard University Press.

Gordon, L. A. (1974). Accounting rate of return vs. economic rate of return. *Journal of Business Finance & Accounting*, *1*(3), 343–356.

Gordon, L. A., & Stark, A. W. (1989). Accounting and economic rates of return: A note on depreciation and other accruals. *Journal of Business Finance & Accounting*, *16*(3), 425–432.

Gow, I., & Reichelstein, S. (2007). Capital budgeting: The role of cost allocations. In K. H. Waldmann & U. M. Stocker (Eds.), *Operations research proceedings 2006*. Berlin: Springer.

Graham, J., & Harvey, C. (2001). The theory and practice of corporate finance: Evidence from the field. *Journal of Financial Economics*, *60*, 187–243.

Graham, J., & Harvey, C. (2002). How do CFOs make capital budgeting and capital structure decisions? *Journal of Applied Corporate Finance*, *15*(1), 8–23.

Grant, E. L. (1930). *Principles of engineering economy* (3rd ed.). New York: Ronald Press Company.

Grant, J. (1996). Foundations of EVA for investment managers. *Journal of Portfolio Management*, 41–48.

Grant, J. (1998). Security analysis using EVA. In F. J. Fabozzi (Ed.), *Handbook of portfolio management*. New Hope: Frank J. Fabozzi Associates.

Grant, R. (1991). The resource-based theory of competitive advantage: Implications for strategy formulation. *California Management Review*, *33*(3), 114–135.

Grant, R., & Robert, N. (1995). *Contemporary strategy analysis: Concept, techniques, applications*. Oxford: Blackwell.

Gray, K. B., & Dewar, R. B. (1971). Axiomatic characterization of the time-weighted rate of return. *Management Science*, *18*(2), 32–35.

Graziani, R., & Veronese, P. (2009). How to compute a mean? The Chisini approach and its applications. *The American Statistician*, *63*(1), 33–36.

Grinblatt, M., & Liu, J. (2008). Debt policy, corporate taxes, and discount rates. *Journal of Economic Theory*, *141*(1), 225–254.

Grinblatt, M., & Titman, S. (2002). *Financial management and corporate strategy* (2nd ed.). Boston: Irwin/McGraw-Hill.

Gronchi, S. (1984/1987). *Tasso di rendimento e valutazione dei progetti: una analisi teorica* [Rate of return and project appraisal]. Università degli Studi di Siena, Collana dell'Istituto di Economia. Reprinted by Franco Angeli, Milano

Gronchi, S. (1986). On investment criteria based on the internal rate of return. *Oxford Economic Papers*, *38*(1), 174–180.

Hajdasinski, M. M. (1987). On relevant and irrelevant rates of return. *The Engineering Economist*, *32*(4), 347–353.

Hajdasinski, M. M. (1995). Meaningful rates of return for project evaluation. In R. J. Singhal et al. (Eds.), *Proceedings of the 4th international symposium on mine planning and equipment selection, Calgary, Canada, October 31/November 3*. Rotterdam: AA Balkema.

Hajdasinski, M. M. (1997). NPV-compatibility, project ranking, and related issues. *The Engineering Economist*, *42*(4), 325–339.

Haley, C. W., & Schall, L. D. (1979). *The theory of financial decisions* (2nd ed.). New York: McGraw-Hill Book Company.

Hamada, R. S. (1969). Portfolio analysis, market equilibrium and corporation finance. *The Journal of Finance*, *24*(1), 13–31.

Hamilton, R. (1777). *An introduction to merchandise*. Edinburgh.

Hansen, P. (1972). *The accounting concept of profit* (2nd ed.). Amsterdam: North-Holland.

Harcourt, G. C. (1965). The accountant in a golden age. *Oxford Economic Papers, 17*(1), 66–80.

Harris, P. H., German, T., & Mills, P. (1996). Children's use of counterfactual thinking in causal reasoning. *Cognition, 61*, 233–259.

Harris, R. S., & Pringle, J. J. (1985). Risk-adjusted discounted rates - extensions from the average-risk case. *The Journal of Financial Research, 8*(3), 237–244.

Hartman, J. C. (2000). On the equivalence of net present value and market value added as measures of a project's economic worth. *The Engineering Economist, 45*(2), 158–165.

Hartman, J. C. (2007). *Engineering economy and the decision-making process*. Upper Saddle River: Pearson.

Hazen, G. B. (2003). A new perspective on multiple internal rates of return. *The Engineering Economist, 48*(1), 31–51.

Hazen, G. B. (2009). An extension of the internal rate of return to stochastic cash flows. *Management Science, 55*, 1030–1034.

Herbst, A. F. (1978). The unique, real internal rate of return: Caveat emptor. *Journal of Financial and Quantitative Analysis, 13*, 363–370.

Herbst, A. F. (2002). *Capital asset investments: Strategy, tactics and tools*. Chichester: Wiley.

Hevener, C. C. (2006). *Alternative financial vehicles: Rotating savings and credit associations (ROSCAs)*. (Community Affairs Discussion Paper 06-01). *Federal Reserve Bank of Philadelphia*. https://ideas.repec.org/p/fip/fedpcd/06-01.html.

Hibbeler, R. C. (2016). *Statics & dynamics* (14th ed.). Hoboken: Pearson.

Hicks, J. R. (1973). *Capital and time. A neo-Austrian theory*. Oxford: Clarendon Press.

Hirshleifer, J. (1958). On the theory of optimal investment decision. *Journal of Political Economy, 66*, 329–352.

Horngren, C. T., Harrison, W. T, Jr., & Oliver, M. S. (2012). *Financial and managerial accounting* (3rd ed.). Upper Saddle River: Pearson.

Hotelling, H. (1925). A general mathematical theory of depreciation. *Journal of the American Statistical Association 20*(151), 340–353.

Howe, K. M. (1991). Perpetuity rate of return analysis. *The Engineering Economist, 36*(3), 248–257.

Hülsmann, J. G. (2003). Facts and counterfactuals in economic law. *Journal of Libertarian Studies, 17*(1), 57–102.

Hartman, J. C., & Schafrick, I. C. (2004). The relevant internal rate of return. *The Engineering Economist, 49*(2), 139–158.

Iman, R. L., & Conover, W. J. (1987). A measure of top-down correlation. *Technometrics, 29*(3), 351–357.

Inselbag, I., & Kaufold, H. (1997). Two DCF approaches for valuing companies under alternative financing strategies (and how to choose between them). *Journal of Applied Corporate Finance, 10*(1), 114–122.

Jagannathan, R., Liberti, J., Liu, B., & Meier, I. (2017). A firm's cost of capital. *Annual Review of Financial Economics, 9*, 259–282.

Jagannathan, R., & Meier, I. (2002). Do we need CAPM for capital budgeting? *Financial Management, 31*(4), 55–77.

Jensen, M. (1986). Agency costs of free cash flow, corporate finance, and takeovers. *American Economic Review, 76*(2), 323–329.

Jensen, M. C., & Meckling, W. H. (1976). Theory of the firm: Managerial behaviour, agency costs and ownership structure. *Journal of Financial Economics, 3*(4), 305–360.

Jean, W. H. (1968). On multiple rates of return. *The Journal of Finance, 23*, 187–191.

Jiang, Y. (2017). Introducing excess return on time-scaled contributions: An intuitive return measure and new solution to the IRR and PME problem. *Journal of Alternative Investments, 19*(4), 77–91.

Jorgensen, D. (1963). Capital theory and investment behavior. *American Economic Review, 53*, 247–259.

Joy, O. M., & Bradley, J. O. (1973). A note on sensitivity analysis of rates of return. *The Journal of Finance*, *28*(5), 1255–1261.

Kahneman, D., & Miller, D. T. (1986). Norm theory: Comparing reality to its alternatives. *Psychological Review*, *93*, 136–153.

Kahneman, D., & Tversky, A. (1982). The simulation heuristic. In D. Kahneman, P. Slovic, & A. Tversky (Eds.), *Judgment under uncertainty: Heuristics and biases*. New York: Cambridge University Press.

Kahneman, D., & Tversky, A. (1984). Choices, values and frames. *American Psychologist*, *39*, 341–350.

Kaplan, S. (1965). A note on a method for precisely determining the uniqueness or non-uniqueness of the internal rate of return for a proposed investment. *The Journal of Industrial Engineering*, *16*, 70–71.

Kaplan, S. (1967). Computer algorithms for finding exact rates of return. *The Journal of Business*, *40*, 389–392.

Kaplan, S. N., & Ruback, R. S. (1995). The valuation of cash flow forecasts: An empirical analysis. *The Journal of Finance*, *50*(4), 1059–1093.

Karmel, P. H. (1959). The marginal efficiency of capital. *Economic Record*, *35*(72), 429–434.

Kay, J. A. (1976). Accountants, too, could be happy in the golden age: The accountants rate of profit and the internal rate of return. *Oxford Economic Papers*, *28*(3), 447–460.

Kay, J. A., & Mayer, C. P. (1986). On the applications of accounting rates of return. *The Economic Journal*, *96*, 199–207.

Keane, S. M. (1979). The internal rate of return and the reinvestment fallacy. *Abacus*, *15*(1), 44–55.

Keef, S. P., & Roush, M. L. (2001). Residual income: A review essay. *Australian Accounting Review*, *11*(24), 8–14.

Keef, S. P., & Roush, M. L. (2001). Discounted cash flow methods and the fallacious reinvestment assumption: a review of recent texts. *Accounting Education: An International Journal*, *10*(1), 105–116.

Kellison, S. G. (2009). *The theory of interest* (3rd ed.). New York: McGraw-Hill/Irwin.

Keynes, J. M. (1936/1967). *The general theory of employment, interest and money*. London: Macmillan.

Khan, Z. A., Siddiquee, A. N., Kumar, B., & Abidi, M. H. (2018). *Principles of engineering economics with applications* (2nd ed.). Cambridge: Cambridge University Press.

Kierulff, H. (2008). MIRR: A better measure. *Business Horizon*, *51*(4), 321–329.

Kimball, R. C. (1998). Economic profit and performance measurement in banking. *New England Economic Review*, 36–53.

Kirshenbaum, P. S. (1965). A resolution of the multiple rate-of-return paradox. *The Engineering Economist*, *10*(1), 11–16.

Kneale, W. (1950). Natural laws and contrary to fact conditionals. *Analysis*, *10*(6), 121–125.

Krugman, P. (1979). A model of balance-of-payments crises. *Journal of Money Credit and Banking*, *11*(3), 311–325.

Kumar, H. A. (2014). Return on Investment (ROI) from a business school library: An Indian perspective. In T. Wales (Ed.), *Business schools libraries in the 21st century*. London: Routledge.

Kumbhakar, S. C. (2011). Estimation of production technology when the objective is to maximize return to the outlay. *European Journal of Operational Research*, *208*(2), 170–176.

Kurtz, D. V. (1973). The rotating credit association: An adaptation to poverty. *Human Organization*, *32*(1), 49–58.

Laland, K. N. (2001). Imitation, social learning, and preparedness as mechanisms of bounded rationality. In G. Gigerenzer, R. Selten (Eds.), *Bounded rationality. The adaptive toolbox*. Cambridge: The MIT Press.

Lander, D. M., & Pinches, G. E. (1998). Challenges to the practical implementation of modeling and valuing real options. *Quarterly Review of Economics and Finance 38*(Special Issue), 537–567.

Laughhunn, D. J., Payne, J. W., & Crum, R. (1980). Managerial risk preferences for below-target returns. *Management Science*, *26*(12), 1238–1249.

Leake, P. D. (1921). *Goodwill. Its history, value and treatment in accounts*. London: Pitman and Sons Ltd.

Leck, C., Upton, D., & Evans, N. (2016). Social return on investment: Valuing health outcomes or promoting economic values? *Journal of Health Psychology*, *21*(7), 1481–1490.

Lee, M. C. (1999). Accounting-based valuation: Impact on business practices and research. *Accounting Horizons*, *13*(4), 413–425.

Lee, T. (1985). *Income and value measurement. Theory and practice* (3rd ed.). Worcester: Billing & Sons.

Levinthal, D. A. (1995). Strategic management and the exploration of diversity. In C. A. Montgomery (Ed.), *Resource-based and evolutionary theories of the firm*. Boston: Kluwer.

Lewellen, W. G., & Emery, D. R. (1986). Corporate debt management and the value of the firm. *Journal of Financial and Quantitative Analysis*, *21*(4), 415–426.

Lewis, D. (1973). Causation. *The Journal of Philosophy*, *70*(17), 556–567.

Li, J., Min, K. J., Otake, T., & Van Voorhis, T. (2008). Inventory and investment in setup and quality operations under return on investment maximization. *European Journal of Operational Research*, *185*(2), 593–605.

Lima e Silva, J., Sobreiro, V. A., & Kimura, H, (2018). Pre-purchasing financing pool: Revealing the IRR problem. *The Engineering Economist*, *63*(2), 193–217.

Lin, S. (1976). The modified internal rate of return and investment criterion. *The Engineering Economist*, *21*(4), 237–247.

Lindblom, T., & Sjögren, S. (2009). Increasing goal congruence in project evaluation by introducing a strict market depreciation schedule. *International Journal of Production Economics*, *121*, 519–532.

Livingstone, J. L., & Salamon, G. L. (1970). Relationship between the accounting and the internal rate of return measures: A synthesis and an analysis. *Journal of Accounting Research*, *8*(2), 199–216.

Lohmann, J. R. (1988). The IRR, NPV and the fallacy of the reinvestment rate assumption. *The Engineering Economist*, *33*(4), 303–330.

Lorie, J., & Savage, L. J. (1955). Three problems in capital budgeting. *The Journal of Business*, *28*, 229–239.

Lücke, W. (1955). Investitionsrechnungen auf der Grundlage von Ausgaben oder Kosten. *Zeitschrift für betriebswirthschaftliche Forschung*, *7*, 310–324.

Luckett, P. F. (1984). ARR vs. IRR: A review and an analysis. *Journal of Business Finance & Accounting*, *11*(2), 213–231.

Lundholm, R., & O'Keefe, T. (2001). Reconciling value estimates from the discounted cash flow model and the residual income model. *Contemporary Accounting Research*, *18*(2), 311–335.

Machlup, F. (1942). Competition, pliopoly and profit Part II. *Economica*, *9*(34), 153–173.

MacMinn, R. D. (2005). *The Fisher model and financial markets*. Singapore: World Scientific Publishing.

Magni, C. A. (2000a). Decomposition of a certain cash flow stream: Differential systemic value and net final value. In *Proceedings of the XXIV AMASES Conference (Association for Mathematics Applied to Economic and Social Sciences), Padenghe, Italy, September*. Available at SSRN. http://ssrn.com/abstract=1096213.

Magni, C. A. (2000b). Tir, Roe e Van: Convergenze formali e concettuali in un approccio sistemico [IRR, ROE and NPV: Formal and conceptual convergences in a systemic approach]. *Finanza Marketing e Produzione* *18*(4), 31–59. Available at SSRN. http://ssrn.com/abstract=1104597.

Magni, C. A. (2000c). *Systemic value added, residual income and decomposition of a cash flow stream* (Working Paper No. 318). Department of Economics, University of Modena and Reggio Emilia. Available at SSRN. http://ssrn.com/abstract=1032011.

Magni, C. A. (2001a). Valore aggiunto sistemico: un'alternativa all'EVA quale indice di sovraprofitto periodale [Systemic value added: an alternative to EVA as a residual income model]. *Budget* *25*(1), 63–71. Available at SSRN. http://ssrn.com/abstract=1103492.

Magni, C. A. (2001b). Scomposizione di sovraprofitti: Economic Value Added e Valore Aggiunto Sistemico [Excess-profit decomposition: economic value added and systemic value added]. *Finanza marketing e produzione 19*(4), 94–119. Available at SSRN. http://ssrn.com/abstract=1140163.

Magni, C. A. (2002). Investment decisions in the theory of finance: Some antinomies and inconsistencies. *European Journal of Operational Research, 137*(1), 206–217.

Magni, C. A. (2003). Decomposition of net final values: Systemic value added and residual income. *Bulletin of Economic Research, 55*(2), 149–176.

Magni, C. A. (2004). Modelling excess profit. *Economic Modelling, 21*(3), 595–617.

Magni, C. A. (2005). On decomposing net final values: EVA SVA and shadow project. *Theory and Decision, 59*(1), 51–95.

Magni, C. A. (2006). Zelig and the art of measuring excess profit. *Frontiers in Finance and Economics, 3*(1), 103–129.

Magni, C. A. (2007a). Project valuation and investment decisions: CAPM versus arbitrage. *Applied Financial Economics Letters, 3*(2), 137–140.

Magni, C. A. (2007b). Project selection and equivalent CAPM-based investment criteria. *Applied Financial Economics Letters, 3*(3), 165–168.

Magni, C. A. (2008a). CAPM-based capital budgeting and nonadditivity. *Journal of Property Investment and Finance, 26*(5), 388–398.

Magni, C. A. (2008b). Economic profit, NPV and CAPM: Biases and violations of Modigliani and Miller's proposition I. *The ICFAI Journal of Applied Finance, 14*(10), 59–72.

Magni, C. A. (2009a). Correct or incorrect application of the CAPM? Correct or incorrect decisions with the CAPM? *European Journal of Operational Research, 192*(2), 549–560.

Magni, C. A. (2009b). Opportunity cost, excess profit, and counterfactual conditionals. *Frontiers in Finance and Economics, 6*(1), 118–154.

Magni, C. A. (2009c). Splitting up value: A critical review of residual income theories. *European Journal of Operational Research, 198*(1), 1–22.

Magni, C. A. (2009d). Investment decisions, net present value and bounded rationality. *Quantitative Finance, 9*(8), 967–979.

Magni, C. A. (2009e). Accounting and economic measures: An integrated theory of capital budgeting (SSRN Working Paper). Available at https://ssrn.com/abstract=1498106.

Magni, C. A. (2010a). Residual income and value creation: An investigation into the lost-capital paradigm. *European Journal of Operational Research, 201*, 505–519.

Magni, C. A. (2010b). Average internal rate of return and investment decisions: A new perspective. *The Engineering Economist, 55*(2), 150–180.

Magni, C. A. (2010c). Relevance or irrelevance of retention for dividend policy irrelevance. *International Review of Applied Financial Issues and Economics, 2*(2), 232–247.

Magni, C. A. (2010d). CAPM and capital budgeting: Present/future, equilibrium/disequilibrium, decision/valuation. *IUP Journal of Financial Economics, 8*(1–2), 7–35.

Magni, C. A. (2010e). *Depreciation classes, return on investment and economic profitability* (SSRN Working Paper). Available at http://ssrn.com/abstract=1718525.

Magni, C. A. (2010f). On the long-standing issue of the internal rate of return: A complete resolution. In *Proceedings of the XXXIV AMASES Conference*, Macerata, September 1–4. Available at SSRN: https://ssrn.com/abstract=1622681.

Magni, C. A. (2011a). Aggregate return on investment and investment decisions: A cash-flow perspective. *The Engineering Economist, 56*(2), 140–169.

Magni, C. A. (2011b). Addendum to "Average internal rate of return and investment decisions: A new perspective". *The Engineering Economist, 56*(2), 181–182.

Magni, C. A. (2011c). Return on equity, internal rate of return and shareholder value creation. In *Proceedings of EAA 2011 annual congress, European Accounting Association*, Rome, April 20–22. Available at SSRN. https://ssrn.com/abstract=1678057.

Magni, C. A. (2011d). L'approccio del reddito residuale e il metodo S&A per la valutazione delle aziende [The residual-income approach and the Sum&Discount method for appraising firms]. *La Valutazione delle Aziende*, *63*, 43–47.

Magni, C. A. (2012). In search of the "lost capital". A theory for valuation, investment decisions, performance measurement. *Frontiers in Finance and Economics*, *9*(1), 87–146.

Magni, C. A. (2013a). The internal rate of return approach and the AIRR paradigm: A refutation and a corroboration. *The Engineering Economist*, *58*(2), 73–111.

Magni, C. A. (2013b). Generalized Makeham's formula and economic profitability. *Insurance: Mathematics and Economics*, *53*(3), 747–756.

Magni, C. A. (2014a). Mathematical analysis of average rates of return and investment decisions: The missing link. *The Engineering Economist*, *59*(3), 175–206.

Magni, C. A. (2014b). Arithmetic returns for investment performance measurement. *Insurance: Mathematics and Economics*, *55*, 291–300.

Magni, C. A. (2014c). *Pseudo-naïve approaches to investment performance measurements* (SSRN Working Paper). Available at https://ssrn.com/abstract=2538496.

Magni, C. A. (2015a). Investment, financing and the role of ROA and WACC in value creation. *European Journal of Operational Research*, *244*(3), 855–866.

Magni, C. A. (2015b). Aggregate return on investment for investments under uncertainty. *International Journal of Production Economics*, *165*, 29–37.

Magni, C. A. (2015c). *ROI and profitability index: A note on managerial performance* (SSRN Working Paper). Available at https://ssrn.com/abstract=2546917.

Magni, C. A. (2016a). An average-based accounting approach to capital asset investment: The case of project finance. *European Accounting Review*, *25*(2), 275–286.

Magni, C. A. (2016b). Capital depreciation and the underdetermination of rate of return: A unifying perspective. *Journal of Mathematical Economics*, *67*, 54–79.

Magni, C. A. (2018). On "Introducing excess return on time-scaled contributions": A clarification. *The Journal of Alternative Investments*, *21*(2), 1–3.

Magni, C. A., & Marchioni, A. (2018). Project appraisal and the intrinsic rate of return. In *Proceeding of the 4th international conference on production economics and project evaluation, ICOPEV*. University of Minho, Guimarães, Portugal, 20–21 September (pp. 225–229). Available at SSRN: https://ssrn.com/abstract=3253956.

Magni, C. A, & Marchioni, A. (2019). The accounting-and-finance of a solar photovoltaic plant: economic efficiency of a replacement project. In *Proceedings of 4th international conference on energy and environment: Bringing together engineering and economics, Guimarães, Portugal, 16–17 May 2019*. Available at SSRN: https://papers.ssrn.com/abstract=3351150.

Magni, C. A., & Martin, J. D. (2017). *The reinvestment rate assumption fallacy for IRR and NPV* (SSRN Working Paper). Available at https://ssrn.com/abstract=3090678.

Magni, C. A., & Martin, J. D. (2018). *Defining a reliable multi-period project rate of return. An alternative to IRR* (SSRN Working Paper). Available at https://ssrn.com/abstract=3253948.

Magni, C. A., & Peasnell, K. V. (2012). *Economic profitability and the accounting rate of return* (SSRN Working Paper). Available at https://ssrn.com/abstract=2027607.

Magni, C. A., & Peasnell, K. V. (2015). *The term structure of capital values: An accounting-based framework for measuring economic profitability* (SSRN Working Paper). Available at https://ssrn.com/abstract=2656651.

Magni, C. A., & Vélez-Pareja, I. (2009). Potential dividends versus actual cash flows in firm valuation. *The ICFAI Journal of Applied Finance*, *15*(7), 51–66.

Magni, C. A., Veronese, P., & Graziani, R. (2018). Chisini means and rational decision-making: Equivalence of investment criteria. *Mathematics and Financial Economics*, *12*(2), 193–217.

Makeham, W. C. (1874). On the solution of problems connected with loans repayable by instalments. *Journal of the Institute of Actuaries*, *18*, 132–143.

Malkiel, B. G. (1962). Expectations, bond prices, and the term structure of interest rates. *The Quarterly Journal of Economics*, *76*(2), 197–218.

Malkiel, B. G. (2003). The efficient market hypothesis and its critics. *Journal of Economic Perspectives*, *17*(1), 59–82.

Marchioni, A., & Magni, C. A. (2016). *Investment decisions and sensitivity analysis: NPV-consistency of straight-line rate of return* (SSRN Working Paper). Available at https://ssrn.com/abstract=2797685.

Marchioni, A., & Magni, C. A. (2018). Investment decisions and sensitivity analysis: NPV-consistency of rates of return. *European Journal of Operational Research*, *268*(1), 361–372.

Marshall, A. (1890). *Principles of economics*. London: MacMillan.

Martin, J. D., & Petty, J. W. (2000). *Value based management. The corporate response to shareholder revolution*. Boston: Harvard Business School.

Martin, J. D., Petty, J. W., & Rich, S. P. (2003). *An analysis of EVA and other measures of firm performance based on residual income* (Hankamer school of business Working Paper). Available at SSRN. http://papers.ssrn.com/sol3/papers.cfm?abstract_id=412122.

Mason, S. P., & Merton, R. C. (1985). The role of contingent claims analysis in corporate finance. In E. Altman & M. Subrahymanyam (Eds.), *Recent advances in corporate finance*. Boston: Irwin.

Massari, M., Roncaglio, F., & Zanetti, L. (2008). On the equivalence between the APV and the WACC approach in a growing leveraged firm. *European Financial Management*, *14*(1), 152–162.

Masters, R., Anwar, E., Collins, B., Cookson, R., & Capewell, S. (2017). Return on investment of public health interventions: A systematic review. *Journal of Epidemiology and Community Health*, *71*, 827–834.

McDonald, R. L. (2000). Real options and rules of thumb in capital budgeting. In M. J. Brennan & R. L. Trigeorgis (Eds.), *Project flexibility agency and competition: New developments in the theory and application of real options*. New York: Oxford University Press.

McMahon, W. W. (2018). The total return to higher education: Is there underinvestment for economic growth and development? *The Quarterly Review of Economics and Finance*, *70*, 90–111.

Mejía-Peláez, F., & Vélez-Pareja, I. (2010). Cost of capital and value without circularity for constant growth perpetuities. (SSRN working paper). Available at https://ssrn.com/abstract=1659446.

Mejía-Peláez, F., & Vélez-Pareja, I. (2011). Analytical solution to the circularity problem in the discounted cash flow valuation framework. *Innovar 21*(42), 55–68. Available at SSRN. https://ssrn.com/abstract=1596426.

Menezes, M. B. C., Kim, S., & Huang, R. (2015). Return-on-investment (ROI) criteria for network design. *European Journal of Operational Research*, *245*(1), 100–108.

Mepham, M. J. (1980). The residual income debate. *Journal of Business Finance & Accounting*, *7*(2), 183–199.

Michaud, T. L., You, W., Wilson, K. E., Su, D., McGuire, T. J., Almeida, F. A., Bayer, A. L., Estabrooks, P. A. (2017). Cost effectiveness and return on investment of a scalable community weight loss intervention. *Preventive Medicine*, *105*, 295–303.

Miles, J. A., & Ezzell, J. R. (1980). The weighted average cost of capital, perfect capital markets, and project life: A clarification. *Journal of Financial and Quantitative Analysis*, *15*(3), 719–730.

Miles, J. A., & Ezzell, J. R. (1985). Reformulating tax shield valuation: A note. *The Journal of Finance*, *40*(5), 1485–1492.

Miller, M. H., & Modigliani, F. (1961). Dividend policy, growth, and the valuation of shares. *The Journal of Business*, *34*(4), 411–433.

Miller, K. D., & Shapira, Z. (2004). An empirical test of heuristics and biases affecting real option valuation. *Strategic Management Journal*, *25*(3), 269–284.

Miller, P., & Napier, C. (1993). Genealogies of calculation. *Accounting, Organizations and Society*, *18*(7/8), 631–647.

Modigliani, F., & Miller, M. H. (1958). The cost of capital, corporation finance and the theory of investment. *American Economic Review*, *48*(3), 261–297.

Modigliani, F., & Miller, M. H. (1963). Corporate income taxes and the cost of capital: A correction. *American Economic Review*, *53*(3), 433–443.

Mohnen, A., & Bareket, M. (2007). Performance measurement for investment decisions under capital constraints. *Review of Accounting Studies, 12*(1), 1–22.

Mørch, O., Fagerholta, K., Pantuso, G., & Rakkec, J. (2017). Maximizing the rate of return on the capital employed in shipping capacity renewal. *Omega, 67*, 42–53.

Mossin, J. (1966). Equilibrium in a capital asset market. *Econometrica, 34*(4), 768–783.

Mossin, J. (1969). Security pricing and investment criteria in competitive markets. *American Economic Review, 59*(5), 749–756.

Mowen, M. M., & Hansen, D. R. (2008). *Cornerstones of managerial accounting* (2nd ed.). Mason: Thomson South-Western.

Muliere, P., & Parmigiani, G. (1993). Utility and means in the 1930s. *Statistical Science, 8*(4), 421–432.

Myers, S. C. (1974). Interactions of corporate financing and investment decisions - implications for capital budgeting. *Journal of Finance, 29*(1), 1–25.

Myers, S. C. (1977). Determinants of corporate borrowing. *Journal of Financial Economics, 5*(2), 147–175.

Myung, Y.-S., Kim, H., & Tcha, D. (1997). A bi-objective uncapacitated facility location problem. *European Journal of Operational Research, 100*(3), 608–616.

Nau, R. F. (1999). *Arbitrage choice theory* (ResearchGate, Working Paper). Available at https://www.researchgate.net/publication/228926280.

Nau, R. F. (2001). De Finetti was right: Probability does not exist. *Theory and Decision, 51*, 89–124.

Nau, R. F., & McCardle, K. F. (1991). Arbitrage, rationality, and equilibrium. *Theory and Decision, 31*, 199–240.

Nauss, R. M. (1988). On the use of internal rate of return in linear and integer programming. *Operations Research Letters, 7*(6), 285–289.

Newnan, D. G., Eschenbach, T. G., & Lavelle, J. P. (2009). *Engineering economic analysis* (International Tenth ed.). Oxford: Oxford University Press.

Norstrøm, C. J. (1970). Uniqueness of the internal rate of return with variable life of investment: A comment. *The Economic Journal, 80*(320), 983–984.

Norstrøm, C. J. (1972). A sufficient condition for a unique non-negative internal rate of return. *Journal of Financial and Quantitative Analysis, 7*(3), 1835–1839.

Nosal, E., & Wang, T. (2004). Arbitrage: The key to pricing options (pp. 1–4). *Federal Reserve Bank of Cleveland*, 1st January 2004.

Nuti, D. M. (1974). On the rates of return on investment. *Kyklos, 27*(2), 345–369.

Oakford, R. V., Bhimjee, S. A., & Jucker, J. V. (1977). The internal rate of return, the pseudo internal rate of return, and the NPV and their use in financial decision-making. *The Engineering Economist, 22*(3), 187–202.

O'Byrne, S. F. (1996). EVA and market value. *Journal of Applied Corporate Finance, 9*(1), 116–125.

O'Byrne, S. F., & Young, D. S. (2006). Incentives and investor expectations. *Journal of Applied Corporate Finance, 18*(2), 98–105.

O'Hanlon, J., & Peasnell, K. V. (1998). Wall Street's contribution to management accounting: The Stern Stewart EVA financial management system. *Management Accounting Research, 9*, 421–444.

O'Hanlon, J., & Peasnell, K. V. (2002). Residual income and value creation: The 'missing link'. *Review of Accounting Studies, 7*(2/3), 229–245.

O'Hanlon, J., & Peasnell, K. V. (2014). Residual income and EVA. In C. L. Cooper (Ed.) *Wiley encyclopedia of management. Volume 1. Accounting* (3rd ed., pp. 1–5). New York: Wiley.

Ohlson, J. A. (1989). Accounting earnings, book values, and dividends: The theory of the clean surplus equation in equity valuation. Unpublished manuscript. Reprinted in: Brief, R., Peasnell, K. V. (Eds.) (1996). *Clean surplus: A link between accounting and finance*. New York: Garland Publishing.

Ohlson, J. A. (1995). Earnings, book values, and dividends in equity valuation. *Contemporary Accounting Research, 11*(2), 661–687.

Ohlson, J. A. (2003). Positive (zero) NPV projects and the behavior of residual earnings. *Journal of Business Finance & Accounting*, *30*(1/2), 7–16.

Osborne, M. (2010). A resolution to the NPV-IRR debate? *The Quarterly Review of Economics and Finance*, *50*(2), 234–239.

Pacioli, L., & Cripps, J. (1494/1994). *Particularis de computis et scripturis. A contemporary translation*. Seattle: Pacioli Society.

Palepu, K. G., Healey, P. M., & Bernard, V. L. (2000). *Business analysis and valuation using financial statements*. Cincinnati: South-Western College Publishing.

Park, C. S. (2011). *Contemporary engineering economics* (5th ed.). Upper Saddle River: Pearson Prentice Hall.

Park, C. S. (2013). *Fundamentals of engineering economics* (International ed.). Harlow: Pearson.

Parker, R. H. (1968). Discounted cash flow in historical perspective. *Journal of Accounting Research*, *6*(1), 58–71.

Pasqual, J., Padilla, E., & Jadotte, E. (2013). Technical note: Equivalence of different profitability criteria with the net present value. *International Journal of Production Economics*, *142*(1), 205–210.

Pattison-Williams, J. K., Yang, W., Liu, Y., & Shane, G. (2017). Riparian wetland conservation: A case study of phosphorous and social return on investment in the Black River watershed. *Ecosystem Services*, *26*, 400–410.

Peasnell, K. V. (1981). On capital budgeting and income measurement. *Abacus*, *17*(1), 52–67.

Peasnell, K. V. (1982a). Some formal connections between economic values and yields and accounting numbers. *Journal of Business Finance & Accounting*, *9*(3), 361–381.

Peasnell, K. V. (1982b). Estimating the internal rate of return from accounting profit rates. *The Investment Analyst*, 26–31.

Peasnell, K. V. (1995a). Analytical properties of earned economic income. *British Accounting Review*, *27*(1), 5–33.

Peasnell, K. V. (1995b). Second thoughts on the analytical properties of earned economic income. *British Accounting Review*, *27*(3), 229–239.

Peasnell, K. V. (1996). Using accounting data to measure the economic performance of firms. *Journal of Accounting and Public Policy*, *15*(4), 291–303.

Peccati, L. (1987). DCF e risultati di periodo [DCF and period results]. In *Proceedings of the XI AMASES conference (Association for mathematics applied to economic and social sciences, Aosta, Italy, 9–11 September)*.

Peccati, L., (1989). Multiperiod analysis of a levered portfolio. *Decisions in Economics and Finance*, 12(1), 157–166. Reprinted. In: J. Spronk, B. Matarazzo (Eds.). (1992). *Modelling for financial decisions*. Berlin: Springer.

Peccati, L. (1991). *Valutazione analitica e sintetica di attività finanziarie* [Analytic and synthetic evaluation of financial activites]. Supplement to *Rivista milanese di economia* 39 (Lug-Set), serie quaderni n. 21. Milano: Laterza

Penman, S. H. (2010). *Financial statement analysis and security valuation* (4th ed.). Boston: Irwin/McGraw-Hill.

Percoco, M., & Borgonovo, E. (2012). A note on the sensitivity analysis of the internal rate of return. *International Journal of Production Economics*, *135*, 526–529.

Pfeiffer, T. (2004). Net present value-consistent investment criteria based on accruals: A generalisation of the residual income-identity. *Journal of Business Finance & Accounting*, *31*(7), 905–926.

Pfeiffer, T., & Velthuis, L. (2005). On the optimality of linear contracts to induce goal-congruent investment behavior. *Applied Economics Letters*, *12*(4), 207–211.

Pianosi, F., Beven, K., Freer, J., Hall, J. W., Rougier, J., Stephenson, D. B., et al. (2016). Sensitivity analysis of environmental models: A systematic review with practical workflow. *Environmental Modelling & Software*, *79*, 214–232.

Pierru, A. (2010). The simple meaning of complex rates of return. *The Engineering Economist*, *55*(2), 105–117.

Pike, R., Neale, B., & Linsey, P. (2012). *Corporate finance and investment: Decisions and strategies* (7th ed.). Harlow: Pearson.

Pinches, G.E. (1982). Myopia, capital budgeting and decision making. *Financial Management*, 11, 6–19.

Pitchford, J. D., & Hagger, A. J. (1958). A note on the marginal efficiency of capital. *The Economic Journal*, *68*, 597–600.

Pope, P., & Wang, P. (2003). Positive (zero) NPV projects and the behavior of residual earnings: Discussion and extension. *Journal of Business Finance & Accounting*, *30*(1–2), 16–22.

Popper, K. R. (1949). A note on natural laws and so-called "contrary-to-fact conditionals". *Mind*, *49*, 62–66.

Poterba, J. M., & Summers, L. H. (1995). A CEO survey of US companies' time horizons and hurdles rates. *Sloan Management Review*, *37*(1), 43–53.

Pratt, J. W., & Hammond, J. S. (1979). Evaluating and comparing projects: Simple detection of false alarms. *The Journal of Finance*, *34*(5), 1231–1242.

Pratt, S. P., Reilly, R. F., & Schweihs, R. P. (1996). *Valuing a business: The analysis and appraisal of closely-held companies* (3rd ed.). New York: McGraw-Hill.

Preinreich, G. (1936). The fair value and yield of common stock. *The Accounting Review*. Reprinted in: R. Brief, K. V. Peasnell, (Eds.). (1996). *Clean surplus: A link between accounting and finance*. New York: Garland Publishing.

Preinreich, G. (1937). Valuation and amortization. *The Accounting Review*, *12*(3), 209–226.

Preinreich, G. (1938). Annual survey of economic theory: The theory of depreciation. *Econometrica*, *6*(3), 219–241.

Pressacco, F., Magni, C. A., & Stucchi, P. (2014). A quasi-IRR for a project without IRR. *Frontiers in Financial Economics*, *11*(2), 1–23.

Pressacco, F., & Stucchi, P. (1997). Su una estensione bidimensionale del teorema di scomposizione di Peccati, [On a two-dimensional extension of Peccati's decomposition theorem]. *Decisions in Economics and Finance*, *20*, 169–185.

Quinn, J. B., & Mintzberg, H. (1996). *The strategy process: Concepts, contexts, and cases*. Englewood Cliffs: Prentice-Hall.

Quine, W. V. (1951). Two dogmas of empiricism. *Philosophy Review*, *60*, 20–43.

Ramsey, J. B. (1970). The marginal efficiency of capital, the internal rate of return, and net present value: An analysis of investment criteria. *Journal of Political Economy*, *78*(5), 1017–1027.

Rao, R. K. S. (1992). *Financial management. Concepts and applications* (2nd ed.). New York: Macmillan.

Rapp, B. (1980). The internal rate of return method – a critical study. *Engineering Costs and Production Economics*, *5*(1), 43–52.

Rappaport, A. (1986). *Creating shareholder value: The new standard for business performance*. New York: The Free Press.

Rappaport, A. (1998). *Creating shareholder value: a guide for managers and investors,* (Revised and update ed.). New York: The Free Press.

Remer, D. S., & Nieto, A. P. (1995a). A compendium and comparison of 25 project evaluation techniques. Part 1: Net present value and rate of return methods. *International Journal of Production Economics*, *42*, 79–96.

Remer, D. S., & Nieto, A. P. (1995b). A compendium and comparison of 25 project evaluation techniques. Part 2: Ratio, payback, and accounting methods. *International Journal of Production Economics*, *42*, 101–129.

Remer, D. S., Stokdyk, S. B., & VanDriel, M. (1993). Survey of project evaluation techniques currently used in industry. *International Journal of Production Economics*, *32*, 103–115.

Rendleman, R. J, Jr. (1978). Ranking errors in CAPM capital budgeting applications. *Financial Management*, *7*(4), 40–44.

Revsine, L., Collins, D. W., & Johnson, W. B. (2005). *Financial reporting and analysis* (3rd ed.). Upper Saddle River: Prentice Hall Education.

Reynolds, L. G. (1963). *Economics: A general introduction*. Homewood: Richard Irwin.

Robichek, A. A., & Myers, S. C. (1965). *Optimal financing decisions*. Englewood Cliffs: Prentice-Hall.

Robichek, A. A., & Myers, S. C. (1966). Conceptual problems in the use of risk-adjusted discount rates. *The Journal of Finance*, *21*(4), 727–730.

Roese, N. J. (1997). Counterfactual thinking. *Psychological Bulletin*, *121*, 133–148.

Roese, N. J., & Olson, J. M. (Eds.). (1995). *What might have been. The social psychology of counterfactual thinking*. Mahwah: Lawrence Erlbaum Associates.

Rogers, M., & Duffy, A. (2012). *Engineering project appraisal. The evaluation of alternative development schemes* (2nd ed.). Chichester: Wiley-Blackwell.

Rogerson, W. P. (1997). Intertemporal cost allocation and managerial investment incentives: A theory explaining the use of economic value added as a performance measure. *Journal of Political Economy*, *105*(4), 770–795.

Ross, S. A. (1978). A simple approach to the valuation of risky streams. *The Journal of Business*, *51*(3), 453–475.

Ross, S. A. (1995). Uses, abuses, and alternatives to the net-present-value rule. *Financial Management*, *24*(3), 96–102.

Ross, S. A., Spatt, C. S., & Dybvig, P. H. (1980). Present values and internal rates of return. *Journal of Economic Theory*, *23*, 66–81.

Ross, S. A., Westerfield, R. W., & Jaffe, J. (2013). *Corporate finance* (7th ed.). New York: McGraw-Hill/Irwin.

Ross, S. A., Westerfield, R. W., & Jordan, B. D. (2011). *Essentials of corporate finance* (7th ed.). New York: McGraw-Hill/Irwin.

Ruback, R. S. (1994). A note on capital cash flow valuation. *Harvard Business School Background Note 295-069*. (Revised January 1995)

Ruback, R. S. (2002). Capital cash flows: A simple approach to value risky cash flows. *Financial Management*, *31*(2), 85–103.

Rubinstein, M. (1973). A mean-variance synthesis of corporate financial theory. *The Journal of Finance*, *28*(1), 167–181.

Rubinstein, M. (2003). Great moments in financial economics: I. Present value. *Journal of Investment Management*, *1*(1), 45–54.

Ryan, P. A., & Ryan, G. P. (2002). Capital budgeting practice of the Fortune 2000: How have things changed? *The Journal of Business Management*, *8*(4), 355–364.

Salamon, G. L. (1985). Accounting rates of return. *American Economic Review*, *75*(3), 495–504.

Saltelli, A., Tarantola, S., Campolongo, F., & Ratto, M. (2004). *Sensitivity analysis in practice. A guide to assessing scientific models*. Chichester: Wiley.

Samuelson, P. A. (1937). Some aspects of the pure theory of capital. *Quarterly Journal of Economics*, *51*(3), 469–496.

Samuelson, P. A. (1947). *Foundations of economic analysis*. Cambridge: Harvard University Press.

Samuels, J. M., Wilkes, F. M., & Brayshaw, R. E. (1990). *Management of company finance* (5th ed.). London: Chapman and Hall.

Sandahl, G., & Sjögren, S. (2003). Capital budgeting methods among Sweden's largest groups of companies. The state of the art and a comparison with earlier studies. *International Journal of Production Economics*, *84*, 51–69.

Sartori, D., Catalano, G., Genco, M., Pancotti, P., Sirtori, E., Vignetti, S., & Del Bo, C. (2014). *Guide to cost-benefit analysis of investment projects. Economic appraisal tool for cohesion policy 2014–2020*. European commission, directorate-general for regional and urban policy. Available at https://ec.europa.eu/inea/sites/inea/files/cba_guide_cohesion_policy.pdf.

Scannella, E. (2012). Project finance in the energy industry: New debt-based financing models. *International Business Research*, *5*(2), 83–93.

Schlick, M. (1931). Die Kausalität in der gegenwärtigen Physik. *Naturwissenschaften 19*(2), 145–162. Reprinted in M. Schlick (1948). *Gesetz, Kausalität und Wahrscheinlichkeit*. Wien: Gerald & Co.

Scorgie, M. B. (1965). Rate of return. *Abacus*, *1*(1), 85–91.

Scorgie, M. E. (1996). Evolution of the application of present value to valuation of non-monetary resources. *Accounting and Business Research*, *26*(3), 237–248.

Sen, A. (1975). Minimal conditions for monotonicity of capital value. *Journal of Economic Theory*, *11*(3), 340–355.

Senbet, L. W., & Thompson, H. E. (1978). The equivalence of alternative mean-variance capital budgeting models. *Journal of Finance*, *33*(2), 395–401.

Sharpe, W. F. (1964). Capital asset prices: A theory of market equilibrium under conditions of risk. *The Journal of Finance*, *19*(3), 425–442.

Shemin, R. (2004). *The learning annex presents making money in real estate: A smarter approach*. Hoboken: Wiley.

Shestopaloff, Y., & Shestopaloff, A. (2013). Choosing the right solution of IRR equation to measure investment success. *The Journal of Performance Measurement*, *18*(1) (Fall), 35–50.

Shiller, R. J. (1981). The use of volatility measures in assessing market efficiency. *The Journal of Finance*, *36*(2), 291–304.

Shrieves, R. E., & Wachowicz, J. M., Jr. (2001). Free cash flow (FCF), economic value added (EVA^{TM}) and net present value (NPV). A reconciliation of discounted-cash-flow (DCF) valuation. *The Engineering Economist*, *46*(1), 33–52.

Simon, H. A. (1955). A behavioral model of rational choice. *Quarterly Journal of Economics*, *69*, 99–118.

Simon, H. A. (1957). *Models of man*. New York: John Wiley.

Simon, H. A. (1981). *The sciences of the artificial*. Cambridge: MIT Press.

Sloan, L. H. (1929). *Corporate profits: A study of their size, variation, use, and distribution in a period of prosperity*. New York: Harper and Brothers Publishers.

Smith, C. W, Jr., & Warner, J. B. (1979). On financial contracting: An analysis of bond covenants. *Journal of Financial Economics*, *7*, 117–161.

Smith, J. E., & Nau, R. F. (1995). Valuing risky projects: Options pricing theory and decision analysis. *Management Science*, *41*, 795–816.

Smithers, A. (2013). *The road to recovery: How and why economic policy must change*. Chichester: Wiley.

Solomon, E. (1956). The arithmetic of capital budgeting decisions. *The Journal of Business*, *29*(2), 124–129.

Solomon, M. B., Jr. (1966). Uncertainty and its effect on capital investment analysis. *Management Science*, *12*(8), Series B:B334–B339.

Solomons, D. (1965). *Divisional performance: Measurement and control*. Homewood: Richard D Irwin.

Soper, C. S. (1959). The marginal efficiency of capital: A further note. *The Economic Journal*, *69*(273), 174–177.

Sosa, E. (Ed.). (1975). *Causation and conditionals*. Oxford: Oxford University Press.

Spaulding, D. (2011). *The handbook of investment performance: A user's guide* (2nd ed.). Somerset: TSG Publishing Inc.

Spearman, C. (1904). The proof and measurement of association between two things. *The American Journal of Psychology*, *15*(1), 72–101.

Stanley, M. T., & Block, S. B. (1984). A survey of multinational capital budgeting. *The Financial Review*, *19*, 36–54.

Stark, A. W. (1989). A note on the aggregation properties of Hotelling depreciation. *The British Accounting Review*, *21*, 69–76.

Stark, A. W. (2004). Estimating economic performance from accounting data - a review and a synthesis. *The British Accounting Review*, *36*(4), 321–343.

Stern, J. M., Stewart, G. B., & Chew, D. H. (1995). The EVA financial management system. *Journal of Applied Corporate Finance*, *8*(2), 32–46.

Stewart, G. B. (1991). *The quest for value: The EVA^{TM} management guide*. New York: HarperCollins Publishers Inc.

Stoughton, N. M., & Zechner, J. (2007). Optimal capital allocation using RAROC^{TM} and EVA^{TM}. *Journal of Financial Intermediation, 16*(3), 312–342.

Sugden, R. (2000). Credible worlds: The status of theoretical models in economics. *Journal of Economic Methodology, 7*(1), 1–31.

Taggart, R. A, Jr. (1991). Consistent valuation and cost of capital expressions with corporate and personal taxes. *Financial Management, 20*(3), 8–20.

Talavera, D. L., Nofuentes, G., & Aguilera, J. (2010). The internal rate of return of photovoltaic grid-connected systems: A comprehensive sensitivity analysis. *Renewable Energy, 35*(1), 101–111.

Teichroew, D., Robichek, A., & Montalbano, M. (1965a). An analysis of criteria for investment and financing decisions under certainty. *Management Science, 12*, 151–179.

Teichroew, D., Robichek, A., & Montalbano, M. (1965b). Mathematical analysis of rates of return under certainty. *Management Science, 11*, 395–403.

Tham, J., & Vélez-Pareja, I. (2004). *Principles of cash flow valuation*. Burlington: Elsevier Academic Press.

Thirlby, G. F. (1946). The subjective theory of value and accounting "cost". *Economica, XIII*, 32–49.

Titman, S., & Martin, J. D. (2011). *Valuation. The art and science of corporate investment decisions* (2nd ed.). Boston: Pearson.

Titman, S., & Martin, J. D. (2016). *Valuation. The art and science of corporate investment decisions* (3rd ed.). Boston: Pearson.

Titman, S., Keown, A. J., & Martin, J. D. (2017). *Financial management. Principles and applications* (13th ed.). Boston: Pearson.

Tobin, J. (1969). A general equilibrium approach to monetary theory. *Journal of Money, Credit and Banking, 1*(1), 15–29.

Tobin, J., & Brainard, W. C. (1977). Asset markets and the cost of capital. In W. Fellner, B. A. Balassa, & R. R. Nelson (Eds.), *Economic progress, private values and public policy, essays in honor of William Fellner*. Amsterdam: North-Holland Publishing Company.

Tomkins, C. (1975a). Another look at residual income. *Journal of Business Finance & Accounting, 2*(1), 39–53.

Tomkins, C. (1975b). Residual income - a rebuttal of Professor Amey's argument. *Journal of Business Finance & Accounting, 2*(2), 161–168.

Tuttle, D. L., & Litzenberger, R. H. (1968). Leverage, diversification and capital market effects on a risk-adjusted capital budgeting framework. *Journal of Finance, 23*(3), 427–443.

Tversky, A., & Kahneman, D. (1981). The framing of decisions and the psychology of choice. *Science, 211*, 453–458.

Van de Mieroop, M. (2005). The invention of interest. Sumerian loans. In W. N. Goetzmann, K.G. Rouwenhorst (Eds.) *The origins of value. The financial innovations that created modern capital markets*. New York: Oxford University Press.

Varian, H. R. (1987). The arbitrage principle in financial economics. *Journal of Economic Perspectives, 1*(2), 5–72.

Vasigh, B., & Gorrjidooz, J. (2017). *Engineering economics for aviation and aerospace*. Abingdon: Routledge.

Vatter, W. J. (1966). Income models, book yield, and the rate of return. *The Accounting Review, 41*(4), 681–698.

Vélez-Pareja I, & Magni, C. A. (2009) Potential dividends and actual cash flows in equity valuation. A critical analysis. *Estudios Gerenciales – Journal of Management and Economics for Iberoamerica, 25*(113), 123–150.

Walker, A., Sibley, F., Carter, A., & Hurley, M. (2017). Social return on investment analysis of a physiotherapy-led service for managing osteoarthritis in primary care. *Lancet, 389*(1), S98.

Weaver, S. C., & Weston, J. F. (2003). *A unifying theory of value based management* (Paper 4-03, November 27). Available at http://repositories.cdlib.org/anderson/fin/4-03/.

Weingartner, H. M. (1966). The generalized rate of return. *Journal of Financial and Quantitative Analysis, 1*(3), 1–29.

Welch, I. (2009). *Corporate finance. An introduction*. Boston: Pearson Education.

Wellington, A. M. (1887). *The economic theory of the location of raiways*. New York: Wiley.

Wells, G. L., Taylor, B. R., & Turtle, J. W. (1987). The undoing of scenarios. *Journal of Personality and Social Psychology*, *53*(3), 421–430.

Weygandt, J. J., Kimmel, P. D., & Kieso, D. E. (2012). *Accounting principles* (10th ed.). New York: Wiley.

Whisler, W. D. (1976). Sensitivity analysis of rates of return. *The Journal of Finance*, *31*(1), 63–69.

White, G. I., Sondhi, A. C., & Fried, D. (1997). *The analysis and use of financial statements* (2nd ed.). New York: Wiley.

White, J. A., Grasma, K. S., Case, K. E., LaScola, Needy K., & Pratt, D. B. (2014). *Fundamentals of engineering economic analysis* (1st ed.). Hoboken: Wiley.

Whittington, G. (1988). The usefulness of accounting data in measuring the economic performance of firms. *Journal of Accounting and Public Policy*, *7*(4), 261–266.

Whittington, G. (2017). *Value and profit. An introduction to measurement in financial reporting*. Cambridge: Cambridge University Press.

Will, F. L. (1947). The contrary-to-fact conditional. *Mind*, *56*, 236–249.

Williams, J. R., Haka, S. F., Bettner, M. S., & Carcello, J. V. (2010). *Financial and managerial accounting. The basis for business decisions* (International ed.). New York: McGraw-Hill.

Wing, G. A. (1965). Capital budgeting, circa 1915. *The Journal of Finance*, *20*(3), 472–479.

Yates, B. T., & Marra, M. (2017). Introduction: Social return on investment (SROI). *Evaluation and Program Planning*, *64*, 95–97.

Yates, J. K. (2017). *Engineering economics*. Boca Raton: CRC Press.

Yee, K. K. (2005). Aggregation, dividend irrelevancy, and earnings-value relations. *Contemporary Accounting Research*, *22*(2), 453–480.

Yoshigawa, H. (1980). On the "q" theory of investment. *American Economic Review*, *70*(4), 739–743.

Young, S. D., & O'Byrne, S. F. (2001). *EVA and value-based management*. New York: McGraw-Hill.

Yung, J. T., & Sherman, L. F. (1995). Investment analysis for loan decision making. *Real Estate Review*, *25*(3), 16–22.

Zhang, D. (2005). A different perspective on using multiple internal rates of return: The IRR parity technique. *The Engineering Economist*, *50*, 327–335.

Symbols and Abbreviations

Symbol	Description
$\begin{array}{\|c\|c\|}\hline C^o & C^d \\ \hline & C^e \\ \hline\end{array}$	Project levered (non-operating assets are absent)
$\begin{array}{\|c\|c\|}\hline C^o & \\ \hline & C^e \\ \hline\end{array}$	Project unlevered (i.e., non-operating assets and debt are absent)
$\begin{array}{\|c\|c\|}\hline & C^d \\ \hline C^l & C^e \\ \hline\end{array}$	Financial transaction (equity-and-debt financed)
$\begin{array}{\|c\|c\|}\hline C^o & C^d \\ \hline C^l & C^e \\ \hline\end{array}$	Project levered (non-operating assets are present)
$\begin{array}{\|c\|c\|}\hline C^o & \\ \hline C^l & C^e \\ \hline\end{array}$	Project equity-financed (non-operating assets are present)
$\begin{array}{\|c\|c\|}\hline & \\ \hline C^l & C^e \\ \hline\end{array}$	Financial transaction (equity-financed)
$\succ$	preferred to
α_t	payout ratio (as a fraction of PD)
β_t	payout ratio (as a fraction of NI and/or PD)
Δ	variation, change

Continued on next page

C. A. Magni, *Investment Decisions and the Logic of Valuation*,
https://doi.org/10.1007/978-3-030-27662-1

Continued from previous page

Symbol	**Description**
ϵ	economic efficiency in the IARR approach (excess IARR)
$\epsilon(K)$	standardized excess IARR
$\epsilon(x)$	excess IARR function
ε	economic efficiency in the AROI approach (excess AROI)
$\varepsilon_{\mathrm{d}}(K)$	standardized excess AROI
$\varepsilon(x)$	excess AROI function
θ	shares owned by existing shareholders at the announcement of a project undertaking
ϑ	net-debt-to-operating-value ratio (in market value)
ξ	economic efficiency in the AIRR approach (excess AIRR)
$\xi(K)$	standardized excess AIRR
$\xi(x)$	excess AIRR function
π_t	risk premium
$\bar{\rho}$	project COC/MARR in the IARR approach
$\bar{\rho}^d$	debt COC/MARR in the IARR approach
$\bar{\rho}^{\mathcal{D}}$	net-debt COC/MARR in the IARR approach
$\bar{\rho}^e$	equity COC/MARR in the IARR approach
$\bar{\rho}^o$	operating COC/MARR in the IARR approach
$\bar{\rho}^l$	non-operating COC/MARR in the IARR approach
$\bar{\rho}^u$	unlevered COC/MARR in the IARR approach
$\bar{\rho}^{uo}$	unlevered operating COC/MARR in the IARR approach
$\bar{\varrho}$	project COC/MARR in the AROI approach
$\bar{\varrho}^d$	debt COC/MARR in the AROI approach
$\bar{\varrho}^{\mathcal{D}}$	net-debt COC/MARR in the AROI approach
$\bar{\varrho}^e$	equity COC/MARR in the AROI approach
$\bar{\varrho}^o$	operating COC/MARR in the AROI approach
$\bar{\varrho}^l$	non-operating COC/MARR in the AROI approach
$\bar{\varrho}^u$	unlevered COC/MARR in the AROI approach
$\bar{\varrho}^{uo}$	unlevered operating COC/MARR in the AROI approach
$\bar{\varrho}_{\mathrm{d}}$	project COC/MARR in the AROI approach (in present terms)
σ	internal rate of return
τ	tax rate
$\phi(x)$	discounted-cash-flow (NPV) function
$\Phi_t^{\mathcal{D}}$	Modified CF$\mathcal{D}$
Φ_t^e	Modified CFE

Continued on next page

Continued from previous page

Symbol	**Description**
AIRR	average internal rate of return
AP_t	accounts payable
APV	adjusted present value
AR_t	accounts receivable
AROI	aggregate return on investment
BAIRR	book AIRR
BAROI	book AROI
BC	benefit-cost ratio
BIARR	book IARR
BO_t	bonds
BOP	beginning-of-period
C_t	BOP capital
C_t^{fin}	capital raised from capital providers (claimholders)
C_t^{inv}	capital invested in the assets
C_t^d	debt
$C_t^{\mathcal{D}}$	net debt (net financial obligations)
C_t^e	equity
C_t^l	liquid, non-operating assets
C_t^o	operating assets
C_t^{pd}	potential (equity) capital
$C_t(\sigma)$	internal (IRR-implied) capital
C_t^{ec}	capital associated with the economic EAIRR
$\mathcal{C}_t$	capital of benchmark portfolio replicating project's cash flows from time 0 to time $n - 1$
C	total capital
C_{d}	total capital discounted from n to 0
C^d	total debt capital
$C^{\mathcal{D}}$	total net-debt capital
C^e	total equity capital
C^l	total liquid assets
C^o	total operating assets
C_B	total capital borrowed in the borrowing periods
C_L	total capital invested in the investment periods
CE_t	certainty equivalent
CCF	capital cash flow (also labeled CFO)
CFA	cash flow from assets
CFD	cash flow to debt
CF$\mathcal{D}$	cash flow to net debt

Continued on next page

Continued from previous page

Symbol	Description
CFE	cash flow to equity
CFL	cash flow from/to liquid assets
CFO	cash flow from operations (also labeled CCF)
CFP	cash flow to capital providers
CM	cash multiple
COC	cost of capital
COGS_t	cost of goods sold
COGS_t°	COGS, including manufacturing depreciation
COM_t	cost of material purchased
$\text{d}_{j,t}$	discount factor for CFAs
$\text{d}_{j,t}^{d}$	discount factor for CFDs
$\text{d}_{j,t}^{\mathcal{D}}$	discount factor for CF$\mathcal{D}$
$\text{d}_{j,t}^{e}$	discount factor for CFEs
$\text{d}_{j,t}^{l}$	discount factor for CFLs
$\text{d}_{j,t}^{o}$	discount factor for CFOs
$\text{d}_{j,t}^{uo}$	discount factor for FCFs
$\text{d}_{j,t}^{u}$	discount factor for asset FCFs
Dep_t	depreciation and amortization expenses
Dep_t^{m}	depreciation for manufacturing assets
Dep_t^{nm}	depreciation for nonmanufacturing assets
DL_t	direct labor costs
DDB	double-declining balance depreciation
DDB-SL	double-declining balance depreciation switching to SL depreciation
E_t	EOP capital
$E_n^{\mathcal{C}}$	terminal capital of benchmark portfolio replicating project's cash flows from time 0 to time $n-1$
E_n^{d}	terminal debt
$E_n^{\mathcal{D}}$	terminal net debt
E_n^{e}	terminal equity
E_n^{l}	terminal non-operating assets
E_n^{o}	terminal operating assets
EAIRR	economic AIRR
EBIT_t	earnings before interest and taxes
EBITDA_t	earnings before interest, taxes, depreciation, and amortization
EBT_t	earnings before taxes
EC_t	excess cash
EOP	end-of-period
F_t	cash flow

Continued on next page

Continued from previous page

Symbol	Description
F_t^{fin}	cash flow to capital providers (claimholders)
F_t^{inv}	cash flow from assets
F_t^{d}	cash flow to debtholders
$F_t^{\mathcal{D}}$	cash flow to net debt
F_t^{e}	cash flow to equity
F_t^{l}	cash flow from/to non-operating assets (e.g., liquid assets or other non-operating assets)
F_t^{o}	cash flow from operations, capital cash flow
F_t^{pd}	potential dividends (cash available for distribution to shareholders)
F_t^{ul}	non-operating free cash flow
F_t^{u}	asset free cash flow (financially unlevered CFA)
F_t^{uo}	operating free cash flow (unlevered CFO)
F_t^{bo}	cash flow paid to bondholders
F_t^{c}	cash from customers
F_t^{capex}	cash for capital expenditures
F_t^{cem}	cash for purchases of manufacturing fixed assets (e.g., plants and machinery)
F_t^{cenm}	cash for purchases of nonmanufacturing fixed assets (e.g., building, office equipment)
F_t^{cogs}	payments for COGS
F_t^{cop}	payments to suppliers
F_t^{ec}	withdrawals from excess cash (net of deposits)
F_t^{fa}	cash from sales of financial assets (net of purchases)
F_t^{lc}	payments to employees
F_t^{lo}	loan instalment
F_t^{nowc}	cash flow from net operating working capital
F_t^{np}	cash flow paid to holders of notes payable
F_t^{opc}	payments for operating costs
F_t^{sga}	payments for SGA
F_t^{T}	payments for taxes
$F_t^{\mathcal{C}}$	cash flow of benchmark portfolio replicating project's cash flows from time 0 to time $n-1$
F_t^{V}	cash flow of benchmark portfolio replicating project's cash flows from time 1 to time n
F_t^{*}	cash flow of benchmark portfolio replicating project's capitals (normal cash flow)
F_t^{+}	inflow
F_t^{-}	outflow
F	total cash flow
F^{d}	total cash flow to debt
F^{e}	total cash flow to equity

Continued on next page

Continued from previous page

Symbol	Description
F^o	total cash flow from operations
F^l	total cash flow from liquid assets
F^+	total inflow
F^-	total outflow
FA_t	marketable securities and other financial assets
FCF	free cash flow (unlevered CFO)
FCFE	free cash flow to equity (also labeled PD)
GP_t	gross profit
GP_t°	gross profit, net of manufacturing depreciation
I_t	income, profit
I_t^{fin}	income to capital providers
I_t^{inv}	income from assets
I_t^d	interest expense
$I_t^{\mathcal{D}}$	net interest
I_t^e	net income
I_t^l	interest income
I_t^o	operating income
I_t^{pd}	potential (net) income
I_t^{uo}	net operating profit after taxes (unlevered operating income)
I_t^u	financially unlevered income
I_t^{bo}	interest on bonds
I_t^{ec}	interest on excess cash (if any)
I_t^{fa}	interest income from financial assets
I_t^{lo}	interest on loans
I_t^{np}	interest on notes payable
$I_t^{\mathcal{C}}$	income of benchmark portfolio replicating project's cash flows from time 0 to time $n-1$
I_t^V	income of benchmark portfolio replicating project's cash flows from time 1 from time n
I_t^*	income of benchmark portfolio replicating project's capitals (normal profit)
I	total income
I^e	total net income
I^d	total interest
I^o	total operating income
I^l	total non-operating income
I_{d}	total income discounted from n to 0
IARR	internal average rate of return
Inv_t	inventory

Continued on next page

Continued from previous page

Symbol	Description
IROR	intrinsic rate of return
IRR	internal rate of return
i_t	income rate, capital growth rate, return on investment, return on capital, return on assets
i_t^{fin}	income rate earned by capital providers
i_t^{inv}	income rate generated by the investments
i_t^d	return on debt
$i_t^{\mathcal{D}}$	return on net debt
i_t^e	return on equity
i_t^l	return on financial assets
i_t^o	return on operating assets
i_t^{pd}	return on potential capital
i_t^{uo}	unlevered (operating) ROI
i_t^u	unlevered asset ROI (financially unlevered ROI)
i_1^{dis}	disequilibrium rate
$\bar{\imath}_0, \bar{\imath}$	average ROI (istantaneous, discrete)
$\bar{\imath}_0^d, \bar{\imath}^d$	average ROD (istantaneous, discrete)
$\bar{\imath}_0^{\mathcal{D}}, \bar{\imath}^{\mathcal{D}}$	average RO$\mathcal{D}$ (istantaneous, discrete)
$\bar{\imath}_0^e, \bar{\imath}^e$	average ROE (istantaneous, discrete)
$\bar{\imath}_0^l, \bar{\imath}^l$	non-operating average ROI (istantaneous, discrete)
$\bar{\imath}_0^o, \bar{\imath}^o$	operating average ROI (istantaneous, discrete)
$\bar{\imath}_0^{uo}, \bar{\imath}^{uo}$	unlevered operating average ROI (istantaneous, discrete)
$\bar{\imath}_0^u, \bar{\imath}^u$	unlevered average ROI (istantaneous, discrete)
$\bar{\imath}(\text{PV}[C^{\text{ec}}])$	economic AIRR
$\bar{\imath}(x)$	AIRR function
$\bar{\imath}(K)$	standardized AIRR
i_B	TRM's lending rate
i_L	TRM's borrowing rate
$\bar{\imath}_0^B, \bar{\imath}^B$	borrowing-period ROI (instantaneous, discrete)
$\bar{\imath}_0^L, \bar{\imath}^L$	lending-period ROI (instantaneous, discrete)
$\bar{\jmath}$	internal average ROI
$\bar{\jmath}^d$	internal average ROD
$\bar{\jmath}^{\mathcal{D}}$	internal average RO$\mathcal{D}$
$\bar{\jmath}^e$	internal average ROE
$\bar{\jmath}^l$	internal non-operating average ROI
$\bar{\jmath}^o$	internal operating average ROI
$\bar{\jmath}^u$	internal (financially) unlevered average ROI

Continued on next page

Continued from previous page

Symbol	Description
$\bar{\jmath}^{uo}$	internal unlevered operating average ROI
$\bar{\jmath}_{\mathrm{d}}$	internal average ROI (in present terms)
$\bar{\jmath}(K)$	standardized IARR
$\bar{\jmath}_{\mathrm{d}}(K)$	standardized AROI
k_t^{uo}	unlevered cost of assets (required return on unlevered assets)
k_t^{u}	financially unlevered cost of assets (required return on financially unlevered assets)
K	standardized capital
LO_t	loans
MAFR	maximum attractive financing rate
MARR	minimum attractive rate of return
MD	modified dietz return
MIRR	modified internal rate of return
MOH_t	manufacturing overhead
n	length of a project
$\mathbb{N}_k^j$	set of all natural numbers from j to k
NFA_t	net fixed assets
NFA_t^m	manufacturing net fixed assets
NFA_t^{nm}	nonmanufacturing net fixed assets
NFV	net future value
NI	net income
NOL_t	net operating liabilities
NOL_t^m	manufacturing liabilities
NOL_t^{nm}	nonmanufacturing liabilities
NOPAT_t	net operating profit after taxes (unlevered operating income)
NOWC_t	net operating working capital
NP_t	notes payable
NPV	net present value
$\mathrm{NPV}^{\mathrm{fin}}$	NPV captured by debtholders and equityholders
$\mathrm{NPV}^{\mathrm{inv}}$	NPV generated by the assets
NPV^d	debt NPV
$\mathrm{NPV}^{\mathcal{D}}$	net-debt NPV
NPV^e	equity NPV
NPV^l	non-operating NPV
NPV^o	operating NPV
NPV^{pd}	potential (equity) NPV
NPV_B	NPV generated in the borrowing periods
NPV_L	NPV generated in the lending periods

Continued on next page

Continued from previous page

Symbol	Description
OpC_t	operating costs
OpC_t°	operating costs, inclusive of depreciation
p_0	firm's share price without the project
$p_0 + \Delta p_0$	firm's share price with the project
PD	potential dividends (cash flow available for distribution to shareholders)
PI	profitability index
PV	present value
$\text{PV}[C]$	present value of capitals (at r)
$\text{PV}[C^d]$	present value of debt (at r^d)
$\text{PV}[C^{\mathcal{D}}]$	present value of net debt (at $r^{\mathcal{D}}$)
$\text{PV}[C^e]$	present value of equity (at r^e)
$\text{PV}[C^l]$	present value of non-operating assets (at r^l)
$\text{PV}[C^o]$	present value of operating assets (at r^o)
$\text{PV}[C^u]$	present value of financially unlevered assets (at r^u)
$\text{PV}[C^{uo}]$	present value of unlevered assets (at r^{uo})
$\text{PV}[C_B]$	present value of capital borrowed in the borrowing periods (at r)
$\text{PV}[C_L]$	present value of capital invested in the lending periods (at r)
$\text{PV}[C(\sigma)]$	present value of internal capital (at r)
$\text{PV}[I]$	present value of incomes (at r)
$\text{PV}[I^d]$	present value of interest expenses (at r^d)
$\text{PV}[I^{\mathcal{D}}]$	present value of net interest expenses (at $r^{\mathcal{D}}$)
$\text{PV}[I^e]$	present value of net incomes (at r^e)
$\text{PV}[I^l]$	present value of non operating incomes (at r^l)
$\text{PV}[I^o]$	present value of operating incomes (at r^o)
$\text{PV}[I^u]$	present value of unlevered incomes (at r^u)
$\text{PV}[I^{uo}]$	present value of NOPATs at r^{uo}
$\text{PV}[I^*]$	present value of normal profits (at r)
r_f, r_{ft}	risk-free rate
r_t	equilibrium market rate, COC, MARR
r_t	cost of assets, asset pre-tax WACC (required return on assets, levered perspective)
r_t^d	cost of debt (required return on debt)
$r^{\mathcal{D}}$	cost of net debt (required return on net debt)
r_t^e	cost of equity (required return on equity)
r_t^l	cost of non-operating assets (required return on non-operating assets)
r_t^o	cost of operating assets, operating pre-tax WACC (required return on operating assets)
r_t^{pd}	cost of potential capital (required return on potential capital)
r_t^{TS}	cost of tax shield, discount rate for tax shields

Continued on next page

Continued from previous page

Symbol	Description
r_t^{uo}	operating after-tax WACC (required return on operating assets, unlevered perspective)
r_t^u	asset after-tax WACC (required return on assets, financially unlevered perspective)
$r_{h,j}$	forward rate for transactions starting at time h and ending at time j
$\bar{r}_0, \bar{r}$	project COC/MARR in the AIRR approach (instantaneous, discrete)
$\bar{r}_0^L, \bar{r}^L$	lending-period MARR (instantaneous, discrete)
$\bar{r}_0^B, \bar{r}^B$	borrowing-period MARR, MAFR (instantaneous, discrete)
$\bar{r}_0^d, \bar{r}^d$	debt COC/MARR in the AIRR approach (instantaneous, discrete)
$\bar{r}_0^{\mathcal{D}}, \bar{r}^{\mathcal{D}}$	net-debt COC/MARR in the AIRR approach (instantaneous, discrete)
$\bar{r}_0^e, \bar{r}^e$	equity COC/MARR in the AIRR approach (instantaneous, discrete)
$\bar{r}_0^l, \bar{r}^l$	non-operating COC/MARR in the AIRR approach (instantaneous, discrete)
$\bar{r}_0^o, \bar{r}^o$	operating COC/MARR in the AIRR approach (instantaneous, discrete)
$\bar{r}_0^u, \bar{r}^u$	unlevered asset COC/MARR in the AIRR approach (instantaneous, discrete)
$\bar{r}_0^{uo}, \bar{r}^{uo}$	unlevered operating COC/MARR in the AIRR approach (instantaneous, discrete)
$\bar{r}_V$	average pre-tax WACC (IARR approach)
$\bar{r}_V^d$	average cost of debt (IARR approach)
$\bar{r}_V^{\mathcal{D}}$	average cost of net debt (IARR approach)
$\bar{r}_V^e$	average cost of equity (IARR approach)
$\bar{r}_V^l$	average cost of non-operating assets (IARR approach)
$\bar{r}_V^o$	average cost of operating assets (IARR approach)
$\bar{r}_V^u$	average asset after-tax WACC (IARR approach)
$\bar{r}_V^{uo}$	average operating after-tax WACC (IARR approach)
RC_t	undistributed cash, retained cash
RI_t	residual income
ROA	return on assets
ROC	return on capital
ROD	return on debt
$\text{RO}\mathcal{D}$	return on net debt
ROE	return on equity
ROI	return on investment
s_t	spot rate
S_t	sales revenues
SFM	sinking fund method
SGA_t	selling, general, and administrative expenses
SGA_t°	operating expenses (SGA, including nonmanufacturing depreciation)
SPV	special purpose vehicle (project financing)
SL	straight-line depreciation

Continued on next page

Continued from previous page

Symbol	Description
SV_n	salvage value
SWP_t	salaries and wages payable
T_t	income taxes
T_B	set of borrowing (financing) periods
T_L	set of lending (investment) periods
TP_t	income taxes payable
TRM	Teichroew-Robichek-Montalbano
$u_{j,t}$	compounding factor
V_t	market value of benchmark portfolio replicating the project's cash flows from time 1 to time n
V_t	project's market/intrinsic/economic value
V_t^d	market/intrinsic/economic value of debt
$V_t^{\mathcal{D}}$	market/intrinsic/economic value of net debt
V_t^e	market/intrinsic/economic value of equity
V_t^l	market/intrinsic/economic value of non-operating (liquid) assets
V_t^o	market/intrinsic/economic value of operating assets
V_t^{pd}	market/intrinsic/economic value of potential dividends
$V_t(\text{RI})$	market/intrinsic/economic value of RI
V_t^{TS}	market/intrinsic/economic value of tax shield
V_t^u	financially unlevered market/intrinsic/economic value of assets
V_t^{uo}	unlevered market/intrinsic/economic value of assets
VA	value added
WACC	weighted average cost of capital
y^{ext}	auxiliary (external) rate
z	gains (> 0) or losses (< 0) on disposal of assets

Index

C. A. Magni, *Investment Decisions and the Logic of Valuation*,
https://doi.org/10.1007/978-3-030-27662-1

D

E

L

S

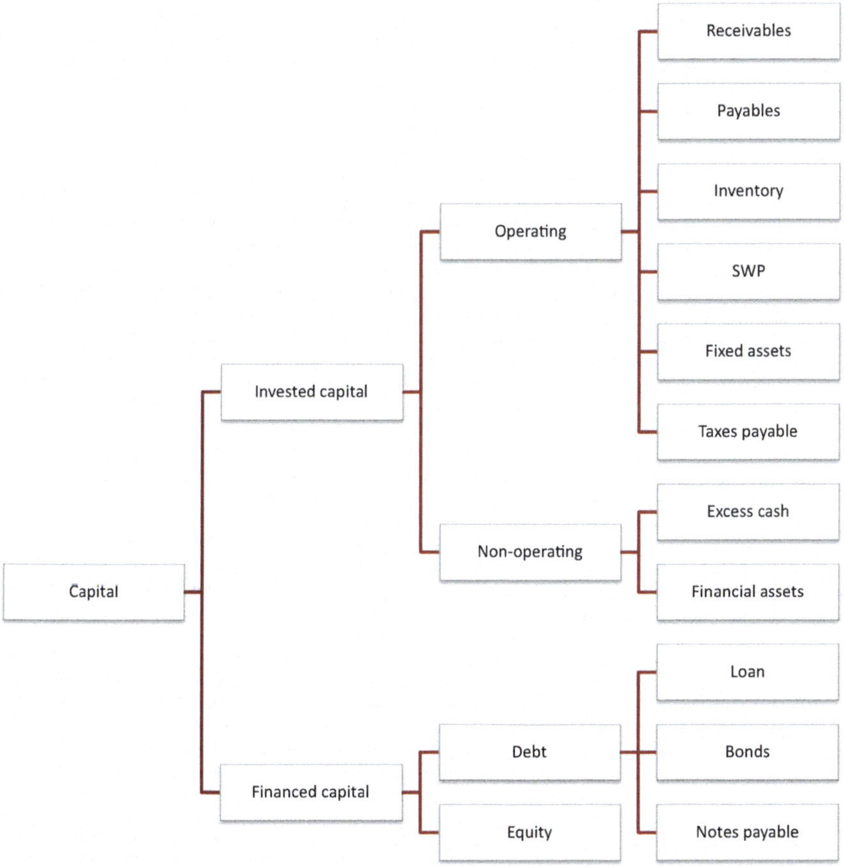
Capital
Invested capital
Financed capital
Operating
Non-operating
Debt
Equity
Receivables
Payables
Inventory
SWP
Fixed assets
Taxes payable
Excess cash
Financial assets
Loan
Bonds
Notes payable

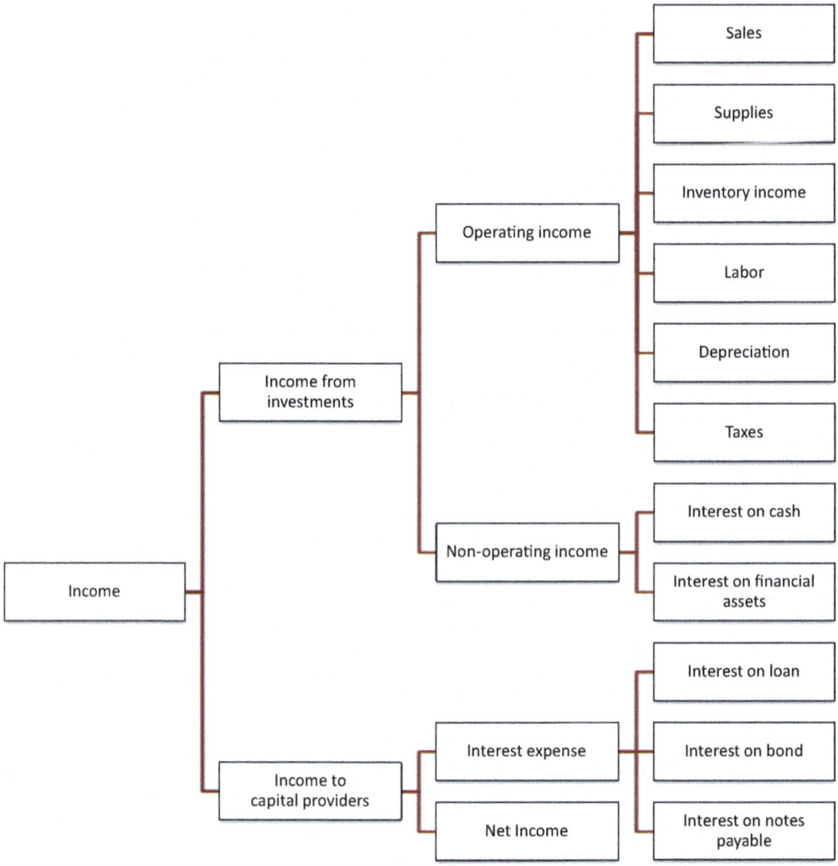

Income
Income from investments
Income to capital providers
Operating income
Non-operating income
Interest expense
Net Income
Sales
Supplies
Inventory income
Labor
Depreciation
Taxes
Interest on cash
Interest on financial assets
Interest on loan
Interest on bond
Interest on notes payable

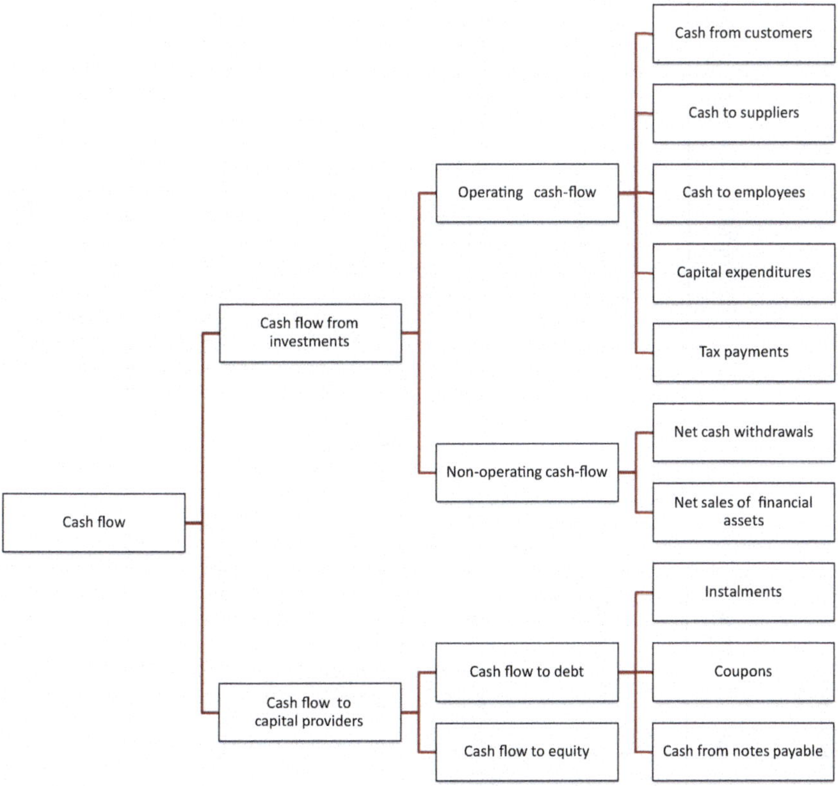

Cash flow
Cash flow from investments
Cash flow to capital providers
Operating cash-flow
Non-operating cash-flow
Cash flow to debt
Cash flow to equity
Cash from customers
Cash to suppliers
Cash to employees
Capital expenditures
Tax payments
Net cash withdrawals
Net sales of financial assets
Instalments
Coupons
Cash from notes payable

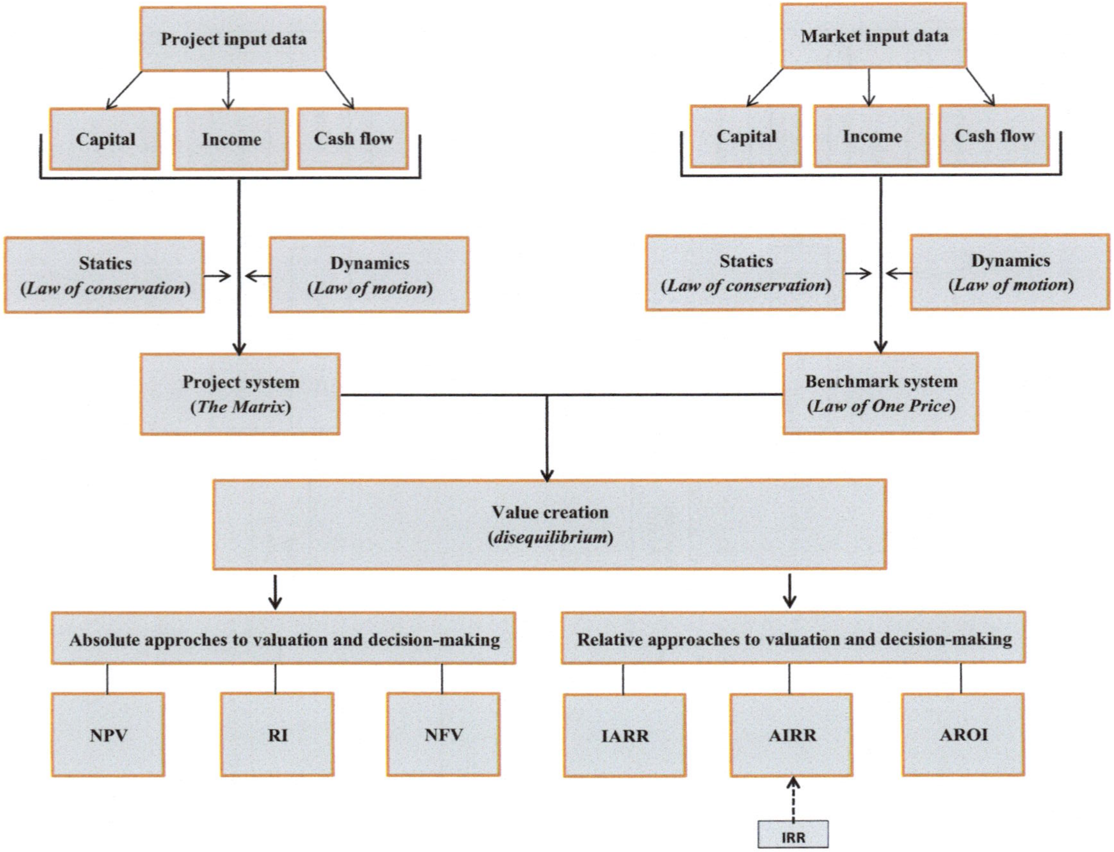
Project input data
Capital
Income
Cash flow
Statics
(Law of conservation)
Dynamics
(Law of motion)
Project system
(The Matrix)
Market input data
Capital
Income
Cash flow
Statics
(Law of conservation)
Dynamics
(Law of motion)
Benchmark system
(Law of One Price)
Value creation
(disequilibrium)
Absolute approches to valuation and decision-making
NPV
RI
NFV
Relative approaches to valuation and decision-making
IARR
AIRR
AROI
IRR

The manufacturer's authorised representative in the EU is Springer Nature Customer Service Centre GmbH, Europaplatz 3, 69115 Heidelberg, Germany. If you have any concerns regarding our products, please contact ProductSafety@springernature.com

Printed and bound by CPI Group (UK) Ltd, Croydon, CR0 4YY
30/09/2026
02225458-0020